Bacterial
Transport

MICROBIOLOGY SERIES

Series Editor
ALLEN I. LASKIN
EXXON Research and Engineering Company
Linden, New Jersey

Volume 1 Bacterial Membranes and Walls
edited by Loretta Leive

Volume 2 Eucaryotic Microbes as Model Developmental Systems
edited by Danton H. O'Day and Paul A. Horgen

Volume 3 Microorganisms and Minerals
edited by Eugene D. Weinberg

Volume 4 Bacterial Transport
edited by Barry P. Rosen

Other Volumes in Preparation

Bacterial Transport

edited by

BARRY P. ROSEN

Department of Biological Chemistry
University of Maryland School of Medicine
Baltimore, Maryland

MARCEL DEKKER, INC.   New York · Basel

Library of Congress Cataloging in Publication Data

Main entry under title:

Bacterial transport.

(Microbiology series ; v. 4)
Includes bibliographies and indexes.
1. Bacterial cell walls. 2. Biological transport.
I. Rosen, Barry P.
QR77.3.B32   589.9'08'75   78-16191
ISBN 0-8247-6670-9

MARCEL DEKKER, INC.

270 Madison Avenue, New York, New York 10016

Current printing (last digit):
10 9 8 7 6 5 4 3 2

PRINTED IN THE UNITED STATES OF AMERICA

PREFACE

This volume, because it is part of the "Microbiology Series," deals exclusively with procaryotic microorganisms. However, there is a more fundamental reason to limit it thusly. At the risk of publicizing a competing text, I would have to admit that Christensen is correct in his "plea for eclecticism" [H. N. Christensen, Biological Transport (2nd ed.), W. A. Benjamin, Reading, Mass., 1975, 514pp.]. There has been too little cross-fertilization between the various subspecialties of biological transport: the mitochondrial, eucaryotic cytoplasmic, plant, and bacterial areas. Sadly, this book cannot correct that defect. The field of bacterial transport encompasses too much information to allow for coverage of the other subspecialties; yet this text is designed for the same audiences, namely graduate and advanced undergraduate students. It should also provide a useful summary for investigators, but it is not designed to be utterly comprehensive.

Knowing that the time lag in publication is too great to allow for a current summary, I have encouraged each contributor to briefly summarize their areas and to provide an intensive look at specific points. In addition, I have requested speculation where possible. Most of that will turn out to be incorrect; by the time publication is achieved our collective score may be known. That type of divination, however, is something to which our students should be exposed.

Finally, I would like to thank Loretta Leive for her advice during the initiation of this project. I must also acknowledge gratefully the gentle pressure to push on to completion applied by Simon Silver and Eva Kashket. Thanks are also due the members of my laboratory for reading and criticizing the various chapters: Lawrence Adler, Jeanne Beck, Robert Brey, Syed Hasan, Tomio Ichikawa, and Erik Sorensen, and especially to Lisa Schwender for much-needed assistance in proofreading and compilation of the Subject Index.

Barry P. Rosen
Baltimore, Maryland

CONTRIBUTORS

Yasuhiro Anraku, Department of Botany, Faculty of Science, University of Tokyo, Hongo, Bunkyo-Ku, Tokyo, Japan

Philip J. Bassford, Jr., [1] Department of Microbiology, University of Virginia School of Medicine, Charlottesville, Virginia

Winfried Boos, Department of Biology, University of Konstanz, Konstanz, Federal Republic of Germany

Thomas Ferenci, Department of Biology, University of Konstanz, Konstanz, Federal Republic of Germany

Masamitsu Futai, [2] Department of Botany, Faculty of Science, University of Tokyo, Hongo, Bunkyo-ku, Tokyo, Japan

Charles Gilvarg, Department of Biochemical Sciences, Frick Chemical Laboratory, Princeton University, Princeton, New Jersey

John B. Hays, Department of Chemistry, University of Maryland, Baltimore County, Catonsville, Maryland

---

Present Address

[1] Department of Microbiology and Molecular Genetics, Harvard Medical School, Boston, Massachusetts

[2] Department of Microbiology, Faculty of Pharmaceutical Sciences, Okayama University, Okayama, Japan

Robert J. Kadner, Department of Microbiology, University of Virginia
School of Medicine, Charlottesville, Virginia

Eva R. Kashket, Department of Microbiology, Boston University School
of Medicine, Boston, Massachusetts

William W. Kay, [3] Department of Biochemistry, University of Saskatchewan,
Saskatoon, Saskatchewan, Canada

Edward G. Moczydlowski, Department of Biology, The John Muir College,
University of California, San Diego, La Jolla, California

John W. Payne, Department of Botany, University of Durham, Durham,
DH1 3LE, England

Barry P. Rosen, Department of Biological Chemistry, University of
Maryland School of Medicine, Baltimore, Maryland

Milton H. Saier, Jr., Department of Biology, The John Muir College,
University of California, San Diego, La Jolla, California

Thomas J. Silhavy, [4] Department of Biology, University of Konstanz,
Konstanz, Federal Republic of Germany

Simon Silver, Department of Biology, Washington University, St. Louis,
Missouri

Jeffrey B. Smith, [5] Department of Biochemistry, Molecular and Cell
Biology, Cornell University, Ithaca, New York

David B. Wilson, Department of Biochemistry, Molecular and Cell
Biology, Cornell University, Ithaca, New York

Present Address
[3] Department of Biochemistry, University of Victoria, Victoria,
British Columbia, Canada
[4] Department of Microbiology and Molecular Genetics, Harvard
Medical School, Boston, Massachusetts
[5] Imperial Cancer Research Fund, London, WC2A 3PX, England

CONTENTS

Preface   iii
Contributors   v

INTRODUCTION, Barry P. Rosen   1

Chapter 1

EXPERIMENTAL SYSTEMS FOR THE STUDY
OF ACTIVE TRANSPORT IN BACTERIA   7
Masamitsu Futai

    I   Introduction   8
   II   Transport into Whole Cells   8
  III   Preparation and Use of Membrane Vesicles   16
  IV   Orientation of Membrane Vesicles   23
   V   Conclusion   32
       References   34

Chapter 2

GROUP TRANSLOCATION TRANSPORT SYSTEMS   43
John B. Hays

    I   Introduction   44
   II   The Phosphoenolpyruvate: Hexose
       Phosphotransferase Systems   46

III   Fatty Acid Transport in <u>Escherichia</u> <u>coli</u>   84
IV   Transport of Purines, Pyrimidines, and
      Nucleosides   89
      References   98

Chapter 3

THE REGULATION OF CARBOHYDRATE TRANSPORT IN
<u>ESCHERICHIA</u> <u>COLI</u> AND <u>SALMONELLA TYPHIMURIUM</u>   103
Milton H. Saier, Jr. and Edward G. Mocydlowski

  I   Introduction   104
 II   Historical Perspectives and Physiological
       Considerations   104
III   Bacterial Carbohydrate Transport Mechanisms   107
 IV   Postulated Mechanisms of Carbohydrate
       Transport Regulation   108
  V   PTS-Mediated Regulation   109
 VI   Regulation by Intracellular Sugar Phosphates   116
VII   Possible Regulation of Methyl-α-Glucoside Transport
       by the Energized Membrane State   118
VIII   Conclusion on a Eucaryotic Note   121
       References   122

Chapter 4

SUGAR TRANSPORT SYSTEMS IN <u>ESCHERICHIA</u> <u>COLI</u>   127
Thomas J. Silhavy, Thomas Ferenci, and Winfried Boos

  I   Introduction   128
 II   Sugar Transport in <u>Escherichia</u> <u>coli</u>   128
III   The Active Sugar Transport Systems of
       <u>Escherichia</u> <u>coli</u>   154
 IV   Epilog   160
       Recent Developments   160
       References   160

Chapter 5

ACTIVE TRANSPORT OF AMINO ACIDS   171
Yasuhiro Anraku

  I   Introduction   172
 II   Historical Remarks   173

CONTENTS

   III Transport Systems for Amino Acids   176
   IV Current Topics in the Study of Transport Systems   194
    V Molecular Apparatus of Active Transport   198
   VI Conclusion   210
       References   212

Chapter 6

TRANSPORT OF CATIONS AND ANIONS 221
Simon Silver

    I Introduction   222
   II Cation Transport Systems   226
   III Inorganic Anion Transport Systems   284
   IV Comparative Aspects   301
    V Conclusion   304
       Recent Developments   305
       References   308

Chapter 7

TRANSPORT OF PEPTIDES IN BACTERIA 325
John W. Payne and Charles Gilvarg

    I Introduction   326
   II Abbreviations and Terminology   327
   III General Observations on Peptide Transport and
       Utilization   327
   IV Methods of Studying Peptide Transport   337
    V Properties of Peptide Transport Systems   340
   VI Structural Requirements for Peptide Transport
       in Microorganisms   347
       References   369

Chapter 8

TRANSPORT OF CARBOXYLIC ACIDS 385
William W. Kay

    I Introduction   386
   II Monocarboxylate Transport   386
   III Dicarboxylate Transport   395
   IV Tricarboxylate Transport   401

V   Summary   407
    Recent Developments   408
    References   408

Chapter 9

THE ROLE OF THE OUTER MEMBRANE IN
ACTIVE TRANSPORT   413
Robert J. Kadner and Philip J. Bassford, Jr.

    I   Introduction   414
    II   Barrier Properties of the Outer Membrane   415
    III Role of the Lambda Receptor in Maltose Transport 426
    IV   Role of the Outer Membrane in Iron Transport   430
    V   Vitamin B_{12} Uptake   442
    VI   Functional Properties of the tonB Product and
         the Possible Role of Membrane Adhesion Sites   447
         References   455

Chapter 10

TRANSPORT OF VITAMINS AND ANTIBIOTICS   463
Robert J. Kadner

    I   Introduction   463
    II   Vitamin Uptake   465
    III Antibiotic Transport   481
         References   487

Chapter 11

BACTERIAL TRANSPORT PROTEINS   495
David B. Wilson and Jeffrey B. Smith

    I   Introduction   495
    II   The Bacterial Proton-Translocating ATPase   498
    III Respiratory Chain Linked Dehydrogenases   521
    IV   Nutrient Transport Proteins   525
    V   Concluding Remarks   546
         Recent Developments   547
         References   548

Chapter 12

ENERGETICS OF ACTIVE TRANSPORT   559
Barry P. Rosen and Eva R. Kashket

I	Introduction	560
II	Development of Early Concepts of Energy Coupling	561
III	Coupling Mechanisms	563
IV	Coupling of Primary Systems to Secondary Active Transport	582
V	Quantitative Aspects of Proton Gradient Coupled Solute Transport	594
VI	Energy Coupling of Transport Systems Sensitive to Osmotic Shock	599
VII	Bacteriocins and Energy Transduction	603
VIII	Avenues of Future Study	608
	Recent Developments	611
	References	612

AUTHOR INDEX   621
SUBJECT INDEX   671

Bacterial Transport

INTRODUCTION

Barry P. Rosen

Why was this book written? A standard question for which there is more-
or-less standard answer: the field has progressed sufficiently rapidly to
justify the appearance of a new text. That is a true and valid answer, but,
rather than simply leaving it at that, I would like to justify the statement
with a brief review of the history of the field. I have divided the development of the thought in the field of bacterial transport into four periods:
(1) discovery (to 1955), (2) physiology (1955 to 1960), (3) biochemistry
(1960 to 1970), and (4) molecular biology (1970 onwards). The time
periods are, of course, approximate, and overlap each other.

Discovery

The third and final edition of Marjory Stephenson's classic text Bacterial
Metabolism [1] appeared in 1949. This text represented a major contribution to the field of microbiology since it demonstrated that the physiology
and metabolism of microorganisms were areas which could be studied in
more than a descriptive fashion. Yet, the subject index contains no entries
on transport or other membrane-related phenomena. There is one short
section (pp. 142-143) on the assimilation of amino acids by bacteria in
which some of the pioneering work on transport is cited, but the point of
permeation as a phenomenon separate from metabolism is missed. As
Gale relates Stephenson's attitude, "...permeability is the last resort of
the biochemist who cannot find any better explanation" [2]. Stephenson's
feelings were not unusual. The first edition of the report of the Carnegie
Institute of Washington on Studies of Biosynthesis in Escherichia coli
concluded "...it does not appear that any mechanism other than simple
diffusion through a highly permeable membrane is involved in the transport
of material across the membrane...the protoplasm may be likened to a
sponge, the cell membrane to a surrounding hair net unable to exclude the
entrance or emergence of small molecules" [3].

1

But within half a decade following the publication of Stephenson's third edition, review articles on the nature of active transport were written by Gale [4] and Mitchell [5]. Further, the conclusions of the Carnegie group were modified in the second edition of their report, a scant 2 years after the first: "Accordingly the simple concept of a freely permeable cell was no longer tenable. The other simple concept of a cell with an impermeable cell wall was equally incompatible with the data ...since it now appears that the cells are neither completely permeable nor completely impermeable, the whole situation must be reconsidered" [3]. By that time there were few who doubted the existence of active transport.

Physiology

The early part second stage is summarized in the reviews by Cohen and Monod [6] and Kepes and Cohen [7]. That period included the work of the Paris school on lactose transport and metabolism. Upon this framework was built a wealth of information about what bacteria transport, including the recognition that a large number of discrete systems exist, each with its own specificity and kinetic characteristics. Within this period also was the conceptual advance of recognizing the genetic basis of transport systems: on the one hand, the isolation of mutants with transport defects, and on the other hand, the ability to induce certain systems cryptic in the absence of inducer. These observations led to the idea that specific proteins, "permeases", were involved in the transport reaction [7].

Biochemistry

Once the thought surfaced that proteins might be components of transport systems, the field expanded rapidly from microbiology to biochemistry. Microbial physiology utilized the intact organism; biochemical characterization required in vitro systems. The major cell-free system for the study of bacterial transport is that devised by Kaback [8]. Without a doubt, the use of isolated bacterial membrane vesicles resulted in a quantum jump in our ability to gather knowledge of bacterial transport. Cross-fertilization has in this case occurred between this area and that of the mammalian transport physiologist: more and more, cytoplasmic membrane vesicle preparations are being utilized for biochemical investigations of mammalian transport systems [9, 10].

Again, with the image of proteins in mind, investigators turned to the isolation of protein components of transport systems. The first major advance was the identification of the M protein as a component of the lactose transport system by Fox et al. [11]. Although the inability to isolate that protein in any sort of active form has been a disappointment, it did provide an impetus to other investigators to search for transport-related proteins.

At about the same time Neu and Heppel demonstrated that an osmotic shock procedure caused the loss of periplasmic proteins [12]. Since such proteins are associated with the cell envelope of Escherichia coli, it was a reasonable next step to look for transport-related proteins in the supernatant fluid from osmotically shocked cells. Piperno and Oxender [13] did just that, discovering that leucine transport activity was reduced by osmotic shock treatment with the simultaneous appearance of a leucine binding protein in the shock fluid. Subsequently, numerous binding proteins related to numerous transport systems have been reported [14]. Even though the exact function of these proteins is still unknown, they were the first proteins definitely known to be components of transport systems to be isolated in an active form. The use of active should be qualified, since it is not possible to assay for an unknown function. As the name applies, however, the binding proteins are active in forming complexes with the substrates of their associated transport systems. None have been found to have associated enzymatic activity, though, and no convincing demonstration of a binding protein functioning in a reconstituted transport system has appeared.

On the other hand, a group of proteins found to comprise an enzymatic complex was subsequently shown to function in a transport capacity. That refers, of course, to the phosphoenolpyruvate-sugar phosphotransferase system discovered by Kundig et. al. [15]. Since the components of the phosphotransferase system could be investigated as enzymes, the system provided a unique opportunity to study a translocating mechanism with biochemical tools. Chapter 2 by Hays discusses the phosphotransferase system in such terms, while Chapter 3 by Saier and Moczydlowski considers it in more physiological terms.

No scientific field is complete without its major guiding light. Thomas Kuhn in his commentary on the evolution of scientific thought [16] suggests that normal science progresses by the accretion of knowledge, leading to the formulation and subsequent acceptance of a paradigm. According to Kuhn, there are two criteria which a hypothesis must meet to be a paradigm: it must be sufficiently unprecedented to attract a following and must be open-ended enough to allow for extensive experimentation, modification, and/or confirmation. Kuhn points out that "...acquisition of a paradigm and of the more esoteric type of research it permits is a sign of maturity in the development of any given scientific field" [16]. The chemiomotic hypothesis proposed by Peter Mitchell [17] is the only major paradigm to have originated in the area of bacterial transport. Little space will be given here to an exposition of the hypothesis, since it is covered in detail in the final chapter of this book.

There are a few points which should be made, though. First, the paradigm was certainly unprecedented as far as the biochemists and microbial physiologists were concerned; so unprecendented, in fact, that it was not taken seriously for a number of years. The idea of forces acting at a distance was vaguely disquieting to the biochemist, who thought in terms of intermolecular interactions. But to the biophysicist and mammalian physiologist, membrane potentials and ion fluxes were everyday events. So from that point of view, Mitchell's formulation was not unprecedented. But the mammalian physiologists were unconcerned with molecules and mechanisms. Mitchell can be credited with a true integration of the purity of thought of the physiologist with the practicality of the biochemist. Although acceptance was slow, especially among the mitochondriologists, Mitchell's ideas were eminently testable and had the tremendous advantage of providing correct predictions. Much of the most recent work in the area of bacterial transport has centered around the ramification of Mitchell's postulates. Even though the hypothesis in itself does not predict molecular mechanisms, the acquisition of the paradigm is right now permitting a more esoteric type of research to be performed. It is certainly a sign of maturity in the development of the field!

Molecular Biology

There are those who would quibble about the use of the term "molecular biology" for an area which does not deal with nucleic acids, protein synthesis, or related matters. But the direction in which current ideas in biological transport are pointing appear to be the understanding of permeation in terms of the molecular mechanisms of the individual components and of the way in which they form supramolecular structures. So, assembly and mechanism will be the major concern in years to come. Already there are numerous laboratories active in the isolation of the intrinsic membrane proteins which are most likely the carriers themselves. The ability to incorporate those proteins into artificial membrane systems provides the means to study both assembly and mechanism. From another point of view, investigations of the energetics in intact cells and membrane vesicles have progressed to the point where meaningful statements can be made in molecular terms.

On the other hand, there is a whole class of transport systems which is not accounted for in Mitchell's terms, namely those linked more intimately with phosphate bond energy, as first described by Berger [18]. It is much too early to think about this type of system in molecular terms; basic biochemical studies are yet to be performed. But it is quite apparent that the future must include a detailed examination of that type of transport system, perhaps leading to the acquisition of a new paradigm of importance equal to that of Mitchell.

In conclusion, this abbreviated guided tour through the development of the basic tenets of the bacterial transport domain was meant to demonstrate the necessity for a text of this sort. Of necessity I have had to pass over many important contributions by many respected colleagues: May they not take offense!

REFERENCES

1. M. Stephenson, Bacterial Metabolism 3rd ed., Longmans, Green and Co., London, 1949, 398 pp.
2. E. F. Gale, J. Gen. Microbiol., 68: 1-14 (1971).
3. R. B. Roberts, P. H. Abelson, D. B. Cowie, E. T. Bolton and R. J. Britten, Studies of Biosynthesis in Escherichia coli (2nd ed.) Carnegie Institution of Washington (Publication 607), Washington, D. C., 1957, 521 pp. (with addendum pages additional).
4. E. F. Gale, in Active Transport and Secretion, Academic Press, Inc., New York, 1954, pp. 242-253.
5. P. Mitchell, in Active Transport and Secretion, Academic Press, Inc., New York 1954, pp. 254-261.
6. G. N. Cohen and J. Monod, Bact. Rev., 21: 169-194 (1957).
7. A. Kepes and G. N. Cohen, in The Bacteria, Vol. 4 (I. C. Gunsalus and R. Y. Stanier, eds.), Academic Press, Inc., New York, 1962, pp. 179-221.
8. H. R. Kaback, Fed. Proc., 19: 130 (1960).
9. J. Hochstadt, D. C. Quinlan, R. L. Rader, C. Li, and D. Dowd, in Methods in Membrane Biology, Vol. 5 (E. D. Korn, ed.), Plenum Press, New York, 1975, pp. 117-162.
10. P. S. Aronson and B. Sacktor, Biochim. Biophys. Acta, 356: 231-243 (1975).
11. C. F. Fox, J. R. Carter, and E. P. Kennedy, Proc. Natl. Acad. Sci. USA 57: 698-705 (1967).
12. H. C. Neu and L. A. Heppel, J. Biol. Chem., 240: 3685-3692 (1965).
13. J. R. Piperno and D. L. Oxender, J. Biol. Chem., 241:5732-5734 (1966).
14. B. P. Rosen and L. A. Heppel, in Bacterial Membranes and Walls (L. Leive, ed.), Marcel Dekker, Inc., New York, 1973, pp. 209-239.
15. W. Kundig, S. Gosh, and S. Roseman, Proc. Natl. Acad. Sci. USA 52: 1067-1074 (1964).
16. T. S. Kuhn, The Structure of Scientific Revolutions, The University of Chicago Press, Chicago, 1962. 172 pp.
17. P. Mitchell, Nature 191: 144-148 (1961).
18. E. A. Berger, Proc. Natl. Acad. Sci. USA 70: 1514-1518 (1973).

Chapter 1

EXPERIMENTAL SYSTEMS FOR THE STUDY OF ACTIVE TRANSPORT IN BACTERIA

Masamitsu Futai*

Department of Botany
Faculty of Science
University of Tokyo
Hongo, Bunkyo-ku, Tokyo, Japan

I.   INTRODUCTION 8
II. TRANSPORT INTO WHOLE CELLS 8
     A. General Remarks 8
     B. Use of Whole Cells in Transport Studies 11
     C. Different Methods of Measuring Transport in Whole Cells. 13
III. PREPARATION AND USE OF MEMBRANE VESICLES 16
     A. General Remarks 16
     B. Membrane Vesicles Prepared by Osmotic Lysis of
        Spheroplasts 16
     C. Membrane Vesicles Prepared by Other Procedures . . . 20
     D. Everted Membrane Vesicles Prepared with a French
        Pressure Chamber 21
IV. ORIENTATION OF MEMBRANE VESICLES 23
     A. Sidedness of Membrane Vesicles Prepared by Osmotic
        Lysis of Spheroplasts 23
     B. Orientation of Other Vesicles Capable of Transport . . . 30
     C. Orientation of Membrane Vesicles Prepared by
        French Pressure Chamber or Sonication 31
V.   CONCLUSION 32
     REFERENCES 34

*Present address: Department of Microbiology, Faculty of Pharmaceutical Sciences, Okayama University, Okayama, Japan

I. INTRODUCTION

Bacterial cells are capable of taking up biologically important molecules
from the surrounding medium. The translocation of solutes across membranes is mediated by transport mechanisms which have been defined as
facilitated diffusion, group translocation, and active transport. The mechanisms of these processes are discussed in previous reviews [1-10] and
the following chapters of this book.
      Translocation of a solute by a specific transport system and the mech-
anism of energy coupling are of major interest. Biochemical studies on
transport began with studies on the uptake of various solutes into whole
cells. Different transport activities and their mechanisms of energy
coupling have been examined with the aid of useful tools such as kinetics,
genetics, and metabolic inhibitors. Isolation of vesicles by Kaback and
his associates has enabled us to study transport phenomena in vitro [11-13].
Vesicles derived from cytoplasmic membranes of bacteria have transport
activities and energy transduction apparati. However, vesicles do not have
all the components of transport reactions and energy production present
in whole cells, as discussed below. Thus, combined studies with membrane vesicles and whole cells are essential for understanding transport
reactions in bacteria.
      In this chapter I will discuss these two experimental systems, espe-
cially their respective advantages. More space is given for discussion of
membrane vesicles, mainly because they are formed from fragments of
plasma membranes of lysed cells. Since Kaback first introduced vesicles
to transport work, various modifications of his procedure and new methods
for preparing vesicles have been developed. Detailed discussion of the
procedures used for fractionation and the orientations of the resulting
vesicles are thus important in analysis of the results obtained using
vesicles. This is also important in considering the structure of membranes. I have tried to write a critical review, and have avoided a comprehensive treatise, citing only limited numbers of references on the
results obtained with each experimental system.

II. TRANSPORT INTO WHOLE CELLS

A. General Remarks

Analysis of the initial rate of uptake and accumulation of solutes into whole
cells has been a logical method for studying transport. The main advantage of whole cells is that their transport components are not damaged or
lost and are present under physiological conditions. Moreover, no special
technique is necessary to obtain whole cells capable of transport. It has
been shown using whole cells that bacteria have developed a variety of

different transport systems: different sugars, amino acids, and ions are transported by different highly specific transport systems. Moreover, a single solute can generally be transported by more than one system. These different transport systems can be distinguished not only kinetically by their substrate specificities and modes of action, but also genetically by isolating mutants defective in one or more transport system [2-10,14,15]. Solute is transported across the cell envelope and accumulates in the cytoplasm. The cell envelopes of bacteria, and especially those of gram-negative bacteria, consist of a complex multilayered structure [16,17]. Escherichia coli and Salmonella typhimurium, which have been widely used in transport studies, have cytoplasmic membranes surrounded by peptidoglycans and outer membranes.

Important progress in studies with this experimental system was achieved using the osmotic shock procedure [3,18-24]. In this procedure gram-negative bacteria are subjected to a sudden change of osmolarity of the medium [18,19]. This results in release of binding proteins into the medium with concomitant loss of specific transport activities [20-22]. These proteins are believed to be released from the periplasm, which is a space between the outer and cytoplasmic membranes [22,24-26]. A large body of evidence suggests that binding proteins have a significant role in active transport [3,9,24]. The interaction of histidine binding protein with component(s) of the cytoplasmic membranes of S. typhimurium has been suggested by the results of genetic studies [27,28]. It would be very helpful if transport systems sensitive to osmotic shock could be reconstituted from binding protein and membrane vesicles. Whole cells are good experimental systems to study transport activities which are sensitive to osmotic shock, because membrane vesicles prepared by the method of Kaback have no binding proteins and shockable transport activities seem to be lost during fractionation [12,13]. Bacteria have developed other active transport activities which are insensitive to osmotic shock. Components of these systems are only found in cytoplasmic membranes [12,13]. Major components of phosphotransferase systems which catalyze group translocation of sugar moieties accompanied with phosphoenol pyruvate-dependent phosphorylation are also found in cytoplasmic membranes [11, 13].

Increasing evidence suggests that the outer membrane also constitutes a barrier to certain kinds of molecules. Moreover, the inward penetration of molecules such as vitamin B_{12} [29], maltose [30,31], and iron [32,33] is mediated by specific transport systems that are located in the outer membranes, whereas most low molecular weight solutes seem to pass through pores in the outer membranes [34,35].

Whole cells have all the cytoplasmic components which themselves are capable of synthesizing phosphate bond energy and transferring the energy to another form through respiration or hydrolysis of ATP. Energy metabolism in the cytoplasm and cytoplasmic membranes can be controlled either biochemically by using cells which have been depleted of energy reservoirs followed by addition of an energy source or genetically by using mutants with defects in energy transformation, such as ATPase-negative strains. Introducing these approaches, Berger and Heppel [36,37] and other workers [38-44] have shown that active transport activities that are sensitive and insensitive to osmotic shock, are coupled to ATP or phosphate bond energy and to an energy-rich membrane state, respectively. These studies were only possible using whole cells, because shockable transport activity and glycolysis with ATP formation are not present in membrane vesicles.

The effect of metabolism on the transport reaction can be largely reduced by rapid assay of transport. However, some metabolic reactions in the cytoplasm may change the rate of transport and modify the solute transported [4,45,46], thus complicating the interpretation of results on the mechanism of transport reactions. To avoid such complications, isolation of transport reactions has been performed genetically using mutants and biochemically using nonmetabolizable analogs. The techniques used for studies on transport of amino acids and sugars have been discussed [4,46]. Measurement of the initial rate of formation of modified solute is also a unique way of assaying transport in whole cells, as discussed below.

It is often desirable to identify and characterize the components of membranes of transport systems studied. An important advance was the characterization by Fox and Kennedy [47] of "M protein" from cells induced for β-galactoside transport. Fox and Kennedy found that this protein contains the binding site of β-galactoside, although they purified it in an inactive form [47]. Identification of transport protein has been facilitated by the recent developments in one- and two-dimensional electrophoreses [48-51]. These procedures have been useful for identification of periplasmic components of the transport systems of glycerol-3-phosphate [52] and histidine [27,28] from shock fluid. The technique of two-dimensional electrophoresis recently developed by O'Farrell [49] and Ames and Nikaido [51] seems to be particularly useful. It involves solubilization of membrane protein with dodecyl sodium sulfate, and then electrophoresis in the first dimension according to charge and in the second dimension according to molecular weight. By this procedure about 150 different proteins from the cell envelope and 1100 proteins from whole cells were separated by Ames and Nikaido [51] and O'Farrell [49], respectively. Thus even the transport component(s) of whole cells may be identified by this procedure.

I have briefly discussed studies on transport using whole cells as experimental systems. The advantages of using whole cells are evident from the results obtained. For details of these results the reader is referred to other chapters of this book. To avoid overlapping with other chapters further discussion here is limited to two fundamental aspects of whole cells as experimental systems: use of whole cells for transport studies and procedures of measuring transport in whole cells.

B. Use of Whole Cells in Transport Studies

1. Energy Depleted Cells

Oxidative phosphorylation and glycolysis are the major mechanisms of energy transduction in cells. However, not all bacteria have both mechanisms. For instance, Streptococcus faecalis [6, 53] and Streptococcus lactis [6] do not have a respiratory chain and cannot carry out oxidative phosphorylation. They have low endogenous energy reserves. Cells harvested from growing medium are depleted almost completely of endogenous energy and are unable to take up solute unless an energy source such as glucose or arginine is added. All reactions dependent on energy ultimately depend on ATP produced via glycolysis or catabolism of arginine to ornithine. Therefore, these cells have been used in studies on the coupling of energy with active transport. Clostridium perfringens probably also lacks a respiratory chain [6, 54]. Anthrobacter pyridinolis is another example of an organism with low energy stores [55], although it has a respiratory chain.

Most bacteria have more complicated patterns of energy metabolism. E. coli, which has been widely used in transport studies, can grow aerobically using energy from oxidative phosphorylation, or anaerobically using energy from either fermentation or redox pathways [6]. Most bacteria also have stores of high energy compounds such as ATP, the concentration of which seems to be controlled within a constant range [56]. The intracellular concentrations of ATP in washed cells of E. coli and S. lactis are about 3 and 0.1 mM, respectively [57]. Therefore, freshly harvested cells of E. coli show a high rate of endogenous transport which tends to mask the effect of added energy sources on transport.

Several procedures have been developed to deplete E. coli cells of energy stores. Koch [58] did this by inducing the glucose permease in the cells and then making them transport α-methylglucoside to exhaustion in the absence of a utilizable energy source and in the presence of azide. α-Methylglucoside and thiomethyl-β-D-galactoside were taken up only slowly by these cells, although they recover the full transport ability with the addition of glucose. However, using this procedure, Berger [36] observed that the inhibitory effect of azide on transport of amino acids could not be completely reversed. In a modification of this procedure azide was replaced by sodium arsenate and sodium cyanide [57]. This treatment reduced the ATP level in the cytoplasm by more than 98% in wild-type and ATPase-negative strains.

Recently, Berger [36] and Berger and Heppel [37] reported another starvation procedure in which cells were incubated with 2,4-dinitrophenol and then washed extensively before assay. This treatment markedly reduced the endogenous rate of active transport and the ATP level, but did not affect glycolysis, respiration, or transport systems for eight amino acids [37]. Thus, these cells took up the amino acids on addition of an energy source at essentially the same rate as untreated cells. These workers also starved cells by aeration in the absence of a carbon source [37,59]. This procedure was not as effective as treatment with dinitrophenol, but gave qualitatively similar results on the sources of energy for active transport [37]. It is noteworthy that ATPase mutants were much more sensitive to dinitrophenol than wild-type cells: the optimal period of treatment with dinitrophenol was 10 hr with the wild type and 1 hr with the ATPase mutants [37]. Thus, optimal conditions may vary for different strains. With minor modification this procedure was used for other transport systems [39-44]. Recently, E. coli cells starved in this way were shown to accumulate β-galactoside [60] and to synthesize ATP [57,61] in response to an artificial proton motive force.

Kobayashi et al. [38] developed a different procedure for starvation: cells were incubated with sodium arsenate in tris(hydroxymethyl)amino-methane (Tris) buffer containing magnesium and then washed extensively with potassium phosphate. Incubation with arsenate for 40 min was enough with both the wild-type and ATPase-negative mutant. The endogenous level of ATP and other respiratory substrates was found to be low enough not to support transport of amino acids. Thus, these cells were also useful systems for the study of energy requirement.

2. Osmotically Modified Cells

The osmotic shock procedure is useful in studying transport systems of gram-negative bacteria [4,20-22,24]. In this technique cells are first suspended in 20% sucrose containing ethylenediamine tetracetic acid (EDTA) and Tris buffer. The suspension is centrifuged and the pelleted cells are rapidly transferred to 0.5 mM $MgCl_2$ solution at 4°C [20,21]. The shocked cells thus obtained have lost solute-specific binding proteins and do not show activities for transport of galactose, leucine, glutamine, cysteine, and arginine [3,4,21,24]. Transport activities of shocked cells can be assayed by essentially the same procedure as that of untreated cells because the shocked cells are resistant to osmotic pressure and viable [20]. It was found that the shocked cells retained activities for transport of proline, phenylalanine, lysine, etc. The osmotic shock procedure and binding proteins have been extensively reviewed [21,23,24,62], and they are also discussed elsewhere in this book. It should be mentioned here that this procedure should not be used uncritically. Proper controls should be used, such as transport systems which are definitely known to be shockable or nonshockable. Release of known periplasmic enzymes is another control. It is also noteworthy that sensitivity to osmotic shock depends on the strain and growth conditions [22,62]. The cystine-specific transport system of E. coli is entirely unaffected by osmotic shock when cells are grown in enriched medium [63]. Similar results have been obtained on lysine transport [64]. These examples suggest that sometimes there is no clear distinction between membrane bound and soluble binding protein.

Gram-negative cells are impermeable to certain hydrophobic inhibitors and nucleotides. So, for transport studies, procedures are required for making the cells permeable to these compounds. Leive [65] developed a useful procedure for doing this: she washed cells with Tris buffer and then treated them with EDTA. After carefully controlled treatment in this way E. coli cells became completely sensitive to actinomycin D, although their activities for transport of proline and methyl-β-glucoside were not affected. This procedure can also be used for Aerobacter aerogenes and S. typhimurium. Weiner and Heppel [59] observed that transport of glutamine became more sensitive to N,N'-dicyclohexylcarbodiimide (DCCD), an inhibitor of ATPase in the membranes, after this treatment, although transport activity itself remained unchanged. Rosen confirmed this observation [66] and also determined that DCCD treatment of an EDTA-treated culture of a proton-permeable Mg^{2+}-ATPase mutant restored the membrane to a proton-impermeable state [67]. It was also reported that cells became permeable to nucleoside triphosphate after this treatment [68,69], although the rate of uptake seemed to be quite low. Nucleoside triphosphate could be introduced into cells by washing them with cold Tris buffer containing these compounds [70,71]. On the other hand, shocked cells did not take up significant amounts of radioactive ATP [72], although these cells became unusually sensitive to actinomycin D [20].

Plasmolysed cells were useful in studies on passive diffusion of non-electrolytes through the outer membrane, since their periplasmic space is wider than that of control cells [34,35]. Spheroplasts have sometimes been used with an osmotic stabilizer [73-76]. Osmometric properties of spheroplasts induced by transported solute have been reported [73,74]. These cells have only nonshockable transport systems, since binding proteins are lost during spheroplast formation [24].

C. Different Methods of Measuring Transport in Whole Cells

Uptake of solute by whole cells has been measured by various procedures, some of which are only applicable to whole cells. The initial rate of uptake of solute is usually measured using radioactive substrate, by determining the amount of label in cells, separated after various times from the medium. Membrane filters have been most widely used for separating the

cells [4]. Usually, uptake of solute is started by addition of radioactive
substrate to the cell suspension, and 15 to 60 sec later samples of the
incubation mixture are rapidly filtered through membrane filters. The
temperature and composition of the washing medium affect retention of
accumulated solute [22, 77]. Usually medium of essentially the same composition and temperature as the incubation medium is used for washing
cells [77]. In assay of passive entry of solute, the membrane filters cannot
be washed. With slight modifications this procedure can be used for
examining uptake by membrane vesicles. Transport can also be measured
by placing cells on the filter and adding labeled substrate to them [78].
Although isotopes have been most widely used, solutes such as potassium
ion have also been assayed quantitatively by flame photometry [81].

The rate of efflux of a solute from cells can be assayed by essentially
the same procedure [45, 77, 79-86]. Usually cells with accumulated radioactive solute are diluted with a large excess of medium, and appropriate-
sized samples are filtered at various times after dilution.

Uptake of solute can be assayed by measuring accumulation of radioactive substrate in cells separated by centrifugation [87, 88]. Nikaido and
his associates [34, 35] determined the size of pores in the outer membranes
of E. coli and S. typhimurium by this procedure. They measured the space
permeable to a series of sugars in plasmolysed cells or isolated outer
membranes and concluded that outer membranes allow passive penetration
of saccharides with molecular weight of less than 666. Centrifugation assays
can be used for more detailed kinetic studies. The uptake of Ca^{2+} and
3-hydroxybutyrate into mitochondria were assayed by separating samples
of mitochondria by a centrifugation procedure which took 20 sec [89]. The
time course of uptake of the solutes longer than 20 sec after their addition
were plotted, and the results showed stoichiometry of transport of both
solutes. The value for radioactivity in the pellet was corrected for that in
extramatrix material, since the pellet was not washed with buffer. Assays
using centrifugation have the advantage that large quantities of cells or
mitochondria can be used, although it takes more time than assays using
membrane filters.

Centrifugal filtration of a reaction mixture through silicon oil may be
used in studies on bacterial transport. This method was devised for the
measurement of the steady-state concentration of internal solutes in
mitochondria [90-93]. Recently, McCarty and his associates [94-98]
modified the original procedure for study of the uptakes of hexylamine,
aniline, and ammonia into chloroplasts. The reaction mixture was placed
in a microcentrifuge tube containing trichloroacetic acid solution overlain
with silicon fluid [95, 98]. The tubes were centrifuged after appropriate
times of incubation. The chloroplasts precipitated into the trichloroacetic
acid solution within 7 to 8 sec [98]. Uptake of [^{14}C]hexylamine determined
by this procedure gave a precise estimation of the light-induced proton concentration gradient [98]. Winkler et al. [99] discussed the advantage of

using membrane filters, standard centrifugation, or centrifugal filtration through silicon oil in assays of the ADP/ATP transporter in mitochondria.

Measurement of metabolite formed from transported solute could be a unique assay procedure for transport in whole cells. The most widely used assay of this criterion is an assay of β-galactoside transport using 0-nitrophenyl-β-D-galactoside (ONPG) [100]. The transported solute can be measured as the absorbancy of nitrophenol liberated from ONPG by β-galactosidase in the cytoplasm. A correction can be made for entry of ONPG into cells via mechanisms not mediated by the lactose transport system by observing the amount of hydrolysis in the presence of 1-thio-β-D-digalactoside [101,102]. Studies on rapid kinetics are also possible using stop-flow apparatus [103]. Transport of ONPG was suggested to be thermodynamically "downhill" since excess intracellular β-galactosidase cleaves ONPG as rapidly as it enters the cell. This downhill cellular hydrolysis of ONPG no longer takes place in cells depleted of an energy reserve [58]. Treatment of energy-depleted cells with a proton conducting agent (m-chlorophenylhydrazone) resulted in stimulation of ONPG transport [104]. Cecchini and Koch [104] called this transport a quasi-facilitated diffusion system consisting of a lactose carrier and a proton conductor. These results suggest that facilitated diffusion carries protons along with ONPG, and this creates a reverse potential that stops the process in energy-depleted cells. Either respiration or action of an uncoupler extrudes protons resulting in the stimulation of ONPG transport.

Usually chloramphenicol is present in the incubation mixture to inhibit protein synthesis during the transport assay. However, the rates of incorporation of some amino acids into protein may be greater than their rates of transport into growing cells, and thus if other metabolic conversions are negligible, the rate of incorporation of amino acids into protein can be used as a measure of their rates of transport. Ames [105] measured transport of histidine and other aromatic amino acids in S. typhimurium in this manner. Other metabolic conversions, such as decarboxylation, can be used for transport assay. Rosen [106] measured evolution of [^{14}C]CO$_2$ from [^{14}C]arginine by decarboxylase.

Flow dialysis [107], recently introduced in this field, has been useful for the study of binding of dansylgalactoside to membrane carriers [108] and measurement of proton gradients generated by membrane vesicles [109].

Measurement of pH change after addition of solute can also be used as an assay of transport in some cases. It was shown that the pH of the medium of an anaerobic suspension of E. coli changed on addition of thiomethyl- -D-galactoside [67,110]. Later stoichiometric studies on initial rate of uptake showed that one proton moved with the transport of one molecule of lactose [111]. Although not discussed here in detail, the rate of transport can also be measured by swelling of cells [112] or spheroplasts [73].

Sometimes it is necessary to examine the percentage of membrane vesicles or cells in the whole population showing active transport. A very good electron microscopic autoradiographic technique using [^{14}C]vinyl-glycolate was recently developed for this purpose [113]. This compound is transported by the lactate transport system and is converted to 2-keto-3-butenoate in membrane vesicles. Since 2-keto-3-butenoate binds covalently to membrane protein, the vesicles which have transported vinylglycolate can be detected by electron microscopic autoradiography.

III. PREPARATION AND USE OF MEMBRANE VESICLES

A. General Remarks

As discussed above transport can be studied using whole cells. Compli-cations due to the presence of cytoplasm can be reduced by using biochemi-cal and genetic procedures. However, cells have to be incubated under rather drastic conditions to deplete their energy reserves. It is also not easy to exclude the effects of catabolism and intracellular pools of solute on the results. To overcome these difficulties Kaback and his colleagues [11-13] prepared cytoplasmic membrane vesicles by disrupting osmotically sensitive spheroplasts in hypotonic buffer. Most of these vesicles have been claimed to have the same orientation as intact cells. These vesicles appear to retain activities of the major active transport systems, com-ponents of phosphotransferase systems, and also to contain electron carriers of the respiratory chain. They respond to an added energy source, and this energy drives transport of amino acids and sugars. These vesicles have been very useful in studying the coupling of energy with solute translocation [11,12]. Sugars are also transported into the vesicles by a phosphotransferase system on addition of soluble components [13]. The preparation of vesicles with similar properties have been reported [114-116]. Everted membrane vesicles with orientations opposite to intact cells have also been prepared [117-119]. The mechanism of secretion of solutes from cells can be studied as uptake into these vesicles [119]. The preparation and uses of different types of membrane vesicles are outlined below.

B. Membrane Vesicles Prepared by Osmotic
Lysis of Spheroplasts

1. Preparation

There are extensive reviews on the preparation [120] and properties [11-13] of membrane vesicles prepared by the method of Kaback. Membrane vesicles can be prepared by the following three steps.   (1) The conversion of the bacteria into osmotically fragile spheroplasts or protoplasts. Treatment of bacteria with EDTA-lysozyme has been used most frequently for this, and organisms in the logarithmic phase seem most suitable. We found that spheroplasts could be obtained reproducibly from physiologically young cells, as reported by Birdsell and Cota-Robles [121]. Penicillin treatment is another means of obtaining osmotically sensitive cells [120]. Lysostaphin has been used for Staphylococcus aureus [122,123]. The procedures used for making spheroplasts and protoplasts have been reviewed [124,125]. (2) In the second step the pellet of spheroplasts is suspended in a minimal volume of potassium phosphate buffer containing 20% sucrose, 20 mM MgCl$_2$, and nucleases using a Teflon or glass homogenizer. This suspension is then diluted with 300 to 500 vol of 0.05 M potassium phosphate and incubated at 37°C. This step is rather difficult, because the pellet of spheroplasts is viscous and hard to suspend and so preparations made in different laboratories may vary. (3) The membrane fraction obtained by removal of whole cells is washed with potassium phosphate buffer containing EDTA. Quantitative removal of whole cells or unlysed spheroplasts by centrifugation through a sucrose gradient has been reported [120], although careful centrifugation at low speed is usually sufficient.

This procedure has been applied to E. coli [126,127], S. typhimurium [122], Bacillus subtilis [122], S. aureus [122,123], Pseudomonas putida [122], Bacillus megaterium [122], Micrococcus denitrificans [122], Azotobacter vinelandii [128], Veillonella alcalescens [129], A. pyridinolis [130], Klebsiella aerogenes [131], Thiobacillus neapolitanus [132], and a thermophilic bacterium [133].

The properties of the membrane vesicles and especially those of E. coli have been studied extensively [11-13] and can be summarized as follows. (1) The vesicles prepared from E. coli ML 308-225 are of simpler composition than whole cells, having lost the outer membranes and cytoplasmic and periplasmic components. Thus, metabolic reactions in them that interfere with attempts to define transport processes are greatly reduced. It is noteworthy that vesicles prepared from other strains of E. coli [120] or similar preparations from S. typhimurium [134] contain significant amounts of outer membrane. (2) The vesicles retain the basic parameter of transport, and are thus capable of active transport and group translocation. The uptake of sugars such as glucose, fructose, and mannose is due to the phosphotransferase system [11], while other sugars, amino acids, and dicarboxyloc acids are taken up by active transport systems [13] with energy supplied either by the respiratory chain or an artificial proton motive force. With vesicles from E. coli, D-lactate is the most effective respiratory substrate for stimulating active transport. (3) The preparations show almost the same specificities for sugar and amino acid transport as the intact cells from which they are derived. For example, membranes from a mutant organism defective in the transport of

a particular solute are also defective in transport of this solute [135,136].
From these properties it is evident that vesicles offer many advantages
over whole cells for use in studying solute transport process.

   The following observations should be taken into account when interpreting results obtained with membrane vesicles. (1) There are still
questions about the sidedness of vesicles. These are discussed in detail
later. (2) Vesicles are capable of at least limited metabolism, such as
formation of phosphatidylserine using ATP [137]. (3) Vesicles contain low
levels of phosphoenolpyruvate and enzyme I [7]. (4) The rate of transport
by vesicles has been reported to be less than that of intact cells [7,14].
Lombardi and Kaback [138] compared the rates of uptake by vesicles and
control cells and found that vesicles retained 42 to 102% of the uptake
activities of control cells for seven amino acids and lactose. However, it
must be noted that control cells treated by the standard procedure for
preparations of membranes except for omission of treatment with lysozyme
had significantly reduced activity of lactose transport. Loss of some
transport systems, such as that for glutamine in the vesicles, can be
explained by the observation that these systems are sensitive to osmotic
shock [59]. (5) Although transport of amino acids and sugars by vesicles
is stimulated by energy from respiration, not all respiratory substrates
are equally effective [11-13]. Furthermore, no correlation was found
between the abilities of substrates to stimulate transport and their rate
of respiration [139]. In this connection Ramos et al. [109] showed recently
that not all respiratory substrates could produce a membrane potential and
proton gradient upon uptake of oxygen.

2. Modification of the Procedure

Many useful modifications of the original procedure for making vesicles have
been developed. In the original procedure a motor-driven plunger or Potter-
Elvejhem homogenizer was used for thorough suspension of the membranes
[120]. However, Short et al. [140] reported that this homogenization procedure caused dislocation of ATPase from the inner surface of membranes
of E. coli to their outside. They suggested that to avoid this, membranes
should be dispersed with a syringe equipped with a hypodermic needle.
The significance of this modification is discussed below. Membrane
vesicles have also been prepared using sodium phosphate instead of
potassium phosphate [136]. The resulting vesicles appeared to lose much
more structural integrity and the capacity to transport on freezing and
thawing [141].

   As discussed above, the procedure includes dilution of spheroplasts
with 300 to 500 vol of hypotonic buffer and extensive washing with buffer
containing EDTA. This method did not give vesicles capable of active
transport from cells grown anaerobically with nitrate or fumarate [142].
An alternative procedure was developed for cells grown anaerobically

[142]: 2 g of spheroplasts were rapidly dispersed in 10 ml of phosphate buffer containing nucleases, and the membrane fraction obtained by removal of unbroken cells had transport activity dependent on glycerol-3-phosphate or D-lactate utilizing nitrate or fumarate as an electron acceptor.

When the original procedure was applied to organisms such as B. subtilis that excrete protease, the membrane vesicles obtained were labile and lost their activity within a few hours [143]. To overcome this difficulty Konings et al. [143] treated organisms with lysozyme and EDTA in hypotonic medium. In the modified procedure membrane vesicles were formed immediately after partial hydrolysis of cell walls. These vesicles were more stable and could transport glutamate at a rate comparable to intact cells. Recently, Hellingwerf et al. [144] prepared vesicles from Rhodopseudomonas spheroides by maintaining the redox potential of the suspension between 0 and 100 mV. Cyclic electron flow drove amino acid uptake in these vesicles. Membrane vesicles were prepared from Pseudomonas aeruginosa after lysis of cells treated with LiCl and lysozyme [145]. The use of EDTA and Tris damaged the transport activity in this organism.

One advantage of vesicles is that the function of transport systems can be modified by agents which affect the composition of the membranes. Treatment of vesicles with phospholipase D and chaotropic agents has been reported [146-148]. Vesicles treated with guanidine-HCl showed increased proton permeability, and this could be restored to the original level with DCCD [145]. Patel and Kaback [149] obtained essentially the same results on a DCCD-resistant mutant, except that vesicles from this strain require a water soluble carbodiimide for reversal. Similar experiments have also been reported on everted vesicles [150] and electron transfer particles from Mycobacterium phlei [151].

3. Membrane Vesicles Containing Incorporated Compounds

For studies on the mechanism of transport or oxidative phosphorylation, it is sometimes desirable to use membrane vesicles containing known compounds. Bacterial cytoplasmic membranes are generally impermeable to nucleotides such as NADH, ADP, and cyclic AMP. Thus it is useful for preparing membrane vesicles containing these compounds. These vesicles can also be used for studies on efflux, if they contain a compound which has a transport system.

Futai [152] prepared membrane vesicles of E. coli which generated NADH inside. For this, spheroplasts were lysed in the presence of NAD and alcohol dehydrogenase, and the resulting membrane vesicles containing these compounds inside were washed with potassium phosphate buffer. The NADH formed inside these vesicles upon addition of ethanol stimulated uptake of oxygen, and this respiration in turn stimulated transport of amino

acids. These results were confirmed by Tucker and Lilligh [153] using
M. denitrificans. Recently, Tsuchiya and Rosen [154] prepared membrane vesicles loaded with ADP by a similar procedure. These vesicles
responded to a proton motive force formed by respiration or an artificial
procedure, and the ADP inside was phosphorylated, although the P:O ratio
was low. These findings suggested that ADP or NADH must be inside the
vesicles to be used properly. Saier et al. [155] preloaded membrane
vesicles with cyclic AMP by a similar procedure. Efflux of this nucleotide from the vesicles was stimulated by DL-lactate and phenazine methosulfate plus ascorbate. This stimulatory effect was largely blocked by an
uncoupler, suggesting that energy was coupled to the efflux. Membrane
vesicles preloaded with potassium ions have been prepared by incubating
vesicles with potassium phosphate at 40° C [136]. This procedure was
based on the observation that vesicles became leaky at higher temperature
[156].

C. Membrane Vesicles Prepared by
    Other Procedures

1. Inner Membrane Vesicles and Outer
    Membrane Vesicles

Miura and Mizushima [157] first described the separation of the cell envelopes of E. coli into two fractions, consisting predominantly of cytoplasmic
and outer membranes. Their techniques involved isopycnic sucrose
gradient centrifugation of crude membranes obtained by lysis of spheroplasts. White et al. [158] reported a similar preparation obtained after
disrupting E. coli in a French pressure chamber. Osborn et al. [134]
developed a modification of the procedure of Miura and Mizushima [157],
which appears to be applicable to most gram-negative bacteria. None of
these workers studied the transport of solutes into the cytoplasmic membrane vesicles obtained in their preparations, except that Mizushima [159]
recently showed uptake of proline in his preparation.

   Recently Yamato et al. [115] modified this procedure and obtained
cytoplasmic membrane vesicles from E. coli K12 capable of taking up
amino acids. They compared the initial rate of transport in their vesicles
with that of vesicles prepared by the method of Kaback [120]. The rate of
uptake of proline by their vesicles was higher than that by Kaback vesicles.*
Contamination of the cytoplasmic membrane fraction with outer membranes
was negligible. They also observed that isoleucine was taken up by their
vesicles but not by Kaback vesicles.

   Outer membranes prepared from S. typhimurium and E. coli have
been used to study the penetration of oligo- and polysaccharides [34, 35].

---

*I. Yamato and Y. Anraku, personal communication.

The procedures for preparation of vesicles discussed so far take time. Joseleau-Petit and Kepes [160] fractionated bacteria cell envelopes by rapid electrophoretic fractionation. It is of interest to know if cytoplasmic membrane vesicles thus obtained have activity to transport amino acids and sugars.

2. Vesicles Prepared from <u>Mycobacterium</u> <u>phlei</u> and <u>Halobacterium halobium</u>

Brodie and his colleagues [161-169] studied oxidative phosphoration and active transport on ghosts and electron transfer particles obtained from M. pheli. They prepared the ghost fraction by lysing osmotically sensitive cells in dilute buffer. These ghosts could transport amino acids on addition of succinate [163], although their activity for oxidative phosphorylation was low, presumably because of low permeability to adenine nucleotides [165].

Electron transfer particles (membrane vesicles) were prepared from M. phlei by sonication and then fractionation by differential centrifugation [161]. The properties of these vesicles were studied extensively and were as follows. (1) Uptake of proline by these vesicles stimulated by ascorbate plus N, N, N', N'-tetramethyl-p-phenylenediamine was about 5% that of ghosts. Succinate did not stimulate transport by electron transfer particles, although it stimulated transport by ghosts as much as did artificial electron donors [114,163]. On the other hand, succinate induced efflux of accumulated proline from the particles [163], but not from ghosts. (2) These particles showed a high rate of phosphorylation coupled to oxidation of NADH [165]. These results suggest that the particles are quite different from ghosts. The properties of these vesicles are discussed later in relation to the question of sidedness of membranes.

Envelope vesicles of H. halobium [116] were prepared from cells disrupted by brief sonication in the presence of nuclease. The membrane fraction was obtained by removal of large fragments and the soluble fraction by differential centrifugation. These vesicles took up leucine upon illumination, suggesting that the proton motive force formed by bacterio-rhodopsin stimulated transport of amino acids.

D. Everted Membrane Vesicles Prepared with a French Pressure Chamber

As discussed above, it is useful to have a procedure for preparing vesicles with the opposite orientation to that of intact cells. Hertzberg and Hinkle [117] reported that vesicles prepared by lysis with a French pressure chamber catalyze oxidative phosphorylation. Moreover, they translocated protons to the inside during respiration or hydrolysis of ATP. These results suggest that vesicles prepared by the French pressure chamber are everted. Futai [118] made similar vesicles and obtained evidence for inversion from studies on the location of enzymes. These vesicles have other vectorial activities consistent with this suggestion. (1) Futai et al. [170] showed that these vesicles catalyzed aerobic or ATP-driven trans-hydrogenase. Both activities were lost on washing the vesicles with dilute EDTA to remove ATPase, but they could be restored by mixing washed vesicles with purified ATPase. These results confirmed the idea that ATPase has a structural role in maintenance of the energized state of membranes [171]. These findings also suggest that EDTA treatment did not change the everted orientation of the vesicles. (2) Rosen and McClees [119] showed that these vesicles accumulated $^{45}Ca^{2+}$ dependent on energy supplied by respiration, hydrolysis of ATP, or an artificially imposed proton motive force. These findings are consistent with the suggestion that E. coli may contain a transport system that is responsible for active extrusion of calcium [172]. ATPase has also been shown to have structural role in this system [173]. (3) These vesicles could not transport proline depending on respiration [118,119]. These studies suggest that membrane vesicles prepared by the French pressure chamber are useful for studies on oxidative phosphorylation and the mechanism of extrusion of solutes from cells.

In preparing the vesicles by passage through a French pressure chamber, the pressure of the chamber is the most critical factor, and it is advisable to titrate the pressure against the activity studied. Tsuchiya and Rosen [174] measured the activity of calcium uptake in vesicles prepared using various pressures and found that at pressure above 4,000 psi the activity of calcium uptake decreased, while recovery of membrane protein per unit wet weight of cells increased. It should be noted that to obtain biologically active vesicles it was necessary to add protective reagents such as glycerol and dithiothreitol [117,119].

This procedure has been used to evert membrane vesicles prepared by the method of Kaback [120]. Mével-Ninio and Yamamoto [175] reported that passage of membrane vesicles of E. coli which catalyzed active transport but not oxidative phosphorylation through a French pressure chamber yielded vesicles which catalyzed oxidative phosphorylation, although the P:O ratio of these vesicles was much lower than that of vesicles obtained by passing whole cells through the pressure chamber. Using essentially the same procedure on vesicles, Houghton et al. [176] obtained vesicles having 8 to 10-fold higher activity than the original vesicles for trans-hydrogenase activity dependent on respiration or hydrolysis of ATP. Again, the pressure seems to be critical in obtaining everted vesicles from right-side-out vesicles, since Kaback and Deuel [177] observed active transport in vesicles which had been passed through the pressure chamber.

IV. ORIENTATION OF MEMBRANE VESICLES

A. Sidedness of Membrane Vesicles Prepared
   by Osmotic Lysis of Spheroplasts

1. Evidence for a Strictly Right-side-out
   Orientation

Kaback and his collaborators claimed that vesicles prepared from E. coli
ML308-225 by the published procedure [120] were strictly right-side-out,
i.e., that they had the same orientation as the cells from which they are
derived. Results in support of this are: (1) by electron microscopy,
vesicles after freeze fracture appeared to have essentially the same
structure as intact cells [12]; (2) Short et al. [113] found by electron
microscopic autoradiography that almost all the vesicles in the preparation
incorporated [^{14}C]vinylglycolate, suggesting that most of the vesicles
catalyzed active transport; and (3) antibody against D-lactate dehydrogenase did not inhibit D-lactate-dependent respiration or transport of proline
[140]. Altendorf and Staehelin [141] confirmed the observation by freeze-
fracture electron microscopy using freshly prepared vesicles from E. coli
ML308-225. However, when vesicles were frozen and thawed as described
by Kaback [120], about 25% of them became everted, although this treatment did not cause appreciable loss of activity for sugar or amino acid
transport. This suggests that the portion of everted vesicles was very
small, and its volume was insignificant in terms of transport. Similar
results have been obtained by Konings et al. [143] with vesicles from
B. subtilis: they reported that only 15% of the vesicles had the opposite
orientation. Recently Futai and Tanaka [178] confirmed that most of the
activity of D-lactate dehydrogenase in the membrane preparation was
inaccessible to antibody against this enzyme.* However, they observed

---

*It seems relevant here to comment on the immunochemical procedures
which have been widely used in studies on the locations of enzymes in membranes. If a certain membrane enzyme is not inhibited by its specific antibody, we are tempted to conclude that it is not on the outside of vesicles.
However suitable controls are needed for interpretation of results. As discussed above, D-lactate dehydrogenase was concluded to be inside membrane vesicles, because the enzyme in inverted vesicles or reconstituted
vesicles was completely accessible by the antibody [140,178]. However, it
is not always possible to obtain suitable antibody. The purified membrane
protein used as antigen is usually in detergent solution. Thus it is quite
possible that resulting antibody is produced against the portion of the enzyme
normally buried in the membrane but exposed by detergent. This antibody
will therefore not react with protein in the membrane, although it will

(Continued on page 24.)

that a small but significant amount (about 15%) of the activity was accessible
to the antibody. This may all be due to everted vesicles, and the difference
in the amount of inverted vesicles may be due to the slightly different conditions used in different laboratories. Thus the orientation of vesicles
claimed by Kaback seems to be correct, at least for most of the vesicles
in a fresh preparation. These vesicles showed movement of protons to the
outside on addition of D-lactate [179]. The direction of this movement was
the same as that in whole cells. However, this cannot be regarded as
proof that vesicles are strictly right-side-out, because the efficiency
(H^+:O) is much lower than that of whole cells.

2. Location of ATPase in Vesicles

Studies on the location of individual membrane enzymes have shown that
question of sidedness is very complex. Harold [5] pointed out in his review
that right-side-out vesicles would not be expected to oxidize NADH or to
hydrolyze ATP, although vesicles prepared by the method of Kaback are
capable of these functions. Oppenheim and Salton [180] studied the location
of ATPase in vesicles prepared from Micrococcus lysodeikticus by lysis of
spheroplasts in dilute magnesium solution. Ferritin-labeled antibody was
shown to become attached to the outside of some vesicles, although it did
not bind to intact spheroplasts. Thus this internal enzyme became detectable on the outside surface of some vesicles. Their procedure for preparing membrane fraction was not exactly the same as that of Kaback [120],
but it is important to note that osmotic lysis of spheroplasts in hypotonic
solution yielded two kinds of vesicles with ATPase in different locations.
Gorneva and Ryabova [181] studied ion uptake and the location of ATPase
in vesicles prepared from M. lysodeikticus and reached the same conclusion as Oppenheim and Salton [180].
     Studies on membrane vesicles prepared from E. coli ML308-225 have
suggested that the location of ATPase is not the same as that in the original
cell membranes. Normally the cytoplasmic membrane is rather impermeable to nucleoside triphosphate, while the outer membrane is readily
permeable to it [182]. This was confirmed by Futai [118]: spheroplasts
which had cytoplasmic membrane exposed showed no detectable ATPase
activity unless they were incubated with toluene to destroy their permeability barriers. ATP may be able to leak into spheroplasts, since cells
treated with EDTA or cold Tris buffer became somewhat permeable to
nucleotides [68-71]. However, this permeability, if any, was too low to
allow penetration of ATP to support the ATPase reaction. Moreover, we

---

react with enzyme in detergent solution. Antibody of this type against
D-lactate dehydrogenase was recently obtained in our laboratory (Y. Tanaka
and M. Futai, manuscript in preparation).

could not demonstrate any significant rate of transport of ATP in membrane vesicles [118]. These findings suggest that cytoplasmic membranes are impermeable to ATP. However van Thienen and Postma [183] detected ATPase activity in membrane vesicles and found that it doubled when the vesicles were treated with a low concentration of Triton X-100. This detergent made the membrane permeable to ATP. Futai [118] obtained similar results in more extensive studies: that is, about 80% of ATPase activity was measurable before treatment with toluene or Triton X-100, and 60% of this measurable activity was accessible to a specific antibody against ATPase. Houghton et al. [176] reported that about 30% of the ATPase activity in the vesicles was measurable (there was about threefold increase in activity on disruption of the vesicles by a French pressure chamber). These findings are consistent with the interpretation that a significant portion of ATPase activity in vesicles is on the outer surface. Recently, Hare et al. [184] fractionated vesicles into two populations with ATPase on the inner and outer surfaces using antibody against ATPase.

It is known that EDTA at low concentration solubilizes ATPase [170, 185]. Thus repeated washing with buffer containing a higher concentration of EDTA during preparation of vesicles may deplete them of ATPase, and Futai [118] and Short et al. [140] actually observed significant loss of ATPase activity during preparation of vesicles. The volume of solution used for washing is not strictly defined, so different amounts of ATPase may remain outside vesicles prepared in different laboratories. The ATPase activity lost during the preparation may be that located on the outside of the vesicles. van Thienen and Postma [183] suggested that on washing vesicles with buffer of low ionic strength, the ATPase activity outside the vesicles was completely lost, whereas that inside was fully retained.

The problem of whether the ATPase outside the vesicles is functionally active is of great interest. Futai [118] observed that the ATPase activities of vesicles prepared by the method of Kaback [120] and of sonicated particles from E. coli ML308-225 were both sensitive to DCCD, although purified ATPase was not. Moreover, vesicles made from an ATPase-negative mutant (DL54) could bind ATPase in a DCCD-sensitive way, whereas spheroplasts made from the same mutant could not. Since membrane component(s) other than ATPase (F_1) is necessary for sensitivity to DCCD [186-188], this suggests that components of the membrane which interact with ATPase moved during lysis. West and Mitchell [189] observed significant movement of protons on addition of ATP to the vesicles, although they observed uptake of more protons after sonication of vesicles. Recently, Adler and Rosen [190] reported uptake of calcium by vesicles when ATP or NADH was added. This uptake was inhibited by D-lactate or phenazine methosulfate plus ascorabte. On the contrary, proline uptake dependent on D-lactate was inhibited by ATP or NADH, and the inhibition by ATP was reversed by antibody against ATPase or azide. From these results they concluded that ATPase or NADH dehydrogenase functions in establishing a proton motive force which is basic and negative outside. Hirata et al. [133] showed that addition of ATP inhibited transport of alanine by vesicles prepared from thermophilic bacteria, suggesting that ATPase outside the vesicles is also functional. These findings together with the earlier observation of van Thienen and Postma [183] that ATP can energize membranes, measured by 9-amino-6-chloro-2-methoxyacridine fluorescence, via this ATPase outside, suggest that the complete complex of ATPase had changed its location in some vesicles. However, the activities of ATP-driven trans-hydrogenase [176] and oxidative phosphorylation [175] were reported to be low in vesicles, although both activities increased about 10-fold after inversion of vesicles (P:O ratio after inversion, 0.006 [175]). On the other hand, vesicles with incorporated ADP were capable of a higher rate of oxidative phosphorylation (P:O ratio of 0.02) [191]. Thus it is quite possible that ATPase complexes outside the vesicles are somehow more damaged than those inside the vesicles. This may partly be due to the impermeability of membranes to agents used for washing which may damage the ATPase complex. These results also support the idea that the cytoplasmic membranes of E. coli are impermeable to adenine nucleotides. It must be noted that the efficiency of oxidative phosphorylation of these vesicles is much less than that of vesicles prepared by passing cells through a French pressure chamber [117]. This may be because factors required for oxidative phosphorylation are damaged more in vesicles prepared by the method of Kaback [120].

In contrast to all these findings, Short et al. [140] showed that ATPase activity at pH 6.6 in the vesicles was inaccessible to antibody unless the vesicles were homogenized vigorously in a motor-driven homogenizer, subjected to sonication, or incubated in Tris buffer, pH 9.0. Thus they concluded that measurable ATPase at pH 6.6 was all on the inside of the vesicles. However, their results must be considered more carefully. They observed 60% inhibition of ATPase at pH 9.0 by antibody against ATPase, confirming previous work [118], but no inhibition of D-lactate-dependent reactions by the antibody against D-lactate dehydrogenase. This excludes the possibility that the vesicles were permeable to antibody molecules at pH 9.0. Other possible explanations are: (1) ATPase molecules inside the vesicles were dislocated to the outside at pH 9.0; (2) ATPase molecules outside were inaccessible to specific antibody at lower pH, but became accessible at pH 9.0 or on homogenization; or (3) ATPase activity at lower pH was not due to a coupling factor of oxidative phosphorylation (BF$_1$). In this regard Rosen and Adler* observed no BF$_1$ activity at pH6.6. They detected ATP hydrolase activity which was resistant to antibody against BF$_1$ and to azide. This hydrolase was also detectable in membrane vesicles prepared from mutants lacking BF$_1$. Short et al. [140] suggested possibility (1). However, it seems unlikely that dislocation of enzyme molecules

_____

*B. P. Rosen and L. W. Adler, personal communication.

occurred simply on suspending vesicles at an alkaline pH, while gross orientation stayed unchanged. We favor possibility (3), although possibility (2) cannot be excluded. I think negative results on the titration of antibody against membrane enzymes should not be regarded as conclusive evidence.

The findings discussed above suggest that ATPase bound to the inner surface of cytoplasmic membranes is dislocated to the outside of membranes during preparation of vesicles, where it is easily lost. The activity of ATPase remaining in the membrane must be carefully evaluated, since loss of the enzyme by washing with dilute EDTA solution or its impairment by mutation may render the vesicles leaky to protons [67,192]. Partial loss of ATPase from membrane vesicles may affect transport, because leakage of protons lowers the efficiency of transport [192]. It would be interesting to prepare membrane vesicles with high ATPase activity and study the efficiency of transport in them.

3. Location of Other Enzymes in Membrane
   Vesicles

Studies on the locations of glycerol-3-phosphate [193], succinate [193], and NADH dehydrogenase [118] also support the view that the orientations of some enzymes are not the same as those in whole cells. Weiner [193] has shown genetically that membrane bound glycerol-3-phosphate dehydrogenase is located on the inner surface of the membranes of intact cells. Further evidence for this was obtained by showing that the activity of this enzyme could not be detected in either intact cells or spheroplasts using ferricyanide as electron acceptor, due to the impermeability of the latter, but that activity could be detected after loss of the permeability barrier by toluene treatment. In preparations of membrane vesicles obtained by Kaback's procedure [120], nearly half the dehydrogenase was accessible to ferricyanide as well as to a nonpermeable inhibitor of the enzyme. Similar results were obtained on NADH-ferricyanide reductase [118]. Recent experiments by Futai [152] support the notion that the membranes of E. coli ML308-225 are impermeable to ferricyanide. Ferricyanide did not inhibit transport by internally generated NADH, whereas it did inhibit transport stimulated by externally added NADH. This suggests that NADH dehydrogenase inside the vesicles is not accessible to ferricyanide, although dehydrogenase outside the vesicles is. Hampton and Freese [194] also suggested from kinetic studies on transport and uptake of oxygen that NADH dehydrogenase may have two locations in membrane vesicles.

The role of NADH oxidation in membrane vesicles should be carefully evaluated, especially when the NADH is added externally. Futai [152] showed that NADH generated inside the vesicles could stimulate transport of proline much more efficiently than that added outside. This suggests that NADH must be oxidized inside to produce a proton motive force which can support transport of proline. However, Stroobant and Kaback [195] recently showed that addition of ubiquinone-1 to vesicles greatly increased coupling between oxidation of NADH and active transport. They concluded that inefficient coupling of NADH oxidation with active transport could not be due to the presence of everted vesicles. It is not obvious why NADH oxidation required ubiquinone-1 to support active transport, since membranes contain ubiquinone-8. Furthermore, inward movement of protons [117], oxidative phosphorylation [117], and uptake of calcium [119] dependent on NADH have been observed in everted vesicles without addition of ubiquinone-1. These results suggest that intact NADH oxidase is capable of producing a proton motive force. Thus the experiment of Stroobant and Kaback [195] may not be relevent in discussing the orientation of NADH oxidase. Recently Adler and Rosen suggested that NADH dehydrogenase and ATPase can establish a proton motive force which is opposite to the formed by D-lactate [190].

It would be desirable to have marker proteins originally located outside the cytoplasmic membrane, since evidence discussed above concerns the enzymes located on the internal surface of cytoplasmic membrane. Wickner [196,197] has shown recently that major coat protein of bacterio-phage M13 was located in the membrane. More specifically, its antigenic site (N-terminus) was located on the outer surface of the cytoplasmic membrane [197].

By immunochemical fractionation using antibody against this protein Wickner [198] has suggested that vesicles are mosaics with respect to the orientation of some membrane proteins and are not a mixture of correctly oriented and everted ones. On the other hand, Hare et al. [184] frac-tionated vesicles with antibody against ATPase into two populations: Only vesicles which were not precipitated by the antibody could transport proline, suggesting that they are correctly orientated. Adler and Rosen [190] recently repeated the experiment of Hare et al. [184] and found that nearly all cytoplasmic membranes were precipitated by the antibody against ATPase. As discussed below, the results of Adler and Rosen [190] and Wickner [198] are more probable considering the available evidence, although the reason for the discrepancies between different laboratories is not clear.

4. Interpretation of Available Evidence

There are two main possibilities to explain why enzymes that are normally on the inner surface of the cell membrane are found on the outer surface of vesicles. (1) Preparations of vesicles may consist of inside-out and right-side-out vesicles. This seems unlikely if one accepts evidence obtained by freeze fracture [12,141], and the findings that D-lactate dehydrogenase is insensitive to its specific antibody [140,178] and that all vesicles can trans-port radioactive substrate [113]. Tsuchiya [191] recently estimated that about 10% of the vesicles were everted by measuring the activity of D-lac-tate-dependent calcium uptake. For the calculation he assumed that the

calcium transport systems in everted and right-side-out vesicles were
equal in amount, and that the location of D-lactate dehydrogenase was the
same as that of other membrane enzymes. His value is consistent with our
observation that a significant portion (about 15%) of the D-lactate dehydrogenase was accessible to its specific antibody [178]. Thus it is highly
probable that vesicle preparations contain significant amounts of everted
vesicles and that the amounts vary in different laboratories. (2) A second
possible explanation is that the gross orientation of the membrane vesicles
is right-side-out, but that the orientation of some of enzymes are reversed.
This explanation is the more probable, if we accept all the evidence so far
published. Moreover, it is highly probable that dislocated enzymes function
in transport and respiration. Weiner [193] showed that mutant cells lacking
transport activity for glycerol-3-phosphate could not transport amino acids
using glycerol-3-phosphate in the medium as an energy source, suggesting
that this compound must be transported into the cells to stimulate the
uptake of another solute. However, he reproducibly observed that membrane vesicles from this mutant showed amino acid transport stimulated
by glycerol-3-phosphate, although the rate of transport was about half that
of vesicles from wild-type cells. Weiner's observation supports the idea
that the location of the dehydrogenase may change during preparation of
vesicles and that the enzyme continues to support transport activity in the
new location, although he did not discuss the possibility. It seems rather
unlikely that only vesicles from mutant cells somehow became leaky to
glycerol-3-phosphate.

The idea that dislocated enzymes can function in transport like those in
reconstituted vesicles is also supported by the following evidence. Transport dependent on D-lactate in deficient vesicles could be reconstituted on
incubating vesicles with a guanidine-HCl extract of wild-type vesicles [199]
or purified D-lactate dehydrogenase [75, 200]. Futai [75] made a similar
observation with glycerol-3-phosphate dehydrogenase. It is known that
enzyme bound to the outside of vesicles supports respiration and transport
[75, 140]. The glycerol-3-phosphate dependent transport of proline in
reconstituted vesicles was almost completely inhibited by ferricyanide [75].
This suggests that ferricyanide competes with an endogenous electron
acceptor in the respiratory chain, and this competition prevents transport
into the reconstituted vesicles. On the other hand, transport into spheroplasts was insensitive to ferricyanide. However, partial inhibition was
also observed with natural vesicles prepared from cells induced for gly-
cerol-3-phosphate dehydrogenase. Results with natural vesicles were
somehow intermediate between those for spheroplasts and for reconstituted
vesicles, so we favor the interpretation that this dehydrogenase moved
during preparation of membranes, and that after dislocation it still
supported transport.

There are also two other possibilities. (3) Vesicles may each consist
of patchwork portions orientated in opposite directions. Such vesicles
might be formed from fusion of vesicles of different orientations. If the

preparation had consisted of such hybrid vesicles, the ratio of the enzymes on the inner surface to the outer surface would be the same for all inner membrane enzymes. However, the ratio differs significantly as discussed above depending on the enzymes tested. Thus this possibility is unlikely. (4) Some vesicles may be leaky so that they show enzyme activity, but do not contribute transport activity. This also seems improbable if one accepts the autoradiographic evidence that practically all vesicles transport radioactive solute [110].

The above discussion is limited largely to E. coli strains. The orientation of membrane vesicles prepared from B. subtilis may be different. From electron microscopic studies [143] and the finding that the respiratory chain was accessible to 5-N-methylphenazonium methylsulfate and trypsin [201], Konings and coworkers suggested that vesicles from B. subtilis are right-side-out. This organism seems to be permeable to NADH, since NADH could support active transport in whole cells [202]. The growth conditions of Paracoccus denitrificans seemed to affect the orientation of membrane vesicles, even if the same fractionation procedure was used [203].

From the findings discussed above it is concluded that in preparations of membrane vesicles from E. coli, most of the vesicles are right-side-out but that some membrane enzymes are dislocated from the inside to the outside of the membrane. It is possible that preparations may contain both undamaged right-side-out vesicles and those with dislocated enzymes. Preparations also contain a small proportion of inside-out vesicles, although the percentage of these seems to vary in preparations from different laboratories.

B. Orientation of Other Vesicles Capable
      of Transport

1. Orientation of Inner Membrane Vesicles
      from E. coli

Recently, Yamato et al.* studied the locations of membrane enzymes in vesicles prepared by their new procedure [115]. Their results can be summarized as follows. (1) About half of the glycerol-3-phosphate ferricyanide reduction was measurable before loss of permeability barriers. (2) D-Lactate-dependent uptake of oxygen and proline uptake were inhibited 70 and 30%, respectively, by antibody against D-lactate dehydrogenase. (3) Glycerol-3-phosphate-dependent oxygen consumption and proline uptake were inhibited 60 and 40%, respectively, by ferricyanide. Thus the dehydrogenase which is accessible to ferricyanide or the specific antibody

---

*I. Yamato, Y. Anraku, and M. Futai, manuscript in preparation.

supported transport of proline into these vesicles. These results suggest that the preparation consisted of a mixture of vesicles with different orientations as discussed above for vesicles prepared by the method of Kaback [120].

2. Orientation of Ghosts and Electron Transfer Particles from M. phlei

Ghosts prepared from M. phlei appear to be oriented in the same way as the intact cells from the following findings. (1) No oxidative phosphorylation was observed, unless the ghosts were either sonicated or preincubated with a high concentration of adenine nucleotide [165]. The resulting phosphorylation was insensitive to antibody against ATPase. (2) Electron microscopy showed that ghosts still contained some cell wall materials so that their shape remained similar to that of intact cells [165]. (3) The ghosts transported proline stimulated by succinate or an artificial electron donor.

The problem of the orientation of electron transfer particles seems to be more complex than that of the ghosts. These particles show highly efficient oxidative phosphorylation dependent on NADH (P:O ratio close to 1) [165], which is sensitive to antibody against ATPase, suggesting that the vesicles are inside-out. However, these vesicles can transport amino acids, and this is stimulated by respiration. Thus these particles consist of at least two kinds of vesicles: right-side-out and inside-out. However, the particles which transported amino acids cannot be regarded as undamaged right-side-out vesicles. Proline which had been accumulated in the particles was rapidly released on addition of succinate [163]. This efflux was only observed with electron transfer particles. Thus it is quite possible that succinate dehydrogenase or other components of the respiratory chain were dislocated during preparation of these particles. More specifically, succinate and an artificial electron donor formed proton motive forces in different directions in the same vesicles.

C. Orientation of Membrane Vesicles Prepared by French Pressure Chamber or Sonication

The localization of membrane enzymes in membrane vesicles prepared by French pressure chamber from E. coli have been studied extensively by biochemical and immunochemical procedures. More than 90% of the ATPase [118] and D-lactate dehydrogenase [178] were accessible to the corresponding antibodies, and the NADH-ferricyanide reductase activity [118] was measurable. The binding sites for ATPase were also shown to be predominantly outside the vesicles [118]. These results suggest that vesicles are completely everted, and this is consistent with the vectorial activities of the vesicles as discussed above. Morphological studies on these vesicles by Altendorf and Staehelin [141] using freeze fracture and negative staining also support this idea: in contrast to membranes prepared by osmotic lysis of spheroplasts, those obtained by passing intact cells through a French pressure chamber were uniformly very small (only 40 to 110 nm in diameter) and approximately 60 to 80% were inside-out. This estimation of the percentage everted particles in the population is different from those made by studies on the locations of membrane enzymes. This may be because biochemical and morphological estimations are essentially different.

    Membrane vesicles prepared by prolonged sonication of E. coli showed essentially the same locations of membrane enzymes as those prepared by a French pressure chamber [118]. This kind of preparation may contain hybrid vesicles of inner and outer membranes, since Tsukagoshi and Fox [204] showed that hybrid vesicles were formed by sonication of membranes of different origins. Most of the vesicles prepared by sonication of M. phlei cells were also shown to be everted as discussed above. Brodie et al. [205] showed that formation of everted vesicles by sonication was dependent on the length of time of the treatment.

    Sonication has been used to prepare submitochondrial particles. These particles are generally believed to be inside-out. However, they contained membrane fragments with dislocated electron carriers [206,207], whereas they are completely everted with respect to mitochondrial ATPase [207]. Eytan et al. [208] could obtain a preparation low in dislocated cytochrome oxidase by immunochemical fractionation of submitochondrial particles. The possibility of dislocation of membrane proteins could not be excluded in bacterial vesicles prepared by sonication or French pressure chamber, although no evidence for it has been presented yet.

    Sonication did not necessarily produce an everted population. Membrane vesicles obtained from H. halobium were suggested to be 85% right-side-out from the following results [116]: (1) NADH-menadione reductase, which is located on the inner surface of the cytoplasmic membranes, was not measurable in membrane vesicles, unless they had been treated with detergents, and (2) these vesicles could transport amino acids, as discussed above.

V. CONCLUSION

I have briefly reviewed the experimental systems widely used in bacterial transport. Both whole cells and membrane vesicles are useful experimental systems for studying bacterial transport. Whole cells will still be useful in future studies, because not all mechanisms of transport are retained by membrane vesicles.

As discussed above, it is questionable whether it is possible to prepare vesicles, especially right-side-out vesicles, without some dislocation of enzymes, contamination with inside-out vesicles, or other damage. Thus it is extremely important to know the exact orientation of vesicles when analyzing the results of transport studies. In studying active transport in particular, knowledge of the orientation of membranes provides valuable information on the direction in which a proton motive force can be formed in the vesicles when a certain energy source is added. Recent studies have suggested that active transport is coupled to a proton motive force, and more specifically that the direction of transport is determined by the direction of the proton motive force generated by the respective energy source used (see Chap. 12). The recent work of Rosen and coworkers [190, 209] is important in this connection. Membrane vesicles prepared by the method of Kaback [120] could not transport calcium when the proton motive force was produced by respiration (positive and acid outside), whereas they could when the proton motive force was formed artificially in the opposite direction (negative and basic outside). As already discussed, hydrolysis of ATP supported uptake of calcium in these vesicles [190]. The direction of the proton motive force formed by hydrolysis of ATP could be inferred to be negative and basic outside, since ATPase shown to be outside as discussed above.

Finally, it must be noted that experimental systems are continuously being modified and new systems not discussed here are being devised in many laboratories. One of the most exciting developments is the solubilization and reconstitution of membrane carrier activities: the D-glucose carrier of erythrocytes [210], ADP/ATP transporter of mitochondria [211], alanine carrier of thermophilic bacteria [212], and proline carrier of E. coli [213] were solubilized. Liposomes obtained by incubating solubilized carrier with phospholipids were capable of taking up the corresponding solute. Developments in this direction are extremely important for understanding the process of translocation of solutes at a molecular level.

ACKNOWLEDGMENTS

I am grateful to Drs Y. Anraku and L. A. Heppel for critical reading of the manuscript. I wish to thank to all those who sent me preprints and reprints and useful suggestions, particularly to Drs. H. R. Kaback, G. F.-L. Ames, W. Epstein, T. A. Krulwich, H. L. Kornberg, A. F. Brodie, D. L. Oxender, Y. Kagawa, H. Hirata, D. B. Wilson and B. P. Rosen. I wish to thank Ms. Elizabeth Ichihara for her patient English corrections.

REFERENCES

1. W. D. Stein, in The Movement of Molecules Across Cell Membranes Academic, New York, 1967.
2. E. C. C. Lin in Structure and Function of Biological Membranes (L. I. Rothfield, ed.), Academic, New York, 1971, p. 285.
3. D. L. Oxender, Ann. Rev. Biochem., 41: 777 (1972).
4. D. L. Oxender, in Metabolic Transport Metabolic Pathways Vol. VI (L. E. Hokin, ed.), Academic, New York, 1972, p. 133.
5. F. M. Harold, Bacteriol. Rev., 36: 172 (1972).
6. F. M. Harold, in Current Topics on Bioenergetics, Vol. 6 (D. R. Sanadi, ed.), New York, Academic (1977) p. 84.
7. S. Roseman, in Metabolic Transport Metabolic Pathways, Vol. VI (L. E. Hokin, ed.), Academic, New York, 1972, p. 41.
8. W. Boos, Ann. Rev. Biochem., 43: 123 (1974).
9. W. Boos, in Current Topics in Membrane and Transport, Vol. 5 (F. Bronner and A. Kleinzeler, eds.), Academic, New York, 1974, p. 51.
10. R. D. Simoni and P. W. Postma, Ann. Rev. Biochem., 44: 523 (1975).
11. H. R. Kaback, in Current Topics in Membrane and Transport (F. Bronner and A. Kleinzeller, eds.) 1: 36 (1970).
12. H. R. Kaback, Biochim. Biophys. Acta, 265: 367 (1972).
13. H. R. Kaback, Science, 186: 882 (1974).
14. Y. S. Halpern, Ann. Rev. Genet., 8: 103 (1974).
15. E. C. C. Lin, Ann. Rev. Genet., 4: 225 (1970).
16. A. Wright and S. Kanegasaki, Physiol. Rev., 51: 748 (1971).
17. M. R. J. Salton, in The Bacterial Cell Wall, Elsevier, Amsterdam, 1964.
18. H. C. Neu and L. A. Heppel, J. Biol. Chem., 240: 3685 (1965).
19. N. G. Nossal and L. A. Heppel, J. Biol. Chem., 241: 3055 (1966).
20. Y. Anraku and L. A. Heppel, J. Biol. Chem., 242: 2561 (1967).
21. L. A. Heppel, Science, 156: 1451 (1967).
22. Y. Anraku, J. Biol. Chem., 243: 3128 (1968).
23. L. A. Heppel, in Structure and Function of Biological Membranes (L. I. Rothfield, ed.), Academic, New York, 1971, p. 223.
24. B. P. Rosen and L. A. Heppel, in Bacterial Membrane and Walls (R. Leive, ed.), Dekker, New York, 1973, p. 209.
25. P. K. Nakane, G. E. Nichoalds, and D. L. Oxender, Science, 161: 182 (1968).
26. A. B. Pardee and K. Watanabe, J. Bacteriol., 96: 1049 (1968).
27. S. G. Kustu and G. F. L. Ames, J. Biol. Chem., 249: 6976 (1974).
28. G. F. L. Ames and E. N. Spudich, Proc. Natl. Acad. Sci. USA, 73: 1877 (1976).
29. D. R. Dimasi, J. C. White, C. A. Schnaitman, and C. Bradbeer, J. Bacteriol., 115: 506 (1973).
30. G. L. Hazelbauer, J. Bacteriol., 124: 119 (1975).
31. S. Szmelcman and M. Hofnung, J. Bacteriol., 124: 112 (1975).

32. K. Hantke and V. Braun, FEBS Lett., 49: 301 (1975).
33. G. E. Frost and H. Rosenberg, J. Bacteriol., 124: 704 (1975).
34. G. Decad, T. Nakae, and H. Nikaido, Fed. Proc., 33: 1240 (1974).
35. T. Nakae and H. Nikaido, J. Biol. Chem., 250: 7359 (1975).
36. E. A. Berger, Proc. Natl. Acad. Sci. USA, 70: 1514 (1973).
37. E. A. Berger and L. A. Heppel, J. Biol. Chem., 249: 7747 (1974).
38. H. Kobayashi, E. Kin, and Y. Anraku, J. Biochem., 76: 251 (1974).
39. D. B. Wilson, J. Bacteriol., 120: 866 (1974).
40. S. J. Curtis, J. Bacteriol., 120: 295 (1974).
41. J. L. Cowell, J. Bacteriol., 120: 139 (1974).
42. J. M. Wood, J. Biol. Chem., 250: 4477 (1975).
43. R. J. Kadner and H. H. Winkler, J. Bacteriol., 123: 985 (1975).
44. D. B. Rhoads and W. Epstein, J. Biol. Chem., 252: 1394 (1977).
45. A. Morikawa, H. Suzuki and Y. Anraku, J. Biochem., 75: 229 (1974).
46. P. J. F. Henderson and H. L. Kornberg, Ciba Foundation Symposium (new series) 31: 243 (1975).
47. C. F. Fox and E. P. Kennedy, Proc. Natl. Acad. Sci. USA, 54: 891 (1965).
48. G. F. L. Ames, J. Biol. Chem., 249: 634 (1974).
49. P. H. O'Farrell, J. Biol. Chem., 250: 4007 (1975).
50. W. C. Johnson, T. J. Silhavy, and W. Boos, Appl. Microbiol., 29: 405 (1975).
51. G. F. L. Ames and K. Nikaido, Biochemistry, 15: 616 (1976).
52. T. J. Silhavy, I. Hartig-Beecken, and W. Boos, J. Bacteriol., 126: 951 (1976).
53. F. M. Harold, Ann. NY Acad. Sci., 227: 297 (1974).
54. V. Riebling, R. K. Thauer, and K. Jungermann, Eur. J. Biochem., 55: 445 (1975).
55. T. A. Krulwich, M. E. Sobel, and E. B. Woldson, Biochem. Biophys. Res. Comm., 53: 258 (1973).
56. Y. Anraku, E. Kin, and Y. Tanaka, J. Biochem., 78: 165 (1975).
57. P. C. Maloney, E. R. Kashket, and T. H. Wilson, Proc. Natl. Acad. Sci. USA, 71: 3896 (1974).
58. A. L. Koch, J. Mol. Biol., 59: 447 (1971).
59. J. H. Weiner and L. A. Heppel, J. Biol. Chem., 246: 6933 (1971).
60. J. L. Flagg and T. H. Wilson, J. Bacteriol., 125: 1235 (1976).
61. D. M. Wilson, J. F. Alderete, P. C. Maloney, and T. H. Wilson, J. Bacteriol., 126: 327 (1976).
62. D. L. Oxender and S. C. Quay, in Methods in Membrane Biology, Vol. VI (E. D. Korn, ed.,), Plenum, New York, 1976, p. 183.
63. E. A. Berger and L. A. Heppel, J. Biol. Chem., 247: 7684 (1972).
64. B. P. Rosen, J. Biol. Chem., 246: 3653 (1971).
65. L. Leive, J. Biol. Chem., 243: 2373 (1969).
66. B. P. Rosen, J. Bacteriol., 116: 1124 (1973).
67. B. P. Rosen, Biochem. Biophys. Res. Commun., 53: 1289 (1973).

68. G. Buttin and A. Kornberg, J. Biol. Chem., 241: 5419 (1966).
69. B. Z. Cavari, Y. Avi-Dor, and N. Grossowicz, Biochem. J., 103: 601 (1967).
70. G. A. Scarborough, M. K. Rumley, and E. P. Kennedy, Proc. Natl. Acad. Sci. USA, 60: 951 (1968).
71. I. C. West, FEBS Lett., 4: 69 (1969).
72. Y. Anraku, H. Kobayashi, H. Amanuma, and A. Yamaguchi, J. Biochem., 74: 1249 (1973).
73. W. R. Sistrom, Biochim. Biophys. Acta, 29: 579 (1958).
74. J. T. Wachsman and R. Stock, J. Bacteriol., 80: 600 (1960).
75. M. Futai, Biochemistry, 13: 2327 (1974).
76. R. C. Essenberg and H. L. Kornberg, J. Biol. Chem., 250: 939 (1975).
77. J. R. Piperno and D. L. Oxender, J. Biol. Chem., 243: 5914 (1968).
78. S. E. Burrous and R. P. DeMoss, Biochim. Biophys. Acta., 73: 623 (1963).
79. H. H. Winkler and T. H. Wilson, J. Biol. Chem., 241: 2200 (1966).
80. Y. Anraku, J. Biol. Chem., 242: 793 (1967).
81. D. B. Rhoads, F. B. Waters, and W. Epstein, J. Gen. Physiol., 67: 325 (1976).
82. J. R. Parnes and W. J. Boos, J. Biol. Chem., 248: 4436 (1973).
83. R. J. Kadner, J. Bacteriol., 117: 232 (1974).
84. Y. S. Halpern and H. L. Ennis, J. Bacteriol., 122: 332 (1975).
85. R. J. Kadner, J. Bacteriol., 122: 110 (1975).
86. D. B. Wilson, J. Bacteriol., 126: 1156 (1976).
87. J. More and E. E. Snell, Biochemistry, 2: 136 (1963).
88. R. E. Marquis and P. Gerhardt, J. Biol. Chem., 239: 3361 (1964).
89. M. D. Brand, C. Chen, and A. L. Lehninger, J. Biol. Chem., 251: 968 (1976).
90. W. C. Werkheiser and W. Bartley, Biochem. J., 66: 79 (1957).
91. J. E. Amoore, Biochem. J., 70: 718 (1958).
92. H. W. Heldt and L. Rapley, FEBS Lett., 7: 139 (1970).
93. E. Pfaff, M. Klinberg, E. Ritt, and W. Vogell, Eur. J. Biochem., 5: 222 (1968).
94. R. E. Gaensslen and R. E. McCarty, Arch. Biochem. Biophys., 147: 55 (1971).
95. R. E. Gaensslen and R. E. McCarty, Anal. Biochem., 48: 504 (1972).
96. A. R. Portis, Jr. and R. E. McCarty, Arch. Biochem. Biophys., 156: 621 (1973).
97. A. R. Portis, Jr. and R. E. McCarty, J. Biol. Chem., 249: 6250 (1974).
98. A. R. Portis, Jr. and R. E. McCarty, J. Biol. Chem., 251: 1610 (1976).
99. H. H. Winkler, F. L. Byrave, and A. L. Lehninger, J. Biol. Chem., 243: 20 (1968).

100. H. V. Rickenberg, G. N. Cohen, G. N. Buttin, and J. Monod, Ann. Inst. Pasteur, 91: 829 (1956).

101. E. P. Kennedy, in The Lactose Operon (J. R. Beckwith and D. Zipser, eds.), Cold Spring Harbor Laboratories, 1970, p. 49.

102. W. D. Nunn and J. E. Cronan, Jr., J. Biol. Chem., 249: 724 (1974).

103. A. L. Koch, J. Bacteriol., 120: 895 (1974).

104. G. Cecchini and A. L. Koch, J. Bacteriol., 123: 187 (1975).

105. G. F. L. Ames, Arch. Biochem. Biophys., 104: 1 (1964).

106. B. P. Rosen, J. Biol. Chem., 248: 1211 (1973).

107. S. P. Colowick and F. C. Womack, J. Biol. Chem., 244: 774 (1969).

108. S. Schuldiner, R. Weil, and H. R. Kaback, Proc. Natl. Acad. Sci. USA, 73: 109 (1976).

109. S. Ramos, S. Schuldiner, and H. R. Kaback, Proc. Natl. Acad. Sci. USA, 73: 1819 (1976).

110. I. C. West. Biochem. Biophys. Res. Commun., 41: 655 (1970).

111. I. C. West and P. Mitchell, Biochem. J., 132: 587 (1973).

112. J. de Gier, J. G. Mandersloot, J. V. Hupkes, R. N. McElhaney, and W. P. van Beek, Biochim. Biophys. Acta, 233: 610 (1971).

113. S. A. Short, H. R. Kaback, G. Kaczorowski, J. Fischer, C. T. Walsh, and S. C. Silverstein, Proc. Natl. Acad. Sci. USA, 71: 5032 (1974).

114. H. Hirata, F. C. Kosmakos, and A. F. Brodie, J. Biol. Chem., 249: 6965 (1974).

115. I. Yamato, Y. Anraku, and K. Hirosawa, J. Biochem., 77: 705 (1975).

116. R. E. MacDonald and J. K. Lanyi, Biochemistry, 14: 2882 (1975).

117. E. L. Hertzberg and P. C. Hinkle, Biochem. Biophys. Res. Commun., 58: 178 (1974).

118. M. Futai, J. Membr. Biol., 15: 15 (1974).

119. B. P. Rosen and J. S. McClees, Proc. Natl. Acad. Sci. USA, 71: 5042 (1974).

120. H. R. Kaback, in Methods in Enzymology, Vol. 22, (W. B. Jacoby, ed.), Academic, New York and London, 1971, p. 99.

121. D. C. Birdsell and E. H. Cota-Robles, J. Bacteriol., 93: 427 (1967).

122. W. N. Konings, E. M. Barnes, Jr., and H. R. Kaback, J. Biol. Chem., 246: 5857 (1971).

123. S. A. Short and H. R. Kaback, J. Biol. Chem., 249: 4275 (1974).

124. J. Spitzen, in Methods in Enzymology, Vol. 5, (S. P. Colowick and N. O. Kaplan, eds.), 1962, p. 122.

125. H. H. Martin, J. Theoret. Biol., 5: 1 (1963).

126. H. R. Kaback and E. R. Stadtman, Proc. Natl. Acad. Sci. USA, 55: 920 (1966).

127. H. R. Kaback, J. Biol. Chem., 243: 3711 (1968).

128. E. M. Barnes, Jr., Arch. Biochem. Biophys., 152: 795 (1972).

129. W. N. Konings, J. Boonstra, and W. deVries, J. Bacteriol., 122:
     245 (1975).
130. E. B. Wolfson, M. E. Sobel, and T. A. Krulwich, Biochim. Biophys.
     Acta, 321: 181 (1973).
131. A. C. L. Johnson, Y.-A. Cha, and J. R. Stern, J. Bacteriol., 121:
     682 (1975).
132. A. Martin, W. N. Konings, J. G. Kuenen, and M. Emmens,
     J. Gen. Microbiol., 83: 311 (1974).
133. H. Hirata, N. Sone, M. Yoshida, and Y. Kagawa, J. Biochem., 79:
     1157 (1976).
134. M. J. Osborn, J. E. Gander, E. Parisi, and J. Carson, J. Biol.
     Chem., 247: 3962 (1972).
135. H. R. Kaback and A. B. Kostellow, J. Biol. Chem., 243: 1384 (1968).
136. H. Hirata, K. Altendorf, and F. M. Harold, J. Biol. Chem., 249: 2939
     (1974).
137. H. Weissbach, E. L. Thomas, and H. R. Kaback, Arch. Biochem.
     Biophys., 147: 249 (1971).
138. F. J. Lombardi and H. R. Kaback, J. Biol. Chem., 247: 7844 (1972).
139. E. M. Barnes, Jr. and H. R. Kaback, J. Biol. Chem., 246: 5518
     (1971).
140. S. A. Short, H. R. Kaback, and L. D. Kohn, J. Biol. Chem., 250:
     4291 (1975).
141. K. H. Altendorf and L. A. Staehelin, J. Bacteriol., 117: 888 (1974).
142. W. N. Konings and H. R. Kaback, Proc. Natl. Acad. Sci. USA, 70:
     3376 (1973).
143. W. N. Konings, A. Bisschop, M. Voenhuis, and C. A. Verneulen,
     J. Bacteriol., 116: 1456 (1973).
144. K. J. Hellingwerf, P. A. M. Michels, J. W. Dorpema, and W. N.
     Konings, Eur. J. Biochem., 55: 397 (1975).
145. J. D. Stinnett, L. F. Guymon, and R. G. Eagon, Biochem. Biophys.
     Res. Commun., 52: 284 (1973).
146. L. S. Milner and H. R. Kaback, Proc. Natl. Acad. Sci. USA, 65:
     683 (1970).
147. R. A. Long and J. C. Dittmer, Biochim. Biophys. Acta, 367: 295
     (1974).
148. L. Patel, S. Schuldiner, and H. R. Kaback, Proc. Natl. Acad. Sci.
     USA, 72: 3387 (1975).
149. L. Patel and H. R. Kaback, Biochemistry, 15: 2741 (1976).
150. S. M. Hasan and B. P. Rosen, Biochim. Biophys. Acta,
     459: 225 (1977).
151. R. Prasad, V. K. Kalra, and A. F. Brodie, J. Biol. Chem., 250:
     3699 (1975).
152. M. Futai, J. Bacteriol., 120: 861 (1974).
153. A. N. Tucker and T. T. Lilligh, J. Bacteriol., 129: 559
     (1977).

154. T. Tsuchiya and B. P. Rosen, Biochem. Biophys. Res. Commun., 68: 497 (1976).
155. M. H. Saier, B. U. Feucht, and M. T. McCann, J. Biol. Chem., 250: 7593 (1975).
156. E. Schechter, T. Gulik-Krzywicki, and H. R. Kaback, Biochim. Biophys. Acta, 274: 466 (1972).
157. T. Miura and S. Mizushima, Biochim. Biophys. Acta., 150: 159 (1968).
158. D. A. White, W. J. Lennarz, and C. A. Schnaitman, J. Bacteriol., 109: 686 (1972).
159. S. Mizushima, Biochim. Biophys. Acta, 419: 261 (1976).
160. D. Joseleau-Petit and A. Kepes, Biochim. Biophys. Acta, 406: 36 (1975).
161. A. F. Brodie, J. Biol. Chem., 234: 398 (1959).
162. H. Hirata, A. Asano, and A. F. Brodie, Biochem. Biophys. Res. Commun., 44: 368 (1971).
163. H. Hirata and A. F. Brodie, Biochem. Biophys. Res. Commun., 47: 633 (1972).
164. A. Asano, H. Hirata, and A. F. Brodie, Biochem. Biophys. Res. Commun., 46: 1340 (1972).
165. A. Asano, N. S. Cohen, R. E. Baker, and A. F. Brodie, J. Biol. Chem., 248: 3386 (1973).
166. T. R. Hinds and A. F. Brodie, Proc. Natl. Acad. Sci. USA, 71: 1202 (1974).
167. F. C. Kosmakos and A. F. Brodie, J. Biol. Chem., 249: 6956 (1974).
168. R. Prasad, V. K. Kalra, and A. F. Brodie, Biochem. Biophys. Res. Commun., 63: 50 (1975).
169. R. Prasad, V. K. Kalra, and A. F. Brodie, J. Biol. Chem., 250: 3699 (1975).
170. M. Futai, P. C. Sternweis, and L. A. Heppel, Proc. Natl. Acad. Sci. USA, 71: 2725 (1974).
171. J. M. Fassenden-Raden, J. Biol. Chem., 244: 6662 (1969).
172. S. Silver and M. U. Kralovic, Biochem. Biophys. Res. Commun., 34: 640 (1969).
173. T. Tsuchiya and B. P. Rosen, J. Biol. Chem., 250: 8409 (1975).
174. T. Tsuchiya and B. P. Rosen, J. Biol. Chem., 250: 7687 (1975).
175. M. Mével-Ninio and T. Yamamoto, Biochim. Biophys. Acta, 357: 63 (1974).
176. R. L. Houghton, R. J. Fisher, and D. R. Sanadi, Biochim. Biophys. Acta, 396: 17 (1975).
177. H. R. Kaback and T. F. Deuel, Arch. Biochem. Biophys., 132: 118 (1969).
178. M. Futai and Y. Tanaka, J. Bacteriol., 124: 470 (1975).
179. J. P. Reeves, Biochem. Biophys. Res. Commun., 45: 931 (1971).
180. J. D. Oppenheim and M. R. J. Salton, Biochim. Biophys. Acta, 298: 297 (1973).

181. G. A. Gorneva and I. D. Ryabova, FEBS Lett., 42: 271 (1974).
182. R. W. Brockman and L. A. Heppel, Biochemistry, 7: 2554 (1968).
183. G. van Thienen and P. W. Postma, Biochim. Biophys. Acta, 323: 429 (1973).
184. J. F. Hare, K. Olden, and E. P. Kennedy, Proc. Natl. Acad. Sci. USA, 71: 4843 (1974).
185. P. L. Davies and P. D. Bragg, Biochim. Biophys. Acta, 266: 273 (1972).
186. R. H. Fillingame, J. Bacteriol., 124: 870 (1975).
187. N. Sone, M. Yoshida, H. Hirata, and Y. Kagawa, J. Biol. Chem., 250: 7917 (1975).
188. R. Serrano, B. I. Kanner, and E. Racker, J. Biol. Chem., 251: 2453 (1976).
189. I. C. West and P. Mitchell, FEBS Lett., 40: 1 (1974).
190. L. W. Adler and B. P. Rosen, J. Bacteriol., 129: 959 (1977).
191. T. Tsuchiya, J. Biol. Chem., 251: 5315 (1976).
192. K. Altendorf, F. M. Harold, and R. D. Simoni, J. Biol. Chem., 249: 4587 (1974).
193. J. H. Weiner, J. Membr. Biol., 15: 1 (1974).
194. M. L. Hampton and E. Freese, J. Bacteriol., 118: 497 (1974).
195. P. Stroobant and H. R. Kaback, Proc. Natl. Acad. Sci. USA, 72: 3970 (1975).
196. W. Wickner, Proc. Natl. Acad. Sci. USA, 72: 4749 (1975).
197. W. Wickner, Proc. Natl., Acad. Sci. USA, 73: 1159 (1976).
198. W. Wickner, J. Bacteriol., 127: 162 (1976).
199. J. P. Reeves, J.-S. Hong, H. R. Kaback, Proc. Natl. Acad. Sci. USA, 70: 1917 (1973).
200. S. A. Short, H. R. Kaback, and L. D. Kohn, Proc. Natl. Acad. Sci. USA, 71: 1461 (1974).
201. W. N. Konings, Arch. Biochem. Biophys., 167: 570 (1975).
202. A. Bisschop, J. Boonstra, H. H. Sips, and W. N. Konings, FEBS Lett., 60: 11 (1975).
203. J. N. Burnell, P. John, and F. R. Whatley, Biochem. J. 150: 527 (1975).
204. N. Tsukagoshi and C. F. Fox, Biochemistry, 10: 3309 (1971).
205. A. F. Brodie, H. Hirata, A. Asano, N. S. Cohen, T. R. Hinds, H. N. Aithol, and V. K. Kalva, in Biological Membranes (C. F. Fox, ed.), Academic, New York, 1972, p. 445.
206. B. Chance, M. Erecinska, and C. P. Lee, Proc. Natl. Acad. Sci. USA, 66: 928 (1970).
207. D. L. Schneider, Y. Kagawa, and E. Racker, J. Biol. Chem., 247: 4047 (1972).
208. G. D. Eytan, R. C. Carrol, G. Schatz, and E. Racker, J. Biol. Chem., 250: 8598 (1975).

209. T. Tsuchiya, J. S. McClees, and B. P. Rosen, in Abstracts of the Annual Meeting of the American Society for Microbiology, Atlantic City, American Society for Microbiology, 1976, p. 154.
210. M. Kasahara and P. C. Hinkle, Proc. Natl. Acad. Sci. USA, 73: 396 (1976).
211. H. G. Shertzer and E. Racker, J. Biol. Chem., 251: 2446 (1976).
212. H. Hirata, N. Sone, M. Yoshida, and Y. Kagawa, Biochem. Biophys. Res. Commun., 69: 665 (1976).
213. H. Amanuma, K. Motojima, A. Yamaguchi, and Y. Anraku, Biochem. Biophys. Res. Commun., 74: 366 (1977).

Chapter 2

GROUP TRANSLOCATION TRANSPORT SYSTEMS

John B. Hays

Department of Chemistry
University of Maryland, Baltimore County
Catonsville, Maryland

I.   INTRODUCTION 44
II.   THE PHOSPHOENOLPYRUVATE:HEXOSE PHOSPHO-
      TRANSFERASE SYSTEMS 46
      A. Reactions and Components of the PTS; General
          Description and Nomenclature 47
      B. Enzymology of Nonspecific PTS Proteins 48
      C. Enzymology of Sugar-specific PTS Proteins 50
      D. Biological Distribution of the PTS... 56
      E. Comparative Physiology and Genetics of the PTS 59
      F. Function of the PTS in Cellular Transport 69
III. FATTY ACID TRANSPORT IN ESCHERICHIA COLI 84
      A. Long-Chain Fatty Acids 85
      B. Short-Chain Fatty Acids 86
IV.   TRANSPORT OF PURINES, PYRIMIDINES, AND
      NUCLEOSIDES 89
      A. Introduction 89
      B. Uptake of Free Bases 90
      C. Uptake of Nucleosides 93
      D. Discussion 96
      REFERENCES 98

I. INTRODUCTION

The name "group translocation" was applied to a special class of transport
mechanisms by Mitchell almost 20 years ago [1,2]. Sometimes this term
has been used synonomously with "group transfer," the class of coupled
metabolic reactions of the following type:

$$DS + C \rightleftharpoons D + CS$$

$$CS + A \rightleftharpoons AS + C. \tag{1}$$

In this case the group S is said to be transferred or translocated from the
donor D to an acceptor A via a carrier C. For example, coupling of the
pyruvate kinase and hexokinase reactions would be described by a scheme
in which DS = P-enolpyruvate, C = ADP, CS = ATP, A = glucose, and AS =
glucose-6-phosphate. The carrier can be a protein, for instance, acyl
carrier protein. In this chapter, however, "group translocation" is
reserved for the processes described by Mitchell, in which the reaction is
vectorial in the macroscopic sense, i.e., the transfer of the group from
donor to acceptor is concomitant with its translocation across a plasma
membrane.

It is important to distinguish between transport by a genuine group
translocation process (Fig. 2-1A) and by a so-called "trapping" mechanism. In the latter, an apparently vectorial process results from an anisotropic arrangement of the system, as diagrammed in Fig. 2-1B. If either
the group transfer enzyme E or the acceptor substance AX (or both) are
present only on the inside of the membrane, and if in addition the membrane is impermeable to AS and DX, then the constituent groups of the
solute DS will be accumulated inside the cell. In this mechanism, however,
the unaltered solute passes through the membrane by diffusion and appears
in the cytoplasm. The diffusion can be either passive or facilitated by a
membrane carrier, but the apparent accumulation of solute moieties against
a concentration gradient depends entirely upon the anisotropy of the
arrangement and the equilibrium constant for the group transfer reaction.
The third class of transport mechanisms, "true" active transport (Fig.
2-1C), is represented by the systems discussed in other chapters. In
these cases the unaltered solute is accumulated against a concentration
gradient. Thus the distinguishing characteristics of a true group translocation transport process are that membrane permeation is concomitant
with a macroscopically vectorial group transfer, and that the presence of
unaltered solute inside the cell is not the immediate result of the transport
process.

Some aspects of specific bacterial group translocation systems are
presented in this chapter. Although the available biochemical data are in
no case sufficient to justify formulation of a complete molecular mechanism, it is useful to list the requisite elements of such a mechanism as a

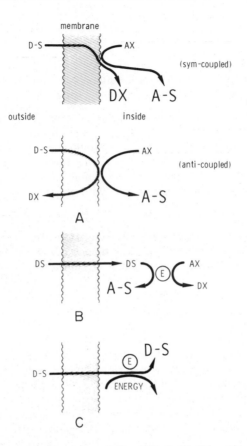

FIG. 2-1. Transport mechanisms (after Mitchell) [2]. (A) Group translocation of S moiety, by either "sym-coupled" or "anti-coupled" mechanisms. (B) Diffusion and trapping of S moiety. (C) Active transport of
free S.

somewhat utopian goal for researchers. The minimum set of events which
must be catalyzed by a group translocation system has been described by
Mitchell [3] and may include the following (not necessarily in the order
listed): replacement of interactions between the solute and its hydration
shell in the external aqueous medium with a set of solute-membrane interactions of approximately the same free energy; breaking of group-donor
bonds; movement of the group across the width (about 7 nm) of the membrane, perhaps by conformational changes in a carrier molecule and/or
diffusion along a stereospecific "pore"; formation of group-acceptor bonds;
and rehydration of the group by the water of the cell interior. In addition
to specification of the nature of these events, mechanistic schemes, should,
according to Mitchell, provide answers to the following questions:

1. How is the translocation coupled to the group transfer? The simplest alternatives are given in Fig. 2-1A.

2. How is the given solute chemically "divided" for the transfer? For example the moiety of a solute G-OH which is actually translocated could be G., G^+, G:$^-$, GO., or GO:$^-$.

3. What is the actual microscopic mechanism of the apparent macroscopic group translocation? For example, facilitated diffusion of the unaltered solute across the membrane into an inner region not accessible from the cell cytoplasm, followed by group transfer mediated by an enzyme directly connected to that region, would be macroscopically indistinguishable from a group translocation transport mechanism.

Descriptions of various bacterial group translocation mechanisms appear in a review by Kaback [4]. The three considered here are those for sugars, fatty acids, and nucleic acid precursors.

II. THE PHOSPHOENOLPYRUVATE:HEXOSE PHOSPHOTRANSFERASE SYSTEMS

A potential advantage of group translocation transport systems is that the group transfers are biochemical reactions, and thus can often be assayed in cell extracts, where transport experiments are impossible. The components of the system can then be purified, isolated, and characterized, and the reaction subjected to at least some of the usual enzymological analyses. The potential for biochemical characterization has been extensively exploited in the case of the bacterial phosphoenolpyruvate: hexose phosphotransferase systems (PTS), particularly the glucose, fructose, and mannose systems of enteric bacteria, and the lactose PTS of Staphylococcus aureus. After a general introduction to the PTS, the detailed enzymology of these particular systems will be presented in the two following sections. With this background the physiological characteristics of the PTS for other sugars and those found in other bacterial genera will be surveyed in Secs. D and E. In the final section, the hazardous transition from in vitro to in vivo phenomena will be attempted, as the functioning of the PTS as a transport process is discussed and mechanistic questions are considered. The PTS has been covered by a number of excellent reviews, which have stressed either physiological and biochemical [5-7] or genetic [8-10] aspects.

A. Reactions and Components of the PTS; General Description and Nomenclature

The overall PTS reaction, first described by Kundig et al. in 1964 [11], is the phosphoenolpyruvate (PEP)-dependent conversion of a hexose to its phosphate ester. In most, but not all, the PTS thus far subjected to thorough biochemical analysis, the conversion is accomplished by a series of phosphoryl transfers, according to the general scheme below:

$$P\text{-Enolpyruvate} + \text{enzyme I} \underset{}{\overset{Mg^{2+}}{\rightleftharpoons}} P\text{-enzyme I} + \text{pyruvate}$$

$$P\text{-Enzyme I} + HPr \rightleftharpoons P\text{-HPr} + \text{enzyme I}$$

$$P\text{-HPr} + \text{enzyme III}^X \rightleftharpoons P\text{-enzyme III}^X + HPr$$

$$P\text{-Enzyme III}^X + \text{sugar}_X \xrightarrow[\text{complex}]{\text{enzyme II}^X} \text{sugar}_X \text{-P} + \text{enzyme III}^X.$$

(2)

Enzyme I* and HPr[†] are nonspecific proteins, and catalyze the phosphorylation of all sugar substrates for the PTS (PTS sugars[††]). (In certain special cases HPr is not obligatory. See Sec. II.E3.) The enzymes II are membrane bound and are usually sugar specific; the enzymes III[§] are also sugar specific, but may be either soluble or membrane bound. These components are denoted by $EIII^{lac}$ and EII^{lac} for a lactose PTS, $EIII^{glc}$ and EII^{glc} for a glucose PTS, and so forth. A brief explanation of PTS nomenclature is helpful here. Enzyme I can be considered as an enzyme in the classical sense, since it catalyzes the transfer of a phosphoryl group from PEP to the protein substrate HPr via a stable enzyme intermediate, P-EI. With equal validity enzyme I can be viewed as a cyclically regenerated phosphoryl carrier protein which catalyzes its own PEP-dependent phosphorylation. The function of HPr is not truly enzymatic, but rather analogous to the acyl carrier protein (ACP) which functions in fatty acid

---

*Enzyme I (EI): nonspecific protein of the PTS.

[†] HPr: nonspecific phosphoryl carrier proteins of the PTS; distinct from enzymes II (membrane bound, glucose-specific, lactose-specific, etc., enzymes of the PTS).

[††] PTS sugars: those transported via the PTS in bacteria.

[§] Enzymes III: lactose-specific, glucose-specific, etc., phosphoryl carrier proteins of the PTS; distinct from enzymes II.

biosynthesis. It is unfortunate that the rather cryptic designation HPr
(Histidine Protein) is so firmly embedded in the literature. Both HPr and
the phosphoryl carrier proteins with which they interact, the enzymes III
in the above scheme, are most logically considered as coenzymes, serving
as substrates for enzyme I and II, respectively. The enzymes III are
soluble in some PTS but are membrane bound in others. The cofactor
nature of the soluble enzymes III is unambiguous, and they are usually
referred to in the literature as "Factors III." Although the role of the
membrane bound enzymes III in the reaction sequence is analogous to that
of the soluble ones, they have usually been given designations like enzyme
II-A^{glc}, etc. This stems from the logical surmise that they form some
sort of complex with their corresponding enzymes II in the membrane. It
is not at all clear whether a membrane bound "enzyme" III should be considered a cofactor, a prosthetic group, or a subunit of the corresponding
enzyme II. In many PTS the existence of a membrane bound enzyme III^X
has not been demonstrated, but only inferred from the absence of a soluble
one. It remains to be proven that the four-component PTS are universal.
In many instances a fractionation like that described in Sec. II.C2a has not
been accomplished, so the term "enzyme II complex"* is used, reflecting
uncertainty as to the actual state of affairs. In summary then, no set of
names ("enzyme", "factor", "phosphoryl carrier protein," etc.) which will
be both accurate and general is possible at this time. For the sake of
conformity with the accompanying chapter on regulation by the PTS, the
designations enzyme I (EI), enzyme II^X (EII^X), enzyme III^X (EIII^X), and
HPr will be used, with apologies for the biochemical inconsistency of the
nomenclature.

B. Enzymology of Nonspecific PTS Proteins

1. Enzyme I

The enzyme I from S. aureus has been partially purified [12,13]; that from
Salmonella typhimurium has been highly purified by the Roseman group [14].
(The Escherichia coli and S. typhimurium enzymes appear to be identical.)
Molecular weights, estimated by analytical gel filtration, were about
70,000 and about 90,000, respectively. In no case has homogeneity been
claimed, probably because of the extreme lability of the enzymes. The
S. typhimurium enzyme appears to contain an unknown number of subunits,
which may or may not be identical. In contrast, the S. aureus enzyme was
reported by Hengstenberg to be a single polypeptide chain, since crude
[^{32}P]enzyme I migrated at a position corresponding to a molecular weight

---

*A membrane bound PTS activity for which existence of a distinct
enzyme III has not been proven or disproven.

of 80,000 during electrophoresis on polyacrylamide gels containing dodecyl sodium sulfate [15]. The uncertainties inherent in molecular weight estimates by gel filtration and gel electrophoresis would allow for two subunits of unequal size in the S. aureus enzyme, however. Phosphorylation of enzyme I, using [^{32}P]PEP as a donor, has been demonstrated in both cases. The reaction requires Mg^{2+}, and the product in both cases is apparently a phosphoryl group linked to a histidine residue, probably at N-3. Formation of [^{32}P]HPr from [^{32}P]EI and HPr in the absence of Mg^{2+} has also been reported for both enzymes. A soluble enzyme I, molecular weight about 150,000, has been identified in Mycoplasma capricolum by Cirillo [16,17].

2. HPr

The second phosphoryl carrier protein in the PTS sequence is the so-called Histidine Protein, or HPr. Apparently identical HPr's were purified to homogeneity from E. coli by Anderson and coworkers [18] and from S. typhimurium. The S. aureus protein was similarly purified and characterized by Simoni et al. [19], and the HPr from M. capricolum has recently been purified by Ullah and Cirillo [17]. These proteins all consist of small single polypeptide chains (molecular weights: S. typhimurium, 9500; S. aureus, 8700; M. capricolum, 9500). They are quite stable and easily purified. Despite their physical similarity, the amino acid compositions of these three HPr's differ considerably. Amino acid sequence studies on S. typhimurium and S. aureus HPr are well along. It will be interesting to see if there is homology between the region of the single histidine of S. aureus HPr and of one of the two histidines in the S. typhimurium protein.

In all cases a single phosphoryl group is transferred to HPr from PEP; the equilibrium constant for the overall reaction

$$\text{PEP} + \text{HPr} \underset{\text{Mg}^{2+}}{\overset{\text{EI}}{\rightleftharpoons}} \text{P-HPr} + \text{pyruvate}$$

has been measured using [^{32}P]PEP and homogeneous HPr [7,19]. The value, about 10 in both S. aureus and S. typhimurium, corresponds to a drop of only 1.4 kcal/mol from the high energy level of the PEP phosphoryl group (ΔG° of hydrolysis, -14.8 kcal/mol [20]). Anderson and coworkers established the nature of the phosphoryl protein linkage by analysis of isolated E. coli [^{32}P]HPr. Alkaline hydrolysis yielded a product with the chromatographic properties of N-1 phosphohistidine, confirming the results of studies of the kinetics of phosphoryl hydrolysis [18]. In the case of S. aureus P-HPr, early hydrolysis experiments by Simoni et al. [19] suggested an N-1 phosphohistidine linkage; Hengstenberg and coworkers subsequently confirmed this assignment by paper electrophoresis of an alkaline hydrolysate and studies of the proton magnetic resonance spectrum of the intact phosphoprotein [15].

In vitro complementation experiments indicate that P-HPr of S. aureus cannot interact functionally with the sugar-specific components (enzymes II, III) of E. coli, and vice versa, but that E. coli enzyme I can (inefficiently) phosphorylate S. aureus HPr, and conversely [12]. There is an interesting asymmetry to the complementation pattern with M. capricolum and E. coli components [17]. A P-HPr generating system (EI + HPr + PEP) from M. capricolum was as effective as one from E. coli when an E. coli membrane fraction was used to catalyze the phosphorylation of methyl-α-glucoside (α-MG), but when M. capricolum membranes were used, the heterologous E. coli P-HPr was only 1/10 as effective as Mycoplasma P-HPr. Experiments with isolated components indicated that E. coli EI catalyzed the phosphorylation of Mycoplasma HPr about one-fourth as well as Mycoplasma EI did, and vice versa.

In E. coli and S. typhimurium, enzyme I and HPr are coded for by the ptsI and ptsH genes, respectively. They are essentially constitutive proteins, but their levels are coordinately elevated three- to five-fold under certain growth conditions [21]. Genetic evidence indicates that these two genes constitute an operon under the control of a promotor-like element proximal to ptsH [22,23]. The term PTS-dependent will be used in this chapter to denote a demonstrated dependence upon PEP, enzyme I, and HPr, unless otherwise indicated.

C. Enzymology of Sugar-specific PTS Proteins

1. Systems with Soluble Enzymes (Factors) III

   (a) The EIII^{lac}/EII^{lac} System of Staphylococcus aureus

   (1) General Description. In S. aureus, the uptake of β-galactosides is concomitant with their phosphorylation. Simoni and coworkers showed that this reaction requires the generation of P-HPr and the action of two lactose-specific proteins, the soluble enzyme III^{lac} (factor III^{lac}) and the membrane bound enzyme II^{lac} [24,25,19]. The cells accumulate lactose phosphate (6-phosphogalactosyl-β,1→4-glucose) rather than the free disaccharide; subsequent hydrolysis (to glucose + galactose-6-phosphate) is catalyzed by a phospho-β-galactosidase. The latter enzyme is inactive with free lactose [26-28]. All three sugar-specific proteins—EIII^{lac}, EII^{lac}, and the phospho-β-galactosidase—are induced if galactose-6-phosphate is added to the medium; presumably this compound (or a metabolic derivative) is the actual inducer during growth on lactose [29]. This staphylococcal lactose system should be clearly distinguished from the familiar E. coli system, which mediates the accumulation of free lactose by a PEP-independent active transport system and its subsequent cleavage by a β-galactosidase

(inactive with lactose phosphate). The reactions involving the S. aureus
lactose PTS proteins may be depicted as:

$$(\text{P-HPr} + \frac{1}{3}\text{EIII}^{\text{lac}}) \rightleftharpoons \text{HPr} + \frac{1}{3}(\text{P}_3\text{-EIII}^{\text{lac}})$$

$$\frac{1}{3}(\text{P}_3\text{-EIII}^{\text{lac}}) + \beta\text{-galactoside} \xrightarrow{\text{EIII}^{\text{lac}}} \frac{1}{3}\text{EIII}^{\text{lac}} + \beta\text{-galactoside-6-P}$$

(3)

(Free galactose is a substrate, but has less affinity for EII^{lac} than any
β-galactoside tested [12].)

(2) Enzyme III^{lac}. The phosphorylcarrier protein EIII^{lac} has been
purified to homogeneity by Hays et al. and extensively characterized [30].
It has a molecular weight of 36,000 and consists of three identical subunits.
The evidence for this novel trimeric structure consists of the following:
molecular weight determinations for the native protein by analytical equilibrium sedimentation at a variety of protein concentrations and centrifuge
speeds (and by gel filtration [101]); determination of the monomer molecular
weight (11,000 to 12,000) in guanidinium hydrochloride solutions by analytical equilibrium sedimentation and gel filtration; amino acid analyses; a
partial amino acid sequence (30 N-terminal residues); phosphorylation
experiments (see below). Neither equilibrium sedimentation nor gel
filtration experiments provided any evidence for further aggregation or
dissociation of the trimer in neutral aqueous buffers. The protein contains
no tryptophan and has three cysteines (one per monomer). When the latter
are reacted with iodoacetamide or other sulfhydryl reagents, the covalently
modified EIII^{lac} retains full enzymatic activity.

If an excess of P-HPr is used, up to three phosphoryl groups can be
directly transferred to EIII^{lac} [30]; neither any additional component nor
Mg^{2+} is required for the transfer. The phosphoryl groups are apparently
attached to histidine residues. Each EIII^{lac} monomer contains four
histidines; presumably a unique residue in each monomer is phosphorylated.
The effects of pH, hydroxylamine, and pyridine on the hydrolysis of the
phospho-EIII^{lac} phosphoryl linkage were characteristic of phosphohistidines;
alkaline hydrolysis of [^{32}P]phospho-EIII^{lac} yielded a product which co-
chromatographed with N-3-phospho [^{14}C]histidine [30]. It is not known
whether the phosphorylation of the EIII^{lac} monomers is independent,
cooperative, or anticooperative. Since P$_3$-EIII^{lac} does not dissociate into
monomers under physiological conditions, it would provide an interesting
system for a study of covalent modification of oligomeric proteins, a process to be contrasted with the reversible binding of ligands to oligomers,
which has been the subject of considerable research. The equilibrium
constant for the phosphoryl transfer from P-HPr to EIII^{lac} is about 0.5
to 1.0, if independent phosphorylation of the monomers is assumed [19].
Thus the "high-energy bond" of the phosphoryl linkage in PEP is still
largely conserved in phospho-EIII^{lac}.

(3) <u>Enzyme II^{lac}</u>. Purification and characterization of enzyme II^{lac}, the membrane bound component of this system, has been hindered by the usual difficulties encountered in work with particulate enzymes. The protein appears to be quite firmly attached to the membrane, since a variety of detergents, chaotropic agents, and organic solvents failed to solubilize EII^{lac} in a form amenable to chromatographic fractionation [12]. A maximal purification of 10-fold was achieved by Simoni and coworkers [12] using a three-stage detergent treatment. The resulting preparation retained less than 10% of the original membrane phospholipids; its enzymatic activity could be stimulated by readdition of lipids. Simpler procedures, yielding "washed membranes," or "deoxycholate-alkali membranes" were used to prepare membrane samples for binding studies. Membrane preparations containing EII^{lac}, either from a strain constitutive for the lactose PTS or from wild-type cells grown on lactose, bound lactose with a K$_D$ of 0.3 μM [19], as measured by a centrifuge assay; enzyme III^{lac} was not required. No binding was detected using membrane preparations from uninduced wild-type cells, or from most mutant derivatives (of the constitutive strain) missing EIII^{lac} activity. Interestingly, a few mutants defective in EII^{lac} activity (as judged by the in vitro phosphorylation assay and transport experiments with whole cells) nevertheless retained substantial lactose binding capability.

In another laboratory, Korte and Hengstenberg [31] "solubilized" EII^{lac} by treatment with the nonionic detergent Triton X-100 and subjected it to ion exchange and gel filtration chromatography. Solubilization was defined by a sedimentation criterion: retention in the supernatant after centrifugation at 155,000 x g for 60 min. The ambiguous nature of solubilization by detergents is illustrated in this instance by the apparent molecular weight of 10^6 for EII^{lac} in the presence of Triton, as judged by chromatography on an agarose gel column. The molecular weight of Triton micelles is about 10^5 [32], so this EII^{lac} preparation may consist of an aggregated complex of membrane protein or lipoprotein with some intercalated lipid. After a treatment which delipidated the EII^{lac}, enzymatic activity was lost; activity could be restored by addition of Triton but not of phospholipids. Gel electrophoresis of the purified preparation in dodecyl sulfate indicated a high degree of purity. Although only a single band was observed, the apparent molecular weight, about 35,000, is characteristic of many membrane proteins, and the result cannot be taken as definitive proof of homogeneity (or as indicative of the native molecular weight of EII^{lac}). Binding studies were performed by Hengstenberg and Weil using the purified preparation and the fluorescent lactose analog 2-(N-dansyl)-aminoethylthio-β-galactoside [15]. This compound, which is reportedly phosphorylated and transported by the lactose PTS, binds to EII^{lac} with a concomitant increase in fluorescence intensity; the K$_D$ is about 10^{-5} M. The presence of lactose or IPTG competitively inhibits binding by the fluorescent substrate; K_D values for these sugars estimated from the competition data are in good agreement with previous estimates from [^{14}C]lactose binding experiments [19].

(b) The High-Affinity $EIII^{glc'}/EII^{glc'}$ System of E. coli and S. typhimurium. Two distinct glucose PTS play major roles in the metabolism of glucose by enteric bacteria. The nomenclature employs a prime to distinguish these systems, whose physiological and genetic characteristics are compared in Sec. II.E2. The biochemically identifiable sugar-specific components of the $EIII^{glc'}/EII^{glc'}$ system, as studied in both E. coli and S. typhimurium, are a soluble phosphoryl carrier protein enzyme $III^{glc'}$ ("factor $III^{glc''}$") and a membrane bound enzyme $II^{glc'}$ ("enzyme IIB'"). Enzyme $III^{glc'}$ has been purified to apparent homogeneity by Kundig [33]; the molecular weight, as estimated by gel filtration, is about 20,000. The native protein is dissociated irreversibly into three or four subunits by heating (70°C, 5 min) or by chelating agents, but the dissociated preparation retains full enzymatic activity. The homogeneity of the purified preparation is put into question by copurification of a potent hexose-6-P phosphatase activity, specific for sugars of the D-gluco and D-manno configurations. Although one or two preparations do not exhibit the phosphatase activity, treatments which dissociate $EIII^{glc'}$ to its subunits do eliminate the latter and not the phosphoryl carrier activity. The independence of the two activities thus remains in doubt.

The direct reversible transfer of a phosphoryl group from [^{32}P]HPr to $EIII^{glc'}$ has been demonstrated by Kundig [33]; the equilibrium favors P-$EIII^{glc'}$. The stoichiometry, one phosphoryl per 20,000 molecular weight, indicates that only one of the subunits (molecular weight, 5000–7000) is phosphorylated. Since dissociated preparations retain full activity, the subunits are either not identical, or are arranged in a highly asymmetric fashion in the oligomer. Genetic evidence identifies a single genetic site for $EIII^{glc'}$ close to or identical with the crr locus (see Chap. 3), but this is not strong evidence that the subunits are identical. This locus is quite close to the genes for enzyme I and HPr [22], but is not a part of the pts operon [23]. The phosphoryl protein linkage of P-$EIII^{glc'}$ appears to be an acyl phosphate, on the basis of hydrolysis studies and chemical identification of the phosphorylated residue [33].

Although the transfer of the phosphoryl group to sugar requires enzyme $II^{glc'}$, it does not involve any further stable phosphoryl protein intermediates. (The phosphatase activity does not require $EIII^{glc'}$.) The specificity of sugar phosphorylation is mainly for the D-gluco and D-manno configurations [33]. The $EII^{glc'}$ protein is particularly refactory to solubilization, but dispersion in Sarkosyl and subsequent partial purification [7] have served to clearly differentiate it from the other membrane bound enzymes II for glucose, EII^{GMF} (see Sec. II.C2c). The phosphorylation activity of the partially purified $EII^{glc'}$ can be stimulated about 40-fold by either phosphatidylglycerol or cardiolipin; neither phosphatidylserine nor phosphatidylethanolamine are effective. The $EII^{glc'}$ protein is extremely sensitive to sulfhydryl reagents, a property which serves to further distinguish it from EII^{GMF}.

2. Systems Lacking Soluble Enzymes (Factors) III

(a) The Low-Affinity GMF System ("IIA/IIB") of E. coli and S. typhimurium. Cells of E. coli strain K235, when grown on a glucose-minimal salts medium, yield membranes which, in the presence of PEP, EI, and HPr, catalyze the conversion of glucose, mannose, fructose, and certain of their analogs to the respective 6-phosphate esters [34]. This system is often referred to as the "constitutive" or "low-affinity" system, and is distinct from the $EIII^{glc'}/EII^{glc'}$ system (the "high-affinity" or "inducible" system). The actual inducibility of these two systems, and their relative physiological importance, are somewhat controversial. These matters are discussed briefly in Sec. II.E2.

Kundig and Roseman [35] found that extraction of membranes from glucose-grown K235 cells with urea and 1-butanol yielded an inactive supernatant fraction ("EIIA") which, when supplemented with the residual pellet ("EIIB") efficiently catalyzed PTS-dependent sugar phosphorylation. Subsequent purification of the supernatant fraction in aqueous buffers yielded a (lipid-free) preparation which was resolved into three protein fractions when subjected to isoelectric focusing. The peaks coincided with peak activities for phosphorylation of fructose, methyl-α-glucoside and N-acetylmannosamine, respectively, as determined by assays in the presence of the pellet fraction, EI, HPr, and PEP. In accordance with the nomenclature adopted in this chapter, I designate these "IIA" fractions $EIII^{fru}$, $EIII^{glu}$, and $EIII^{man}$, although there is considerable overlap in specificity (for example, the fraction designated $EIII^{glc}$ has some activity with sugars of the manno configuration [35]). (Note that the cytoplasmic $EIII^{glc'}$ and the $EIII^{glc}$ extracted from membranes into a water-soluble form are entirely different proteins.) Each of the three isoelectric focusing fractions exhibited a single protein band upon conventional disc gel electrophoresis and appeared to be homogeneous.

The residual pellet from the urea-butanol extraction, suspended in buffers containing deoxycholate, was further purified by Kundig and Roseman. The lipid-free preparation obtained, designated "EIIB", aggregated in the absence of detergent, but gel electrophoresis under denaturing conditions (dodecyl sulfate) gave a single protein band, corresponding to a molecular weight of about 36,000 [35]. This was not considered sufficient evidence for a definite conclusion about the homogeneity of this protein, which was active with $EIII^{glc}$, $EIII^{fru}$, and $EIII^{man}$ in reconstitution experiments. Although recent genetic evidence [7,10] is consistent with the notion that the "IIB protein" is the product of a single gene, it is not yet proven that the preparation described above is a single protein with multiple specificity rather than several similar proteins with the same polypeptide chain weight. In this chapter the designation enzyme II^{GMF} is used to indicate this uncertainty. Note again that EII^{GMF} ("EIIB") is entirely distinct from $EII^{glc'}$ ("EIIB'"), although both are membrane proteins catalyzing glucose phosphorylation by PTS. If the EII^{GMF} preparation is indeed a single protein, the yield from the purification procedure suggests that it is a major component of the membrane.

Reconstitution of an active complex from EII^{GMF} and EII^{glc} (or $EIII^{fru}$ or $EIII^{man}$) required, in addition, phospholipid and a divalent cation. Interestingly, good reconstitution of activity depended on mixing in the following order: EII^{GMF}, divalent cation, lipid, EII^{glc}. Either Mg^{2+} or Ca^{2+} could be used, but the only effective lipid was phosphatidylglycerol. Since this lipid is only a minor component of the E. coli membrane, this result suggests caution in attempting to correlate changes in the physical properties of the bulk membrane lipids with effects upon individual membrane bound enzymes. Optimal reconstitution of activity was achieved at a lipid:protein ratio corresponding to about 50 phosphatidylglycerol molecules per EII^{GMF} polypeptide chain.

The nomenclature used here implies some biochemical analogy between the soluble enzymes III discussed in the preceding sections and the $EIII^{glc}$, $EIII^{fru}$, and $EIII^{man}$ proteins, which are extracted into a water-soluble form by a relatively mild treatment of the membrane. The analogy is strengthened by the demonstration by Kundig that purified $EIII^{glc}$ is directly phosphorylated by P-HPr in the absence of added EII^{GMF} or lipid [33]. Subsequent transfer to sugar requires EII^{GMF} and lipid. In another experiment small membrane vesicles (containing both EII^{GMF} and the EIII proteins) were phosphorylated using a phosphoryl generating system (EI + HPr + [32P]PEP, and the [32P]membranes isolated by sedimentation in a glycerol density gradient [33]. Addition of glucose to the isolated fraction resulted in the rapid formation of glucose-6-[32P]phosphate.

(b) The Inducible β-Glucoside PTS and Other "Unresolved" Systems. Strains of E. coli which can use methyl-β-glucoside as a carbon source for growth are able to induce an arylphospho-β-glucosidase [36] (analogous to the phospho-β-galactosidase of S. aureus) and a β-glucoside transport system. The latter was shown by Fox and coworkers to be PTS-dependent [37], and partial purification of its enzyme II $^{\beta-glc}$ complex was achieved [38]. The purified preparation contained some phospholipid (but could be stimulated by phosphatidylglycerol) and exhibited several bands when subjected to gel electrophoresis in dodecyl sulfate. The important fact is that the procedure did not resolve the EII $^{\beta-glc}$ complex into fractions analogous to those described in Section II.C2a (EII^{GMF} + EII^{glc}, EII^{man}, EII^{fru}), nor was there any evidence for a soluble factor in addition to enzyme I and

HPr. This result puts in question the generality of the four-component systems described thus far. It cannot be concluded in this case, however, whether the EII β-glc complex was simply not resolved, or was, in fact, a single protein species, since the final preparation was impure. The authors did not report whether their final preparation contained any elements of the GMF or EIIglc' systems. The question of how many components are in the EII β-glc complex thus remains open. This situation prevails for most of the PTS described in later sections of this chapter, since often enzymological characterization has not progressed beyond identification of membrane bound activities (enzyme II complexes), and demonstration of requirements for PEP, EI, and HPr. The only systems definitely known to require three soluble factors are the inducible lactose and mannitol/sorbitol systems of S. aureus [24], the E. coli EIIglc' system, and the inducible fructose and rhamnose system of Arthrobacter pyridinolis described by Krulwich and coworkers [39]. In no case, however, has a negative result excluding a second membrane bound component been unequivocally established.

D.   Biological Distribution of the PTS

Although PEP-dependent hexose phosphotransferase systems are found in a wide variety of bacterial genera, a number of investigations have failed to provide firm evidence for such systems in other procaryotic organisms or in eucaroytes [6]. It seems a little surprising that the strategy of coupling the transport step to the first step of metabolism is not more general. The distribution within the bacterial kingdom has been the subject of considerable study. The most extensive survey was performed by Romano and coworkers for glucose phosphotransferases; they grew cells on glucose-salts media and assayed crude extracts for phosphorylation of the nonmetabolizable analog 2-deoxyglucose [40]. The results of these and other surveys are given in Table 2-1. The general pattern is the following: (1) strict aerobes do not have glucose PTS (but note fructose and rhamnose PTS in A. pyridinolis); (2) most facultative anaerobes have a glucose PTS, if and only if they metabolize hexoses by the glycolytic (Embden-Meyerhof) pathway, which yields two molecules of PEP for each hexose metabolized; (3) heterofermentative genera such as Lactobacillus and Leuconostoc, which use the phosphoketolase pathway, or Zymonas mobilis, which uses the anaerobic Entner-Doudoroff pathway, do not have a PTS, presumably because these pathways only yield one PEP per hexose, and PEP-dependent transport would result in no net energy gain [41]; and (4) although few anaerobes have been tested, they would be expected to have PTS (but see Clostridium thermocellum for glucose). In Table 2-1, in cases where the available data for glucose is negative or not known, known PTS for other sugars in that organism have been noted. Thus the list is not necessarily inclusive for sugars other than glucose.

TABLE 2-1 Distribution of PTS's Among Bacterial Genera

Strain	Glucose analogs[a] phosphorylated	Other sugars[b] phosphorylated	Reference
Escherichia coli	αMG, 2dGlc	(Many)	7
Salmonella typhimurium	αMG, 2dGlc	(Many)	7
Streptococcus lactis	?[c]	β-galactosides	73
Streptococcus mutans	2dGlc	Mannitol, Sorbitol	51, 72
Bacillus subtilis	αMG	?[c]	98
Bacillus cereus	2dGlc	?[c]	40
Bacillus megaterium	2dGlc	?[c]	40
Achromobacter pavulus	2dGlc	?[c]	40
Corynebacterium ulcerans	2dGlc	?[c]	40
Clostridium thermocellum	None	Fructose, Mannitol	65
Rhodospirillum rubrum	None	Fructose	66
Mycoplasma (various species)	αMG	?[c]	99
Arthrobacter pyridinolis	None	Fructose, Rhamnose	39, 64
Clostridia (various species)	None	Fructose	100
Lactobacillus arabinosus	(Glucose)	Mannose	11

TABLE 2-1 Distribution of PTS's Among Bacterial Genera (Cont'd)

Strain	Glucose analogs[a] phosphorylated	Other sugars[b] phosphorylated	Reference
Staphylococcus aureus	αMG, 2dGlc	(Many)	80, 46
Spirochaeta aurantia	None	Mannitol	95
Lactobacillus (heterofermentative species)	None	(Probably none)	41
Leuconostoc (heterofermentative species)	None	(Probably none)	41
Zymonas mobilis	None	(Probably none)	41
Bacterioides fragilis	None	(Probably none)	41

[a]Nonmetabolizable glucose analogs phosphorylated indicated where possible; in many cases glucose was positive as well.

[b]Other sugars indicated for species where glucose is not a PTS sugar; entries in this column are not necessarily complete.

[c]?, information not available.

E. Comparative Physiology and Genetics
   of the PTS

1. Introduction: Problems Inherent
   in Comparative Studies

   (a) Quantitative Assay of PTS Components and Parameters. Most
bacterial PTS have been only partially characterized. Typically, genetic
identification of a given system (by isolation of mutants unable to grow on
a given sugar) is verified by in vitro phosphorylation assays. Sometimes
these assays are performed using whole cells made permeable by toluene
or ether, sometimes a separation of membrane and soluble fractions is
made, and sometimes further subdivision of the soluble fraction is
achieved, so that a dependence on HPr-like and EI-like components can be
demonstrated. Often apparent Michaelis constants (K_m) are estimated.
The objective of the sections below is to compare with one another the
phosphotransferase systems employed for a given sugar by various bacterial species. (Sometimes a given species uses two different systems for
the same sugar.) Items of interest include the nature of the components of
the systems, their inducibility or constitutivity, and their relative affinities
for their substrates (as judged by K_m values). Crucial to these comparisons are accurate and reproducible measurements of the relative levels of
each of the PTS components in cells grown under various conditions, and
K_m measurements that genuinely reflect enzyme-substrate affinities. The
experimental complexities inherent in such measurements have not always
been appreciated. In fact, the complexity of the experimental situation is
such that direct quantitative comparisons between in vitro and in vivo
results are virtually hopeless, and comparisons between different in vitro
situations are hazardous at best. The result has been considerable confusion and controversy whenever data from different laboratories are compared. If systems as complicated as the PTS are ever to be understood at
the molecular level, it is essential that the requirements for quantitatively
meaningful measurements be recognized. It seems worthwhile, therefore,
to discuss these factors in some detail. The four-component lactose PTS
of S. aureus (see Sec. II.C1a) will be used as a model. The reactions can
be written in the simplified form below, ignoring the facts that the first
reaction proceeds via a stable P-EI intermediate, and that $EIII^{lac}$ is
trimeric.

$$PEP + HPr \xrightarrow{\text{EI}} P\text{-}HPr + pyruvate \quad \text{"enzyme I reaction"} \quad (4)$$

$$P\text{-}HPr + EIII \rightleftharpoons P\text{-}EIII + HPr \quad (5)$$

$$P\text{-}EIII + S \xrightarrow{\text{EII}} EIII + S\text{-}P \quad \text{"enzyme II reaction"} \quad (6)$$

In systems for which no EIII has been demonstrated, reactions (5) and (6) must be combined, i.e., P-HPr is considered to be a substrate of the EII complex. The conditions required for accurate estimation of the level of each component in a reaction mixture are not the same and are described separately below.

Quantitative measurement of HPr levels requires the following:

1. The amount of HPr, considered as a substrate of EI, must be low enough to be in the first-order range of the EI reaction.

2. The PEP level should be saturating. If physiological conditions do not correspond to saturating PEP, and the apparent K_m for HPr is markedly PEP-dependent, then this problem must be considered when in vivo and in vitro results are compared.

3. The steady-state level of P-HPr must be negligible, so that the amount of HPr added corresponds to the actual steady-state level, and so that the EI reaction does not run in reverse. This requirement demands that reaction (5) not be at all rate-limiting, which in turn requires a high level of enzyme III.

4. A high level of enzyme III requires not only addition of a large amount of this protein to the reaction mixture, but also a steady-state EIII:P-EIII ratio so high that reversal of reaction (5) is not significant.

5. The high levels of EIII required are likely to result in levels of P-EIII which saturate EII, if the latter is present in only catalytic amounts. Thus there will be a significant steady-state concentration of free P-EIII, and reaction (5) will be driven in reverse. Such a situation can be avoided only by using enough EII to bind nearly all of the P-EIII, i.e., stoichiometric amounts. Such a situation may or may not prevail in vivo. If high amounts of a soluble EIII are added to an in vitro assay mixture, the EII level must be increased accordingly. (In the case of a membrane bound EIII-EII complex, the need for approximate stoichiometry presumably is satisfied automatically.)

It should be clear from the foregoing that even if HPr assays are "linear" in two different experiments, it does not necessarily follow that the actual levels are being measured with the same efficiency in both cases. It is necessary to demonstrate that further increases in enzyme II levels do not increase the rate of sugar phosphorylation. Assays for HPr which are not linked to sugar phosphorylation, e.g., pyruvate formation, may offer significant advantages.

Quantitative assay of enzyme I requires the following:

1. Saturation with PEP.

2. Saturating levels of HPr.

3. A steady-state ratio P-HPr:HPr low enough that the reverse reaction is not significant.

4. Conditions (3) above is the same as (3) for HPr measurement, so conditions (4) and (5) concerning EIII and EII levels apply to EI assays as well.

Quantitative assay of enzyme III requires the following:

1. A sufficiently potent phosphoryl-generating system (PEP, enzyme I, HPr) to maintain all of the EIII completely phosphorylated in the steady state.

2. A level of PIII sufficiently low for it to be in the linear range as a substrate in the enzyme II reaction.

3. Saturation of enzyme II with sugar.

Requirements for quantitative assays of enzyme II are

1. Saturation with sugar.

2. Saturation with P-EIII. This requires large amount of purified EIII protein, and a potent phosphoryl-generating system. If EIII and EII are in a membrane bound complex, the concept of saturation is meaningless; the "natural" EIII-EII stoichiometry will presumably be preserved unless the membranes are subjected to drastic treatments.

3. Reproducible membrane preparation procedures. Cell breakage efficiencies are variable, and activities must be expressed as units per milligram of membrane protein. The most effective cell disruption techniques are not necessary optimal for preparation of active membranes. Furthermore, variations in the extent of denaturation, oxidation, and proportion of vesicles oriented wrong-side-out vs right-side-out are likely. Sometimes dispersion of the membranes in nonionic detergent improves the reproducibility of membrane enzyme assays.

The requisite conditions for accurate measurement of Michaelis constants (K_m) and other kinetic parameters include the following:

1. Appropriate conditions to make the desired reaction (enzyme I or enzyme II) rate limiting.

2. Levels of the other substrate(s) that are either saturating, or reproducibly the same in all experiments. Ideally, apparent K_m values for one substrate should be extrapolated to saturating levels of the other substrate.

3. Recognition of the possibility that the components of one reaction may indirectly affect the apparent kinetic parameters of the rate-limiting reaction, e.g., altered enzyme I properties in the presence of membranes.

(b) Comparisons of In Vivo and In Vitro Kinetic Parameters. In general, agreement between the apparent sugar K_m values for the in vitro phosphorylation reaction with those for uptake by whole cells is not to be expected. Some of the reasons have been discussed in detail by Schachter [42] and will only be summarized here.

First, the actual conformations of a given enzyme II in whole cells, vesicles, and disrupted membrane fragments may not be the same.

Second, since the apparent kinetic parameters for the enzyme II reaction depend on the P-EIII level, they will be altered when EIII is subsaturating, or when the ratio EIII:P EIII is not negligible. Although this problem can be overcome in vitro by using enough EIII and a potent phosphoryl-generating system, the situation inside intact cells is much more difficult to manipulate.

Third, the kinetic equations for transport and phosphorylation are not the same, even for the simplest mechanisms and in the absence of other complicating factors [42]. For example, if we assume that the membrane translocation steps are equivalent to isomerizations of enzyme-substrate complexes, we can write the mechanism as:

$$EII_{Outside} + S_{Out} \rightleftharpoons (EII \cdot S)_{Out}$$

$$(EII \cdot S)_{Out} \rightleftharpoons (EII \cdot S)_{Inside} \qquad (*)$$

$$(EII \cdot S)_{In} + P\text{-}EIII_{In} \rightleftharpoons (EII \cdot S \cdot P\text{-}EIII)_{In}$$

$$(EII \cdot S \cdot P\text{-}EIII)_{In} \rightleftharpoons (EII \cdot S\text{-}P \cdot EIII)_{In} \qquad (*) \qquad\qquad (7)$$

$$(EII \cdot S\text{-}P \cdot EIII)_{In} \rightleftharpoons (EII \cdot EIII)_{In} + S\text{-}P_{In}$$

$$(EII \cdot EIII)_{In} \rightleftharpoons EII_{In} + EIII_{In}$$

$$EII_{In} \rightleftharpoons EII_{Out} \qquad (*)$$

As Cleland [43] and others have pointed out, such isomerization steps (*) in a kinetic mechanism may not alter the form of the equations and will therefore not be detectable as such by the methods of steady-state enzyme kinetics. The rate constants for these steps will necessarily appear in the various terms of the kinetic equations, however, and will thus affect the apparent values for the kinetic parameters. It seems obvious that even if there are analogous isomerization steps in the in vitro phosphorylation reaction, the rate constants will be different from those for the transport process.

The K_m values for a series of lactose analogs were determined by Simoni and Roseman [44] for both in vitro phosphorylation and in vivo transport by the lactose PTS of S. aureus. Similarly, the inhibition constants for tosyl galactosides (dead-end inhibitors of the lactose PTS) were determined by Hays and Sussman [45]. In both instances the in vitro values were as much as an order of magnitude lower (more apparent affinity) than the in vivo transport values. Such a result is not surprising, in view of the preceding discussion.

(c) Sugar Specificity of PTS. Comparisons of the levels of PTS components in various strains is complicated by the overlapping sugar specificities of some of the systems. For example, in the "low-affinity" $EIII^{GMF}$ system of E. coli and S. typhimurium, the factors $EIII^{glc}$, $EIII^{fru}$ and $EIII^{man}$ have significant secondary specificities; $EIII^{glc}$ catalyzes mannose phosphorylation fairly efficiently (See for instance Ref. 35). In S. aureus, recent work by Friedman and Hays indicates that mannitol and sorbitol induce the same soluble EIII and membrane bound EII [46]. Even galactose, a sugar not considered to be a PTS substrate ("non-PTS sugar")* in enteric bacteria, is phosphorylated by the EII^{GMF} system in certain mutant strains of S. typhimurium [47]. Fortunately, two systems with apparent specificity for the same sugar will often have different affinities for various analogs of the primary substrate. For example, the $EII^{glc'}$ system works well for αMG ($K_m < 20\ \mu M$), whereas the EII^{GMF} system works poorly for αMG (K_m, 25 mM) and well for 2-deoxyglucose (K_m, 0.2 mM). Thus a judicious choice of substrates and concentrations permits reasonable estimates of the relative levels of the two systems in unfractionated extracts [7].

(d) Constitutivity and Inducibility of PTS Components. In view of the problems discussed above, it is not surprising that much of the literature concerning the constitutivity of PTS components is confusing. Nevertheless, a few generalizations are possible.

First, most PTS for sugars other than the "main-line" substrates (glucose, fructose, and mannose) display the "classical" induction pattern: very low constitutive enzyme levels when cells are grown on the corres-

---

*Sugars which bacteria transport by other systems.

ponding sugar. Good examples are the lactose and mannitol systems of
S. aureus [24, 46] and the β-glucoside PTS of E. coli [37]. Second, the
constitutive levels of PTS activity for some non-main-line sugars in enteric
bacteria can be appreciable, for example, the mannitol PTS in some S.
typhimurium strains [48]. Third, some sugars can be phosphorylated by
both an inducible and a constitutive system. Generally the former have
higher affinities than the latter. The fructose systems in enteric bacteria
are good examples [49, 50]. Fourth, even proteins thought to be constitutive,
such as enzyme I, HPr, EII^{GMF}, and $EII^{glc'}$ and the corresponding
enzyme III, fluctuate markedly in their levels (three- to fivefold and more).
The levels depend on stage of growth, concentration of sugar substrate in
the medium, and other factors [7]. Fifth, an apparently constitutive
system, the EII^{GMF}, is totally missing in some strains of E. coli and
S. typhimurium [7].

2. Glucose and Mannose Phosphotransferases

Although PTS for glucose are widely distributed among bacterial genera
(see Sec. II. D), only the E. coli and S. typhimurium systems have been
extensively fractionated and biochemically characterized. Thus it is not
known whether analogs of the high affinity ($EIII^{glc'}$ + $EIIglc'$) and low
affinity ($EIII^{glc}$, $EIII^{fru}$, $EIII^{man}$ + EII^{GMF}) systems exist in other genera.
Partial characterization of the Mycoplasma [16, 17] and Streptococcus
mutans glucose systems has been reported [51, 52]. No inducible mannose-
specific systems have been reported.
    A major question is whether the two-system scheme is general for
most strains of E. coli and S. typhimurium, and whether it can be reconciled with existing genetic analyses. Two distinct classes of mutants which
grow "poorly" on glucose and are deficient in membrane bound PTS activity
for glucose and its analogs have been isolated using a variety of selection
techniques. The mutations, designated gpt by Curtis and Epstein [53], umg
by Kornberg and Smith [54], and tgl by Bourd and coworkers [55], are all
localized at 23 to 24 min on the E. coli map, as is the cat mutant described
by Tyler and collaborators [56] several years ago. Apparently distinct
from this class are the mpt mutations of Curtis and Epstein [53] and the
ptsX lesions of Jones-Mortimer and Kornberg [57]; all of these map at 35.5
min. Curtis and Epstein physiologically characterized these two classes of
mutants, and proposed that rather than corresponding to two distinct
glucose PTS, the genes coded for an enzyme II^{glc} (gpt, umg, etc.) and an
enzyme II^{man} (mpt and ptsX), the latter possessing a broader specificity
and therefore able to phosphorylate glucose as well. However, the data
are also consistent with the assumption that gpt mutants are missing
$EIIglc'$ and that mpt mutants have a deficiency in one or more of the membrane components of the GMF system. Resolution of the conflict requires

careful biochemical characterization of the mutants, including the following:
assays performed under conditions which clearly distinguish the two
systems; fractionation of the membrane protein by some variant of the
Kundig procedure [35] sufficient either to resolve the EII^{GMF} and $(EIII^{glc}$
$EIII^{fru} + EIII^{man})$ fractions in these strains or to demonstrate that most
"EII complexes" cannot be subdivided into EII and EIII components. Assay
conditions which take advantage of the much greater affinity of $EII^{glc'}$ for
methyl-α-glucoside, its requirement for added $EIII^{glc'}$, and its sensitivity
to sulfhydryl-blocking reagents, have been employed to distinguish $EII^{glc'}$
from EII^{GMF} [7,33]. Cordaro [10] used E. coli cells that had high levels
of both systems (even when grown on lactate) and showed that mutants of
the gpt class lacked the $EIII^{glc'}$-dependent membrane enzyme (high affinity
for αMG), but retained the low-affinity ($EIII^{glc'}$-independent) activities
for αMG, mannose, and fructose. In other experiments, similar mutants,
selected by Melton [58] for resistance to 2-deoxy-2-fluoro-glucose,
displayed comparable genetic properties, but the biochemical properties
were not unequivocally characterized. Cordaro further demonstrated that
mpt mutants retained the high-affinity αMG phosphorylation activity, but
were deficient in the low-affinity phosphorylation activities for glucose,
mannose and fructose. The simplest hypothesis is that gpt and similar
lesions are in the gene for $EIII^{glc'}$, whereas the mpt and ptsX mutations
affect the GMF system, and may well be in EII^{GMF} itself. In a recent
review [10], Cordaro proposed a systematic nomenclature encompassing
all of the known PTS genes.

3. Fructose Phosphotransferases

Anderson and coworkers found two fructose PTS in Aerobacter aerogenes
[59-61]. The high-affinity system, routinely assayed at 0.1 mM fructose,
is induced by growth on fructose; the product is fructose-1-phosphate. A
novel feature of this system is its absolute requirement for an inducible
soluble protein, originally designated "K_m factor." This protein actually
replaces HPr, so that the system still has only two soluble components.
A somewhat analogous HPr-independent PTS is characteristic of a class of
S. typhimurium strains, phenotypic revertants of ptsH mutants selected
for their ability to grow on fructose. Some of these revertants remained
HPr-negative but had greatly increased levels of a "pseudo-HPr complex"
[10], which replaced HPr in the PTS-dependent phosphorylation of a
number of sugars. The other fructose PTS in A. aerogenes is constitutive,
has a low affinity for sugar (assay is at 90 mM fructose), requires HPr
(K_m factor has no effect), and catalyzes the formation of fructose-6-
phosphate.
    Similar systems have been described in E. coli [40,50]. The activity
of the inducible, high-affinity system for fructose-1-phosphate formation
is missing from the membranes of the ptsF mutants of Ferenci and

Kornberg [50,62]. No evidence for an induced fructose-specific soluble factor has been reported, nor have attempts at fractionation of the EII^{fru} complex been described. The low-affinity constitutive PTS which phosphorylates fructose at carbon-6 (presumably the GMF system) is missing in ptsX mutants [55].

In S. aureus, lactate-grown cells possess a substantial fructose PTS activity, which is elevated about 10-fold in cells grown on fructose [46]. Preliminary experiments seem to rule out any soluble enzyme III^{fru}.

A genuine four-component PTS for fructose in the obligate aerobe A. pyridinolis has been characterized by Krulwich and coworkers [39]. A large number of fructose-negative mutants of that organism were classified into groups by in vitro complementation studies of the PEP-dependent formation of fructose-1-phosphate. Most of the mutants were rather leaky, but four complementation classes were obtained: a membrane bound component induced by growth on fructose, a soluble induced component, and two constitutive soluble components. These were designated "enzyme II" (EII^{fru}), "factor III" ($EIII^{fru}$), "phosphocarrier protein" (HPr), and enzyme I in a recent report. In this organism, there is no PTS for glucose, which is transported instead by a respiration-dependent active transport system [13]. A similar respiration-dependent system provides a second pathway for fructose uptake [63] ; initial uptake of fructose by this latter system is a prerequisite for induction of the fructose PTS. The only other PTS activity reported in this bacterium is for rhamnose [64].

Although no glucose PTS has been reported for C. thermocellum, this organism metabolizes fructose (and mannitol) by inducible PTS [65]. The photosynthetic bacteria Rhodospirillum rubrum and Rhodopseudomonas spheroides similarly lack glucose PTS but do possess a novel PEP-dependent system for the formation of fructose-1-phosphate [66]. The only two components in this fructose PTS are a firmly membrane bound enzyme and a soluble protein which adheres to the membrane but can be removed at low ionic strength. The latter was partially purified and had a molecular weight of about 200,000; it thus does not correspond to any known EI, HPr or EIII.

4. Mannitol, Sorbitol (Glucitol), and
   Galactitol Phosphotransferases

Inducible systems for these hexitols are widespread; galacitol systems are least common. Lengeler [67,68] has thoroughly described the genetic and physiological complexities of the E. coli systems. There are three parallel but distinct operons, each with a primary specificity for a given hexitol. The operons, mtl (mannitol), gut [glucitol(sorbitol)] and gat (galactitol), map at three different genetic loci. Each operon contains a cis-dominant control element (mtlC, gutC, gatC), an inducible EII (mtlA,

etc.), and an inducible hexitol phosphate dehydrogenase (mtlD, etc.). No
evidence for sugar-specific soluble components was obtained, and no
attempts were made to further subdivide the EII complexes. Considerable
overlap in specificity was observed. The EII^{mtl} had a low-level glucitol
activity, as did EII^{gat}; EII^{gut} had a good mannitol activity. Only EII^{gat} was
active with galactitol. The three systems are evidently the only ones for
hexitol uptake in E. coli, since mtlA⁻-gutA⁻-gatA⁻ triple mutants could
neither grow on hexitols nor transport them. The E. coli mannitol PTS
was first characterized by Solomon and Lin [69], who concluded that free
mannitol was probably the actual inducer of the mtl operon, since induction
was successful in an EI mutant. Evidence relevant to the question of
PTS-mediated facilitated diffusion of mannitol obtained by the Lin group
is discussed in Sec. II.E2. The first identification of a mannitol PTS was
made by Tanaka and Lin in A. aerogenes [70]; induction of this system in
EI mutants was also demonstrated. The sorbitol PTS of this organism was
described by Kelker and Anderson [71].

Inducible PTS for sorbitol and mannitol in S. mutans have been
described by Maryanski and Wittenberger [72]. Membrane bound inducible
components and soluble factors analogous to EI and HPr were identified,
but there was no evidence for an EIII-like factor. These systems are much
more sugar-specific than those in E. coli; sorbitol-grown cells had only
slight mannitol activity, and vice versa. Interestingly, the activity apparently corresponding to EII^{mtl} remained high in the soluble fraction obtained
from sonicated extracts, despite repeated ultracentrifugation. Perhaps the
EII^{mtl} in this organism has an unusually loose attachment to the membrane.

In S. aureus, a PTS containing both a soluble $EIII^{mtl}$ and a membrane
bound EII^{mtl} activity is induced by growth on mannitol [24]. Purification
and characterization of EII^{mtl} is in progress in the author's laboratory,
and preliminary experiments indicate several interesting properties. There
is a salt-dependent reversible aggregation of the protein, which elutes
from Sephadex G-100 in 50 mM KC1 as a broad peak (molecular weight
20,000-30,000) with a very broad shoulder of higher molecular weight
material. In 0.5 M KC1, the activity elutes as a single peak corresponding
to a molecular weight of about 15,000. The mannitol-induced $EIII^{mtl}$ and
EII^{mtl} also catalyze sorbitol phosphorylation [46]. The sorbitol reaction
requires PEP, HPr, and EI, and has a K_m comparable to that for the
mannitol reaction but only about one-fourth the maximal velocity. The
mannitol-induced system is apparently specific for hexitols; neither
$EIII^{mtl}$ nor EII^{mtl} can replace $EIII^{lac}$ and EII^{lac}, and vice versa [24].
When cells are grown for several generations on nutrient media with added
sorbitol, an EII-EIII pair with substantial hexitol PTS activity and the
same fourfold preference for mannitol over sorbitol is induced. When the
soluble fractions of extracts from cells grown with either mannitol or
sorbitol are further fractionated, the $EIII^{mtl}$ and $EIII^{stl}$ activities co-
chromatograph in every case, and with the same 4:1 activity ratio. In

contrast to the measurements of sorbitol PTS activity by the in vitro phosphorylation reaction are the results of whole cell transport experiments. Cells previously grown in the presence of sorbitol took up [^{14}C]sorbitol at a rate which was about 0.1% of the rate of mannitol uptake by mannitol-grown cells, and was almost indistinguishable from the rate for lactate-grown cells. Although addition of mannitol to a trypticase-vitamin medium greatly stimulated cell growth, sorbitol had little or no effect. Thus it appears that there is a single broad specificity PTS for the phosphorylation of these two hexitols and that the apparent sorbitol activity in vitro is physiologically irrelevant, a sobering thought for biochemists.

5. Lactose Phosphotransferases

In most bacteria which utilize lactose, metabolism proceeds via active transport of the free sugar and its subsequent cleavage to glucose and galactose by a β-galactosidase. The lac operon of E. coli is a well-known example (see Chap. 4). The lactose PTS of S. aureus described biochemically in Sec. II.C1a is thus an exception. A similar system, which also accumulates lactose phosphate and uses a phospho-β-galactosidase has been reported in certain strains of Streptococcus lactis by McKay and coworkers [73]. Crude extracts of these strains hydrolyze orthonitrophenyl-β-galactoside only when supplemented with PEP. Other strains of this species, however, have instead a non-PTS lac system analogous to that of E. coli.

6. Galactose Phosphotransferases

In S. aureus, galactose is a secondary substrate for the lactose PTS, with a K_m value about 500 times greater than that for lactose [12]. In enteric bacteria, however, galactose is not a PTS sugar, but is taken up by at least four distinct active transport systems [9] (see Chap. 4). Two recent reports suggest that in the absence of the active transport system, some E. coli and S. typhimurium strains can take up free galactose via a glucose enzyme II, without a need for EI or HPr. One interpretation of these results is that the EII mediates facilitated diffusion of the galactose. The biochemical bases for these phenomena are not well understood, however; these results will be discussed further in Sec. II.F2.

7. Phosphotransferases for Other Sugars

Maltose is not a PTS sugar in enteric bacteria. In. S. aureus, Button and coworkers found that the glucose PTS appeared to be involved indirectly.

After entry of maltose, apparently by facilitated diffusion, and cleavage by a cellular maltase, the released glucose is metabolized (primarily) by the PTS [74].

Egan and Morse found that sucrose was accumulated by S. aureus to very high levels (up to 3 M); it is at least partially in a derivatized form in sucrase-negative cells [75]. Preliminary experiments in the author's laboratory indicate that there is a substantial PTS-dependent activity which converts uniformly labeled sucrose to [^{14}C]sugar phosphate, and that glucose and fructose cause partial (but additive) inhibitions. (The glucose PTS and some fructose PTS activity are constitutive in these cells.) In A. aerogenes, the fructose generated internally by sucrose hydrolysis is normally metabolized via an inducible fructokinase. However, Kelker and coworkers found that mutants lacking this enzyme can still utilize the fructose by an alternate pathway which depends on the presence of the inducible fructose PTS [76]. Later experiments indicated that the process involved uptake and hydrolysis of sucrose, excretion of fructose, and reuptake via the fructose PTS [60] (see also Sec. II.F3). Sorbose can be used as a sole source of energy by A. aerogenes; an inducible PTS converts it to sorbose-1-phosphate [77]. Bacillus papillae grows on trehalose in its natural host, the Japanese beetle; the metabolic pathway involves PTS-dependent conversion to trehalose-6-phosphate and subsequent cleavage to glucose and glucose-6-phosphate by a phosphotrehalase [78]. No trehalase was detected.

F. Function of the PTS in Cellular
   Transport

Most of the enzymology described in Sec. II.B and II.C was elucidated by extensive study of the in vitro phosphorylation reaction; much of the physiological information presented in Sec. II.E was obtained in the same way. Correlation of this information with that obtained by in vivo sugar uptake studies has provided a pattern of circumstantial evidence strongly implicating the PTS in the actual transport process. In this section the emphasis is on the functioning of the PTS in vivo and in the "semi-in vivo" vesicle system devised by Kaback [79]. In particular, the following questions will be discussed. (1) Are the phosphorylation and membrane translocation steps obligatorily coupled? (2) Does the PTS mediate the facilitated diffusion of free sugars across the membrane in any (or all) circumstances? (3) Can the PTS catalyze the phosphorylation of internal free sugars? (4) Can the PTS catalyze the transfer of phosphoryl groups between sugars? (5) What can be said about transport mechanisms? Most of these questions are dealt with in the review by Postma and Roseman [7], and summaries of their arguments occur in several places in the discussion that follows.

1. Evidence for the Direct Role of the
    PTS in Sugar Transport

The first solid evidence for a major role of the PTS in transport was the
discovery by Egan and Morse that (enzyme I) mutants failed to transport
a number of sugars [80, 81] and that S. aureus accumulated most sugars as
the respective phosphate esters [75]. Similar results have been obtained
in other genera, but in many cases substantial internal free sugar is found
as well, probably because of hydrolysis by cellular phosphatases. Indirect
but compelling evidence is provided by correlations between the enzymatic
defects of mutants lacking PTS activity and their transport deficiencies.
Such correlations, obtained in a wide variety of bacterial species, are
particularly well exemplified by the results reported by Simoni and Roseman
[44] for S. aureus, probably because there are fewer secondary transport
systems in this bacterium. In this organism mutants lacking enzyme I
could not take up either lactose analogs (TMG) or glucose analogs (αMG).
The concentrations of radioactive sugar within the cells remained well
below their levels in the outside medium for at least 20 min, i.e., sugar
did not even enter these mutants by a diffusive process. Mutants deficient
only in factor III^{lac} or enzyme II^{lac} transported αMG at normal rates,
although TMG uptake was totally blocked.

     The membrane vesicles originally obtained from E. coli by Kaback [79]
have been successfully employed for the study of a variety of transport
systems in this and other organisms. These plasma membrane vesicles
lack most of the normal cytoplasmic constituents and the components
associated with the outer membrane and cell wall layers. They can take
up glucose and its analogs at a high rate if supplied with PEP, presumably
because soluble PTS components adhere to the membrane during preparation of the vesicles. This phenomenon was exploited by Kaback in a penetrating study of the uptake process [82]. He found that the accumulation
product was exclusively αMG-phosphate, which leaked into the external
medium only after its accumulation to high internal levels. The transport
process was absolutely and exclusively dependent on PEP, which was con-
sumed in 1:1 stoichiometry with αMG phosphate formation. Vesicles
prepared from enzyme I mutants failed to accumulate the αMG moiety. In
a crucial experiment, he preloaded vesicles with free [^{14}C]glucose under
nonphosphorylating conditions and transferred them to a medium containing
[^{3}H]glucose and PEP. The vesicles phosphorylated the external [^{3}H]glucose
much more rapidly than that present in the internal pool. This observation
thus provides strong evidence for a tight coupling of the translocation and
phosphorylation processes. This conclusion is strengthened by the results
obtained by Kelker and Anderson [60] using mutants of A. aerogenes which
lacked the inducible fructokinase. They found that when fructose was generated inside intact cells by hydrolysis of sucrose, it was used for growth

only after it had been excreted into the medium and then taken up again by the fructose PTS.

All the experiments discussed in this section strongly suggest that the PTS is a true group translocation system, mediating a PEP-dependent vectorial phosphorylation:

$$\text{Sugar (outside)} \xrightarrow[\text{PTS}]{\text{PEP}} \text{sugar phosphate (inside)} \qquad (8)$$

A schematic of a model PEP-mediated process is depicted at the top of Fig. 2. A number of reports have suggested that this view of the PTS is too simple, and that in some cases it can catalyze the facilitated diffusion of free sugars across the membrane, phosphorylate sugars presented to it internally, and exchange internal sugar phosphate for external free sugar. One or more of these activities have been invoked as part of physiological models for the transport process. The experimental situations which prompt these models are usually complex, and the results as well as the interpretations placed on them by various laboratories are sometimes contradictory. Although these hypothesized processes are clearly interrelated, it is easier to discuss them one at a time, and this is done in the next three sections. Before doing so, however, it is useful to have in mind just what sorts of molecular events are under consideration. A minimum set of postulated molecular processes is presented below and schematically indicated in Fig. 2; these postulates will be used as a basis for the subsequent discussions of the complex physiological data.

Postulate I: Free sugars, still complexed to enzyme II, are in physical contact with the cytosol while still unphosphorylated. (This implies that translocation precedes phosphorylation and is not directly coupled to it.)

Postulate II: Free sugar dissociates from its complex with EII directly into the cytosol without prior phosphorylation.

Postulate III: Enzyme II sugar sites on the interior of the membrane are available to free sugars in the cytosol.

Postulate IV: Free sugars which bind to internal EII sites can be translocated to the outside medium.

Postulate V: Free sugars from the cytosol which bind to internal EII sites can be phorphorylated in situ by the PTS.

For example, EII-mediate entry of free sugars by facilitated diffusion implies Postulates I and II, exit of free sugars via the PTS requires III and IV, and phosphorylation of internal sugars demands III and V. Note that II implies I, and IV and V imply III, but not conversely.

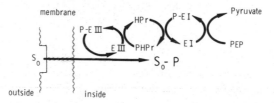

VECTORIAL PHOSPHORYLATION

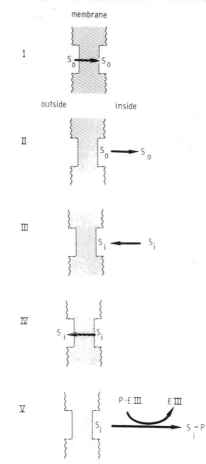

POSTULATED EVENTS IN ALTERNATIVE PROCESSES

FIG. 2-2 Vectorial phosphorylation of sugars by the PTS and postulated alternate PTS-mediated events. Top, coupling of PEP to vectorial phosphorylation by a typical PTS sequence. I-V, postulated events in other processes proposed to involve PTS components (see Sec. I. F1). S_0, sugar from outside medium, interacting with the PTS; S_i, sugar from cell cytoplasm, interacting with the PTS.

2. Does the PTS Catalyze Facilitated
   Diffusion?

Postma and Roseman [7] examined the evidence from a large number of
experiments and thoroughly discussed this question. Their conclusions
will be summarized here and supplemented with some additional material.
Facilitated diffusion seems to demand that PTS action include Postulates
I and II (Fig. 2-2) and the experimental evidence should be considered in
this light. Two studies provide strong evidence against facilitated diffusion.

The results of Simoni and Roseman [44] with S. aureus EI mutants
were described in Sec. II.F1. The essential point is that negligible quantities of the radioactive moieties of the sugars tested were taken up by
these mutants.

In his studies of PTS-mediated uptake of αMG by membrane vesicles,
Kaback [82] found that when the external concentrations exceeded 10^{-5} M,
a small but significant amount of free sugar appeared inside. However, the
rate of entry of free sugar was directly proportional to the external concentration over a million-fold range, whereas (PEP-dependent) sugar
phosphate accumulation displayed "saturation kinetics" (hyperbolic rate-
concentration curve) with an apparent K_m of 4×10^{-6} M. In this case,
then, the rate-limiting step in diffusive entry does not involve a saturable
carrier, i.e., an enzyme II. The levels of the internal pool of free sugar
were always less than one-third of the external level, and there was no
evidence for rapid exchange of that internal pool with external sugar. The
initial product accumulated was exclusively αMG-phosphate; free sugar
appeared only at later times.

Although the experiments described above, which utilized relatively
straightforward experimental systems, provide direct evidence against
facilitated diffusion via PTS enzyme II, a number of studies of more
complex situations have yielded considerable indirect evidence which can
be interpreted to the contrary.

Saier and coworkers found that unlike S. aureus enzyme I mutants,
those from S. typhimurium equilibrated free glucose across the membrane
fairly rapidly (about 3 min) and equilibrated mannose and fructose more
slowly (about 10 min [83]). Neither mannitol nor N-acetylglucosamine
could be equilibrated in 30 min, but the membranes of these cells could
phosphorylate the latter two sugars as well. It would thus appear that
either the apparent facilitated diffusion is limited to certain enzymes II, or
other hexose transport systems can facilitate a slow entry of the first three
sugars. It is known, in fact, that glucose and αMG can be taken up by at
least one of the non-PTS galactose transport systems [77]. A more recent
study of this phenomenon by Stock and Roseman (described in [7]) employed
methyl-α-glucoside and two PTS mutants of S. typhimurium [7]. Both a
tight ptsI mutant and a strain defective in the $EIII^{glc'} / EII^{glc'}$ system (the
principal pathway for αMG under the experimental conditions) equilibrated

αMG in 10 to 15 min. Gachelin found that the rate of PEP-dependent phosphoryl-ation by toluene-treated cells approached the rate of uptake by whole cells, but was extremely sensitive to N-ethylmaleimide (NEM) and other sulfydryl-blocking reagents [84,85] (both EI and EIII$^{glc'}$ are quite NEM-sensitive). Although NEM-treated intact cells equilibrated αMG with the external medium, the process was inhibited by galactose rather than glucose. It should be further noted that the strain used was constitutive for the lactose operon. Although the results are somewhat contradictory, the weight of the evidence suggests that the phenomena described in this paragraph are most simply ascribed to other carriers than PTS enzymes II.

A process closely related to facilitated diffusion is that of facilitated exchange. A typical experiment is to preload cells with radioactive sugar, transfer them to a medium containing an excess of unlabeled sugar, and then determine whether the decrease in internal sugar-specific activity (if any) appears to be carrier-mediated. A facilitated exchange mechanism could be constructed using the steps proposed in Postulates I, II, III, and IV (Fig. 2-2). However, the same phenomenon could result from PTS-mediated group translocation of sugar and subsequent hydrolysis of internal sugar phosphate, followed by efflux of free sugar. Such an efflux could in principle be mediated by a PTS EII (Postulates III and IV) or by another hexose permease, or could simply be the result of passive leakage. Al-though there have been several reports describing apparent facilitated exchange of methyl-α-glucoside by E. coli cells, interpretation of the results is complicated by the high rate of glucoside efflux characteristic of these cells (50% of the total internal pool of accumulated sugar moiety exited in 20 min [86]). After a detailed study of these events Haguenauer and Kepes [86] concluded that the principal route of exchange was in fact entry by group translocation, hydrolysis by cellular phosphatase, and efflux by an undetermined mechanism. In the experiments of Stock and Roseman (described in [7]), cells preloaded with radioactive αMG rapidly ex-changed their internal pool of αMG-phosphate when transferred to an unlabeled sugar solution, but exhibited little change in their pool of intracellular radio-active free sugar [7]. The sugar-sugar phosphate exchange, an interesting phenomenon in its own right, will be discussed in Sec. II.F4e. The exchange studies, taken as a whole, suggest that PTS enzymes II mediate neither influx of free sugars (Postulate II) nor direct efflux of free sugars (not coupled to exchange with incoming sugar phosphates) (Postulate IV).

In apparent contradiction to the results above are those obtained by Solomon and associates [87] in an ingenious experiment. They incubated E. coli cells constitutive for the lac operon with galactosyl-[^{3}H]mannitol, which was taken up by the lac permease and hydrolyzed internally by β-galactosidase. Radioactive mannitol was thus accumulated by the cells, but if the PTS enzyme II^{mtl} had previously been induced by prior growth on mannitol, then the steady-state level of the [^{3}H]mannitol moiety was about one-fifth the level of that in cells not induced for EII^{mtl}. The lower level of accumulation was attributed to an enhanced rate of [^{3}H]mannitol efflux mediated by the EII^{mtl}. It was not experimentally feasible to observe the mannitol efflux directly, however, and from the time course of accumulation shown, it is not possible to definitely conclude that the lower level of accumulation by the cells induced for EII^{mtl} was not actually the result of a lower initial rate of galactosyl-[^{3}H]mannitol uptake. The authors addressed this point indirectly by showing that the rate of ortho-nitrophenyl-β-galactoside uptake was unaffected by induction of the manni-tol PTS. This latter result, however, does not rule out an effect of EII^{mtl} on uptake of galactosyl-mannitol by the lac permease, perhaps by binding the disaccharide in an unproductive complex.

Two recent studies have shown that certain bacterial strains, appar-ently unable to take up galactose by any of the four transport systems normally utilized by that sugar, can nevertheless grow on galactose. The mechanism proposed was diffusion facilitated by a PTS enzyme II and sub-sequent phosphorylation by an (inducible) galactokinase. It was already known that the PTS does catalyze the phosphorylation of galactose in vitro, but the product, galactose-6-phosphate, is the substrate of no pathway known in enteric bacteria. In the first study, Postma [88] utilized a strain of S. typhimurium carrying the pleiotropic mem-1 mutation of Cordaro et al. [89]. This strain is defective in a number of membrane activities, including the PTS EII^{GMF} system and all three of the normal galactose transport systems ("melibiose," "methyl β-galactoside," and "galactose permease"). (Like most Salmonella strains, it has no lac operon.) The mutation seems to result in faulty insertion of membrane bound enzymes. All derivatives of this mem-1 strain selected for their ability to grow on mannose had, as expected, recovered PTS activity for αMG, mannose, and fructose, but they had also acquired the ability to grow on galactose, and to phosphorylate this sugar by the PTS in vitro [88]. The strains remained genotypically mem-1. Introduction of a ptsI mutation into these phenotypic revertants resulted in the loss of their ability to grow on mannose, but in a slight enhancement of their growth rate on galactose. It was concluded that entry of galactose into these revertants was mediated by a PTS enzyme II, but that the process was not coupled to the phosphorylation reaction. These revertants had not reacquired the ability to transport known substrates of the three "normal" galactose systems. Furthermore, αMG inhibited the galactose-dependent uptake of oxygen in these revertants of mem-1, but did not inhibit the process in wild-type strains possessing the normal galactose systems. This work thus provides indirect proof that a presumably mutant enzyme II complex can mediate facilitated diffusion of galactose, but two cautionary points should be made. First, the molecular basis for the reversion phenomenon is not known. The author's conclusions seem to require either that the membrane has been further altered by mutation in a gene coding for another component so that it now accomodates a PTS enzyme II but none of the galactose transport proteins, or that a mutant enzyme II has overcome the insertion problem and in every case simultaneously acquired an enhanced affinity for galactose. Neither hypothesis seems entirely plausible, and faulty insertion of a galactose transport protein so that its substrate range is altered cannot be ruled out by the data presented. Second, the putative role of the PTS enzyme II would be the result of considerable selective pressure upon the strain carrying this complex and poorly understood mutation, so that the relevance of these results to the physiology of normal cells is uncertain. More complete biochemical characterizations of the relevant strains is clearly required.

Similar phenomena exhibited by certain E. coli Hfr strains have been studied in considerable detail by Kornberg and Riordan [90]. These strains could grow on galactose only at high sugar concentrations, and did not appear to have either the methyl-β-galactoside or galactose permeases; the endogenous galactokinase effectively prevented induction of the lac operon by galactose. The absence of the melibiose system was inferred from previous reports that this system does not operate at 37°C in K12 strains of E. coli [91], but no experimental verification of this assumption was presented. Derivatives of this strain with mutations at the umg locus (EII$^{glc'}$) were identified by their inability to blacken film exposed to plates containing [^{14}C]αMG. These mutants simultaneously lost the ability to grow on galactose. Revertants of these galactose-negative mutants (selected for αMG uptake ability) and similarly selected umg$^+$ transductants had regained the ability to grow on galactose. Similarly, revertants of the umg strains selected for their ability to again grow on galactose had regained EII$^{glc'}$ activity (or had become constitutive for the lac operon). Although EII$^{glc'}$ was thus implicated in the low affinity system used for galactose by the Hfr strains, PTS-mediated group translocation was not involved. This conclusion follows from the observation that when ptsI mutations were introduced into the parental strains, the low-affinity utilization of galactose was not affected. However, several observations suggest that the model proposed, facilitated diffusion of free galactose via EII$^{glc'}$, may be too simple. First, although the umg (EII^{glc}-negative) mutants grew more slowly on galactose, their apparent "growth K_m" (galactose concentration for half-maximal growth rate) was identical with that of the parental strain. This suggests that a rate-limiting step involving recognition of galactose by some protein other than EII$^{glc'}$ may be involved. Second, the behavior of strain KR163, a galactose-positive revertant of a umg mutant is puzzling. This strain grows well on galactose, but still grows poorly on glucose, and PTS activity for αMG in toluenized cells is low. Although 5 mM glucose did interfere with the ability of this revertant to grow on 10 mM galactose, no direct support for the inference of competitive inhibition was presented. The complexity of these in vivo phenomena makes interpretation of the results difficult. It would appear that these strains have evolved an alternative to the usual galactose transport systems. This alternative could involve EII$^{glc'}$ directly, or a membrane factor, essential for the action of both the low-affinity galactose system and $EII^{glc'}$, which depends somehow on the presence of the $\underline{umg}^+$ allele. Further biochemical characterization of these strains and transport experiments using membrane vesicles would be helpful in clarifying these phenomena. It is noteworthy that the E. coli results were interpreted as evidence for galactose diffusion facilitated by $EII^{glc'}$ whereas the S. typhimurium study suggested a similar role for the EII^{GMF} system.

In summary it would appear that no firm conclusion about facilitated diffusion via the PTS can be drawn. On the one hand, the simpler systems, intact S. aureus cells and E. coli membrane vesicles, provide direct evidence against such a role for the PTS. On the other hand, the more complex experiments with intact cells of E. coli and S. typhimurium (which possess a greater variety of transport systems) provide indirect evidence on both sides of the question. Many of the results cannot be simply interpreted in terms of molecular events. It does appear that under selective pressure, some strains have evolved membrane components which permit galactose utilization independent of enzyme I and probably without the involvement of the "normal" galactose uptake systems. These components could either be an enzyme II "co-opted" for such a role or another membrane protein involved indirectly in EII PTS activity. The evidence does not appear sufficient at this time to directly implicate enzyme II as the carrier in these special cases, nor to justify generalization of these conclusions to all PTS-mediated transport processes.

3. Can the PTS Phosphorylate Internal
    Free Sugars?

The available evidence on this matter is also conflicting. The double-label experiments with vesicles (see Sec. II.F1) argue strongly against it, as do similar experiments with intact cells. For example, Kelker and Anderson [60] showed that $[^{14}C]$fructose formed by internal hydrolysis of sucrose was excreted by cells of A. aerogenes which were simultaneously utilizing external (unlabeled) fructose via the PTS. (The system which mediated the fructose exit was not determined.) There is both direct and indirect evidence in favor of PTS-mediated phosphorylation of internal sugars, however. Kashket and Wilson [92] studied an E. coli mutant which rapidly accumulated excessive quantities of TMG; internal TMG-phosphate subsequently appeared with a time course which suggested phosphorylation of the internal TMG pool. A likely explanation of the results is that the mutant had an $EIII^{glc'}/EII^{glc'}$ system with an enhanced affinity for TMG, normally a poor PTS substrate for the in vitro phosphorylation reaction. A double-label experiment was not performed, however, so that parallel accumulation of TMG via the lac permease and conversion of external TMG to internal TMG-phosphate by the PTS cannot be ruled out. Two other

studies dealt with the fate of potential PTS substrates released internally
by cleavage of disaccharides. In one case Curtis and Epstein [53] suggested
that the growth of galactose-negative, glucokinase-negative mutants on
lactose was a result of PTS-mediated internal phosphorylation of glucose
released by the action of β-galactosidase. However, efflux of free glucose
and subsequent uptake by the PTS is also possible. Such a process might
account for the slightly lower (20%) growth rate of this double mutant on
lactose. A similar proposal was made by Button and coworkers concerning
the pathway of maltose metabolism in S. aureus [74]. The postulated
sequence was entry of the disaccharide by diffusion, hydrolysis by the
known cellular maltase [75], and internal phosphorylation by the PTS. Exit
of free glucose was observed in ptsI mutants, however. The principal
evidence against metabolism of glucose (generated from maltose) by efflux
and subsequent PTS-mediated reentry was the absence of detectable gluconic
acid (or glucose) in the medium during an 18-hr fermentation in the presence
of an external "glucose trap" (glucose oxidase plus catalase). It was not
shown, however, that such a trap competed effectively with the PTS EII^{glc}
for effluxing glucose. Note that the studies with A. aerogenes described
earlier in this section led to an opposite conclusion about the fate of a
monosaccharide generated internally from a disaccharide [60]. Although
there seems to be fairly good evidence for PTS action on "internal" sugar
in at least one special case (the E. coli "hyper-TMG" mutant of Kashket
and Wilson [92]), no firm conclusions about the generality of the phenomenon
or its physiological relevance can be made.

4. Do Sugar Phosphates Interact with
   PTS Enzymes II?

Sugar phosphates do not inhibit PTS-mediated sugar phosphorylation in
vitro [12], but a number of experiments suggest that the enzymes II do
interact with these compounds; the interactions may play an important
regulatory role.

Regulation of the PTS by sugar phosphates was first proposed by
Kaback, on the basis of earlier work by Ward and Glaser [93] and his own
experiments with membrane vesicles [94]. He found that uptake/phosphoryl-
ation of glucose and αMG by membrane vesicles was strongly inhibited by
glucose-6-phosphate and glucose-1-phosphate; other related hexose phosphates inhibited to a considerably lesser extent. The inhibition was noncompetitive, and was not enhanced by initial loading of the vesicles with
glucose phosphates by a preliminary "shock treatment." Evidence was
presented that the sites for glucose-1-P and glucose-6-P were separate
and distinct, but that the presence of glucose-1-P antagonized inhibition by
glucose-6-P (and vice versa) in a complex and asymmetric manner. The
conditions under which the cells were grown had a considerable effect on
the susceptibility to sugar phosphates of the membrane prepared from them.

The phenomenon termed "transphosphorylation" has been observed in
several instances. When S. typhimurium cells, preloaded with radioactive
αMG-P, were transferred to a medium containing an excess of unlabeled
glucoside, the internal αMG-P pool exchanged with external sugar at a
rate comparable to the initial sugar uptake rate [7]. (The small internal
pool of free sugar did not exchange; see Sec. II.F2d.) A simple mechanism
for such an exchange is diagrammed in Fig. 2-3. The process would require
two interacting sites on the internal surface of the enzyme II. The shift
of a phosphoryl group from a sugar phosphate, S_B-P, at one site, to a free
sugar, S_A, bound initially to another site, must be concomitant with the
translocation of free S_B to the external medium. In the experiment described above, S_A and S_B were both αMG. However Egan and Morse [75]
found that when S. aureus cells which had accumulated one sugar phosphate
species (S_B-P) were transferred to a medium containing a different sugar,
S_A, an efflux of free S_A was concomitant with the accumulation of S_B-P
inside the cells.

The first evidence of transphosphorylation in vitro has recently been
obtained by Saier and Newman [95]. They found that sugar phosphates can
apparently serve as phosphoryl donors in the PTS enzyme II reaction. In
contrast to the reaction when the ultimate source of the phosphoryl is PEP,
no soluble PTS phosphoryl carrier proteins are required. The specificity
of the enzyme II for the sugar phosphate donor parallels that for the free
sugar in the PEP-dependent reaction; only mannitol-1-phosphate is an
effective donor for mannitol transphosphorylation by the EII^{mtl} of
Spirochaeta aurantia; similar specificities were found for glucose and αMG
transphosphorylation by S. typhimurium extracts, and for an analogous
process mediated by the S. aureus enzyme II^{lac}. Extracts of mutants lacking
the respective PEP-dependent EII activities were similarly deficient in the
transphosphorylation activity.

Although the S. aureus EII^{lac} is not inhibited by galactose-6-phosphate,
Sussman and Hays found that this compound enhanced the rate at which
EII^{lac} is inactivated by mild heating in the presence of the nonionic detergent Triton X-100 [96]; glucose-6-phosphate had no effect. In contrast,
free β-galactosides markedly depressed the inactivation rate. Indirect
evidence for a possible second sugar-specific site, distinct from the
primary substrate site, is provided by the kinetics of inhibition of EII^{lac}

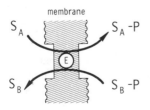

FIG. 2-3   Exchange transphosphorylation of
sugars.

by the dead-end substrate analog, 6-0-(p-toluenesulfonyl)galactose, reported
by Hays and Sussman [45]. Inhibition persists at saturating substrate concentrations, i.e., the inhibition is noncompetitive.

The results outlined above, taken as a whole, make a strong case for
physiologically important interactions of sugar phosphates with the respective enzyme II. Plausible roles for this reaction would include prevention
of needless dissipation of PEP when the internal sugar phosphate level is
high, and detoxification of the cell by lowering the level of nonmetabolizable
sugar phosphates. The latter role would involve a $(S_B-P)-(S_A)$ exchange
like that described in previous paragraphs. It is difficult to imagine a
mechanism for this process compatible with simple models for the transport process. This matter is considered further in the following section.

5. Mechanisms for PTS-Mediated
Transport

Although the ultimate goal of studies of the PTS is an understanding of the
molecular events of the transport process, many of the standard experimental techniques are only applicable to the in vitro phosphorylation
reaction. The PTS is a complex system, even in vitro, and the dangers
of extrapolating results obtained using purified components to complicated
in vivo processes are obvious. The futility of comparing the values for
the kinetic parameters of the in vitro phosphorylation and in vivo transport
processes were discussed in Sec. II.E1b. Nevertheless the kinetic pattern
of the phosphorylation reaction can provide important clues, and sometimes
one can at least rule out transport mechanisms on the basis of what an EII
apparently cannot do. For simplicity, we restrict ourselves here to the
lactose PTS of S. aureus; transport mechanisms will thus be discussed in
terms of the reaction

$$P-EIII^{lac} + Sugar \xrightarrow{EII^{lac}} EIII^{lac} + sugar-P \qquad (9)$$

The first kinetic analyses of this reaction by Simoni and coworkers
[19, 25] employed the usual plots of (initial velocity)$^{-1}$ versus (sugar)$^{-1}$
at changing fixed levels of $EIII^{lac}$. The lines were intersecting rather
than parallel; it therefore appears that the reaction proceeds via a ternary
$EII^{lac} \cdot S \cdot P-EIII^{lac}$ complex ("sequential" mechanism) rather than via a
stable $P-EII^{lac}$ intermediate ("ping-pong" mechanism). All efforts to
isolate a $P-EII^{lac}$ complex were unsuccessful. Since [^{14}C]lactose bound
tightly to EII^{lac} preparations in the absence of other PTS components, a
sugar-first sequential mechanism was at least possible. However, sub-
strate-initial rate studies can never permit unambiguous determination of
the reaction order. Subsequent kinetic analyses by Hays and Sussman [45]
employed tosyl-β-galactosides as dead end inhibitors. The unexpectedly

high affinity for EII^{lac} displayed by these compounds required both the polar and nonpolar moieties of the tosyl group; α anomers were completely inactive [97]. The obvious structural analogies between these inhibitors and a plausible intermediate in a sequential reaction scheme (see Fig. 2-4) prompted the proposal that the tosyl galactosides acted like transition state analogs. In principle, dead end inhibitors can be used to determine reaction orders, since the slopes of (initial velocity)$^{-1}$ versus (variable substrate)$^{-1}$ are affected by inhibitor concentrations if and only if the inhibitor binds <u>before</u> the variable substrate in the reaction sequence [43]. Experiments using both sugar and $P-EIII^{lac}$ indicated that <u>both</u> reaction sequences were allowed, but that in the presence of detergent (0.1% Triton X-100), the principle reaction pathway was $P-EIII^{lac}$-first. The inhibition by tosyl galactosides was not affected by the $P-EIII^{lac}$ concentration. This latter result indicates that the sites to which sugar and $P-EIII^{lac}$ <u>initially</u> bind are well removed (by at least the distance of the bulky tosyl group), although close juxtaposition of sugar and phosphoprotein is eventually required if the phosphoryl transfer is to be accomplished without a $P-EII^{lac}$ intermediate. If one recalls the obligatory coupling of transport and phos- phorylation demonstrated in <u>S. aureus</u> [44], it is tempting to speculate that it is the translocation process itself which results in the required juxta- position of the two substrates, and that the driving force for the process derives from the large difference between the energy level of the phosphoryls in $P-EIII^{lac}$ and in sugar phosphates (about 9 kcal/mol). A model incor- porating this notion is depicted in Fig. 2-5(A). The mechanism proposed

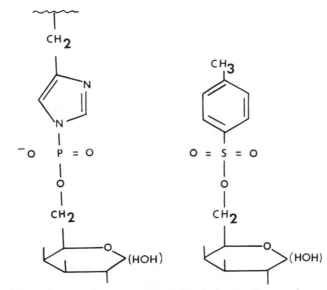

FIG. 2-4 Comparison of structures of tosyl galactose and proposed inter- mediate in PTS-mediated vectorial phosphorylation.

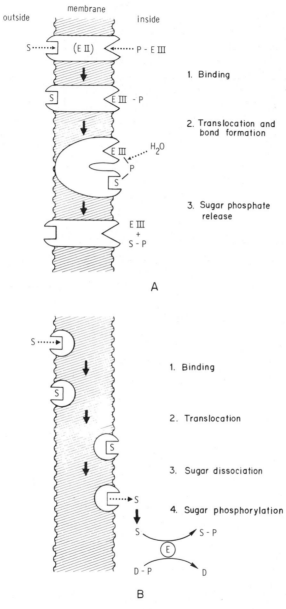

FIG. 2-5

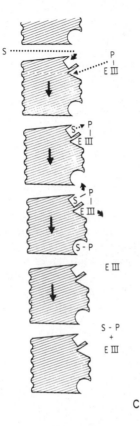

1. Binding and translocation

2. Bond formation

3. Shift to sugar phosphate site

4. Sugar phosphate release

C

FIG. 2-5 Models for PTS-mediated sugar transport. (A) Vectorial phosphorylation in which translocation occurs mainly as a result of a conformational change in enzyme II. (B) Apparent vectorial phosphorylation resulting from facilitated diffusion plus trapping, as proposed by Kornberg and Riordan [90]. (C) Vectorial phosphorylation in which translocation occurs mainly as a result of movement by sugar and sugar phosphate thru a stereospecific pore, as proposed by Hengstenberg [15] and Saier and Newman [95].

here is a true group translocation and is thus to be differentiated from the alternative of facilitated diffusion plus trapping (Fig. 2-5B). A hybrid of the two mechanisms, depicted in Fig. 2-5(C), has been proposed by Hengstenberg [15] and by Saier and Newman [95]; it involves translocation by a sugar-specific pore, and a vectorial phosphorylation at the inner membrane surface.

A major difficulty with the simple mechanisms proposed above is they do not account for the effects of sugar phosphates. It cannot be that these compounds occupy the $P-EIII^{lac}$ site, since they do not inhibit the reaction. One must propose a separate sugar phosphate binding site, for which there is at present no physical evidence. The mechanism of transphosphorylation is another problem. Although an uncatalyzed (sugar)-(sugar phosphate) exchange seems implausible, there is no evidence for a $P-EII^{lac}$ inter-mediate. Even if there is a transitory covalent linkage of the phosphoryl to some EII^{lac} amino acid side chain, it is hard to see how $P-EIII^{lac}$ and S-P, bound at different sites, can transfer their phosphoryls via the same catalytic side chain. One is required to postulate an enzyme with at least three active sites, two catalytic centers, and a capability for massive but precise conformational changes. The mechanisms in Fig. 2-5B and C have essentially the same difficulties. There must be simpler schemes, but more experimental clues are needed.

The mechanisms above attempt to describe the transport events only in macromolecular terms, but even here they are unsatisfying. We are certainly very far from understanding the process at the level discussed by Mitchell (see the beginning of this chapter). Assuming that some β-galactosyl moiety is indeed translocated, what precisely is the moiety, and what happens to the rest of the molecule? Is $RCH_2-O:^-$ the sugar moiety translocated, with H^+ remaining behind, or do they both enter in a sym-coupled mechanism? Is RCH_2OH divided in a different way instead? The only encouraging thing about our ignorance of the molecular events of this transport process is the incentive it offers for future investigations.

III. FATTY ACID TRANSPORT IN
      ESCHERICHIA COLI

Early work indicated that the enzymes for fatty acid catabolism are induced to elevated levels in E. coli cells when long-chain fatty acids are introduced into the growth medium [102-105]. This observation stimulated work by several laboratories aimed at elucidating the uptake system involved. Experiments by Overath and his collaborators [106] subsequently demonstrated that the uptake system itself was inducible, that uptake was tightly coupled to the subsequent catabolism of the fatty acids, and that the oleate-induced system exhibited a marked preference for long-

chain substrates (eight or more carbon atoms). Work by several groups
has revealed that short-chain fatty acids are transported by a different
system [107-109].

A. Long-Chain Fatty Acids

An acyl:Coenzyme A (acyl:CoA)-synthetase and the enzymes for fatty acid
degradation are coordinately induced in E. coli [102,103]; regulatory
mutants constitutive for the degradation enzymes are constitutive for the
synthetase as well. Klein and coworkers [106] found that the uptake
system induced by oleate had a marked preference for long-chain fatty
acids. Uptake rates (relative to that for oleate) were: dodecanoate, 94%;
decanoate, 80%; octanoate, 63%; and hexanoate, 3%. The same preference
was observed when the oxidation of various [^{14}C]fatty acids to [^{14}C]CO_2
was measured. The uptake was saturable, suggesting that it was carrier-
mediated. The trend of apparent K_m values for transport (oleate, 15 μM;
octanoate, 90 μM) was consistent with the long-chain specificity. It should
be noted, however, that like most transport studies with long-chain fatty
acids, these experiments were performed using a fairly high (0.5 to 1.0%)
concentration of the nonionic detergent Triton X-100. Under the experimental conditions the detergent:lipid ratios were in excess of 50:1. Thus
the fatty acids were almost certainly incorporated into mixed micelles
containing mostly detergent and a few fatty acid molecules similar to those
described by Dennis [110]. Since the micelle properties may depend in a
complex way on fatty acid chain length and concentration, the conventional
kinetic analysis used for single-phase systems may not be strictly
applicable.

A vital role for the acyl:CoA-synthetase in transport was suggested by
the properties of mutants lacking the enzyme. These synthetase mutants,
isolated in three different laboratories by penicillin selection [108,109],
or by selection for resistance to "suicide" by uptake of [^{3}H]oleate [111],
were in all cases deficient in transport as well. Furthermore, Frerman
and Bennett [112] found that 4-pentenoate inhibited oleate uptake, presumably by trapping free CoA, and making it unavailable to the synthetase.
The effects of other inhibitors suggested that the uptake required cellular
ATP and oxidative metabolism [112].

Transport of fatty acids appears to be tightly coupled to subsequent
degradation steps. Mutant strains deficient in one or more of the degradative enzymes (thiolase, β-hydroxyacyl-CoA-dehydrogenase, enoyl-CoA
hydrase, isomerase) exhibited depressed rates of fatty acid oxidation in
vivo (measured by [^{14}C]CO_2 release), and were correspondingly deficient
in uptake activity [106,112]. Since degradation enzyme mutants exhibited
partial loss of uptake activity but synthetase mutants were totally trans-

port deficient, a key role for the latter enzyme in uptake, and some sort of feedback control of it by degradation products, were proposed [112].

The alcohol-extractable products of [³H]oleate uptake were sufficiently characterized by Frerman and Bennett to show that there was no free fatty acid or oleyl-CoA inside the cells [112]. Furthermore, there was no detectable efflux of material from preloaded cells. All of the experimental evidence suggested that a true group translocation was involved, and the term "vectorial acylation" was used for the proposed mechanism [112]. Although the process seems to be mediated by the acyl:CoA synthetase, involvement of another component in a prior translocation step could not be ruled out.

B. Short-Chain Fatty Acids

Strains of E. coli K-12 which had acquired the ability to use butyrate as a carbon source were characterized in several laboratories [107, 108, 109]. They proved to be double regulatory mutants, having become constitutive both for the enzymes of the general fatty acid degradation pathway (fad operon) and for the two enzymes which comprise the short-chain specific ato pathway characterized by Pauli and Overath [109]. One of these latter enzymes (atoA), an acetyl coenzyme A:butyrate (acetoacetate) CoA-transferase, is obligatory for the uptake of butyrate (and acetoacetate) [107, 109]; the second (atoB) is a short-chain specific thiolase, which can be replaced (with lower efficiency) by the fad thiolase [109]. Although the presence of butyrate in the growth medium cannot derepress the ato operon, acetoacetate is an effective inducer. Thus wild-type cells can take up acetoacetate and use it as a carbon source. (The fad enzymes are not required in this case; the atoA and atoB gene products suffice to convert acetoacetate into acetyl CoA.) The same enzyme has been designated both an acetyl CoA:acetoacetate CoA-transferase and an acetyl CoA:butyrate CoA-transferase in the literature.

Although the absolute dependence of uptake upon the CoA-transferase was noted, early workers were reluctant to assign a major role in translocation to this enzyme, which is mainly soluble rather than membrane bound. In a subsequent investigation, Frerman prepared membrane vesicles from constitutive(ato^C fad^C) cells and used them for transport experiments [113]. About 5 to 10% of the original cellular CoA-transferase activity remained associated with the membranes. [Under certain conditions, vesicles retaining 50% of the transferase activity can be prepared (F. Frerman, personal communication).] Although the vesicles had also retained significant amounts of some of the degradative enzymes, the level of butyryl CoA-dehydrogenase was negligible, so that fatty acid uptake was essentially uncoupled from metabolism in the vesicles. Efficient uptake of butyrate by the vesicles required the addition of acetyl CoA (or ATP + Coenzyme A); neither reducing compounds (e.g., lactate, NADH), nor "high-energy" phosphate compounds were stimulatory. Evidence implicating the CoA-transferase in the uptake process included the following: vesicles prepared from wild-type cells were inactive if they had not been induced for the ato enzymes; the uptake process and the transferase displayed similar kinetic parameters and the same pattern of response to pH and sulfhydryl blocking reagents; and known product inhibitors of the transferase reaction (butyryl CoA, acetate) inhibited uptake by the vesicles. Uptake by a group translocation mechanism ("vectorial acylation") was proposed, citing the requirement for acetyl CoA, the presence of butyryl CoA rather than free butyrate inside the vesicles, and the absence of a demonstrable rapid efflux process. The level of butyryl CoA inside the vesicles never exceeded that in the external medium, for reasons which are probably inherent in the nature of the transferase and the transfer reaction (see below).

Subsequent work has been focused on the nature of the transferase enzyme, which is usually assayed as an acetyl CoA:acetoacetate CoA-transferase. It has been purified to homogeneity and thoroughly characterized by Sramek and Frerman [114]. The native protein consists of α chains (molecular weight, 26,000) and β chains (molecular weight, 23,000) arranged as an $\alpha_2 \beta_2$ tetramer. It displays the expected preference for short fatty acid chains. The following reaction scheme was proposed, based on extensive biochemical and kinetic studies [115,116] with the purified transferase

$$\text{Acetyl CoA} + \text{enzyme} \rightleftharpoons \text{acetyl-enzyme} + \text{CoA}$$

$$K_{eq} = 50$$

$$\text{Acetyl-enzyme} + \text{acetoacetate} \rightleftharpoons \text{acetoacetyl CoA} + \text{acetate}$$

$$K_{eq} = 0.025$$

(The equilibrium constants were calculated from kinetic data [116].) Direct evidence for the acetyl-enzyme intermediate includes the following [115]. (1) The enzyme was inactivated by treatment with borohydride if and only if acetyl CoA was present as well; treatment of the acetylated enzyme with $NaB[^3H]H_4$ permitted identification of a glutamyl residue in the β subunit as the acylation site. (2) The rate of $[^{14}C]$acetate exchange into acetyl CoA was comparable to the overall reaction rate. (3) A somewhat unstable enzyme-CoA intermediate could be isolated by gel filtration subsequent to reaction of the transferase with a fluorescent acetyl CoA analog; no intermediate could be isolated if acetate was present as well. The steady-state kinetic patterns provide further support for the proposed reaction scheme [116]. Double reciprocal plots of the data from (initial velocity)-(substrate concentration) experiments have the parallel pattern characteristic of ping-pong mechanisms. The product inhibition patterns for inhibition of the forward reaction by acetate and inhibition of the reverse reaction by acetoacetate are also consistent with this mechanism. Acetoacetyl CoA is an inhibitor of the reverse reaction. These inhibition properties may be involved in the coupling of the uptake process to subsequent degradative steps.

In view of the tight coupling of uptake and subsequent fatty acid metabolism seen in vivo, it is instructive to compare the parameters for the transferase reaction with those recently determined by Duncombe and Frerman for the thiolase [117]:

$$\text{Acetoacetate} + \text{acetyl CoA} \xrightleftharpoons{\text{Transferase}} \text{acetoacetyl CoA} + \text{acetate}$$

$$K_{eq} = 0.13$$

$$\text{Acetoacetyl CoA} + \text{acetyl CoA} \xrightarrow{\text{Thiolase}} 2 \text{ acetyl CoA}$$

$$K_{eq} = 10^6$$

The Michaelis constants for acetoacetyl CoA are about 20 μM for the transferase reaction (in reverse) and for the thiolase reaction, and there seem to be more molecules of the former enzyme present in the cell [114, 117]. It thus appears that the transferase reaction may be near equilibrium in vivo. The unfavorable K_{eq} for this reaction would mean that acetoacetyl groups can never be concentrated inside the cell, a prediction consistent with the inability of membrane vesicles to concentrate butyryl CoA (see above). The flux of net acetyl CoA units to cellular metabolism via the thiolase reaction is thus determined by the equilibrium position of the transferase reaction.

The parallelism between the general process of PTS-mediated sugar transport ("vectorial phosphorylation") and the fatty acid systems ("vectorial acylation") is close, but the differences are interesting. Despite its apparent involvement in the transport process, it is difficult to view the CoA-transferase as the only protein involved in recognition and translocation of short-chain fatty acids. It is essentially a soluble protein, and binds to membranes only when charged with CoA (F. Frerman, personal communication). In this respect, then, it resembles a phospho-EIII of the PTS more than an enzyme II. Nevertheless, the transfer of the CoA moiety to fatty acid is mediated by the transferase, so that it plays an enzyme II-like role as well. The entire transfer reaction requires only the β subunit (F. Frerman, personal communication); the role of α subunit remains unknown. Perhaps it is involved in recognition of some membrane component involved in the initial translocation process. It would be interesting to see if any uptake-deficient mutants proved to have defective α chains. Finally, the striking difference in the energetics of the sugar uptake and fatty acid uptake processes should be noted. The phosphoryl donors of the PTS are phosphohistidyl proteins, and the equilibrium constants for transfer of phosphoryl to sugar are on the order of 10^8. In contrast, the equilibrium constant for transfer of the two-carbon fragment from acetyl CoA to acetoacetyl CoA is slightly unfavorable, so that the only compound whose accumulation is thermodynamically favored is acetyl CoA, the product of the subsequent thiolase reaction.

IV. TRANSPORT OF PURINES, PYRIMIDINES, AND NUCLEOSIDES

A. Introduction

The fundamental nature of the processes by which free bases and nucleosides are transported into bacterial cells is the subject of considerable controversy. It is uncertain whether group translocations are the general rule. In some instances, not only the conclusions reached but also the experimental data obtained by different laboratories seem to be in direct conflict. It is worthwhile, therefore, to first consider possible reasons for the apparent contradictions. A fundamental cause is the complex nature of base, nucleoside, and nucleotide metabolism, which includes conversions of bases to nucleosides and nucleotides, e.g.,

$$\text{Adenine} + \text{PRPP} \xrightarrow{\text{phosphoribosyltransferase}} \text{AMP} + \text{PP}_i$$

and

$$\text{Uridine} + \text{ATP} \xrightarrow{\text{kinase}} \text{UMP} + \text{PP}_i$$

and vice versa, e.g.,

$$\text{Adenosine} + \text{P}_i \xrightarrow{\text{phosphorylase}} \text{adenine} + \text{ribose-1-P}$$

and conversions of one base of nucleoside to another, e.g.,

$$\text{Adenosine} \xrightarrow{\text{deaminase}} \text{inosine.}$$

Furthermore, the precise location of many of these enzymes in enteric bacteria (the subject of most of the studies discussed here) is somewhat undefined. A number of these enzymes appear to be external to the cell cytoplasm, either in the so-called periplasmic space between the inner and outer bacterial membranes (see Chap. 11), or more or less attached to the external surface of the inner membrane. It seems likely that many as yet unidentified components of the transport apparatus will prove to be similarly located. Both the number and the "nativeness" of these peripheral proteins are susceptible to alteration by preparative treatments. Drastic procedures, such as conversion of cells to spheroplasts or to membrane vesicles, are likely to release many loosely bound "periplasmic" components. Less extreme treatments, such as osmotic shock, may result in partial release, and it is possible that even relatively mild perturbations, such as ice-cold washes or prolonged starvation, can cause some loss of these components. Vesicle preparations, although extremely useful in transport studies, are subject to variations in wrongsidedness, leakiness, and loss of peripheral components, and thus in their metabolic and transport capacities. Other problems include differences in the catalytic properties of membrane bound and detached enzymes, difficulties in measuring true initial rates of uptake, and uncertainty as to which moieties of complex substrates are actually transported. Since there is little agreement about even the basic mechanism(s) of purine, pyrimidine, and nucleoside transport, a comprehensive review of all systems does not seem in order here. Instead, only the most recent work will be discussed in detail, and the emphasis will be on situations in which the experimental problems have been minimized.

B. Uptake of Free Bases

1. Adenine

In an important early study, Berlin and Stadtman [118] carefully measured, at various times, the specific radioactivity of the internal pools of the various compounds accumulated by Bacillus subtilis cells exposed to [^{14}C]adenine. They found that external adenine was converted directly to internal AMP, apparently bypassing the internal pool of free base.

Extensive investigations of adenine transport in E. coli K12 were described by Hochstadt-Ozer and Stadtman in a series of papers [119-121]. Motivated by evidence implicating the adenine phosphoribosyltransferase (adenine PRTase) in transport, they purified the enzyme to apparent homogeneity from cells hyperderepressed for purine PRTase activity. The E. coli enzyme, which is specific for the reaction

$$\text{Adenine + PRPP} \xrightarrow{\hspace{3cm}} \text{AMP + PP}_i$$

had a molecular weight of about 4×10^4 (analytical gel filtration), but could be disaggregated into catalytically active subunits [119]. The pattern of response of the purified enzyme to putative regulatory inhibitors was studied carefully, in order to provide a reference to which the effects of these compounds on the actual transport process could be compared. The 5'-nucleotides (especially those containing adenine) were effective inhibitors, but adenosine and free adenine were not. All nucleoside triphosphates and purine nuceloside diphosphates were potent inhibitors. The pattern of inhibition was determined for a few compounds by double reciprocal plots of initial velocity vs substrate at a single inhibitor concentration [119].

The inhibition appeared to be competitive, but no rigorous determination of linearity of inhibition of inhibition constants was made. Membrane vesicles, prepared by the method of Kaback [79], were used to study the transport process. Uptake of adenine was stimulated by PRPP and inhibited by the negative effectors of the PRTase listed above. Vesicles which lost adenine uptake activity after freeze-thaw treatments proved to be correspondingly lacking in vesicle bound adenine PRTase, and the predominant product accumulated by the vesicles was AMP. These observations led the authors to propose group translocation by the PRTase as the mechanism for adenine transport. The process in this case would be a "vectorial phosphoribosylation." The vesicles exhibited a basal level of uptake in the absence of added PRPP, which could reflect either endogenous PRPP or a second transport system. Freshly harvested cells took up adenine at a low rate which could not be stimulated by glucose (or other energy sources). Instead, cells starved in the cold (for 6 hr or more) were used. Uptake was higher, and was enhanced in the presence of iodoacetamide, which presumably inhibits glycolytic metabolism in the cells. The kinetic parameters and pattern of response to nucleotide inhibitors for adenine uptake by starved cells was similar to that for uptake by vesicles. Although the primary product accumulated by the cells was not AMP, this may be the result of intracellular metabolism of the initial transport product.

Results differing considerably from those discussed above have recently been reported by Roy-Burman and Visser [122]. They found that uptake of adenine (and other bases) by freshly harvested cells of E. coli strains B and K12 was stimulated about 20-fold by glucose. Under these conditions the kinetics of adenine uptake corresponded to the action of a single saturable system, $K_m = 1.2 \times 10^{-7}$ M; the K_m for the adenine PRTase is about 10^{-5} M. Although PRPP did not stimulate uptake, this compound may not get into the cells. There was considerable accumulation of free adenine, as well as other compounds. The uptake process was strongly inhibited by KCN. On the basis of their results, these workers proposed that adenine was taken up as the free base by an active transport process. A mutant strain, resistant to the antibiotic showdomycin and deficient in nucleoside uptake (NUC$^-$; see Sec. IV. C), transported free adenine normally.

2. Guanine, Xanthine, Hypoxanthine

The transport of 6-OH purines into freshly harvested cells was studied by Roy-Burman and Visser in parallel with the adenine experiments [122]. The essential features (stimulation by glucose but not PRPP, inhibition by KCN, accumulation of free bases as well as derivatives, normal uptake by NUC$^-$ mutations) were the same. Transport of these bases also appeared to be by single saturable systems; K_m values were 1/10 to 1/100 those reported for the corresponding phosphoribosyltransferases. Uptake of 6-OH purines was not inhibited by adenine or pyrimidines, which are presumably transported by different systems. The only pair of substrates exhibiting strong mutual competitive inhibition of uptake were guanine and hypoxanthine. Thus there seems to be a common guanine-hypoxanthine system and a distinct xanthine system in E. coli B. This specificity pattern differs from that exhibited in vitro by the phosphoribosyltransferases of this organism, which is instead consistent with the existence of distinct hypoxanthine and guanine-xanthine enzymes [122]. These results are thus inconsistent with the notion that the PRTases mediate the rate-limiting step in free base uptake.

Different results were reported in a recent study by Jackman and Hochstadt of guanine and hypoxanthine transport by membrane vesicles prepared from S. typhimurium [123]. Vesicles from a strain (TR119) having phosphoribosyltransferase activity for guanine and for hypoxanthine took up both substrates about three to five times faster in the presence of PRPP, whereas those prepared from a mutant strain (proAB 47), lacking the guanine PRTase, exhibited PRPP stimulation only for hypoxanthine transport. The pattern of inhibition of PRPP-dependent uptake by nucleotide effectors (GTP, ITP, GDP, IDP) was similar in most respects to that of the phosphoribosylation catalyzed by vesicle bound PRTases. Release of PRTase enzymes from the vesicles resulted in enhanced total activity (two- to threefold), and what is more important, the appearance of significant guanine PRTase activity in the guanine-negative mutant strain proAB 47. The latter result suggests that the apparent specificities of enzymes which have been released from membranes do not necessarily correspond to the specificities of the same enzymes in their "natural" environment, and could be used to explain some of the differences between transport specificity and PRTase specificity seen by Roy-Burman and Visser [122] (see above). Jackman and Hochstadt observed significant transport activity for both guanine and hypoxanthine in both strains in the absence of PRPP. Further evidence for two types of uptake systems was provided by the observation that N-ethylmaleimide inhibits the PRPP-independent uptake activity in both strains, but not the PRPP-dependent hypoxanthine uptake (or any guanine transport). The pattern of apparent product accumulation was quite complex and is difficult to generalize concisely, but the following should be noted: GMP and IMP accumulation is significant in the presence of PRPP, if and only if the strain has the appropriate PRTase; free guanine is significant in the absence of PRPP; and inosine (and IMP) appear in the absence of PRPP, rather than free hypoxanthine.

3. Uracil

Roy-Burman and Visser found that transport of uracil by freshly harvested E. coli B cells displayed essentially the same characteristics as that of purine bases [122]. Uptake is not inhibited by free purines, but is reduced in the presence of 5-fluorouracil. It is noteworthy that no uracil phosphoribosyltransferase activity can be detected in extracts of these cells. On the other hand, Beck and coworkers [124] and Bulman and Hochstadt [125] found that mutant strains lacking the uracil PRTase could not take up free uracil, and were completely resistant to fluorouracil.

C. Uptake of Nucleosides

In order to avoid some of the complications inherent in analyzing the results of experiments where deaminations and other degradative steps may be important, I will emphasize in this section the transport of deoxyadenosine, which is degraded less than adenosine (and other purine ribonucleosides), and of uridine (cytidine is usually degraded to uridine).

1. Deoxyadenosine and Adenosine

Roy-Burman and Visser [122] compared the uptake of [8-^{14}C]deoxyadenosine by freshly harvested E. coli B cells to that by a mutant derivative (NUC⁻), selected for its resistance to showdomycin, an antibiotic analog of uridine. Although deoxyadenosine transport by wild-type cells was stimulated about sixfold by glucose, there was substantial uptake in its absence under both conditions. The dependence of uptake rate upon nucleoside concentration suggested a single saturable process, K_m about 2 μM. In the mutant NUC⁻ cells, the uptake rate was reduced about 20-fold, and was nonsaturable (to 1 mM deoxyadenosine). Transport of deoxyadenosine into wild-type cells was not significantly inhibited by any heterologous nucleoside (including adenosine), except for uridine and deoxyuridine; free adenine was slightly inhibitory. Deoxyadenosine uptake was reduced in the presence of KCN, but less so than the transport of free bases. The products accumulated were primarily nucleotides and polymeric material. Although adenosine is rapidly degraded in the periplasmic space, its uptake could still be measured. The apparent adenosine uptake was saturable (and reduced more than 15-fold) in the NUC⁻ mutant. Curiously, deoxyadenosine was a potent inhibitor of adenosine uptake, although the converse is not true, a result possibly due to periplasmic degradation of the ribonucleoside. The failure of free adenine to significantly inhibit deoxyadenosine uptake was considered evidence for the independence of the nucleoside and free base uptake systems.

Earlier work by Hochstadt-Ozer using membrane vesicles led to the proposal of a two-step mechanism for nucleoside transport [126]. Their model was based in part on a previous report by Roberts and coworkers [127]. These latter workers had found that during the transport of uniformly labeled adenosine by intact cells, the adenine and ribose moieties

accumulated intracellularly at different specific activities, suggesting
cleavage of the nucleoside before the uptake step. Interpretation of the
vesicle results was hampered by the extensive deamination of adenosine to
inosine. However, citing the observed similarities in kinetics, response
to inhibitors, and PRPP stimulation between base and nucleoside uptake,
as well as other evidence, Hochstadt-Ozer and Stadtman proposed that
adenosine (and inosine) were transported according to the model below,
and not as intact nucleosides:

$$\text{Purine nucleoside (outside)} + P_i \xrightarrow[\text{nucleoside phosphorylase}]{\text{purine}}$$

$$\text{purine (periplasmic)} + \text{ribose-1-phosphate}$$

$$\text{Purine (periplasmic)} + \text{PRPP base} \xrightarrow[\text{PRTase}]{\text{membrane}}$$

$$\text{purine nucleotide (inside)} + PP_i.$$

Indirect support for this model has been provided by Yagil and Beacham
[128], who found that mutants of S. typhimurium defective in purine
nucleoside phosphorylase (Pup⁻) were deficient in the uptake of [8-^{14}C]aden-
osine, but transported free adenine normally. There was a significant
residual level of adenosine transport in Pup⁻ cells, which could reflect
either a second uptake system or leakiness of the pup mutation. It should
be noted that S. typhimurium strains which cannot synthesize purines and
are also Pup⁻ cannot satisfy their purine requirements using adenosine or
deoxyadenosine.

Komatsu [129] studied adenosine transport using E. coli vesicles, and
concluded that there were three adenosine transport systems, one of which
involved prior cleavage by a purine phosphorylase. No evidence implicating
a phosphoribosyltransferase was obtained. One difficulty with these studies
is that adenosine was rapidly deaminated to inosine, so that the actual
substrates and products of the transport process(es) are not known.

2. Uridine

Analysis of a large number of mutants of S. typhimurium, deficient in
nucleoside and nucleotide metabolism, led Beck and coworkers [124] to
propose that there were three distinct pathways for uridine metabolism,
one of which seemed to involve both uridine phosphorylase and uridine
phosphoribosyltransferase.

Hochstadt and Bulman [125] compared the uptake of uridine by membrane vesicles prepared using E. coli strain LW2, which lacks the uracil
PRTase, to that by wild-type vesicles. Although vesicles of the former
strain could still accumulate radioactivity from uniformly labeled uridine

([U-^{14}C]uridine), there was little or no uptake when [2-^{14}C]uridine was the substrate. Rader and Hochstadt subsequently employed vesicles prepared from S. typhimurium by a new method. The uracil PRTase is removed by this procedure [130]. Ribose-1-phosphate was accumulated in the vesicles, while uracil appeared in the external medium. Since accumulation of the ribosyl moiety was not inhibited by added ribose-1-P or ribose, cleavage by an external phosphorylase prior to the actual transport step did not seem likely. Instead, group translocation by a trans-membrane phosphorylase was proposed:

$$\text{Uridine (outside)} + P_i \xrightarrow{\text{phosphorylase}} \text{ribose-1-P (inside)}$$

$$+ \text{ uracil (outside)}$$

Such a process could be termed a "vectorial phosphorolysis." The uracil remaining outside would then be taken up by a free base system. The apparent uptake of the ribosyl moiety by the vesicles was stimulated by electron transport substrates (D- and L-lactate, succinate). It was inhibited by adenosine. Reversal of the process was observed when excess free base was added externally. Under these conditions an efflux of ribose-1-phosphate was mediated by the phosphorylase, with concomitant formation of de novo nucleosides.

An important study of uridine transport into intact E. coli cells by Leung and Visser has been the subject of a recent preliminary report [131]. The pattern of uptake by freshly harvested cells of E. coli B and certain mutant derivatives was as follows. The NUC$^-$ (showdomycin-resistant) strain transported free uracil normally, but did not appreciably accumulate either the uracil or the ribosyl moiety of uridine. A fluorouracil-resistant mutant U$^-$ could not transport free uracil; it retained normal activity for the ribosyl moiety of uridine and 75% of normal for the uracil moiety. A derivative of the U$^-$ strain U$^-$UR$^-$ could not take up free uracil or the uracil moiety of uridine, but nevertheless retained 40% of the normal transport activity for the ribosyl moiety. The authors proposed that NUC$^-$ cells lack some component of a uridine recognition element, and that this recognition is rate limiting for all utilization of uridine, that U$^-$ cells lack the transport system for free uracil but can take up intact uridine by a distinct second system and its ribosyl moiety by a third system, and that the U$^-$UR$^-$ mutant has lost the second system but retained the third one. When the rapid intracellular oxidation of the ribosyl moiety to [^{14}C]CO$_2$ was taken into account, the release of uracil from [U-^{14}C]uridine to the medium by the U$^-$UR$^-$ cells was found to be nearly stoichiometric with the accumulation of the ribosyl moiety, and was subsequently used as a measure of the latter process. The observed failure of ribose and ribose-1-P to inhibit uptake of the ribosyl moiety seems to rule out prior cleavage by an external phosphorylase, but vectorial phosphorolysis by a membrane-associated enzyme seems quite possible. The uptake of both moieties of uridine was unaffected by 6-OH purine nucleosides, moderately inhibited by deoxy-uridine and deoxycytidine (the latter in the presence of the deaminase inhibitor tetrahydrouridine), and strongly inhibited by adenosine and deoxyadenosine. This intriguing uridine-adeonsine dualism seems to be a recurring feature of all of the nucleoside transport studies. The relationship between the postulated systems for the transport of intact uridine and for the vectorial phosphorolysis of the ribosyl moiety is not yet clear. The existence of a rate-determining step in common is suggested by the absence of uptake activity for both moieties in the NUC$^-$ mutant, and by the similar K_m values and nearly identical responses to adenosine, deoxyadenosine, and other inhibitors found for the two processes. In the U$^-$UR$^-$ mutant, which doesn't take up the uracil moiety, uptake of the ribosyl moiety is still inhibited by adenosine and deoxyadenosine. Conversely, the existence of two distinct systems is indicated by the absence of uracil moiety uptake by U$^-$UR$^-$ cells, and by the observation that whereas accumulation of the uracil moiety by wild-type cells does not occur in the absence of glucose, ribosyl uptake is only slightly stimulated by this energy source.

D. Discussion

There is apparently a major conflict between the bulk of the data for free base uptake by membrane vesicles and starved cells and that for transport into freshly harvested cells. Whereas the latter data are most consistent with the notion that unaltered bases are taken up by active transport, the former implicates phosphoribosyltransferases in a group translocation mechanism. Some of the differences are probably inherent in the experimental systems: differing degrees of interconversion of bases, nucleosides, and nucleotides; different permeabilities to charged compounds like PRPP; differential leakiness. The properties of guanine and hypoxanthine phosphoribosyltransferases apparently change when they are released from a membrane-associated into a free form. If this is a general phenomenon, it could explain some of the differences between the substrate specificities of transport processes and those of the PRTases postulated to mediate the transport. It should be emphasized again that in general, agreement between the kinetic parameters for transport and for analogous reactions in vitro is not to be expected [42]. Nevertheless some of the conflicts are difficult to explain. For instance, extracts of cells of E. coli B do not contain a detectable uracil PRTase, but the cells take up uracil perfectly well. In contrast, however, when mutants lacking this enzyme are isolated from other strains, uracil transport activity is lost concomitantly. One must in this case postulate a unique instability for the E. coli B PRTase such that it is inactivated by extract preparation.

Another apparent conflict is between the kinetic pattern for base uptake by freshly harvested cells, which suggests a single saturable system, and that obtained using vesicles, which is consistent with the existence of both PRPP-dependent and independent transport processes. Hochstadt (personal communication) has suggested that cells do contain two systems, but that the PRPP-dependent one, being more subject to control by nucleic acid precursor pools, is repressed in rapidly growing cells. The system is apparently derepressed in starved cells, where it may be more important than the PRPP-independent system. In vesicles, both systems would presumably be released from feedback control. They would thus both be detectable, as long as they were not lost during the vesicle preparation procedure, as was the uracil PRTase in the experiments of Rader and Hochstadt cited in Sec. IV.C2 [130]. Such selective losses during vesicle preparation, if made reasonably quantitative, exploited in a systematic way, and cautiously interpreted, may help to unravel some of the complexities of these multisystem processes.

The data for nucleosides obtained by different groups seem more compatible than those for free bases, but there is no agreement as to the number of distinct systems or their biochemical nature.

Clearly, considerable further research will be necessary before even the general nature of base and nucleoside uptake processes can be unequivocally established. It would be helpful if the confusion resulting from metabolism of these compounds in the periplasmic space and in the cell could be reduced by using nonmetabolizable analogs. This will require considerable effort directed towards organic synthesis, but the results obtained using nonmetabolizable sugars suggest that it would be time well spent. It is essential that the nature of the products initially accumulated be determined by the appropriate double-label experiments. Nucleoside uptake studies would profit by the availability of differentially labeled substrates, for example, $[^{14}C]$uracil-$[^{3}H]$ribose for uridine studies.

Mutant strains deficient in transport components and/or key metabolic enzymes are obviously desirable. For instance, mutants whose PRTases have reduced affinity for bases should have abnormally high K_m values for transport, if those enzymes are directly involved in the process, and the trends should be detectable whether or not the absolute values of transport K_m values agree with the corresponding PRTase parameters. Conditional lethal mutations in key enzymes, e.g., temperature-sensitive PRTases, would be extremely useful. Finally, it is important to be able to compare studies in which different groups use the same systems.

To an outsider it seems worthwhile to consider whether in fact the same step is rate limiting in both cellular and vesicular transport of bases and nucleosides. For instance, if in intact cells there is a rate-limiting, high-affinity concentrative step in the periplasmic space, followed by a rapid low-affinity group translocation mediated by a PRTase, then many apparent conflicts could be explained. Presumably the vesicles would have lost the "concentrating" system, and the rate-limiting steps would now exhibit lower uptake rates, higher K_m values, PRPP dependence, and altered substrate and inhibitor specificities. One testable consequence of such a hypothesis would be the presence in the periplasmic space of binding proteins with high affinities for bases and nucleosides.

ACKNOWLEDGMENTS

I am grateful to the large number of investigators who kindly supplied reprints, preprints, and discussions of their work, and to Drs. J. C. Cordaro and M. H. Saier, who read a draft version of this manuscript. Work in the author's laboratory was supported by Grant No. GM 20211 from the National Institute of General Medical Sciences, National Institutes of Health.

REFERENCES

1. P. Mitchell, Nature, 180: 134-136 (1957).
2. P. Mitchell, in The Structures and Function of the Membranes and Surfaces of Cells (D. Bell and J. Grant, eds.), University Press, Cambridge, England, 1963.
3. P. Mitchell, Bioenergetics, 4: 63-91 (1973).
4. H. Kaback, Science, 182: 882-892 (1974).
5. H. Kaback, Ann. Rev. Biochem., 39: 561-598 (1970).
6. W. Kundig, in The Enzymes of Biological Membranes, Vol. 3, Membrane Transport (A Martonosi, ed.), Plenum, New York, 1976.
7. P. Postma and S. Roseman, Biochim. Biophys. Acta, 457:213-257 (1976).
8. H. Kornberg, Symp. Soc. Exp. Biol., 27: 175-193 (1973).
9. E. Lin, Ann. Rev. Genet., 4: 225 (1970).
10. C. Cordaro, Ann. Rev. Genet., 10: 341-359 (1976).
11. W. Kundig, S. Ghosh, and S. Roseman, Proc. Natl. Acad. Sci. USA 52: 1067-1074 (1964).
12. R. Simoni, T. Nakazawa, J. Hays, and S. Roseman, J. Biol. Chem., 248: 932-940 (1973).
13. R. Stein, O. Schrecker, H. F. Lauppe, and H. Hengstenberg, FEBS Lett., 42: 98-100 (1974).
14. N. Weigel, E. Waygood, A. Nakazawa, M. Kukuruzinska, and S. Roseman, Proc. Can. Fed. Biol. Soc., 20: 54 (1977).
15. W. Hengstenberg, in Proceedings of the 3rd International Symposium on Staphylococci and Staphylococcal Infections, Verlag, Stuttgart, 1976.

16. A. Ullah and V. Cirillo, J. Bacteriol., 131: 988-996 (1977).
17. A. Ullah and V. Cirillo, J. Bacteriol., 127: 1298-1306 (1976).
18. B. Anderson, N. Weigel, W. Kundig, and S. Roseman, J. Biol. Chem., 246:7023-7033 (1971).
19. R. Simoni, J. Hays, T. Nakazawa, and S. Roseman, J. Biol. Chem., 248: 957-965 (1973).
20. A. Lehninger, Biochemistry, Worth, New York, 1975, p. 398.
21. M. Saier, R. Simoni, and S. Roseman, J. Biol. Chem., 245: 5870-5873 (1970).
22. J. Cordaro and S. Roseman, J. Bacteriol., 112: 17-29 (1972).
23. J. Cordaro, R. Anderson, E. Grogan, D. Wenzel, M. Engler, and S. Roseman, J. Bacteriol., 120: 245-252 (1974).
24. R. Simoni, M. Smith, and S. Roseman, Biochem. Biophys. Res. Commun., 31: 804-811 (1968).
25. T. Nakazawa, R. Simoni, J. Hays, and S. Roseman, Biochem. Biophys. Res. Commun., 42: 836-843 (1971).
26. E. Kennedy and G. Scarborough, Proc. Natl. Acad. Sci. USA, 58: 225-228 (1967).
27. W. Hengstenberg, J. Egan, and M. Morse, Proc. Nat. Acad. Sci. USA 58: 274-279 (1967).
28. W. Hengstenberg, J. Egan, and M. Morse, Eur. J. Biochem., 14: 27-32 (1970).
29. M. Morse, K. Hill, J. Egan, and W. Hengstenberg, J. Bacteriol., 95: 2270-2274 (1968).
30. J. Hays, R. Simoni, and S. Roseman, J. Biol. Chem., 248: 941-956 (1973).
31. T. Korte and W. Hengstenberg, Eur. J. Biochem., 23: 295-302 (1971).
32. O. Helenius and K. Simons, Biochim. Biophys. Acta, 415: 29-79 (1976).
33. W. Kundig, J. Supramol. Struct., 2: 695-714 (1974).
34. W. Kundig and S. Roseman, J. Biol. Chem. 246: 1393-1406 (1971).
35. W. Kundig and S. Roseman, J. Biol. Chem., 246: 1407-1418 (1971).
36. S. Schaeffer, J. Bacteriol. 93: 254-263 (1967).
37. C. Fox and G. Wilson, Proc. Nat. Acad. Sci. USA 59: 988-995 (1968).
38. S. Rose and C. Fox, J. Supramol. Struct., 1: 565-587 (1973).
39. E. Wolfson, M. Sobel, and T. Krulwich, Biochim. Biophys. Acta, 321: 181-188 (1973).
40. A. Romano, S. Eberhard, S. Dingle, and F. McDowell, J. Bacteriol., 104: 808-813 (1970).
41. A. Romano and J. Trifore, Abstracts of the Annual Meeting of the American Society for Microbiology, Abstract K103 Washington, D.C., Am. Soc. Microbiol., (1970).
42. H. Schachter, J. Biol. Chem., 248: 974-976 (1973).

43. W. Cleland, in The Enzymes, Vol. II (P. Boyer, ed.), Academic, New York, 1970, p. 4.
44. R. Simoni and S. Roseman, J. Biol. Chem., 248: 966-976 (1973).
45. J. Hays and M. Sussman, Biochim. Biophys. Acta, 443: 267-283 (1976).
46. S. Friedman and J. Hays, J. Bacteriol., 130: 991-999 (1977).
47. B. Anderson, PhD. Thesis, University of Michigan (1968).
48. C. Cordaro, Ann. Rev. Genet., 10: 341-359, see Table 1 (1976).
49. D. Fraenkel, J. Biol. Chem., 243: 6458-6463 (1968).
50. T. Ferenci and H. Kornberg, FEBS Lett., 13: 127-130 (1971).
51. C. Schachtele and J. Mayo, J. Dent. Res., 52: 1209-1215 (1973).
52. C. Schachtele, J. Dent. Res., 54: 330-338 (1975).
53. S. Curtis and W. Epstein, J. Bacteriol., 122: 1189-1199 (1975).
54. H. Kornberg and J. Smith, FEBS Lett., 20: 270-272 (1972).
55. G. Bourd, S. Andeeva, V. Shabolenko, and V. Gershanovitch, Mol. Biol., 2: 89-94 (1968).
56. B. Tyler, R. Wishnow, W. Loomis, and B. Magasanik, J. Bacteriol., 100: 809-816 (1969).
57. M. Jones-Mortimer and H. Kornberg, Proc. Roy. Soc., Series B, 187: 121-131 (1974).
58. T. Melton, PhD. Thesis, The Johns Hopkins University (1976).
59. T. Hanson and R. Anderson, Proc. Natl. Acad. Sci. USA, 61: 269-276 (1968).
60. N. Kelker and R. Anderson, J. Bacteriol., 112: 1441-1443 (1972).
61. R. Walter and R. Anderson, Biochem. Biophys. Res. Commun., 52: 93-97 (1973).
62. T. Ferenci and H. Kornberg, Biochem. J., 132: 341-347 (1973).
63. P. Wolfson and T. Krulwich, J. Bacteriol., 112: 356-364 (1972).
64. S. Levinson and T. Krulwich, Arch. Biochem. Biophys., 160: 445-450 (1974).
65. N. Patni and J. Alexander, J. Bacteriol., 105: 226-231 (1971).
66. M. Saier, B. Feucht, and S. Roseman, J. Biol. Chem., 246: 7819-7821 (1971).
67. J. Lengeler, J. Bacteriol., 124: 26-38 (1976).
68. J. Lengeler, J. Bacteriol., 124: 39-47 (1976).
69. E. Solomon and E. Lin, J. Bacteriol., 111: 566-574 (1976).
70. S. Tanaka and E. Lin, Proc. Natl. Acad. Sci. USA, 57: 913-919 (1967).
71. N. Kelker and R. Anderson, J. Bacteriol., 105: 160-164 (1971).
72. J. Maryanski and C. Wittenberger, J. Bacteriol., 124: 1475-1481 (1975).
73. L. McKay, L. Walter, W. Sandine, and P. Elliker, J. Bacteriol., 99: 603-610 (1969).
74. D. Button, J. Egan, W. Hengstenberg, and M. Morse, Biochem. Biophys. Res. Commun., 52: 850-855 (1973).

75. J. Egan and M. Morse, Biochim. Biophys. Acta, 112: 63-73 (1966).
76. N. Kelker, T. Hanson, and R. Anderson, J. Biol. Chem., 245: 2060-2065 (1970).
77. N. Kelker, R. Simkins, and R. Anderson, J. Biol. Chem., 247: 1479-1483 (1972).
78. A. Bhumiratana, R. Anderson, and R. Costilow, J. Bacteriol., 119: 484-493 (1974).
79. H. R. Kaback, in Current Topics in Membranes and Transport (A. Kleinzeller and F. Bronner, eds.), Academic, New York, 1970.
80. J. Egan and M. Morse, Biochim. Biophys. Acta, 97: 310-319 (1965).
81. J. Egan and M. Morse, Biochim. Biophys. Acta, 109: 172-183 (1965).
82. H. Kaback, J. Biol. Chem., 243: 3711-3724 (1968).
83. M. Saier, W. Young, and S. Roseman, J. Biol. Chem., 246: 5838-5840 (1971).
84. G. Gachelin, Eur. J. Biochem., 16: 342-357 (1970).
85. G. Gachelin, Ann. Inst. Pasteur, 122: 1099-1116 (1972).
86. R. Haguenauer and A. Kepes, Biochemie, 53: 99-107 (1971).
87. E. Solomon, K. Miyai, and E. Lin, J. Bacteriol., 114: 723-728 (1973).
88. P. Postma, FEBS Lett., 61: 49-53 (1976).
89. J. Cordaro, P. Postma, and S. Roseman, Fed. Proc., 33: 1326 (1974).
90. H. Kornberg and C. Riordan, J. Gen. Microbiol., 94: 75-89 (1976).
91. B. Rotman, A. Ganesan, and R. Guzman, J. Mol. Biol., 36: 247-260 (1968).
92. E. Kashket and F. Wilson, Biochim. Biophys. Acta, 193: 294-307 (1964).
93. J. Ward and L. Glaser (unpublished work), quoted in H. Kaback, Ann. Rev. Biochem., 39: 561-598 (1970).
94. H. Kaback, Proc. Natl. Acad. Sci. USA, 63: 724-731 (1969).
95. M. Saier and M. Newman, J. Biol. Chem., 251: 3834-3837 (1976).
96. M. Sussman and J. Hays, Biochim. Biophys. Acta, 465: 559-570 (1977).
97. J. Hays, M. Sussman, and T. Glass, J. Biol. Chem., 250: 8834-8839 (1975).
98. E. Freese, A. Klofat, and E. Galliers, Biochim. Biophys. Acta, 222: 265-289 (1970).
99. V. Cirillo and S. Razin, J. Bacteriol., 113: 212-217 (1972).
100. H. von Hugo and G. Gottschalk, FEBS Lett., 46: 106-108 (1974).
101. O. Shrecker and W. Hengstenberg, FEBS Lett., 13: 209-212 (1971).
102. P. Overath, E. Raufuss, W. Stoffel, and W. Ecker, Biochem. Biophys. Res. Commun., 29: 28-33 (1967).
103. P. Overath, G. Pauli, and H. Schairer, Eur. J. Biochem., 7: 559-574 (1969).
104. G. Weeks, M. Shapiro, R. Burns, and S. Wakil, J. Bacteriol., 97: 827-836 (1969).

105. D. Samuel and G. Ailhaud, FEBS Lett., 2: 213-216 (1969).
106. K. Klein, R. Steinberg, B. Fiethen, and P. Overath, Eur. J. Biochem., 19: 442-450 (1971).
107. E. Vanderwinkel, M. DeVlieghere, and J. Meersche, Eur. J. Biochem., 22: 115-120 (1971).
108. J. Salanitro and W. Wegener, J. Bacteriol., 108: 893-901 (1971).
109. G. Pauli and P. Overath, Eur. J. Biochem., 29: 553-562 (1972).
110. E. Dennis, J. Supramol. Struct., 2: 682-694 (1974).
111. F. Hill and D. Angelmaier, Molec. Gen. Genet., 117: 143-152 (1972).
112. F. Frerman and W. Bennett, Arch. Biochem. Biophys., 159: 434-443 (1973).
113. F. Frerman, Arch. Biochem. Biophys., 159: 444-452 (1973).
114. S. Sramek and F. Frerman, Arch. Biochem. Biophys., 171: 14-26 (1975).
115. S. Sramek and F. Frerman, Arch. Biochem. Biophys., 171: 27-35 (1975).
116. S. Sramek and F. Frerman, Arch. Biochem. Biophys., 181: 178-184 (1977).
117. G. Duncombe and F. Frerman, Arch. Biochem. Biophys., 176: 159-170 (1976).
118. R. Berlin and E. Stadtman, J. Biol. Chem., 241: 2679-2686 (1966).
119. J. Hochstadt-Ozer and E. Stadtman, J. Biol. Chem., 246: 5294-5303 (1971).
120. J. Hochstadt-Ozer and E. Stadtman, J. Biol. Chem., 246: 5304-5311 (1971).
121. J. Hochstadt-Ozer and E. Stadtman, J. Biol. Chem., 246: 5312-5320 (1971).
122. S. Roy-Burman and D. Visser, J. Biol. Chem., 250: 9270-9275 (1975).
123. L. Jackman and J. Hochstadt, J. Bacteriol., 126: 312-326 (1976).
124. C. Beck, J. Ingraham, J. Newhard, and E. Thomassen, J. Bacteriol., 110: 219-228 (1972).
125. Bulman and Hochstadt (unpublished work), quoted in J. Hochstadt, CRC Crit. Rev. Biochem., 2: 259-310 (1974).
126. J. Hochstadt-Ozer, J. Biol. Chem., 247: 2419-2426 (1972).
127. R. Roberts, R. Abelson, D. Cowie, E. Bolton, and R. Button, Carnegie Institute of Washington Publication No. 607 (1955).
128. E. Yagil and S. Beacham, J. Bacteriol., 121: 401-405 (1975).
129. Y. Komatsu, Biochim. Biophys. Acta, 330: 206-221 (1973).
130. R. Rader and J. Hochstadt, J. Bacteriol., 128: 290-301 (1976).
131. K. Leung and D. Visser, Fed. Proc., 35: 1586 (1976).

Chapter 3

THE REGULATION OF CARBOHYDRATE TRANSPORT IN
ESCHERICHIA COLI AND SALMONELLA TYPHIMURIUM

Milton H. Saier, Jr. and Edward G. Moczydlowski

Department of Biology
The John Muir College
University of California, San Diego
La Jolla, California

I.   INTRODUCTION 104
II. HISTORICAL PERSPECTIVES AND PHYSIOLOGICAL
     CONSIDERATIONS 104
III. BACTERIAL CARBOHYDRATE TRANSPORT
     MECHANISMS 107
IV. POSTULATED MECHANISMS OF CARBOHYDRATE
     TRANSPORT REGULATION 108
V.   PTS-MEDIATED REGULATION 109
VI. REGULATION BY INTRACELLULAR SUGAR
     PHOSPHATES 116
VII. POSSIBLE REGULATION OF METHYL-α-GLUCOSIDE
     TRANSPORT BY THE ENERGIZED MEMBRANE STATE . 118
VIII. CONCLUSION ON A EUCARYOTIC NOTE 121
     REFERENCES 122

Wonder not if such my transports were; for in all things I saw
one life, and felt that it was joy.

William Wordsworth, The Prelude

I. INTRODUCTION

The recent yield of information from diverse biological fields bears directly on the problems encountered in the study of transport regulation. The intimate structure of biological membranes and the properties of several proteins which normally reside in the phospholipid bilayer have been revealed in a variety of studies [1]. Considerable progress has been made regarding the mechanisms by which energy is coupled to the translocation of substrates across the cell membrane (see Chap. 12) [2,3] . Also, the basic principles governing the regulation of soluble enzymes have been elucidated [4]. In light of these advances, the study of transport regulation may be approached with several criteria in mind. First, it is important to establish that a particular regulatory phenomenon occurs at the level of transmembrane solute transport and is not a consequence of altered metabolism or protein synthesis. The activity of a transport protein must be shown to be modified directly. Second, we must aim to gain insight into the molecular details of regulatory processes. Models involving allosteric activation or inhibition, chemical modification, or alteration in the electrodynamic or fluid properties of the membrane may be considered. Third, the physiological consequences may include modified rates of enzyme synthesis, nutrient catabolism, and/or growth. Finally the relatedness of regulatory mechanisms in evolutionarily divergent organisms must be ascertained. The elucidation of the molecular details of a regulatory process in one organism is likely to provide valuable insight into corresponding processes in other biological systems.

The measurement of transport and its regulation involve special topological problems imposed by the membrane which are not encountered in studies of soluble enzymes. Proof of a proposed mechanism ultimately requires reconstitution of the regulatory process in a chemically defined membrane system.

II. HISTORICAL PERSPECTIVES AND PHYSIOLOGICAL CONSIDERATIONS

Even before the turn of the century, it was known that when microorganisms were exposed to two different carbon sources, one was utilized preferentially. Figure 1 shows an example of the biphasic growth curves that are sometimes observed under these conditions. The growth behavior which is illustrated has been termed diauxie. Only after one sugar (glucose in Fig. 3-1) is completely metabolized during the first phase of growth does metabolism of the other sugar (maltose in Fig. 3-1) begin. Monod [5] and Gale [6] were among the early workers who initiated the characteri-

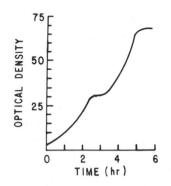

FIG. 3-1 An example of diauxie, the biphasic growth behavior which results when E. coli cells are grown in minimal salts medium containing two carbon sources. The two carbon sources, glucose and maltose, were both present at 100 μg/ml. Glucose was utilized exclusively in the first phase of growth, while maltose was utilized in the second phase. The curve was redrawn from data included in Ref. 5.

zation of these regulatory phenomena. It was found that glucose had the effect of reducing the concentrations of certain catabolic enzymes. This phenomenon became known as the glucose effect.

In 1961 and again in 1970, Magasanik published reviews [7,8] in which he attributed the glucose effect to three distinct processes, termed cata- bolite repression, transient repression, and inducer exclusion. Catabolite repression referred to the permanent inhibition of enzyme synthesis which occurred when cells were grown in the presence of glucose or another repressing sugar. It was though to be caused by catabolites of the repressing sugar present in the cell, but these catabolites have never been identified. Transient repression was distinguished kinetically. It was more intense than catabolite repression, but of short duration. Cer- tain nonmetabolizable sugars were found to be as effective as metabolizable ones in causing the transient inhibition of enzyme synthesis. The phenom- enon termed inducer exclusion occurred when the intracellular concen- tration of an inducer for a given catabolic enzyme system was reduced upon addition of a second sugar to the extracellular medium. This phenom- enon is a direct consequence of transport regulation (see Sec. IV). Al- though the molecular mechanisms responsible for these three processes were unknown, the permeases for the respective sugars were implicated both in transient repression and in inducer exclusion [8-10].

An involvement of the phosphoenolpyruvate:sugar phosphotransferase system (PTS) in the repression of catabolic enzyme synthesis
was first suggested by the observation that Escherichia coli mutants
deficient in either HPr or enzyme I of the PTS (ptsH or ptsI mutants,
respectively, see below) were defective for β-galactosidase induction
[11-12].* Pastan and Perlman studied this phenomenon, showing that
β-galactosidase synthesis was hypersensitive to repression in pts mutants
as compared with the parental strains [13-14]. They showed that either
cyclic AMP or isopropyl-β-thiogalactoside restored β-galactosidase and
lac specific mRNA synthetic rates, leading to the suggestion that repression
was due to depressed cellular levels of the cyclic nucleotide and/or inducer.
Several recent investigations have focused on the involvement of the PTS in
the regulation of cellular cyclic AMP levels [13-19].

Another approach to the regulation of sugar utilization was taken by
McGinnis and Paigen [20]. These investigators measured the rates at
which E. coli cells utilized a number of carbon sources, and they determined the responses of this organism to the addition of a second carbon
source. Both production of [^{14}C]CO$_2$ and incorporation of label into acid
insoluble material were measured with cells growing on the following
^{14}C-labeled carbon sources: maltose, lactose, galactose, mannose,
xylose, L-arabinose, and glycerol. A general phenomenon was observed:
glucose caused the immediate and reversible inhibition of carbohydrate
utilization (usually greater than 50% inhibition). An example of this
behavior is shown in Fig. 3-2. The inhibitory effect was different from
catabolite repression since the activity rather than the synthesis of a
cellular constituent was the target of regulation.

A number of sugars other than glucose were tested as inhibitors, but
only glucose, and to a lesser extent glucose-6-phosphate, caused the
immediate decline in the rate of carbohydrate utilization. The effect of
induction of the catabolic enzymes and transport systems specific for the
inhibitory sugars was not determined. In a subsequent report McGinnis
and Paigen proposed that inhibition of carbohydrate utilization was a
manifestation of transport regulation [21].

In a few cases, these phenomena can be explained by simple competition
between substrate and glucose at the extracellular recognition site of a
transport system. It is known, for example, that the enzyme II^{man} of
the PTS [22,23], the galactose specific permease [24,25], and the β-methyl-
galactoside permease [24,26] exhibit high affinity for D-glucose. However,
the transport systems specific for the other carbohyrates studied do not
possess appreciable affinity for glucose, and, therefore, indirect inhibitory
mechanisms must be proposed.

---

*The abbreviations used are as described in Chapter 2.

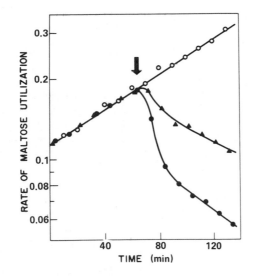

FIG. 3-2 Inhibition of the utilization of uniformly labeled [14C]maltose. Open circles show data for a control culture. Closed symbols show corresponding data for parallel cultures to which glucose (●) or glucose-6-phosphate (▲) was added at the time indicated by the arrow. The units are micrograms of sugar per minute per milliliter of bacterial culture and are plotted on a logarithmic scale. The data are reproduced from Ref. 20 with permission.

III. BACTERIAL CARBOHYDRATE TRANSPORT MECHANISMS

Recent studies have revealed that carbohydrates enter bacterial cells by a variety of different mechanisms. Glycerol, for example, is thought to cross the cytoplasmic membrane by facilitated diffusion, an energy independent process [27,28]. By contrast, the disaccharides maltose, lactose, and melibiose, pass through the gram-negative bacterial cell membrane by several distinct energy-dependent active transport processes. The maltose transport system is a member of the binding protein class of systems [29] which have been postulated to depend on a chemical form of energy such as ATP [30,31]. Lactose and melibiose, on the other hand, are transported by two distinct permease systems, neither of which requires chemical energy or a periplasmic binding protein for activity.

These two transport systems appear to utilize the energy of the electrochemical potential gradient of different ions, where the ion is a proton in
the case of the lactose transport system [32,33] and a sodium ion in the
case of melibiose [34].

Still another mechanism by which transmembrane sugar translocation
can be coupled to metabolic energy involves phosphorylation of the sugar
concomitant with permeation. This process is catalyzed by the PTS [35,
36]. Transfer of phosphate from phosphoenolpyruvate to a given sugar via
this enzyme system initially requires the sequential transfer of phosphate
to each of several energy coupling proteins of the PTS. The reactions
catalyzed by the system are as follows:

$$\text{Phosphoenolpyruvate + enzyme I} \underset{}{\overset{Mg^{++}}{\rightleftharpoons}} \text{P-enzyme I + pyruvate}$$

$$\text{P-Enzyme I + HPr} \rightleftharpoons \text{P-HPr + enzyme I}$$

$$\text{P-HPr + enzyme III} \rightleftharpoons \text{P-enzyme III + HPr}$$

$$\text{P-Enzyme III + sugar} \xrightarrow{\text{enzyme II complex}} \text{sugar-P + enzyme III}$$

Enzyme I and HPr are general proteins of the PTS, common to all sugars
phosphorylated by the system. The enzymes II and III, on the other hand,
are sugar-specific. While the enzyme II for each sugar is an integral
component of the cytoplasmic membrane which interacts directly with the
sugar and may serve as a carrier for sugar transport [37], enzyme III,
which may or may not be membrane-associated, appears to serve as a
sugar-specific energy coupling protein [38]. The enzymology of the PTS
has been reviewed in Chap. 2.

IV. POSTULATED MECHANISMS OF
     CARBOHYDRATE TRANSPORT
     REGULATION

Just as a multiplicity of transmembrane carbohydrate transport mechanisms have evolved in bacterial cells, a variety of mechanisms responsible
for the regulation of these activities appears to have evolved in parallel.
Thus, recent preliminary evidence indicates that at least three distinct
mechanisms are operative in the control of sugar uptake. One mechanism,
termed PTS-mediated regulation, depends on the catalytic activities of the
protein components of the PTS. This type of regulation seems to be operative only with certain non-PTS sugar transport systems. A mechanism
involving allosteric inhibition by a regulatory protein which is itself subject
to phosphorylation has been proposed [36]. A second possible mechanism
involves the direct binding of intracellular sugar phosphates to a regulatory
site on the cytoplasmic face of the transport protein. The proposed inter-

action would result in an inhibition of uptake activity [39-41]. Both PTS and certain non-PTS transport systems may be subject to this type of inhibition. A third postulated mechanism [42,43] presupposes that the conformations (and therefore the activities) of certain permease proteins are influenced by the transmembrane electrical potential which, according to the chemiosmotic theory, may correspond in part to the energized membrane state of oxidative phosphorylation. In this case, an enhanced cellular level of chemiosmotic energy may either depress or increase transport activity. While the glucose enzyme II complex of the E. coli phosphotransferase system may be subject to negative control by such a mechanism, the activities of permease systems responsible for the uptake of several carbon sources (lactose, 3-phosphoglycerate, citrate) [44-46] may require an appreciable membrane potential (negative inside), even for facilitated diffusion. In this last case, the permease system would be subject to positive control by the membrane potential.

It is teleologically satisfying to note that all three of the proposed regulatory mechanisms serve to sense the availability of utilizable carbohydrate and energy: PTS-mediated inhibition of transport results when the extracellular energy level (in the form of an extracellular PTS sugar) is high; the extent of inhibition by intracellular sugar phosphate is determined by the levels of intracellular metabolites; and inhibition of the activity of a permease protein by the imposition of a membrane potential generated by O_2-dependent electron transport or ATP hydrolysis via the Ca^{2+}, Mg^{2+}-ATPase, would provide sensitivity to the cellular level of chemiosmotic energy. All three mechanisms may function to prevent the consumption of carbon and energy in excess of the needs of the bacterial cell. In the succeeding sections of this chapter, the experimental evidence which led to the proposal of these three regulatory mechanisms in bacteria is evaluated, and transport regulatory phenomena that may be operative in eucaryotic cells are discussed from a comparative standpoint.

V. PTS-MEDIATED REGULATION

The lactose permease of E. coli was the first system in which the noncompetitive inhibition of transport by a sugar substrate of the PTS was clearly demonstrated. Winkler and Wilson [47] studied lactose permease activity using the nonmetabolizable lactose analog methyl-β-thiogalactoside, and they measured the activity of the glucose permease (the enzyme II^{glc} of the PTS) employing the nonmetabolizable glucose analog methyl-α-glucoside (αMG). When the bacteria had been grown under conditions which resulted in the simultaneous induction of the activities of both permeases, αMG was shown to reversibly inhibit the initial rate of methyl-β-thiogalactoside uptake as well as the steady-state level of accumulation. Methyl-β-thiogalactoside did not appear to exert a reciprocal effect on αMG uptake. The in vivo hydrolysis of o-nitrophenyl-β-galactoside, which had been shown to be a measure of lactose permease function, was also strongly inhibited by αMG.

Boniface and Koch conducted a kinetic study of this phenomenon, demonstrating that αMG depressed the maximal rate of o-nitrophenyl-β-galactoside entry (V_{max}) without altering the apparent affinity of the transport system for its substrate (K_m). A slight lag period for inhibition was noted in some experiments. The apparent binding constant for the inhibitory sugar (K_i) was estimated at 10 μM, lower than the apparent K_m measured for αMG uptake [48-50].

More recent experiments employing E. coli and S. typhimurium strains have established the involvement of the PTS in the regulation of the activities of the lactose permease and of other transport systems. Additionally, these experiments have provided important evidence regarding the mechanism of transport regulation and the significance of inducer exclusion to the repression of enzyme synthesis.

S. typhimurium cells which completely lacked enzyme I of the PTS could not be induced for the synthesis of the catabolic enzymes involved in the utilization of maltose, melibiose, or glycerol under normal conditions. Synthesis of these enzymes could be induced in "leaky" enzyme I mutants, but was sensitive to repression by extremely low concentrations of any sugar substrate of the PTS [51,52]. The sensitivity of a strain to repression was inversely related to the amount of residual enzyme I (or HPr) in the cell. A kinetic analysis of α-galactosidase synthesis in the presence or absence of methyl-α-glucoside in a leaky enzyme I mutant (ptsI17) and in the parental strain is reproduced in Fig. 3-3. The top figures reproduce growth rates, while the bottom figures show α-galactosidase induction. Data for the leaky enzyme I mutant, strain ptsI17, are shown in parts A, B, D, and E, while the results obtained with the wild type parent, strain LT-2, are reproduced in parts C and F. In parts A and D the only source of carbon in the growth medium was melibiose. Growth was therefore dependent on induction of the melibiose catabolic enzymes. It can be seen that methyl-α-glucoside had a strong repressive effect on α-galactosidase induction, which was reflected by depressed growth rates.

Figures 3-3B and E show the effect of methyl-α-glucoside on growth and induction of α-galactosidase synthesis when the medium contained, in addition to melibiose, a second carbon source, potassium lactate. The repressive effect of methyl-α-glucoside on α-galactosidase induction was more intense in the presence of lactate (compare Figs. 3-3D and E). However, growth was only slightly affected. The cells could utilize potassium lactate as the primary source of carbon, and the growth rate on this carbon source was only slightly depressed by inclusion of the glucoside. As shown in Fig. 3-3F, αMG exerted little or no repressive effect in the wild-type strain LT-2 under the conditions employed. If the enzyme II^{glc} had been induced prior to the initiation of the induction experiment, strong

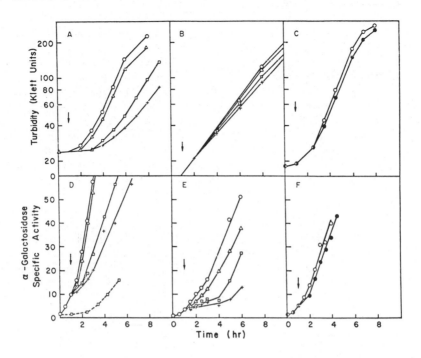

FIG. 3-3 Kinetics of the effect of methyl-α-glucoside on growth and α-galactosidase synthesis in a leaky enzyme I mutant, strain ptsI17, and the wild-type parent, strain LT-2. Washed inocula, grown in lactate minimal medium, were transferred to Medium 63 containing 0.5% melibiose (120 ml) in 500 ml Erlenmeyer flasks, fitted with side arms for turbidimetric determinations. The medium had been preincubated at 37°C before addition of cells at time zero. Methyl-α-glucoside (free of glucose) was added at the time indicated by the arrows; $\bigcirc$, 0 M; $\triangle$, 2 x 10^{-5} M; $\square$, 2x 10^{-4} M; +, 2 x 10^{-3} M; $\bullet$, 10^{-2} M methyl-α-glucoside. Aliquots were periodically removed and immediately chilled to 0°C. Cells were washed and assayed for α-galactosidase activity [52]. (A) Growth of ptsI17 cells in Medium 63 containing 0.5% melibiose as the sole source of carbon. (B) Growth of ptsI17 cells in Medium 63 containing 0.5% melibiose and 1% potassium lactate. (C) Growth of strain LT-2 in Medium 63 containing 0.5% melibiose as the sole source of carbon. (D) Induction of α-galactosidase synthesis in ptsI17 cells growing as in (A). (E) Induction of α-galactosidase synthesis in ptsI17 cells growing as in (B). (F) Induction of α-galactosidase synthesis in strain LT-2 growing as in (C). The dashed line in part D shows α-galactosidase induction when 2 x 10^{-4} M methyl-α-glucoside was added simultaneously with inducer.

repression of α-galactosidase synthesis by αMG could be demonstrated in this wild-type strain [23].

The degree of repression exerted by αMG on α-galactosidase synthesis was dependent on the extent of induction of the melibiose operon. This can be seen from the data in Fig. 3-3D (dashed line). When the glucose analog was added simultaneously with inducer, induction of α-galactosidase synthesis was prevented for hours. The same concentration of methyl-α-glucoside produced only brief repression when added after appreciable induced synthesis of the melibiose-specific proteins had occurred. When noninduced cells were transferred to minimal medium containing a single metabolizable sugar (such as melibiose) in the presence of methyl-α-glucoside, growth did not occur because the requisite catabolic enzymes could not be induced.

The data reproduced in Fig. 3-3 and results obtained in other experiments suggested that repression of α-galactosidase synthesis was most intense when the concentration of the inducing sugar was low, the concentration of the repressing sugar was high, and when the repressing sugar was added before appreciable induction of the melibiose catabolic enzyme system had occurred. These results were consistent with the suggestion that exclusion of the inducer from the cell by the repressing sugar was partly responsible for repression of α-galactosidase synthesis. This conclusion was substantiated by direct measurement of uptake rates [53-56]. Transport experiments showed that the melibiose, maltose, and glycerol permease systems in S. typhimurium strains were sensitive to PTS-mediated regulation. Representative results for the inhibition of melibiitol uptake by αMG are illustrated in Fig. 3-4. Employing cells which were not induced for the melibiose or glucose transport systems, only weak inhibition of melibiitol uptake by αMG was observed in the wild-type strain (LT-2, Fig. 3-4A). By contrast, strong inhibition was observed in an isogenic strain which was greatly deficient for, but not completely lacking, enzyme I of the PTS (ptsI17, Fig. 3-4B). No uptake of [^{3}H]melibiitol was observed in a "tight" enzyme I mutant (ptsI18, [Fig. 3-4(D)]). Figure 3-4(C) shows that the introduction of a secondary crr mutation (see below) into the leaky enzyme I mutant abolished the inhibition of melibiitol uptake by αMG. A similar mutation allowed uptake of melibiitol in strain ptsI18 (Fig. 3-4E).

Genetic analyses indicated that inhibition of the activities of the non-PTS sugar permeases was promoted by an interaction of a PTS sugar with the enzyme II complex specific for that sugar. Regulation was at the level of sugar uptake, rather than sugar efflux, and the maximum velocity of transport (V_{max}) was depressed. This last fact, plus the observation that exceedingly low concentrations of the inhibiting sugar were effective, showed that competition between substrate and inhibitor for a site on the transport carrier was not responsible for inhibition. This conclusion was in agreement with the observations of Winkler and Wilson, and of Koch and Boniface [47-50].

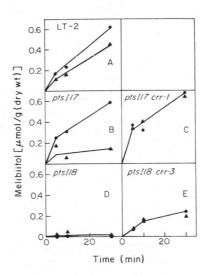

FIG. 3-4 Uptake of 1-[³H]melibiitol by S. typhimurium strains grown in
potassium lactate minimal medium. Washed cells suspended in minimal
salts Medium 63 containing an energy source, 10 mM NaCl, and 50 μg/ml
of chloramphenicol were employed for the uptake experiment. Transport
was conducted at 32°C in the presence (▲) or absence (●) of methyl-α-
glucoside as described previously [51,52]. See text for description of
bacterial strains.

    The following specific mutations were found to abolish regulation by
the PTS: (1) enzyme II⁻ mutations, (2) crr mutations, and (3) permease-
specific mutations.

1. Mutants which lacked an enzyme II activity exhibited reduced
   sensitivity to repression by the sugar(s) for which that enzyme II
   was specific. In all such mutants, resistance to repression correlated with resistance to inhibition of inducer uptake [23,53,54].

2. Carbohydrate repression resistance (crr) mutations rendered a
   strain completely resistant to PTS-mediated repression of catabolic enzyme synthesis as well as to PTS-mediated inhibition of
   non-PTS sugar permease systems (see Figs. 3-4C and E). Genetic
   experiments showed that all of several crr mutations tested in
   S. typhimurium were closely linked to the genes which coded for
   HPr and enzyme I (the ptsH and ptsI genes, respectively). All crr
   mutants tested contained depressed activity of a particular sugar-
   specific protein of the PTS involved in the phosphorylation of glucose and methyl-α-glucoside, an enzyme III^{glc} [53,55]. This protein has been purified and studied by Kundig [56]. It appears to be

a tetrameric protein with a molecular weight of 20,000. The protein can be phosphorylated by transfer of phosphate from phospho-HPr. Although it seems reasonable that the crr gene is the structural gene for this protein, this possibility has not been unequivocably established [19].

3. A permease–specific mutation in an S. typhimurium ptsI mutant rendered synthesis of the catabolic enzyme system involved in the metabolism of a single non-PTS sugar resistant to PTS-mediated repression [54,55]. Maltose, melibiose, and glycerol permease-specific mutants were isolated. In each of these three types of mutants a single transport system became resistant to inhibition by all sugar substrates of the PTS, but the other permease systems retained normal sensitivity to inhibition. Each of these three regulatory genes was mapped on the bacterial chromosome and was found to map within or very near the gene(s) which coded for the respective permease protein(s). An analogous mutation specifically abolished PTS-mediated regulation of the lactose permease in E. coli (unpublished results). These important observations led to the suggestion that some permeases are allosteric regulatory proteins, and that they can be indirectly modulated by the PTS. Further experiments, described below, led to the formulation of a model by which the PTS may regulate the activities of sensitive transport systems.

The mutant analyses described above suggested that regulation of transport activity by the PTS depended on several gene products as shown in Fig. 3-5. Regulation by a particular sugar substrate of the PTS is presumed to depend first, on the enzyme II complex specific for that sugar; second, on the products of the crr gene; and third, on a permease–specific gene product. This last gene product may correspond to a regulatory site (or subunit) of the target transport system.

Salmonella strains (in contrast to E. coli) possess a phosphoglycerate permease system which can transport phosphoenolpyruvate, 2-phosphogly-cerate, and 3-phosphoglycerate intact across the membrane. This transport system has been shown to be inducible, specific for the above mentioned phosphorylated compounds, and distinct from previously described transport systems [45]. It was employed to supply lysozyme-treated energy-depleted bacteria with a continuous intracellular source of phosphoenolpyruvate. When the cells had been induced to high levels of the phosphoglycerate transport systems, a low extracellular concentration of phosphoenolpyruvate (0.1 mM) half maximally stimulated uptake of methyl-α-glucoside via the PTS. If the phosphoglycerate transport system had not been induced, 100 times this concentration of phosphoenolpyruvate was required for half maximal stimulation. Phosphoenolpyruvate could not be replaced by other energy sources if potassium fluoride (an inhibitor of enolase) was present. Phosphoenolpyruvate also relieved inhibition of

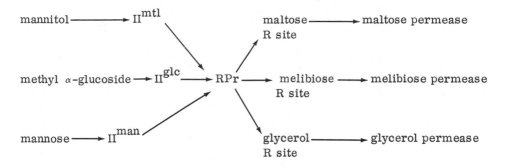

FIG. 3-5. Schematic representation of a hypothesis for the mechanism of PTS-mediated inhibition of transport activity in S. typhimurium. The hypothesis assumes that the three classes of mutations which render cells resistant to repression are due to loss of specific proteins (see text). II represents an enzyme II complex of the PTS. RPr is presumed to be the product of the crr gene and may be the enzyme III^{glc}. The sugar-specific R site is assumed to be an allosteric regulatory component of each of the respective transport systems. This regulatory site may or may not be part of the same polypeptide chain which comprises the transport carrier.

[^{14}C]glycerol uptake. The concentration of this phosphate donor which half maximally stimulated methyl-α-glucoside accumulation half maximally counteracted the inhibitory effect of the sugar. In the presence of potassium fluoride, other energy sources such as ATP, D-lactate, and 3-phosphoglycerate were without effect (M. H. Saier, Jr., unpublished results).

These data were interpreted in terms of a mechanism in which a protein (here referred to as RPr, possibly the product of the crr gene) could physically interact with the allosteric regulatory site of the sensitive transport system. It was assumed that RPr could be phosphorylated by transfer of a phosphoryl group from phospho-HPr with the formation of a high energy phosphoprotein. According to this hypothesis, only the free (unphosphorylated) form of RPr would bind to the allosteric binding site of the permease system and depress transport activity. This simple model accounts for several observations.

1. In the absence of a PTS sugar, RPr should be largely phosphorylated so that transport activity would be uninhibited. Addition of a PTS sugar to the cell suspension should drain phosphate from RPr to sugar. RPr would then be expected to inhibit the activities of the sensitive transport systems. This accounts for the PTS-mediated inhibition of transport activity observed in the wild-type bacterial cell.

2. If a cell is deficient for either enzyme I or HPr, the rate at which RPr can be phosphorylated should be depressed. Therefore, a

lower concentration of the PTS sugar should be effective in dephos-
phorylating RPr and inhibiting transport activity. This possibility
can explain the hypersensitivity of transport systems to inhibition
by PTS sugars in leaky enzyme I and HPr mutants.

3. An energy-depleted cell should be hypersensitive to inhibition by a
   PTS sugar because the cellular level of phosphoenolpyruvate would
   be insufficient to maintain rapid RPr phosphorylation. Enhancing
   the intracellular concentration of this phosphate donor should
   enhance the amount of phospho-RPr, thereby decreasing the concentration of the free protein and sensitivity of the PTS-controlled
   transport systems to inhibition. Again, this prediction is borne
   out by the observations noted above for energy-depleted cells.

If this hypothesis is correct, the regulatory role of the PTS derives
from its role as a "protein kinase" system. An interesting analogy may
exist between transport regulation in bacteria and the involvement of protein
kinases in the regulation of animal cell physiology (see below).

VI. REGULATION BY INTRACELLULAR SUGAR PHOSPHATES

The possibility that intracellular hexose phosphates may be inhibitory
modulators of carbohydrate transport proteins has been advanced by
several investigators [39-41, 58-60]. H. L. Kornberg and his coworkers
have pursued genetic and physiological approaches to the problem of transport regulation [39,40,58]. Employing an E. coli strain which synthesized
the enzyme II^{glc} constitutively, growth on 10 mM lactose or galactose was
shown to be prevented by the presence of 5 mM αMG in the culture medium.
Only weak inhibition was observed when a strain inducible for the enzyme
II^{glc} was employed, and the genetic loss of this enzyme II complex resulted
in complete abolition of the noncompetitive regulation.

Extensive studies on the inhibitory effects of glucose analogs on
fructose metabolism were also conducted. Fructose, like glucose, is
transported via the PTS. When wild type E. coli cells were grown in
minimal medium containing both glucose and fructose, biphasic sugar
utilization, reminiscent of the classic maltose plus glucose diauxie
documented by Monod (Fig. 3-1), was observed. Additionally, fructose-
grown E. coli cells, which synthesized the enzyme II^{glc} constitutively,
removed glucose from the medium eight times more rapidly than they
removed fructose when both sugars were present in the medium at the
same concentration. Fructose utilization was not inhibited by glucose
analogs in mutants deficient for the enzyme II^{glc}.

More recently, Amaral and Kornberg [40] described a mutant in which
fructose utilization was partially insensitive to inhibition by glucose analogs.

This mutant differed from previously studied mutants since the activity of the enzyme II^{glc} appeared to be normal, and the inhibitory effects of glucose analogs on lactose and glycerol utilization were not relieved. Transductional analyses revealed that the genetic defect associated with the regulatory deficiency mapped within or near the gene which coded for the enzyme II^{fru}. Thus, the mutation may have altered a regulatory site on the enzyme II^{fru} protein.

Amaral and Kornberg proposed that the inhibition of fructose utilization may be mediated by intracellular sugar phosphate. The following evidence was presented in support of this contention: (1) mutations which restricted the catabolism of glucose-6-phosphate (i.e., impairment of phosphoglucose isomerase, glucose-6-phosphate dehydrogenase and 6-phosphofructokinase activities) increased the degree of inhibition of fructose uptake exerted by exogenously added glucose; (2) in such strains, genetic loss of the enzyme II^{glc} resulted in the loss of sensitivity to inhibition of fructose utilization by glucose; (3) addition of glucose-6-phosphate to the growth medium of cells constitutive for the hexose phosphate transport system resulted in inhibition of fructose utilization; genetic loss of the enzyme II^{glc} did not relieve this inhibitory effect; and (4) the enzyme II^{fru}-specific genetic lesion discussed above, which reduced the inhibitory effect of glucose on fructose uptake, also diminished the inhibitory effect of exogenously added glucose-6-phosphate. These observations argue in favor of a mechanism involving intracellular sugar phosphate.

Related to this hypothesis are the observations of Kaback concerning the inhibitory effects of hexose phosphates on sugar uptake into bacterial membrane vesicles [41,59]. Measurement of the rates of uptake of radiolabeled glucose, αMG, fructose, or galactose in the presence of various sugar phosphates led to the suggestion that there are two classes of inhibitor binding sites, one for hexose-6-phosphates and another for the 1-phosphate esters. The data indicated that both glucose-6-phosphate and glucose-1-phosphate noncompetitively inhibited glucose uptake.

At present, there appears to be little evidence for the existence of regulatory sugar phosphate binding sites on the extracellular faces of transport proteins in intact bacterial cells. However, it has recently been suggested that the enzymes II of the PTS may possess sugar phosphate binding sites on their intracellular surfaces [61,62]. Support for this notion derives from the observation that the enzymes II (in the presence or absence of the soluble components of the PTS) apparently catalyze phosphoryl transfer from sugar phosphate to [^{14}C]sugar. This transphosphorylation reaction was demonstrated both in vitro and in bacterial membrane vesicles. The reaction in membrane vesicles required that the sugar phosphate be present intravesicularly and that the [^{14}C]sugar be present in the external medium [62]. These results can be explained if the enzymes II are assumed to possess two substrate binding sites: one on the outer surface of the membrane which binds free sugar, and one on the cytoplasmic surface which can bind sugar phosphate.

An important question concerns the possibility that intracellular sugar phosphates may regulate the uptake of carbohydrates other than fructose. Evidence has been presented that in Staphylococcus aureus and Bacillus subtilis intracellular sugar phosphates inhibit the uptake of several carbohydrates [60]. Another study provided evidence that the accumulation of intracellular glucose-6-phosphate inhibits glucose utilization in E. coli [63]. On the other hand, Winkler [64], Gachelin [65], and Haguenauer and Kepes [66] each independently reported that the initial rate of uptake of αMG into glycerol-grown E. coli cells was not significantly inhibited by intracellular methyl-α-glucoside-6-phosphate. Recently, however, this experiment was repeated with glucose-grown cells, and just the opposite result was obtained [67]: the initial rate of [^{14}C]αMG uptake was drastically reduced following preincubation of the cells with nonradioactive αMG. This effect was less apparent in cells grown on carbon sources other than glucose, in agreement with the negative results reported in previous investigations. These results indicate that intracellular methyl-α-glucoside-6-phosphate may act as a feedback inhibitor of αMG uptake in glucose-grown cells, although partial depletion of the phospho-HPr pool by preincubation with substrate is another possible explanation.

The available evidence supports the notion that intracellular sugar phosphates inhibit the uptake of several sugars into a variety of bacterial cells. More detailed studies will be required to definitively establish that the inhibitory effects result from the direct modulation of the activities of carbohydrate transport proteins.

VII. POSSIBLE REGULATION OF METHYL-α-GLUCOSIDE TRANS-PORT BY THE ENERGIZED MEMBRANE STATE

An unusual aspect of αMG transport in E. coli and S. typhimurium is its behavior with respect to certain inhibitors of oxidative phosphorylation. The first extensive study of this problem was undertaken in the early 1960s by Hoffee and Englesberg [68,69]. These investigators observed a several-fold increase in the steady-state level of αMG accumulation upon preincubation of cells with sodium azide or 2,4-dinitrophenol. Likewise, anaerobiosis resulted in an increase in the steady-state level of αMG accumulated by cells grown aerobically on fructose. In contrast, preincubation of cells with fructose or another metabolizable carbon source caused a decrease in the internal level of αMG accumulated compared to the level reached in washed cells deprived of an energy source. It appeared that inhibition of αMG accumulation was observed under conditions of rapid metabolism, and a stimulation of uptake resulted upon energy starvation. Hoffee and Englesberg [68,69] and others [70,71] showed that exit of preloaded αMG was affected by these treatments. Energy sources stimulated exit of αMG while sodium azide <u>inhibited</u> exit.

Gachelin [72] studied the effect of 50 mM azide on the kinetic parameters of αMG uptake and found that this energy inhibitor shifted the K_m of αMG from 200 to 70 μM without altering the maximal uptake rate (V_{max}). Incubation with azide caused a decrease in the intracellular concentration of phosphoenolpyruvate from 1.2 to 0.4 μmol/g, dry weight of cells. This latter observation ruled out the possibility that azide exerted its stimulatory effect by increasing the level of cellular phosphoenolpyruvate.

Since the PTS is dependent on an intracellular supply of phosphoenolpyruvate, it might be expected that fluoride, an inhibitor of enolase, would decrease the rate of αMG uptake. Haguenauer and Kepes [66] found that NaF alone had no effect on αMG uptake or phosphorylation at pH 7.0. On the other hand, when cells were preincubated with azide plus fluoride at the same pH, the azide stimulation was not only prevented, but accumulation was totally blocked. The inhibitory synergism of azide plus fluoride, as well as an inhibitory effect of fluoride alone, was found to be pH dependent. At pH 7, NaF had little effect on the uptake and phosphorylation of αMG, but at pH 6, uptake in the presence of fluoride was abolished. Similarly, strong inhibition by azide plus fluoride was noted at pH 6 or 7, but not at pH 8. Although the authors proposed an explanation for their observations, the unexplained effects of fluoride on αMG uptake and phosphorylation in bacterial vesicles reported by Kaback [59] serve to emphasize the complexity of the phenomena at hand and to discourage simple interpretation of the data.

Since the PTS functions via a cascade of phosphoprotein intermediates where phosphate is originally derived from phosphoenolpyruvate, one might expect that arsenate treatment would abolish PTS function. Klein and Boyer showed that exposure of <u>E. coli</u> cells to arsenate in the absence of phosphate severely reduced the cellular levels of ATP and phosphoenolpyruvate [57]. Although the concentration of phosphoenolpyruvate was reduced to 1% of its original value, the initial rate of αMG uptake was inhibited only 50%. The intracellular pool of glucose-6-phosphate was largely retained, leading to the precocious suggestion that this compound might serve as the phosphoryl donor for αMG uptake [57]. This possibility has recently been substantially strengthened by the observation that the enzymes II of the PTS apparently catalyze phosphoryl transfer from sugar phosphate to sugar in the absence of phosphoenolpyruvate [61,62]. The significance of these observations with respect to azide stimulation of αMG uptake is not yet clear.

It appears difficult to explain the azide effect in terms of a change in the level of some intracellular metabolite. Instead, the glucose transport protein (the enzyme II^{glc}) may be sensitive to the transmembrane electrical potential. Hernandez-Asensio et al. [73] showed that any substrate which stimulated cell respiration also inhibited the initial rate of αMG uptake. Furthermore, the uncoupler of oxidative phosphorylation, CCCP, which appeared to dissipate the membrane potential without blocking electron flow, prevented the inhibition of αMG uptake by respirable carbon sources, and sodium azide had the same effect. In both cases the principal effect was on the K_m rather than the V_{max} for uptake, in agreement with the data of Gachelin [72].

In a second series of experiments carried out by the same investigators [43] mutants defective for the Ca^{2+}, Mg^{2+} -ATPase were employed. It was found that the inhibition of αMG uptake produced by a respirable carbon source was more pronounced in the mutants than in the wild type. CCCP was able to completely reverse this inhibition, and direct measurements of the ATP pools in both mutant and parental strains failed to show any correlation between αMG uptake rate and the level of cellular ATP.

In an independent analysis of αMG uptake in ATPase mutants, Moczydlowski and Wilson [67] found that αMG uptake was inhibited when the energized membrane state was formed as a result of electron transport. Proline and methyl-β-thiogalactoside uptake were assayed to monitor the membrane potential [2, 3], and mannitol uptake was measured to monitor the energy source for the PTS. It was found that ATPase mutants starved by incubation in the absence of a carbon source and in the presence of 2, 4-dinitrophenol had greatly reduced levels of endogenous methyl-β-thiogalactoside and proline uptake, while uptake of αMG and mannitol was largely unaffected. This starvation procedure, developed by Berger [30, 31], is known to deplete the cellular ATP pool. When the energized membrane state was restored by the addition of an oxidizable substrate or the artificial electron donor system, phenazine methosulfate plus ascorbate, the rates of proline and methyl-β-thiogalactoside uptake were restored to normal, but αMG uptake was inhibited. While anaerobiosis inhibited uptake of proline and methyl-β-thiogalactoside, αMG transport was stimulated. Additional evidence came from the study of Gilchrist and Konisky [74] who showed that colicin Ia, which appeared to inhibit the formation of the energized membrane state of oxidative phosphorylation, inhibited proline and methyl-β-thiogalactoside uptake while stimulating αMG uptake. Also, Singh and Bragg [75] have recently contributed further evidence with respect to this problem. They found that a cytochrome-deficient mutant of E. coli exhibited a strong inhibition of αMG uptake when an energized state of the membrane was formed from ATP hydrolysis by the Ca^{2+}, Mg^{2+} -ATPase. This inhibition was reversed by inhibitors of the ATPase, such as dicyclohexylcarbodiimide (DCCD) and azide. The uncouplers 2, 4-dinitrophenol and carbonylcyanide-m-chlorophenylhydrazone also reversed the inhibition under appropriate conditions. Although the proposal that the energized membrane state directly inhibits the activity of the enzyme II^{glc} seems to be a tenable explanation for the results summarized above, quantitative studies will be required to further substantiate this hypothesis.

VIII. CONCLUSION ON A EUCARYOTIC NOTE

Extensive studies in many laboratories have established that the regulation of solute transport is a universal cellular phenomenon common to the eucaryotic, as well as the procaryotic kingdom. Holley [76] has put forth the postulate that nutrient transport regulation serves as a primary mode of growth regulation in animal cells. He has also proposed that the membrane changes associated with malignancy may account for the loss of growth regulation and the uncontrolled proliferative abilities of transformed cells.

In the eucaryotic cell, hormones are fundamental regulatory mediators. Christensen [77] has published an extensive list of examples where hormones are known to affect the transport of various compounds. The molecular events by which hormones bring about these alterations in transport activities are almost completely unknown. The effects of many hormones, however, have been found to correlate with changes in the levels of second messengers such as the cyclic nucleotides.

Hormones may modulate transport activities by mechanisms which involve cyclic AMP-responsive protein kinases. These enzymes have been shown to phosphorylate proteins which are associated with the animal cell plasma and microsomal membranes [78]. In several instances, consequential regulation of ion transport has been proposed, and correlative studies [78, 79] provide evidence for a relationship between membrane phosphorylation and transport modulation. In one interesting example, protein phosphorylation appeared to account for certain temporal changes in the postsynaptic electrical potential in neurons, possibly as a consequence of the modification of the activities of transmembrane ion channels [79].

Evidence for the regulation of solute transport in eucaryotes by a mechanism involving intracellular phosphorylated metabolites is still very preliminary. It has been suggested [60] that intracellular sugar phosphates inhibit the uptake of glycerol into cells of the yeast Saccharomyces cerevisiae, but this suggestion requires further substantiation.

Elbrink and Bihler [80] have recently reviewed the literature concerned with glucose transport in various animal tissue types and have discussed the conditions under which glucose transport can be considered to be the rate-limiting step in glycolysis. In several tissues where glucose transport displays the characteristics typical of facilitated diffusion (muscle, adipose tissue, and avian erythrocytes), glucose transport has been shown to be stimulated by anoxia and inhibitors of oxidative phosphorylation. These are conditions which result in the depletion of cellular energy. Possibly the increase in glucose transport during oxygen depletion is partially responsible for the Pasteur effect [80]. These authors and other workers [81,82] in the eucaryotic field are currently entertaining the possibility that the regulation of glucose transport may be mediated by Ca^{2+} ions, a suggestion that should be explored by workers in the procaryotic field as well. In relation to this problem and the effect of the membrane potential on transport, it appears that under certain conditions the ATP-ADP exchange transport system in the inner mitochondrial membrane can be stimulated by uncouplers [83]. This observation is suggestive of a regulatory phenomenon involving the transmembrane electrical potential.

The results summarized above provide preliminary evidence that in both eucaryotes and procaryotes of divergent evolutionary origin, a variety of mechanisms have evolved for the regulation of transmembrane solute permeation. In this review, experiments supporting three postulated regulatory mechanisms, involving protein phosphorylation, intracellular metabolites, and the transmembrane electrical potential, have been discussed. Nature appears to have kept an open mind during the evolution of complex eucaryotic life forms, conserving in principle the useful regulatory mechanisms of simpler procaryotic cells. Students of transport regulation would do well to consider all of the various models proposed in studies of dissimilar biological systems.

ACKNOWLEDGMENTS

The original research in the authors' laboratory reported in this chapter was supported by grant BMS73-06802 A01 from the National Science Foundation and Public Health Service Grant #1 RO1 CA165521-01A1 MBY. M.H.S. was supported by NIH Research Career Development Award 5 KO4 CA00138-02.

REFERENCES

1. S. J. Singer, Ann. Rev. Biochem., 43: 805-833 (1974).
2. F. M. Harold, Bacteriol. Rev., 36: 172-230 (1972).
3. R. D. Simoni, and P. W. Postma, Ann. Rev. Biochem., 44: 523-554 (1975).
4. D. E. Koshland and K. E. Neet, Ann. Rev. Biochem., 37: 359-410 (1968).
5. J. Monod, Growth 11: 223-289 (1947).
6. E. F. Gale, Bacteriol. Rev., 1: 139-173 (1943).
7. B. Magasanik, Cold Spring Harbor Symp. Quant. Biol., 26: 249-256 (1961).
8. B. Magasanik, in The Lactose Operon (J. Beckwith, and D. Zipser, eds.), Cold Spring Harbor Laboratories, Cold Spring Harbor, New York, 1970, pp. 189-219.

9. K. Paigen, and B. Williams, Adv. Microbial Physiol., 4: 251-324 (1970).
10. B. Tyler and B. Magasanik, J. Bacteriol., 102: 411-422 (1970).
11. V. N. Gershanovitch, G. I. Bourd, N. V. Jorovitz-Kaya, A. G. Shavronskaya, V. V. Klyutchova, and V. P. Shabolenko, Biochim. Biophys. Acta, 134: 188-190 (1967).
12. C. F. Fox and G. Wilson, Proc. Natl. Acad. Sci. USA, 59: 988-995 (1968).
13. I. Pastan and R. Perlman, J. Biol. Chem., 244: 5836-5842 (1969).
14. I. Pastan and R. Perlman, Science, 169: 339-344 (1970).
15. P. K. Wayne and O. M. Rosen, Proc. Natl. Acad. Sci. USA, 71: 1436-1440 (1974).
16. W. Epstein, L. B. Rothman-Denes, and J. Hesse, Proc. Natl. Acad. Sci. USA, 72: 2300-2304 (1975).
17. H. V. Rickenberg, Ann. Rev. Microbiol., 28: 353-369 (1974).
18. A. Peterkofsky and C. Gazdar, Proc. Natl. Acad. Sci. USA, 72: 2920-2924 (1975).
19. M. H. Saier, Jr. and B. U. Feucht, J. Biol. Chem., 250: 7078-7080 (1975).
20. J. F. McGinnis and K. Paigen, J. Bacteriol., 100: 902-913 (1969).
21. J. F. McGinnis and K. Paigen, J. Bacteriol., 114: 885-887 (1973).
22. S. J. Curtis and W. Epstein, J. Bacteriol., 122: 1189-1199 (1974).
23. M. H. Saier, Jr., B. U. Freucht, and L. J. Hofstadter, J. Biol. Chem., 251: 883-892 (1976).
24. B. Rotman, A. K. Ganesan, and R. Guzman, J. Mol. Biol., 36: 247-260 (1968).
25. M. H. Saier, Jr., F. G. Bromberg, and S. Roseman, J. Bacteriol., 113: 512-514 (1973).
26. W. Boos, J. Biol. Chem., 247: 5414-5424 (1972).
27. S. Hayashi and E. C. C. Lin, Biochim. Biophys. Acta, 94: 479-487 (1965).
28. E. C. C. Lin, Ann. Rev. Genet., 4: 225-262 (1970).
29. B. P. Rosen and L. A. Heppel, in Bacterial Membranes and Walls (L. Leive, ed.), Marcel Dekker, 1973, pp. 209-239.
30. E. A. Berger, Proc. Natl. Acad. Sci. USA, 70: 1514-1518 (1973).
31. E. A. Berger and L. A. Heppel, J. Biol. Chem., 249: 7747-7755 (1974).
32. I. West and P. Mitchell, Bioenergetics, 3: 445-462 (1972).
33. I. West and P. Mitchell, Biochem. J., 132: 587-592 (1973).
34. J. Stock and S. Roseman, Biochem. Biophys Res. Commun., 44: 132-138 (1971).
35. S. Roseman, in Metabolic Pathways, Vol. VI, Ed. 3, (L. E. Hokin, ed.), Academic, New York, 1972, pp. 41-89.
36. M. H. Saier, Jr. and C. D. Stiles, Molecular Dynamics in Biological Membranes, Springer Verlag, New York, 1975.

124   M. A. SAIER AND E. G. MOCZYDLOWSKI

37. R. D. Simoni and S. Roseman, J. Biol. Chem., 248: 966-976 (1973).
38. R. D. Simoni, J. B. Hays, T. Nakazawa, and S. Roseman, J. Biol. Chem., 248: 957-965 (1973).
39. H. L. Kornberg, Symp. Soc. Exp. Biol., 27: 175-193 (1973).
40. D. Amaral and H. L. Kornberg, J. Gen. Microbiol., 90: 157-168 (1975).
41. H. R. Kaback, Proc. Natl. Acad. Sci. USA, 63: 724-731 (1969).
42. E. G. Moczydlowski and D. B. Wilson, Fed. Proc., 34: 491 (1975).
43. F. F. del Campo, M. Hernandez-Asensio, and J. M. Ramirez, Biochem. Biophys. Res. Commun., 63: 1099-1105 (1975).
44. A. L. Koch, J. Mol. Biol., 59: 447-459 (1971).
45. M. H. Saier, Jr., D. L. Wenzel, B. U. Feucht, and J. J. Judice, J. Biol. Chem., 250: 5089-5096 (1975).
46. S. Schuldiner, R. Weil, and H. R. Kaback, Proc. Natl. Acad. Sci. USA, 73: 109-112 (1976).
47. H. H. Winkler and T. H. Wilson, Biochim. Biophys. Acta, 135: 1030-1051 (1967).
48. J. Boniface and A. L. Koch, Biochim. Biophys. Acta, 135: 756-770 (1967).
49. A. L. Koch and J. Boniface, Biochim. Biophys. Acta, 225: 239-247 (1971).
50. A. L. Koch, Biochim. Biophys. Acta, 249: 197-215 (1971).
51. M. H. Saier, Jr., R. D. Simoni, and S. Roseman, J. Biol. Chem., 245: 5870-5873 (1970).
52. M. H. Saier, Jr., R. D. Simoni, and S. Roseman, J. Biol. Chem., 251: 6584-6597 (1976).
53. M. H. Saier, Jr. and S. Roseman, J. Biol. Chem., 247: 972-975 (1972).
54. M. H. Saier, Jr. and S. Roseman, J. Biol Chem., 251: 6598-6605 (1976).
55. M. H. Saier, Jr. and S. Roseman, J. Biol. Chem., 251: 6606-6615 (1976).
56. W. Kundig, J. Supramol. Struct., 2: 695-714 (1974).
57. W. L. Klein and P. D. Boyer, J. Biol. Chem., 247: 7257-7265 (1972).
58. H. L. Kornberg, in The Molecular Basis of Biological Transport (J. F. Woessner, Jr. and F. Huijing, eds.), Academic, New York, 1972, pp. 157-180.
59. H. R. Kaback, in Current Topics in Membranes and Transport, (A. Kleinzeller, ed.), Academic, New York, 1970, pp. 35-99.
60. M. H. Saier, Jr. and R. D. Simoni, J. Biol. Chem., 251: 893-894 (1976).
61. M. H. Saier, Jr. and M. J. Newman, J. Biol. Chem., 251: 3834-3837 (1976).
62. M. H. Saier, Jr., B. U. Feucht, and W. K. Mora, J. Biol. Chem., 252: 8899-8907 (1977).

63. D. N. Dietzler, M. P. Leckie, J. L. Magnani, M. J. Sughrue, and P. E. Bergstein, J. Biol. Chem., 250: 7194-7203 (1975).
64. H. H. Winkler, J. Bacteriol., 106: 362-368 (1971).
65. G. Gachelin, Eur. J. Biochem., 16: 342-357 (1970).
66. R. Haguenauer and A. Kepes, Biochemie, 53: 99-107 (1971).
67. E. G. Moczydlowski and D. B. Wilson, unpublished results.
68. P. Hoffee and E. Englesberg, Proc. Natl. Acad. Sci. USA, 48: 1759-1765 (1962).
69. P. Hoffee, E. Englesberg, and F. Lamy, Biochim. Biophys. Acta, 79: 337-350 (1964).
70. Y. S. Halpern and M. Lupo, Biochim. Biophys. Acta, 126: 163-167 (1966).
71. H. Hagihira, T. H. Wilson, and E. C. C. Lin, Biochim. Biophys. Acta, 78: 505-515 (1963).
72. G. Gachelin, Ann. Inst. Pasteur, 122: 1099-1116 (1972).
73. M. Hernandez-Asensio, J. M. Ramirez, and F. F. del Campo, Arch. Microbiol., 103: 155-162 (1975).
74. M. J. R. Gilchrist and J. Konisky, J. Biol. Chem., 250: 2457-2462 (1975).
75. A. P. Singh and P. D. Bragg, FEBS Lett., 64: 169-172 (1976).
76. R. W. Holley, Proc. Natl. Acad. Sci. USA, 69: 2840-2841 (1972).
77. H. N. Christensen, Biological Transport, Benjamin, 1975, Chap. 12.
78. C. S. Rubin and O. M. Rosen, Ann. Rev. Biochem., 44: 831-887 (1975).
79. P. Greengard, Nature, 260: 101-108 (1976).
80. J. Elbrink and I. Bihler, Science, 188: 1177-1184 (1975).
81. J. P. Reeves, J. Biol. Chem., 250: 9413-9420 (1975).
82. J. P. Reeves, J. Biol. Chem., 250: 9428-9430 (1975).
83. M. Klingenberg, in Mitochondria: Bioenergetics, Biogenesis, and Membrane Structure (L. Packer and A. Gomez-Puyou, eds.), Academic, New York, 1976, pp. 127-149.

Chapter 4

SUGAR TRANSPORT SYSTEMS IN ESCHERICHIA COLI

Thomas J. Silhavy,* Thomas Ferenci, and Winfried Boos

Department of Biology
University of Konstanz
Konstanz, Federal Republic of Germany

Dedicated to Herman M. Kalckar

I. INTRODUCTION 128
II. SUGAR TRANSPORT IN ESCHERICHIA COLI 128
    A. Facilitated Diffusion Transport Systems 130
    B. Group Translocation Systems 131
    C. Shock-Resistant Active Transport Systems 131
    D. Shock-Sensitive Active Transport Systems 142
    E. Transport Systems Containing Outer Membrane
        Components 149
    F. Systems that Are Transported in Membrane Vesicles
        but Contain Periplasmic Components 149
    G. The Exit Reaction 150
III. THE ACTIVE SUGAR TRANSPORT SYSTEMS OF
     ESCHERICHIA COLI 154
IV. EPILOG. 160
    RECENT DEVELOPMENTS 160
    REFERENCES 160

*Present address: Department of Microbiology and Molecular Genetics,
Harvard Medical School, Boston, Massachusetts

I. INTRODUCTION

Transport studies in microorganisms have been greatly stimulated in recent years by the pioneering studies of Kaback and his collaborators who introduced the system of membrane vesicles. These vesicles are osmotically active sacs composed of correctly oriented cytoplasmic membranes capable of exhibiting active transport. Since these membrane vesicles were freed of most of the cytoplasmic constituents as well as the complex structure of cell wall and outer membrane, this system could be regarded as an in vitro transport system more suitable for biochemical studies than whole cells.

However, it has also become obvious that vesicles studies are suitable only for a certain type of transport systems but not for others. For example, several systems that operate via periplasmic binding proteins cannot be studied in membrane vesicles since these proteins are apparently lost during vesicle preparation. In addition, it has become apparent that the outer membrane, at least in gram-negative bacteria, can play an important role as a permeability barrier for the molecules to be transported, particularly when their size exceeds that of a disaccharide. Obviously, the characteristics of these transport systems cannot be described in membrane vesicles either. Moreover, the problem of energy coupling to active transport cannot be fully described in membrane vesicles. Recently there have been strong indications that besides PEP-dependent substrate phosphorylation via the phosphotransferase system and the protonmotive force as the driving force for systems active in membrane vesicles, there are other modes of energy coupling. These modes involve direct participation of phosphate bound energy, possibly ATP itself.

Thus, despite the progress which has been made with the membrane vesicle studies, experiments with whole cells are still meaningful and necessary. In the following we will discuss the different modes of sugar transport. Since energy coupling of active transport as well as the mechanism of the sugar phosphotransferase system will be covered elsewhere in this book (see Chapters 2 and 12), these areas will not be discussed here in detail. As a model organism mainly Escherichia coli will be discussed, since most studies have been done in this organism.

II. SUGAR TRANSPORT IN ESCHERICHIA COLI

The cell envelope of E. coli is a complex structure composed of three distinct layers: the outer membrane, the cell wall or peptidoglycan, and the inner or cytoplasmic membrane. The periplasm [1], then, is the space between the inner and outer membrane. Despite the presence of this multilayered permeability barrier, these bacteria are quite capable of selectively transporting a number of substrates through this barrier to the cytoplasm.

Furthermore, they can carry out this process actively, i.e., they can transport against and maintain concentration gradients (inside/out) of as high as 10^5 [2]. How the bacteria can transport many hydrophilic compounds through the hydrophobic permeability barrier, where the cell derives the energy needed to maintain such large concentration gradients, and how this energy is coupled to the transport process, are the subjects of a considerable amount of current research in this field of membrane function. At present these questions remain largely unanswered; however, the research carried out to date clearly indicates that bacteria utilize a variety of different transport mechanisms and immediate energy sources. In order to facilitate a discussion of these transport mechanisms and to provide a framework of reference, the known sugar transport systems have been divided into categories based primarily upon what is known concerning their mechanism. These categories can be summarized as follows:

1. Facilitated diffusion systems

2. Group translocation systems (the phosphotransferase system)

3. Active transport systems

   a. Shock-resistant, and functioning in membrane vesicles.

   b. Shock-sensitive, and binding protein mediated.

   c. Systems containing outer membrane components.

   d. Systems that transport in membrane vesicles but contain periplasmic components.

Such a division into different transport systems is based on the type and location of the proteins involved, on the type of energy coupling and type of mechanism by which translocation is accomplished. It is quite clear that not all bacteria contain all the different types of transport systems, and it is equally clear that certain sugars are not transported in all bacteria by the same way. In addition, in many organisms certain sugars are often transported by more than one system. The phosphotransferase system (PTS) has been found to catalyze the uptake of sugars such as glucose, fructose, sorbose, and mannose, and sugar alcohols such as glucitol, mannitol, and galactitol in many types of anaerobic and facultative anaerobic as well as photosynthetic bacteria [3-11], and limited to fructose in strictly aerobic organisms [12,13]. Staphylococcus aureus prefers to transport most sugars via the PTS, [14] even though few sugars are still transported via facilitated diffusion [15] or even active transport [16]. In strictly aerobic organisms sugars are transported via active transport systems. This is true for glucose transport in Azotobacter vinelandii [17], in Brucella abortus [18] or in Pseudomonas aeruginosa [19]. Active transport of the same substrate in different organisms is not neces-

sarily mediated by similar mechanisms. While transport of glucose-1-
phosphate in E. coli is mediated via a typical shock-resistant transport
system [20], in Agrobacterium tumelaciens transport appears to be dependent on a periplasmic binding factor [21]. Similarly, transport of myoino-
sitol in Klebsiella aerogenes [22] was not dependent on binding
protein [23] whereas it involved a binding protein in Pseudomonas
fluorescens [24]. Active maltose transport in E. coli is mediated via the
periplasmic maltose binding protein [25]. In S. aureus it is mediated via
facilitated diffusion [15] and in P. fluorescens it is mediated via a shock-
resistant active transport system [26]. As a general rule, periplasmic
binding protein dependent systems are found in gram-negative but not gram-
positive organisms. Few transport studies have been done in Mycoplasma
even though these organisms should be quite suitable for transport studies.
They have limited biosynthetic capabilities and their membrane is free of
cell walls and external polysaccharides [27]. Both types of transport systems have been recognized in these organisms, PTS [9] and active transport systems [28].

The discussion which follows will not be a compendium of all the known
sugar transport systems. At least one example from each of the above
categories will be presented. In categories containing more than one
transport system, only the most thoroughly studied will be discussed.
Others will be mentioned only when necessary for discussion or if they
provide a unique insight. The classification into different types of transport systems is only meant as a superficial guideline in order to facilitate
a systematic approach. Variations within the above classes may become
evident when more details on the particular systems are available.

A. Facilitated Diffusion Transport
    Systems

For a number of years it was believed that glycerol entered cells by a nonspecific mechanism. This belief was supported by the fact that the K_m
for growth on glycerol was found to be identical with the K_m of glycerol
for the enzyme glycerolkinase. Also, glycerol entered the cells so rapidly
that no measurements could be made [29]. Recently, however, it became
apparent that the process of glycerol entry is carrier mediated. This
carrier, though, does not concentrate glycerol inside the cell [30]. The
carrier, called the glycerol facilitator, was detected in the following
manner. If a hypertonic NaCl solution is added to a bacterial suspension,
an immediate increase of optical density (OD) is seen to be due to plasmolysis of the cell. If glycerol is used instead of NaCl, the bacteria do
not remain plasmolyzed, but return to their normal shape as the glycerol
equilibrates across the membrane. This can be followed by the simultaneous decrease in optical density. Sanno et al. [31] found that cells pregrown

on glycerol equilibrated glycerol across the cell envelope much more rapidly than cells pregrown on glucose. Furthermore, it was found that this increased permeability to glycerol was constitutive in glpR cells, i.e., cells which produce the glycerol catabolic enzymes constitutively. Further investigation of this phenomenon revealed that a gene, glpF, located in the operon which contains the gene for the enzyme glycerokinase, coded for the facilitator [32]. The physiological importance of this gene was demonstrated by showing that glpF cells grow much slower on millimolar glycerol than do the corresponding wild-type cells [33].*

A special form of facilitated diffusion systems appears to be some enzyme II systems of the phosphotransferase system. Thus, Kornberg and Riordan [35] could demonstrate that mutants of E. coli that are defective in all known active transport systems for galactose could still transport galactose via facilitated diffusion catalyzed by an enzyme II for glucose. Transport of galactose occured in the absence of phosphorylation. Similar observations had been made for enzyme II-mediated facilitated diffusion of galactose in Salmonella typhimurium [36]. Furthermore, from exit studies of endogeneously-produced mannitol, facilitated diffusion properties of enzyme II for mannitol has been concluded [37]. A similar observation has been made for fructose exit in Bacillus subtilis [7].

B. Group Translocation Systems

It is known that bacteria transport a number of sugars by a process which liberates sugar phosphates in the cytoplasm. In other words, these sugars are phosphorylated during transport. These transport systems are called phosphotransferase systems (PTS), and to date they provide the single most characterized and understood transport mechanism. All the components of several PTS transport systems have been purified, and the actual phosphorylation reaction has been demonstrated in vitro. In addition, fully induced and "operating" enzymes II of PTS have been found to interfere with the transport activity of other non-PTS sugar transport systems [38]. These and other properties of PTS that concern its participation in gene regulation are dealt with in other chapters of this book and are therefore not further discussed here (see Chapters 2 and 3).

C. Shock-Resistant Active Transport
    Systems

Shock-resistant transport systems are largely unaffected by the classical cold osmotic shock procedure of Neu and Heppel [39] and accordingly, none

---

*A more detailed picture on the dissimilation of glycerol in bacteria can be found in a recent review by Lin [34].

of these systems contains or is dependent upon periplasmic binding proteins that are released by this procedure. The best studied of these systems, the lactose transport system, contains only a single component, which is presumed to be tightly bound to the inner or cytoplasmic membrane. Membrane vesicles exhibit transport activity for these shock resistant systems, and the data available on the model of energy coupling suggest that most, possibly all, of these systems utilize the proton motive force as the driving force for the active transport, while translocation occurs via a proton symport. Again, these questions of energy coupling are discussed in detail in another chapter of this book (see Chapter 12).

1. The Lactose Transport System

So far as is known, the lactose transport system is composed of a single protein, namely the product of the lacY gene. This gene, of course, is a component of the lactose operon, and accordingly the regulation of this transport system has been thoroughly studied and is the classic example of an operon under negative control [40]. In addition to being under the control of the lactose repressor, the operon is subject to catabolite, transient, and carbohydrate repression (see Chap. 3). The substrate specificity and the kinetics of this transport system have been thoroughly reviewed [41-43] and will not be discussed here. Rather, this section will deal with biochemical studies of the carrier protein and how these studies related to the various transport systems.

Despite considerable efforts, the lactose carrier protein remains largely uncharacterized. Many investigators have searched for this carrier in whole cells, membrane fractions, shock fluids, and even cytoplasmic fractions. They have used a variety of different electrophoresis techniques, numerous purification procedures, super-producing mutants, and various differential labeling techniques to maximize the difference between induced and uninduced cells. To date, no one has yet succeeded in even unequivocally identifying the native carrier protein, let alone purifying it. There have been, however, several notable successes. In what has become a classic experiment, Fox and Kennedy [44] succeeded in covalently labeling a membrane protein which exhibits the properties of the lactose carrier. They have called this protein the "M protein" and this name will be used hereafter. These authors took advantage of the known sensitivity of the transport process to sulfhydryl reagents such as N-ethylmaleimide (NEM). It was found that β, β'-thiodigalactoside (TDG), a substrate of the lactose transport system, could protect against the inhibitory effects of NEM, presumably by binding to the M protein and preventing the approach of NEM. Therefore, if whole cells or a membrane fraction were treated with NEM in the presence of TDG, all available sulfhydryl groups, with the exception of the TDG-protected group in the M protein, will be covalently modified. Subsequent removal of excess NEM and TDG will leave only the protected sulfhydryl groups available for the following reaction with [14]C- or ^{3}H-labeled NEM. To further enhance the sensitivity of this method Fox and Kennedy [44] differentially labeled induced and uninduced cultures. The cultures were then mixed, and subsequent fractionation revealed a protein with a high [14]C:[3]H ratio present in the membrane particulate fraction. The authors were able to solubilize this protein with Triton, and a partial purification was reported. In a subsequent report, Fox et al. [45] provided more conclusive evidence that the M protein was the product of the lacY gene by showing that no labeled protein could be detected in a variety of lacY mutants, including a temperature-sensitive strain. In addition it was shown that the M protein is a constitutent of the cytoplasmic membrane. The authors estimate 9000 copies of M protein per cell, a number which is consistent with data concerning the other enzymes of the operon. The authors also reported the curious fact that, although TDG is effective in protecting the M protein from covalent modification by NEM, other substrates, such as lactose are not. This phenomenon was investigated in more detail [46], and it was found that the substrates of the lactose transport system fall into two classes: those that protect against NEM, and those that do not. On this basis it was proposed that the M protein contains two distinct binding sites (see below). To date, the M protein has not been purified. Jones and Kennedy [47] succeeded in solubilizing the protein with dodecyl sodium sulfate (SDS), and by following the radioactive label they could estimate a molecular weight of 30,000. The protein proved to be very difficult to handle because of its hydrophobic nature, and irreversible aggregation of the protein upon removal of detergent prevented any more detailed studies. The reader is referred to the review of Kennedy [41] for a more thorough discussion of this work.

One of the most intriguing conclusions of the work on the M protein was the proposal of two different substrate binding sites [46]. Additional evidence for the existence of a second site (called site II; substrates which bind here i.e., TDG, protect against NEM labeling) has been recently provided by the use of a direct binding assay [48]. This procedure utilizes ^{14}C-labeled sugar and [^{32}P]phosphate. Binding is detected as an increase in sugar radioactivity over the [32]P- background when membrane fractions are centrifuged out of a solution containing both. Using this procedure it was possible to show that TDG and certain α-galactosides bind to the M protein at a site which was not competitively inhibited by TMG (methyl-β-thiogalactoside) and other β-galactosides. This provides rather convincing evidence for the existence of a second site. However, it should be carefully noted that TDG can inhibit TMG transport. The K_i for this process is 160 μM [49], and this is not too different from the binding constant of 67 μM measured by Kennedy et al. [48]. Unfortunately, it was not determined if TMG bound to a site which was not inhibited by TDG. Therefore, it is not clear if TDG has an affinity for both site I and site II. In view of the similarity of the aforementioned K_i and K_d this does not seem likely. How TDG inhibits TMG transport without binding at the same site is not clear, but certainly it adds another complication to any carrier molecule model. It should be noted here that attempts to offer genetic proof for the existence of two sites have so far been unsuccessful. Messer [50] searched for mutants in which lactose transport was not subject to TDG inhibition. Such mutants were found; however, the defect was not in lacY but in araC, the reason being that lactose is a poor substrate of the araF transport system.

Returning to the NEM labeling technique, the possibility has been raised [51] that the sulfhydryl group in question is not a part of the carrier molecule but is a part of a different but closely related protein, for example, a protein involved in the energy coupling process. According to this view, TDG binding to the carrier would cause the sulfhydryl group on the adjacent protein to become buried, possibly as the complex translocates through the membrane. This adjacent protein may always be present, and conceivably it could function with more than one transport system. A certain amount of circumstantial evidence can be cited to support such a hypothesis. First, substrates which protect against NEM are translocated slowly when compared with substrates which do not protect [52]. Also, energy uncoupled mutants of the lactose transport system exhibit an increased sensitivity to sulfhydryl reagents concomitant with an increased V_{max} of transport [53, 54].

Other investigators have reported the successful labeling of the lactose carrier protein. Yariv et al. [55] report that N-bromoacetyl-β-D-galactosamine can be used as a labeling agent. They find that both lactose and melibiose can inhibit the incorporation of label. This is interesting since lactose is a site I sugar while melibiose binds to site II [46].

Even more intriguing results have been provided recently by Rudnick et al. [56], who succeeded in labeling the carrier protein with a new class of labeling agents, photoaffinity labels. They employed the ONPG (ortho-nitrophenyl-β-D-galactoside) analog 2-nitro-4-azidophenyl-1-thio-β-D-galactoside (APG), and their results can be summarized as follows:

1. APG is actively transported by membrane vesicles of strains containing an induced lactose transport system, K_m 75 μM.

2. APG is a competitive inhibitor of lactose uptake, K_i 75 μM.

3. APG inactivates the lactose transport system with a binding constant of 77 μM, and this inactivation is light-dependent.

4. Lactose (50-fold excess) protects against photoinactivation by APG.

5. Other transport systems were not photoinactivated by APG.

6. Photoinactivation by APG required the presence of an energy source such as D-lactate.

These results clearly indicate a specific labeling of the carrier protein. However, it should be pointed out that the amount of light necessary to decompose the label also inactivates many transport systems. Rudnick et al. [56] report that the initial rate of lactose transport in membrane vesicles are inhibited by 18 to 55% during a 10 min exposure to light in the absence of APG. If APG is added, then transport is inhibited by 90 to 100%. It should be mentioned here that other investigators have attempted to use azidophenylgalactosides as photoaffinity labels for the lactose transport system [57, 58]. They found, in contrast to Rudnick et al. [56], that many transport systems, including lactose, were so inhibited by light (85% lower initial rates of uptake after 10 min) that the use of a photoaffinity label would not be possible [57, 58].

In subsequent reports Rudnick et al. [59, 60] describe another photoaffinity label 2'-N-(2-nitro-4-azidophenyl)-aminoethyl-β-D-galactoside (APG$_2$). This compound is also a competitive inhibitor of lactose transport, K_i 30 to 40 μM, and when irradiated with visible light, it also rapidly inactivates the transport system in vesicles, provided D-lactate is present as an energy source, K_m 35 μM. However APG$_2$, unlike APG, is not transported [60]. It was found that, although the photoinactivation was D-lactate dependent, artificially imposed ion gradients could also serve in place of D-lactate as energy source. This is consistent with the chemiosmotic theory of energy coupling to active transport. In addition, the sulfhydryl reagent p-chloromercuribenzenesulfonate (PCMBS) did not prevent photoinactivation by APG$_2$. Even more surprising, it was found that APG$_2$ could inactivate the transport system in the absence of energy, provided PCMBS was present. This evidence suggested to the authors [59] that the critical sulfhydryl group was not present in the binding site of the lactose carrier protein. It was suggested that PCMBS altered the carrier so that the binding site became exposed to the external media even in the absence of energy.

Besides the results with photoaffinity labels, other lines of experimental evidence support this view. Schuldiner et al. [61] have reported that a number of dansylgalactosides are capable of inhibiting lactose transport in a competitive fashion by binding to the active site [62] without the compounds themselves being transported. The fluorescent dansyl moiety of these compounds makes them particularly useful, since specific binding to the hydrophobic carrier can be detected by an increase in the fluorescence of these compounds [63]. As with the photoaffinity labels, no effect is seen in the absence of energy.

The recent results obtained with the photoaffinity labels as well as with the fluorescent substrates are not entirely consistent with the earlier work on the M protein:

1. Binding of dansylgalactoside or inactivation by the photo affinity labels does not occur in the absence of energy, yet TDG binding to sonicated membranes in the absence of an added energy source has become a routine test for the M protein [48]. For example, Struve and McConnell [64] have used a similar assay to show that a spin-labeled β-galactoside binds to sonicated membrane fractions in a manner which is competitively inhibited by lactose. A possible explanation would be that sonicated membranes and "Kaback vesicles" differ in their membrane orientation. One would then have to conclude that the binding site of the M protein which faces the cytoplasm is accessible to substrate in the absence of energy. Energy-independent efflux of lactose from vesicles [65,61] supports this view. Another possibility offered by Schuldiner et al. [62] is residual specific binding even in the absence of energy that is seen in the experiments of Kennedy et al. [48].

As pointed out by Kaback and his collaborators, it is indeed critical for their proposed model [61] that the "uptake" of dansylgalactosides by energized membranes reflects binding of the substrate to, but not transport through, the membrane. Generally, the lower the apparent K_m of translocation for a given substrate of the lactose transport system, the slower it is transported [52]. Since the dansylgalactosides have an unusually low "K_m of uptake," one might interpret this uptake of the dansylgalactosides not as binding but transport and accumulation. Similarly, the observed fluorescence increase of the dansylgalactosides after energization of the membranes could be explained by the slow translocation of this substrate through the membrane resulting in the simultaneous exposure to the hydrophobic environment of the membrane. However, the comparison of the amount of dansylgalactosides taken up by the energized membranes (1.1 nmol/mg membrane protein [62] agrees well with the amount of M protein estimated to be present in the E. coli membranes (3% of the total membrane protein with a molecular weight of 30,000 [47]). In contrast, substrates that are accumulated by energized membrane vesicles take up amounts that are in the order of 30 to 50 nmol/mg membrane protein [66].

2. The idea that the accessibility of the substrate to its transport site is dependent on energy input is obviously in variance with the phenomenon of facilitated diffusion in energy uncoupled whole cells described and formulated by the classic work of Wong, Winkler, and Wilson [67,68]. In addition, facilitated diffusion as well as energy-independent efflux can also be demonstrated in membrane vesicles of both types, right-side-out and inside-out [69]. This could indicate that the accessibility of the M protein for substrate on the external site of the membrane is dependent on the loading state of the binding site on the internal site. In this respect it is interesting to note that efflux of lactose from preloaded membrane vesicles promotes binding of dansylgalactosides to the external binding sites even in the absence of energy input [61].

3. Photoaffinity labels inactivate the lactose carrier in the presence of PCMBS, even in the absence of energy. According to Rudnick et al. [59] PCMBS reacts with a sulfhydryl group which is not present in the binding site, and this interaction causes the carrier to assume a conformation which exposes the binding site to the external media. This seems plausible, and it is consistent with the observation that sulfhydryl reagents, unlike energy uncouplers of the proton conductive type, prevent substrate efflux after preloading [65,70]. However, this is difficult to reconcile with the view that the M protein has a reactive sulfhydryl group in one of the substrate binding sites [46].

4. Sulfhydryl reagents appear to have differential effects with different substrates. PCMBS was found to prevent the binding of dansylgalactosides [70] as well as to release bound dansylgalactosides [62,70], yet APG_2 inactivates the lactose carrier protein in the presence of PCMBS [59]. This was observed even though neither the dansylgalactosides nor APG_2 is transported in vesicles. This finding could be interpreted by the two-site model of the M protein [46]. Accordingly, APG_2 would bind to site I which does not contain a reactive sulfhydryl group, while the dansylgalactosides would bind to site II which is sensitive to sulfhydryl reagents. Although the site I sugars lactose and TMG inhibit dansylgalactoside binding, this does not rule out the latter compound binding to site II. Under the experimental conditions, the added lactose or TMG is being actively transported not simply bound. Thus, inhibition will be exerted not by binding to site II, but on the translocation level, i.e., by removing available carrier by lactose or TMG transport.

Clearly more biochemistry on the lactose carrier is needed to understand its complex properties. Recent reports on the reconstitution of vesicles of lacY strains by cell-free preparations of the lactose carrier are promising steps in this direction: aprotic solvents such as hexamethylphosphoric triamide are able to solubilize the lactose carrier protein from membrane vesicles of a wild-type strain. The addition of this extract to an aqueous suspension of mutant membrane vesicles was able to restore the energy-dependent uptake of lactose in these vesicles [71]. This successful solubilization of the M protein in a potentially native state and the availability of a suitable assay might be the first step in the purification and in vitro characterization of this elusive protein.

2. Galactose Transport: The Multiplicity
   of Transport Systems

It would seem that E. coli has a particular affinity for the sugar galactose, for there are certainly a large number of transport systems with specificity for this sugar. Besides the galactose transport system there is the lactose (TMG permease I) transport system and the melibiose (TMG permease II) transport system. In addition, galactose is a substrate for two different arabinose transport systems, the MeGal (methyl-β-D-galactoside) transport system, and possibly even a PTS system [72-75]. This overlapping specificity has greatly hindered mutant isolation and complicated any kinetic analysis of galactose transport. The study of the galactose transport system in particular has suffered because of this. There are no known specific substrates, no well-characterized mutants. In fact, the very existence of a unique galactose transport system remained in doubt until recently [76].

Horecker and coworkers [77,78] were the first to study the transport of galactose in E. coli. This was done with a strain which carried a defect in galactokinase (galK), thus enabling them to demonstrate that substrate was accumulated in an unaltered form. They found that galactose transport was highly specific, with an apparent K_m of 10 μM, and thus clearly distinct from other known transport systems. Several observations were made which remained a source of confusion for several years. First, they found that the galactose transport in the galK mutant was constitutive despite the fact that the parental strain exhibited a normally inducible phenotype. Second, they found galactose transport to be much more efficient at low rather than at high external galactose concentrations, i.e., at external galactose concentrations of 1 μM more than half of the galactose was taken up, generating a very considerable concentration gradient (in/out). They also found that the sugars L-arabinose and MeGal produced significant competitive inhibition at low but not at high external galactose concentrations.

A closer look at the kinetics of galactose transport [79] revealed two K_m values. This suggested the involvement of more than a single system, and subsequently Ganesan and Rotman [80] demonstrated the existence of two galactose transport systems. One was called the galactose transport system, the other the MeGal transport system since it also transported MeGal [81]. This latter transport system will be discussed in detail later. However, it should be noted at this point that the MeGal transport system is a high affinity galactose transport system and it has also affinity for L-arabinose. This explains why Horecker et al. [77,78] observed the somewhat different transport properties at low and high external galactose concentrations. The fact that galK mutants are constitutive for these two galactose transport systems is due to the retention of endogenously produced inducer by the MeGal transport system [82]. The galactose transport system does not transport MeGal.

Ganesan and Rotman [80] had isolated a MeGal-negative mutant (mgl). This mutation was present in a lacY background and therefore if the strains were grown at 37°C to prevent the synthesis of the temperature-sensitive TMG permease II, the galactose transport system could be studied without significant complications arising due to multiple transport systems [73]. The galactose transport system was found to be highly specific for galactose (the only substrate), with a K_m of 0.14 mM.

Recently, these studies have been verified in several different E. coli K-12 strains by Wilson [76]. He also found that galactose is the only substrate transported significantly (K_m 0.17 mM); however, fucose appears to be a poor substrate as well. The compounds MeGal and glyceryl galactoside were found to be not transported by the galactose transport system, although both are substrates of the MeGal transport system [83, 84]. According to Wilson, cells carrying a lacY mutation grown on galactose at 37°C (to prevent induction of TMG permease II) contain only two galactose transport systems. By adding saturating amounts of glyceryl galactoside to inhibit the MeGal transport system, he found that the galactose transport remaining was due solely to the galactose transport system. Wilson also found, in agreement with Buttin [85], that galR strains synthesize the galactose transport system constitutively, thus demonstrating that it is under the same control as the galactose operon. The same conclusion has been reached for the galactose transport system in S. typhimurium [86].

Mutants lacking the galactose transport system but still retaining the MeGal transport system have only recently been isolated. From the analysis of these mutants it is clear that the MeGal and the galactose transport systems are indeed two different systems. While the galactose transport system is a proton symporter, the MeGal transport system is not [87, 88]. The gene locus for the galactose transport system was found to be located between the markers fda and lys [89] at 61.5 min of the E. coli genetic map [90]. The galactose transport system is clearly not dependent on the galactose binding protein, which is a component of the MeGal transport system (see below). However, it does appear that these two galactose transport systems interact in some complex way, such that, at high external galactose concentrations, the MeGal transport system is inhibited. Wilson [76] suggests that this effect may be similar to the previously mentioned effect of PTS on certain other sugar transport systems. Whatever the mechanism, it is clear that galactose transport at high external galactose concentrations is not simply the expected sum of the activity of the MeGal and galactose transport systems [82].

At present it is difficult to clearly classify the galactose transport system. The system has been reported to be shock-sensitive and dependent on phosphate bond energy, in analogy to other binding protein-dependent systems [220]. However, so far no binding protein specific for this system has been found. Moreover, transport of galactose, most likely mediated via the galactose transport system, can still be observed in membrane vesicles [91]. In addition, transport occurs in whole cells concomitant with proton translocation [87].

Beside the TMG permease I and II, the MeGal, the galactose, and the arabinose transport systems, it appears that there is yet another system identified as recognizing galactose. This system has been found as a facilitated diffusion system most likely identical with enzyme II for glucose of the phosphotransferase system [35]. This is also true for EII-mediated translocation of galactose in S. typhimurium [36].

3. Other Shock-Resistant Systems

(a) The Melibiose Transport System. Because of the similarity of
substrate specificity to the lactose transport system (TMG permease I), the
melibiose transport system has been termed TMG permease II [92]. The
discovery of this transport system was based on the observation [93] that
galactinol (α-galactosyl inositol) did induce transport activity for TMG
without the concurrent formation of β-galactosidase. What first appeared
to be an exception to the operon model, was later suggested by Rotman [94]
and finally demonstrated by Prestidge and Pardee [92] to be due to a second
transport system for TMG. The early experiments on the melibiose transport system were done in E. coli B. Attempts to reproduce these experiments in E. coli K-12 were uniformly negative. Beckwith [95] observed
that various K-12 strains that were defective in the lactose transport system could grow on melibiose at 25°C but not at 37°C. Further investigation
made it clear that it is the inducible synthesis of the gene products coded
for by the melibiose operon that is temperature-sensitive [92, 96, 97]. The
melibiose transport system differs from the lactose transport system in
that it is not inducible by β-thioisopropylgalactoside (IPTG), but it can be
induced by melibiose and galactinol. Lactose is not transported while TMG
and TDG are [92, 73]. The melibiose transport system appears to be absent
in ML strains [92], temperature-sensitive in K-12 strains [92, 96], but not
temperature sensitive in E. coli B strains [93]. Salmonella typhimurium
also contains a non-temperature sensitive melibiose transport system.
Stock and Roseman [98] have studied melibiose transport in this organism.
They conclude that this system is a Na^+-symporter. Not only is TMG transport dependent on the presence of Na^+ in the medium, but the reverse is
also true. Na^+ transport can be stimulated by TMG. Although stoichiometric measurements were complicated by the very rapid extrusion of Na^+,
it would appear that the TMG:Na^+ ratio is 1. This provided as yet the only,
but most compelling evidence that Na^+ gradients can be involved in the
coupling of active sugar transport in bacteria.

(b) The Hexose Phosphate Transport System. Lin et al. [99] found that
mutants lacking alkaline phosphatase could still grow on glucose-6-phos-
phate (glucose-6-P) and fructose-6-phosphate (fructose-6-P). Mutants that,
in addition to the defect in alkaline phosphatase, were defective in glucose
transport also were still able to grow on glucose-6-P [100]. Direct evidence for a transport system was provided by Fraenkel et al. [101]. They
demonstrated that a mutant ptsI (of the phosphotransferase system), also
lacking glucokinase would still grow on glucose-6-P but not on glucose.
This transport system was shown to be active: it accumulated its substrate
against the concentration gradient [102]. In these studies [^{32}P]2-deoxy-D-
glucose-6 phosphate was used. Despite the fact that bacteria cannot use this compound as a carbon source, a fraction of the label did not stay with this compound after it was accumulated. These results have been verified by Dietz
and Heppel [103], and it was shown that the 2-deoxy-D-glucose portion of

the molecule is not the part that is metabolized. Recent studies by Grover and Winkler [104] that employed a double mutant lacking phosphoglucoiso-merase (pgi) and glucose-6-P dehydrogenase (zwf) conclusively demon-strated that glucose-6-P is transported unaltered. Mutants selected for their failure to grow on glucose-6-P could also not grow on fructose-6-P and mannose-6-phosphate. Revertants could again grow on all three sugars [105]. All three sugars were found to be inducers of the transport system. Glucose-1-P and fructose-1-P, even though transported by the system, do not act as inducers [20,106]. Similarly, D-arabinose-5-phosphate and sedoheptulose-7-phosphate are also transported by this system in S. typhimurium, although they do not cause induction [107].

Mutations in the hexose phosphate transport system of E. coli, con-ferring either negativity (uhpT) or constitutivity (uhpR), map near the pyrE marker on the E. coli chromosome [108,106,109]. It is interesting to note that mutants selected for resistance to phosphonomycin can be negative in the hexose phosphate transport system [110,111].

Despite reports that the hexose phosphate transport system is inhibited by osmotic shock [112], it seems clear that this system can be classified as shock-resistant, since several investigators have observed it to be operating in membrane vesicles [113,114]; in addition, cotransport with protons as immediate driving forces has been demonstrated [115].

The hexose phosphate transport system mediates the translocation of several common metabolic intermediates such as glucose-6-P and fructose-6-P. In nature these compounds are almost always present in higher con-centrations in the cell than in the outside medium. Since some of the sub-strates are also inducers of the system, one would assume that the hexose phosphate transport system is continually induced. However this is not the case; instead, induction is dependent exclusively on the inducer being present in the medium. This type of regulation has been termed exogenous induction, and this phenomenon is one of the most interesting aspects of the hexose phosphate transport system. Dietz and Heppel [116] showed that a pgi/zwf double mutant accumulates up to 60 mM glucose-6-P when it is given glucose, yet no induction of the hexose phosphate transport system takes place. However, glucose-6-P in the external medium at concentra-tions as low as 0.5 mM caused induction. Growth of pgi/zwf double mutants is sensitive to the accumulated glucose-6-P. This sensitivity can be over-come by adding glucose-6-P to the external medium. Apparently, induction of the transport system allows the toxic glucose-6-P to exit from the cell. Winkler [117] made the same observation and further demonstrated that fructose-6-P was not an inducer in a pgi mutant. Apparently, fructose-6-P must first be converted to glucose-6-P, then exit from the cell before induction can take place. Another explanation has been provided by Fried-berg [118], who demonstrated that phosphoglucoisomerase, the enzyme responsible for the conversion of fructose-6-P to glucose-6-P, is partially localized on the cell surface and can be released by osmotic shock. Therefore, fructose-6-P can be converted to glucose-6-P externally to the cytoplasmic membrane, and it is not necessary to postulate leakage of internally produced glucose-6-P. In a further study by Winkler [19] it was demonstrated that the recognition sites of the hexose phosphate transport system and the exogenous recognition site for the induction of the transport system are different: fructose-6-P had no effect on the induction of the transport system by glucose-6-P, even though it is a potent inhibitor of glucose-6-P transport. A factor in the regulation of hexose phosphate transport that has not been fully clarified is the role of catabolite repression. The hexose phosphate transport system is definitely catabolite repressible; glucose prevents the induction of the system at low external glucose-6-P concentrations [117], and the level of activity of mutants constitutive for hexose phosphate transport is lower in glucose-6-P or glucose-grown cells than in those grown on glycerol [106,120]. Additionally, the concentration dependence of induction by glucose-6-P is complicated by a drop in the level of induction at external glucose-6-P concentrations above millimolar range [119]. These observations suggest that glucose-6-P, directly or indirectly, may act both as inducer and repressor of the transport system. This naturally complicates studies of induction of the system, especially when large internal concentrations of glucose-6-P are present.

Despite these complicating factors it is clear that exogenous induction of the transport system can be initiated in the absence of transport. This represents one of the most intriguing problems in regulation. Obviously, the signal of recognition of the inducer on the outside of the cytoplasmic membrane must be transferred to the intracellular control mechanism. This is reminiscent of the phenomenon of chemotaxis: there, the attractant is recognized by the chemoreceptor which can be a periplasmic binding protein and is localized outside the cytoplasmic membrane. This signal of recognition can then be transferred again to the tumble generator without translocation of the attractant through the cytoplasmic membrane [121].

D. Shock-Sensitive Transport Systems

The activity of shock-sensitive transport systems is almost completely abolished by the cold osmotic shock procedure [39], as this results in the specific removal from the cell of essential components, the periplasmic binding proteins [51]. Since membrane vesicles contain no periplasmic proteins, the transport activity of these systems cannot be studied in membrane vesicles but only in whole cells.

1. The β-Methylgalactoside Transport System

This transport system was first identified by Rotman [81] who showed that it differed from both TMG permease I and II in that it transported MeGal but not TMG; hence the name MeGal transport system. Ironically, MeGal is a rather poor substrate of this transport system. Other sugars such as galactose or glyceryl galactoside are transported with a much lower K_m. Even though attempts have been made to change the name of this transport system accordingly [72], the original name remains the most commonly used. As was previously mentioned, the existence of this system separate from the galactose transport system was not recognized until Rotman and Radojkovic [79] showed by kinetic analysis the existence of two separate transport systems for galactose. Later, Ganesan and Rotman [80] were able to show that the gene(s) (mgl) coding for this transport system mapped near the his region on the E. coli chromosome.

Along other lines, Anraku [122] found that galactose transport was reduced by about 50% when cells were subjected to a cold osmotic shock procedure. Concomitant with the loss of transport activity a factor was released from the cell which could bind galactose. In a further study, Anraku [123] purified this factor and found it to be a protein apparently free of enzyme activity which specifically and reversibly bound galactose and glucose with a K_d of about 1 μM. Anraku suggested that this binding protein was involved in galactose transport. However, since no well-characterized mutants were available, and since no other galactose or galactoside derivatives such as MeGal were checked for binding, it was not possible to tell which transport system required this galactose binding protein (GBP). Boos [124], utilizing the mgl mutant W4345 of Ganesan and Rotman [80], was able to show that this strain contained no GBP. Subsequent mating studies were performed with strain W4345, and it was found that the gene coding for GBP was closely linked, if not identical to, the gene(s) coding for the MeGal transport system [125]. In addition, it could later be shown [126] that the synthesis of GBP and the MeGal system was coregulated, and independent of galR (see also Ref. 76). Although these studies suggested a close relationship between GBP and the MeGal transport system, they also introduced another complicating factor. A survey of E. coli strains showed that all mgl$^+$ strains tested produced GBP. However, the reverse was found not to be the case. The mgl strains fell into two categories: those which contain no detectable GBP and those which are still able to synthesize it in normal amounts. Apparently, then, the MeGal transport system is coded for by more than a single gene [125].

Fluorescence studies with purified GBP showed that galactose and other substrates produced a characteristic increase in the intrinsic fluorescence of the protein [127]. This proved to be a valuable tool. A mutant was found, EH 3039, which produced a defective but immunologically cross-reactive binding protein [128]. When GBP was purified from the strain and examined by fluorescence, it too showed a characteristic increase in the emission spectra. However, in this case, approximately 7000 times more substrate was required in order to see this fluorescence increase. This mutant strain was reverted to a transport-positive phenotype. Subsequent examination of the purified GBP from this revertant showed that it had re-acquired the fluorescence properties of the wild-type protein. This simultaneous alteration in the structure and activity of GBP upon mutation to transport-negative and reversion to transport-positive phenotype clearly demonstrates the essential role of GBP in the MeGal transport system. These results have recently been verified by Ordal and Adler [129,130]. These investigators have concluded from a genetic study that three genes are necessary to code to the MeGal transport: mglA, mglB, and mglC. The gene mglB codes for GBP. The proteins coded for by mglA and mglC have not yet been conclusively identified. However, several proteins have been located in a membrane fraction which appear to be related to the MeGal transport system, and it is possible that one or more of these proteins may be the mglA or mglC gene products [131,132]. All three mgl genes appear to be part of an operon [133].

2. Periplasmic Sugar Binding Proteins

There are at present four sugar periplasmic binding proteins known in E. coli, namely the galactose, arabinose, ribose, and maltose binding proteins. Each of these proteins is required for the active transport of their respective substrates [128,74,134,25]. In addition, all of the proteins except the arabinose binding protein [135] also function as receptors for chemotaxis [136-138]. These proteins are all remarkably similar. They bind substrate with high affinity ($K_d \leq 10^{-6}$M), their molecular weights are 30,000 to 40,000, all are isolated as monomers,* and all exhibit remarkable stability. At least with galactose binding protein and arabinose binding protein, the protein structure appears to be β-conformation with very little α helix [127,139]. In addition, the galactose, arabinose, and maltose binding proteins exhibit changes in the fluorescence emission spectrum upon substrate binding [127,139,140]. With the galactose and the arabinose binding proteins, evidence has been presented which indicates the presence of a tryptophan residue in the binding site [139,141]. Fluorescence studies performed with the maltose binding protein also suggest similar tryptophan involvement [140]. Despite the fact that these fluorescence changes suggest a conformational change, in cases where this has been looked for neither the arabinose [139] nor the galactose binding proteins [127,141] appear to undergo any major alteration in protein structure upon substrate binding. Both galactose and maltose binding proteins had been found to exhibit heterogeneous binding behavior at protein concentrations of 0.5 mg/ml [127, 142,2 5,143,144]. However recently Zukin et al. [145] reported that the

---

*Storage of GBP as precipitate in saturated ammonium sulfate results in the formation of dimers that are stable and active in buffer solutions. They only dissociate slowly in boiling SDS [254] .

E. coli and S. typhimurium galactose binding protein have only one binding
site and homogeneous binding characteristics. They concluded that the
earlier observed heterogeneous binding characteristics [127,142] were
artifactual, due to the presence of a radioactive compound contaminating
the galactose preparation used as substrate in the binding assays. Similarly,
the previously reported heterogenic binding behavior of the maltose binding
protein [25,143,144] has been reexamined by Schwartz et al. [146]. They
now also conclude that the maltose binding protein has homogeneous binding
characteristics with only one binding site. Binding experiments with the
ribose binding protein [137,147,134] and the arabinose binding protein [139]
gave linear binding characteristics and only one binding site.

   (a) Retention Phenomenon of Binding Proteins. When a solution of a
binding protein and its ligand is dialyzed against a larger volume of buffer,
the rate of disappearance of the ligand from the dialysis bag will be much
smaller than it would be from a bag containing no protein (retention phenomenon). This has often been interpreted to mean that the ligand dissociation rate from the binding protein is very slow [134,25,148]. However
this is not the case, and fluorescence measurements have shown that ligand
dissociation from both the galactose [149] and the maltose binding proteins
[140] is a rapid process.
   The consideration of the mass law for the retention phenomenon leads to
an equation that can describe the loss of ligand $d[L]/dt$ into the ligand-free
medium. It will be valid when the total ligand concentration [L] is low in
comparison to the total concentration of binding sites [P].

$$\frac{d[L]}{dt} = \frac{[L]\alpha}{1 + [P]/K_d}$$

In this equation α is a constant depending on the diffusion properties of the
molecule and on the experimental conditions, such as the porosity of the
membrane and the geometry of the dialysis system.
   The validity of this equation has been verified using the maltose binding
protein from E. coli [150]. Results similar to those obtained with the
maltose binding protein were observed with the ribose binding protein [134].
   When a protein endowed with a high affinity for a given ligand exists at
a high local concentration in a living cell, the retention effect should occur
and may have a biological role. The periplasm of gram-negative bacteria
provide examples of such a situation. The value of K_d for these proteins in
the range of 10^{-7} to 10^{-6} M. If the protein molecules are evenly distributed
in the periplasm, their concentration in this compartment is in the order of
10^{-3} to 10^{-2} M. In this case, $[P]/K_d$ ranges between 10^3 and 10^5, so that
the retention of ligand should be considerable. This means that ligand
release from the periplasm may be 10^4 to 10^5 times slower than entry. If
one assumes a diffusion rate of 1 cm/sec and a periplasmic space of 10^{-5}

cm thickness, then this corresponds to a time delay of 0.1 to 1 sec. Since the speed of a bacteria swimming in a straight line is about 20 body lengths per second [151], then this translates to an average distance of about 10 μm over which this delay could be important [152]. This means the bacteria would move 10 μm without sensing a decrease in attractant concentration. Increases, on the other hand, would be sensed immediately. In view of the fact that attractant concentrations may fluctuate wildly in the immediate vicinity of a bacteria, this phenomenon would be of particular value for a sensory mechanism which causes the bacteria to move up a gradient via a "biased random walk."

(b) The Role of Binding Proteins in Translocation. Despite considerable knowledge on the properties of binding proteins, their precise role in the transport process is not understood. However, several recent experiments have provided some important clues. Parnes and Boos [153] studied substrate exit through the MeGal transport system. They found that the K_m of galactose exit could not be exactly determined but must be higher than 0.01 M. This is vastly different from the binding constants (0.1 and 10 μM) exhibited by galactose binding protein for galactose, and accordingly they concluded that it could not be the recognition site for substrate exit on the inside of the cytoplasmic membrane. It was also found that a mutant defective in galactose binding protein structure exhibited the same rate of substrate exit as the wild-type parent even though entry in the mutants was only 6% of that observed in the parent. From these results it was concluded that the galactose binding protein did not participate in substrate exit. It follows then that it must not participate in substrate translocation per se as had previously been proposed [154]. Although similar studies have not been performed with the other three sugar binding protein mediated systems, it seems unlikely in view of the hydrophilic nature of these binding proteins that they participate in translocation.

One proposal for the function of periplasmic substrate binding proteins was that of a K_m factor. The transport system would operate continuously, but addition of periplasmic binding protein would greatly increase the system's affinity for substrate [155-157]. This conclusion seemed to be corroborated by the finding that membrane vesicles of E. coli also exhibited transport activity for leucine-isoleucine-valine, histidine, glutamic acid, lysine [158], galactose [91], and arabinose [114], compounds for which binding proteins had been isolated [159-164, 123, 148]. Uptake of these compounds occurs in membrane vesicles, even though the periplasmic binding proteins supposedly were removed during their preparation [114]. Yet for all of the above compounds more than one transport system has been reported in whole cells [165-168, 72-74]. Accordingly, it is probable that the transport seen in vesicles is due to these other transport systems. Therefore, any argument for an auxiliary function of periplasmic binding proteins for membrane bound systems has to be restricted to homogeneous systems. At the present time this is true only for the uptake of glutamine [169], diaminopimelic acid via the cystine general system [170], arginine via the arginine specific system [163], and maltose [25]. Indeed, no transport activity for these compounds can be found in membrane vesicles [158; Kaback, personal communication].

These observations strongly suggest that the binding protein mediated transport of glutamine, diaminopimelic acid, arginine, and maltose is entirely different and independent from the transport systems observed in membrane vesicles. It is highly likely that this is also true for other binding protein related systems such as the MeGal, arabinose, and ribose systems. Indeed, membrane vesicles of a lacY strain which still take up galactose do not transport MeGal [91], which in this mutant is only transported by the galactose binding protein mediated MeGal system.

It would seem then that removal of the binding protein prevents transport and translocation via shock-sensitive transport systems. This agrees well with the fact that mutants that have lost the maltose binding protein (including known nonsense mutants) can no longer utilize maltose [25]. These results indicate that binding proteins are not simply K_m factors, but are essential for translocation to occur even though they are most likely not involved in the translocation step itself.

Recently, however, Robbins and Rotman [156] and Robbins et al. [157] have provided evidence that substrate translocation can occur in whole cells in the absence of functional galactose binding protein. They find that mutants carrying a defect in mglB (structural gene of the galactose binding protein) could grow on MeGal at concentrations of 1 mM (mgl+ parents can grow at MeGal concentrations as low as 50 μM). In contrast, mglA or mglC mutants could not grow on MeGal at concentrations as high as 5 mM. Evidence was presented which ruled out the participation of the PTS, lactose, or either arabinose transport system, and therefore it was concluded the MeGal could enter the cell via the mglA and mglC gene products without a functional galactose binding protein.

All five of the mglB mutants tested were able to grow on 1 mM MeGal. Unfortunately, none of these mutants have been carefully characterized, i.e., there are not known nonsense or deletion mutations. Accordingly, it is possible that these mglB mutants can grow because they still have a poorly functioning galactose binding protein.

Studies on the growth inhibiting properties of glyceryl galactoside in a glpD/galE background are also not in agreement with the conclusions of Robbins and Rotman. mglB Mutants are not more inhibited in their growth in the presence of glyceryl galactoside than mutants defective in all three mgl genes, indicating that substrates can only be translocated in the presence of a functional galactose binding protein [84,173].

Thus, we feel that it is the interaction of the binding protein-substrate complex and not free substrate that has to interact with the membrane bound components of the system in order to accomplish translocation of the substrate.

3. The Remaining Components of Shock
    Sensitive Transport Systems

Since the periplasmic binding proteins are not likely to be involved in the
actual translocation step per se, it follows that such transport systems
must have additional components. Indeed, genetic analysis of the MeGal
[129,130] and maltose systems [171] , and the histidine transport system
in S. typhimurium [165] have proven this to be the case. Moreover, two of
the genetically defined components must interact functionally, as has been
shown by the observed suppression of a mutation in the hisJ protein by a
mutation in hisP [172]. No additional components have been recognized
with either the ribose or araF transport system, but in all probability the
same will prove true for these systems as well. By the process of elimin-
ation we must assign to the components the two most important processes
in transport, namely substrate translocation and energy coupling. In view
of their probable role, it is assumed that these components exist tightly
bound to the inner or cytoplasmic membrane. It is interesting to note that
with both the maltose and MeGal systems there are two additional components, namely malF and malK and mglA and mglC. Accordingly, one might
think that the processes of translocation and energy coupling are separated.
If this is true, then mutants which lack only the energy coupling component
do not carry out facilitated diffusion. This is clear since malK and malF
mutants cannot use maltose, and mglA and mglC mutants are not sensitive
to glyceryl galactoside (in a glpD/galE/lacY background), nor can they use
MeGal [156,157] as a source of carbon. In view of the fact that these
transport systems probably utilize phosphate bound energy as an immediate
driving force, this need not be surprising.

The remaining two components of these shock-sensitive transport
systems have yet to be characterized. In fact, it has not yet been conclusively shown that these components are membrane bound, although the
available evidence would strongly indicate it. Using a two-dimensional
polyacrylamide gel electrophoresis system we have at least tentatively
identified proteins in the particulate membrane fraction of E. coli that might
be these as yet unknown components of the MeGal transport system [131,132].
Similar experiments with the maltose [173] and with the histidine transport
system in S. typhimurium [174] have been unsuccessful.

Further insight into this problem has been obtained by studying protein
fusions between β-galactosidase and the malF gene product. Using techniques developed by Casadaban [175], strains were constructed which produced a hybrid protein. The N-terminal portion of this hybrid protein is
coded for by a portion of the malF gene while the C-terminal portion is
β-galactosidase. This hybrid protein is induced with maltose and still
retains β-galactosidase activity. Subsequent fractionation of the cell producing this hybrid protein revealed that the β-galactosidase activity was
strongly bound to the cytoplasmic membrane [176]. Hopefully, experiments

such as this may provide a method for purifying the malF gene product and demonstrate its location in the cytoplasmic membrane.

E. Transport Systems Containing Outer
    Membrane Components

It is becoming increasingly apparent that the outer membrane of gram-negative organisms is a significant permeability barrier. This evidence can be summarized as follows

1. Mutants carrying a lipopolysaccharide (LPS) defect are more sensitive to various dyes and antibiotics [177,178].

2. Short exposure to ethylendiaminetetraacetic acid (EDTA) causes the release of about 50% of the LPS and causes a concomitant increase in antibiotic sensitivity [179].

3. Kinetics of substrate exit through the lactose transport system suggest the existence of a permeability barrier outside of the cytoplasmic membrane [180].

4. Isolated outer membrane vesicles are essentially impermeable to polysaccharides with molecular weight over 1000 [181]; substances below a molecular weight of 1000 require a certain protein that might function as an unspecific pore [182].

In view of this outermost permeability barrier, it is not surprising that certain transport systems for larger molecules require the presence of an outer membrane component for normal function. For example, maltose [183,140], vitamin B_{12} [184,185], and several types of iron transport systems [186-188] have been shown to require the presence of outer membrane components. An interesting observation is that so far all known outer membrane transport components also function as colicin and/or bacteriophage receptors. This aspect is the subject of a separate chapter in this book (see Chapter 9).

F. Systems that Are Transported in
    Membrane Vesicles but Contain
    Periplasmic Components

It is generally assumed that phosphorylated compounds do not diffuse through the cytoplasmic membrane [189]. A possible exception to this was the compound glycerol-3-phosphate (G-3-P), since it was observed that K. aerogenes could grow on this compound but not on β-glycerophosphate [190]. These results were verified by Lin et al. [99] in E. coli by using a mutant lacking alkaline phosphatase. These authors could find no specific phosphatase in cell extracts. Furthermore, they found that a mutant devoid of both alkaline phosphatase and glycerol kinase could still grow on G-3-P but not on glycerol. They concluded that G-3-P enters the cell without prior hydrolysis. It did not seem likely that the cell membrane would be permeable to this metabolic intermediate, and the existence of a specific transport system was postulated. This prediction was verified by Hayashi et al. [191], as they found that a mutant lacking both alkaline phosphatase and the necessary catabolic enzyme, G-3-P dehydrogenase actively transported G-3-P. It would seem that the transport system is coded for by a single gene, and the chromosomal location of this gene (glpT) has been determined [192]. The gene glpT is a constituent of the glp regulon, and as such it is under the control of a repressor which is coded for by glpR. Besides this form of regulation this regulon (which also contains the previously mentioned glycerol facilitator) is also subject to catabolite and respiratory regulation. The different operons in this regulon exhibit varying degrees of sensitivity to each of these controls [193]. It has been reported that glpT mutants can be isolated as phosphonomycin-resistant clones [194,195,111]. However, there are several other ways in which bacteria can become resistant to this antibiotic as well [110,111].

The observation that glycerol phosphate cannot only energize membrane vesicles but can also be transported by these vesicles [196-198] would suggest that the glycerol phosphate transport system is a typical membrane bound system analogous to the lactose transport system. However, two-dimensional electrophoresis of proteins released from E. coli by the cold osmotic shock procedure of Neu and Heppel [39] revealed the presence of a protein that is under the regulation of glpR, the regulatory gene of the glp regulon. Mutant analysis of the glpT-type showed that this protein is likely to be the gene product of the glpT gene [199].

This protein has no binding activity toward glycerol phosphate and exhibits quaternary structure. It is composed of four identical subunits [198]. Despite the fact that glycerol phosphate is transported in membrane vesicles, transport in whole cells is severely reduced by osmotic shock [199]. Therefore, it appears likely that the periplasmic component of this system is only necessary in the intact cells but not in membrane vesicles. Similar observations have been made with a periplasmic component for a dicarboxylic acid transport system [200]. This phenomenon is at present not understood; it might indicate that these proteins are not classical binding proteins but are at least functionally involved in overcoming the diffusion barrier of the outer membrane.

G. The Exit Reaction

When mutants can be isolated that have a block in the first degradative enzyme for the substrate, or when substrate analogs are available that are not metabolized, the "active nature" of a transport system can be easily recognized. In these cases the substrate is accumulated to a point where the exit reaction becomes equal to the entry reaction, and net uptake of substrate no longer takes place. This situation can be characterized as an equilibrium state of accumulation, since in bacterial sugar transport systems that have been examined exchange reactions were demonstrated [153, 67, 201, 202]. The observation that a dynamic equilibrium exists between the entry and exit reaction does not prove that both reactions are mediated via the same carrier or even the same system. It only says that the substrate concentration dependence of both fluxes is different. Generally, in bacterial sugar transport systems the entry reaction is a saturable process with a certain apparent K_m, while the exit reaction has a "very high" K_m or is strictly first order, i.e., linearly dependent on the substrate concentration [67, 153, 203, 204]. This is the reason why, in systems that actively transport their substrates without metabolizing it, the K_m of the entry reaction can be measured by two methods. First, one can determine the concentration dependence of initial rate of entry. Under these conditions the exit reaction is still considered insignificant. Second, one can determine the substrate concentration dependent steady-state level of accumulation. Since the exit reaction can be considered first order, this steady-state level becomes dependent on the K_m of entry. This, of course, is the reason why "accumulation ratios" are always highest at the lowest possible substrate concentration that can be measured, but approach 1 when the substrate concentration is increased orders of magnitude above the apparent K_m of the entry reaction.

From the classical work on the glucose carrier in red blood cells [205] the phenomenon of counterflow has been used to characterize the properties of the mobile carrier. This facilitated diffusion system is seemingly able to concentrate radioactively labeled substrate that is added in low concentrations to cells that are preloaded with the unlabeled substrates. As soon as the labeled substrate enters the cell it is being diluted by the unlabeled substrate and prevented from exit by competitive inhibition. Both the labeled and the unlabeled substrates will eventually leave the cell according to their chemical gradients across the membrane. In order to demonstrate the counterflow experiment in bacterial systems, active transport systems have to be altered into facilitated diffusion systems. This is usually done by the use of energy uncouplers such as dinitrophenol or carbonyl cyanide-m-chlorophenylhydrazone (CCCP). A prerequisite of these experiments is that only the accumulation of substrate against the concentration gradient is abolished without influencing the rate of the fluxes of entry and exit. In the E. coli lactose transport system counterflow has been observed [206, 67, 68], even though energy uncouplers have some inhibiting effect on the rate of entry [206, 207]. Similarly, galactose transport in E. coli was found to exhibit the phenomenon of counterflow [202]. Even though these experiments are now somewhat difficult to interpret because of the multiplicities of galactose transport systems, it seems

clear that counterflow can be attributed to the galactose transport system.
This can be concluded from the observation that MeGal did not exhibit the
phenomenon of counterflow [202]. Counterflow has also been observed with
the 2-keto-3-deoxy-D-gluconate transport system [204]. As stated before,
the main requirement for the observation of counterflow in a bacterial sugar
transport system is that the rate of the respective fluxes are not altered by
the uncoupler. One might then ask why uncouplers abolish active transport,
or more basically, what is the molecular mechanism by which the steady-
state level of accumulation is maintained. The recent advances that have
been made in the understanding of energy coupling in systems that still
operate in membrane vesicles can be very briefly summarized in the
following way. The substrate carriers are obligatory proton symporters,
and translocation of substrate occurs in response to a pH gradient, to a
membrane potential, or both (for a detailed discussion see Chap. 12). Therefore, it is the maintenance of the protonmotive force ($\Delta \tilde{\mu}_{H^+}$)

$$\Delta \tilde{\mu}_{H^+} = \Delta \psi - \frac{2.3 \ RT}{F} \ \Delta pH$$

that maintains the level of accumulation of substrate in a steady-state
situation. Winkler and Wilson in their classical work on the counterflow
phenomenon of the lactose transport system in whole cells of E. coli have
demonstrated that neither the V_{max} nor the K_m of the entry reaction is
altered when energy uncouplers are employed [67]. More important, they
conclude that the K_m of the exit reaction has been reduced by the action of
energy uncouplers from a very high value to the same K_m as that of the
entry reaction. Knowing that the lactose transport carrier is a proton
symporter, one would have to conclude that protonated and unprotonated
carriers differ greatly in their affinity for the substrate, and that it simply
is the energy-dependent maintenance of the protonated and unprotonated
forms of the carrier on the opposite site of the membrane that brings about
active transport.

Energy-uncoupled mutants for the lactose transport system in E. coli
have been isolated [53, 54]. These mutants exhibit a strongly reduced
ability to accumulate substrate even though they exhibit a normal counterflow phenomenon. The mutation of this uncoupling effect is apparently
located within the polypeptide of the transport carrier [208], and results in
an uncoupling of proton-substrate cotransport [209]. Very similar but less
complex is the situation of TMG transport in the facultative anaerobic
organism Streptococcus lactis. This organism does not exhibit oxidative
phosphorylation and does not possess significant utilizable endogenous
substrate. Thus, accumulation of TMG only occurs after addition of a
metabolizable carbon source [210].

Translocation of TMG occurs via obligatory protonsymport, and
accumulation is dependent on the protonmotive force available to the cell
[211, 212]. The system shows the phenomenon of counterflow in the absence

of energy sources or in the presence of uncoupling agents of the proton conductive type [210]. Again, there is no inhibitory effect of the energy uncoupler on the translocation of substrate per se. In contrast, in a situation where no endogenous energy is available (in E. coli by depletion, in S. lactis by the absence of energy sources), these energy uncouplers even facilitate the transmembrane movement of substrate since they dissipate the concomitant buildup of electrical charges [207].

It is very likely that the entry and exit reactions of lactose transport in E. coli are mediated via the same carrier protein: TMG exit from lacY mutants occurs only at the rate of passive diffusion [201], and the extent of counterflow is dependent on the amount of lacY gene product present [68]. The result of energy coupling is to reduce the K_m of exit [67], and the ability to undergo this reaction must reside in the polypeptide chain of the lacY gene product [208]. In addition, it has been shown in membrane vesicles of the right-side-out and the inside-out types that the entry and exit processes are in fact equivalent [69]. All these experiments clearly indicate that entry and exit in the lactose transport system are mediated by the same carrier protein. By extension, one is inclined to conclude that systems that operate via proton symport and that have been shown to be dependent on the protonmotive force, are, in fact, carriers of the classical type mediating both entry and exit.

The situation becomes quite different when systems are considered that involve periplasmic binding proteins. Using galactose as substrate in a lacY strain it was demonstrated that the galactose binding protein does not participate significantly in an exit reaction [153]. Moreover, Wilson reported recently that exit of MeGal is independent of any mgl gene and equal to the exit observed in the wild type [203]. However, he used strains that still contained uninduced levels of the lactose transport system. This system is clearly the main pathway of galactose exit [213], most likely identical with the inducible galactose exit reported by Horecker et al. [78]. In addition, since the K_m of MeGal entry and its K_i for inhibition of galactose entry via the MeGal transport system are not identical [203], MeGal is most likely not transported only via the MeGal transport system. Therefore, it seems that MeGal exit [203] as well as galactose exit [153] are not mediated via the MeGal transport system.

Thus it seems clear that the main function of the galactose binding protein mediated MeGal transport system is entry and not exit. Attempts to demonstrate counterflow by the MeGal transport system were negative [153]. The reason for the lack of counterflow is clearly the effect of the energy uncouplers. In contrast to the lactose transport system, where energy uncouplers abolish accumulation but not translocation of substrate, the entry reaction mediated by the MeGal transport system is severely reduced by uncouplers [153]. Thus it is clear that energy uncouplers affect the two different types of transport systems in an entirely different way.

Therefore, it appears in retrospect somewhat naive to elucidate the mechanism of energy coupling in binding protein-mediated systems [214, 218] by studying the effects of energy uncouplers on these systems without dissecting their influence on the different parameters: the entry, the exit, and the accumulation.

The exit reaction in particular can be a source of unexpected surprises. Maltose transport in E. coli is a typical binding protein-mediated system that transports maltose actively and is inhibited by energy uncouplers [25, 215]. However, by studying the different processes separately, it was found that the exit reaction is practically abolished by energy uncouplers of the proton conductive type [216]. The observation that glycolytic carbon sources are able to stimulate binding-protein mediated sugar transport systems in mutants defective in the Ca^{2+}-, Mg^{2+}-dependent ATPase clearly indicates that phosphate bound energy is the immediate driving force in these systems [214, 217, 218]. However, they still are subjected to alterations in the protonmotive force brought about by the addition of respiratory substrates of by energy uncouplers of the proton conductive type. In addition, the effect of colicin K on the transport of binding protein-dependent systems in an ATPase mutant clearly demonstrates the importance of the protonmotive force for these systems [219]. Similarly, the isolation of mutants defective in coupling of energy to active transport that affect both shock-resistant and shock-sensitive systems alike [220] indicates the participation of the protonmotive force. The lack of proton cotransport in the binding protein dependent systems [87, 88] indicates that the protonmotive force is not the immediate energy source. Possibly, a high protonmotive force is required to keep the supposed membrane bound components of these transport systems in an operating state. But clearly more experiments with these systems are necessary to describe their mode of action.

III. THE ACTIVE SUGAR TRANSPORT SYSTEMS OF ESCHERICHIA COLI

Table 4-1 is an attempt to briefly summarize the sugar transport systems in E. coli. Listed are the names, the substrates, mode of energy coupling, if known, participation of periplasmic binding proteins, and whether they were observed to operate in membrane vesicles. In addition, pecularities are mentioned that are unique for some of these systems. In order to complete this table several simplifications had to be made. For instance, $\Delta\tilde{\mu}_{H+}$ was listed as the driving force for a system that is operating in membrane vesicles and can be stimulated aerobically with substrates of the respiratory chain. In most of these cases it is not clear if both components of $\Delta\tilde{\mu}_{H+}$, the electrical potential $\Delta\psi$, and the pH gradient $(-2.3\ RT/F)\ (\Delta pH)$ are effective for energy coupling as has been demonstrated for the lactose transport system.

TABLE 4-1 Active Sugar Transport in Escherichia Coli

Name	Substrates	Genetic Loci	Properties
Arabinose	L-arabinose [221] D-fucose β-methyl-L-arabinoside D-xylose D-galactose	araE [222]	Operating in membrane vesicles; energy coupling; $\Delta\tilde{\mu}_{H^+}$; proton symporter [87]; present in E. coli B/r and E. coli K-12 [225]
Arabinose	L-arabinose [221] D-fucose [223] D-galactose [50] β-methyl-L-arabinoside D-xylose β-methyl-D-galactoside lactose	araF [221,142]	Dependent on the periplasmic arabinose binding protein; present in E. coli B/r [139,148] and K-12 [164]
Fucose	L-fucose [224]	fuc [90]	The properties of this system have not been studied as yet [224]
Galactose	D-galactose [80, 73] D-fucose [76] D-glucose	galP [89]	Operating in membrane vesicles; energy coupling: $H+$ [91] ; proton symporter [87], opposite view on energy coupling (phosphate bound energy) [218] . The analogous system is present in S. typhimurium [86].

TABLE 4-1 (Con't)

Name	Substrates	Genetic Loci	Properties
Gluconate	D-gluconate [225, 226]	gntM; gntR [224-229]	Operating in membrane vesicles and stimulated by electron donors [228]; gluconate transport results in alkalinization of the external medium [227].
Glycerol phosphate	sn-glycerol-3-phosphate inorg. phosphate glyceraldehyde-3-phosphate [111, 230]	glpT [192]	Operating in membrane vesicles; energy coupling: $\Delta\tilde{\mu}_{H^+}$ [197,198]; related to a periplasmic protein [199] that is not necessary for transport in membrane vesicles [198]
Hexose phosphate	glucose-6-phosphate fructose-6-phosphate mannose-6-phosphate 2-deoxyglucose-6-phosphate glucose-1-phosphate fructose-1-phosphate [105,103,20,106] (In S. typhimurium also D-arabinose-5-phosphate and sedoheptulose-7-phosphate [107])	uhpT, uhpR [106,108,109]	Operating in membrane vesicles; energy coupling: $\Delta\tilde{\mu}_{H^+}$ [113]; in intact cells, hexose phosphate transport is accompanied by proton influx [115]; unusually, extracellular but not internally generated glucose-6-phosphate was found to induce this transport system [116,117]
Hexuronate	D-glucuronate [231] D-galacturonate [232, 233]	exuT [234, 235]	

2-keto-3-deoxy-D-gluconate	2-keto-3-deoxy-D-gluconate [236] D-glucuronate	KdgT, kdgP, kdgR [237, 238]	Operating in membrane vesicles, is stimulated by electron donors and inhibited by proton conductors [239]; the transport system is not present in wild-type cells and found only in derepressed mutants [237]
Lactose TMG permease I	a. lactose and most O-β-D-galacto-pyranosides; the aglyconic moiety can be a monosaccharide such as glucose or galactose; an alcohol such as methanol, ethanol, butanol, glycerol; a sugar alcohol such as mannitol; or a aromatic alcohol such as ortho-nitrophenol b. thio-β-methylgalactoside and most thio-β-D-galacto-pyranosides; again the aglyconic moiety can be monosaccharide, alcohol, or phenol c. melibiose and some other α-D-galactopyranosides d. galactose [41, 52, 240, 37, 49, 73, 92]	lacY [49]	Mediated via the membrane bound M protein [41], two different types of substrate recognition sites [46], operating in membrane vesicles [241]; proton symporter [243], system exhibits counterflow in whole cells [67]; energy coupling: $\Delta \tilde{\mu}_{H^+}$ [242]; $\Delta \tilde{\mu}_{H^+}$ is required for the recognition of substrate in membrane vesicles [61, 62]; existence of energy uncoupled mutants [244]

TABLE 4-1 (Con't)

Name	Genetic Loci	Properties	Substrates
Maltose	malB region [171] [246] consisting of malE, malF, malK, lamB	Dependent on periplasmic maltose binding protein [25]; the λ receptor, an outer membrane protein [247] is part of the maltose transport system [183, 140]; energy coupling: exit requires $\Delta\tilde{\mu}_{H^+}$ [216]; the maltose binding protein is also the chemoreceptor for maltose chemotaxis [138]	maltose [215, 245] maltotriose and higher maltodextrin
MeGal	mglA, mglB, mglC [80, 125,	Dependent on the periplasmic galactose binding protein [124]; energy coupling; phosphate bound energy [218]; the galactose binding protein is also the chemoreceptor for galactose chemotaxis [136]	D-glucose D-galactose β-glycerolgalactoside L-arabinose 6-deoxy-D-glucose D-fucose β-methylgalactoside xylose [80, 73, 83, 76, 153]
Melibiose TMG permease II	melB [97]	Synthesis of this system is absent in E. coli ML [92] and temperature-sensitive in E. coli K12 [92, 96]; the system is also present in S. typhimurium where it has been found to be a Na$^+$ symporter [98]	melibiose [73, 92, 94, 248] galactinol thio-β-methylgalactoside galactose

Raffinose	raffinose [249]	rafB [249]	Specified by a transmissible plasmid [249], no other properties known so far
Ribose	ribose possibly ribulose in E. coli and D-allose in S. typhimurium [134, 250]	rbsP [251, 252]	Dependent on a periplasmic ribose binding protein in E. coli [134] and S. typhimurium [137, 147]; some strains of E. coli seem to have two ribose transport systems [252], others only one [251, 253]; energy coupling: phosphate bound energy [214]; the ribose binding protein is also the chemoreceptor for ribose chemotaxis [137, 147]

IV. EPILOG

During the past several years a considerable amount of progress has been made in the field of transport. It has become clear that bacteria such as E. coli possess a surprising diversity of transport mechanisms. Also, it is now clear that a variety of energy coupling mechanisms are employed. Despite this progress though, the membrane carrier proteins remain a virtual enigma. Since these are the proteins that carry out the actual transport process, our knowledge of this important biological phenomenon is correspondingly vague. Clearly, a better understanding of the translocating and energy coupling mechanisms must await the isolation and purification of these components. At present though, even identifying these components and localizing them in the cytoplasmic membrane has proven to be exceedingly difficult.

RECENT DEVELOPMENTS

II C. 1   There is now considerable genetic evidence [255] that the lactose transport system in fact consists of only one gene product, i.e. a single protein. In addition, these studies are at variance with the two-binding site model of the M-protein [46].

II C. 1   Microcalorimetric studies of TDG binding to the M-protein indicated that the substrate bound to deenergized vesicles with the same affinity as found for the K_m of transport when energized [256].

II C. 2   The melibiose transport system of E. coli as well as that of S. typhimurium has also been shown to exhibit Na-cotransport [257]. Membrane vesicle studies are consistent with a mechanism of Na-symport [258].

ACKNOWLEDGMENTS

We would like to express our gratitude to Mrs. Siebert who patiently typed and retyped this manuscript. To all our colleagues in the field of transport that we misinterpreted or even did not mention we would like to extend our sincere apologies. Thanks to those who let us in on their latest news.

REFERENCES

1. P. Mitchell, in Biological Structure and Function, Vol. 2, (T. W. Goodwin and O. Lindberg, eds.), Academic, New York, 1961, p. 581.
2. J. Vorisek and A. Kepes, Eur. J. Biochem., 28: 364 (1972).
3. S. Roseman, in The Molecular Basis of Biological Transport (J. F. Woessner and F. Huijing, eds.), Academic, New York, 1972, p. 181.
4. J. Lengeler, J. Bacteriol., 124: 39 (1975).
5. D. J. Groves and A. F. Gronlund, J. Bacteriol., 100: 1256 (1969).

6. P. Harris, and H. L. Kornberg, Proc. Roy. Soc., Series B, 182: 159 (1972).
7. A. Delobbe, H. Chalumean, J. M. Claverie, and P. Gay, Eur. J. Biochem., 66: 485 (1976).
8. M. H. Saier, B. Feucht, and S. Roseman, J. Biol. Chem., 246: 7819 (1971).
9. V. P. Cirillo, and S. Razin, J. Bacteriol., 113: 212 (1973).
10. J. H. Maryanski and C. L. Wittenberger, J. Bacteriol., 124: 1475 (1975).
11. N. J. Patni and J. K. Alexander, J. Bacteriol., 105: 226 (1971).
12. M. E. Sobel and T. A. Krulwich, J. Bacteriol., 113: 907 (1973).
13. M. H. Sawyer, P. Baumann, L. Baumann, S. M. Berman, J. L. Canovas, and R. H. Berman, Arch. Microbiol., 112: 49 (1977).
14. J. B. Egan and M. L. Morse, Biochim. Biophys. Acta, 109: 172 (1965).
15. D. K. Button, J. B. Egan, W. Hengstenberg, and M. L. Morse, Biochem. Biophys. Res. Commun., 52: 850 (1973).
16. H. H. Winkler, J. Bacteriol., 116: 1079 (1973).
17. E. M. Barnes, J. Biol. Chem., 248: 8120 (1973).
18. R. F. Rest and D. C. Robertson, J. Bacteriol., 118: 250 (1974).
19. R. G. Eagon and P. V. Phibbs, Can. J. Biochem., 49: 1031 (1971).
20. G. W. Dietz and L. A. Heppel, J. Biol. Chem., 246: 2891 (1971).
21. S. Fukui and S. Miyairi, J. Bacteriol., 101: 685 (1970).
22. J. Deshusses and G. Reber, Biochim. Biophys. Acta, 274: 598 (1972).
23. J. Deshusses and G. Reber, Eur. J. Biochem., 72: 87 (1977).
24. G. Reber, M. Belet, and J. Deshusses, J. Bacteriol., 131: 872 (1977).
25. O. Kellermann and S. Szmelcman, Eur. J. Biochem., 47: 139 (1974).
26. A. A. Euffanti and W. A. Corpe, Arch. Microbiol., 108: 75 (1976).
27. S. Razin, in Advances in Microbial Physiology, Vol. 10 (H. Rose and D. W. Tempest, eds.), Academic, New York, 1973.
28. M. A. Tarshis, A. G. Bekkonzjin, V. G. Ladygina, and L. F. Panchenko, J. Bacteriol., 125: 1 (1976).
29. S. Hayashi and E. C. C. Lin, Biochim. Biophys. Acta, 94: 479 (1965).
30. E. C. C. Lin, Ann. Rev. Genet., 4: 225 (1970).
31. Y. Sanno, T. H. Wilson, and E. C. C. Lin, Biochem. Biophys. Res. Commun., 32: 344 (1968).
32. M. Berman-Kurtz, E. C. C. Lin, and D. P. Richey, J. Bacteriol., 106: 724 (1971).
33. D. P. Richey and E. C. C. Lin, J. Bacteriol., 112: 784 (1972).
34. E. C. C. Lin, Ann. Rev. Microbiol., 30: 535 (1976).
35. H. L. Kornberg and C. Riordan, J. Gen. Microbiol., 94: 75 (1976).

36. P. W. Postma, FEBS Lett., 61: 49 (1976).
37. E. Solomon, K. Miyai, and E. C. C. Lin, J. Bacteriol., 114: 723 (1973).
38. M. H. Saier, Jr., B. U. Feucht, and L. J. Hofstädter, J. Biol. Chem., 251: 883 (1976).
39. H. C. Neu and L. A. Heppel, J. Biol. Chem., 240: 3685 (1965).
40. F. Jacob and J. Monod, "On the Regulation of Gene Activity," Cold Spring Harbor Symp. Quant. Biol., 26: 193 (1961).
41. E. P. Kennedy, "The Lactose Permease System of Escherichia coli," in The Lactose Operon (J. R. Beckwith and D. Zipser, eds.), Cold Spring Harbor Laboratories, Cold Spring Harbor, New York, 1970 p. 49.
42. E. C. C. Lin, "The Molecular Basis of Membrane Transport Systems" in Structure and Function of Biological Membranes (L. I. Rothfield, ed.), Academic, New York, 1971, p. 285.
43. A. Kepes, J. Membr. Biol., 4: 87 (1971).
44. C. F. Fox and E. P. Kennedy, Proc. Natl. Acad. Sci. USA, 54: 891 (1965).
45. C. F. Fox, J. R. Carter, and E. P. Kennedy, Proc. Natl. Acad. Sci. USA, 57: 698 (1967).
46. J. R. Carter, C. F. Fox, and E. P. Kennedy, Proc. Natl. Acad. Sci. USA, 60: 725 (1968).
47. T. H. D. Jones and E. P. Kennedy, J. Biol. Chem., 244: 5981 (1969).
48. E. P. Kennedy, M. K. Rumley, and J. B. Armstrong, J. Biol. Chem., 249: 33 (1974).
49. H. V. Rickenberg, G. N. Cohen, G. Buttin, and J. Monod, Ann. Inst. Pasteur, 91: 829 (1956).
50. A. Messer, J. Bacteriol., 120: 266 (1974).
51. W. Boos, Ann. Rev. Biochem., 43: 123 (1974).
52. A. Kepes and G. N. Cohen, "Permeation," in The Bacteria, Vol. IV (I. C. Gunsalus and R. Y. Stanier, eds.), Academic, New York, 1962, p. 179.
53. P. T. S. Wong, E. R. Kashket, and T. H. Wilson, Proc. Natl. Acad. Sci. USA, 65: 63 (1970).
54. T. H. Wilson and M. Kusch, Biochim. Biophys. Acta, 255: 786 (1971).
55. J. Yariv, A. J. Kalb, and M. Yariv, FEBS Lett., 27: 27 (1972).
56. G. Rudnick, H. R. Kaback, and R. Weil, J. Biol. Chem., 250: 1371 (1975).
57. J. Y. D'Aoust, J. Giroux, L. R. Barran, H. Schneider, and W. G. Martin, J. Bacteriol., 120: 799 (1974).
58. L. R. Barran, J. Y. D'Aoust, J. L. Labelle, W. G. Martin, and H. Schneider, Biochem. Biophys. Res. Commun., 56: 522 (1974).
59. G. Rudnick, R. Weil, and H. R. Kaback, Fed. Proc., 34: 491 (1975).
60. G. Rudnick, H. R. Kaback, and R. Weil, J. Biol Chem., 250: 6847 (1975).

61. S. Schuldiner, G. K. Kerwar, H. R. Kaback, and R. Weil, J. Biol. Chem., 250: 1361 (1975).
62. S. Schuldiner, R. Weil, and H. R. Kaback, Proc. Natl. Acad. Sci. USA, 73: 109 (1976).
63. S. Schuldiner, R. D. Spencer, G. Weber, R. Weil, and H. R. Kaback, J. Biol. Chem., 250: 8893 (1975).
64. W. G. Struve and H. M. McConnell, Biochem. Biophys. Res. Commun., 49: 1631 (1972).
65. H. R. Kaback and E. M. Barnes, J. Biol. Chem., 246: 5523 (1971).
66. E. M. Barnes and H. R. Kaback, J. Biol. Chem., 246: 5518 (1971).
67. H. H. Winkler and T. H. Wilson, J. Biol. Chem., 241: 2200 (1966).
68. P. T. S. Wong and T. H. Wilson, Biochim. Biophys. Acta, 196: 336 (1970).
69. E. M. Teather, O. Hamelin, H. Schwarz, and P. Overath, Biochim. Biophys. Acta, 467: 386 (1977).
70. S. Schuldiner, S. F. Kung, H. R. Kaback, and R. Weil, J. Biol. Chem., 250: 3679 (1975).
71. K. Altendorf, C. R. Müller, and H. Sandermann, Jr., Eur. J. Biochem., 73: 545 (1977).
72. H. M. Kalckar, Science, 174: 557 (1971).
73. B. Rotman, A. K. Ganesan, and R. Guzman, J. Mol. Biol., 26: 247 (1968).
74. C. E. Brown and R. W. Hogg, J. Bacteriol., 111: 606 (1972).
75. D. L. Oxender, Ann. Rev. Biochem., 41: 777 (1972).
76. D. B. Wilson, J. Biol. Chem., 249: 553 (1974).
77. B. L. Horecker, J. Thomas, and J. Monod, J. Biol. Chem., 235: 1580 (1960).
78. B. L. Horecker, J. Thomas, and J. Monod, J. Biol. Chem., 235: 1586 (1960).
79. B. Rotman and J. Radojkovic, J. Biol. Chem., 239: 3153 (1964).
80. A. K. Ganesan and B. Rotman, J. Mol. Biol., 16: 42 (1966).
81. B. Rotman, Biochim. Biophys. Acta, 32: 599 (1959).
82. H. C. P. Wu, J. Mol. Biol., 24: 213 (1967).
83. T. J. Silhavy and W. Boos, J. Biol. Chem., 248: 6571 (1973).
84. T. J. Silhavy and W. Boos, J. Bacteriol., 120: 424 (1974).
85. G. Buttin, J. Mol. Biol., 7: 183 (1963).
86. M. Saier, F. G. Bromberg, and S. Roseman, J. Bacteriol., 113: 512 (1973).
87. P. J. F. Henderson, "Application of the Chemiosmotic Theory to the Transport of Lactose, D-Galactose, and L-Arabinose by Escherichia coli," in Comparative Biochemistry and Physiology of Transport (L. Bolis, K. Bloch, S. E. Luris, and F. Lynen, eds.), North Holland Publ., Amsterdam, 1974, p. 409.
88. P. J. F. Henderson, R. A. Giddens, and M. C. Jones-Mortimer, Biochem. J., 162: 309 (1977).

89. C. Riordan and H. L. Kornberg, Proc. Roy. Soc. Series B, 198: 401 (1977).
90. B. J. Bachmann, K. B. Low, and A. C. Taylor, Bacteriol. Rev., 40: 116 (1976).
91. G. Kerwar, A. S. Gordon, and H. R. Kaback, J. Biol. Chem., 247: 291 (1972).
92. L. S. Prestidge and A. B. Pardee, Biochim. Biophys. Acta, 100: 591 (1965).
93. A. B. Pardee, J. Bacteriol., 73: 376 (1957).
94. B. Rotman, Biochim. Biophys. Acta, 32: 599 (1959).
95. J. R. Beckwith, J. Mol. Biol., 8: 427 (1964).
96. G. Buttin, Adv. Enzymol., 30: 81 (1968).
97. R. Schmitt, J. Bacteriol., 96: 462 (1968).
98. J. Stock and S. Roseman, Biochem. Biophys. Res. Commun., 44: 132 (1971).
99. E. C. C. Lin, J. P. Koch, T. M. Chused, and S. E. Jorgensen, Proc. Natl. Acad. Sci. USA, 48: 2145 (1962).
100. H. Hagihira, T. H. Wilson, and E. C. C. Lin, Biochim. Biophys. Acta, 78: 505 (1963).
101. D. G. Fraenkel, F. Falcoz-Kelly, and B. L. Horecker, Proc. Natl. Acad. Sci. USA, 52: 1207 (1964).
102. B. M. Pogell, B. R. Maity, S. Frumkin, and S. Shapiro, Arch. Biochem. Biophys., 116: 406 (1966).
103. G. W. Dietz and L. A. Heppel, J. Biol. Chem., 246: 2881 (1971).
104. W. H. Grover and H. H. Winkler, Biochim. Biophys. Acta, 363: 428 (1974).
105. H. H. Winkler, Biochim. Biophys. Acta, 117: 231 (1966).
106. T. Ferenci, H. L. Kornberg, and J. Smith, FEBS Lett., 13: 133 (1971).
107. L. Eidels, P. D. Rick, N. P. Stimler, and M. J. Osborn, J. Bacteriol., 119: 138 (1974).
108. H. L. Kornberg and J. Smith, Nature (London), 224: 1261 (1969).
109. R. J. Kadner, J. Bacteriol., 116: 764 (1973).
110. R. J. Kadner and H. H. Winkler, J. Bacteriol., 113: 895 (1973).
111. P. S. Venkateswaran and H. C. Wu, J. Bacteriol., 110: 935 (1972).
112. G. Dietz, Y. Anraku, and L. A. Heppel, Fed. Proc., 27: 831 (abstract 3480) (1968).
113. G. W. Dietz, J. Biol. Chem., 247: 4561 (1972).
114. H. R. Kaback, Biochim. Biophys. Acta, 265: 367 (1972).
115. R. C. Essenberg and H. L. Kornberg, J. Biol. Chem., 250: 939 (1975).
116. G. W. Dietz and L. A. Heppel, J. Biol. Chem., 246: 2885 (1971).
117. H. H. Winkler, J. Bacteriol., 101: 470 (1970).
118. I. Friedberg, J. Bacteriol., 112: 1201 (1972).
119. H. H. Winkler, J. Bacteriol., 107: 74 (1971).

120. T. Ferenci, unpublished observations.
121. J. Adler, Ann. Rev. Biochem., 44: 341 (1975).
122. Y. Anraku, J. Biol. Chem., 242: 793 (1967).
123. Y. Anraku, J. Biol. Chem., 243: 3116 (1968).
124. W. Boos, Eur. J. Biochem., 10: 66 (1969).
125. W. Boos and M. O. Sarvas, Eur. J. Biochem., 13: 526 (1970).
126. J. Lengeler, K. O. Hermann, H. J. Unsöld, and W. Boos, Eur. J. Biochem., 19: 457 (1971).
127. W. Boos, A. S. Gordon, R. E. Hall, and H. D. Price, J. Biol. Chem., 247: 917 (1972).
128. W. Boos, J. Biol. Chem., 247: 5414 (1972).
129. G. W. Ordal and J. Adler, J. Bacteriol., 117: 509 (1974).
130. G. W. Ordal and J. Adler, J. Bacteriol., 117: 517 (1974).
131. W. C. Johnson, T. J. Silhavy, and W. Boos, Appl. Microbiol., 29: 405 (1975).
132. T. J. Silhavy, W. Boos, and H. M. Kalckar, in Biochemistry of Sensory Functions, the 25th Mosbacher Colloquium (L. Jaenicke, ed.), Springer-Verlag, Berlin, 1974, p. 165.
133. A. R. Robbins, J. Bacteriol., 123: 69 (1975).
134. R. C. Willis and C. E. Furlong, J. Biol. Chem., 249: 6926 (1974).
135. J. Adler, G. L. Hazelbauer, and M. M. Dhal, J. Bacteriol., 115: 824 (1973).
136. G. L. Hazelbauer and J. Adler, Nature New Biol., 230: 101 (1971).
137. R. R. Aksamit and D. E. Koshland, Jr., Biochemistry, 13: 4473 (1974).
138. G. L. Hazelbauer, J. Bacteriol., 122: 206 (1975).
139. R. G. Parsons and R. W. Hogg, J. Biol. Chem., 249: 3602 (1974).
140. S. Szmelcman, M. Schwartz, T. J. Silhavy, and W. Boos, Eur. J. Biochem., 65: 13 (1976).
141. E. B. McGowan, T. J. Silhavy, and W. Boos, Biochemistry, 13: 993 (1974).
142. R. G. Parsons and R. W. Hogg, J. Biol. Chem., 249: 3608 (1974).
143. G. L. Hazelbauer, J. Bacteriol., 122: 206 (1975).
144. G. L. Hazelbauer, Eur. J. Biochem., 60: 445 (1975).
145. R. S. Zukin, P. G. Strange, L. R. Heavy, and D. E. Koshland, Biochemistry, 16: 381 (1977).
146. M. Schwartz, O. Kellerman, S. Szmelcman, and G. L. Hazelbauer, Eur. J. Biochem., 71: 167 (1976).
147. R. Aksamit and D. E. Koshland, Biochem. Biophys. Res. Commun., 48: 1348 (1972).
148. R. W. Hogg and E. Englesberg, J. Bacteriol., 100: 423 (1969).
149. T. J. Silhavy and W. Boos, Eur. J. Biochem., 54: 163 (1975).
150. T. J. Silhavy, S. Szmelcman, W. Boos, and M. Schwartz, Proc. Natl. Acad. Sci. USA, 72: 2120 (1975).
151. H. C. Berg and D. A. Brown, Nature, 239: 500 (1972).

152. H. M. Kalckar, Biochimie, 58: 81 (1976).
153. J. R. Parnes and W. Boos, J. Biol. Chem., 248: 4436 (1973).
154. B. Rotman and J. H. Ellis, J. Bacteriol., 111: 791 (1972).
155. K. M. Miner and L. Frank, J. Bacteriol., 117: 1093 (1974).
156. A. R. Robbins and B. Rotman, Proc. Natl. Acad. Sci. USA, 72: 423 (1975).
157. A. R. Robbins, R. Guzman, and B. Rotman, J. Biol. Chem., 251: 3112 (1976).
158. F. J. Lombardi, and H. R. Kaback, J. Biol. Chem., 247: 7844 (1972).
159. W. R. Penrose, G. E. Nochoalds, J. R. Piperno, and D. L. Oxender, J. Biol. Chem., 243: 5921 (1968).
160. J. E. Lever, J. Biol. Chem., 247: 4317 (1972).
161. B. P. Rosen and F. D. Vasington, J. Biol. Chem., 246: 5351 (1971).
162. H. Barash and Y. S. Halpern, Biochem. Biophys. Res. Commun., 45: 681 (1971).
163. B. P. Rosen, J. Biol. Chem., 248: 1211 (1973).
164. R. Schleif, J. Mol. Biol., 46: 185 (1969).
165. G. F.-L. Ames and J. E. Lever, Proc. Natl. Acad. Sci. USA, 66: 1096 (1970).
166. B. P. Rosen, J. Biol. Chem., 246: 3653 (1971).
167. M. Rahmanian, D. R. Claus, and D. L. Oxender, J. Bacteriol., 116: 1258 (1973).
168. Y. S. Halpern and A. Even-Shoshan, J. Bacteriol., 93: 1009 (1967).
169. J. H. Weiner and L. A. Heppel, J. Biol. Chem., 246: 6933 (1971).
170. E. A. Berger and L. A. Heppel, J. Biol. Chem., 247: 7684 (1972).
171. M. Hofnung, Genetics, 76: 169 (1974).
172. G. F.-L. Ames and E. N. Spudich, Proc. Natl. Acad. Sci. USA, 73: 1877 (1976).
173. T. J. Silhavy, Ph.D. Thesis, Harvard University, Cambridge, Mass. (1975).
174. G. F.-L. Ames, J. Biol. Chem., 249: 634 (1974).
175. M. J. Casadaban, J. Mol. Biol., 104: 54 (1976).
176. T. J. Silhavy, M. J. Casabadan, H. A. Shuman, and J. R. Beckwith, Proc. Natl. Acad. Sci. USA, 73: 3423 (1976).
177. S. Tamaki, T. Sato, and M. Matsuhashi, J. Bacteriol., 105: 968 (1971).
178. P. Gustafsson, K. Nordström, and S. Normark, J. Bacteriol., 116: 893 (1973).
179. L. Leive, J. Biol. Chem., 243: 2373 (1968).
180. J. P. Robbie and T. H. Wilson, Biochim. Biophys. Acta, 173: 234 (1969).
181. T. Nakae and H. Nikaido, J. Biol. Chem., 250: 7359 (1975).
182. T. Nakae, J. Biol. Chem., 251: 2176 (1976).
183. S. Szmelcman and M. Hofnung, J. Bacteriol., 124: 112 (1975).

184. D. R. DiMasi, J. C. White, C. A. Schnaitman, and C. Bradbeer, J. Bacteriol., 115: 506 (1973).
185. C. Bradbeer, M. L. Woodrow, and L. I. Khalifah, J. Bacteriol., 125: 1032 (1976).
186. C. C. Wang and A. Newton, J. Biol. Chem., 246: 2147 (1971).
187. K. Hantke and V. Braun, FEBS Lett., 49: 301 (1975).
188. G. E. Frost, H. Rosenberg, J. Bacteriol., 124: 704 (1975).
189. B. D. Davis, Arch. Biochem. Biophys., 78: 497 (1958).
190. B. Magasanik, M. S. Brooke, and D. Karibian, J. Bacteriol., 66: 611 (1953).
191. S. Hayashi, J. P. Koch, and E. C. C. Lin, J. Biol. Chem., 239: 3098 (1964).
192. N. R. Cozzarelli, W. B. Freedberg, and E. C. C. Lin, J. Mol. Biol., 31: 371 (1968).
193. W. B. Freedberg and E. C. C. Lin, J. Bacteriol., 115: 816 (1973).
194. E. C. C. Lin, Ann. Rev. Genet., 4: 225 (1970).
195. D. Hendlin, E. O. Stapley, M. Jackson, H. Wallick, A. K. Miller, F. J. Wolf, T. W. Miller, L. Chaiet, F. M. Kahan, E. L. Foltz, H. B. Woodruff, J. M. Mata, S. Hernandez, and S. Mocholes, Science, 166: 122 (1969).
196. A. L. Koch, J. Mol. Biol., 59: 447 (1971).
197. L. A. Heppel, B. P. Rosen, I. Friedberg, E. A. Berger, and J. H. Weiner, in The Molecular Basis of Biological Transport (J. F. Woessner and F. Huijing, eds.), Academic, New York, 1972, p. 133.
198. W. Boos, I. Hartig-Beecken, and K. Altendorf, Eur. J. Biochem., 72: 571 (1977).
199. T. J. Silhavy, I. Hartig-Beecken, and W. Boos, J. Bacteriol., 126: 951 (1976).
200. T. C. Y. Lo and B. D. Sanwal, J. Biol. Chem., 250: 1600 (1975).
201. A. Kepes, Biochim. Biophys. Acta, 40: 70 (1960).
202. M. J. Osborn, W. L. McLellan, and B. L. Horecker, J. Biol. Chem., 236 (2585 (1961).
203. D. Wilson, J. Bacteriol., 126: 1156 (1976).
204. A. Lagarde and F. R. Stoeber, Eur. J. Biochem., 55: 343 (1975).
205. T. Rosenberg and W. Wilbrandt, J. Gen. Physiol., 41: 289 (1957).
206. A. L. Koch, Biochim. Biophys. Acta, 79: 177 (1964).
207. G. Cecchini and A. L. Koch, J. Bacteriol., 123: 187 (1975).
208. T. H. Wilson, E. R. Kashket, and M. Kusch, in The Molecular Basis of Biological Transport (J. F. Woessner and F. Huijing, eds.), Academic, New York, 1972.
209. I. C. West and T. H. Wilson, Biochem. Biophys. Res. Commun., 50: 551 (1973).
210. E. R. Kashket and T. H. Wilson, J. Bacteriol., 109: 784 (1972).
211. E. R. Kashket and T. H. Wilson, Biochem. Biophys. Res. Commun., 49: 615 (1972).

212. E. R. Kashket and T. H. Wilson, Proc. Natl. Acad. Sci. USA, 70: 2866 (1973).

213. H. C. P. Wu, Ph.D. Thesis, Harvard University (1967).

214. S. J. Curtis, J. Bacteriol., 120: 295 (1974).

215. H. Wiesmeyer and M. Cohn, Biochim. Biophys. Acta, 39: 440 (1960).

216. T. Ferenci, W. Boos, M. Schwartz, and S. Szmelcman, Eur. J. Biochem., 75: 187 (1977).

217. E. A. Berger and L. A. Heppel, J. Biol. Chem., 249: 7747 (1974).

218. D. B. Wilson, J. Bacteriol., 120: 866 (1974).

219. C. A. Plate, J. L. Suit, A. M. Jehen, and S. E. Luria, J. Biol. Chem., 249: 6138 (1974).

220. M. A. Lieberman and J. S. Hong, Arch. Biochem. Biophys., 172: 312 (1976).

221. C. E. Brown and R. W. Hogg, J. Bacteriol., 111: 606 (1972).

222. E. Englesberg, G. Irr, J. Power, and N. Lee, J. Bacteriol., 90: 946 (1965).

223. R. Schleif, J. Mol. Biol., 46: 185 (1969).

224. A. J. Hacking and E. C. C. Lin, J. Bacteriol., 126: 1166 (1976).

225. A. Robin and A. Kepes, FEBS Lett., 36: 133 (1973).

226. J. M. Pouysségur, P. Faik, and H. L. Kornberg, Biochem. J., 140: 193 (1974).

227. P. Faik and H. L. Kornberg, FEBS Lett., 32: 260 (1973).

228. R. Nagel de Zwaig, N. Zwaig, T. Isturiz, and R. S. Sanchez, J. Bacteriol., 114: 463 (1973).

229. N. Zwaig, R. Nagel de Zwaig, T. Isturiz, and M. Wecksler, J. Bacteriol., 114: 469 (1973).

230. S. Hayashi, J. P. Koch, and E. C. C. Lin, J. Biol. Chem., 239: 3098 (1964).

231. J. Jimeno-Abendano and A. Kepes, Biochem. Biophys. Res. Commun., 54: 1342 (1973).

232. F. Stoeber, C. R.Hebd.SéancesAcad.Sci.(Paris), 244: 1091 (1957).

233. F. Antissier and A. Kepes, Biochimié, 54: 93 (1972).

234. G. Nemoz, J. Robert-Baudouy, and F. Stoeber, C. R. Acad. Sci. Paris, Ser. D, 278: 675 (1974).

235. G. Nemoz, J. Robert-Baudouy, and F. Stoeber, J. Bacteriol., 127: 706 (1976).

236. A. Lagarde and F. Stoeber, Eur. J. Biochem., 36: 328 (1973).

237. J. Pouysségur and A. Lagarde, Mol. Gen. Genet., 121: 163 (1973).

238. J. Pouysségur and F. Stoeber, J. Bacteriol., 117: 641 (1974).

239. A. Lagarde and F. Stober, Eur. J. Biochem., 43: 197 (1974).

240. W. Boos, P. Schaedel, and K. Wallenfels, Eur. J. Biochem., 1: 382 (1967).

241. E. M. Barnes and H. R. Kaback, Proc. Natl. Acad. Sci. USA, 66: 1190 (1970).

242. S. Ramos, S. Schuldiner, and H. R. Kaback, Proc. Natl. Acad. Sci. USA, 73: 1892 (1976).

243. I. West, Biochem. Biophys. Res. Commun., 41: 655 (1970).

244. P. T. S. Wong, E. R. Kashket, and T. H. Wilson, Proc. Natl. Acad. Sci. USA, 65: 63 (1970).

245. M. Schwartz, C.R.Hebd.Séances Acad.Sci (Paris), 260: 2613 (1965).

246. M. Schwartz, J. Bacteriol., 92: 1083 (1966).

247. L. Randall-Hazelbauer and M. Schwartz, J. Bacteriol., 116: 1436 (1973).

248. I. G. Leder and J. W. Perry, J. Biol. Chem., 242: 457 (1967).

249. K. Schmid and R. Schmitt, Eur. J. Biochem., 67: 95 (1976).

250. J. David and H. Wiesmeyer, Biochim. Biophys. Acta, 208: 45 (1970).

251. A. Anderson and R. A. Cooper, J. Gen. Microbiol., 62: 335 (1970).

252. M. Abu-Sabe and J. Richman, Mol. Gen. Genet., 122: 303 (1973).

253. M. Abu-Sabe and J. Richman, Mol. Gen. Genet., 122: 291 (1973).

254. I. Rashed, H. Shuman, and W. Boos, Eur. J. Biochem., 69: 545 (1976).

255. A. C. Hobson, D. Gho, and B. Müller-Hill, J. Bacteriol., 131: 830 (1977).

256. A. Belaich, P. Simonpietri, and J.-P. Belaich, J. Biol. Chem., 251: 6735 (1976).

257. T. Tsuchiya, J. Raven, and T. H. Wilson, Biochem. Biophys. Res. Commun., 76: 26 (1977).

258. H. Tokuda and H. R. Kaback, Biochemistry, 16: 2130 (1977).

Chapter 5

ACTIVE TRANSPORT OF AMINO ACIDS

Yasuhiro Anraku

Department of Botany
Faculty of Science
University of Tokyo
Hongo, Bunkyo-Ku, Tokyo, Japan

I. INTRODUCTION 172
II. HISTORICAL REMARKS 173
    A. Permease Concept 173
    B. Breakthroughs in the Study of Transport Proteins 174
    C. Definition of the Molecular Apparatus of Active Transport . . 175
III. TRANSPORT SYSTEMS FOR AMINO ACIDS 176
    A. General Remarks 176
    B. Transport Systems in Escherichia coli 177
    C. Transport Systems in Other Bacteria 187
    D. Transport Systems Studied with Membrane Vesicles 190
IV. CURRENT TOPICS IN THE STUDY OF TRANSPORT
    SYSTEMS 194
    A. Osmotic Shock 194
    B. Retention Effect of Binding Protein 195
    C. Kinetic Evaluation of Transport Reactions 195
    D. Studies of Transport Carrier Proteins 197
V. MOLECULAR APPARATUS OF ACTIVE TRANSPORT 198
    A. Transport Systems for Branched-Chain Amino Acids
       of E. coli 198
    B. Roles of Binding Proteins in Active Transport 201
    C. Explanation for Multiple Transport Systems for Branched-
       Chain Amino Acids from Various Perspectives 202

D. Molecular Apparatus of Transport and Possible
    Roles of Binding Proteins 207
VI. CONCLUSION 210
    REFERENCES 212

I. INTRODUCTION

Modern biological studies on active transport of nutrients in bacteria have
their origin in 1957. Cohen and Monod [1] presented a novel concept on
permeation systems for carbohydrates and amino acids in living organisms
by using a generic term, permease, which stereospecifically catalyzes
transport of nutrients across the cytoplasmic membranes, and is independent of the metabolic processes in the cells. Synthesis and regulation of
various bacterial permeases, including those specific for amino acids,
were studied in the light of this benchmark concept, and various of their
functional properties were characterized kinetically [1,2].

Biochemical studies on the nature of the transport systems have been
initiated through the following several important breakthroughs in technique
and ideas, which came one after another from 1965 through 1967. Fox and
Kennedy [3] had their first success in demonstrating the existence of a
membrane protein which is functionally active for the β-galactoside transport in Escherichia coli. Kundig and coworkers discovered the phosphoenol
pyruvate-dependent sugar phosphotransferase [4] and demonstrated that this
enzyme complex could be responsible for the processes of group translocation of sugar moieties accompanied by phosphorylation [5]. Neu and
Heppel established the procedure of osmotic shock [6] and showed that this
mild modification of the cell envelope of gram-negative bacteria gave a
useful system for studying the active transport [7,8]. Kaback and Stadtman
[9] reported a procedure for preparing membrane vesicles after osmotic
lysis of spheroplasts and described the properties of proline uptake in the
vesicles. Undoubtedly these works are monuments in studies on active
transport and have invited forthcoming advances in the past 10 years. Although not discussed in detail in this chapter, important concepts on the
mechanism of energy coupling in active transport appeared in the early 1960s
[10-12].

This chapter deals first with historical aspects of studies on the active
transport systems for amino acids in bacteria and summarizes their kinetic
and genetic properties. Then, the molecular apparatus for the active transport is discussed, with special reference to recent advances in the study of
transport carriers in association with periplasmic binding protein. Readers
are referred to other chapters of this book and comprehensive reviews for
recent advances in the studies of bacterial active transport [13-20].

II. HISTORICAL REMARKS

A. Permease Concept

The notion that bacterial cells have a permeability barrier for certain sugars was demonstrated first by Doudoroff et al. [21]. They showed that a mutant of E. coli could not grow on maltose while the cell-free extract of the same strain could ferment it. However, this kind of selective crypticity for fermenting sugars was found not to be unique in bacteria, and studies on specific processes in the permeation of nutrients were developed greatly by the investigators at the Pasteur Institute. They demonstrated clearly that the process was highly stereospecific, with a catalytic nature, and catalyzed by a protein component distinct from a metabolic enzyme [22, 23]. The protein component was originally named permease [1] from the following criteria: (1) transfer by a permease includes the transitory formation of a specific complex between the protein of permease and the substrate, and (2) the permease is a functionally specialized system and is not involved in the intracellular metabolism itself.

It seems that in the original model, the enzyme-like nature of permease in the catalytic transfer of a substrate was emphasized. However, the definition of the mechanism of action of permease as an enzyme was not warranted and was put aside until evidence could be obtained for the protein nature of permease. Some critical arguments against this enzyme-like nomenclature appeared in the early literature [24, 25]. In addition, the definition cited above did not mention the process in which permease catalyzes the transfer of substrate against a concentration gradient, i.e., active transport. Although the energy requirement for β-galactoside uptake was studied in the original [1] and more carefully in the extended models [26, 27], the most brilliant contribution of the permease concept was that it enhanced investigations on the mechanism of induction and physiological roles of permease action.

It should perhaps be mentioned here that the term permease was given by bacteriologists independent of the studies on carbohydrate transport in animal cells. Many animal physiologists used the word carrier for the major component responsible for osmotic work transferring substrate across the membrane barrier. The most distinguishing feature of the carrier seemed to be its ability to induce the uphill movement of a solute across the membrane against a concentration gradient when coupled to metabolic energy and, therefore, the carrier itself is thermodynamically mobile. In several papers [28, 29] and a review [30], the thermodynamic aspect of the carrier function was discussed critically.

Winkler and Wilson [12], using mutants for β-galactoside uptake, demonstrated clearly that the product of lacY can facilitate the substrate transfer depending on chemical concentration gradient and that, therefore, the lactose permease itself shows a carrier function. This same protein

can couple with metabolic energy, so that the uphill movement of a substrate is catalyzed under proper conditions. This finding was quite
important because it added the concept of energy coupling to the original
permease model.

B. Breakthroughs in the Study of
     Transport Proteins

Studies on the biochemical nature of transport proteins were quite difficult,
and no suitable experimental method of treating these proteins has been
available for long. The main reasons for this are: (1) these proteins are
located in membranes and difficult to solubilize; (2) since their function is
to transfer a substrate across the membrane without chemical modification,
classical techniques in enzymology are not directly applicable for in vitro
analysis, and assays of binding activity have not been used until recently;
(3) the notion that the function of transport proteins requires energy had
not been recognized for a long time, and there were no useful subcellular
assay systems; and (4) there has been no relevant collection of mutants
for various transport systems other than lactose.

Since 1965 valuable new techniques have been introduced, and studies
on bacterial active transport have developed into an exciting new phase.
First, Fox and Kennedy [3] succeeded in identifying the membrane component having a high binding affinity for β-galactosides by an elegant,
preferential labeling method using ^{14}C-labeled N-ethylmaleimide. Inci-
dentally, their work gave the first biochemical demonstration that this
membrane component, the lacY protein, is the only component involved in
the active transport system for β-galactoside [31].

In 1965 Neu and Heppel, who worked on nucleic acid metabolism in
E. coli, demonstrated the selective release of a group of hydrolytic
enzymes when intact cells were converted to spheroplasts or subjected to
osmotic shock [6]. The procedure of osmotic shock was established [7, 32,
33], and the optimal conditions for the release of enzymes were elucidated
[8, 34]. Then Anraku and Heppel found that the shocked cells, despite the
loss of periplasmic proteins were fully viable on nutrient agar plate but
pleiotropically lose the ability to adapt to certain sugars when incubated in
a minimal salt medium [8]. Consistent with these observations, it was found
that the shocked cells had decreased transport activities for various nutrients,
and that there were corresponding binding proteins in shock fluid [7, 35].

The discovery of many binding proteins from gram-negative bacteria,
and the contribution of these findings to advances in the study of active
transport have been already documented in several reviews [36-39]. It
should perhaps be mentioned that in the early work on the isolation and
purification of binding proteins, investigators were interested chiefly in
their substrate specificities. Close correlation between the substrate
specificity of binding protein and that of the corresponding transport

reaction received much attention. These observations were adapted into
an idea that binding protein was the main protein component determining the
substrate specificity of active transport. However, observations against
this soon accumulated [14,40]. In addition, recent findings [41,42] on the
characteristic binding nature of binding protein in vitro have called attention
to their physiological role. More recently, two laboratories in Japan reported their success in solubilizing transport carrier proteins from membranes and reconstituting their activities in liposomes with energy added
in the form of a membrane potential introduced by K^+-diffusion via
valinomycin [43,44].

The great contribution to the development of an experimental system for
examining the mechanism of energy coupling to active transport has come
from the early work of Kaback and Stadtman [9] and Kaback [45]. Cell-free,
closed membrane vesicles prepared carefully after osmotic lysis of spheroplasts were recognized as useful preparations for analyzing energy coupling
in the active transport reaction [14,45]. By the use of this membrane
preparation, studies on energy transducing components such as the respiratory chain and ATPase (a coupling factor of oxidative phosphorylation)
have progressed greatly. Recent advances in the knowledge of various
membrane vesicles and energy coupling have been reviewed [15,46-48] and
are also discussed in this book. At present, preparations of membrane
vesicles serve as important experimental systems for analyzing energy
transducing reactions in vitro. However, it is also noted that studies with
membrane vesicles alone cannot answer every question about the mechanism
of active transport reactions in vivo.

C. Definition of the Molecular Apparatus
of Active Transport

The mechanism of active transport can be studied with membrane vesicles
when the molecular apparatus for solute translocation remains intact. This
notion was best examined in the case of β-galactoside transport. The
genetically defined and functionally active protein (the lacY protein) is in
the membrane [3] and facilitates the stereospecific transfer of a substrate
across the membrane barrier [49,50]. The protein was found to bind the
substrate stereospecifically when tested by either indirect [3] or direct
methods [51-53], and this same component translocates the substrate when
chemical concentration gradient [12] or a proton motive force [54,55] is
given. The observations that in the active transport of β-galactoside the
binding of substrate to the lacY protein requires "energization" of membranes [56], and that the substrate transfer by the protein is accompanied
concomitantly with proton uptake in the ratio of 1 to substrate [55], seem to
be important in elucidating the molecular mechanism of energy coupling. As
is well known, this functional component has been called permease [1],
carrier [12], and porter [55], or in a special case, M protein [3].

To understand the nature of the molecular apparatus for active transport in general, an example of the β-galactoside transport is the simplest
apparatus to consider. It is obvious that the essential feature of active
transport is the vectorial transfer of a compound from one compartment
to the other against an osmotic barrier, dependent on energy and without any
chemical modification of the molecule. Thus, use of the term carrier is
recommended as the simplest molecular apparatus catalyzing this type of
a biological process for the following reasons: (1) The binding of substrate
to carrier initiates only the transfer of the bound substrate and does not
cause chemical modification of the molecule; (2) ability of the carrier to
bind and release the substrate depends on a substrate concentration; and
(3) concomitant binding of a substrate and a proton or other cations to the
carrier appears to change its affinity for the substrate and to induce an
uphill flux of the molecule against a concentration gradient.

These reasons indicate clearly that although the carrier binds the
substrate stereospecifically and forms a transitory complex with it, the
thermodynamic aspect of energy transfer by the complex is different from
that in enzyme action. That is to say, the carrier catalyzes transduction
of osmotic energy while the enzyme catalyzes conversion of a chemical bond
energy. Therefore, the carrier is not an enzyme in the regular sense of
the word. The enzyme-like nature of the catalytic reaction must be
expressed by the term carrier which stresses its inherent catalytic mechanism of performing the vectorial transfer of a substrate against an
osmotic barrier.

Accordingly, the carrier should be defined as a protein which (1) is in
the membrane as an intrinsic protein, (2) binds substrate stereospecifically,
(3) facilitates transfer of the bound substrate across the membrane depending on chemical concentration gradient, and (4) mediates an uphill flux
of the molecule against a concentration gradient when chemiosmotic energy
is added to the system.

In the following sections, critical examination will be made as to what
extent the criteria defined above have been demonstrated in the case
of active transport systems for amino acids in bacteria.

III. TRANSPORT SYSTEMS FOR
      AMINO ACIDS

A. General Remarks

Active transport for amino acids in bacteria can be characterized as systems having a high affinity for substrates with K_t values of entry on the
order of 10^{-7} to 10^{-8} M and strict substrate specificity without significant
overlapping. These properties are different from those of eukaryotes such
as Ehrlich ascites cells. Comparative aspects of these systems have

been reviewed [57-59]. These properties of high affinity and strict specificity should enable us to isolate relevant mutants uncomplicated by
metabolic effects. However, these special advantages have not been
fully appreciated until recently, and the genetic background for the isolation of transport mutants has not been sufficiently considered.

This section summarizes the properties of transport systems for
amino acids in bacteria, studied with intact cells and membrane vesicles,
referring especially to their substrate specificity and genetics. Readers
are also referred to Chapters 11 and 12 for information on other topics
such as binding proteins and energy coupling in the active transport.

B.   Transport Systems in Escherichia
coli

There are at least 12 separate, distinct transport systems in E. coli cells:
glycine-alanine, threonine-serine, leucine-isoleucine-valine, phenylalanine-tyrosine-tryptophan, methionine, proline, lysine-arginine-ornithine,
cystine, asparagine, glutamine, aspartate, and glutamate. Some of these
transport systems include subgroups which in general show narrow specificity for one of the corresponding substrate amino acids with different
affinity. Transport systems for histidine and cysteine have not been studied
in E. coli.

1. Glycine-Alanine

Existence of this transport system was first indicated by the inhibition of
growth of E. coli by D-serine and its recovery by alanine and glycine [60].
Schwartz and coworkers isolated mutants resistant to D-serine and found
that they are defective in the uptake activity for glycine, alanine, and D-
serine [61]. Lubin and his associates studied glycine transport by selecting
mutants resistant to D-serine or D-cycloserine, and a mutant requiring a
high concentration of glycine for growth [62,63]. Kinetic studies on the
uptake of glycine and other related amino acids indicated that glycine, D-
alanine, and D-cycloserine are transported by the same system (designated
as cycA), which also seemed to transport alanine and D-serine [63]. The
cycA gene was located at 94 min on the E. coli genetic map by three groups
of investigators [64-66].

Independently, Cosloy [67] isolated a mutant EM1302, which is derived
from a D-serine deaminaseless strain and is resistant to D-serine. This
mutant was found to be defective in the uptake of D-serine, D-alanine, and
glycine; and the dagA mutation was mapped at 94 min [67,68], a location
similar to that of cycA (see above). Therefore, in E. coli cells a common
transport system exists which is specific for glycine and alanine, and shares
D-alanine and D-serine uptake. As shown in Table 5-1, this common trans-

TABLE 5-1 Transport Systems for Amino Acids in Escherichia coli

Transport System	Substrate and K_t (μM)	Competitive Inhibitor and K_i (μM)	Ref.	Gene Symbol	Map position[a] (min)	Phenotype Trait Affected	Ref.
I. Glycine–alanine	Glycine(2.8–4.8) Alanine(27)	D–Alanine	68	cycA	94	Transport of D–alanine, D–serine, and glycine	64–66
		D–Serine		dagA	94	Transport of D–alanine, D–serine, and glycine	67, 68
Ia. Alanine	Alanine(2)	Leucine(0.2) Serine Threonine	68				
II. Threonine–serine	Threonine(2)	Serine(1.4)	71				
IIa. Threonine	Threonine (0.39)	Alanine(8.2) Leucine(0.55) Isoleucine(0.37) Valine(0.78)	68				
III. Leucine–iso-leucine–valine	(For details, see Secs. III.B and V.C)			brnQ	8	Transport system 1 for leucine, iso-leucine, and valine	86, 87
				brnR	8	Component of transport systems 1 and 2 for leucine, iso-leucine, and valine	87
				brnS	1	Transport system 2 for leucine, iso-leucine, and valine	87
				hrbA	9	Transport carrier for leucine, iso-leucine, and valine	72

	Name	Substrate (constant)	Ref.	Gene	Ref.	Function	Ref.
				hrbR	83	Utilization of transported leucine, isoleucine, and valine	72
				lstR	20	Derepression of leucine-specific transport system	88
				livR	20	Derepression of both leucine-specific and common transport systems	88
IV.	Phenylalanine–tyrosine–tryptophan	Phenylalanine (0.47) Tyrosine (0.57) Tryptophan (0.40)	95	aroP	2	Transport of phenylalanine, tyrosine, and tryptophan	95–97
IVa.	Phenylalanine	Phenylalanine (2.0)	95				
IVb.	Tyrosine	Tyrosine (2.2)	95				
IVc.	Tryptophan	Tryptophan (3.0)	95	tryP	27	Transport of tryptophan	98
V.	Methionine-1	Methionine (0.075) L-Ethionine(23) D-Methionine(600)	69	metD	5	Transport of D-, and L-methionine	99–101
	Methionine-2	Methionine(40) Selenomethionine	99				
VI.	Proline	Proline(0.44; 40) L-3,4-Dehydroproline(2.6) L-Azetidine-2-carboxylate(24)	102; 104,105				
VII.	Lysine–arginine–ornithine	Lysine(0.5) Ornithine(1.4) Arginine D,L-Canavanine Citrulline	110,111	argP	62	Transport of lysine, arginine, and ornithine	107–109
VIIa.	Lysine	Lysine(10)	110				
VIIb.		Arginine(0.026)	110				

TABLE 5-1 (Cont.)

Transport System	Substrate and K_t (μM)	Competitive Inhibitor and K_i (μM)	Ref.	Gene Symbol	Map[a] position (min)	Phenotypic Trait Affected	Ref.
VIII. Cystine-1	Cystine(0.02)	Selenocystine	115				
Cystine-2	Cystine(0.3)	Diaminopimelic acid (14)	115				
IX. Asparagine-1	Asparagine (3.5)	Aspartate (50), 5-diazo-4-oxo-norvaline (4.6), β-hydroxylamyl-L-aspartate (10)	116				
Asparagine-2	Asparagine(80)						
X. Glutamine-1	Glutamine (0.15)	α-Glutamylhydra-zice(75) α-Glutamylhydro-xamate	117, 118				
Glutamine-2	Glutamine(2)		119				
XI. Aspartate-1	Aspartate(3.7)	Glutamate(640) Glutamine(840) D-Aspartate(200)	120–122				
Aspartate-2	Aspartate(39)	Succinate Fumarate D, L-Malate oxaloacetate	121				
XII. Glutamate	Glutamate in presence of 40 mM NaCl (0.7)	Aspartate(60:non-competitive)	126	gltS	81	Transport of gluta-mate	134
	Glutamate in the absence of Na^+ (10)	Aspartate(5) Aspartate(7.5–10: noncompetitive) D-Glutamate(245)		gltR	91	Regulation of gluta-mate transport	134

[a]See Ref. 222.

port system has K_t values of 2.8 to 4.8 μM of entry for glycine and 27 μM for alanine [68].

Existence of a separate transport system specific for alanine but not for glycine was suggested by several investigators from kinetic and genetic studies [63, 66, 69, 70]. This transport system was found to have a K_t value of 2 μM for alanine and to be inhibited by leucine competitively (K_i value of 0.2 μM) and by serine and threonine [68]. Based on these findings, Oxender and Oxender suggested that this system might be a LIV-I system (see below).

2. Threonine-Serine

A specific transport system for threonine was reported by Templeton and Savagean [71], showing the K_t value of 2 μM. This inhibited competitively by serine (K_i value of 1.4 μM, see Table 5-1). Yamato and coworkers [72] recently isolated transport mutants of E. coli which are defective in the uptake of leucine, isoleucine, and valine, but normal in that of threonine.

On the other hand, Smulson and coworkers [73] suggested that threonine could be transported in the cells via a branched-chain amino acid transport system, judging from a study on the inhibition of growth by O-methyl-D, L-threonine. Robbins and Oxender [68] described kinetic parameters of threonine uptake via a LIV-system and its inhibition by competitive inhibitors (see Table 5-1).

3. Leucine-Isoleucine-Valine

Among various active transport systems known in E. coli K12, the multiple transport systems for branched-chain amino acids have been most extensively studied biochemically and genetically. Thus, the properties of the systems with respect to their specificities [23, 69, 74-77], binding proteins [42, 78-83], energy coupling [77, 84, 85], genetics [72, 86-88], membrane vesicles [89, 90], and restoration and regulation [35, 90-94] have been reported. At present these particular systems seem to provide the best example for discussing molecular models of active transport reaction which exhibits multiplicity in substrate specificity and which require membrane transport carrier in functional association with periplasmic binding proteins. In this section the genetic analysis of the transport systems will be described; detailed biochemical properties and current explanations for the multiplicity of the systems are fully discussed in Sec. V.

Genetic analyses of the transport systems in a broader aspect have been developed by Iaccarino and his coworkers [86, 87]. They divided the transport systems into three groups by determining the apparent K_t values of entry for three substrates and by testing inhibition by analog compounds. First, a very high-affinity system was noted which takes up leucine, iso-leucine, and valine with K_t values in the order of 10^{-8} M and which is inhibited by methionine, threonine, and alanine. This system was missing in strains carrying the brnR-6^{am} mutation and in those carrying the brn-8 mutation [87]. Second, a high-affinity system was shown to have specifi-city for the three branched-chain amino acids, with apparent K_t values of 2×10^{-6} M. Later this system was divided into two subgroups. The high-affinity system 1 is coded by the brnQ gene and is insensitive to threonine, while high-affinity system 2 is coded by the brnS gene, and its activity is inhibited by threonine [87]. Finally, there seem to be three different low-affinity systems with apparent K_t values in the order of 10^{-4} to 10^{-5}M, each of which was reported to be specific only for one of the three branched-chain amino acids [86,87].

Various types of mutations were obtained after selection of mutants of E. coli K12 for sensitivity to valine and/or glycylvaline for growth. Four genes were thus identified, and three of them were mapped by transduction with phage P1kc [86,87]. The brnQ gene is located close to phoA at 8 min, and the brnS gene is linked to pdxA at 1 min of the E. coli genetic map. The brnR gene is located close to lac at 8 min, and the product of this gene seemed to be required for the activities of high-affinity systems 1 and 2 (see Table 5-1). Biochemical examination has not been made of the three gene products.

Recently, two laboratories have independently reported evidence indi-cating the presence of new genes for the transport systems [72,88]. Anraku and his associates [72] investigated the transport systems using E. coli K12, strain W1-1 (Leu⁻). Mutants were isolated which require high concentrations of branched-chain amino acids for growth. The mutation activity was affecting transport identified and designated as hrbA. The hrbA codes for the transport carrier, since a lesion in this gene results in the loss of transport activity as assayed with cytoplasmic membrane vesicles [90]. The hrbR gene appeared to regulate the ef-ficient utilization of branched-chain amino acids taken up via ATP-depend-ent transport systems [85]. The hrbA gene is mapped at 9 min, close to proC, and the hrbR gene is at 83 min on the ilv region of the E. coli genetic map [72].

Oxender and his coworkers isolated D-leucine-utilizing mutants from E. coli K12, strain EO 0301 [74], and analyzed them genetically [88]. Two mutant loci resulting in derepression of the leucine-specific trans-port system (lstR), and both the leucine-specific system and a common transport system for leucine, isoleucine, and valine (livR) were mapped by conjugation and transduction. Both livR and lstR were found to be closely linked to aroA at 20 min on the E. coli genetic map. It was ob-served that lesion of the lstR gene resulted in a striking increase of leucine-specific binding protein, whereas the livR gene derepressed the leucine-isoleucine-valine binding protein [88].

4. Phenylalanine-Tyrosine-Tryptophan

A transport system common for phenylalanine, tyrosine, and trypto-phan, called the general aromatic amino acid transport system, has been studied extensively by Brown [95,96]. The apparent K_t values for these three amino acids were determined to be in the order of 10^{-7} M (Table 5-1). High concentrations of histidine, leucine, and methionine inhibit the activity. Mutants defective in this general aromatic amino acid transport system were obtained by selecting the cells resistant to β-thienylalanine. The aroP gene was mapped at 2 min by transduction with phage Plkc [95-97]. It was also noted from competition studies that there are three additional transport systems in E. coli K12, each of which is specific for a different aromatic amino acid, with affinity for its respective substrate in the order of 10^{-6} M. A tryptophan-specific transport system was described, and the tryP gene was mapped at 27 min on the E. coli genetic map [98].

5. Methionine

Kadner [99] and Kadner and Watson [100] reported methionine transport systems in E. coli K12. One of the systems has a high affinity for meth-ionine (K_t value of about 1 x 10^{-7} M) and is inhibited by methionyl peptides. The other system has a low affinity (K_t value of 40 μM) and is inhibited by selenomethionine [99]. Mutants defective in the gene metD lack the high-affinity system and are unable to utilize D-methionine as a methionine source. This indicates that the metD gene, which maps at 5 min on the E. coli genetic map [99-101], codes for this high-affinity system which is also responsible for D-methionine uptake.

A mutant RK4201 (metP, metD), was isolated which not only lacks the high-affinity system but which also exhibits decreased activity of the low-affinity system [100]. The mutation in the metP gene was detectable only in the presence of a metD mutation. The metP gene thus appears to code the low-affinity system.

6. Proline

Morikawa and coworkers described the properties of a specific transport system for proline in E. coli W3092 [102]. The transport reaction was characterized as representing a biphasic Lineweaver-Burk plot, and this kinetic property was explained by a mechanism of negative homotropic cooperativity [103]. The apparent K_t values were estimated as 0.44 and 40 μM for the high-affinity and low-affinity transport processes, respectively [102]. Evidence that a single mutation is responsible for alteration in both high- and low-affinity transport processes has recently been obtained (Moto-jima, Yamato, and Anraku, manuscript in preparation).

Substrate specificity of the transport system has been studied by several laboratories [102,104-106]. Tristram and Neale [105] reported a K_t value for high-affinity proline transport of 0.46 μM and a K_i value of 24 μM for L-azetidine-2-carboxylate.

7. Lysine-Arginine-Ornithine

Schwartz and associates suggested the existence of a basic amino acid transport system common for lysine, arginine, and ornithine by studying mutants resistant to canavanine [61]. Then two laboratories examined this system independently and found that, although the mutation selected by canavanine resistance is pleiotropic for arginine biosynthesis and transport, resistance to a canavanine in a group of mutants is accompanied by a decrease in the transport of the three basic amino acids [107-112]. Mutation of the argP gene was mapped at 62 min on the E. coli genetic map by Maas and his coworkers [107-109]. This common transport system shows K_t values of 0.5 and 1.4 μM, respectively, for lysine and ornithine [110, 111]. Canavanine is transported via this system with a K_t value of 400 μM. The data suggested, however, that arginine is not transported by this system. [110,111].

Rosen found that arginine transport in E. coli K12, strain 7, showed a K_t value of 0.026 μM and was not inhibited by lysine or ornithine even at a 500-fold molar excess of the inhibitor, suggesting the existence of an arginine-specific transport system [110]. In addition, a plot of initial rates of lysine uptake was found to be curvilinear in the Lineweaver-Burk plot. The resulting low-affinity component (apparent K_t value of 10 μM) was not inhibited by a large excess of arginine or ornithine [110]. It was thus indicated that lysine is transported in the cells via two routes; one by the common transport system for basic amino acids with high affinity (see Table 5-1) and the other by a second, low-affinity system specific for lysine alone. It was also noted that the latter activity was resistant to osmotic shock.

8. Cystine

A cystine transport system which shares transport of diaminopimelic acid as a common substrate was first described by Leive and Davis [113,114]. They also reported the existence of a cystine-specific transport system. Berger and Heppel [115] reexamined these systems and obtained the following results: the common transport system is sensitive to osmotic shock and shows a K_t value of 0.3 μM for cystine. The K_i value of diaminopimelic acid for cystine uptake is 14 μM. The K_t value for cystine via the cystine-specific transport system is 0.02 μM, and this activity is inhibited by selenocystine.

9. Asparagine

Willis and Woolfok [116] reported two distinct transport systems for
asparagine in E. coli K12; the high-affinity system shows a K_t value of
3.5 μM and the low-affinity system, 80 μM. The former system is
inhibited competitively by the following structurally related compounds:
5-diazo-4-oxo-L-norvaline (K_i, 4.6 μM), β-hydroxyamyl-L-aspartic acid
(K_i, 10 μM), and L-aspartic acid (K_i, 50 μM). The activity of this high-
affinity system was repressed by 1 mM asparagine. It was found that the
low-affinity system was necessary for utilization of asparagine as a nitrogen source.

10. Glutamine

Two transport systems for glutamine were investigated by Heppel and his
coworkers [117,118]. The first, high-affinity system shows its K_t value
of 0.15 μM for glutamine in E. coli K12, strain 7. The K_d value of
glutamine binding protein, purified from the same strain, was determined
to be 0.3 μM [118]. α-Glutamylhydrazide and α-glutamylhydroxamate are
competitive inhibitors for transport and binding. It was noted that this
high-affinity system was derepressed when the cells were cultured in nitro-
gen-limited conditions and repressed about 20-fold when they were grown
in a glutamine-depleted nutrient broth [119]. The second, low-affinity
system for glutamine (K_t value larger than 2 μM) was observed only in
these repressed cells.

11. Aspartate

Transport of aspartate is catalyzed by two distinct systems and has been
studied by Kay and Kornberg biochemically and genetically [120-122]. A
high-affinity system with a K_t value of 3.7 μM is inhibited by β-hydroxy-
aspartate but affected only slightly by glutamate, glutamine, and asparagine
(see Table 5-1). An aspartate transport negative mutant selected for
resistance to inhibition of growth by DL-threo-β-hydroxyaspartic acid was
found to be defective in this high-affinity system. A second, low-affinity
system with a K_t value of 39 μM seemed to share general transport systems for dicarboxylic acids such as succinate, fumarate, DL-malate, and
oxaloacetate [121]; thus, this low-affinity system is lost in dct mutants. The
gene dctA is mapped at 79 min on the E. coli genetic map [120,123].

12. Glutamate

Glutamate transport in E. coli K12 is interesting because the uptake
activity appears to be controlled by a sodium ion [124-126]. This system

has long been studied genetically by Halpern [18]. Recently, information about this particular system, including especially the properties of binding protein, uptake activity in membrane vesicles, and genetic control has been obtained [125-129]. Interestingly, evidence so far obtained suggests a similarity in the molecular apparatus of this transport system to that of the branched-chain amino acid transport in E. coli (see below).

Early studies on the glutamate transport system were carried out using mutants capable of utilizing this amino acid as a sole source of carbon [130,131]. These mutants, which are derived from E. coli W and K12, were found to have several-fold higher uptake activity for glutamate than their parents [131-133]. The transport system is highly stereospecific for glutamate, and its structurally related derivatives such as α-methylglutamate, β-hydroxyglutamate, and α-methylglutamate are competitive inhibitors [132].

The transport reaction of glutamate often showed a unique shape in the substrate saturation curve [132]. Frank and Hopkins [124] investigated this system in detail and found that sodium ion potentiates the uptake activity for glutamate as well as growth inhibition by analogs of glutamate. The cotransport mechanism of glutamate uptake which is coupled with the presence of optimal sodium and potassium ions has now been described by several laboratories. Willis and Furlong investigated the glutamate transport of E. coli K12, strain W3092, grown in a medium with succinate as a carbon source [126]. They found an apparent K_t value of 10 μM for glutamate in the absence of sodium ion and 0.7 μM in the presence of an optimal level of sodium ion.

Halpern and his collaborators examined the regulation of glutamate transport activities and discovered mutants having different genotypes [131,133,134]. The gltS gene, which codes a structural component in the transport system, was located at 81 min on the E. coli genetic map by interrupted mating and by transduction with phage Plkc [134]. The regulatory gene gltR, which regulates the synthesis of the glutamate transport system, was mapped at 91 min [134].

Barash and Halpern [135] and Miner and Frank [125] found the presence of a periplasmic binding protein specific for glutamate. This protein component was recently purified to homogeneity as a glutamate-aspartate binding protein [127]. As noted above, early studies [135] implicated a close correlation of this binding protein with the glutamate transport system. However, recent studies, particularly on the regulation of the transport activity, demonstrated a need for more caution in deciding the physiological role of this binding protein. For example, the effects of growth conditions on the glutamate transport activities, both in intact cells and membrane vesicles, and on the level of glutamate-aspartate binding protein were examined independently by Halpern and his coworkers [128,129] and Willis and Furlong [126,127]. It was found that when E. coli K12, strain

CS101, was grown with aspartate as the sole source of carbon or nitrogen, the glutamate transport activities in intact cells and membrane vesicles increased several fold, while the synthesis of the binding protein was not affected [128]. Moreover, growth of strain CS101 in a rich broth greatly reduced the level of the binding protein but did not appreciably affect the transport activities of intact cells and membrane vesicles [128].

Similarly, Willis and Furlong [126] found that membrane vesicles prepared from E. coli K12, strain D_2W, retained activity for D-lactate-dependent aspartate uptake but not glutamate, although this membrane preparation had lost glutamate-aspartate binding protein. The glutamate-aspartate binding protein was repressed two- to threefold when the cells were grown in the presence of glucose while the glutamate transport activity did not change essentially under the same growth conditions [126]. Based on these results, the authors suggested that, although the glutamate-aspartate binding protein may be a component of the glutamate transport system [129], it is conditionally required for the system, depending on the prior history of growth conditions of the cells [126].

C. Transport Systems in Other Bacteria

1. Glycine-Alanine

Lactobacillus casei has two distinct transport systems for glycine and alanine; one specific for L- and D-alanine and the other for glycine [136]. Two groups of investigators examined the transport systems for these amino acids in Streptococcus faecalis and obtained essentially the same results, namely that the organism has a single transport system specific for L- and D-alanine and glycine [137,138]. Transport systems for these amino acids were also studied using cells of Streptococcus strain Challis [139] and of Bacillus megaterium KM [140].

2. Leucine-Isoleucine-Valine

Kiritani studied the genetics of the transport system for branched-chain amino acids in Salmonella typhimurium and found that the ilvP gene, which codes the transport activity is closely linked to proC on the S. typhimurium genetic map [141]. On the other hand, Thorne and Corwin [142] reported evidence that a gene coding leucine transport in S. typhimurium maps on the side of the chr locus distal to the try operon.

3. Phenylalanine-Tyrosine-Tryptophan

Transport systems for these three amino acids have been studied using Pseudomonas aeruginosa and Bacillus subtilis. Kay and Gronlund [143] reported the existence of two separable transport systems specific for three aromatic amino acids in P. aeruginosa. In order of uptake rate, System I takes up phenylalanine, tyrosine, and tryptophan, while System II takes up tryptophan, phenylalanine, and tyrosine. Rosenfeld and Feigelson [144] described a highly specific transport system for tryptophan in Pseudomonas acidovorans. Aromatic amino acid transport in Pseudomonas sp. 1129a was studied by Guroff and coworkers [145]. The transport activity was induced 5- to 15-fold by phenylalanine. Induced cells released phenylalanine binding protein upon osmotic shock, the amount of which was found to be twofold that of the control cells [146].

Jensen and his coworkers reported properties of a transport system specific for tyrosine and phenylalanine in B. subtillis and claimed that this activity is regulated by a mechanism of negative cooperativity [147,148].

4. Methionine

Ayling and Bridgeland [149] examined the methionine transport system in S. typhimurium. The K_t value was found to be about 1 to 2 x 10^{-7} M. The activity was inhibited competitively by D, L-ethionine and other methionine analogs. They isolated transport mutants by selecting for resistance to α-methyl-DL-methionine and D, L-methionine sulfoximine and found two genetic loci governing the transport system. The metP gene-760 was mapped at about 7 min, between leu and proB, and the metP gene-761 near the gal locus on the S. typhimurium genetic map. The two genes are not linked to any known structural or regulatory gene of the methionine biosynthetic pathway.

5. Proline

A highly specific transport system for proline in P. aeruginosa was studied by Kay and Gronlund [150,151]. Its specificity and kinetic profile appear to be essentially the same as those of E. coli mentioned above. The transport activity of this organism [150] and of Pseudomonas putida [152] are inducible and regulated by the mechanism of catabolite repression.

6. Lysine-Arginine-Ornithine

Rodwell and his collaborators [153,154] described three different transport systems for basic amino acids in P. putida. A common transport system for the three amino acids is inducible in the presence of lysine or D, L-pipecolic acid, and the K_t values for the substrates were determined to be on the order of 10^{-6} M. A second transport system takes up lysine and ornithine with K_t values on the order of 10^{-7} M. A third, inducible transport system specific only for arginine (K_t value, 5×10^{-8} M) was observed in the cells grown in the presence of arginine.

7. Glutamate-Aspartate

Transport systems sharing glutamate and aspartate as a common substrate are present in Lactobacillus plantarum [155], S. faecalis [155-157], Mycobacterium avium [158], and Mycobacterium smegmatis [159]. The specific requirement for sodium for the transport activity has been studied using Halobacterium salinarium [160] and Marine pseudomonad [161].

8. Histidine

The histidine transport of S. typhimurium, like the branched-chain amino acid transport of E. coli, is an interesting system for evaluating the role of a binding protein in active transport. This important system was studied by Ames and her coworkers genetically and biochemically [162-168]. In the most recent model for histidine transport, Ames and Spudich [168] proposed that one of the histidine binding proteins, hisJ protein, is essential for the system. Rosen and Vasington have also investigated the role of the histidine binding protein in the active transport of histidine in S. typhimurium [169,170].

Histidine transport in S. typhimurium is regulated by at least three genes. The hisJ gene determines hisJ protein, a periplasmic binding protein having a K_d value on the order of 10^{-8} M. The hisP gene is a structure gene for the hisP protein which is presumably located in the cytoplasmic membrane. The dhuA gene is a regulatory gene for both proteins mentioned above. All these genes are included in an operon in the order of dhuA-hisJ-hisP [168]. There is one more protein, the hisK protein, in S. typhimurium, which is a minor periplasmic binding protein specific for histidine [164]. Genetic analysis of this protein has not been reported yet.

S. typhimurium has a high-affinity histidine-specific transport system with a K_t of entry on the order of 10^{-8} M and a low-affinity aromatic amino acid transport system which shares histidine and others as substrates [162]. It is known also that hisP protein is required for arginine uptake of low affinity when the cells are grown in the presence of arginine as a sole source of nitrogen [163,166]. Genetic and biochemical evidence indicate that the hisJ protein is a component which only determines the K_t value and substrate specificity of the high-affinity transport system, since mutation in the binding activity of hisJ protein results in the loss of the transport activity, with the low-affinity system remaining unchanged [164,165,168]. Thus, the function of the hisJ protein is dominant over hisP protein. Translocation of the substrate across the membrane is suggested to be mediated by a

hisJ-hisP complex. Evidence supporting this idea has been obtained as
follows. (1) hisJ protein has two distinct active sites in the molecule;
one is the binding site for the substrate and the other the interaction site
for hisP protein [167]. (2) A histidine transport mutant, TA308, produces a hisJ protein defective in the interaction site but normal in the
binding site [168]. (3) Strain TA342, a pseudorevertant of TA308, was
isolated by a sophisticated two-step selection method and showed an
improved transport activity, although it produces a hisJ protein identical
to that in the parent strain [168]. Based on these and other lines of
evidence, Ames and Spudich [168] concluded that the mutation occurring
in strain TA342 can be reverted to the modification of hisP protein in such
a way that it can combine with the altered interaction site of hisJ protein,
thereby compensating for the defective transport activity.

   This evidence, therefore, indicates that (1) the high affinity histidine
transport system of S. typhimurium requires hisJ protein as an essential
component which determines its substrate specificity and affinity, and
(2) hisP protein recognizes the interaction site of hisJ protein, and the
complex formed mediates substrate translocation across the membrane.
Although the precise mechanisms of translocation (for example, the mechanisms of formation and dissociation of a transitory ternary complex of
hisP-hisJ-histidine), and the energy requirement for this process have not been
elucidated, this system provides a novel reaction model for active transport. In this connection, characterization of hisP protein is especially
required with respect to whether it has a binding site for a substrate.

   Ames and her coworkers have not done an experiment using membrane
vesicles of S. typhimurium. Restoration of the uptake activity of shocked
cells by histidine binding protein was reported to be unsuccessful. Further
studies are required for the biochemical properties of this hisP protein.
As Ames and Spudich [168] stated, physiological and immunochemical
evidence showing direct interaction of hisJ protein with hisP protein are of
particular importance.

D. Transport Systems Studied with
     Membrane Vesicles

1. "Kaback Vesicles" Prepared from
     E. coli ML308-225

Lombardi and Kaback [89] examined specificities of amino acid transport
in membrane vesicles of E. coli ML308-225, prepared after osmotic lysis
of sphcroplasts. The properties of this membrane preparation are given
in several studies [14,15,45] and also discussed in Chapter 1. It was noted
that this membrane preparation takes up 16 amino acids actively, depending
on respiratory substrate D-lactate and the artificial electron donor
ascorbate-phenazine methosulfate, through nine kinetically different trans-

port systems [89]. Table 5-2 summarizes these results. The following are comments on these transport activities in comparison with those observed in intact cells (see Table 5-1).

(a) <u>Glycine-Alanine.</u> This system is specific for glycine and alanine and inhibited by D-serine but not by L-serine. The observed specificity of the system and its K_t values for substrates are nearly equal to those of intact cells.

(b) <u>Threonine-Serine.</u> The specificity of this system seems to be the same as that of intact cells.

TABLE 5-2 Transport Systems for Amino Acids Detectable in Membrane Vesicles Prepared from <u>E. coli</u> ML308-225 [89]

System	K_t (μM)	Inhibitor
Glycine-alanine	Glycine (1.6) Alanine (8.4)	D-Serine
Threonine-serine	Threonine (2.6) Serine (5.4)	D-Serine
Leucine-isoleucine-valine	Leucine (1.1, 18) Isoleucine (1.7, 21) Valine (2.0, 29)	L-Methionine L-Cysteine
Phenylalanine-tyrosine-tryptophan	Phenylalanine (0.42) Tyrosine (0.68) Tryptophan (0.33)	L-Histidine
Cysteine	Cysteine (38)	L-Leucine L-Isoleucine L-Valine
Proline	Proline (1.0)	
Lysine	Lysine (0.94)	L-Thiosine
Histidine	Histidine (0.15, 4.0)	L-Phenylalanine L-Tyrosine L-Tryptophan
Glutamate-aspartate	Glutamate (1.6) Aspartate (8.4)	

(c) Leucine-Isoleucine-Valine. Kinetics of transport reactions for these three amino acids show the presence of two saturable components in the membrane vesicles. Two apparent K_t values for each of the substrates were obtained by visual inspection of the Lineweaver-Burk plots. In contrast to the results obtained with intact cells, cysteine was found to be inhibitory for uptake activities. In addition, methionine inhibited these activities considerably, although this amino acid was not taken up at all by the membranes.

(d) Phenylalanine-Tyrosine-Tryptophan. These three amino acids seem to be taken up through a general transport system common for aromatic amino acids, as judged by competition experiments and kinetic parameters. Any specific transport system for each of the three substrates was not detected or was missing in this membrane preparation.

(e) Lysine-Arginine-Ornithine. The membrane preparation is defective in this transport system since it cannot take up arginine and ornithine. The membranes take up lysine via the thiosine-inhibitable lysine specific system [110].

(f) Glutamate-Aspartate. The membrane vesicles show the presence of a common transport system for glutamate and aspartate as judged by a mutual competition in the uptake activities. This result is different from that in intact cells. An effect of sodium ion on uptake activity was not reported.

(g) Other Comments. The membrane vesicles take up histidine and cysteine but not methionine, cystine, glutamine, and asparagine, although intact cells of strain ML308-225 do. In a few cases V_{max} values of transport reactions were reduced considerably in membrane vesicles. For example, V_{max} values for histidine and leucine were reduced to almost one-tenth of those in intact cells [89]. These results suggest that certain transport systems are unstable in the procedure for preparing membrane vesicles and underwent inactivation of transport carriers or of the coupling system with energy.

2. Cytoplasmic Membrane Vesicles
   Prepared from E. coli K12

Anraku and associates [171] published a new procedure for preparing cytoplasmic membrane vesicles from E. coli K12. This procedure involves the disruption of spheroplasts in 20% (w/w) sucrose by a French pressure chamber and dialysis of crude membranes against 3 mM EDTA at 4°C. The dialyzed membrane fraction is then subjected to centrifugation

on 43% (w/w)sucrose, which results in the complete separation of cytoplasmic membranes from outer membranes. The membrane preparation
thus obtained contains right-side-out closed vesicles of 70% of the total
membrane population [172]. Morphology and orientation of the cytoplasmic
membrane vesicles have been described [172,173].

The cytoplasmic membrane vesicles are of high purity as judged by
specific contents of cytochromes and specific activities for D-lactate
dehydrogenase or succinate oxidase [171], and show high transport activities
for isoleucine, valine, threonine, lysine, glutamine, and cystine [90]. As
is well known, when E. coli K12 was converted to membranes by the
procedure of Kaback [45], the resulting vesicles showed little uptake
activity for certain amino acids [174,175]. For instance, Cox and coworkers
reported that uptake activities for proline and leucine in the vesicles were
about 0.06 and 0.03 nmol/min/mg protein, respectively [175]. In contrast,
the uptake activities for proline and isoleucine in cytoplasmic membrane
vesicles prepared by the method of Yamato et al. [171] were 3.4 and 0.6
nmol/min/mg protein, respectively [90].

These results suggest that different procedures affect the uptake
activities in the resulting membrane preparations, and therefore much
caution is required in preparing membrane vesicles from different strains
of E. coli for transport studies.

3. Membrane Vesicles Prepared from
    Other Bacteria

A number of membrane vesicles capable of transporting various amino
acids have been prepared from bacteria such as E. coli, S. typhimurium,
Micrococcus denitrificans, Arthrobacter pyridinolis, Mycobacterium phlei,
Marine pseudomonas B-16, P. putida, Proteus mirabilis, Staphylococcus
aureus, B. subtilis, and B. megaterium (see ref. 15), and also Veillonella
alcalescens [176], Halobacterium halobium [177-179], and thermophilic
bacteria PS-3 [180].

Detailed studies on the substrate specificity of the transport systems
in membrane vesicles prepared from B. subtilis [50,181] and S. aureus
[182,183] have been reported. These results are summarized below. The
readers are referred to Table 5-2 for comparison. B. subtilis has 10
transport systems for glycine-alanine, threonine-serine, leucine-isoleucine-
valine, phenylalanine-tyrosine, methionine, cysteine, proline, lysine-
arginine, asparagine-glutamine, and aspartate-glutamate. S. aureus has
12 transport systems for glycine-alanine, threonine-serine, leucine-
isoleucine-valine, phenylalanine-tyrosine-tryptophan, methionine, cysteine,
proline, lysine, arginine, histidine, asparagine-glutamine, and aspartate-
glutamate.

IV. CURRENT TOPICS IN THE STUDY
OF TRANSPORT SYSTEMS

A. Osmotic Shock

As mentioned earlier, biochemical studies using the procedure of osmotic shock have contributed greatly to recent developments in the field of active transport for amino acids. In 1966, this method was first employed for sulfate transport of S. typhimurium by Pardee and his coworkers [184, 185], who described the release of a sulfate binding protein in the shock fluid. In the study of galactose transport of E. coli, Anraku found that the galactose uptake activity was reduced by osmotic shock, and that the reduced uptake activity of shocked cells was restored to near the original level by incubating them with dialyzed shock fluid [7, 35]. Reduction of transport activities for a number of amino acids was found with concomitant release of corresponding binding proteins [35, 186]. Cytochemical studies on the localization in the periplasm of a group of proteins released by osmotic shock have also been reported [187-189]. Accordingly, in the early phase of development, biochemical studies on amino acid transport systems have been carried out with close attention to binding protein and also with close correlation between release of binding protein and reduction of apparent transport activity. Summary and evaluation of these earlier results were described in several reviews [36-39].

Berger and Heppel [190] compared the effect of osmotic shock on various transport activities in E. coli ML308-225 and observed the presence of two types of systems; (1) the shock-sensitive systems which are associated with periplasmic binding proteins and which are absent in isolated membrane vesicles, such as uptake for glutamine, diaminopimelic acid, arginine, histidine, and ornithine, and (2) the shock-resistant systems, whose carrier proteins are not released by osmotic shock and which are active in membrane vesicles, like transport for proline, phenylalanine, serine, glycine, and cysteine. Transport systems in these two categories are also different in their sources of energy for the uptake [190].

However, it should be noted that the effect of osmotic shock is pleiotropic for physiological activities in the cells, not only for modification of the transport systems [8]. There are several examples in which reduction of transport activities for leucine [85], glucose-6-phosphate [191], and glycylglycine [192] took place after osmotic shock without release of the binding proteins or of detectable binding activities. A preliminary suggestion that restoration of reduced transport activities in shocked cells required a minor protein component other than the binding protein has also been reported [92]. As have been reported, conditions for osmotic shock [8] and for restoration of reduced activities by the binding proteins [35, 92] are critical. In this respect, it is also interesting to note that modified osmotic procedures of the original method failed to restore the reduced transport activities (personal communications from Drs. D. L. Oxender, J. Wood and W. Boos).

B. Retention Effect of Binding Protein

On some occasions, when the substrate binding reactions of periplasmic binding proteins were examined by a conventional equilibrium dialysis method using radioactive substrates, the Lineweaver-Burk plots of the reactions were concave downward [42,193-195], and the specific activities of the proteins were significantly altered by the protein concentration [42].

These phenomena have been extensively studied by Amanuma and coworkers [42], who demonstrated chemically that purified, homogeneous leucine-isoleucine-valine-threonine binding protein (LIVT-binding protein) contains about an equimolar amount of its substrate, bound noncovalently to the molecule. The mechanism of the retention of substrate by the binding protein was studied biochemically and theoretically. They concluded that the substrate binding reaction is a simple association-dissociation reaction with dissociation constants of around 10^{-7} M for its substrates, and that owing to these small dissociation constants, the protein can retain the substrates in solution in a dynamic equilibrium. Several investigators have examined similar problems with different binding proteins. Richarme and Kepes [196] showed the release of glucose from purified galactose binding protein upon addition of galactose. On the contrary, Silhavy and Boos [197], and Hazelbauer [198] reported that their purified preparations of galactose and maltose binding proteins, respectively, did not contain the corresponding substrates.

Based on this and other evidence, Amanuma et al. [42] postulated that binding protein may have dual physiological roles in retaining the substrate in the periplasmic space and in stimulating the transport carrier activity. In other words, binding proteins conserve the substrate concentration at a homeostasic level in the space inside the outer membrane layer without the use of metabolic energy, owing to their inherent ability to retain substrate in a dynamic equilibrium in solution [199].

Recently, Silhavy and associates [41] reported that, theoretically, binding proteins having small dissociation constants for substrates are able to retain the substrates in a solution. The results of Amanuma et al. [42] give a unique biochemical example for this theory.

C. Kinetic Evaluation of Transport Reactions

Active transport reactions, with respect to substrate specificity, affinity, and energy coupling can be studied by kinetics using radioactive substrates.

Kinetic studies are powerful methods, and much useful information about
the nature of transport reactions is obtained easily. For a long time,
bacterial active transport reactions were described by the Michaelis-Menten
equation. However, careful analyses have shown that most transport
reactions for amino acids of intact cells are not described by a single
set of the Michaelis constants. For example, the Lineweaver-Burk
plots of certain transport reactions are biphasic rather than linear,
suggesting that the intact cells have heterogeneous transport systems [69,
74,77,162,170] and homogeneous carrier with negative cooperativity [75,
102,103,147,148].

Mathematical methods for analyzing these kinetic data were reported. Neal
[200] described the method for two independent carriers and Borst-Pauwels
[201] for two binding sites in a single carrier. To elucidate a transport
reaction showing a concave downward Lineweaver-Burk plot and negative
homotropic cooperativity, Awazu and associates [103] investigated the kinetic
model shown in Fig. 5-1.

Figure 5-1 shows a transport model representing sequential binding and
translocation of a substrate by a single transport carrier with three binding
sites for substrate. Assuming that equilibration in the binding of a substrate outside the cells with various forms of carriers (C, CS, and CSS in
Fig. 1) is rapid, and the dissociation of carrier-substrate complexes,
$CSn \longrightarrow CSn-1 + S$ (n = 1, 2, and 3), is rate limiting in the translocation of a

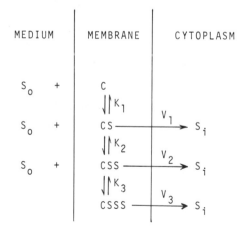

FIG. 5-1 Sequential binding and membrane translocation of substrate
mediated by a single carrier with three binding sites for substrate. K_1,
K_2, and K_3 are dissociation constants and V_1, V_2, and V_3 are maximal
velocities. S_o and S_i are substrate outside and inside the cells, respectively.

substrate across the cytoplasmic membrane (see Fig. 5-1), the following equation is obtained, which describes the initial rate of entry of a substrate mediated by a single carrier with three binding sites.

$$v = \frac{\dfrac{V_1}{K_1} S + \dfrac{V_2}{K_1 K_2} S^2 + \dfrac{V_3}{K_1 K_2 K_3} S^3}{1 + \dfrac{S}{K_1} + \dfrac{S^2}{K_1 K_2} + \dfrac{S^3}{K_1 K_2 K_3}}$$

Awazu et al. [103] tested the fit of this general model with experimental data on the initial rates of uptake for branched-chain amino acids by an iterative computation using the nonlinear least squares method and concluded that these reactions are catalyzed by a mechanism in which two of the three carrier-substrate complexes, CS and CSSS, are capable of mediating the translocation of a substrate. Biochemical validity of this model has not been convincing, but the postulated model is important because it suggests that a carrier has its own ability to regulate the transport activity by negative cooperativity induced by sequential binding of a substrate to the binding sites of the carrier.

It is obvious that in the future kinetic data will have to be computed according to relevant models, and thus the reliability of parameters must be given with respect to criteria such as the minimum sum of squares, standard deviation, and weighing of data. Kinetic analysis of the transport reaction is useful but not conclusive as to its molecular mechanism. Investigations in combination with genetic and biochemical analyses undoubtedly potentiate the present level of kinetic studies on active transport.

D. Studies of Transport Carrier Proteins

To examine transport carrier proteins directly, using solubilized preparations uncomplicated by translocation reactions, has long been an attractive problem for many investigators. Gordon and coworkers [202] once observed that proline binding activity was solubilized by the detergent Brij 36T from membrane vesicles having D-lactate-dependent amino acid uptake systems of E. coli. They separated the proline binding activity of a low molecular weight by column chromatography and showed that this activity was strictly specific for proline and sensitive to sulfhydryl reagents, as is proline transport in the membrane vesicles.

Recently, two very important contributions on the nature of carrier proteins in membranes have been reported. Hirata and associates [43, 180] reported properties of membrane vesicles prepared from thermophilic bacteria PS-3 capable of taking up alanine, and succeeded in solubilizing the protein component(s) from the vesicles with the detergents cholate and deoxycholate. The solubilized protein fraction was found to be reconstitutively active for alanine uptake in liposomes when the transport energy was supplied as a membrane potential introduced by K^+ diffusion via valinomycin [43].

Anraku and his coworkers [44] also demonstrated that the treatment of sonic membranes of wild strain E. coli with an organic solvent, acidic butanol, resulted in the solubilization of a protein fraction active for the uptake of several amino acids, including proline, glutamate, and cysteine, when reconstituted in liposomes with E. coli phospholipids and with energy supplied as a membrane potential. By using this procedure and proline transport defective mutants, they showed that the reconstituted vesicles with the protein fraction from mutant cells took up glutamate but not proline.

Amanuma and associates [203] carefully measured the proline binding activity of membranes by a procedure uncomplicated by the uptake, using sonic membranes prepared from wild strain and a mutant of E. coli defective in proline transport. The membranes prepared from wild cells showed a stereospecific binding of [^{14}C]proline. The binding activity was not affected by the presence of D-lactate, with or without a proton conducting uncoupler, SF6847 (3,5-di-tert-butyl-4-hydroxybenzylidenemalononitrile) [204-206]. The activity was inhibited by Hg^{2+}. All these observations were consistent with the properties of proline transport of membrane vesicles [89,174]. Under the same conditions, sonic membranes prepared from the mutant defective in proline transport did not show the binding activity. Interestingly the K_d value of the binding depended on pH of the medium without change in the maximal number of the binding sites. Kinetic examination of the binding data indicated that the proton first binds to the membrane carrier, and this causes change in the affinity of the binding sites for proline by lowering the K_d value. Therefore, these results suggest the mechanism of cobinding of proton to carrier for proline in the membranes.

V. MOLECULAR APPARATUS OF
   ACTIVE TRANSPORT

A. Transport Systems for Branched-
   Chain Amino Acids of E. coli

As mentioned in the preceding sections, certain transport systems for amino acids in bacteria seem to consist of membrane bound transport carriers in functional association with periplasmic binding proteins. Although studies on the nature of carrier proteins are as yet incomplete, it would be of considerable value to discuss the possible molecular mechanisms of transport reactions based on presently available evidence. Since the active

transport systems for branched-chain amino acids in E. coli K12 have been
studied most extensively, these systems provide the best example for this
purpose. In the following sections are pertinent findings representing the
kinetic and biochemical characteristics of these transport systems.

1. Specificity

Cohen and Rickenberg [23] first showed that leucine uptake activity is inhibited competitively by isoleucine and valine, and proposed that E. coli
has a common transport system for branched-chain amino acids. Oxender
and his coworkers further studied this system and concluded that the
common transport system has a strict specificity for L-isomers of three
branched-chain amino acids [69]. For example, the K_t value for leucine
uptake was 1 μM, and the K_i values of isoleucine and valine were 1 and 5
μM, respectively. On the other hand, the K_i values of D-leucine and
cycloleucine were 27 and 400 μM, respectively. They claimed that this
common transport system shares the uptake of threonine and alanine, based
on the results of inhibition kinetics of leucine transport [74].

It has been found by several investigators that kinetics of entry of all
the three branched-chain amino acids cannot be described by a single set
of Michaelis-Menten constants, since the double reciprocal plots of the
initial rate of uptake versus substrate concentration give biphasic curves
[69,74-77]. Further investigations on substrate specificities have indicated that there are multiple systems catalyzing the uptake of a group of
branched-chain amino acids in E. coli K12 [74,76,77,82]. One is a
common transport system as mentioned above. The second system is
specific only for leucine. Furlong and Weiner [82] found this system,
which is sensitive to trifluoroleucine but not to isoleucine. The third
system is described for the transport activities in cytoplasmic membrane
vesicles and takes up three branched-chain amino acids [85, 90].
This system is sensitive to norleucine but not to threonine [90].

2. Binding Proteins

Three distinct species of binding proteins are obtained from E. coli K12,
strain W3092. One is the LIVT binding protein, which binds leucine, isoleucine, valine, and threonine [83]. This protein was initially called
leucine binding protein [77,186]. The identification of the leucine binding
protein as LIVT binding protein was made by the observation that shock
fluid contained a threonine binding activity and all the activity was eluted
with leucine binding activity from a DEAE-Sephadex column [207]. After
careful purification it was demonstrated clearly that the threonine binding
activity can be copurified with the leucine binding protein, which is homo-

geneous as judged by determination of the N-terminal amino acid of the preparation [42, 83].

The second binding protein (Ls binding protein) which binds leucine only, was found by Furlong and Weiner [82], and confirmed by several laboratories [74, 83]. The third binding protein was recently purified by Amanuma et al. [42] and named sLIVT (second LIVT) binding protein. This protein has exactly the same substrate specificity as LIVT binding protein, but its chromatographic and electrophoretic properties are different from the known binding proteins. In E. coli W3092, LIVT, Ls, and sLIVT binding proteins are present in the ratio of 0.85:0.13:0.02 [42]. There have been papers reporting other binding proteins which bind isoleucine preferentially [74, 208], but the substrate specificity of these components has not been studied in detail.

3. Substrate Binding of Binding Proteins

Inhibition of [^{14}C]leucine binding to the binding proteins by analog compounds was studied [42, 83]. It was concluded that the binding proteins require a free α-hydrogen atom in the substrate molecule but not necessarily a free α-carboxyl residue. The substrate specificity depends on the aliphatic residue of the substrate amino acids. Isoleucine, valine, and threonine inhibit competitively the binding of [^{14}C]leucine to LIVT binding protein in this decreasing order of magnitude. Unnatural amino acids such as L-α-aminobutyrate and L-alloisoleucine were found to inhibit the activity competitively. It is important to note that L-norleucine does not inhibit the binding reaction [42]. Ls binding protein shows a strict specificity for leucine, and its activity is inhibited only by trifluoroleucine [82].

Formerly, the K_d value of LIVT binding protein for leucine was reported to be of the order of 10^{-6} M [78, 82, 186]. However, Amanuma et al. [42] recently demonstrated that LIVT binding protein contained bound substrate amino acids in the molar ratio of 1 to protein, even in the homogeneous and dialyzed preparation. This bound substrate interferes with the binding kinetics using labeled substrate under certain experimental conditions. The bound substrate can be removed from LIVT binding protein by treating the protein with 6 M urea and dialysis. The resulting protein, which is free from bound substrate, shows the true K_d value of 0.19 μM for leucine [42].

4. Membrane Vesicles

Yamato and Anraku [90] analyzed D-lactate-dependent uptake of isoleucine by the cytoplasmic membrane vesicles and found that the uptake reaction shows a K_t of entry of 6 μM and is inhibited competitively by norleucine but not by threonine. The reaction in the membrane vesicles depends on a proton motive force, in agreement with reported observations [209,210]. They also found that the initial rate of isoleucine uptake is enhanced significantly by the addition of LIVT binding protein but not by Ls binding protein, without a significant change in the K_t value of entry. The cytoplasmic membrane vesicles prepared from a mutant defective in the transport of branched-chain amino acids show neither uptake activity for isoleucine nor stimulation of the activity by LIVT binding protein. These results indicate that partial restoration of osmotic shock-sensitive transport is obtained with this membrane preparation in the presence of a binding protein.

B. Roles of Binding Proteins in Active
Transport

Since the demonstration by Anraku [7,35] that osmotic shock reduced galactose transport activity of E. coli, and that this reduced activity was restored to nearly the original level by incubating shocked cells with protein components in the shock fluid, positive evidence showing the specific role of binding proteins in active transport has been obtained as listed below.

1. Transport activities for various solutes such as amino acids, carbohydrates, and phosphate ions are shock-sensitive. Few of these reduced activities are restored upon incubation with respective binding proteins and other protein components [35,92,135,211-213].

2. Repression of transport activities is accompanied by repression of the synthesis of binding proteins [35,75,81], and mutants which synthesize binding proteins in high amounts show high transport activities [74,93,163,164,214].

3. Mutation in genes responsible for the synthesis and function of binding proteins affects transport activity greatly. Thus, cells which produce binding proteins defective in substrate binding lose the transport activity completely or show only a little activity [165,168,215].

4. Isoleucine uptake activity by cytoplasmic membrane vesicles is stimulated by the addition of LIVT binding protein but not Ls binding protein [90].

These observations, particularly those by Ames and her coworkers [165,168], suggest strongly that the binding proteins may play a significant role in the entire process of active transport, especially for determining its substrate specificity. However, there have been other observations that suggest that the binding proteins may not be involved in the essential part of the active transport. These are summarized as follows.

1. Membrane vesicles take up various amino acids and carbohydrates irrespective of the release of corresponding binding proteins after formation of spheroplasts [89, 90].

2. Leucine auxotroph mutants, whose synthesis of binding proteins for branched-chain amino acids is repressed by leucine, show their respective transport activities when measured using intact cells and cytoplasmic membrane vesicles [85, 90].

3. When added to cytoplasmic membrane vesicles prepared from a mutant defective in branched-chain amino acid transport, LIVT binding protein does not catalyze the respiration-dependent uptake of the substrates [90].

This evidence indicates that the binding protein itself is neither a carrier for transport nor solely a component catalyzing the rate-limiting process of the translocation of substrate. Therefore, it seems most probable that binding proteins are not essential components in the above cases, but stimulate the transport carrier activities of the membranes, either obligatorily or additionally.

C. Explanation for Multiple Transport
   Systems for Branched-Chain Amino
   Acids from Various Perspectives

1. Classification of the Systems by
   Oxender and his Coworkers

Rahmanian and associates [74] and Wood [77] seem to take the point of view that binding proteins are essential components capable of determining the substrate specificities of the transport systems. After having done careful inhibition kinetic studies and experiments on the repression of transport activities and of binding proteins, Rahmanian et al. [74] concluded that there are three distinct transport systems, LIV-I, LIV-II, and L systems, in E. coli K12, as shown in Table 5-3A. The three systems are markedly different in their specificities. The LIV-I system corresponds to a major and common transport system for branched-chain amino acids. They also claimed that this system shares threonine and alanine transport because these amino acids inhibited the leucine uptake via this system, and that the LIVT binding protein seemed to have a similar broad specificity. This system shows a high affinity for the three branched-chain amino acids and is sensitive to osmotic shock. Osmotic shock and growth of cells with leucine were found to lower the amount of the LIVT binding protein in the periplasm, thus greatly reducing the uptake activity. Taking these observations into account, they concluded that the LIVT binding protein is

essential for the LIV-I system and that its specificity is determined by
the specificity of the binding protein.

The L system is specific only for leucine and is not inhibited by isoleucine [82]. However, the L system is difficult to describe kinetically
because its activity is minor and its affinity for leucine is essentially the
same as that of the LIV-I system. Trifluoroleucine is a specific inhibitor
for the activity of Ls binding protein as well as for that of leucine uptake
via the L-system. The D-leucine-utilizing mutant shows a significant increase in the activity of the L system and in the level of the Ls binding
protein [74,88]. These facts once again demonstrate a close correlation
between the transport activity of the L system and the amount of the Ls
binding protein.

Rahmanian et al. [74] thus concluded that the LIV-I and L systems are
independent of each other and that their specificities are determined by
the respective binding proteins as essential components. So far as the
available evidence is concerned, this conclusion may be valid. But the
questions arise: How do binding proteins translocate the substrate? Are
binding proteins the only components capable of determining substrate
specificity of the systems?

The LIV-II system is minor and has a low affinity for the three
branched-chain amino acids. The substrate specificity of the system is
strict and not inhibited by threonine and alanine (Table 5-3A). This system does not require binding proteins because it is resistant to osmotic
shock and is active in repressed cells. Thus it is clear that the LIV-II
system is unique and distinct from the LIV-I and L systems. Rahmanian
et al. [74] have suggested that this LIV-II system corresponds to the transport activity measured in membrane vesicles by Lombardi and Kaback [89].
This suggestion was later confirmed by Wood [77]. A question concerning this
conclusion may be: What component other than the binding protein does
promote this distinct system?

2. Classification of the Systems by
    Anraku and His Coworkers

The questions raised here by the author are critical in order to understand
the molecular apparatus of transport and the mechanism for determining
the substrate specificity. An important idea for answering these questions
has already been described: Anraku [35] claimed that the restoration of
reduced galactose transport activity of shocked cells was achieved through
an interaction of the binding protein with a limited number of membrane
sites because the effect of the binding protein was saturable and dependent
on metabolic energy. In addition, it is now certain that the cytoplasmic
membrane has carriers capable of translocating solutes in the absence
of binding proteins [44,90].

TABLE 5-3 Multiple Transport Systems for Branched-Chain Amino Acids of E. coli and Possible Roles of Binding Proteins in the Systems

Transport System[a]	Substrate[b]	K_t[c]	Energy	Osmotic Shock	Inhibitor[f]			Repression by Leucine[g]	Proposed Roles of Binding Protein
					Thr	Norleu	Others		
CLASSIFICATION A									
LIV-I	LIVTA	0.2	--[d]	Sensitive	Yes	No	Alanine	LIVT-BP,[h] repressed	LIVT-BP, essential
LIV-II	LIV	2	Δp[e]	Resistant	No	Yes	--	Activity, not repressed	LIVT- and Ls-BPs, not required
L	L	0.2	--	Sensitive	No	No	Trifluoro-leucine	Ls-BP,[h] repressed	Ls-BP, essential
CLASSIFICATION B									
LIV-1	LIV	0.4	ATP	Sensitive	No	No	--	Activity, not repressed[i]	LIVT- and Ls-BPs, obligatorily stimulative
LIV-2	LIV	6	ATP	Resistant	No	Yes	--	Activity, repressed[i]	LIVT- and Ls-BPs, additionally stimulative
LIV-3	LIV	6	Δp	Resistant	No	Yes	--	Activity, not repressed	LIVT- and Ls-BPs, additionally stimulative

[a] Classification A reported by Rahmanian et al. [74] and Classification B by Yamato et al. [85].

[b] Abbreviations used are: L, leucine; I, isoleucine; V, valine; T, threonine; A, alanine.

[c] K_t value of entry (in μM) for leucine (Classification A) and for isoleucine (Classification B).

[d] Not determined.

[e] Studied by Wood [77].

[f] Abbreviations used are: Thr, threonine; Norleu, norleucine.

[g] Repression of transport activity and synthesis of binding proteins by leucine during the growth of cells.

[h] Abbreviations used are: LIVT-BP, LIVT binding protein; Ls-BP, Ls binding protein.

[i] Unpublished observation of I. Yamato and Y. Anraku.

Anraku and his coworkers [72, 85, 90] investigated the nature of carrier activities of intact and shocked cells, and of cytoplasmic membrane vesicles under conditions uncomplicated by the function of binding proteins. When E. coli K12, strain W1-1 (Leu⁻) was grown in minimal salt medium containing leucine (80 μg/ml), synthesis of binding proteins was repressed completely. The kinetics of isoleucine uptake by intact cells showed a biphasic curve in the Lineweaver-Burk plot, resulting in high-affinity (LIV-1) and low-affinity (LIV-2 and LIV-3) systems, as shown in Table 5-3B. It should be noted that the LIV-1 and LIV-2 systems do not correspond to the LIV-I and LIV-II systems of Table 5-3A, respectively, because the classification of A and B in Table 5-3 is conventional and is based on different aspects and experimental approaches.

All the systems are shown to have affinity for leucine, isoleucine, and valine and not to be inhibited by threonine, since the cells have no LIVT binding protein. However, the LIV-1 system is lost by osmotic shock and the other two systems are inhibited by norleucine (Table 5-3B). The LIV-1 system was also demonstrated in E. coli K12, strain NR-70, having a normal level of LIVT binding protein [84]. In this case threonine inhibited competitively the LIV-1 system which was also shock sensitive. Although the source of energy for driving the LIV-1 and LIV-2 systems apparently depended on the cellular concentration of ATP [85, 90], the former activity was found not to be repressed by leucine. These results indicate that the LIV-1 system is distinct from the LIV-2 system (Table 5-3B).

Therefore, it is suggested that osmotic shock causes the release of binding protein as well as some unknown changes in energy coupling, and that both biochemical modifications result in the loss of the LIV-1 system. These findings also suggest that the binding protein may not be essential for either the LIV-1 or the LIV-2 system. So far as their genetics are concerned, the mutants examined are all defective in the uptake of the three branched-chain amino acids but normal in threonine uptake. No mutant has yet been isolated in which the LIV-I and/or the L systems are present [72].

The cytoplasmic membrane vesicles prepared from strain W1-1 showed isoleucine uptake (LIV-3 system), which was kinetically indistinguishable from that of the LIV-2 system. However, this system is driven by a proton motive force and its activity is not repressed by leucine (Table 5-3B). Genetic examination showed that the hrbA gene is required for this LIV-3 system [72].

The properties of these systems were determined using cells and cytoplasmic membrane vesicles in which binding proteins had been repressed and lost, respectively. Therefore, the observed activities were considered to depend on membrane components, possibly a carrier. These results also suggest that there are three functionally independent carriers which are coded by different genes but which show identical substrate specificity for

substrate amino acids. However, their apparent specificities are different
and are affected by the presence of binding proteins, as evidenced by the binding
protein-dependent stimulation of branched-chain amino acid transport of
shocked cells and cytoplasmic membrane vesicles [35,90,92].

In this line of consideration, Yamato et al. [85] postulated the roles of
branched chain amino acid binding proteins as <u>obligatorily stimulative</u> for
the LIV-1 system, because they greatly enhance norleucine-insensitive
activity, while <u>additionally stimulative</u> for the LIV-2 and LIV-3 systems because they slightly stimulate norleucine-sensitive activities (Table 5-3B).
Note that both LIVT and Ls binding proteins are not inhibited by norleucine.

3. Evaluation

Two independent attempts to elucidate the biochemical background of multiple transport systems for branched-chain amino acids can be summarized
as above. These results indicate that both explanations may be valid if
relevant membrane components are present and function in collaboration
with binding proteins. Further evaluation of these explanations has to await
studies on the biochemical nature of relevant membrane transport components such as a <u>carrier</u>, a <u>pseudocarrier</u>, and an <u>apocarrier</u>.

For the reader's convenience, it seems worthwhile to correlate the
two classifications made from different standpoints. First, the LIV-3
system in Table 5-3B may be the same as the LIV-II system in Table 5-3A
because both are measurable with membrane vesicles, driven by a proton
motive force, and sensitive to norleucine. Genetic evidence that both
activities are not subjected to repression by leucine supports this. Second,
the LIV-2 system is not mentioned by Rahmanian et al. [74] or by Wood
[77]. Finally, the author would like to propose that the LIV-1 system
serves as a common carrier for both the LIV-I and L systems. Observed
differences in their apparent substrate specificity can be explained by
assuming that LIVT and Ls binding proteins can stimulate their respective
systems obligatorily as discussed above.

D. <u>Molecular Apparatus of Transport</u>
    <u>and Possible Roles of Binding</u>
    <u>Proteins</u>

To discuss further and define the molecular apparatus of transport involving
periplasmic binding protein, it may be of general value to construct several
transport models theoretically based on the evidence mentioned above.
Figure 5-2 schematically illustrates the representative models for the
molecular apparatus of transport and the possible roles of periplasmic
binding proteins.

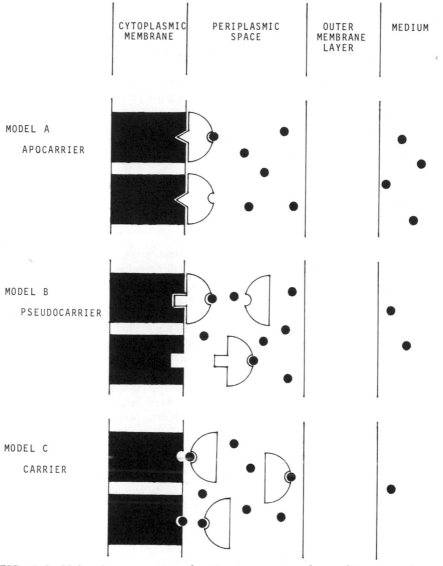

| CYTOPLASMIC MEMBRANE | PERIPLASMIC SPACE | OUTER MEMBRANE LAYER | MEDIUM |

MODEL A
APOCARRIER

MODEL B
PSEUDOCARRIER

MODEL C
CARRIER

FIG. 5-2 Molecular apparatus of active transport and possible roles of
binding proteins. Black squares indicate membrane bound transport proteins and semicircles, periplasmic binding proteins having different properties. The black squares with triangular and square indentations are apocarriers (Model A) and pseudocarriers (Model B), respectively. These indentations represent interaction sites (accepting) for binding proteins. The
black squares with semicircular indentations are carrier (Model C), and the
indentation represents a binding site for a substrate. Binding proteins also
have a binding site for substrate (circular indentations) and closed circles
represent a substrate. In Model A binding protein having an interaction site

Model A shows that an apocarrier exists in the cytoplasmic membrane, and the stable complex of this protein with the binding protein forms the molecular apparatus for solute translocation. The apocarrier is defined here as (1) having an interaction site for the binding protein and no binding site for the substrate, and (2) the complex is stable and no dissociation of the complex takes place during translocation of the bound substrate.

Thus the binding protein determines the specificity and affinity of the reaction and is essential, i.e., absolutely necessary. When no binding protein is present, the apocarrier may promote nonspecific diffusion of a solute across the membrane, since it has no binding site.

Model B represents an idea in which the binding protein is mandatory. A pseudocarrier is unique in the following points. (1) It can form a transitory complex with the binding protein-substrate complex during the translocation reaction. (2) It interacts only with the complex and not with the free binding protein.

Accordingly, in Model B, a conformational change of the binding protein upon binding of the substrate is required in order to produce an interaction site for the pseudocarrier. Thus the pseudocarrier in this category may be a unique enzyme in the sense that it recognizes the substrate to be translocated only through the interaction of a binding protein. Obviously, the specificity and affinity of solute translocation via this apparatus are determined by the mandatory effect of the binding protein. Although periplasmic binding proteins were found to undergo some conformational changes upon binding of substrates in vitro [118, 193, 216-218], whether these changes are responsible for the formation of an interaction site has not been investigated.

In the absence of a binding protein, the pseudocarrier itself may bind free substrate with very low affinity and broad specificity.

Model C shows a typical carrier function in combination with a binding protein in the periplasmic space. The carrier is defined as mentioned in Sec. II.C and translocates substrate by its own function. The binding protein resides in the periplasmic space and affects the carrier activity due to its inherent retention ability as previously discussed. Therefore, interaction of a carrier molecule with the binding protein is not necessarily required, and stereospecific stimulation of the initial rate of translocation by the presence of a binding protein may be obligatory or additional to the carrier

---

(donating) forms a stable complex with an apocarrier, while in a Model B binding protein first undergoes a conformation change upon binding of substrate to produce an interaction site, and the resulting complex is accessible to a pseudocarrier to form a transitory, ternary complex. In Model C, a substrate is retained in dynamic equilibrium in a region where a binding protein exists. (For further details, see text.)

activity. Efforts to demonstrate the binding of periplasmic binding proteins to membranes have been made by several laboratories [17, 90, 92], but the results so far obtained are negative.

The postulated role of the binding protein in Model C suggests that a group of binding proteins is physiologically required in the periplasm, a pseudo milieu intérieur, in order to conserve the substrate concentration from the surrounding medium [42, 199]. The repression control for the synthesis of various amino acid binding proteins is consistent with this idea.

Obviously, evidence presented by Ames and Spudich [168] is applicable to Models A and B, most preferably to Model B, although they prefer a model like Model A. On the other hand, the results obtained by Anraku and his coworkers [85] and by Halpern and his collaborators [129] appear to support Model C rather than the others. To prove whether these models are physiologically valid, we need more biochemical and genetic studies on the respective transport proteins. Some relevant arguments on the molecular models for solute translocation have been given by several investigators [219-221].

Finally, it should be mentioned that the lactose transport of E. coli is a unique example of Model C in which the binding protein may not be required physiologically. E. coli is able to ferment lactose, and the rate-limiting step in the utilization is its translocation across the membrane. The translocated substrate is hydrolyzed rapidly to form glucose and galactose by β-galactosidase. E. coli is an organism which has no biosynthetic ability to form lactose. Therefore, it can be said that the physiological role of the lactose operon is to perform unidirectional catalysis of lactose translocation and hydrolysis, and that accumulation of the intact lactose molecule against a concentration gradient is not an essential part of the transport apparatus. The fact that the lactose operon is under the control of induction and catabolite repression is consistent with its physiological role mentioned above.

VI. CONCLUSION

The major contribution of this chapter is to put forth a survey and evaluation of the present status of studies on the active transport of amino acids in bacteria, with special reference to evidence showing the existence and properties of membrane bound transport carriers. The evidence so far obtained is insufficient and preliminary. However, rapid progress in studies with membrane vesicles and recent development in genetics enable us to examine the carrier activity directly. Moreover, the highly exciting evidence that bacterial transport activities for alanine and proline are able to be reconstituted in liposomes with a protein fraction extracted from the membranes and phospholipids seems to invite a hopeful future for the study of transport.

I have discussed the definition and function of a carrier in the active transport. When the periplasmic binding proteins are essential or mandatory for transport activity, functional membrane proteins such as apocarriers and pseudocarriers active for translocation of a solute, may well have unique properties, as shown in Fig. 5-2. I did not mention the structure of these membrane proteins and the mechanism of energy coupling because relevant evidence has not been available. However, it is interesting to assume that these membrane proteins should exist in the membrane in aggregated form rather than as a single polypeptide monomer. The validity of these models for the molecular apparatus of active transport should be examined biochemically using reconstituted membrane vesicles.

As pointed out repeatedly, the physiological role of periplasmic binding proteins appears to be diverse. They function in active transport as well as in chemotaxis, possibly due to their ability to retain a substrate in a solution in which they are present. This property may not be unique to the homeostatic regulation of organic solutes in the cellular pool of such a gram-negative bacteria as E. coli. This notion implies that the periplasmic space containing a group of binding proteins and hydrolases may be an important biosphere outside the cytoplasmic membrane.

I have pointed out several contradictions and discrepancies presently appearing in the literature. These should be corrected. It is desirable that great care should be taken in determining transport activities for amino acids in membrane vesicles prepared from different strains. Particular caution is also required for osmotic shock and evaluation of heterogeneous transport kinetics.

Finally, I did not mention energy coupling to carrier molecule in the process of active transport. The mechanism of energy coupling has been extensively studied in the past several years and is one of the major topics of this book. The thermodynamic aspect of energy coupling in active transport is now accepted widely, and several models showing uniports as well as proton symports and antiports are proposed. In conclusion, studies on the molecular mechanisms of energy coupling appear to be approaching a new era which will undoubtedly begin with a further biochemical knowledge of membrane bound transport proteins such as the carrier, pseudocarrier, and apocarrier.

ACKNOWLEDGMENTS

I am grateful to Drs. H. Amanuma and I. Yamato for collaboration and helpful comments and to Dr. M. Futai for critical reading of the manuscript. I wish to thank Dr. U. Mizoguchi for English correction of the manuscript. The original work carried out in the laboratory of the author, discussed in this article, was supported by grants (C758034, C858069,

048095, and 148319) from the Ministry of Education, Science, and Culture of Japan and the Naito Science Foundation, Japan.

REFERENCES

1. G. N. Cohen and J. Monod, Bacteriol. Rev., 21: 169 (1957).
2. A. Kepes and G. N. Cohen, in The Bacteria, Vol. 4, (I. C. Gunsalus and R. Y. Stanier, eds.), Academic, New York, 1962, p. 179.
3. C. F. Fox and E. P. Kennedy, Proc. Natl. Acad. Sci. USA, 54: 891 (1965).
4. W. Kundig, S. Ghosh, and S. Roseman, Proc. Natl. Acad. Sci. USA, 52: 1067 (1964).
5. W. Kundig, F. D. Kundig, B. Anderson, and S. Roseman, J. Biol. Chem., 241: 3243 (1966).
6. H. C. Neu and L. A. Heppel, J. Biol. Chem., 240: 3685 (1965).
7. Y. Anraku, J. Biol. Chem., 242: 793 (1967).
8. Y. Anraku and L. A. Heppel, J. Biol. Chem., 242: 2561 (1967).
9. H. R. Kaback and E. R. Stadtman, Proc. Natl. Acad. Sci. USA, 55: 920 (1966).
10. P. Mitchell, Nature (London), 191: 144 (1961).
11. P. Mitchell, Biochem. Soc. Symp., 22: 142 (1962).
12. H. H. Winkler and T. H. Wilson, J. Biol. Chem., 241: 2200 (1966).
13. E. C. C. Lin, in Structure and Function of Biological Membranes, (L. I. Rothfield, ed.), Academic, New York, 1971, p. 285.
14. H. R. Kaback, Biochim. Biophys. Acta, 265: 367 (1972).
15. H. R. Kaback, Science, 186: 882 (1974).
16. D. L. Oxender, Ann. Rev. Biochem., 41: 777 (1972).
17. B. P. Rosen and L. A. Heppel, in Bacterial Membranes and Walls (L. Leive, ed.), Dekker, New York, 1973, p. 209.
18. Y. S. Halpern, Ann. Rev. Genet., 8: 103 (1974).
19. W. Boos, Ann. Rev. Biochem., 43: 123 (1974).
20. D. L. Oxender and S. C. Quay, in Methods in Membrane Biology, Vol. 6, (E. D. Korn, ed.), Plenum, New York, 1976, p. 183.
21. M. Doudoroff, W. Z. Hassid, E. W. Putnam, A. L. Potter, and J. Lederberg, J. Biol. Chem., 179: 921 (1949).
22. H. V. Rickenberg, G. N. Cohen, G. Buttin, and J. Monod, Ann. Inst. Pasteur, 91: 829 (1956).
23. G. N. Cohen and H. V. Rickenberg, Ann. Inst. Pasteur, 91: 693 (1956).
24. H. N. Christensen, Adv. Protein Chem., 15: 234 (1960).
25. A. Kleinzeller and A. Kotyk (eds.), Membrane Transport and Metabolism, Academic, New York, 1961.
26. A. Kepes, Biochim. Biophys. Acta, 40: 70 (1960).
27. A. L. Koch, Biochim. Biophys. Acta, 79: 177 (1964).

28. T. Rosenberg and W. Wilbrandt, J. Gen. Physiol., 41: 289 (1957).
29. W. F. Widdas, J. Physiol., 118: 23 (1952).
30. W. Wilbrandt and T. Rosenberg, Pharmacol. Rev., 13: 109 (1961).
31. E. P. Kennedy, in The Lactose Operon (J. R. Beckwith and D. Zipser, eds.), Cold Spring Harbor Laboratory, Cold Spring Harbor, New York, 1970, p. 49.
32. N. G. Nossal and L. A. Heppel, J. Biol. Chem., 241: 3055 (1966).
33. H. C. Neu and G. Chou, J. Bacteriol., 94: 1934 (1967).
34. L. A. Heppel, Science, 156: 1451 (1967).
35. Y. Anraku, J. Biol. Chem., 243: 3128 (1968).
36. A. B. Pardee, J. Gen. Physiol., 54(supple.): 279 (1969).
37. Y. Anraku, in Protein, Nucleic Acid, Enzyme, 14: 627 (1969) [Kyoritsu Publisher, Tokyo, (Japanese)].
38. L. A. Heppel, in Structure and Function of Biological Membranes (L. I. Rothfield, ed.),Vol. 6, Academic, New York, 1971, p. 223.
39. D. L. Oxender, in Metabolic Pathways, Vol. 6 (L. E. Hokin, ed.), Academic, New York, 1972, p. 133.
40. H. R. Kaback, Ann. Rev. Biochem., 39: 561 (1970).
41. T. J. Silhavy, S. Szmelcman, W. Boos, and M. Schwartz, Proc. Natl. Acad. Sci. USA, 72: 2120 (1975).
42. H. Amanuma, J. Itol, and Y. Anraku, J. Biochem., 79: 1167 (1976).
43. H. Hirata, N. Sone, M. Yoshida, and Y. Kagawa, Biochem. Biophys. Res. Commun., 69: 665 (1976).
44. H. Amanuma, K. Motojima, A. Yamaguchi, and Y. Anraku, Biochem. Biophys. Res. Commun., 74: 366 (1977).
45. H. R. Kaback, in Methods in Enzymology, Vol. 22, (W. B. Jakoby, ed.), Academic, New York, 1971 p. 99.
46. F. M. Harold, Bacteriol. Rev., 36: 172 (1972).
47. F. M. Harold, in Current Topics in Bioenerg., Vol. 6, Academic, New York, 1977, p. 83.
48. R. D. Simoni and F. W. Postma, Ann. Rev. Biochem., 44: 523 (1975).
49. E. M. Barnes, Jr. and H. R. Kaback, Proc. Natl. Acad. Sci. USA, 66: 1190 (1970).
50. W. N. Konings, E. M. Barnes, Jr., and H. R. Kaback, J. Biol. Chem., 246: 5857 (1971).
51. E. P. Kennedy, M. K. Rumley, and G. B. Armstrong, J. Biol. Chem., 249: 33 (1974).
52. J. P. Reeves, E. Shechter, R. Weil, and H. R. Kaback, Proc. Natl. Acad. Sci. USA, 70: 2722 (1973).
53. S. Schuldiner, G. K. Kerwar, R. Weil, and H. R. Kaback, J. Biol. Chem., 250: 1361 (1975).
54. I. C. West and P. Mitchell, J. Bioenerg., 3: 445 (1972).
55. I. C. West and P. Mitchell, Biochem. J., 132: 587 (1973).
56. S. Schuldiner, G. Rudnick, R. Weil, and H. R. Kaback, TIBS, 1: 41 (1976).

57. A. Kotyk, and K. Janáček, Cell Membrane Transport, Principles and Techniques, Plenum, New York, 1970.
58. Z. Böszörményi, E. Cseh, G. Gárdos, and P. Kertai, Transport Processes in Living Organisms, Akadémiai Kiadó, Budapest, 1972.
59. Y. Anraku, in Modern Biological Sciences, Vol. 5 (R. Sato and Y. Nishizuka, eds.), Iwanami Shoten, Tokyo, 1975, p. 17 [Japanese].
60. W. K. Maas and B. D. Davis, J. Bacteriol., 60: 733 (1950).
61. J. H. Schwartz, W. K. Maas, and E. J.Simon, Biochim. Biophys. Acta, 32: 582 (1959).
62. M. Lubin, D. H. Kessel, A. Budreau, and J. D. Gross, Biochim. Biophys. Acta, 42: 535 (1960).
63. D. Kessel and M. Lubin, Biochemistry, 4: 561 (1965).
64. R. Curtiss, L. J. Charmella, III, C. M. Berg, and P. E. Harris, J. Bacteriol., 90: 1238 (1965).
65. R. P. B. Russel, J. Bacteriol., 111: 622 (1972).
66. R. J. Wargel, C. A. Shadur, and F. C. Neuhaus, J. Bacteriol., 105: 1028 (1971).
67. S. D. Cosloy, J. Bacteriol., 114: 679 (1973).
68. J. C. Robbins and D. L. Oxender, J. Bacteriol., 116: 12 (1973).
69. J. R. Piperno and D. L. Oxender, J. Biol. Chem., 243: 5914 (1968).
70. R. J. Wargel, C. A. Shadur, and F. C. Neuhaus, J. Bacteriol., 103, 778 (1970).
71. B. A. Templeton and M. A. Savagean, J. Bacteriol., 117: 1002 (1974).
72. I. Yamato, Y. Anraku, and M. Ohki, Seikagaku, 48: 768 [Abstract, Japanese]. (1976).
73. M. E. Smulson, M. Rabinowitz, and T. R. Breitman, J. Bacteriol., 94: 1890 (1967).
74. M. Rahmanian, D. R. Claus, and D. L. Oxender, J. Bacteriol., 116 : 1258 (1973).
75. Y. Anraku, T. Naraki, and S. Kanzaki, J. Biochem., 73: 1149 (1973).
76. J. Guardiola, M. De Felice, T. Klopotowski, and M. Iaccarino, J. Bacteriol., 117: 382 (1974).
77. J. M. Wood, J. Biol. Chem., 250: 4477 (1975).
78. Y. Anraku, J. Biol. Chem., 243: 3116 (1968).
79. Y. Anraku, J. Biol. Chem., 243: 3123 (1968).
80. W. R. Penrose, G. E. Nichoalds, J. R. Piperno, and D. L. Oxender, J. Biol. Chem., 243: 5921 (1968).
81. W. R. Penrose, R. Zand, and D. L. Oxender, J. Biol. Chem., 245: 1423 (1970).
82. C. E. Furlong and J. H. Weiner, Biochem. Biophys. Res. Commun., 38: 1076 (1970).
83. H. Amanuma and Y. Anraku, J. Biochem., 76: 1165 (1974).
84. H. Kobayashi, E. Kin, and Y. Anraku, J. Biochem., 76: 251 (1974).

85. I. Yamato, H. Amanuma, and Y. Anraku, Abst. Japan Bioenerg. Group, 1: 19 (1975).

86. J. Guardiola and M. Iaccarino, J. Bacteriol., 108: 1034 (1971).

87. J. Guardiola, M. De Felice, T. Klopotowski, and M. Iaccarino, J. Bacteriol., 117: 393 (1974).

88. J. J. Anderson, S. C. Quay, and D. L. Oxender, J. Bacteriol., 126: 80 (1976).

89. F. J. Lombardi and H. R. Kaback, J. Biol. Chem., 247: 7844 (1972).

90. I. Yamato and Y. Anraku, J. Biochem., 81: 1517 (1977).

91. S. Kanzaki, and Y. Anraku, J. Biochem., 70: 215 (1971).

92. Y. Anraku, H. Kobayashi, H. Amanuma, and A. Yamaguchi, J. Biochem., 74: 1249 (1973).

93. M. Rahmanian, and D. L. Oxender, J. Supramol. Struct., 1: 55 (1972).

94. S. C. Quay, W. L. Kline, and D. L. Oxender, Proc. Natl. Acad. Sci. USA, 72: 3921 (1975).

95. K. D. Brown, J. Bacteriol., 104: 177 (1970).

96. K. D. Brown, J. Bacteriol., 106: 70 (1971).

97. J. R. Guest, J. Gen. Microbiol., 80: 523 (1974).

98. J. Kuhn, and R. L. Somerville, Proc. Natl. Acad. Sci. USA, 68: 2482 (1971).

99. R. J. Kadner, J. Bacteriol., 117: 232 (1974).

100. R. J. Kadner and W. J. Watson, J. Bacteriol., 119: 401 (1974).

101. S. Cooper, J. Bacteriol., 92: 328 (1966).

102. A. Morikawa, H. Suzuki, and Y. Anraku, J. Biochem., 75: 229 (1974).

103. S. Awazu, H. Amanuma, A. Morikawa, and Y. Anraku, J. Biochem., 78: 1047 (1975).

104. D. Kessel and M. Lubin, Biochim. Biophys. Acta, 57: 32 (1962).

105. H. Tristram and S. Neale, J. Gen. Microbiol., 50: 121 (1968).

106. I. Rowland and H. Tristram, J. Bacteriol., 123: 871 (1974).

107. W. K. Maas, Fed. Proc., 24: 1239 (1965).

108. W. K. Maas, Mol. Gen. Genet., 119: 1 (1972).

109. T. F. R. Celis, H. J. Rosenfeld, W. K. Maas, J. Bacteriol., 116: 619 (1973).

110. B. P. Rosen, J. Biol. Chem., 246: 3653 (1971).

111. B. P. Rosen, J. Biol. Chem., 248: 1211 (1973).

112. B. P. Rosen, J. Bacteriol., 116: 627 (1973).

113. L. Leive and B. D. Davis, J. Biol. Chem., 240: 4362 (1965).

114. L. Leive and B. D. Davis, J. Biol. Chem., 240: 4370 (1965).

115. E. A. Berger and L. A. Heppel, J. Biol. Chem., 247: 7684 (1972).

116. R. C. Willis and C. A. Woolfolk, J. Bacteriol., 123: 937 (1975).

117. J. H. Weiner, C. E. Furlong, and L. A. Heppel, Arch. Biochem. Biophys., 142: 715 (1971).

216   Y. ANRAKU

118. J. H. Weiner and L. A. Heppel, J. Biol. Chem., 246: 6933 (1971).
119. R. C. Willis, K. K. Iwata, and C. E. Furlong, J. Bacteriol., 122: 1032 (1975).
120. W. W. Kay and H. L. Kornberg, FEBS Lett., 3: 93 (1969).
121. W. W. Kay and H. L. Kornberg, Eur. J. Biochem., 18: 274 (1971).
122. W. W. Kay, J. Biol. Chem., 246: 7373 (1971).
123. T. C. Y. Lo, M. K. Rayman, and B. D. Sanwal, J. Biol. Chem., 247: 6323 (1972).
124. L. Frank and I. Hopkins, J. Bacteriol., 100: 329 (1969).
125. K. M. Miner and L. Frank, J. Bacteriol., 117: 1093 (1974).
126. R. C. Willis and C. E. Furlong, J. Biol. Chem., 250: 2581 (1975).
127. R. C. Willis and C. E. Furlong, J. Biol. Chem., 250: 2574 (1975).
128. S. Kahane, M. Marcus, E. Metzer, and Y. S. Halpern, J. Bacteriol., 125: 762 (1976).
129. S. Kahane, M. Marcus, E. Metzer, and Y. S. Halpern, J. Bacteriol., 125: 770 (1976).
130. Y. S. Halpern and H. E. Umbarger, J. Gen. Microbiol., 26: 175 (1961).
131. Y. S. Halpern and M. Lupo, J. Bacteriol., 90: 1288 (1965).
132. Y. S. Halpern and A. Even-Shoshan, J. Bacteriol., 93: 1009 (1967).
133. M. Marcus and Y. S. Halpern, J. Bacteriol., 93: 1409 (1967).
134. M. Marcus and Y. S. Halpern, J. Bacteriol., 97: 1118 (1969).
135. H. Barash and Y. S. Halpern, Biochem. Biophys. Res. Commun., 45: 681 (1971).
136. R. F. Leach and E. E. Snell, J. Biol. Chem., 235: 3523 (1960).
137. J. Mora and E. E. Snell, Biochemistry, 2: 136 (1963).
138. S. S. Ashgar, E. Levin, and F. M. Harold, J. Biol. Chem., 248: 5225 (1973).
139. R. H. Reitz, H. D. Slade, and F. C. Neuhaus, Biochemistry, 6: 2561 (1967).
140. R. E. Marquis and P. Gerhardt, J. Biol. Chem., 239: 3361 (1964).
141. K. Kiritani, J. Bacteriol., 120: 1093 (1974).
142. G. M. Thorne and L. M. Corwin, J. Bacteriol., 110: 784 (1972).
143. W. W. Kay and A. F. Gronlund, J. Bacteriol., 105: 1039 (1971).
144. H. Rosenfeld and P. Feigelson, J. Bacteriol., 97: 705 (1969).
145. G. Guroff, K. Bromwell, and A. Abramowitz, Arch. Biochem. Biophys., 131: 543 (1969).
146. G. Guroff and K. Bromwell, Arch. Biochem. Biophys., 137: 379 (1970).
147. S. M. D'Ambrosio, G. I. Glover, S. O. Nelson, and R. A. Jensen, J. Bacteriol., 115: 673 (1973).
148. G. I. Glover, S. M. D'Ambrosio, and R. A. Jensen, Proc. Natl. Acad. Sci. USA, 72: 814 (1975).
149. P. D. Ayling and E. S. Bridgeland, J. Gen. Microbiol., 73: 127 (1972).

150. W. W. Kay and A. F. Gronlund, Biochim. Biophys. Acta, 193: 444 (1969).
151. W. W. Kay and A. F. Gronlund, J. Bacteriol., 98: 116 (1969).
152. R. M. Gryder and E. Adams, J. Bacteriol., 101: 948 (1970).
153. D. L. Miller and V. W. Rodwell, J. Biol. Chem., 246: 1765 (1971).
154. C. L. Fan, D. L. Miller, and V. W. Rodwell, J. Biol. Chem., 247: 2283 (1972).
155. J. T. Holden, J. N. A. van Balgooy, and J. S. Kittredge, J. Bacteriol., 96: 950 (1968).
156. K. G. Reid, N. M. Utech, and J. T. Holden, J. Biol. Chem., 245: 5261 (1970).
157. N. M. Utech, K. G. Reid, and J. T. Holden, J. Biol. Chem., 245: 5273 (1970).
158. K. Yabu, Biochim. Biophys. Acta, 135: 181 (1967).
159. K. Yabu, Jap. J. Microbiol., 15: 449 (1971).
160. J. Stevenson, Biochem. J., 99: 257 (1966).
161. G. Sprott and R. A. MacLeod, Biochem. Biophys. Res. Commun., 47: 838 (1972).
162. G. F-L. Ames, Arch. Biochem. Biophys., 104: 1 (1964).
163. G. F-L. Ames and J. H. Roth, J. Bacteriol., 96: 1742 (1968).
164. G. F-L. Ames and J. Lever, Proc. Natl. Acad. Sci. USA, 66: 1096 (1970).
165. G. F-L. Ames and J. Lever, J. Biol. Chem., 247: 4309 (1972).
166. S. G. Kustu and G. F-L. Ames, J. Bacteriol., 116:107 (1973).
167. S. G. Kustu and G. F-L. Ames, J. Biol. Chem., 249: 6976 (1974).
168. G. F-L. Ames and E. N. Spudich, Proc. Natl. Acad. Sci. USA, 73: 1877 (1976).
169. B. P. Rosen and F. D. Vasington, Fed. Proc., 29: 342 (1970).
170. B. P. Rosen and F. D. Vasington, J. Biol. Chem., 246: 5351 (1971).
171. I. Yamato, Y. Anraku, and K. Hirosawa, J. Biochem., 77: 705 (1975).
172. M. Futai, Y. Tanaka, I. Yamato, H. Kimura, and Y. Anraku, Abst. Japan Bioenerg. Group, 1: 22 (1975).
173. I. Yamato, M. Futai, Y. Anraku, and Y. Nonomura, J. Biochem., 83: 117 (1978).
174. M. Kasahara and Y. Anraku, J. Biochem., 76: 977 (1974).
175. G. S. Cox, H. R. Kaback, and H. Weissbach, Arch. Biochem. Biophys., 161: 610 (1974).
176. W. N. Konings, J. Boonstra, and W. deVries, J. Bacteriol., 122: 245 (1975).
177. R. E. MacDonald and J. K. Lanyi, Biochemistry, 14: 2882 (1975).
178. J. K. Lanyi, V. Yearwood-Drayton, and R. E. MacDonald, Biochemistry, 15: 1595 (1976).
179. J. K. Lanyi, R. Renthal, and R. E. MacDonald, Biochemistry, 15: 1603 (1976).

180. H. Hirata, N. Sone, M. Yoshida, and Y. Kagawa, J. Biochem., 79: 1157 (1976).
181. W. N. Konings and E. Freese, J. Biol. Chem., 247: 2408 (1972).
182. S. A. Short, D. C. White, and H. R. Kaback, J. Biol. Chem., 247: 298 (1972).
183. S. A. Short, D. C. White, and H. R. Kaback, J. Biol. Chem., 247: 7452 (1972).
184. A. B. Pardee and L. S. Prestidge, Proc. Natl. Acad. Sci. USA, 55: 189 (1966).
185. A. B. Pardee, L. S. Prestidge, M. B. Whipple, and J. Dreyfuss, J. Biol. Chem., 241: 3962 (1966).
186. J. R. Piperno and D. L. Oxender, J. Biol. Chem., 241: 5732 (1966).
187. A. B. Pardee and K. Watanabe, J. Bacteriol., 96: 1049 (1968).
188. P. K. Nakane, G. E. Nichoalds, and D. L. Oxender, Science, 161: 182 (1968).
189. B. K. Wetzel, S. S. Spicer, H. F. Dvoraki, and L. A. Heppel, J. Bacteriol., 104: 529 (1970).
190. E. A. Berger and L. A. Heppel, J. Biol. Chem., 249: 7747 (1974).
191. G. Dietz, Y. Anraku, and L. A. Heppel, Fed. Proc., 27: 831 (1968).
192. J. L. Cowell, J. Bacteriol., 120: 139 (1974).
193. V. K. Antonov, S. L. Alexandrov, and T. I. Vorotyntseva, in Advances in Enzyme Regulation, Vol. 14 (G. Weber, ed.), Pergamon, New York, 1975, p. 269.
194. W. Boos, A. S. Gordon, R. E. Hall, and H. D. Price, J. Biol. Chem., 247: 917 (1972).
195. O. Kellermann and S. Szmelcman, Eur. J. Biochem., 47: 139 (1974).
196. G. Richarme and A. Kepes, Eur. J. Biochem., 45: 127 (1974).
197. T. J. Silhavy and W. Boos, Eur. J. Biochem., 54: 163 (1975).
198. G. L. Hazelbauer, Eur. J. Biochem., 60: 445 (1975).
199. Y. Anraku, in Proc. 1st Intersectional Congr., IAMS, Vol. 1, (T. Hasegawa, ed.), Science Council of Japan, 1975, p. 563.
200. J. L. Neal, J. Theor. Biol., 35: 113 (1972).
201. G. W. F. H. Borst-Pauwels, J. Theor. Biol., 40: 19 (1973).
202. A. S. Gordon, F. J. Lombardi, and H. R. Kaback, Proc. Natl. Acad. Sci. USA, 69: 358 (1972).
203. H. Amanuma, J. Itoh, and Y. Anraku, FEBS Lett., 78: 173 (1977).
204. S. Muraoka, and H. Terada, Biochim. Biophys. Acta., 275: 271 (1972).
205. H. Terada, and K. van Dam, Biochim. Biophys. Acta, 387: 507 (1975).
206. H. Kobayashi, M. Maeda, and Y. Anraku, J. Biochem., 81: 1071 (1977).
207. Y. Anraku, J. Biochem., 69: 243 (1971).
208. C. E. Furlong, C. Cirakoglu, R. C. Willis, and P. A. Santy, Anal. Biochem., 5: 297 (1973).

209. H. Hirata, K. Altendorf, and F. M. Harold, Proc. Natl. Acad. Sci. USA, 70: 1804 (1973).
210. H. Hirata, K. Altendorf, and F. M. Harold, J. Biol. Chem., 249: 2939 (1974).
211. Y. Anraku, J. Biochem., 70: 855 (1971).
212. H. Medveczky and H. Rosenberg, Biochim. Biophys. Acta,, 211: 158 (1970).
213. O. H. Wilson and J. T. Holden, J. Biol. Chem., 244: 2743 (1969).
214. W. Boos and M. O. Sarvas, Eur. J. Biochem., 13: 526 (1970).
215. W. Boos, J. Biol. Chem., 247: 5414 (1972).
216. W. Boos and A. S. Gordon, J. Biol. Chem., 246: 621 (1971).
217. E. B. McGowan, T. J. Silhavy, and W. Boos, Biochemistry, 13: 993 (1974).
218. S. Szmelcman, M. Schwartz, T. J. Silhavy, and W. Boos, Eur. J. Biochem., 65: 13 (1976).
219. B. Rotman and J. H. Ellis, J. Bacteriol., 111: 791 (1972).
220. W. Boos, in Curr. Topics in Membrane Transport, Academic, New York, 1974, p. 51.
221. S. J. Singer, Ann. Rev. Biochem., 43: 805 (1974).
222. B. J. Bachmann, K. B. Low, and A. L. Taylor, Bacteriol. Rev., 40: 116 (1976).

Chapter 6

TRANSPORT OF CATIONS AND ANIONS

Simon Silver

Department of Biology
Washington University
St. Louis, Missouri

I. INTRODUCTION 222
II. CATION TRANSPORT SYSTEMS 226
    A. Potassium Transport Systems 226
    B. Magnesium Transport Systems 239
    C. Iron Transport Systems 243
    D. Manganese Transport 253
    E. Zinc Transport 262
    F. Other Trace Cations 265
    G. Calcium Transport 265
    H. Sodium Transport 274
    I. Proton Transport 277
    J. Ammonium Transport 281
III. INORGANIC ANION TRANSPORT SYSTEMS 284
    A. Phosphate Transport Systems 284
    B. Sulfate Transport Systems 290
    C. Nitrate and Nitrite Transport 293
    D. Chloride Transport 295
    E. Molybdate Transport 299
    F. Other Anion Systems 300
IV. COMPARATIVE ASPECTS 301
    A. Uptake vs Extrusion Pumps 301
    B. Microrequirements vs Macrorequirements and the
       Possibility of Passive Diffusion 303
V. CONCLUSION 304
    RECENT DEVELOPMENTS 305
    REFERENCES 308

I. INTRODUCTION

Bacterial transport has developed into a mature science in the 20 years
since the classic review on the mechanism of lactose transport by Cohen
and Monod [39] opened the contemporary era of studies on how microbial
cells accumulate materials for growth and energy. This maturity is indicated by the definitive tone of many of the chapters of this book. The general properties of microbial sugar and amino acid transport systems are
well understood; there has been a small degree of progress toward isolating
and characterizing the molecular components of some; and finally, in the
last few years the mechanisms of coupling of energy metabolism to active
transport are becoming basically clear. In this picture of marvelous progress during the last two decades, cation and anion transport has been a
relative backwater. This chapter represents a first attempt to cover the
range of inorganic cations and anions for which microbial cells must have
transport systems, and to consider some of the common properties of these
systems. The unifying hypothesis is: For each and every inorganic min-
eral cation or anion that cells need for growth, separate highly specific
transport systems exist. The properties of these systems will fit the
cellular nutritional needs and will range from high capacities for major
intracellular cations such as K^+ and Mg^{2+} and anions such as PO_4^{3-} and SO_4^{2-},
to very low capacities (V_{max}) for trace nutrient systems. In addition to
the dozens (?) of anticipated accumulation systems for needed cellular ions,
microbial cells have evolved systems for excreting ions that are prevalent
in the normal environment but are either toxic or utilized for regulatory
or other purposes. Na^+ and Ca^{2+} egress systems appear to fall into this
class. I will predict the existence of some cation and anion transport systems
that have not yet been reported. This will be on the basis of known intracellular roles and the premise that essentially no ionic species can pass
into or out from cells by passive leakage. I shall surely make mistakes
and would welcome information from microbiologists who know facts that
have escaped my efforts and from researchers who will reap a rich harvest
by discovering and defining new and important systems.

Not all microbial cation and anion transport falls into the realm of the
unknown—or else it would be silly to attempt writing about the subject.
Some cations and anions are well enough understood to justify the rather
cursory coverage we intend. Others are just about ripe for full chapter
coverage in monographs of more restrictive scope than this. We have
just finished "first ever" reviews of microbial magnesium [109], calcium
[189], and manganese [191] transport systems. Bacterial potassium transport is overdue for detailed consideration. Studies of bacterial iron transport systems and their component extracellular iron chelates essentially
required monograph-length treatment when reviewed several years ago
[146]. With the thoroughly-studied systems, my sins will be superficiality,
and limiting myself only to transport in Escherichia coli and one or two

additional organisms. With the never-looked-for systems, my fault will be the opposite—that of stating a hypothesis in such terms that it may be taken as fact.

Before starting our ion by ion survey through the cations and anions, we need to introduce the concept of the transmembrane potential and how this affects cation and anion transport. The chemiosmotic coupling hypothesis of Mitchell [144] is considered in detail in Chap. 11 because the chemiosmotic potential is no longer a hypothesis but clearly a primary mechanism of coupling cellular energy metabolism to energy requiring active transport systems. Here we want to consider the proton as another type of cation that is transported into and out from bacterial cells, although a highly unusual cation that can be "created" and can disappear. We also need to consider the electrical component of the chemiosmotic potential, since this will often amount to 100 to 200 mV potential (internal negative) across a fully energized bacterial membrane. The requisite quantitative relationships are summarized by the Nernst equation [36]:

$$\Delta \psi = \frac{2.3\,RT}{ZF}\ \log\ \frac{[\text{cation}]_{in}}{[\text{cation}]_{out}} = \frac{2.3\,RT}{ZF}\ \log\ \frac{[\text{anion}]_{out}}{[\text{anion}]_{in}}$$

where $\Delta \psi$ is the potential difference across the membrane; RT/F has its usual meaning and is numerically equal to 59 mV at 25°C and to 61 mV at 35°C. Z is the charge on the ionic species; and $[\text{cation}]_{out}$, $[\text{cation}]_{in}$, $[\text{anion}]_{out}$, and $[\text{anion}]_{in}$ are the respective "activities" of particular ions inside or outside of the cell, and the "activities" range from the total concentration present for an osmotically free cation or anion down toward 0 if the ion is bound or otherwise immobilized [121,132]. The Nernst potential equation relates ion concentration ratio to membrane potential based on the assumption of free mobility of the cation or anion across the membrane. The mobility needs to be sufficient for Nernst equilibrium to be reached, since the equation describes the equilibrium ratios. Then, no additional energy beyond that required to create and maintain the transmembrane potential is needed to establish the Nernst equilibrium ion gradients. We will argue in separate sections below how Nernst equilibrium does not represent the situation in actively metabolizing cells, but that energy metabolism is required for movement of cations and anions in both directions, in and out, independent of the equilibrium predictions from the Nernst equation. The inapplicability of this equation can be demonstrated with a few examples, assuming for the moment that -120 mV is a typical membrane potential. Then the ratio of $[K^+]_{in}:[K^+]_{out}$ according to the equation is 100:1, which is reasonably close to what is found with (for example) E. coli growing in about 1 mM K^+ and with an internal concentration about 100 mM K^+. However, E. coli growing in the same medium with 10 or 100 mM K^+ will have essentially the same 100 mM K^+ internal concentration, considerably less than the Nernst equilibrium prediction.

E. coli growing or suspended in an external K^+ of near 10^{-6} M will retain
the 100 mM internal K^+ [116], and in this case, a Nernst potential equilibrium would require a membrane potential of -300 mV, considerably
greater than has been found. With the divalent cations, Ca^{2+}, Mg^{2+}, and
Mn^{2+}, the Nernst prediction for -120 mV membrane potential is for a
concentration ratio of 10^4 inside to 1 outside. Yet a normal magnesium
ratio is about 2 to 5 mM Mg^{2+} inside with less than 1 mM Mg^{2+} in the
medium, and the 10^4:1 ratio is essentially never seen. Magnesium uptake
does require energy (see below), yet active extrusion of Mg^{2+} against the
10^4:1 Nernst prediction is what would be required if there were free
permeability to Mg^{2+}. A similar situation is found with Mn^{2+}, for which
concentration ratios are often about 10^2:1 instead of 10^4:1; with regard to
Ca^{2+}, our best estimates are that internal Ca^{2+} is perhaps one-tenth that in
the external medium. The Nernst equilibrium just does not apply. Similar
arguments occur with regard to anions: $[Cl^-]_{in}$:$[Cl^-]_{out}$ with free permeability and -120 mV across the membrane should be 1:100, whereas most
experimental values are closer to 1:10. This would require active pumping
inward of Cl^- against the Nernst potential, whereas we believe that Cl^-
experiences carrier-mediated egress from the cells. Both SO_4^{2-} and
PO_4^{3-} are accumulated by the cells by energy-dependent processes, yet
the Nernst equation predictions are of an equilibrium with a $[SO_4^{2-}]_{in}$:
$[SO_4^{2-}]_{out}$ of 1:10^4 and a $[PO_4^{3-}]_{in}$:$[PO_4^{3-}]_{out}$ ratio of 1:10^6. Thus, the
Nernst equation equilibrium is a useful concept, but generally does not
describe the distribution of cations and anions across bacterial membranes.

Genetic methods of isolation and characterization of mutants altered
or defective in specific components of cation or anion transport systems
are the singular most important methodologies in studies of ion transport
as in studies of other types of transport discussed in the remaining chapters of this book. The emphasis on genetics places a premium on studies
in E. coli strain K12, since it is the most thoroughly studied organism from
a genetic point of view. Cation by cation (and anion by anion), we will
introduce the many known mutants and the understanding of transport that
they have provided. Before we begin, however, we provide in Fig. 6-1 a
genetic map of E. coli K12, listing only those genes affecting ion transport
and metabolism, and in Table 6-1 the mnemonics and brief descriptions of
the genes in Fig. 6-1. This is an ion-only subsection of the amazing list
of transport mutants complied by Slayman [200] about 5 years ago and is,
of course, updated to include all new mutations available at the time of
writing. Note that only about 50 of the 700 or so currently known genes of
E. coli have to do with ion metabolism. About 70 of the 650 genes listed in
ref. 7 concern transport of small molecules. With the 14 different ions
listed in the Contents of this chapter as a minimum estimate, and predicting
an average of two separate systems per ion and perhaps four genes per
system (simply loose estimates based on the already studied situations with
transport of iron, potassium, magnesium and phosphate), we will estimate

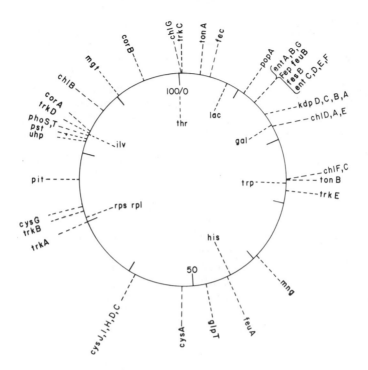

FIG. 6-1. Genetic map of genes determining ion transport and assimilation in E. coli. The standard 100-min genetic map from Ref. 7 is marked within the circle with representative well-placed marker genes and on the outside of the circle with genes concerning ion transport and assimilation as described in Table 6-1.

an anticipated 150 or so genes concerned with cation or anion transport, out of the estimated 2500 or so genes for which E. coli has sufficient DNA to code. This estimate that 6% of the total genes are involved with ion transport does not seem unreasonable, considering the importance for cell growth and survival of keeping out and bringing in specific ion species.

To use Racker's analogy, cells are like countries. Soon after their initial establishment by evolutionary processes, in this case by precellular evolution, they found it necessary to define specific borders (the cell membrane) in order to prevent accidental movement into or out from the cells. Cells also went heavily into the import-export trade and developed highly specific systems to govern movements across the borders, and the equivalent of tariffs in terms of energy that are proportional to the importance of the traded good and its relative supply in the environment and demand in the cell. With this overall rationale for the chapter on cation and anion transport, we will now turn to specific consideration of what is known about ionic transport systems in bacterial cells.

TABLE 6-1   Genes Involved in Ion Transport and Assimilation in E. coli[a]

Gene Symbol	Mnemonic and Phenotype
chlA-G	Chlorate resistance; formate-nitrate reductase
corA and corB	Cobalt resistance; missing or regulation of magnesium transport system I
cysA	Cysteine requirement; sulfate permease
cysJIHDC	Cysteine requirement; $SO_4^{2-} \longrightarrow H_2S$ reduction
entA-G	Enterochelin synthesis (different enzymatic steps in; see Fig. 6-6)
fec	Ferric (citrate) uptake
fep	Ferric enterochelin permease; iron transport
fesB	Enterochelin esterase component B (see Fig. 6-6)
feuA and feuB	Ferric enterochelin uptake; outer membrane components required for
glpT	L-α-Glycerophosphate permease; inorganic phosphate as alternative substrate
kdpA-D	Potassium dependent; high affinity potassium transport system
mgt	Magnesium transport system II
mng .	Manganese resistance; change in the magnesium transport system
phoS and phoT	Phosphate uptake; regulation of alkaline phosphatase
pit	Pi transport, arsenate sensitivity
popA	Porphyrin synthesis; ferrochelatase (see Fig. 6-6)
pst	Phosphate-specific transport; arsenate resistant
tonA	Resistance to phage T1 (one); see also Table 6-5
tonB	Resistance to phage T1; see also Table 6-5
trkA-E	Transport of K^+
uhpT	Uptake of hexose phosphate; alternative system for inorganic phosphate

[a]References to all genes are to be found in Bachmann et al. [7] except for feu (Refs. 85, 87, and Hancock, personal communication), cor and mgt [161], entG [236], pit and pst [10, 177, 230], and fec (Ref. 70; Gibson and Woodrow, personal communication).

II. CATION TRANSPORT SYSTEMS

A. Potassium Transport Systems

Potassium transport in bacterial cells has been more widely studied than that of any other inorganic nutrient (except possibly iron). Nevertheless, little in the way of mechanistic understanding has been accomplished. I will limit consideration here to studies with only three organisms: E. coli, Streptococcus faecalis, and Rhodopseudomonas capsulata. As stated about 4 years ago, consideration of results from other bacteria "are so scattered that their coherent discussion proved infeasible but leave the impression that the basic features of K^+ transport...are universal" [90]. The degree of scatter and impression of universality of basic processes continue today. It is less appropriate for our general survey of ion transport to attempt comprehensiveness than was the case in Harold and Altendorf's [90] more restricted review, which is, as far as I am aware, the sole attempt at a comprehensive review of mechanisms of monovalent cation transport. E. coli is considered because it is so thoroughly studied with regard to the genetic control of potassium transport; S. faecalis is included because Harold and coworkers have used this organism extensively in studies concerning energy coupling and cation exchange reactions; and R. capsulata has recently [108] been the organism with which the specificity of the K^+ transport system has been most thoroughly studied. This short list leaves our own studies of changes in K^+ transport rates during the sporulation-spore germination cycle of Bacillus subtilis [49, 52, 185, 196] beyond our purview.

1. Escherichia coli

Isolation of high K^+-requiring mutants in nine distinct genes has enabled Epstein and coworkers [58, 59, 170, 171] to analyze the properties of three distinct K^+ transport systems in E. coli K12. The kinetic parameters of these systems are given in Table 6-2. The Kdp system (so called because it is missing in K^+-dependent mutants in any of the four closely linked kdpA-D genes) is a high affinity trace K^+ scavenging system. With its K_m of 2 μM K^+, it enables wild-type E. coli to grow readily at submicromolar external K^+ concentrations. When functioning under those conditions, the Kdp system allows the maintenance of well over 100 mM internal K^+ in E. coli when the external K^+ has been reduced to perhaps 0.1 μM, for a concentration gradient of 10^6:1 inside to out. The Nernst potential equilibrium would require a potential of -360 mV (internal negative), nearly twice the values generally estimated for fully "energized" E. coli cells. Unless one wished to convert the membrane potential from its general role in the chemiosmotic hypothesis to a highly special role that varies with available potassium, one might conclude that K^+ is being transported actively with a requirement for direct energy input. K^+ is not responding electrophoretically to the Nernst potential. Further suggestive of active transport (in the strict sense, meaning against an electrochemical gradient) for the Kdp system comes from recent findings of an involvement of ATP availability in the functioning of the Kdp system (see below). These four

TABLE 6-2 Kinetic Parameters of Bacterial Potassium Transport Systems[a]

	Cation	K_m (mM)	V_{max} (μmol min^{-1}g^{-1})	Competitive Inhibitors (mM)
E. coli	K$^+$ TrkA	1.5	550	
	TrkD	0.50	40	Rb$^+$
	Kdp	0.002	150	
R. capsulata	K$^+$	0.20	8.0	$K_i = 0.56$ Rb$^+$ $K_i = 2.57$ Cs$^+$
	Rb$^+$	0.52	5.9	$K_i = 0.2$ K$^+$ $K_i = 2.6$ Cs$^+$
	Cs$^+$	3	2	

[a]From Refs. 108 and 171.

genes have attributes of an "operon" in E. coli, and indeed, the Kdp system is repressible by growth in high potassium. We can speculate as to whether the products of the four kdp genes are components of a multiprotein membrane transport complex or instead, enzymes that sequentially lead to synthesis of a small Kdp substance that functions in turn in transport. Rhoads et al. [171] favor the multicomponent system model. They suggest from unpublished experiments that the kdpD gene may be a regulatory gene (since revertants of kdpD are sometimes partially constitutive for synthesis of the Kdp system) and that the kdpB appears to code for an osmotic shock released periplasmic protein, as is thought to be characteristic of transport systems that are not linked to H$^+$ circulation (see Chaps. 10 and 11).

The TrkA K$^+$ transport system (missing in mutants in the trkA gene that affects transport of K$^+$) is the major potassium transport system of E. coli in that its maximum rate of functioning is almost four times that of the Kdp system. trkA Mutants of E. coli K12 could be isolated only after the Kdp system had been previously eliminated by mutation. However, since the TrkA system is synthesized constitutively, whereas the Kdp system is repressible, the TrkA system dominates the bacterial potassium economy at moderate or relatively high external potassium concentrations. The third (and minor with regard to rate) potassium transport system of E. coli K12 is missing in mutants in still another (trkD) gene. The TrkD system has a relatively low affinity for potassium as well as a relatively low transport rate (Table 6-2). It, like the TrkA system, is synthesized constitutively in high as well as in low potassium growth media.

Finally, triple mutants missing all three Kdp, TrkA, and TrkD systems can grow at concentrations of K^+ well above 10 mM. K^+ uptake by these triple mutants is also energy-dependent but does not follow Michaelis-Menten kinetics—at least there was no sign of saturation of rate with as high as 105 mM K^+ [171]. Two alternatives are possible to explain the growth of these bacteria: either there is still a fourth K^+ transport system (!), called TrkF by Epstein, although there are no mutants to provide a genetic basis for this hypothesis; or alternatively, K^+ can enter the cell nonspecifically, perhaps through other channels but with such low affinity that saturation is never observed.

The description of the three K^+ transport systems still leaves three additional classes of high-K^+ requiring mutants, those in the trkB, trkC, and trkE genes. trkB and trkC mutants are normal with regard to K^+ uptake but defective with regard to K^+ retention [171]. Such mutants, in contrast to wild-type cells, are unable to retain intracellular potassium in the absence of extracellular potassium. Furthermore, the rate of exchange of intracellular K^+ by these cells is generally some five times that found with wild-type cells. The accelerated K^+ loss causes an elevated growth requirement for external potassium so that uptake can balance loss. With the uptake mutants, the wild-type phenotype was dominant over the mutant nonfunctional form (allowing, for example, determination of four kdp genes by complementation analysis [58]). Similarly, the "leakiness" of the trkB and trkC strains appeared recessive to being K^+ "tight" [59]. The defect in a trkC mutant has been extensively studied (e.g., Refs. 81 and 165). While we do not know whether the trkB and trkC genes affect components of a system that functions normally in K^+ turnover or egress, or whether these genes affect a system that is normally responsible for Na^+ egress so that it no longer discriminates adequately against K^+, it is clear that the rate of K^+ loss by the trkC mutant is more rapid in high external Na^+ [81,165]. It is known that the K^+ leaky mutants are not leaky for other cations; in particular they show normal retention of both Mg^{2+} and Mn^{2+} (Silver et al., unpublished data). Furthermore, retention mutants have not been found among the classes of mutants requiring high levels of Mg^{2+} or Fe^{3+} for growth. Apparently similar K^+ retention mutants have been isolated in other E. coli strains and in B. subtilis [228]. Only a single mutation has been found in the last of the K^+ transport genes, trkE. The function of the trkE gene is the least understood of the nine. It may be a regulatory gene and appears to affect the functioning of uptake by the two low affinity K^+ transport systems TrkA and TrkD, since studies with kdp$^-$, trkE$^-$ strains gave variable but low rates of K^+ transport [171]. trkE does not appear to determine a separate fourth system.

The kinetic studies with the K^+ transport mutants were run under conditions of net K^+ accumulation following resuspension of cells depleted of

K^+ by exposure to the uncoupler dinitrophenol [12,171]. Slightly different
kinetics for K^+ transport were found under two alternative "step-up" conditions: suspension in fresh medium of stationary phase cells that had lost
much cellular K^+ during post-log phase exposure, and suspension of log
phase cells with steady state K^+ potassium levels in medium of higher
osmotic pressure. Early studies of K^+/K^+ exchange under conditions of no
net change of cellular K^+ [60,61] showed no effect of varying external K^+
concentration on exchange rate. More recent experiments of Rhoads and
Epstein (personal communication) demonstrate saturation of the K^+ exchange
function by the TrkA and TrkD systems, however. While complex regulatory
control of K^+ flux made conventional kinetic experiments of K^+/K^+ exchange
infeasible with E. coli K-12, K^+/K^+ exchange followed simple Michaelis-
Menten kinetics with both B. subtilis [49] and R. capsulata [108]. Internal
concentrations of K^+ are regulated in response to total osmotic strength of
the suspension medium, from 150 mM K^+ in 80 mosM medium to nearly
600 mM K^+ in 1,200 mosM medium [60,116]. This regulation occurs
essentially independently of the external K^+ available, as long as there is
K^+ available, internal K^+ rises less than a factor of 2 between 10^{-6} M
external K^+ and 10^{-3} M external K^+ in constant osmotic strength medium
[116]. Recent experiments have begun to define the regulation of K^+ flux
rates during net uptake [116]. With K^+-depleted cells exposed to extracellular K^+, an initially high rate of K^+ uptake occurred with essentially
no discernible efflux of K^+. After the cells had reached half of the
equilibrium K^+ level (5 min), the rate of influx declined while the rate
of K^+ efflux increased dramatically; and an equilibrium with equal influx
and efflux rates was obtained within 15 min. When net K^+ uptake was
induced instead by an osmotic step-up, the K^+ efflux rate remained constant
while a transient increase in influx was found, leading to the new internal
equilibrium with exchange rates (K^+/K^+ exchange) in 600 mosM medium with
sucrose the same as had been the case in 300 mosM medium before
addition of sucrose [116]. Influx and efflux appeared to be independently
regulated by "feedback" in response to osmotic pressure of the cell membrane against the cell wall. K^+/K^+ exchange required cellular energy [116],
although this is not theoretically necessary for a process that results in no
net work. Further studies on regulation of cellular K^+ in E. coli K12 will
add greatly to this understanding. The experiments just described were
under conditions where the TrkA system would be responsible for most K^+
uptake [116]. Regulatory interactions between the potassium transport
systems of E. coli were found in experiments with mutant strains. Although
uptake rates via the Kdp and TrkA systems were additive [171], the possession of a functioning Kdp system appeared to regulate and inactivate the
functioning of the TrkD system [171]. This result is now attributed to the
use of an unusual and unstable TrkD strain (Epstein, personal communication). K^+ transport was regulated during the growth of a single cell from
cell division to cell division: with synchronously dividing cells, the cellular

content of K^+ increased smoothly with cell growth [122,123], but the rate of K^+ transport per cell remained constant throughout most of the cell cycle and then doubled abruptly at about the time of cell division [122].

Studies of K^+ transport with subcellular isolated membranes from E. coli, and especially with membranes isolated from the transport mutants, adds another dimension to our understanding of K^+ transport but also (at the moment) adds an unresolved conflict. Bhattacharyya et al. [16] found that membrane vesicles prepared from E. coli cells could not accumulate K^+ although they retained numerous amino acid transport capacities [111], and even the capacity for active transport of the divalent cation Mn^{2+} [13]. The addition of valinomycin or similar potassium mobilizing ionophores restored energy-dependent K^+ transport to these vesicles, a result that has since been shown to be dependent on the transmembrane potential [3,111,126]. Two alternatives were possible: either valinomycin was dissolving into membrane lipid bilayer regions and transporting K^+ in response to the electrical potential or, alternatively, valinomycin was localized in the membranes in a more specific position (associated with specific proteins) and was replacing an ionophore-like component lost during the osmotic shock procedures of membrane preparation. We [16] favored the more specific role for valinomycin on the basis of experiments with membranes from K^+ transport mutants. Membranes from a kdp⁻ mutant strain responded to valinomycin much as wild-type membranes, suggesting that the valinomycin was replacing a product of the Kdp system. However, the membranes were prepared from cells grown in high K^+, conditions where we now know that the Kdp system is repressed. Mutants from a strain with both a kdp⁻ mutation and an additional trkA⁻ mutation responded very poorly to valinomycin, suggesting a residual defect in potassium transport in the membranes; and membranes from a trkB retention mutant showed normal valinomycin-dependent K^+ uptake but accelerated loss of K^+ consistent with the membranes retaining the defect in retention of the intact cells [16]. These experiments have been questioned by some [e.g. 89]; and Altendorf (personal communication, 1976) was unable to obtain reproducible activities or reproducible differences with vesicles prepared from the K^+ transport mutants.

Two results of Lombardi et al. [126], from their detailed study of the mechanism of valinomycin-induced Rb^+ uptake, are basically inconsistent with a model of free valinomycin diffusion across a lipid bilayer. First, the Rb^+ uptake rate was a saturable function of valinomycin concentration with a half-maximum uptake rate at 0.4 nmol valinomycin/mg membrane protein. [Harold (personal communication, 1976) would explain this away by saying that valinomycin is insoluble in water and precipitates at higher concentrations.] Adding valinomycin at levels more than five times the half-saturation concentration caused progressive decreases in the initial rate of Rb^+ uptake [126]. We would like to ascribe the initial stimulatory effect as being due to valinomycin binding to highly specific high-affinity membrane receptors leading to energy-dependent uptake, and the secondary inhibitory effect as being due to passive facilitated diffusion via valinomycin dissolved in bilayer membrane regions.

The second result suggestive of a specific site for valinomycin was the finding that no saturation of the rate of Rb^+ efflux was seen at intravesicular concentrations up to 100 mM Rb^+ [126] [accumulation stimulated by valino-mycin and efflux "triggered" by addition of the uncoupler m-chlorophenyl-carbonylcyanide hydrazone (CCCP)]; whereas the K_m for uptake with valin-omycin was 0.9 mM Rb^+. Passive facilitated diffusion should have shown similar half-saturation values in both directions. Furthermore, addition of the sulfhydryl reagents p-chloromercuribenzenesulfonate and N-ethylmale-imide inhibited valinomycin-stimulated Rb^+ uptake, presumedly by inhibiting energy metabolism, but did not cause efflux of Rb^+ accumulated against a concentration gradient [126] (see also Ref. 3) as might have been expected with a freely diffusible valinomycin shuttle. Whether valinomycin functions as part of a "physiological" K^+ transport system in these membranes is of importance with regard to two questions. It leads directly to the idea of "ionophores" as components of physiologically normal transport systems [79], and it is directly pertinent to questions of energy coupling to K^+ trans-port, as will be discussed next.

The last apsect of K^+ transport in E. coli to consider is the mechanism of coupling of metabolic energy to the process of K^+ uptake. I have already indicated that passive movement to Nernst potential equilibrium, even through highly K^+-specific channels, is probably not sufficient to account for the $10^6:1$ K^+ ratio sometimes found with wild-type E. coli under con-ditions of K^+ starvation. In addition, the sensitivity of K^+ transport to osmotic shock treatment [16,171] is suggestive of nonchemiosmotic coupling of energy to transport [12]. Rhoads and Epstein [170] recently studied energy coupling to K^+ transport with their well-defined K^+ transport mutants. The high affinity Kdp K^+ transport system had the (now almost standard) properties of directly ATP-coupled systems since it was shown to be (a) sensitive to osmotic shock, (b) highly sensitive to inhibition by arsenate, and (c) rather resistant to inhibition by the proton-conducting uncoupler dinitrophenol when in an uncA⁻ genetic background (that elimin-ates the turnover of ATP via the membrane ATPase missing in uncA⁻ strains). The other tested system, TrkA, gave more ambiguous results: (a) the TrkA-mediated potassium transport was fully as osmotic shock-resistant as was the proline transport system, the standard for the chemi-osmotic systems found in membrane vesicles from the cells [12,105,111]. Yet (b) trkA⁺ membrane vesicles were nonfunctional for potassium trans-port except on the addition of a potassium ionophore such as valinomycin [16], indicating the loss of some factor during osmotic lysis in membrane preparation. (c) K^+ transport via the TrkA system was highly sensitive to arsenate inhibition; but (d) TrkA transport was also sensitive to dini-trophenol in the uncA⁻ strain. These results suggest that the Kdp system is like the transport systems for glutamine and other substrates [12] that are directly dependent upon the availability of ATP in the cell, but not on the membrane potential. However, the TrkA system has properties of both types of transport systems and appears to require both ATP and a membrane potential. The sum of these results can be put together in a tentative working model: Since the Kdp system functions under K^+ starvation conditions and against concentration gradients beyond what the proton motive potential ($\Delta\Psi$ + ΔpH) will support, it must be coupled to another more direct source of metabolic energy. A special membrane ATPase may function in this role and utilize a periplasmic (kdpB?) protein to facilitate movement of the substrate to the membrane carrier. Indeed, other periplasmic binding proteins appear to function in this fashion rather than directly as membrane embedded carrier molecules (see Chap. 10). The TrkA system functions with a low affinity and generally under conditions where the intracellular K^+ is equal to or less than that expected from Nernst potential equilibrium. Electrogenic movement of K^+ down the membrane potential to equilibrium would bring in potassium but not provide for cellular control over internal potassium. If instead of electrogenic "uniport" movement of K^+ across the membrane, one hypothesizes K^+-H^+ symport, then the process would still be electrogenic but the mechanism would provide for an added control by separate regulation of the ΔpH component of the chemiosmotic potential. K^+-H^+ symport could also result in the 10^6:1 concentration ratio found under K^+ starvation conditions. An arsenate-sensitive involvement of a phosphorylated intermediate, perhaps the membrane carrier protein itself, might provide regulation over the rate of K^+ movement through the TrkA system. If we visualize the basic carrier molecule (a trkA protein?) as forming a highly K^+-specific channel through the membrane (Fig. 6-2) and if we imagine that the specificity of this channel is determined by placement of oxygen atoms lining the channel, as is the case with the model ionophore gramicidin A [218], then we can continue our speculation by introducing the possibility of "gating" of the channel by phosphorylation-dephosphorylation processes (an open channel in the phosphorylated form and closed in the other). With gramicidin A channels in artificial membranes, it has been proposed that the membrane potential opens and closes the "gate" by aligning the dipole moment of the peptide channel [217,218]. Here we propose that the potential serves only for "electrophoresis" of K^+ through open TrkA channels and that the gating is provided by phosphorylation. Since this model is already far beyond the experimental data supporting it, let me not try to justify its details but only to warn readers of its unsubstantial foundation. One point should be added for readers familiar with the mechanism of K^+ transport in mammalian cells, which is in fact rather different from what occurs in bacteria. There is no evidence whatsoever for a K^+-ATPase in bacterial membranes. Furthermore, despite initial reports to the contrary, it is now very clear that K^+ transport in bacterial cells is completely resistant to specific inhibitors such as ouabain that act on the animal cell Na^+-K^+-transport ATPase [90]. Na^+ efflux in bacteria differs from that in animal cells in that it is not directly coupled with K^+ uptake (see Sec. II.H).

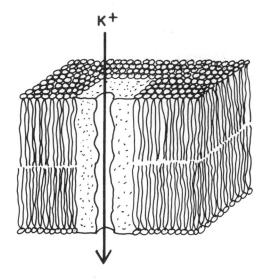

FIG. 2. Hypothetical model for a "channel" containing a membrane spanning transport carrier. A cut-away view through the carrier protein is shown. This model is no less hypothetical than the more conventional "permease" models that generally suggest substrate transport via rotation of the carrier in the membrane as an alternative to a through channel in a relatively fixed carrier.

2. Streptococcus faecalis

The transport of K^+ by S. faecalis has been extensively studied over the past decade by Harold and coworkers. A recent review [90] summarizes the results and compares S. faecalis K^+ transport with that of E. coli and other organisms. The basic mechanisms appear extraordinarily similar. There are some differences, however: S. faecalis has proven much more useful for studies of effects of ionophores and uncouplers on K^+ transport [93, 94, 96]. Although mutants defective in K^+ transport have been isolated in S. faecalis [91], the absence of a system for formal genetic studies in S. faecalis has limited the genetic approach.

Energy-dependent uptake of K^+ by S. faecalis appears to be in response to the membrane potential generated by electrogenic expulsion of protons via the membrane bound ATPase (which in S. faecalis is very similar to the energy-transduction reversible ATPase of the E. coli membrane [88-90]). Glycolysis provides ATP to generate a membrane potential approaching -200 mV, which can account for the $2000:1 \ K^+_{in}:K^+_{out}$ ratio achieved by the cells. This is consistent with the energetically secondary movement of K^+ in response to the potential gradient. No direct involvement of energy processes is required. Consistent with this interpretation were numerous studies with antibiotic compounds that mediated electrogenic K^+ movement (e.g., valinomycin), electrogenic H^+ movement [e.g., the uncouplers such as dinitrophenol, CCCP, and carbonyl cyanide p-trifluoromethoxyphenyl-hydrazone (FCCP)] and electroneutral K^+/Na^+ or K^+/H^+ exchange (e.g., nigericin). Valinomycin or other antibiotics that increased membrane permeability to K^+ did not inhibit net K^+ uptake, whereas uncouplers that discharged the membrane potential by accelerating electrogenic proton uptake, or nigericin which discharged the proton gradient by K^+/H^+ exchange, completely inhibited net potassium uptake. Surprisingly, removal of an energy source or the addition of uncouplers or other poisons that caused dissipation of the membrane potential of S. faecalis did not cause loss of preaccumulated K^+ from the cells, which would be the expected result for freely K^+ permeable membranes responding to the potential. Also unanticipated was the finding that K^+/K^+ exchange required an energized membrane and did not occur with starved cells. Since no net movement is involved in such "autologous" exchange, a membrane potential should not be required. Uncouplers (proton conductors) inhibited K^+/K^+ exchange, whereas dicyclohexylcarbodiimide (DCCD) and other agents that inhibit the membrane ATPase were without effect on K^+/K^+ exchange but completely inhibited net K^+ uptake [90]. Harold and Altendorf [90] were led by these results to the notion of a "gated" potassium carrier, strengthened by analogies with the potential-gated channel formed across artificial membranes by gramicidin A [217,218]. Whereas in the absence of a membrane potential, the K^+ gate apparently remains closed in S. faecalis, transport systems for sugars and amino acids remain "open" and can carry out facilitated diffusion of substrate and also exchange diffusion.

The kinetic parameters of K^+ transport in S. faecalis were not studied in as great detail as has been done with E. coli or R. capsulata (see below). $^{86}Rb^+$ exchange for cellular Rb^+ followed Michaelis-Menten kinetics with a K_m of 0.3 mM Rb^+, and all of the cellular Rb^+ (or K^+) appeared kinetically to be in a single "pool".

Three classes of mutants have been found in S. faecalis that require high K^+ for growth [90]. The first class, represented by the Tr_{K8}^- mutant (for transport of K^+ at pH 8), may be directly affected in a K^+ transport system. K^+ uptake in this strain was abnormally sensitive to inhibition by Na^+, which is a competitive inhibitor of K^+ transport in wild-type cells (K_i, 17 mM). With the Tr_{K8}^- strain, the inhibition of K^+ uptake by Na^+ was complex, showing both an increase in K_m indicative of a K_i for Na^+ of perhaps less than 10 mM and a decrease of the V_{max} for K^+ uptake by 50% at 50 mM Na^+ [91]. K^+ uptake and K^+/K^+ exchange were equally Na^+-sensitive. The presumptive binding site for Na^+ on the K^+ transport carrier is not related to the site used in Na^+/Na^+ exchange or Na^+/H^+ exchange (Sec. II.H), since neither of these processes were affected by the Tr_{K8}^- mutation [90,91]. The inhibition of growth of Tr_{K8}^- cells by Na^+ occurred at pH 7.5 or 8, but not at pH 6.0, and required as much as 300 mM Na^+, even with this Na^+-sensitive mutant. No defect in K^+ retention occurred in the mutant cells, but loss of $^{86}Rb^+$ from the mutant cells was relatively more rapid than from wild-type cells [91]. Studies with mutant class Tr_{K8}^- have not been extended in recent years, but the early results seem reasonably consistent with a change in the substrate recognition site for a basically K^+ uptake system with altered specificities for Rb^+ and for Na^+. Even in the mutant strain, Na^+ was not a substrate for the system.

Mutant class Cn_{K6}^- (for defective in concentration of K^+ at pH 6) represented by strains 325B (now lost) and 687A [96, 98], were first described as K^+ retention mutants that were particularly defective at acid pH [98]. Rb^+ uptake by Cn_{K6}^- mutants was normal, but Rb^+/Na^+ or K^+/Na^+ heterologous exchange was greatly accelerated [98]. Rb^+/Rb^+, Rb^+/K^+, and K^+/K^+ exchange were unaffected in this mutant strain. The Cn_{K6}^- defect was only discernible at acid pH, the opposite result from that with Tr_{K8}^- mutants that were K^+ transport defective only at alkaline pH. Optimum growth of Cn_{K6}^- mutants required only 1 to 2 mM Rb^+ at pH 8 but 100 mM Rb^+ at pH 6 [98]. In the absence of detailed comparisons, it is not possible to tell how similar the Cn_{K6}^- class mutants are to either the trkB or trkC mutants of E. coli [171]. There is some similarity, however, in that the Cn_{K6}^- mutants of S. faecalis showed accelerated loss of Rb^+ in the presence of Li^+ or Na^+ but not with Mg^{2+} or Tris as the sole extracellular cation [98]. With the three E. coli K^+ uptake systems removed by mutation, $kdp^-, trkA^-, trkD^-$ triple mutants showed more rapid loss of K^+ in the presence of Na^+ than with Tris as the extracellular cation (Ref. 171; Epstein, personal communication). Na^+ affected the loss of K^+ from a trkC mutant [81, 165]. Whereas initially S. faecalis Cn_{K6}^- mutants were considered defective in K^+ retention [98], subsequent experiments led to the conclusion that the primary block was in K^+/H^+ exchange [96] and later to the conclusion that the primary block was in H^+ extrusion [90, 100]. Cn_{K6}^- mutant cells preloaded with Na^+ were able to glycolyze glucose actively and utilize the energy in K^+/Na^+ net exchange. However, cells low in Na^+ but with high H^+ as well as high K^+ showed inhibition of glycolysis and neither lost internal K^+ nor internal protons [96]. Not only was K^+ accumulation defective in Cn_{K6}^- mutants, but that of phosphate and alanine was also affected [96]. With the evidence for a primary block in proton extrusion, we will defer further discussion of the K^+ retention mutants to Sec. II.K on H^+ transport.

The third class of high K^+-requiring mutant in S. faecalis has never been assigned a mnemonic and is represented by the mutant strain 7683 [90]. At alkaline pH, strain 7683 required 150 mM K^+ for most rapid growth; at pH 6 it grew at half wild-type rate with only 5 mM K^+ present [96]. This mutant is now thought to lack the normal Na^+/H^+ exchange system and only secondarily to be defective in K^+ retention [90]. If indirect, the effect on K^+ is still relatively specific since transport of phosphate and alanine is normal in strain 7683 [96]. Mutant 7683 was not defective in K^+ uptake and in fact showed somewhat more rapid K^+/K^+ exchange than wild-type cells. Mutant 7683 differed from the Tr_{K8}^- mutants in that Na^+ did not affect K^+ uptake rate. These cells were, however, severely defective in the ability to carry out net K^+ uptake in exchange for Na^+ [90]. Because the defect in strain 7683 is primarily in Na^+ egress, we will defer further consideration of this mutant strain to Sec. II.H on Na^+ transport.

Comparing the result of the use of mutants in S. faecalis with that in E. coli, it is extraordinary that from a similar beginning point, Harold [88-90] was led by mutants defective in H^+ and Na^+/H^+ movement to his elegant development of experimental approaches establishing Mitchell's [144] chemiosmotic coupling model for energy-dependent transport in S. faecalis, and then also in other bacteria. Epstein [58,59,170,171] was led by his mutants to the almost baroque dissection of the properties of three K^+ uptake systems in E. coli. The basic mechanisms appear similar if not precisely the same in the two organisms; the same range of mutants may eventually be found in both. However, the results from studies of K^+ metabolism with these two organisms have been complementary and not very similar during the last decade. A summary of current models for electrogenic proton and potassium transport and electroneutral Na^+/H^+ antiport is shown in Fig. 6-3.

3. Rhodopseudomonas capsulata

We introduce K^+ transport in R. capsulata here because the specificity and kinetic parameters of the K^+ transport system have been more thoroughly studied in this organism than in any other bacterium. Unfortunately, no K^+ transport mutants have been isolated in R. capsulata. A second reason to

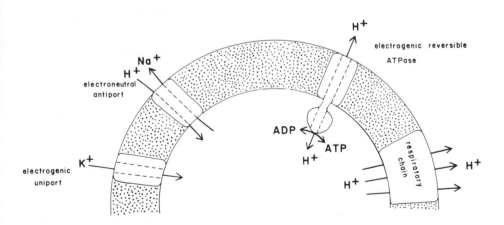

FIG. 6-3 Proposed monovalent cation carriers (modified from Ref. 90).

introduce R. capsulata is to add the dimension of changes in K^+ transport: first, changes during the usual growth cycle from lag phase to log phase to stationary phase growth, and second, changes in the properties of the K^+ transport system during developmental transitions between aerobic respiration-driven growth and anaerobic growth supported by photophosphorylation. Both anaerobic and aerobic R. capsulata behaved kinetically as if they contained a single K^+ transport system [108]. The kinetic parameters in the phototrophic bacteria were studied more thoroughly and are given in Table 6-2. R. capsulata cells had a K_m of 0.20 mM K^+ for uptake during photoheterotrophic growth (predominantly exchange kinetics) for an intermediate affinity between the trace Kdp and the major TrkA systems of E. coli. The K_m for K^+ for aerobic dark-grown cells was virtually identical to that found for phototrophically-grown cells. Rb^+ and Cs^+ were alternative but poorer substrates for the R. capsulata K^+ transport system. The affinity constant (K_m) for Rb^+ was three times that for K^+, and the K_m for Cs^+ was approximately 15 times that for K^+ (Table 6-2). The maximum rates (V_{max}) for K^+ and for Rb^+ uptake were not significantly different; however, that for Cs^+ uptake was three to four times lower. Jasper [108] determined these parameters both by use of the different radioisotopes $^{42}K^+$, $^{86}Rb^+$, and $^{137}Cs^+$, and also by determining the competitive inhibition constants (K_i values) for each monovalent cation against the radioisotopes of the others. Even a modest difference in the kinetic parameters of Rb^+ versus K^+ for transport can severely affect conclusions that can be drawn from experiments using the more convenient $^{86}Rb^+$ radioisotope as a K^+ analog. For example, with E. coli we have found that only about 10% of the equivalent radioactive counts of $^{86}Rb^+$ are accumulated during growth on a mixture of Rb^+ and K^+ (both below millimolar levels) when compared with the radioactive counts of $^{42}K^+$ found in the cells (unpublished data). Harold (personal communication) found a similar preference for K^+ over Rb^+ with S. faecalis, especially at alkaline pH. If the cells are exchanging intracellular K^+ for extracellular K^+ five times per generation, then only a 50% selection for K^+ per round of exchange would lead to nearly 10:1 accumulated $K^+:Rb^+$ during growth ($1.5^5 = 7.6$). This problem does not appear to have been adequately considered in long-term growth experiments using $^{86}Rb^+$.

Jasper [108] also demonstrated that Rb^+ could quantitatively replace K^+ as the monovalent intracellular cation in R. capsulata: cells growing anaerobically on standard (15 mM K^+) growth medium had an internal K^+ content corresponding to 110 mM K^+; when grown in 10 mM Rb^+ and no deliberately added K^+, the intracellular values were 116 mM Rb^+ (99.7%) and only 0.4 mM K^+ (0.35% of the total monovalent cations). Without special precautions, her growth medium contained 5 μM K^+ as a contaminant from other ingredients. R. capsulata, like S. faecalis, showed no evidence for the high affinity trace Kdp system of E. coli and left 2 to 3 μM K^+ in the external growth medium after K^+-limited growth [108].

Although K^+ (and Rb^+) uptake in R. capsulata behaved kinetically as a single system with a constant K_m, the V_{max} for K^+ varied with growth conditions. Dark aerobically growing cells had a 10-times lower V_{max} during early log phase growth (when K^+ uptake rate was roughly equivalent to the growth rate and K^+/K^+ exchange was minimum) when compared with the highest V_{max} found during the beginning stationary phase with light-grown anaerobic cells (when the calculated turnover time for cellular K^+ was as short as 12 min!) [108]. Since both dark-grown and light-grown cells had essentially equivalent intracellular K^+ throughout the growth cycle (from 150 to 250 μmol/g dry weight cells), changes in uptake rate resulted in changes in exchange rates for cellular K^+. The changes in V_{max} for K^+ were established primarily during growth, since switching light-grown cells into dark aerobic conditions (and the reverse) had little effect on K^+ transport, indicating that the cells had both respiratory and photoautotrophic apparatus in sufficient quantity for driving K^+ transport.

B. Magnesium Transport Systems

Magnesium is the major intracellular divalent cation in all living cells. In bacterial cells, the intracellular Mg^{2+} content is equivalent to 20 to 40 mM Mg^{2+} [131, 190] which compares to K^+ values that usually range from 100 to 500 mM. Most probably, more than 95% of intracellular Mg^{2+} is bound (largely to the intracellular polynucleotides), so that it is not osmotically free but relatively readily exchangeable. The movement of Mg^{2+} across the cellular membrane is governed by transport systems, as would be expected. We have just completed a comprehensive treatment of Mg^{2+} transport and metabolism in bacteria [109], and readers are referred to this source for a greater depth of consideration than what follows.

The existence of energy-dependent concentrative uptake of Mg^{2+} by E. coli was first reported in 1969 from experiments using the short half-life radioisotope $^{28}Mg^{2+}$ [130, 188]. The early studies established two primary characteristics of bacterial Mg^{2+} transport: an energy requirement and a high degree of cation specificity. Energy requirements were demonstrated by inhibition of Mg^{2+} uptake by a variety of inhibitors of energy metabolism, including cyanide and uncouplers of oxidative phosphorylation. Mg^{2+} transport was also temperature-dependent: more rapid at 37°C than at 20°C and essentially zero at 0°C. (Arrhenius plots indicated an energy requirement of 16 kcal/mol for Mg^{2+} uptake.) The magnesium transport system is highly substrate specific. An early experiment shown in Fig. 6-4 [190] showed that the addition of 1 mM Mn^{2+}, Ca^{2+}, or Sr^{2+} was essentially without effect on cells accumulating 0.01 mM $^{28}Mg^{2+}$ (Fig. 6-4). One millimolar nonradioactive Mg^{2+} was strongly inhibitory. An extensive series of experiments showed that Mg^{2+} uptake in E. coli and many other types of microbial cells always followed Michaelis-Menten saturation kinetics with

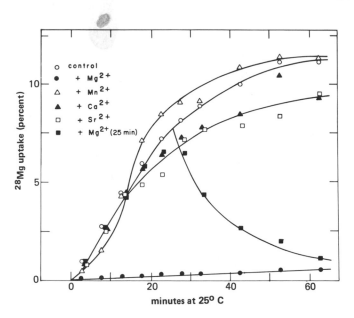

FIG. 6-4 Specificity of magnesium transport in E. coli. Cells were grown and resuspended in a Tris-glucose medium. Accumulation of 0.01 mM $^{28}Mg^{2+}$ in the absence or presence of 1.0 mM chloride salts was measured. To one aliquot, 1.0 mM $MgCl_2$ was added 25 min into the experiment and sampling continued. From Silver and Clark [190] with permission.

defined K_m and V_{max} values. A summary of these results in Table 6-3 shows the similarities with regard to the systems for magnesium uptake of the bacterial cells, the green protist Euglena, and human cells in culture. Although highly specific for magnesium, the accumulation of magnesium by these transport systems was competitively inhibited by Co^{2+}, Mn^{2+}, Ni^{2+} Fe^{2+}, and probably Zn^{2+}. The competitive inhibitors had lower affinities than Mg^{2+}: generally the K_i values for inhibition were some 4 to 10 times higher than the K_m for Mg^{2+} (Table 6-3).

The competitive inhibitors, especially Mn^{2+} and Co^{2+}, are toxic to growing cells. Co^{2+} or Mn^{2+} were themselves accumulated and caused the displacement and loss of 75% or more of the cellular Mg^{2+} (principally from ribosomes). This enabled selection of mutants of E. coli that were resistant to Co^{2+} or Mn^{2+} by virtue of heritable changes in the magnesium transport system, resulting in lesser Mn^{2+} uptake or total loss of Co^{2+} uptake. Co^{2+}-resistant cells were, however, still Mn^{2+}-sensitive, and Mn^{2+}-resistant cells Co^{2+}-sensitive. The genetic loci of these resistances were determined along with another gene locus resulting in loss of a magnesium transport system (Fig. 6-1 and Table 6-1). There are currently four genes known to affect Mg^{2+} transport. Each of these genes maps at a separate position from the others.

TABLE 6-3 Kinetic Parameters of Magnesium Transport in Different
Organisms

Species	K_m (μM)	V_{max} (μmol min^{-1}g^{-1})	Competitive Inhibitors (mM)	References
E. coli	31 (broth) 18 (Tris)	8.0	$K_i = 2$ Mn^{2+} $K_i = 0.4$ Co^{2+}	147,148,161, 190,193
R. capsulata	55	1.8	$K_i = 0.2$ Co^{2+} $K_i = 0.3$ Mn^{2+} $K_i = 0.4$ Fe^{2+}	110
B. subtilis	250	4.4	$K_i = 0.5$ Mn^{2+}	184
B. subtilis (citrate system)	450	123 (citrate inducible)		227
S. aureus	70	9.6		Weiss & Silver, unpub.
Euglena gracilis	300	1.0 nmol/hr per 10^5 cells		Kohl & Silver, unpub.
Human KB cells	100	0.4 nmol/hr per 10^5 cells		8

The use of the mutants allowed Nelson and Kennedy [148] and more
recently Park et al. [161] to show that E. coli has two separate transport
systems, whose basic properties are summarized in Table 6-4. The
cobalt-resistance (COR) system is synthesized constitutively during growth
in low or high Mg^{2+}. It accumulates Mg^{2+} as primary substrate but can
also accumulate Co^{2+}, Ni^{2+}, and Mn^{2+}. (It might be noted here that all
cells that we have looked at have an additional and quite separate "trace
nutrient" manganese transport system that functions at very low Mn^{2+} concentrations and, in general, shows a K_m below 10^{-6} M Mn^{2+}. This trace
Mn^{2+} system is considered in Sec. II.D.) The COR system is missing in
the corA mutants and altered in specificity in the mng mutants. The cobalt-
resistant corB mutants have a regulatory defect so that the COR system is
no longer constitutively synthesized. In corB mutants, the synthesis of the
COR system is induced by the presence of high (e.g., 10 mM) Mg^{2+}. The
other Mg^{2+} transport system is, as far as we are aware, completely

TABLE 6-4 Characteristics of the Two Mg^{2+} Transport Systems of
E. coli K-12[a]

COR System	Constitutive
	Substrates Mg^{2+}, Co^{2+}, Mn^{2+} and Ni^{2+}
	K_m = 30 μM Mg^{2+}; V_{max} = 11 μmol $min^{-1} g^{-1}$ protein
	Missing in corA mutants; altered in mng mutants (?)
MGT System	Repressible by growth in 10 mM Mg^{2+}
	Mg^{2+} substrate specific
	K_m = 30 μM Mg^{2+}; V_{max} = 8 μmol $min^{-1} g^{-1}$ protein
	Missing in mgt mutants

[a]Primarily from Park et al. [161].

specific for Mg^{2+} as a substrate. The kinetic parameters of the MGT
system are rather similar to those of the COR system, and only the
existence of mutants lacking one, the other, or both permit the clear-cut
definition of the two systems [161]. The MGT magnesium transport system
is repressible by growth in high Mg^{2+}, the opposite situation from the
COR system in corB mutants. The MGT system is missing in mgt mutants
which were first isolated as corA, mgt double mutants that are missing both
magnesium transport systems and can grow only in the presence of very
high concentrations of Mg^{2+}, about 10 mM. With the availability of mutants
in four genes affecting magnesium transport, we can anticipate detailed
studies of cellular regulation of magnesium and perhaps the isolation of
the products of some of the genes involved.

Among the various studies carried out with regard to magnesium
transport are some pertaining to changes in the rates and properties of
the magnesium transport system of B. subtilis during the developmental
cycle of bacterial spore germination, growth, and formation, again, of
dormant spores. We [185,196] followed Mg^{2+} transport and also the transport of potassium, manganese, and calcium during this cycle. Each cation
plays highly specific cellular roles. Each is transported by one or more
highly specific transport system(s) whose properties change during the
cycle, and the changes in transport of each are separately regulated, as one
would expect for the regulatory control of cellular components as critical
as these cations. The dormant bacterial spores have no discernible metabolism and no functioning cation transport systems. The earliest metabolically active stage in spore germination is the "turning on" of preexisting
but previously dormant transport systems for Mg^{2+} and for K^+ [52]. These
systems require glucose metabolism for function and "turn on" prior in
time to, and independent of, de novo RNA and protein synthesis. Roughly

concomitant with the turning on of the transport system for the accumulation of Mg^{2+}, the germinating spore loses the massive amounts of Ca^{2+} that had been accumulated days earlier during sporulation. This is an example in bacteria of movement of Mg^{2+} and Ca^{2+} in opposite directions, which is a familiar situation in mammalian cell physiology. During growth, each of the cation transport systems is regulated so as to maintain a relatively constant intracellular level. The rate of Mg^{2+} transport is high during growth and decreases rapidly during the post-growth phase developmental formation of spores [184]. The changes in rate of transport of other cations (K^+, Mn^{2+}, and Ca^{2+}) during sporulation all reflect changes in V_{max} without any changes in K_m, similar to the situation with the R. capsulata K^+ system described above. It is as though variable numbers of cation carrier molecules were functioning, but no changes in the basic affinities of each cation carrier occurred. The exception to this rule is Mg^{2+}. With Mg^{2+}, we found changes in both K_m and V_{max} for transport during sporulation. By 4.5 hr into the sporulation process, by which time the V_{max} for Mg^{2+} transport had declined about 80% [185], the residual transport function was fully saturated at the lowest testable substrate concentration (about 34 μM). The residual transport function in the sporulating cells was indicative of a second very high affinity (K_m less than 10 μM Mg^{2+}) system that persists throughout sporulation, although perhaps declining in rate (V_{max}), and which meets the limited Mg^{2+} requirements of the sporulating cells. This is not unreasonable, since the accumulation of Mg^{2+} during growth is sufficient for the nongrowing developmental cycle, and the cells do not lose Mg^{2+} but rather retain all that is available. Under conditions of limiting Mg^{2+}, the high affinity system would function effectively.

C. Iron Transport Systems

The extraordinary breadth of understanding of microbial iron metabolism recently required a separate monograph [146] for a comprehensive treatment. Transport of iron, especially, under iron starvation conditions, involves a series of high-affinity iron binding chelates that are excreted into the microbial growth medium in the desferri-form (lacking Fe). After binding of iron, the ferrichelate is taken up into the microbial cells by highly specific transport systems [56]. Microbial iron chelates fall into two classes, the catechols and the hydroxamic acids [56], but in both cases a hexadentate cage of six oxygens surrounds the iron atom in the ferri-form of the chelate. In general, bacteria can accumulate iron both with self-made chelates and with chelates that they themselves do not synthesize (see below). The ability to utilize another microbe's iron chelates is obviously of great selective advantage during iron-limited growth in the presence of a variety of competing microorganisms. Three separate chapters on these chelates and transport of iron are to be found in Ref. 146.

As usual I will try to limit our consideration to iron transport in E. coli: (1) because a discussion of E. coli systems will bring out the basic properties common also to other organisms, and (2) because the E. coli systems have been most thoroughly studied biochemically and genetically.

But first, Fig. 6-5 gives the structures of a very limited selection of the known microbial iron chelates. Enterochelin (Fig. 6-5A) is a catechol and is the high-affinity primary iron chelator synthesized by E. coli that will be described in detail below. It was also named enterobactin when isolated from S. typhimurium by Neilands' group [146,178]. Ferrichrome (Fig. 6-5B) was first described by Neilands over 20 years ago and has a radically different structure, being a trihydroxamic acid with three δ-N-acetyl-L- δ-N-hydroxyornithine residues in the structure. The remainder of the covalently closed ring of ferrichrome consists of three contiguous glycine residues, and the ring structure is formed by peptide bonds between the six amino acid residues. In this respect, ferrichrome differs from enterochelin, where the three amino groups are not involved in the ring structure, which is formed by ester linkages between the three constituent 2,3-dihydroxybenzoylserine residues. Ferrichrome is not synthesized by bacteria but rather by a variety of fungi including Ustilago sphaerogena. The extraordinary importance of iron sequestration in nature is attested to by the finding that both gram-positive and gram-negative bacteria possess highly specific energy-dependent transport systems for uptake of Fe-ferrichrome [31,86,128]. Aerobactin (Fig. 6-5C) is also a hydroxamate formed from two residues of ε-N-acetyl- ε-N-hydroxylysine with a single citric acid in between. Aerobactin is synthesized by a bacterium rather closely related to E. coli, Aerobacter aerogenes; this organism also synthesizes the catechol enterochelin. The next trihydroxamic acid listed in Fig. 6-5D is desferal, which is kept on hospital shelves as an antidote for acute iron poisoning in children [146]. Desferal is synthesized by Streptomyces pilosus, but it will function as an iron chelate, stimulating growth of both Bacillus megaterium [31] and Salmonella typhimurium [128]. Desferal is less effective, however, in stimulating growth and in mediating rapid iron uptake in B. megaterium [31] than is the "natural" hydroxamate synthesized by B. megaterium, that has been called "schizokinen" (Fig. 6-5E), since its growth-stimulatory properties were known before its structure and mechanism of action were elucidated. Byers [31] has summarized the extensive range of studies of iron uptake with schizokinen and B. megaterium. This hydroxamate will also stimulate growth of iron-starved S. typhimurium [128]. The last of the hydroxamic acids in Fig. 6-5F is rhodotorulic acid, a cyclic dipeptide made by a number of yeasts including Rhodotorula pilimanae [56]. The "half molecules" in rhodotorulic acid are δ-N-acetyl-δ-N-hydroxyornithine, the same hydroxamate that occurs in ferrichrome. Rhodotorulic acid is still another hydroxamate acid that can be utilized by S. typhimurium [128], E. coli [86], and probably B. megaterium [31], in what constitutes a

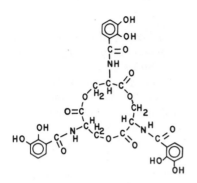

A   Enterochelin

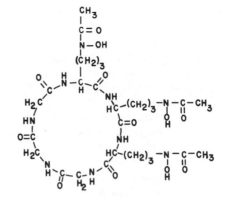

B   Ferrichrome

$$CH_3C-N(CH_2)_4CHNHCOCH_2COHCH_2CONHCH(CH_2)_4N-CCH_3$$

C   Aerobactin

$$NH_2(CH_2)_5N-C(CH_2)_2CONH(CH_2)_5N-C(CH_2)_2CONH(CH_2)_5N-CCH_3$$

D   Desferal (deferrioxamine B)

$$CH_3C-N(CH_2)_3NHCOCH_2COHCH_2CONH(CH_2)_3N-CCH_3$$

E   Schizokinen

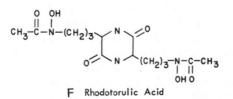

F   Rhodotorulic Acid

FIG. 6-5 Structures of microbial iron binding chelates.

general hydroxymate-iron transport system (see below). Our list of chelates
is very incomplete, and additional microbial hydroxamic acids are described
in detail by Emery [56] and Snow [201] and catechols by Byers [31].

E. coli has four known iron transport mechanisms that can be studied
independently with the use of mutants and exogenous chelates. These are:
(1) the enterochelin-iron transport system, (2) the hydroxamate-iron
transport system(s), (3) the citrate-iron transport system, and (4) a low-
affinity iron transport mechanism with no known chelate [178]. Since the
enterochelin system is the primary system involving an endogenously
synthesized chelate, we will consider it first. The complete cycle for the
synthesis and utilization of enterochelin iron has been thoroughly studied
primarily by Gibson and coworkers (recent summary by Rosenberg and
Young [178]). Much parallel information from studies with S. typhimurium
has come from Neilands' laboratory, and the initial studies of the enzymatic
biosynthesis of enterochelin were carried out by Bryce and Brot [27]. All
these results are diagrammed in Fig. 6-6. The conversion of the central
aromatic intermediate, chorismic acid, to 2,3-dihydroxybenzoic acid (DHB)
involves three sequential enzymatic steps carried out by three soluble
enzymes that are determined by the genes entA, entB, and entC [178,237].
Complementation in vivo and in vitro, and isolation and characterization of
the intermediates have been accomplished. The next step in the synthesis
of enterochelin involves the attachment of L-serine to a protein bound sulfhydryl group in an exchange process involving hydrolysis of an ATP molecule. 2,3-Dihydroxybenzoate is attached to a second protein component
concomitant with hydrolysis of another ATP, and the dihydroxybenzoate is
transferred to the protein-attached serine to form a thiol-attached dihydroxybenzoylserine (DBS) with the release of AMP into solution [27,178]. Free
soluble dihydroxybenzoylserine is not released, and added soluble dihydroxybenzoylserine cannot participate in the subsequent steps in enterochelin
synthesis [178]. There is little experimental data for the next steps, but
presumably dihydroxybenzoylserine dimers and trimers are formed through
ester linkages, and the trimer is finally cyclized with an identical bond and
released from the biosynthetic protein complex. Four polypeptides are
involved in these steps, and four genes entD-G determining the four proteins
are known [129,178,235,236]. All four components were needed for enterochelin synthesis from dihydroxybenzoic acid [235]. Different gene-deter-
mined polypeptides initially bind L-serine (ent F) and dihydroxybenzoic acid
(ent E) in ATP-dependent stages [27,235]. The completed enterochelin
(which was presumably thiol-bound to the protein as a DBS trimer) would
have to be released from the protein as the final serine hydroxyl becomes
incorporated during cyclization. There is no information on the next step of
the cycle, but enterochelin appears rapidly in the extracellular fluid and
does not accumulate intracellularly. (Some may remain localized between
the inner and outer cell membranes and participate in uptake of iron [71]
that has moved into that position by other means.) Once extracellular,

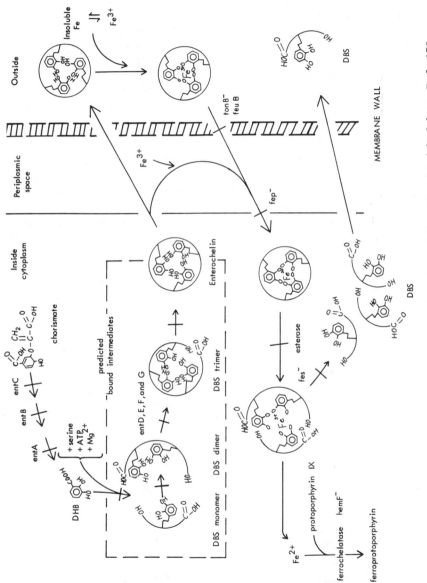

FIG. 6-6. The enterochelin-iron transport system of E. coli. Modified from Ref. 178.

enterochelin solubilizes iron from polynuclear complexes [221] by virtue of
its high binding affinity. Fe^{2+} is oxidized to Fe^{3+} in the process at the
expense of the solvent.

The enterochelin-iron chelate does not cross the outer membrane of
E. coli readily if mutations exist in any of several genes (Table 6-5; see
also Chap. 9). The tonB gene product has not been identified or definitely
located, but is thought likely to be an outer membrane protein. The tonB
mutants were shown by Wang and Newton [221] to have an increased K_m for
iron uptake, from 0.36 μM for tonB$^+$ to 3.6 μM Fe^{3+} for tonB$^-$ with no

TABLE 6-5 Outer Membrane Functions Altered in Mutants Affecting Iron-
Chelate Uptake[a]

| Gene | Outer Membrane Protein MW | Iron Transport System | | | | Resistance to Phage, Colicins, and Antibiotics |
		Entero-chelin	Ferri-chrome	Citrate	Low Affinity	
fec[b]	81K	+	+	−	+	
feuA (cir)	74K	+	+	+	+	colI, colV
feuB	81K	−	+	+	+	colB, colI
tonA	78K	+	−	+	+	T1, T5, Ø80, colM, albo-mycin
tonB[c] (exbA)		−	−	−	+	T1, Ø80, colM, colI, colB, colV, albomycin
Repres-sible by high Fe^{3+}		+	+	+		

[a]Refs. 21, 70, 71, 85-87, 166, 222.

[b]Open questions: fec mutants may or may not determine the citrate-
inducible 81K protein. Ferrichrome-mediated transport has not been
tested with the fec mutants.

[c]The tonB protein has not been identified or localized in the outer
membranes as yet.

change in V_{max}. The genetic background may influence the effect of tonB on iron-enterochelin uptake, since Frost and Rosenberg [71] found no uptake at all with the same tonB mutation as used by Wang and Newton but transduced into another E. coli K12 strain. At the same time as the K_m for Fe^{3+} increased in certain tonB⁻ strains, the K_i for Cr^{3+} as a competitive inhibitor of net Fe^{3+} uptake was reduced from 8 μM Cr^{3+} in the tonB⁺ E. coli to 0.1 μM Cr^{3+} in the tonB⁻ strains [221], providing an explanation for the unusual Cr^{3+}-sensitivity of growth of tonB mutants. The tonB mutants are also defective in the other highly specific transport systems for iron-hydroxamate complexes and for iron-citrate [86]. The other gene that affects iron-enterochelin uptake at the outer membrane level is called feuB for Fe^{3+} enterochelin-uptake. The tonB mutants are resistant to a variety of bacteriophages and colicins in addition to the hydroxamate antibiotic albomycin; the feuB mutants are resistant to a more limited range of colicins (Table 6-5). Mutants in genes affecting iron metabolism are missing specific outer membrane proteins (normally found in derepressed high levels under iron starvation conditions). These proteins have molecular weights of 74,000 daltons (feuA) and 81,000 daltons (feuB) (Table 6-5) [21,85,87]. The initial paper reported that both feuA and feuB mutants were defective in Fe-enterochelin uptake [87]. However, recent experiments with purified Fe-enterochelin showed no such defect (R. E. W. Hancock, personal communication, 1976), and it now appears from experiments in Konisky's, Boos', and Braun's laboratories that feuA is identical with the cir gene (determining the colicin I receptor) and maps close to the his complex (Fig. 6-1). Although the feuA (cir) protein is derepressed on iron starvation conditions, its role in iron metabolism is unclear.

Nothing is known about how these proteins facilitate the movement of iron-enterochelin across the outer membrane, usually a barrier to movement of molecules larger than about 700 daltons [87]. (Iron-enterochelin has a molecular weight of 719 daltons; Fe-ferrichrome's molecular weight is 740 daltons.) In another recent study, four classes of colicin B-resistant mutants were reported to be defective in chelator-mediated uptake of iron [166]. One class was mutant in the tonB gene, but the other three classes (exbB, exbC [82], and cbt) appeared to be different from and in addition to the feuA and feuB mutants in Table 6-5 [166]. The nomenclature of these genes has not yet been standardized. Chapter 9 deals in detail with the outer membrane protein influencing iron uptake.

Once through the outer membrane, passage through the inner (cytoplasmic) membrane requires the functioning of a highly specific (probably energy-requiring) active transport system that is under the control of still another gene called fep (for Fe-permease) [178]. tonB and feuB mutants cannot respond to added extracellular enterochelin because of the outer membrane barrier. tonB strains can, however, carry out iron uptake mediated by internally synthesized enterochelin via the "periplasmic" inner cycle diagrammed in Fig. 6-6. The fep mutants cannot utilize either added or internally synthesized enterochelin. Once within the cell, the problems of providing the cell with usable iron are not over. The tight binding of iron by enterochelin precludes the utilization of the iron in cellular metabolism, and the enterochelin molecule itself must be destroyed (by hydrolysis) in order to make the iron available for cellular functions. An iron-enterochelin esterase, consisting of two components (mutants defective in synthesis of the "B-polypeptide" chain have been reported [178]), first hydrolyzes the enterochelin to the trimer and then to dimeric and monomeric DBS. This sequential hydrolysis has been demonstrated in vitro as well [178]. The dihydroxybenzoylserine monomers and dimers are not reutilized in the biosynthesis of enterochelin. DBS monomers and dimers are secreted from the cell where they can serve, themselves, as low-affinity iron transport chelators by still another system. The iron associated internally with the DBS monomers and dimers is reduced from Fe^{3+} to Fe^{2+}, and much of it is incorporated into heme precursors by the enzyme ferrochelatase, for which once again mutants are available, this time in the hemF gene (Fig. 6-6). All seven ent genes plus the fes and fep genes map together at about 13 min on the E. coli genetic map, while tonA, tonB, and feuB map at a distance from that locus (Fig. 6-1). Preliminary studies with newly isolated mu-phage induced mutants [236] suggest the presence of at least three regulatory units or operons, one consisting of entE(ABG) with the transcription sequence beginning at the entE end, another containing entF, and the third containing the fep gene [235]. The organization of the entC, entD, and fes genes in this region is currently not known. Still another class of iron-dependent mutants was missing all of the components of the enterochelin system. These have mutations, designated fer (for ferric regulation), that map in the same general region as the ent genes [235]. Since these genes have not been ordered by conventional three- and four-factor genetic mapping, the conclusions about the organization of this region of at least nine genes concerned with enterochelin-iron metabolism must be considered tentative. However, early two-factor transductional data [237] suggested separation of the gene for the first enzyme in the pathway, entC, from the next two, entB and entA, with the fep gene in between.

I have presented the iron enterochelin cycle in such detail because of several heuristic advantages. The details of the enterochelin cycle, with at least 13 genes involved and several more that can be postulated, sets a standard for studies of systems involving small chelates or ionophores as components of iron transport systems. Understanding of no other transport system approaches this degree of detail. The investment of cellular protein and energy (ATP) into enterochelin synthesis demonstrates aptly the important role of iron limitation in governing bacterial growth and the lengths to which bacterial cells will go to obtain what little iron is available in the environment.

The second highly specific iron chelate transport mechanism in E. coli is responsible for uptake of iron chelated with hydroxamic acids produced by other microorganisms. This is truly a scavenge pathway for competitive survival! The iron hydroxamate transport system has been studied less well with E. coli than with S. typhimurium, where a wide range of hydroxamic acids were shown to stimulate growth under iron starvation conditions. Just how many Fe-hydroxamate transport systems exist and how they are related are not known, but Leong and Neilands [124] recently provided evidence for two mechanisms. In the first, Fe^{3+} bound to hydroxamate chelator was reduced at the cell surface and taken up while the hydroxamate remained exterior to the cell. The second mechanism involved transport of the complex of Fe-hydroxamate into the cell, followed subsequently by Fe^{3+} reduction and hydroxamate excretion. The mechanisms were distinguished by comparing uptake of $[^{55}Fe^{3+}]$- and $[^{3}H]$ ferriferrichrome and the Cr^{3+}-$[^{3}H]$ferrichrome complex. E. coli K12 accumulated Cr^{3+}-desferrioxamine (Fig. 6-5D), as well as Cr^{3+}-ferrichrome, whereas S. typhimurium LT-2 accumulated Cr^{3+}-ferrichrome but not Cr^{3+}-desferrioxamine [124]. The basis for this difference is not known.

At least 12 phenotypic classes of differing mutants with reduced ranges of hydroxamate utilization exist [128]. These were isolated on the basis of resistance to the ferrichrome-related antibiotic albomycin and are likely to include the equivalent of the tonA and tonB albomycin-resistant mutants in E. coli. Mutations in eight of these classes were mapped by transduction to a map position in S. typhimurium that corresponds to tonA in E. coli. However, mutations in another class map very close to ent, and at least two other classes do not cotransduce with tested marker genes [128]. Both tonA and tonB mutants of E. coli are missing the iron-hydroxamate transport activity (Table 6-5; Refs. 87, 222). Competitive binding experiments between hydroxamates, colicins, and phage to common outer membrane receptors are discussed in Chap. 9.

One can anticipate detailed studies of the hydroxamate-iron cycle in E. coli analogous with those being done with enterochelin. One major difference will be that the hydroxamates are not destroyed by intracellular enzymes in order to remove the Fe^{3+}, but rather that the iron is removed and the intact hydroxamic acids excreted again into the medium for another round of iron uptake [e.g., 124]. This is also the case with hydroxamate-iron uptake with other organisms [31, 56]. Reduction from Fe^{3+} to Fe^{2+} may be a critical enzymatic step in release of iron from hydroxamates. An NADH- or NADPH-dependent reductase has been identified [26] that acts on an iron-hydroxamate from Mycobacterium smegmatis, Fe^{3+}-mycobactin. Fe^{2+} is much more readily released from mycobactin than is Fe^{3+}. Oxidation of Fe^{2+} to Fe^{3+} accompanies the addition of Fe^{2+} to enterochelin (at the expense of the solvent), and Fe^{3+}-enterochelin is 10^{40} times more stable than would be Fe^{2+}-enterochelin [149]. Thus it appears as if enzymatic reduction will suffice to remove iron from hydroxamates (and also may be a step in the release of Fe^{2+} from monomeric DBS), but reduction is not feasible as a mechanism of iron removal from intact enterochelin [149]. It seems reasonable for a cell utilizing iron chelates synthesized by foreign organisms to reutilize these in a cyclic fashion, rather than with the conspicuous consumption of energy involved in the one-time-only use of enterochelin.

The third iron transport system found in E. coli utilizes citric acid as the iron chelate and transport intermediate [178]. Although found in many E. coli strains and in A. aerogenes (in addition to the enterochelin and hydroxamate systems), the citrate-iron system is somewhat less widespread in nature and does not occur, for example, in strains E. coli W or S. typhimurium LT-2, which have been widely used in enterochelin studies [178]. The citrate-iron transport system is strictly inducible and synthesized only in the presence of citrate. This system is also "repressible" by added 2,3-dihydroxybenzoate in an entA mutant unable to make this enterochelin precursor. This repression is presumably due to the intracellular iron made available by the enterochelin-iron transport system, since high Fe^{3+} also represses synthesis of the citrate system and since synthesis of the enterochelin biosynthetic apparatus is also repressible by high intracellular iron [178]. The citrate-iron uptake system had an apparent K_m for Fe^{3+} of 0.2 μM (in the presence of 1 mM citrate) [70]. This is coincidentally very close to the affinity of the enterochelin system (K_m for Fe^{3+} of 0.1 μM); the V_{max} values of both systems are under complex regulatory control but are also approximately comparable: for example, 85 (enterochelin system) and 66 (citrate system) natoms Fe^{3+} per gram dry weight per minute, with one strain grown on low iron plus citrate [70]. Frost [69] studied a mutant (fec) missing the citrate-iron transport system. The fec mutant cotransduced with argF (at about 7 min on the genetic map), and therefore was not in the ent region (G. Woodrow and F. Gibson, personal communication). However, a more precise location of this iron transport gene (near tonA?) must await further studies. It is not known whether mutants in the fec gene are defective in synthesis of the outer membrane protein that is inducible by growth in citrate [85]. The citrate-iron system of E. coli [69,70,178] differs strikingly from the citrate-Mg^{2+} system in B. subtilis [150,227] in that there was no evidence for citrate being taken up by the cells concomitant with Fe^{3+} uptake; nor was citrate metabolized by E. coli [69,70]. Experiments with [^{14}C]citrate and $^{55}Fe^{3+}$ (analogous to those with $^{55}Fe^{3+}$ and [^{3}H]hydroxamates [124]) were suggestive of "delivery" of Fe^{3+} to the cell's surface as a citrate-iron chelate with subsequent movement across the membrane only of the Fe^{3+}.

There is at least a fourth mechanism for Fe^{3+} uptake by E. coli cells that can be measured only when the hydroxamate, enterochelin, and citrate systems are not functional due to mutation and or lack of substrate. This is a non-energy-dependent "low affinity" Fe^{3+} uptake system that appears to involve a membrane carrier but has not been saturable under normal assay conditions [71]. Nitrilotriacetate (NTA) binds Fe^{3+} tightly enough to make it unavailable for the low affinity system, without affecting the activities of the citrate or enterochelin systems [69,70,178].

D. Manganese Transport

Manganese, like iron, is apparently an essential trace element for all cell types. However, the relatively low levels of Mn^{2+} required for optimum growth, coupled with the existence of high-affinity transport systems for scavenging traces of available Mn^{2+}, has made it difficult to demonstrate manganese requirements for growth in many cases. For example, in E. coli where we first reported the existence of a specialized Mn^{2+} transport system, we have never been able to demonstrate a Mn^{2+} requirement for growth. The less than 10^{-7} M Mn^{2+} present in our media is more than sufficient for maximal growth. With B. subtilis we cannot demonstrate a Mn^{2+} requirement for growth but only for sporulation, because at least 10 times more Mn^{2+} is required for this developmental process than for growth (see below). A few bacteria do show a readily measured Mn^{2+} growth requirement [191]. In this section I will summarize the current status of Mn^{2+} transport studies from my laboratory. As far as I am aware, no other laboratory has published comparable studies with the same or other organisms, although occasionally results confirming the existence of these systems have been reported (e.g., in ref. 161). We recently completed a comprehensive review of Mn^{2+} transport, functions, and metabolism [191]; readers are referred to this for more detailed coverage.

We have studied Mn^{2+} transport in four very different bacterial types. In each case, energy-dependent uptake was highly specific for Mn^{2+} and followed Michaelis-Menten kinetics. Co^{2+}, Fe^{2+}, and (where tested) Cd^{2+} were competitive inhibitors of Mn^{2+} uptake. The kinetic parameters of these systems are summarized in Table 6-6. These transport systems are in addition to the active transport of manganese as an alternative low-affinity substrate for the cellular magnesium transport systems (Sec. II.B). The high-specificity Mn^{2+} transport systems enable organisms to concentrate Mn^{2+}, even in the presence of much higher levels of calcium and magnesium. In each case, the K_m is lower than those reported for Mg^{2+} or K^+ transport systems, but is instead more comparable to the Fe^{3+} system K_m values given in the preceding section. Each microorganism has been studied with a different thrust, and therefore we shall consider the results from each in turn.

1. Escherichia coli

The E. coli manganese transport system has a high affinity (K_m of 0.2 μM) and relatively low rate of accumulation (V_{max} of 4 to 16 nmol min^{-1} g^{-1} dry

TABLE 6-6   Kinetic Parameters of Bacterial Mn^{2+} Transport Systems

Organism	K_m (μM)	V_{max} (μmol min^{-1} g^{-1} dry weight)	Competitive inhibitors K_i	μM
E. coli	0.2	0.02	Co^{2+} Fe^{2+}	20 μM 50 μM
B. subtilis	1.0	0.01 (repressed) 5.0 (derepressed)		
R. capsulata	0.5	0.02	Co^{2+} Fe^{2+}	about 100 μM about 100 μM
S. aureus	2.0	0.06	Cd^{2+}	1-6 μM (sensitive) > 100 μM (resistant)

[a]Refs. 50, 110, 185, 192, 225.

weight) in different experiments [192]. While the K_m was reproducible, this range of V_{max} values probably reflects changes during the growth cycle (as described above for K^+ on R. capsulata and as we later studied in detail with the Mn^{2+} system in B. subtilis). It is likely that the lack of reports of specific transport systems for other essential micronutrient cations is primarily due to the presence in normal growth media of contaminants at transport system saturating levels. We might have missed the manganese transport system if tryptone broth had had more than the low (less than 0.2 μM) level of Mn^{2+} that it does. The E. coli system is highly specific for manganese and was unaffected by 100,000-fold molar excess of the related macronutrients magnesium and calcium [192]. Ni^{2+}, Cu^+, Cu^{2+}, and Zn^{2+} were not alternative high affinity substrates for the Mn^{2+} transport system of E. coli. Although Fe^{2+} and Co^{2+} appeared to be competitive inhibitors of Mn^{2+} uptake and perhaps alternative substrates for the system, the K_i values for this inhibition are some 100 times greater than the K_m for Mn^{2+} (Table 6-6). The system is thus essentially specific for manganese. In addition to saturation kinetics and substrate specificity, the manganese transport system of E. coli showed the other characteristics standard for bacterial active transport systems: temperature dependence [194] and inhibition by poisons of energy-dependent processes, such as cyanide and the uncouplers dinitrophenol and CCCP [192]. The manganese transport system is unusual among the several active transport systems that we have studied in that inhibition by the respiratory chain inhibitor cyanide was only partial (about 50%). This relative resistance of Mn^{2+} transport to cyanide inhibition was also found with B. subtilis [50] and aerobic R. capsulata [110], which raises questions about the energy coupling analogous to those we have discussed with regard to K^+ transport in Sec. II.A and to Ca^{2+} transport in Sec. II.G. The critical experiments concerning the coupling of Mn^{2+} transport to proton movement and membrane potential [12, 90] have not been done.

The manganese which is accumulated by E. coli to a level of about 15 nmol/ml cell water is probably not osmotically free, since Mn^{2+} will bind to intracellular polyanions in a way similar to intracellular magnesium. However, at least 85% of this accumulated Mn^{2+} was exchangeable for added extracellular manganese [192], and more than 80% of it was released by toluene in the absence of visible cell lysis [192]. The exchange of intra-cellular for extracellular manganese was also inhibited by uncouplers, suggesting that manganese could not freely leak from the cells. Formalde-hyde, an agent thought to "seal" the cell surface and inactivate membrane transport systems, sealed in more than 80% of the accumulated manganese so that it could no longer exchange with extracellular manganese [192]. These characteristics of the E. coli manganese transport system are just those expected for any micronutrient transport system, and the only sur-prise was that E. coli, an organism without a demonstrable Mn^{2+} require-ment, has such a system.

Under the usual conditions for growth of E. coli and for our studies of the system (low manganese concentrations), the high-affinity, high-specifi-city Mn^{2+} system provides most of the cellular Mn^{2+}. At high external Mn^{2+} concentrations, however, the uptake of Mn^{2+} by the Mg^{2+} system (K_m of about 2 mM Mn^{2+}, i.e., the K_i for inhibition of Mg^{2+} transport, Table 6-3) will be greater than that by the micronutrient system. Assuming equal V_{max} values for Mn^{2+} and Mg^{2+} transport by the Mg^{2+} system, we calculated that below 40 μM Mn^{2+} the high affinity Mn^{2+} system would predominate and that above this level the Mg^{2+} system would be the principal source of cellu-lar Mn^{2+} [192]. The extracellular level below which the Mn^{2+}-specific sys-tem predominates varies with growth medium and with bacterial species, but the principle that the highly specific Mn^{2+} system provides cellular Mn^{2+} required at low external concentrations remains.

2. Bacillus subtilis

We turned to B. subtilis because of its well-known requirement for Mn^{2+} for sporulation [191]. The manganese transport system in B. subtilis shares the basic properties of specificity and energy dependence with those of other organisms. Mn^{2+} uptake by B. subtilis was unaffected by 100,000-fold molar excess Mg^{2+} or Ca^{2+} [50] but was a saturable function of the Mn^{2+} concentration with a K_m of 1.0 μM and a V_{max} that varied with the growth conditions from less than 0.1 μmol min^{-1} g^{-1} dry weight of cells to more than 5.0 μmol min^{-1} g^{-1} dry weight [66,185]. Mn^{2+} uptake was temperature dependent and inhibited by uncouplers [50]. The manganese accumulated by log phase cells or nonsporulating stationary phase cells could be exchanged for extracellular Mn^{2+} and was released by toluene treatment or lysis with lysozyme [50]. During sporulation, however, cellular Mn^{2+} was converted to a toluene- and lysozyme-insensitive form synchronously with the packaging of Ca^{2+} into the developing spore. The two positions of cellular Mn^{2+} were associated with two phases of Mn^{2+} accumulation during sporulation. First, Mn^{2+} was accumulated by B. subtilis cells during log phase growth. This was followed by a period of equilibrium during which relatively little Mn^{2+} was accumulated by the cells, although exchange of extracellular with intracellular Mn^{2+} occurred. At T$_3$ (3 hr into the sporulation cycle), when massive amounts of Ca^{2+} were accumulated, the remaining available Mn^{2+} was also taken up by the sporulating cells, and all of the Mn^{2+} was eventually transferred into the developing spore [50].

Because 1 μM or more Mn^{2+} is specifically required for sporulation in species of Bacillus, we measured the level of cell Mn^{2+} to determine if a movement of Mn^{2+} is associated with early stages of commitment to sporulation. No such movement was found. At very low Mn^{2+} concentrations (below 0.5 μM), cellular Mn^{2+} was limited by the available Mn^{2+}, and the cells did not sporulate. With inadequate medium Mn^{2+}, the intracellular Mn^{2+} was lowered by post-log phase cell growth to levels below that needed for sporulation. However, for cells grown in concentrations of Mn^{2+} sufficient for sporulation (1 to 10 μM Mn^{2+}); the cellular level of Mn^{2+} was essentially constant during log phase growth and early sporulation stages at 200 nmol/ml cell water [50]. This does not imply that all of the intracellular Mn^{2+} was free in solution, but simply that a high and constant Mn^{2+} level was maintained. The Mn^{2+} requirement for spore formation appeared to involve the maintenance of this high level, rather than an extra increase in Mn^{2+} content during early sporulation.

The apparent V$_{max}$ for Mn^{2+} accumulation by B. subtilis cells was regulated as a function of extracellular Mn^{2+} by both "derepression" to high potential transport rates during Mn^{2+} starvation and by inactivation of existing transport function upon the addition of high (1 μM or more) Mn^{2+} to derepressed cells [66,185,196]. No changes in affinity (K$_m$) were observed during regulation of Mn^{2+} transport [185]. Such regulation occurred in both sporulating and nonsporulating cells, and we have studied these processes in some detail. Early post-log phase cells grown in 1 or 10 μM Mn^{2+} had a low Mn^{2+} transport rate. The V$_{max}$ for Mn^{2+} transport increased only after Mn^{2+} was depleted from the growth medium. The V$_{max}$ for Mn^{2+} uptake varied over a 50-fold range (Table 6-6) from 0.1 μmol min^{-1} g^{-1} for cells growing in 10 μM Mn^{2+} to 5.0 μmol min^{-1} g^{-1} for Mn^{2+}-starved cells (either during late sporulation on 1 μM Mn^{2+} or under nonsporulating stationary phase conditions because of the absence of

Mn^{2+} supplements to the medium). The process of derepression of Mn^{2+} transport function was reversible, and 15 min after the adddition of 10 μM Mn^{2+} to the cells sporulating in 1.0 μM Mn^{2+}, the rate of Mn^{2+} transport had decreased some 70% [185,196]. Protein synthesis was required for the derepression process, since inhibitors such as chloramphenicol or rifamycin prevented the rise in rate. However, since the rise in rate occurred slowly over hours in cells depleted of Mn^{2+} by accumulation during growth, and since such cells have a rather low rate of protein synthesis, we have not pursued the mechanism of the derepression process.

We were able to study the decrease in Mn^{2+} transport function in more detail since it occurred over a brief time (10 to 60 min, depending on conditions) and also required both RNA and protein synthesis. Fig. 6-7 depicts our current working model for the Mn^{2+} transport cycle. The V_{max} for Mn^{2+} transport increases during Mn^{2+} starvation in the absence of a change in K_m; we interpret this as being due to the synthesis of additional Mn^{2+}

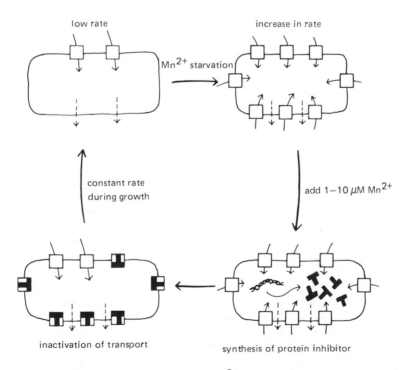

FIG. 6-7. Model for the regulation of Mn^{2+} transport function in B. subtilis during the cycle of manganese starvation followed by addition of exogenous manganese.

transport carriers in the cell membrane (Fig. 6-7). The equilibrium Mn^{2+} levels in the cells [50] followed an absorption isotherm

$$[Mn^{2+}]_{in} = \frac{[Mn^{2+}]_{in}^{max} \, [Mn^{2+}]_{out}}{K_m + [Mn^{2+}]_{out}}$$

The K_m in this equation has the same meaning as that in the Michaelis-Menten equation for Mn^{2+} uptake

$$\frac{d[Mn^{2+}]_{in}}{dt} = \frac{V_{max} \, [Mn^{2+}]_{out}}{K_m + [Mn^{2+}]_{out}}$$

Since we have never observed saturation of the egress of Mn^{2+}, this process can be represented as $[Mn]_t/[Mn]_0 = e^{-k_{exit}t}$, and by setting the rate of uptake equal to the rate of egress, one calculates the equilibrium maximum cellular Mn^{2+} to be $[Mn^{2+}]_{in}^{max} = V_{max}$ for uptake/k_{exit}. In our experiments [50,66], all of these numbers were constants ($K_m = 1 \ \mu M$; $k_{exit} = 0.1 \, min^{-1}$, and $[Mn^{2+}]_{in}^{max} = 200 \, nmol/ml$ cell water), except for the regulated variable, the V_{max} for uptake. The first-order kinetics for Mn^{2+} efflux do not mean that the egress process is by passive diffusion, since efflux was temperature dependent and coupled to uptake in an energy-dependent manner. It means only that system saturating levels of internal Mn^{2+} were never established. The exit rate remained the same throughout a regulatory cycle, during which the internal Mn^{2+} varied by about 25-fold [66].

Since we observed a 90% lowering of the V_{max} for uptake with no concomitant change in K_m and no change in the exit rate, there are several possibilities for the relationship between the Mn^{2+} uptake system and the Mn^{2+} egress system. The first possibility is that Mn^{2+} uptake and Mn^{2+} egress occur by physically different systems (as diagrammed in Fig. 6-7). Alternatively, if the same system is functioning in both directions and if the V_{max} for exit is reduced in proportion to the change in entrance rate, then the K_m for exit must be reduced so as to give constant rates over a regulatory cycle. A third alternative is that a single system is operative in both directions but that the regulatory protein shown in Fig. 6-7 only affects entrance rate. Available data do not allow us to distinguish among the possibilities, but they clearly indicate that a protein regulator is involved in lowering the V_{max} for uptake during a regulatory cycle. Since inhibitors of RNA and protein synthesis prevented the reduction in Mn^{2+}-transport V_{max} [66], the de novo synthesis of a protein inhibitor of Mn^{2+} transport was suggested (Fig. 6-7). This protein inhibitor was specific for the Mn^{2+} transport system and was without effect on the functioning of the Mg^{2+} transport system of the same cell [66]. Two alternatives come to mind as to the functional nature of the protein inhibitor: it might bind one-to-one with part of the Mn^{2+}-transport system, inactivating the transport function in the process, or alternatively, the inhibitor may be a highly specific protease destroying part of the Mn^{2+} transport system. Model systems for regulation by highly specific binding proteins and by protease action are discussed in Ref. 191. There are a few precedents for highly specific inactivation of B. subtilis enzymes by proteases, but the basic mechanisms differ in each case [208]. Switzer [208] has also reviewed the inactivation of specific systems, including some transport systems, in eucaryotic microbes. The results of one type of experiment argue in favor of an inhibitor rather than a protease in reducing Mn^{2+} transport rate. Inhibitors of RNA and protein synthesis added during the cycle of accumulation and release immediately froze the intracellular Mn^{2+} content at the level of the time of addition [66], demonstrating that continuous protein synthesis was required for inactivation of Mn^{2+} transport function. It is surprising that RNA synthesis inhibitors acted so rapidly, showing no period of messenger RNA function followed by decay. If a protease were involved, then it must be destroyed as rapidly as it functions.

The isolation of a mutant that is temperature sensitive in regulation of Mn^{2+} transport function promises another route to studying the inhibitory protein [185]. This mutant was selected as being inhibited in growth in the presence of high Mn^{2+} at 42°C but was normal in its response at 30°C. The mutant showed wild-type derepression of Mn^{2+} transport function on Mn^{2+} starvation and a normal release of accumulated Mn^{2+} (i.e., lowering of the uptake rate) at 30°C. At 42°C, however, the accumulated Mn^{2+} was maintained, indicative of a continued high rate of Mn^{2+} uptake [185]. Upon shifting the mutant from 42 to 30°C, the net loss of accumulated Mn^{2+} started within minutes [185]. Whether this was due to de novo synthesis of the temperature-sensitive inhibitor or reversible inactivation of the inhibitor at 42°C remains to be determined.

Early experiments with derepressed Mn^{2+} transport were often carried out in fresh growth media. More recently, careful measurements have shown a difference in the rate of Mn^{2+} uptake by derepressed cells in fresh or in spent growth medium (medium in which cells had previously been grown to stationary phase). This has led to the characterization of a low molecular weight Mn^{2+} transport stimulator that is excreted by derepressed cells and functions to facilitate Mn^{2+} uptake by derepressed, not repressed, B. subtilis cells. Only a preliminary report of this work has appeared [185]. The stimulator is synthesized during Mn^{2+} starvation, both under sporulating and under nonsporulating conditions. It can be assayed in vitro by its ability to carry radioactive $^{54}Mn^{2+}$ through a gel filtration (G-15 or G-25) column. The rate of movement through the column was indicative of a molecular weight on the order of 1,500 [185]. The Mn^{2+} binding activity was unaffected by excess Mg^{2+} or Ca^{2+} and therefore is highly specific for Mn^{2+}, just as is the cellular transport system. The chemical nature of this material has not been determined.

3. Subcellular Membrane Vesicles

The Mn^{2+} transport system was the first cation transport system shown to
function in the subcellular membrane vesicles isolated and characterized
by Kaback and coworkers [111]. Bhattacharyya studied energy-dependent
Mn^{2+} accumulation by vesicles prepared from both E. coli [13] and B.
subtilis [14]. Basically, the properties of energy dependence and substrate
specificity in the vesicles resembled the properties of the Mn^{2+} transport
systems of the intact cells, and the general properties were very similar
to those of respiratory energy-dependent amino acid transport in the same
vesicles [111]. The K_m values for the vesicle Mn^{2+} systems were, however,
13 (B. subtilis) [14] or 40 (E. coli) [13] times higher than the affinity constants for the intact cells. The most striking difference between the intact
cells and the subcellular membranes was a requirement for high Ca^{2+} levels
for Mn^{2+} uptake with E. coli membranes [13]. Levels near 10 mM were
optimal, and Sr^{2+} but not Mg^{2+} could substitute for Ca^{2+}. This calcium
requirement was not seen with B. subtilis membranes [14].

4. Rhodopseudomonas capsulata

Studies with bacteria other than E. coli and B. subtilis have been initiated
both to establish the range of organisms with Mn^{2+} transport systems (all
that we have tested have high affinity Mn^{2+} transport systems) and to
address specific problems unique to other organisms.
     Manganese is required for the growth of O_2-evolving photosynthetic
organisms. Although Rhodopseudomonas and other photosynthetic bacteria
do not evolve O_2, photosynthetic bacteria do accumulate Mn^{2+} during growth
(discussed in Ref. 191). The R. capsulata Mn^{2+} transport system functions
both in aerobic heterotrophically growing cells and in anaerobic photosynthetically growing cells [110]. Photosynthetically growing R. capsulata
showed a K_m of 0.5 μM Mn^{2+} and a V_{max} of 20 nmol/min/g dry weight of
cells. These kinetic parameters are very similar to those found for Mn^{2+}
transport in E. coli (Table 6-6). The specificity of the highly specific Mn^{2+}
transport system in R. capsulata [110] was similar to that in E. coli.
Toluene caused the release of Mn^{2+}, uncouplers prevented uptake, and
cyanide inhibited Mn^{2+} uptake by aerobically but not by anaerobically
growing R. capsulata [110].

5. Staphylococcus aureus

We looked at Mn^{2+} transport in S. aureus in the process of determining
which transport system was responsible for Cd^{2+} sensitivity in this organism and for Cd^{2+} resistance in plasmid bearing strains [225]. About 60%
of the antibiotic resistance plasmids in S. aureus have genes for Cd^{2+}

resistance. Since the mechanism of Cd^{2+} resistance is a much lowered V_{max} for the uptake of Cd^{2+} by a saturable, energy-dependent process [225], we asked what might be the physiologically normal substrate for the transport system that results in Cd^{2+} accumulation (assuming that there would not be a special transport system for a toxic material). We could find no difference in $^{65}Zn^{2+}$ uptake or turnover between Cd^{2+}-sensitive and resistant S. aureus; there was also no difference with regard to the K^+ or Mg^{2+} transport systems. However, Mn^{2+} transport was specifically inhibited by Cd^{2+} in the sensitive, but not in the resistant strains, and Cd^{2+} specifically exchanged for (chased) intracellular Mn^{2+} in sensitive, but not in resistant cells. Kinetic analysis of both Cd^{2+} and Mn^{2+} transport in S. aureus indicate that Cd^{2+} is a competitive inhibitor of Mn^{2+} uptake in the sensitive cells, as shown in the Lineweaver-Burk plot in Fig. 6-8, and that Cd^{2+} is essentially without effect on Mn^{2+} transport in the resistant cells (Fig. 6-8; Table 6-6).

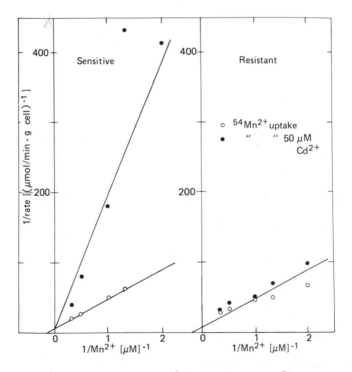

FIG. 6-8   Lineweaver-Burk plot of Cd^{2+} inhibition of Mn^{2+} transport in S. aureus [225].

E. Zinc Transport

Zinc transport, accumulation, and function in microbial cells has just been comprehensively reviewed for the first time [62]. I recommend this paper for more thorough (and sometimes speculative) coverage. Since there is only a single publication that I am aware of concerning zinc transport in bacteria [28], I will first describe the results of that paper and then pass on to the more extensive literature on zinc uptake in eucaryotic microbes. Our working hypothesis (from Sec. I) is that all living cells must have highly specific zinc transport systems and that the only reason why this has not been established is that the systems are generally saturated at the sub-micromolar zinc concentrations that contaminate even highly defined growth media. Eucaryotic microbes generally require some 10 times more zinc for optimum growth than do procaryotes [62] and therefore provide more readily studied material.

Bucheder and Broda [28] reported energy-dependent uptake of $^{65}Zn^+$ by E. coli cells in buffer with all of the standard characteristics of active transport. The process was energy dependent as evidenced by a requirement for exogenous glucose, air, and by temperature dependence (Q_{10} of about 10). Michaelis-Menten kinetics were observed with a K_m of 20 μM and a V_{max} of 2.7 μmol min^{-1} g^{-1} (dry weight) cells [28]. Note that these values are comparable to those for Mg^{2+} transport given in Table 6-3 and much higher than the values for Mn^{2+} transport in Table 6-6. Zn^{2+} uptake also showed specificity in that high Na^+ or Ca^{2+} did not affect Zn^{2+} uptake and in that the addition of 10-fold excess Cd^{2+} (10 μM ^{65}Zn and 100 μM Cd^{2+}) inhibited zinc uptake by only 50% [28]. Mg^{2+} was never present in the assay buffers. There are two alternative explanations for the results of Bucheder and Broda [28]. Either the reported Zn^{2+} uptake was occurring via a specific zinc transport system such as has been hypothesized, or alternatively, Bucheder and Broda were studying the uptake of zinc by the Mg^{2+} transport system for which Zn^{2+} is an alternative substrate [223]. We have been unable to demonstrate the high level of ^{65}Zn uptake (20% of 10 μM ^{65}Zn by suspensions of 10^8 E. coli cells per milliliter) found by Bucheder and Broda in analogous experiments in media containing Mg^{2+}. Generally one observes about 1/50 that level of zinc accumulation (Silver and Weiss, unpublished data with E. coli, B. subtilis, and S. aureus; M. Failla, personal communication). Kung et al. [123] found only 5% of this level, or 1 $\mu mol/10^{12}$ cells for E. coli growing in medium with 12 μM Zn^{2+} and 600 μM Mg^{2+}. The K_m and V_{max} values in Bucheder and Broda's experiments also appear to be too high for a "trace nutrient" system and inconsistent with the vanishingly small zinc requirement for growth of bacterial cells [62,211]. Kung and Glaser (personal communication, 1976) found in preliminary experiments an apparent K_m of 4 μM and a V_{max} only 1/10 of Bucheder and Broda's value. We look forward to an experimental rather than argumentative resolution of this problem, and the two alternatives of a special trace nutrient zinc transport system or zinc uptake into bacterial cells only via broad specificity systems (i.e., the Mg^{2+} system) remain possible.

There arc three additional studies of zinc uptake (but not of transport) in bacteria that should be mentioned here. Torriani [211] controlled the formation of the periplasmic zinc metalloenzyme alkaline phosphatase by reducing available zinc in her medium to about 2×10^{-7} M, which was sufficient for unrestricted growth during 12 successive subculturings. The alkaline phosphatase level of these cells was reduced by two-thirds. Of the available zinc, 30% was accumulated by the cells, with one-fifth of that zinc incorporated into the single enzyme alkaline phosphatase [211] that exists outside of the cytoplasmic membrane of the cells. Kung et al. [123] measured the zinc content of synchronously dividing E. coli cells and found that cellular zinc increased in a step-like fashion 10 to 15 min following cell division, whereas cell content of other cations (K^+, Mg^{2+}, and Ca^{2+}) increased gradually throughout the cell cycle. The differing kinetics for Zn^{2+} and for Mg^{2+} uptake in these experiments argue (albeit indirectly) for separate transport and control mechanisms for these two divalent cations. The uptake and exchange of ^{65}Zn by Salmonella enteritidis was measured [34] with results consistent with cell surface binding, but since these experiments were performed at high (10^{-5} M) Zn^{2+} concentrations, cell surface binding would be expected to mask transport system activity, saturating at perhaps 100 times lower concentrations.

The existence of highly specific zinc transport systems appears firmly established with a variety of fungal forms, and the evidence for this has been reviewed by Failla [62]. The energy-dependent uptake of zinc by mycelia of Neocosmospora vasinfecta was measured with properties characteristic of active transport [163], but again the experimental conditions and results give rise to doubt as to whether Zn^{2+} was taken up by a special zinc transport system or was a low affinity (K_m 200 μM Zn^{2+}) substrate for a basically Mg^{2+} transport system, assayed in the absence of added Mg^{2+}. Zn^{2+} uptake under these buffer assay conditions was inhibited both by Mg^{2+} and by Mn^{2+} (competitively with a K_i of about 250 μM Mn^{2+}) [163].

Finally we come to the only series of experiments with which I am familiar that concern a highly specific trace nutrient zinc transport system [63,64]. Candida utilis accumulates ^{65}Zn by an energy- and temperature-dependent process following saturation kinetics, with a high affinity (K_m of 1.8 μM) and low capacity (V_{max} of 0.22 μmol Zn^{2+} per minute per gram cells) [63,64]. These kinetic parameters were determined in medium containing 1 mM Mg^{2+}, and the rate of Zn^{2+} uptake was unaffected by added 0.1 mM Ca^{2+}, Mn^{2+}, Co^{2+}, or Ni^{2+}. Only Ag^+, Hg^+, and Cd^{2+} inhibited Zn^{2+} uptake [63]. Zn^{2+} uptake by C. utilis was unidirectional. Neither exchange nor net loss of cellular zinc was observed under physiological conditions, although Zn^{2+} was released by membrane active agents such as toluene and nystatin. Evidence was found [62,63] for intracellular zinc storage associated with a binding protein analogous to metallothione, a cysteine-rich protein from animal cells. Even high intracellular Zn^{2+} was not growth inhibitory [64].

Zn^{2+} uptake rates were regulated by C. utilis by a mechanism apparently similar to that described in Sec. II.D for Mn^{2+} in B. subtilis. When zinc-starved stationary phase C. utilis were added to fresh medium, there was a rapid phase of massive zinc uptake, followed by a period of no appreciable uptake of zinc as cell mass increased during mid-log phase of growth [64]. Zinc uptake was irreversible in C. utilis, and lowering of cellular Zn^{2+} was accomplished only be dilution upon increase of cell mass (Fig. 6-9). In late log phase when the cellular Zn^{2+} concentration had been reduced by cell growth, the Zn^{2+} uptake system appeared to be "derepressed" to a high level (Fig. 6-9). Cells exposed to higher zinc concentrations 3 to 6 hr into the growth cycle did not show this derepression phenomenon (Fig. 6-9) [64]. This derepressed Zn^{2+} uptake was specific, as the accumulation of Fe^{3+} by C. utilis smoothly followed growth. The changes in accumulation of Zn^{2+} shown in Fig. 6-9 reflect changes in initial

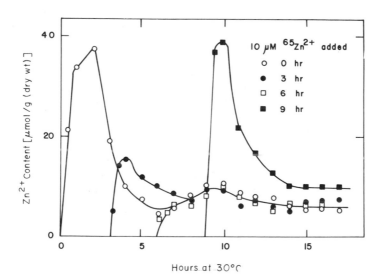

Hours at 30°C

FIG. 6-9 Regulation of zinc uptake in C. utilis [64]. Starting with a zinc-starved inoculant diluted into fresh growth medium containing only 0.1 μM Zn^{2+} (as a medium contaminant), 10 μM Zn^{2+} (^{65}Zn) was added to aliquots periodically throughout the 18-hr growth cycle (2-hr lag phase; log phase growth from 3 to 13 hr) and zinc content per gram cells determined. For the 9 hr addition time, all of the 10 μM ^{65}Zn was accumulated by hour 10, and the "decay" in the figure is dilution of this zinc by subsequent growth.

rate of "uptake", which on a per cell basis decreased more than 90% from early lag phase into mid-log growth phase [64]. The changes in rate were determined by changes in V_{max} (from 0.2 μmol min^{-1} g^{-1} cells in early log growth phase to 3.7 μmol min^{-1} g^{-1} cells in late log phase after derepression) with no change in K_m (1.8 μM Zn^{2+}) for this transport system. The "turning off" of zinc uptake in C. utilis differed in mechanism from the turning off of Mn^{2+} transport in B. subtilis: the same maximum cellular zinc level (120 μmol/g; 0.8% of total dry weight) was obtained over a range of external zinc concentrations and in the presence of or in the absence of protein synthesis [64], indicative of regulation altering the properties of existing molecules and not requiring de novo macromolecular synthesis. Other yeast transport systems also show regulatory inactivation of transport function [208]. Although the C. utilis zinc system is not a bacterial transport system, it currently provides the best standard for related bacterial studies.

F. Other Trace Cations

Additional elements which appear to be essential for some or all microbes include copper, nickel, chromium, and cobalt. There is no evidence for transport systems for inorganic cations of these or even for the oxidation state at the time of cellular uptake. Cobalt is an essential trace nutrient for many bacteria that utilize cobalamin in methyl transfer reactions. Although there is currently no evidence for a highly specific Co^{2+} transport system, Co^{2+} is an alternative substrate for constitutive COR Mg^{2+} transport system [161], and that pathway for Co^{2+} uptake may indeed suffice except under conditions of high Mg^{2+} and low Co^{2+}. Cyanocobalamin (vitamin B$_{12}$), is an alternative source of cellular cobalt. Although E. coli neither synthesizes nor requires vitamin B$_{12}$, it can utilize cobalamin in methyl group transfer reactions, and some methionine auxotrophs will grow on vitamin B$_{12}$ as a substitute for methionine. E. coli has a high affinity transport system for cyanocobalamin, with a K_m for uptake of less than 10^{-9} M [44]. There are less than 200 vitamin B$_{12}$ binding sites per cell [44], and mutants exist [112] that are defective in either an outer membrane binding protein or an inner membrane energy-dependent transport system for cyanocobalamin (discussed further in Chap. 9). Although I know of no studies of copper uptake by bacteria at micronutrient levels, there are a few bacterial copper proteins that appear to serve in electron transport roles. One example of a blue copper protein was recently isolated from the respiratory chain of Thiobacillus ferroxidans where it serves as an intermediate in the oxidation of Fe^{2+} by O$_2$ [37].

G. Calcium Transport

Over the last couple of years, calcium transport has become the most thoroughly studied example of an energy-dependent bacterial egress system, that is, one oriented in the cell membrane so as to reduce the intracellular concentration of Ca^{2+} to a level below that in the outside medium. Since the Nernst potential equilibrium would call for an appreciably higher intracellular concentration, Ca^{2+} egress is active transport against an electrochemical gradient [although it is not apparently a "primary" system directly utilizing energy, but a "secondary" system coupled to the proton-motive potential (Chap. 11)]. I recently reviewed microbial calcium transport and function in considerable detail [189] and will more briefly summarize the characteristics of bacterial calcium transport systems in this section. All bacteria, indeed all cells, appear to have outwardly oriented energy-dependent calcium transport systems.

1. Calcium Transport in Intact
Bacterial Cells

Silver and Kralovic [194] first suggested that bacterial cells might have a transport system for the active extrusion of calcium on the basis of experiments showing three to four times greater accumulation of ^{45}Ca by E. coli incubated below 5°C (as compared with cells in the more physiologically normal ranges of 15 to 37°C). More recently, these experiments have been expanded not only in detail but also with additional bacteria such as B. subtilis [197], B. megaterium [24], R. capsulata [110], and S. faecalis (Harold, personal communication). B. subtilis showed the same patterns as E. coli, but this gram-positive bacterium accumulated quantitatively six to eight times more ^{45}Ca below 5°C, thus facilitating such studies. The accumulation at low temperature showed substrate specificity for Ca^{2+} greater than for Sr^{2+} or Mn^{2+}. Mg^{2+} and monovalent cations were without effect. Calcium accumulation also showed saturation kinetics characteristic of a carrier-mediated process as opposed to a passive diffusion process. The accumulation was reversible, in that more than 80% of the accumulated calcium was extruded upon warming the cells from 0 to 20°C; this extrusion process was energy dependent in that it was inhibited by cyanide [197].

Low temperature was not the only means of inducing calcium accumulation by intact bacterial cells. The addition of uncouplers (proton conductors) stimulated calcium accumulation by E. coli and B. subtilis cells. The optimum concentration of uncoupler was proportional to its potency in affecting other energy-dependent processes: FCCP stimulated calcium accumulation at lower concentrations than CCCP, which was in turn more potent than pentachlorophenol, tetrachlorosalicylanilide, or dinitrophenol [197]. Again, the uptake stimulated by uncouplers displayed substrate specificity and saturation kinetics.

Bronner et al. [24] found an alternative method of studying calcium accumulation in intact cells of B. megaterium. Washed cells grown on limiting glucose accumulated ^{45}Ca. The addition of an electron transport substrate, reduced phenazine methosulfate, inhibited the accumulation of ^{45}Ca and caused the release of previously accumulated calcium [24]. Similarly, reduced phenazine methosulfate caused the release of some of the calcium accumulated by B. megaterium during sporulation [24]. Although the relationship between calcium uptake in energy-starved cells and in sporulating bacteria (see below) requires much further work, this encouraging new approach allows direct experimentation on the question. In all, studies with a variety of intact cells provide plausible evidence for a calcium extrusion system as a general characteristic of bacterial cells.

2. Studies with Subcellular Membranes

(a) Right-side-out and Inside-out Membranes of E. coli. By far the most clear and elegant evidence on calcium accumulation by E. coli comes from the work of Rosen, Tsuchiya, and McClees [176,213-215], who measured calcium and amino acid accumulation by membranes prepared by two methods and inferred that the membranes had two orientations. Membranes prepared by osmotic lysis of E. coli cells by the method of Kaback [111] appeared to have the normal right-side-out orientation of the original cells. These right-side-out membranes accumulated amino acids by an energy-dependent process (for example, proline with energy from reduced phenazine methosulfate in Fig. 6-10D). Right-side-out membranes from E. coli did not accumulate calcium with or without an energy source (Fig. 6-10C). The alternative method of preparing cell membranes from E. coli cells, by explosive decompression in a French pressure chamber, yielded vesicles apparently with the opposite inside-out orientation. These vesicles accumulated calcium in an energy-dependent process (Fig. 6-10A) and were unable to accumulate proline under any conditions (Fig. 6-10B). This experiment was designed in such a manner as to directly suggest an energy-dependent calcium transport system with an orientation in the cell membrane opposite to that of the proline transport system [176]. However, the perennial question of whether the vesicles prepared by osmotic shock treatment are completely right-side-out [111] has cropped up again [1]. The alternative hypothesis of functionally mosaic membrane vesicles is discussed in Chap. 1. Although the models we currently use do not depend on the detailed characteristics of right-side-out versus mosaic membranes, specific questions in the near future could hinge on this problem.

In addition to determining the polarity of Ca^{2+} transport in E. coli membranes, Rosen and McClees [176] demonstrated that calcium uptake by everted membranes was energy-dependent, requiring NADH, D-lactate, succinate, or ATP, and was subject to inhibition by uncouplers, by cyanide (when respiration driven) [214], and by DCCD (when ATP driven) [176]. Since either respiratory substrates or ATP can drive calcium accumulation, this places the energy coupling for the calcium transport system in E. coli

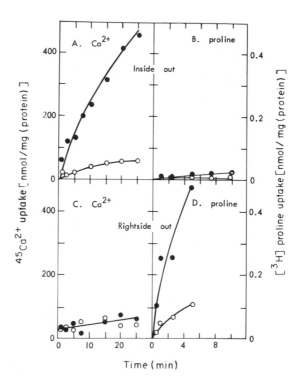

Time (min)

FIG. 6-10 Calcium uptake by everted membranes and proline transport by right-side-out membranes. Redrawn from Rosen and McClees [176], with permission. o, no energy source; ●, energy from reduced phenazine methosulfate.

in the class of proton motive force driven systems [89, 90, 144] and not in the class of transport systems that appear to be more directly coupled to ATP as an energy source [12].

Tsuchiya and Rosen [214] measured the basic characteristics of calcium uptake by everted E. coli membranes in considerable detail. The kinetics of calcium accumulation by E. coli membranes suggested that two systems were operating: (1) a high affinity system with a K_m of 4.5 μM (and a V_{max} of 2 μmol min^{-1} g^{-1} protein), and (2) a low affinity system with a K_m of 340 μM (and a V_{max} of 85 μmol min^{-1} g^{-1} protein). The low affinity system had a similar K_m to those reported for energy-inhibited (360 μM) and sporulating (380 μM) B. subtilis, while the high-affinity system might be more closely related to the 9 μM K_m system of calcium uptake found with everted membranes from B. megaterium [78]. Monovalent cations, especially K$^+$, stimulated calcium uptake when added above 100 mM. Divalent cations were without effect at 0.1 mM, but at 1 mM levels,

Mg^{2+} and Sr^{2+} inhibited calcium accumulation by about half, and inexplicably Mn^{2+} stimulated calcium accumulation (cf, the Ca^{2+} requirement for Mn^{2+} uptake in E. coli membranes, Sec. II.D2).

Given that the normal orientation of the calcium transport system is outward, this raises the interesting questions of (1) coupling between calcium and proton transport, and (2) whether electroneutrality is preserved (two H^+ for one Ca^{2+} antiport), or whether the movement of calcium is electrogenic (one H^+ in for one Ca^{2+} out) and anion movement accompanies calcium efflux. In its simplest form the Mitchell [89,144] hypothesis proposes that only proton extrusion will be outwardly electrogenic and will generate the transmembrane potential, internal negative. All other transport systems are predicted to be either electroneutral or to run down the potential gradient. However, Tsuchiya and Rosen [215] favor an electrogenic 1:1 calcium/proton antiport on the basis of the following indirect evidence: (1) coaccumulation of the neutralizing anion phosphate [215]; (2) the lack of effect, or small stimulation, of calcium accumulation by valinomycin (which would be expected to dissipate the unfavorable internal positive potential built up by electrogenic calcium accumulation) [15,215]; and (3) an experiment showing that valinomycin-induced potassium efflux, which generates an internal negative potential, can drive calcium uptake (Rosen and Tsuchiya, personal communication). We consider the question of electrogenic or electroneutral calcium egress from cells unanswered and of prime importance. Two noncompelling arguments against electrogenic calcium movement in intact cells are: (1) it would have to work against the normal membrane potential that is highly internal negative; and (2) the comovement of phosphate is not the overall situation in intact cells which accumulate phosphate (Sec. III.A) while expelling calcium, both by energy-requiring mechanisms. It is likely that the Ca^{2+}, H^+ antiport mechanism explains the unusual requirement for an alkaline external pH for calcium accumulation both with everted membranes [176,214,215] and with uncoupler-treated intact cells (Scribner, unpublished data). A pH difference of 1.0 across the membrane would be able to maintain a calcium gradient of 100:1.

(b) Membranes from B. subtilis, B. megaterium, and Azotobacter vinelandii. Three other laboratories have reported experiments with calcium uptake in subcellular bacterial membranes. Silver et al. [197] studied calcium uptake by aged right-side-out vesicles of E. coli and B. subtilis prepared by osmotic lysis procedures. With everted E. coli membranes, Rosen and McClees [176] found calcium transport activity was lost within a few hours, even at 4°C, although activity was stable upon storage at -70°C with 50% glycerol present [214]. With aged right-side-out membranes, highly specific calcium uptake was found. Calcium accumulation was not energy dependent, although these membranes retained the energy-dependent proline transport also found in fresh membranes. The relationship between the energy-dependent uptake of calcium by everted membranes and the non-energy-dependent uptake by aged osmotic lysis membranes remains in doubt. Silver et al. [197] found that the rate of calcium uptake by aged B. subtilis membranes was about eight times that with aged E. coli membranes, similar to the ratio of relative rates obtained with intact cells with proton conducting uncouplers (above).

Golub and Bronner [78] found energy-dependent calcium accumulation with membrane vesicles from B. megaterium, even though the membranes had been prepared by osmotic lysis procedures. However, by varying the buffer in which osmotic lysis occurred, Bronner et al. [24] found that the relative rate of calcium uptake by the membranes was inversely related to the relative rate of uptake by a "normally" oriented transport system, in this case for glutamate. Thus they were able to conclude that the calcium uptake by freshly prepared osmotically lysed membranes was due to a variable fraction of membranes with everted orientation. Golub and Bronner [78] observed a loss of energy-dependent calcium uptake by B. megaterium membranes over a few days storage and a concomitant rise in the rate of non-energy-dependent calcium accumulation. Whether a single system is changing from active transport in one orientation to facilitated diffusion in the other direction is not known.

Bhattacharyya and Barnes [15] reported the energy-dependent uptake of ^{45}Ca by membrane vesicles from A. vinelandii prepared by osmotic lysis procedures. These membranes have many properties in common with those produced by Bronner et al. [24] and Golub and Bronner [78] and probably have an everted orientation [15]. The Azotobacter membranes concentrated calcium with substrate specificity, Michaelis-Menten kinetics (K_m, 48 μM; V_{max}, 2 μmol min^{-1} g^{-1}) and strict dependence upon respiratory energy or ATP [15]. The calcium accumulated by the Azotobacter membranes in the absence of phosphate was rapidly released when an uncoupler, exogenous Ca^{2+} or EGTA, was added [15] (unlike calcium accumulated by everted E. coli membranes in the presence of phosphate buffer). In the absence of high phosphate the small amount of calcium accumulated by E. coli membranes was also readily released and exchanged [176]. These results are consistent with the calcium being retained within the membrane by continued calcium/proton exchange rather than being bound in insoluble phosphate salts. Calcium/proton exchange in everted membranes would be consistent with the inhibition of calcium uptake by nigericin or monensin and the lack of an effect of valinomycin with Azotobacter membranes [15]. It appears as if Ca^{2+} uptake with Azotobacter membranes is in response to the ΔpH across the membrane without a potential ($\Delta \psi$) component [15].

(c) Intracellular Membranes. The osmotic lysis and explosive decompression prepared membranes just described were derived from the cellular "cytoplasmic" membrane. Some microbes, including photosynthetic bacteria such as R. capsulata and nitrogen fixing bacteria such as A. vinelandii, also contain small intracellular extensions of the cell membrane with the

photosynthetic or respiratory apparatus of the cells, respectively. These
intracellular membranes are formed by invagination from the cellular
membranes. They are therefore thought to have the reversed polarity or
sidedness of the cytoplasmic membrane [182] (Fig. 6-11A). Jasper [110]
has found energy-dependent uptake of ^{45}Ca by chromatophore membranes
from R. capsulata consistent with the model we have been developing for
oriented calcium transport as a general characteristic of microbial membranes.

3. Calcium Accumulation During
   Sporulation

Massive calcium accumulation occurs during sporulation in members of
the genus Bacillus, along with the accumulation of other divalent cations
(reviewed in Ref. 189). Generally, calcium is found associated with
dipicolinic acid (DPA) in a 1:1 complex that can constitute 20% of the dry
weight of the eventual spore. Since calcium accumulation begins an hour
or two earlier in sporulation than does DPA synthesis and can be separated
with inhibitors from DPA synthesis, the question arose as to whether calcium was passively flowing into the cell where it was bound by a chelate
"sink" such as DPA, or whether the entry to the cell was via a transmembrane transport system. The relative rate of calcium uptake increases
from 0 about 3 hr after sporulation begins, peaks about 6 hr later, and then
rapidly declines [185,196], suggesting the synthesis and regulation of a
specific transport system.

More direct studies have shown that calcium uptake during sporulation
is due to the functioning of a membrane-associated carrier-mediated transport system. Bronner et al. [23,24] suggested the existence of a calcium
"pump" based on studies of the timing of calcium accumulation, and
attempted to isolate the cellular calcium binding component that will be
described below. Eisenstadt and Silver [51] studied the specificity and
saturation kinetics of calcium uptake during sporulation, both of which
were indicative of carrier-mediated uptake. Our current picture of the
uptake and sequestering of calcium is summarized in Fig. 6-11B [189].
The first step is the accumulation of calcium by the sporulating cell by
means of a membrane-located transport system [51]. Whether this transport is by a calcium/proton antiport mechanism as shown in Fig. 6-11B is
not known. Since data are not readily available about the energy requirement for calcium accumulation during sporulation [51], uptake during
sporulation may be either by facilitated diffusion or by an energy-dependent
process. Once within the mother cell cytoplasm, to reach its ultimate
location within the developing spore cytoplasm, or "core", the calcium
must move across two additional membranes (Fig. 6-11B). These are
the outer forespore membrane, which originates by invagination from the

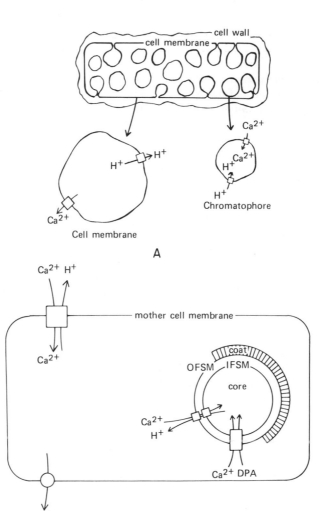

FIG. 6-11 Orientation of bacterial calcium transport systems. (A) <u>R. capsulata</u> cell and chromatophore membranes. (B) <u>B. subtilis</u> scheme for calcium uptake during sporulation including Ca^{2+}/H^+ antiport at the mother cell membrane and either Ca^{2+}/H^+ antiport or Ca^{2+}/dipicolinic acid (DPA) symport across the outer and inner forespore membranes (OFSM and IFSM) [189].

mother cell cytoplasmic membrane and which therefore has the opposite
polarity of the mother cell membrane, and the inner forespore membrane,
that originates as the cellular membrane of the small "protospore" and
which therefore has the normal polarity. Two hypothetical alternatives
are available. Electroneutrality could be maintained either by a calcium/
proton antiport mechanism or by symport with the dicarboxylic acid dipicolinic acid (DPA). Both alternatives are diagrammed in Fig. 6-11B. There
is additionally the question of whether movement across the outer and inner
forespore membranes involves membrane transport systems (as we have
diagrammed) or occurs by passive diffusion through a more or less freely
permeable membrane. At least one membrane transport system is likely
to be required in order to establish the partitioning of components and the
formation of the unique internal structure of the developing spore.

Given the model, what are the data, however incomplete, that led to
its construction? By measuring exchangeability with added nonradioactive
calcium and release by agents such as toluene and lysozyme, we were able
to distinguish four positions or stages of calcium movement during sporulation [57]: (1) Ca^{2+} in solution outside the cells; (2) Ca^{2+} in solution
within the mother cell cytoplasm (exchangeable and releasable by toluene,
which disrupts membrane permeability barriers); (3) Ca^{2+} bound within
the forespore with DPA (no longer subject to release by toluene but still
extractable by lysozyme, which disrupts the cell wall leading to osmotic
lysis); and (4) Ca^{2+} bound in a stable, completed spore (no longer released
by either toluene or lysozyme treatment). The conversion from stage (3) to
stage (4) involves synthesis of the spore coat that protects the otherwise
lysozyme-sensitive structure of the spore cell wall or "cortex". Both the
DPA and the Ca^{2+} are found within the spore cytoplasm (core) (Fig. 6-11B).
This has only become clear during the last few years on the basis of experiments reviewed in Ref. 189.

The mechanism of calcium movement from the mother cell compartment into the forespore has been approached by isolating intact forespores
from sporulation stages during which active calcium accumulation occurs
[53]. Apparently DPA is synthesized exclusively in the mother cell compartment and then rapidly moves into the forespore (Fig. 6-11B). Protoplasts of B. megaterium have a K_m for calcium uptake below 50 μM, whereas the forespores of B. megaterium accumulate appreciable calcium only
when the level is raised to 10 mM or so [53]. With 10 mM external calcium,
the forespores accumulate as much calcium in 2 hr as the protoplasts from
which they had been isolated did from 0.1 mM Ca^{2+} [53]. At 0.1 mM Ca^{2+},
the forespores show no detectable uptake [53] . As with Rosen and Mc-
Clees' [176] studies with everted E. coli membranes, accumulation of
calcium by forespores of B. megaterium required a counter ion for continued accumulation but not for the initial transport event. In this case,
some 20 to 30 times more ^{45}Ca was accumulated by forespores in the
presence of 2 mM PO_4^{3-} buffer in addition to the 10 mM Ca^{2+} [53]. Added

DPA was ineffective in these experiments. The calcium accumulated by forespores in the presence of PO_4^{3-} differed from that accumulated within the mother-cell protoplast in that the in vitro accumulated calcium would exchange with added nonradioactive calcium, whereas in vivo accumulated calcium was not exchangeable, presumably because it was sequestered with DPA.

A serious question is that of the relationship of the sporulating cell's inwardly directed calcium transport system to the growing cell's outwardly directed system: is a new sporulation-specific system synthesized by the cells after the inactivation of the previously outwardly oriented calcium system [176,197], or does the same system function in different orientations during different stages of the sporulation cycle? According to the chemiosmotic hypothesis [88-90,144], the outwardly oriented calcium movement is likely to be coupled to the inward movement of protons, down the pH gradient (and electroneutral), whereas the coupling for inwardly directed calcium movement would be electrogenic and driven by the potential gradient (internal negative).

4. Calcium Binding Factors

The isolation and preliminary characterization of a low molecular weight calcium binding factor was reported by Bronner and Freund [23]. It was synthesized by B. megaterium after the end of log phase growth and early in the sporulation cycle. In fact, the calcium binding activity peaked during sporulation just before net calcium accumulation began, and the level of the chelate then declined during the period of massive calcium accumulation [23]. This material could be precipitated by ethanol and run on a Sephadex G-25 or Biogel P2 column at a position indicative of a molecular weight of a few hundred for the major activity peak. The calcium binding factor from B. megaterium was clearly not dipicolinic acid, however, as determined by comparing chromatographic mobilities and by direct chemical analysis of the active fractions for dipicolinic acid [23,24].

H. Sodium Transport

The properties of outwardly oriented energy-dependent sodium transport in bacteria do not appear to have been studied directly but rather as peripheral measurements in studies directed at K^+ or H^+ transport. Nevertheless, a fair amount of data on Na^+ movements exists, and we can try to "shape out" what the properties of the putative sodium transport system might be. Harold and Altendorf [90], in their recent review of mechanisms of monovalent cation transport, mentioned Na^+ in the title, but no section of the paper was given over to consideration of Na^+ movements themselves.

Bacterial cells excrete Na^+ and under physiologically active conditions maintain a Na^+_{in} to Na^+_{out} gradient between 1:3 to 1:50, perhaps depending upon the organism and the osmotic pressure of the growth medium. The Na^+ extrusion process is <u>not</u> related to that of the Na^+-K^+ ATPase of animal cells. Bacteria do not have a comparable enzyme. In bacterial cells Na^+ extrusion is not tightly coupled to K^+ uptake. K^+ movement is more generally coupled with H^+ movements, and Na^+ egress is currently thought to be coupled antiport to H^+ [90,226].

The most direct study of Na^+ movements was that of West and Mitchell [226] who took anaerobic unenergized <u>E. coli</u> cells and by giving them a "pulse" of O_2 could observe a pulse of H^+ movement to the cellular exterior. The protons gradually moved back into the cells in a process that could be accelerated by the addition of 5 to 10 mM Na^+. This was interpreted as Na^+/Na^+ exchange blocking the normal Na^+/H^+ exchange channel [226]. In another experiment the addition of a "pulse" of 0.5 mM Na^+ drove H^+ efflux, but the addition of choline was without effect. The medium contained 150 mM K^+ [226]. This Na^+-induced H^+ flux was indicative of the Na^+/H^+ antiport functioning in the reverse of a physiologically normal direction.

In their detailed studies of the mechanism of valinomycin-mediated Rb^+ uptake by subcellular membranes, Lombardi et al. [126] found that Rb^+ uptake was associated with $^{22}Na^+$ efflux from the membranes. As much as 50% of the equilibrium (inside:outside) Na^+ initially in the membranes was displaced. The membrane vesicles lacked a potassium carrier and therefore required the addition of valinomycin. However, this experiment showed that the vesicles retained the Na^+ carrier. Since the membranes were energized by formation of a chemiosmotic potential [111], it is reasonable to conclude that Na^+ efflux was antiport to proton uptake.

Membrane vesicles prepared from a <u>trkC</u> mutant strain showed a defect in retention of K^+ [16], and as I will now hypothesize, this could be due to an alteration of substrate specificity for the Na^+/H^+ system. The "potassium-retention" mutants (<u>trkB</u> and <u>trkC</u>, Sec. II.A1) have never been tested for defects in Na^+ extrusion [171], and experiments with these strains generally have been conducted in high Na^+ medium. The extracellular requirements for retention of K^+ and conversely for the rapid loss of cellular K^+ were studied with a <u>trkC</u>-like mutant of another <u>E. coli</u> strain, B [81,165]. Intracellular Na^+ increased in parallel with the loss of K^+ [81]. When the extracellular Na^+ level was increased, the intracellular K^+ decreased and Na^+ increased [81,165]. Whereas with wild-type cells the intracellular Na^+ level was relatively independent of extracellular K^+, with the mutant strain increasing extracellular K^+ lowered the intracellular Na^+ [81]. These mutual interactions between internal and external Na^+ and K^+ levels in the mutant strain were interpreted as indicative of changes in affinity of a common transport carrier for Na^+ and for K^+ in the mutant strain [81,165]. It was not specified whether the transport carrier was primarily a K^+ or a Na^+ carrier, but the idea of a specific Na^+ carrier had not occurred during the time when these experiments were being performed.

Extensive measurements of Na^+ movements have been made with cells of S. faecalis, especially with regard to the properties of a mutant (7683) originally isolated as requiring high K^+ for growth [96] and now thought to be defective in the Na^+/H^+ antiport system [96,100]. Although extrusion of both H^+ and Na^+ occurs with S. faecalis, evidence from studies with lipid-soluble cations led to the conclusion that there need not be an "electrogenic Na^+ pump," but that Na^+ egress could be electroneutrally coupled antiport with H^+ uptake [100]. Wild-type S. faecalis cells suspended in Na^+ buffer retained their cellular K^+ and did not show energy-dependent K^+/Na^+ exchange [98]. Since the addition of valinomycin induced K^+_{out}/Na^+_{in} exchange [90], the Na^+ carrier must be functionally present but inoperative until the net negative membrane potential elicited by the valinomycin-induced potassium loss provided a driving force. $^{22}Na^+/Na^+$ exchange did not occur with unenergized S. faecalis cells but could be stimulated by addition of glucose and was inhibited by DCCD [90,96], leading to the conclusion that an energized membrane state is required even for exchange. K^+ did not inhibit Na^+/Na^+ exchange, indicating that a separate carrier was involved from that of the K^+ transport system [90,96]. It is apparently the Na^+/H^+ transporter that can function also in a Na^+/Na^+ exchange. Both DCCD and proton conductors inhibited Na^+ efflux [96,100], showing that a proton gradient was required for Na^+ egress. Large Na^+ concentration gradients (inside high) were maintained by Na^+ loaded cells unless H^+ movements inward were allowed to balance Na^+ outward movement. Starting with high Na^+, low H^+ (and negligible K^+) cells suspended in water, a direct increase in cellular H^+ concomitant with a lowering of cellular Na^+ was measured [100]. The mutant defective in the Na^+/H^+ antiport mechanism (below) did not show H^+_{in}/Na^+_{out} exchange under these conditions, although these mutant cells were able to extrude protons in exchange for K^+ [100].

Mutant S. faecalis strain 7683, apparently defective in the Na^+/H^+ exchange system, tends to lose cellular potassium, but apparently its K^+ carrier is intact since its K^+/K^+ exchange rate was high [96,100]. Na^+/H^+ antiport movements with the mutant could not be detected, but could be established upon addition of the antibiotic monensin, which can function in lipid bilayers as a Na^+/H^+ exchange carrier [90,100]. Surprisingly, $^{22}Na^+/Na^+$ exchange was normal with the mutant strain [96]. Whether still another carrier is involved in this latter process [90] or whether "gating" and regulation of all of the cation transport carriers occurs in addition to the more simple considerations of "driving energy" is not known. The requirement for glycolytic energy for Na^+/Na^+ and Na^+/H^+ exchange suggests a process regulated beyond the level that occurs with ionophores freely diffusible in lipid bilayer regions.

The question of Na^+ transport is at least peripherally related to the question of Na^+-dependent amino acid transport. Several bacterial amino acid transport systems require sodium for activity, including the glutamate system in E. coli [67,113] and that for α-aminoisobutyric acid transport in a marine pseudomonad [210]. Na$^+$ specificity affected glutamate transport by reducing the K_m for uptake. Rb$^+$, Li$^+$, Cs$^+$, and choline could not replace Na$^+$ in transport stimulation [67]. Extracellular Na$^+$ also reduced the K_m for α-aminoisobutyrate uptake in the marine pseudomonad. It was the presence of extracellular Na$^+$ and not a Na$^+$ gradient (extracellular:intracellular) that was required for amino acid transport [210]. Unaltered transport in the absence of a Na$^+$ gradient excluded the possibility that the Na$^+$ gradient is utilized to "drive" bacterial amino acid transport as is the case with animal cells. Na$^+$ was required for glutamate uptake by subcellular membrane vesicles as well as by cells of E. coli [113]. With the membrane preparations, again, the effect was to increase the affinity for glutamate (lower the K_m) without affecting the V_{max} for this transport system. The stimulation of glutamate transport required high Na$^+$, with a half-maximum stimulation at about 10 mM Na$^+$, whereas the K_m for glutamate was in the range of 10 μM [113]. Kahane et al. [113] were unable to detect ^{22}Na$^+$/glutamate symport, but with a Na$^+$ to glutamate ratio of about 1000:1, this result is not surprising.

I. Proton Transport

The outward translocation of protons across membranes is the primary energy-requiring step according to Mitchell's chemiosmotic hypothesis [88-90,144]. Other energy-requiring processes, including ATP synthesis in oxidative and photophosphorylation, active transport of nutrients, motility, etc., are "coupled" secondarily to proton translocation. This is primarily the topic of Chap. 11. Here we will briefly consider only the initial process of proton movement from the inside to the outside of the cell as an example of an outwardly oriented transport system. Basically, several membrane bound systems show proton transporting activity. These include the membrane respiratory chains, the membrane bound ATPase (involved in oxidative phosphorylation in aerobes), bacterial rhodopsin, and specialized systems for removing metabolic waste products such as lactic acid during glycolysis. The subject of proton coupling to other metabolic processes has been repeatedly reviewed (most recently for bacteria by Harold [89; The Bacteria, 2nd ed., in press) and therefore I will only touch on the wealth of experimental data available concerning that aspect of interest here: the outward translocation of protons.

Bacteriorhodopsin is the simplest known proton pump, consisting of a single protein. It has been isolated and reconstituted into artificial membranes where the light-induced proton transport was studied ([114,168]; reviewed in ref. 89). The bacteriorhodopsin protein has a molecular weight of 26,000 and contains a single retinal per molecule, bound in a Schiff base to the ε-amino group of a specific lysine of the opsin protein [151,153]:

$$\text{Retinal-}C{\overset{\displaystyle \nearrow O}{\underset{\searrow H}{}}} + NH_2\text{-lysine} \rightleftharpoons \text{retinal-}HC=N^+H\text{-lysine}$$

This Schiff base is successively protonated (pigmented) and deprotonated (bleached) upon illumination [125,127], with the protons coming from the cytoplasmic side of the membrane and being released on the external boundary. Several intermediates in the deprotonation-protonation cycle have been identified spectroscopically and with regard to time constants [127,151]. In addition to the protonation cycle, the retinal undergoes a cycle of 13-cis to all trans and back [127,151].

The three-dimensional structure of the protein within the purple membrane has been studied in detail by a combination of x-ray diffraction and electron microscopy [17,103,104]. The bacteriorhodopsin molecules are situated in groups of three around a central axis. Each 26,000 molecular weight molecule is folded into seven α-helical sections, each of which bridges the 45-Å membrane essentially completely [17]. The proton channel must be either enclosed within a single rhodopsin molecule or it is formed by the ring of the nine (three times three) α-helical chains surrounding each 20-Å diameter central hole. This region may be occupied by lipid molecules in a bilayer configuration [17,103]. The current level of resolution did not allow placement of the retinal molecules, which from spectroscopic measurements appeared parallel to the plane of the membrane (i.e., perpendicular to the α-helical structures [103]).

Bacteriorhodopsin is the only protein found in the purple membranes of Halobacterium, which can occupy up to 50% of the total cell surface. On illumination of these membranes with visible light, protons are rapidly released from the membranes and the bacteriorhodopsin is bleached. The membrane pigment regenerates in milliseconds with the reabsorption of protons [152]. With ether-treated membranes only a single proton was released per bacteriorhodopsin molecule on illumination. However, with intact cells, up to 20 protons per molecule could be released on illumination [152], indicative of the proton "pump" role.

The purple membrane patches contain 75% bacteriorhodopsin and 25% lipids by weight [153]. When added to mitochondrial or soybean phospholipids or to purified phospholipids, membrane vesicles formed with the bacteriorhodopsin oriented so that illumination resulted in extensive proton uptake [114,167]. The uptake of protons by the reconstituted vesicles was electrogenic, and the establishment of a potential gradient could be followed by the uptake of lipid-soluble anions [114,167]. If the membrane bound mitochondrial ATPase (F_1-F_0) was included in the reconstitution mixture, these membrane vesicles showed light-dependent ATP synthesis [168] with the proton pump from the Halobacterium providing the proton gradient coupled with the ATPase from beef heart mitochondria! The membrane vesicles containing bound bacteriorhodopsin could associate with planar membranes separating two small chambers, and illumination-generated proton movement was directly measured as an electrical potential across the planar membrane [45]. The potential generated upon illumination of such preparations was as high as 150 mV. The final type of preparation utilized to study the light-dependent proton movement with bacteriorhodopsin consisted of purple membrane added to a water phase associated with an octane phase. Some purple membrane apparently oriented at the interphase, and upon illumination protons were transferred onto octane-soluble anions generating, again, a voltage that could be directly measured [18]. In summary, the bacteriorhodopsin molecule is both the simplest and the most thoroughly characterized proton "pump". What remains to be learned with this system is the exact location and pathway of the proton channel.

The proton tranduscing ATPase is far more complicated than the bacteriorhodopsin molecule. It consists of two parts: a dissociable ATPase "knob" (BF_1) and a residual membrane "patch" (BF_0) [40]. Its structure is considered in detail in Chap. 10. The dissociable ATPase is analogous to the mitochondrial coupling factor (MF_1) and appears to play the same role in bacteria as coupling factors do in oxidative and photophosphorylation in higher forms [40, 209, 220]. The residual membrane region corresponds to the F_0 of the mitochondrial membrane. Perhaps as many as a dozen polypeptides are involved, and therefore a dozen genes to specify their amino acid sequences. Since it is not clear whether the proton channel passes through the ATPase coupling factor [145] or whether the ATPase generates a proton channel in the F_0, we will not consider the polypeptide composition of the ATPase (see Chap. 10). The proton channel in principle must pass through the F_0 region of the membrane [145]. Because this region is embedded in the inner cell membrane, it is not easy to specify its protein components. Nearly 10 polypeptides are found in the membranes containing this region after removal of the coupling factor (e.g., Refs. 65, 198; see Chap. 10). Not all of these polypeptides need be part of the F_0 complex but it is difficult to determine directly which are and which are not. Several "functions" of the residual membrane have been approached by means of isolation of mutants defective in them. These include the dicyclohexylcarbodiimide (DCCD) binding protein of molecular weight about 10,000 that is missing or altered in DCCD-resistant [65] or uncoupled [4] mutant strains. DCCD inhibits the proton translocating function of the wild-type membrane [88-90, 95] and also inhibits the ATPase activity when the coupling factor is associated with the membrane. However, DCCD can restore the essential proton impermeability of some mutants [2, 174, 175] that cannot maintain an energized membrane apparently due to a "leaky" unregulated proton channel through the F_0. The polypeptide defect in these mutants has not been localized. The mutants in F_0 just listed occur in E. coli strains. In addition, the S. faecalis mutant class Cn_{K6}^- appears defective in proton extrusion [96, 100] and may also be F_0 mutants. Alanine and phosphate transport were both defective with these strains, defects that might be secondary to the primary change in proton efflux and K^+ retention [96].

Although the current state of knowledge of the polypeptides involved and their roles is very much in a state of flux (Chap. 10), it is clear that the ATPase aggregate does function as a "reversible proton pump" as envisaged by Mitchell [144,145]. The ATPase aggregate can generate a proton gradient by expelling protons during ATP hyrolysis, or it can mediate ATP synthesis with the energy provided by the proton motive gradient and concomitant with proton uptake.

The respiratory chain also appears to generate a proton motive force by expelling H^+ from bacterial cells. Each of the "coupling sites" for oxidative phosphorylation, which were once considered physical sites of association between the respiratory chain and the ATPase coupling factor, is now considered to be a site of proton translocation [88-90,144,172] , with the coupling factor at a distance. This subject too is in a state of rapid flux with highly speculative models. However, an encouraging change in the last couple of years has been the introduction of very detailed models specifying the physical components thought to be responsible for the movement of protons across the respiratory chain membrane [89,172]. A thorough review of the evidence for proton movements associated with the respiratory chain is beyond the scope of this review. Furthermore, this topic has been lucidly reviewed every year or two by Harold [89,90]. I will end this section with another prejudice for the reader to accept or reject: The Mitchell chemiosmotic hypothesis [144], postulating a critical and central role for proton movements in energy metabolism and active transport, has moved beyond the stage of hypothesis to the stage of being a well-established paradigm for understanding essentially the entire body of data on the subject. It is unfortunate that the original phrasing of the hypothesis was in such apparently nonbiochemical terms as to delay understanding by most of us. It is also unfortunate that polemics clouded discussions of the topic from all directions. But the time of hypothesis and polemics appears certainly over in 1978.

The need to expel protons generated by intracellular metabolism--such as during lactic acid production by glycolysis--is the last example of proton transport systems that we will consider. Harold and Levin [99] studied this problem in S. faecalis. Using evidence from pH shifts and proton- and cation-conducting compounds, they concluded that lactate is expelled with protons as lactic acid [99]. The lactic acid carrier could be studied functioning "backwards" and carried out electroneutral symport of lactate with H^+ but would not facilitate transport of the lactate anion [99]. ^{14}C-lactate accumulation was "driven" by pH gradients (alkaline interior). Normally during glycolysis lactic acid efflux was "driven" by the lactate concentration gradient itself (glycolytically generated lactate could build up to as much as 30 mM within the cells [99]). Lactic acid uptake showed saturation kinetics with a K_m of 2 mM lactate and competitive inhbition by pyruvate with a K_i of 10 mM [99].

J. Ammonium Transport

There are only three reports of which I am aware that suggest a highly specific high affinity NH_4^+ transport system in bacteria. Strenkoski and DeCicco [206] isolated mutants of the Gram negative facultative chemotroph Alkaligenes (Hydrogenomonas) eutropha with growth properties indicative of a defect in retention of NH_4^+ and with a requirement for high NH_3 concentrations (high pH and/or NH_4^+; $NH_4^+ \leftrightarrow NH_3 + H^+$ pK of 9.25) for growth. The mutants had no defect in the ammonium assimilating enzyme glutamate dehydrogenase. Both wild type A. eutropha and the mutants excreted comparable amounts of NH_4^+ (2.5 mM) from amino acids such as alanine (that required deamination and subsequent reassimilation of ammonium by glutamate dehydrogenase). However, while the wild type strains could reassimilate the NH_4^+ and grow on alanine as sole nitrogen source, the mutants could not. This work does not appear to have been pursued during the last 6 to 8 years. Without the chance isolation of the NH_4^+ assimilation defective mutants, Strenkoski and DeCicco [206] would not have found need to postulate the high affinity NH_4^+ transport process.

The ability of A. vinelandii to scavenge its environment for traces of ammonium before derepressing the nitrogenase system for N_2 fixation led Kleiner [119] to study intracellular and extracellular NH_4^+ levels. Azotobacter maintained an intracellular NH_4^+ near 600 μM while growing in continuous culture in media with 1 to 8 μM NH_4^+, for a concentration gradient sometimes of 100:1. Comparable NH_4^+ gradients from other studies with other N_2 fixing bacteria are listed in Ref. 119. The conclusion was that the membrane of Azotobacter is relatively NH_4^+-NH_3 impermeable, and therefore one must postulate an NH_4^+ transport system. The basic impermeability of the Azotobacter membrane to NH_4^+ was previously reported from experiments measuring plasmolysis by light scattering [219]. Because of differences in the immediate rate of uptake of NH_4^+ with different salts of ammonium (lactate, acetate, and phosphate more rapid than with chloride, sulfate, or citrate), Kleiner [119] hypothesized a series of inducible NH_4^+-X^- symport systems. However, the indirectness of the data does not eliminate alternative models involving NH_4^+ uniport plus the requirement for balancing of charges with intracellular anions.

Recently a class of S. typhimurium mutants, nit (for nitrogen utilization), were isolated that were unable to grow on a wide variety of nitrogen sources including low NH_4^+ concentrations [22]. The absence of enzymatic defects in glutamate or glutamine metabolism suggested a defect in NH_4^+ assimilation rather than in subsequent organic nitrogen metabolism [22]. Transport and regulatory defects were the two working hypotheses to explain the properties of these mutants.

Despite the existence of these two studies (and those with eucaryotic microbes described below), the general assumption of workers concerned with bacterial nitrogen metabolism has been that assimilation follows passive movement of NH_3 through the cell membrane, without the involvement of transport carriers. Such passive diffusion occurs in plant and animal materials [see 206 and 238 for references] which are "buffered" with high ammonium levels and against low extracellular pH. Some of the primary experiments demonstrating passive NH_3 movements with S. faecalis were conducted by Zarlengo and Abrams [238], who followed the uptake of NH_3 and of mono-, di-, and trimethylamine indirectly with a pH-stat by determining the number of protons released concomitant with free NH_3 uptake by the cells. Uptake appeared passive and essentially instantaneous. The S. faecalis cells used, however, were grown in a nitrogen-rich medium under conditions that would have repressed the fungal ammonium transport systems that will be described below. The concentrations of NH_4^+ used in the S. faecalis studies were also rather high, generally in the millimolar range.. I do not doubt that NH_3 uptake was passive under these conditions, but rather urge studies of active NH_4^+ transport under conditions predicted to elicit such a system as a trace nitrogen scavenging mechanism.

In another study with the nitrifying bacterium Nitrosomonas europaea, the similar effect of pH on the rate of oxidation of ammonia by intact cells and by cell-free enzyme preparations led to the conclusions that NH_3 rather than NH_4^+ was the substrate for this system, and that NH_3 was moving passively across the bacterial cell membrane to the enzyme site [207].

In addition to systems for assimilation of NH_4^+ and of NO_3^- and/or NO_2^- (Sec. III.C), nitrogen fixing bacteria have a multienzyme complex nitrogenase, for the reduction of N_2 to NH_4^+ (e.g., Ref. 239). Nothing is known about the movement of N_2 into the intracellular membrane bound nitrogenase complex. Since the nitrogenase contains bound molybdenum, we will consider it further in Sec. III.E. Under conditions where the assimilation of NH_4^+ into the amino acid pool is blocked, NH_4^+ produced from N_2 reduction was excreted from Klebsiella mutants up to levels as high as 5 mM [186]. The mechanism for this excretion has not been studied.

In striking contrast to the indirectness of approach and the paucity of data on NH_4^+ transport with procaryotes, in eucaryotic microbes there have been several studies of NH_4^+ transport systems, utilizing the radioactive ammonium analog $^{14}CH_3NH_3^+$, and studies of regulatory control using mutants to characterize the transport systems. In keeping with our goal of postulating bacterial systems where they have not been directly sought, I will first describe these NH_4^+ transport studies in some detail and then introduce preliminary results from experiments that have grown out of the process of writing this review.

The ammonium transport system of Penicillium chrysogenum was synthesized (derepressed) only under conditions of complete nitrogen starvation [83] and therefore would not have been found by workers using the usual range of nitrogen sources. With high NH_4Cl or with most amino acids as nitrogen sources, the formation of this system was completely prevented. KNO_3 or the two amino acids methionine and leucine allowed formation of the ammonium transport system in Penicillium at somewhat reduced levels [83]. The ammonium transport system of <u>Aspergillus nidulans</u> was synthesized during growth on a variety of amino acids but repressed by growth on high NH_4^+ or urea [162]. The rate of $CH_3NH_3^+$ transport by <u>Saccharomyces cerevisiae</u> was not derepressed by nitrogen starvation and showed relatively little change in rate with different nitrogen growth sources [173]. All of these systems carried out energy requiring concentrative uptake of NH_4^+ or analogs. A series of extensive kinetic experiments led to the conclusion that the substrate specificity for the <u>P. chrysogenum</u> system was:

$$NH_4^+ > CH_3NH_3^+ > CH_3CH_2NH_3^+$$

with K_m values of 0.25 μM, 10 μM, and 100 μM, respectively, and with V_{max} values essentially identical at 10 $\mu mol\ min^{-1}\ g^{-1}$ at 25°C [83]. The <u>A. nidulans</u> system was rather similar with a K_m for $CH_3NH_3^+$ of 20 μM and a V_{max} of 11 $\mu mol\ min^{-1}\ g^{-1}$ [162]. With <u>S. cerevisiae</u> the V_{max} was about the same at 17 $\mu mol\ min^{-1}\ g^{-1}$ but the K_m for $CH_3NH_3^+$ was higher at 220 μM [173]. Since NH_4^+ was again a strong competitive inhibitor of $CH_3NH_3^+$ uptake with a K_i of less than 20 μM [173], a background level of NH_4^+ may have been responsible for this difference. The $CH_3NH_3^+$ transport systems have a pH optimum around 6 to 7 [83,173], far below the pK for $CH_3NH_3^+ \leftrightarrow CH_3NH_2$ of 10.6. With a K_m for NH_4^+ uptake as low as 2.5 x $10^{-7}\ M\ NH_4^+$, it was further argued [83] that NH_4^+ and not NH_3 was the likely substrate, since the NH_3 concentration at pH 6.5 would be vanishingly small at this K_m value: 4.5 x $10^{-10}\ M\ NH_3$. $CH_3NH_3^+$ was accumulated to concentration gradients of 120 [162] to 1,000 [173]. $CH_3NH_3^+$ was not subsequently metabolized in Aspergillis [162], but could be used as sole nitrogen source by Penicillium [83].

Regulation of NH_4^+ transport occurred by both repression-type inhibition of formation of the system and also by feedback inhibition or inactivation of existing transport function [83,162]. Repressed <u>P. chrysogenum</u> accumulated [14C]methylamine at rates 1000 times lower than did derepressed mycelia, and this lower uptake did not show saturation at external $CH_3NH_3^+$ levels up to 1 mM, or 100 times the K_m for the active transport system.

Mutants of both <u>A. nidulans</u> and <u>S. cerevisiae</u> that were defective in NH_4^+ transport were selected on the bases of poor growth on NH_4^+ or of resistance to growth inhibition by high levels of methylamine [162,173]. Several types of mutants exist, with some affecting the K_m and others affecting regulation of or the V_{max} for $CH_3NH_3^+$ transport [162]. Decreased NH_4^+ and $CH_3NH_3^+$ uptake was a recessive trait in both organisms.

Stevenson recently started studies of [14C]$CH_3NH_3^+$ transport with <u>E. coli</u> in order to find the hypothetical NH_4^+ transport system of bacteria. With <u>E. coli</u> strain ML grown on limited glutamate or NH_4^+ as nitrogen source, $CH_3NH_3^+$ uptake was seen with nitrogen-starved cells. Temperature and pH dependencies and saturation kinetics were consistent with one (or two) energy-dependent transport systems for NH_4^+ in E. coli. High affinity $CH_3NH_3^+$ uptake by E. coli showed a K_m near 3 μM (preliminary data). E. coli was not sensitive to growth inhibition by $CH_3NH_3^+$ under any conditions tested to date, and no mutants in this system are currently available.

III. INORGANIC ANION TRANSPORT
SYSTEMS

There has been relatively less effort directed toward studies of inorganic anion transport as compared with cation transport studies, and almost none at all except for studies of phosphate and sulfate transport. I will start with these two and then, as with the cations, move to systems that are very poorly understood and studied.

A. Phosphate Transport Systems

Phosphate transport mechanisms have been more thoroughly studied than those for any other anion. The use of mutants has played a prominent role both with E. coli and with S. faecalis. Nevertheless, understanding of phosphate transport is still in a state of flux, and the picture presented here is rather tentative.

1. E. coli Phosphate Transport

There is a problem that concerns phosphate transport in E. coli to a greater degree than any other section of this review: conflicting and essentially irreconcilable results from different laboratories. Much of the difficulty comes from use of strains mutant in any of the half-dozen genes that affect phosphate transport, and often without full understanding of the geno-type of the test organism. This problem is compounded by the amount of available information that has not been formally published and by changes in the conclusions of some of the workers in the field over the last several years. Therefore, the picture presented for E. coli phosphate transport is only my tentative attempt to select the facts to fit a rational pattern. The detail is justified by the wealth of information that does exist and the multiplicity of transport systems.

Four separate transport systems may move inorganic orthophosphate (Pi) across the cell membrane of E. coli. Their properties are summarized in Table 6-7. There is a constitutive Pi transport system (Pit) and a second Pi transport system (Pst) that functions at significant levels constitutively [10,230] but that functions at five times higher levels after

TABLE 6-7 Phosphate Transport Systems of E. coli

| System | Orthophosphate parameters | |
	K_m (μM)	V_{max} μmol min^{-1} g^{-1}
Pst phoS$^+$	0.2	69
Pst phoS$^-$	0.3	5
Pst phoT$^-$	383	--
Pit	25	60
GlpT	244 (12 μM αGP)	5.6 nmol/min per OD unit
Uhp	N.D. (200 μM G-6-P)	N.D.

Refs. [10,11,177,229-231,233]. N.D., not determined; αGP, K_m for α-glycerol phosphate; G-6-P, K_m for glucose-6-phosphate.

phosphate starvation [177]. Furthermore, two inducible organophosphate transport systems also utilize inorganic phosphate as an alternative substrate (Uhp and GlpT) [10]. Each of the four systems is named for the gene which appears to determine a structural component of the system, and mutations in these genes lead to loss of one or another of the four systems. The Pst transport system is referred to as phosphate specific since it does not accumulate radioactive arsenate (Asi) [10,229-231]. However, there is now a question of whether this lack of arsenate transport is absolute or rather a forty times lower ratio of the (Asi:Pi) substrate affinities of the Pst system than for the Pit system [177]. Mutations in two additional genes affect the functioning of the Pst system. The phoS gene determines the formation of the periplasmic phosphate binding protein [73, 230,231] and is involved in regulation of synthesis of alkaline phosphatase. Mutations in this separate gene led to altered specificity for the Pst system so that this transport system then accepted arsenate as an alternative substrate under some conditions; therefore pst$^+$, phoS$^-$ strains were arsenate sensitive while pst$^+$, phoS$^+$ strains were arsenate-resistant [229, 230]. Rosenberg et al. [177] did not find a change in Asi sensitivity in phoS mutants using previously phosphate-starved cells. There is also a conflict in the results between different laboratories on the question of whether the K_m for the Pst system is altered in phoS mutants. With cells grown on 1 mM Pi, pst$^+$, phoS$^-$ cells showed a greatly reduced affinity for phosphate with a transport K_m of 0.4 μM Pi for the wild-type phoS$^+$, pst$^+$ and only 18 μM Pi for the mutant phoS$^-$ pst$^+$ [230]. The K_m values for both phoS$^+$ and phoS$^-$ cells obtained by using phosphate-starved cells were between 0.2 and 0.4 μM Pi [177]. Mutations in the phoT gene, which is closely linked to phoS and also affects regulation of alkaline phosphatase synthesis, led essentially to loss of inorganic orthophosphate transport by the Pst system [177,229] (Table 6-7).

In an independent study, however, Kida [118] obtained rather different results with the phoS⁻ and phoT⁻ mutant strains and reported as little as 5% of normal Pi uptake in phoS⁻ mutants and normal Pi uptake in phoT⁻ strains. This was, of course, the opposite pattern to that found by Willsky et al. [229] and Rosenberg et al. [177].

The periplasmic phosphate binding protein determined by the phoS gene is the only gene product associated with phosphate transport that has been purified and characterized [73]. As with other periplamsic binding proteins [157], it can be removed from the cells by osmotic shock and cells that have been shocked have diminished capacity for phosphate transport. Synthesis of this protein was partially repressed by growth in 10 mM phosphate [73,231], and the Pst system with which it is involved is 20% constitutive and 80% repressible [10,177,229,231]. Binding protein synthesis was completely constitutive in phoT⁻ mutants [229] and 100-fold repressed in wild type strains by growth in 1 mM Pi [231], which only affected the Pst transport level fivefold [177]. The binding protein was purified starting with an affinity chromatography step on a phosphorylated Sepharose resin. The protein had an affinity constant for phosphate in the region of 10^{-6} M [73]. The homogeneous purified protein had a molecular mass of 38,000, the largest of the purified binding proteins, and contained neither cysteine nor methionine [73]. It was calculated that each cell contained 25,000 molecules of this protein between its inner and outer cellular membranes.

The second constitutive inorganic phosphate transport system is called Pit (for "Pi transport"), and it is generally responsible for the arsenate sensitivity of wild-type E. coli cells. The Pit system has a lower affinity for phosphate than does the Pst system (Table 6-7). The Pit system was missing due to mutation in several lines of the frequently studied E. coli K12 strain [177,203,229], somewhat simplifying study of the remaining Pst system. Experiments on the energy coupling mode of these two Pi transport systems suggests a basic difference in mechanism [10,11,74, 169,177,229,230]. The Pst system has the basic properties of an ATP-driven system, including the utilization of a shock releasable binding protein [74] and resistance to inhibition by uncouplers that collapse the chemiosmotic gradient [177]. Only the Pit system works normally in spheroplasts, although the addition of concentrated phoS binding protein restored Pst activity to spheroplasts [74]. The Pit system, on the contrary, shows the basic attributes of systems energy coupled by the proton motive gradient [12,89], including lack of involvement of a binding protein and sensitivity to uncouplers [177]. Whereas Pi uptake and exchange occurred via the Pit system, uptake via the Pst system was essentially unidirectional in strains having only that system [177].

Phosphate uptake in E. coli required H^+ (for symport with $H_2PO_4^-$; Harold, Gerdes, and Rosenberg, in preparation; reported in Ref. 89) and extracellular K^+ [224]. In neither case has the coupling of phosphate transport to H^+ or K^+ been adequately specified. For example, phosphate is taken up by the E. coli Pst system symport with protons. Although this phosphate uptake is driven by "direct" ATP coupling [74, 177], the ΔpH across the membrane (when present) could provide some energy to the system [89].

In addition to the two constitutive orthophosphate transport systems, two inducible organophosphate transport systems have been reported to transport inorganic phosphate as well [10]. These are the glucose-6-phosphate inducible system for the uptake of hexose phosphates (Uhp) that is missing in mutants of the uhp gene and the glycerol phosphate inducible system that is missing in glpT mutants [10, 203]. There is a question as to whether the glpT or Uhp systems can transport sufficient inorganic phosphate for growth, however, since $GlpT^+, Pst^-, Pit^-$ strains either could [10] or could not [203] grow on inorganic phosphate (P_i) as sole phosphate source in different studies with different isolates. Similarly one uhp-constitutive mutant still required organophosphate for growth [203]. When present and induced both the GlpT and the Uhp systems led to arsenate sensitivity [10], suggesting that arsenate is an alternative substrate for both systems. In fact, the arsenate-resistant mutants selected on glycerol containing media were found to be $GlpT^-$ [10, 229].

Glucose-6-phosphate inhibited P_i and arsenate uptake via the Uhp system and stimulated rapid loss of P_i and arsenate from cells containing both the Uhp and the Pst systems [10]. Mutants selected for arsenate resistance in the presence of glucose-6-phosphate fell into two classes, those that could no longer utilize glucose-6-phosphate for growth (uhp^-) and those that were highly arsenate resistant but still able to grow on glucose-6-phosphate as sole carbon and phosphate source (presumedly due to altered specificity of the Uhp system) [10]. In a mutant strain that had accumulated P_i via the Uhp system only, both glucose-6-phosphate and arsenate chased the P_i from the cells, but glycerol phosphate did not ($glpT^-$ mutant) [10]. Among the unresolved questions of energy coupling is that of why P_i uptake by the Uhp and GlpT systems was completely cyanide resistant [10] whereas organophosphate uptake by these systems was energy dependent and cyanide sensitive [43, 233]. One possible difference may be that P_i uptake was assayed under conditions of P_i/P_i exchange without net uptake, whereas under conditions of net concentrative uptake organophosphate uptake was found to be cyanide sensitive [10] . The Uhp system functioned as a facilitated diffusion system in the presence of energy poisons [233]. Differences in chemiosmotic or nonchemiosmotic energy coupling may also account for these results.

As if four transport systems (with six chromosomal genes so far) were not sufficient, gene(s) conferring arsenate resistance are known to occur on a wide range of antibiotic-resistance plasmids, and there is good evidence that the mechanism of plasmid-determined arsenate resistance also involves changes in specificity of cellular phosphate transport systems (Willsky and Malamy, personal communication).

2. S. faecalis Phosphate Transport

S. faecalis phosphate transport shows many similarities to the Pst system in E. coli. Uptake of phosphate is energy dependent and essentially unidirectional [97]. Stimulation of P_i uptake by K^+ [97] is now considered due to a $[K^+]in/[H^+]out$ exchange process that maintains the intracellular pH that would otherwise be lowered by the symport process of H^+ $H_2PO_4^-$ uptake [101]. The isolation of a mutant that required high phosphate for growth [97] allowed the characterization of two phosphate transport systems in S. faecalis [92, 97]. Both carry out transport of arsenate as well [92]. The major system missing in the mutant strain PT-1 showed a relatively broad flat pH optimum from 6 to 9 and had the following kinetic parameters: K_m, 10 to 20 μM Pi (K_i, 3 μM Asi); K_m, 5 to 10 μM Asi; V_{max} 20 μmol min^{-1} g^{-1} P_i essentially independent of pH [92, 97]. The "minor" system that remains in the mutant strain had a pH peak of activity around 5.5 to 6 and was essentially nonfunctional above pH 7. The K_m of this system rose from 20 μM P_i or 50 μM Asi at pH 5, to 500 μM P_i or Asi at pH 6, to above 3 mM P_i at pH 7.4, while the V_{max} was approximately constant at low pH, but significant transport could not be measured at high pH values [92, 97].

Recently, Harold and Spitz [101] proposed a detailed mechanism for the energy requirement for phosphate-arsenate transport in S. faecalis analogous to the ATP-linked Pst system in E. coli. It is thought that the S. faecalis phosphate transport systems are directly coupled to ATP utilization and not driven by the proton motive force [89, 101]. $H_2PO_4^-$ is accumulated by an electroneutral symport with a proton. The pH gradient of approximately one pH unit was insufficient to drive the 100 or 1000:1 inside:outside gradients of P_i and of Asi that were measured [101]. The evidence for a directly ATP-coupled energy input for phosphate (arsenate) uptake in S. faecalis came from experiments showing a lack of effect of agents that dissipate the membrane potential and pH gradient [101]. These agents included DCCD, which prevents maintenance of the gradient by inhibiting the membrane ATPase, uncouplers such as CCCP, and cation exchanging antibiotics such as nigericin. All these reagents strongly inhibit proton-coupled sugar and amino acid transport systems in the same cells under the same conditions [101]. Furthermore, artificially induced membrane potentials and pH gradients did not induce Pi or Asi uptake, whereas these did induce uptake by the proton-coupled lactose and amino acid systems [101]. Experiments with a wide variety of monovalent cations to facilitate exchange indicated a very striking requirement for an intracellular pH greater than 6.8 for functioning of the Asi uptake process.

This requirement provided an explanation for the inhibition of Asi and P_i uptake by uncouplers and antibiotics under specialized conditions where these agents brought about a reduction of the cellular pH below 6.9 [101].

Regulation of phosphate uptake was governed in addition by a type of feedback mechanism both by glycerol, which would be utilized to deplete the intracellular phosphate pool and by intracellular phosphate itself [101]. The details of these regulatory processes remain uncertain.

3. Other Bacteria

Phosphate transport has been studied in a wide variety of additional bacteria. I will make no attempt to cover reports of this, except for two cases where additional or different properties were found. Mitchell [143], in what may have been the earliest series of studies of phosphate transport in bacteria, demonstrated a system for P_i/P_i exchange across the membranes of unenergized "resting cells" of S. aureus. Phosphate transport later studied in energized E. coli and S. faecalis was generally unidirectional, but Pit system-mediated exchange was found with lactate-grown cells [177]. The S. aureus exchange system showed a low affinity for P_i (K_m of 1.6 mM P_i and V_{max} of 9 μmol min^{-1} g^{-1} (at 25°C). Although unaffected by cyanide, the exchange process in these essentially nonmetabolizing cells was still uncoupler sensitive and showed a temperature dependence with an activation energy of 38 kcal/mol P_i exchange [143]. The pH profile of this exchange process, with a relatively sharp peak near pH 7, differed strikingly from those found subsequently for net P_i transport in S. faecalis.

Recently, Burnell, John, and Whatley [29, 30] demonstrated phosphate transport in subcellular membrane vesicles prepared from Paracoccus denitrificans. This was the first successful measurement of inorganic anion transport in membrane vesicles after many unsuccessful attempts with vesicles from other bacteria [111]. The properties of phosphate transport in P. denitrificans vesicles are consistent with electroneutral symport of H^+ and $H_2PO_4^-$ driven by physiologically or artificially generated pH gradients with the direction of this reversible transport determined solely by the direction of the pH gradient [29]. Right-side-out vesicles accumulated P_i when energized by either reduced phenazine methosulfate or succinate, substrates for the respiratory chain. Inside-out vesicles could not be induced to accumulate P_i by respiratory substrates. The phosphate accumulated by the energized right-side-out vesicles was released rapidly upon addition of uncoupler, which would dissipate the chemiosmotic potential and pH gradients. Transient and much lower levels of phosphate accumulation were achieved with both right-side-out and inside-out membranes when pH gradients were artificially formed by addition of nigericin in the presence of high external K^+ (K^+/H^+ exchange mediated by the nigericin would generate a pH gradient) or by the addition of high external NH_4Cl (NH_3 would diffuse across the membranes and cause a pH gradient, internal alkaline, by forming NH_4^+ within the vesicles). Valinomycin, which would generate a potential gradient ($\Delta\Psi$) under high K^+ conditions, did not induce P_i uptake [29]. The phosphate transport system in these membranes was specific for phosphate in that sulfate did not affect P_i uptake, although the vesicles also showed pH-driven sulfate transport via a separate repressible system [30]. There thus appears to be a fundamental difference in the energy coupling mode between the E. coli Pst system and S. faecalis (directly ATP driven) and the E. coli Pit system and P. denitrificans (pH gradient driven, like mitochondria) with regard to phosphate transport.

B. Sulfate Transport Systems

Bacterial sulfate transport has been most thoroughly studied with S. typhimurium [46,47,154,158,160]. Although there is strong genetic evidence implicating a sulfate binding protein to a role in sulfate transport [154], as with the phosphate binding protein [74,230], the role seems secondary, and the binding protein is not part of the transmembrane transport system itself (see Chap. 10).

The initial report of a SO_4^{2-}, $S_2O_3^{2-}$ transport system in S. typhimurium [46] described most of the basic properties of this system, and studies over the last 14 years have added only a little to our understanding. Dreyfuss [46] found temperature-dependent, energy-dependent uptake of SO_4^{2-} by a system that was missing in cysA mutants and subject to repression by growth on cysteine. L-Djenkolic acid provided a sulfur source that allowed growth of S. typhimurium and derepression of synthesis of the sulfate transport system. Dreyfuss [46] devised transport assay conditions that were rather different from those commonly used in transport studies with other substrates. He used very dense suspensions of cells (10 to 25 mg cell protein per milliliter) and measured the $^{35}SO_4^{2-}$ remaining in solution after filtering off the cells following short time exposures, generally 1 to 3 min at 37°C. CysA mutants defective in the sulfate transport system would grow with sulfite, sulfide, or cysteine as sulfur sources but not with sulfate or thiosulfate, both of which supported growth of wild-type cells. Another mutant blocked in the cysC and cysD genes could accumulate SO_4^{2-} but not metabolize intracellular SO_4^{2-}. Use of this cysC-D deletion mutant allowed studies of SO_4^{2-} transport without the complication of intracellular conversions. SO_4^{2-} transport without the complication of intracellular conversions. SO_4^{2-} followed Michaelis-Menten kinetics, with a K_m of 36 μM SO_4^{2-} and a V_{max} of around 6 μmol min^{-1} g^{-1} cell protein [46]. The uptake of SO_4^{2-} was temperature dependent and dependent upon metabolic energy. The ability of cells to take up SO_4^{2-} or $S_2O_3^{2-}$ was completely missing in all cysA mutants tested and completely repressed by growth of the cells with cysteine. Cysteine-grown cells transferred to djenkolic acid-containing medium showed derepressed transport activity after a lag of only 10 min at 37°C, indicative of little if any intracellular storage of sulfur-containing intermediates [46]. Although cysteine repressed formation of the SO_4^{2-} transport system, cysteine did not inhibit its functioning

[46]. Sulfide was without affect on SO_4^{2-} transport in S. typhimurium, but both thiosulfate and sulfite were inhibitory [46]. A most unusual aspect of SO_4^{2-} uptake was the time course with linear uptake for perhaps 1 min at 37°C followed by net loss of as much as 80% of the accumulated SO_4^{2-}, followed by another smaller burst of uptake and loss [46]. These "overshoot" kinetics were indicative of feedback regulatory control of intracellular SO_4^{2-} level in the absence of SO_4^{2-} metabolism, and the mechanism of this regulation was subsequently studied in greater detail. The overshoot phenomenon occurred only when net sulfate uptake increased the internal sulfate concentration to about 10^{-4} M; lower amounts of sulfate were taken up and retained [47]. Regulation of the SO_4^{2-} level did not involve protein synthesis, and the cycle was repeatable, i.e., cells that had undergone an overshoot cycle could be washed free of significant intracellular SO_4^{2-} and would immediately demonstrate another cycle [47]. The functioning of this cycle was energy dependent in that it could be slowed by use of energy-starved cells and low glucose concentrations and by use of lower temperatures and dense semianaerobic cultures. Studies with mutants blocked in early steps of SO_4^{2-} reduction to H_2S suggested that an early metabolic product of SO_4^{2-} and/or sulfate itself regulated the rate of uptake by "feedback inhibition."

The identification [159] and subsequent purification [155] and crystallization [156] of a sulfate binding protein from the periplasmic space between the inner and outer cell membranes [160], along with the finding that this protein was absent in many cysA mutants [154,159] gave rise to great hope for the actual isolation of a transport "permease". Detailed genetic studies [154], however, showed that while many "transport mutants," selected on the basis of a cysteine requirement or on the basis of chromate resistance, had reduced levels of the binding protein, some of these cysA mutants (including some amber mutants in the gene) had significant amounts of apparently normal sulfate binding protein. These findings relegated the binding protein to a secondary role in SO_4^{2-} transport [154]. Nevertheless the role and properties of this protein are of interest. The SO_4^{2-} binding protein is a single polypeptide with a molecular weight of 32,000 and typical amino acid composition, except for the total absence of sulfur-containing amino acids [155]. Up to 10,000 copies of the SO_4^{2-} binding protein were found per cell [159]. This protein bound specifically a single SO_4^{2-} per molecule with a low dissociation constant that varied with the assay buffer from 4 μM [159] down to 0.02 μM [155], compared with the transport K_m of 36 μM [46]. The substrate specificity of the binding protein differed somewhat from that of SO_4^{2-} uptake. Chromate was a strong inhibitor of SO_4^{2-} binding and transport [155,159]. Thiosulfate did not inhibit SO_4^{2-} binding [155,159], although it was a more effective inhibitor of $^{35}SO_4^{2-}$ uptake than was an equivalent level of $^{32}SO_4^{2-}$ itself [46,159]. Sulfite inhibited both transport and binding, when added at 50 times the $^{35}SO_4^{2-}$ concentration. MoO_4^{2-}, VO_4^{2-}, WoO_4^{2-}, S^{2-}, and cysteine were essentially without effect on either process [159]. With the cysA locus assigned the role of determining a SO_4^{2-} permease, it is rather unusual that three (rather than one) complementation groups have been found at that position [154]. Nevertheless, these groups are referred to as cysAa, cysAb, and cysAc rather than as separate genes [154]. The cysA region cannot include the structural gene for the binding protein, since mutants in all three complementation groups produce binding protein that is immuno-logically and by heat sensitivity tests identical with wild-type binding protein. These binding protein-positive, transport-negative mutants include both amber and deletion mutants [154].

Growth in djenkolate-containing medium derepressed both transport activity and binding protein activity; growth in cysteine repressed both. With wild-type S. typhimurium in SO_4^{2-}-containing medium, however, a difference in regulation was found: production of sulfate binding protein was totally repressed, while SO_4^{2-} transport activity was at least 50% of the derepressed level [154].

Surprisingly little attention has been paid to sulfate transport in E. coli. A 6-year-old note constitutes the only report of energy-dependent uptake of $^{35}SO_4^{2-}$ in E. coli K12. Saturation kinetics (K_m about 50 μM SO_4^{2-}) were found, and SeO_4^{2-} strongly inhibited SO_4^{2-} uptake [204].

With membrane vesicles of P. denitrificans, Burnell et al. [30] found reversible pH driven SO_4^{2-} uptake by a system very similar to but separate from the P_i transport system of these vesicles described in Sec. III.A3. SeO_4^{2-} decreased the rates of uptake and loss of SO_4^{2-} strikingly [30]. Respiratory substrates stimulated SO_4^{2-} uptake by right-side-out but not by inside-out vesicles. This uptake was inhibited by uncouplers that also caused release of accumulated SO_4^{2-} [30]. Much lower levels of SO_4^{2-} uptake could be induced with both inside-out and right-side-out membranes by artificially formed pH gradients, and nigericin plus high K^+ could drive SO_4^{2-} uptake as described before for P_i uptake. Valinomycin plus high K^+ did not induce SO_4^{2-} uptake, and valinomycin inhibited the nigericin-driven transport. Uncouplers were also inhibitory of the nigericin-K^+ driven uptake and that driven by the pH gradient formed by addition of high external NH_4^+ [30]. These results are consistent with a mechanism of electroneutral symport of $2H^+$ and SO_4^{2-} across the membranes, driven by a pH gradient but not by the membrane potential [30].

The sulfate transport systems of filamentous fungi should be considered here both because they are well studied and because there are important differences in substrate specificity between the bacterial and the fungal systems. Although Penicillium has a single sulfate transport system as determined by kinetic studies and by the absence of SO_4^{2-} transport in a mutant [216], Neurospora has two distinct SO_4^{2-} transport systems that are lost separately on mutation [133] and regulated separately in response to organosulfur source and stages of development [134]. Whereas the bacterial system transports SO_4^{2-} and $S_2O_3^{2-}$, and probably SO_3^{2-} and CrO_4^{2-} but not SeO_4^{2-}, MoO_4^{2-}, VO_4^{2-}, or WoO_4^{2-} [46,159], the fungal transport system is broader in its specificity, and SeO_4^{2-} and MoO_4^{2-} are also substrates [135,216]. SO_3^{2-}, however, has a separate transport system and S^{2-} still another [216]. Net uptake of sulfate was an energy-dependent process [135, 216] leading to intracellular SO_4^{2-} as high as 15 to 20 mM [20,135]. Under normal growth conditions uptake was essentially unidirectional [20,135], although non-energy-dependent SO_4^{2-}/SO_4^{2-} exchange could be demonstrated in Neurospora [135] but not in Penicillium [20]. Sulfate transport in fungi is regulated, as in S. typhimurium, both by repression of synthesis of the system in cells grown with methionine as a sulfur source and by feedback inhibition by SO_4^{2-} or early products of sulfate assimilation [20,135]. The Penicillium sulfate system appeared to be coupled symport 1 SO_4^{2-}:1 H^+: 1 Ca^{2+} [41]. The net charge of the substrate-loaded carrier would then be positive, and movement of the anion (plus cations) would be electrogenic and driven by the membrane potential. Experimentally, SO_4^{2-} uptake in Penicillium required any one of several divalent cations, and concomitant $^{45}Ca^{2+}$ and $^{35}SO_4^{2-}$ uptake was measured because of the convenient calcium radioisotope [41].

C. Nitrate and Nitrite Transport

As far as I am aware, there have been little in the way of published reports of nitrate (or nitrite) uptake by bacterial cells and no reports of transport systems. I will summarize the indirect results that might lead one to hypothesize the existence of such systems and then will discuss the comparable systems of eucaryote microbes, especially Neurospora, where evidence for a specific transport system for NO_3^- preceding intracellular NO_3^- reduction [38] has been obtained. In principle, one might expect two very different types of transport systems for the two biological roles that nitrate plays in microorganisms: there is the assimilatory system that provides nitrogen for intermediate metabolism starting with reduction of NO_3^- to the level of NH_4^+, and there is the dissimilatory nitrate reductase that serves as a terminal electron acceptor during anaerobic respiration [164]. The assimilatory system must involve transport so that the NO_3^- or its reduced product can enter intracellular metabolism. The respiratory system functions at much higher rates when serving, in general, energy metabolism, and in this case, either the nitrate must be reduced at the cell surface without entering the cytoplasm, or otherwise there must be rapid egress of the reduced product, NO_2^-, or further reduced forms.

1. Escherichia coli

Garland et al. [72] indirectly determined the movement of NO_3^- (and NO_2^-) across E. coli spheroplast membranes by measuring osmotic swelling in the presence of combinations of the cation conducting materials, nigericin, valinomycin, and CCCP. They could infer that NO_3^- movement occurred and was rate limiting for swelling under their conditions. NO_3^- moved passively downhill and electrogenically, i.e., neither NO_3^-/H^+ symport nor NO_3^-/OH^- antiport was seen. Un-ionized HNO_2 could pass across the membranes and dissociate into H^+ and NO_2^- intracellularly [72]. There was no evidence of NO_3^-/NO_2^- exchange processes in these osmotic swelling experiments. NO_3^- uptake was assumed to occur by passive diffusion and not to be carrier mediated since saturation of rate of uptake was not found at external NO_3^- concentrations up to 0.25 M NO_3^-, although the rate of swelling did not increase linearly with NO_3^- above 0.15 M NO_3^- [72]. The time constant for swelling by NO_3^- was about 0.5 min at 30°C. Garland et al. [72] calculated from the osmotic swelling assays that the rate of nitrate uptake at 0.2 mM external NO_3^- (the K_m for the nitrate reductase enzyme complex of these cells) was 1.0 μmol min^{-1} g^{-1} cell protein. This was only 1/1000 of the rate of nitrate reductase activity at 0.2 mM NO_3^-. Therefore the NO_3^- movement measured in the swelling assays could not account for the rate of NO_3^- reduction by the E. coli spheroplasts, and it was concluded that respiratory NO_3^- reduction occurred at the outer surface of the cell [72]. Experiments on the competitive inhibition of nitrate reduction by azide with E. coli cells as a function of pH also were consistent with interaction between azide and nitrate at a site on the outer surface of bacteria. Garland et al. [72] concluded that the respiratory nitrate reductase of E. coli is oriented vectorially across the cell membrane (see also Ref. 19), so that nitrate reduction occurs on the outer surface, concomitant with the movement of two protons from the inner to the outer surface of the membrane (see Sec. II.I.).

Many bacteria, including E. coli, can utilize NO_3^- as sole nitrogen source by means of an assimilatory nitrate reductase [115,164]. As far as I am aware, no measurements of transport of NO_3^- during this process have been reported, although it is evident that movement from extracellular NO_3^- must accompany or precede the reduction steps to intracellular NH_4^+. The assimilatory systems differ in many respects from the dissimilatory systems; for example, only the former are usually repressed by NH_4^+ and are synthesized both under aerobic and anaerobic growth conditions in the presence of NO_3^- [164]. While the dissimilatory NO_3^- reductase is membrane bound and also capable of reducing chlorate (hence the isolation of chlorate-resistant mutants defective in this enzyme complex), the assimilatory enzyme is soluble, cytoplasmic, and unable to reduce ClO_3^-. Studies with a marine pseudomonad showed a similar half-saturation concentration for net cellular NO_3^- uptake (0.26 mM) and for the assimilatory NO_3^- reductase of the organism (0.29 mM) [25]. However, it is difficult to picture how sufficient intracellular NO_3^- levels might be obtained in the absence of a specific transport system when the Nernst equilibrium ratio at -120 mV potential across the membrane (a reasonable estimate) would be 1/100 of the external NO_3^- level.

2. Neurospora

In the absence of bacterial NO_3^- transport studies, we turn to eucaryotic microbes where transport systems have been studied that might serve as
models for future bacterial work. Both nitrate and nitrite transport systems have been reported in Neurospora [38,180,181]. The nitrate transport system was inducible by either NO_3^- or by NO_2^-. It carried out the
energy-requiring uptake of NO_3^- with a K_m 0f 0.25 mM NO_3^- and a V_{max}
of 1.2 μmol g^{-1} min^{-1} at 25°C [180]. NH_4^+ and NO_2^- were noncompetitive
inhibitors of NO_3^- uptake with K_i values of 0.13 mM NH_4^+ and 0.17 mM NO_2^-.
NH_4^+ did not repress the formation of this assimilatory nitrate transport
system but casamino acids did. NH_4^+ did repress synthesis of the nitrate
reductase in Neurospora. In Penicillium, NH_4^+ caused rapid loss of NO_3^-
transport activity by a process requiring protein synthesis [77]. The
Neurospora nitrate transport system was unaffected by mutations in the
nit-1, nit-2, and nit-3 genes which determine the nitrate reductase enzyme
complex, and therefore transport was separable from subsequent metabolism
of NO_3^- [180]. The internal NO_3^- level in Neurospora built to levels approaching 25 mM NO_3^- in the absence of a functional nitrate reductase and when
the external concentration was 20 mM NO_3^-. However, since there is a
large internal negative membrane potential in Neurospora [199], uptake to
chemical equilibrium required energy for movement against the Nernst
potential equilibrium.

Neurospora has a separate NO_2^- transport system [181] with a K_m of
0.09 mM NO_2^- and a V_{max} of 1.7 μmol g^{-1} min^{-1} at 25°C. This system
was also inducible by either NO_3^- or NO_2^-, but differed from the nitrate
transport system in that it was neither repressed nor inhibited by NO_3^-,
NH_4^+, or amino acids [181]. Whereas wild-type Neurospora and nit-4 and
nit-5 mutants took up nitrite, the nit-2 mutant did not [38]. Since nit-2
mutants are also missing the nitrate reductase, the function of this gene
is likely to be at the regulatory level rather than in determining a common
protein for both nitrite transport and nitrate reduction [38]. Nitrate transport was not affected by the nit-2 mutation [180]. The equilibrium level of
NO_2^- uptake by wild-type mycelium at 30°C was about 1.5 mM NO_2^-, about
equal to the external concentration. At 0°C, the mycelium took up three
times as much NO_2^- as at 30°C [38], possibly because the membrane
potential was dissipated.

D. Chloride Transport

Surprisingly, there have been no studies of bacterial chloride transport.
We therefore hypothesize that bacterial cells must have highly specific
chloride transport systems, generally inwardly oriented, and present the
limited data from bacteria that are consistent with this hypothesis. There

is also limited information about chloride transport systems in eucaryotic microbes and in mammalian red blood cells.

In the absence of data on chloride transport and in the absence of reported chloride requirements, we are left with data comparing chloride content of bacterial cells with the predictions from the Nernst potential equilibrium. The Nernst potential equilibrium with perhaps -120 mV potential (internal negative) would lead to an equilibrium with an internal chloride content of about 1% that of the medium level. Damadian [42] has reported very limited chloride data for E. coli growing in a medium containing no added Cl^- and with less than 10^{-4} M Cl^- (my estimate) as contaminant from casamino acids; the cells contained 1 mM Cl^- during growth and less than that during stationary phase. The Cl^- was freely removable by washing. Schultz et al. [183] measured the intracellular chloride content of E. coli in media containing from 25 to 100 mM Cl^- and estimated the transmembrane potential from the Nernst equation, assuming free permeability of the membrane to Cl^-. Log phase cells showed $Cl^-_{in}:Cl^-_{out}$ ratios of 0.3 regardless of external Cl^- concentration (and for comparison, $Na^+_{in}:Na^+_{out}$ ratios of 0.55 and $K^+_{in}:K^+_{out}$ ratios of 46); stationary phase cells had approximately equal internal and external Cl^- and Na^+ concentrations, and the K^+ ratio had dropped to 2:1 [183]. Christian and Waltho [35] measured Cl^- content for six bacterial species growing in high chloride media and obtained results with nonhalophilic bacteria (S. aureus and Staphylococcus oranienberg; Table 6-8) indicating cellular Cl^- as a few percent of the medium level. With moderate halophilic bacteria and with extreme halophiles, the intracellular chloride levels were about 1/10 or nearly equal to the extracellular chloride levels, respectively (Table 6-8). Assuming that the halophilic bacteria maintain a membrane potential comparable to that found with other bacteria [88-90], then we can conclude that Cl^- is being accumulated by the halophilic bacteria to levels approximately 100 times above electrochemical equilibrium. This must involve an energy-dependent transport process.

Rapid changes in Cl^- that are involved in response to osmotic "upshocks" are also most readily understood with the hypothesis of a Cl^- transport system responsible for energy-dependent movements to govern osmotic equilibrium. "Deplasmolysis" following osmotic plasmolysis of bacterial cells required metabolism, and K^+ was the likely cation involved in setting up the new osmotic equilibrium [60,116]. The question then becomes, what was the balancing anion? PO_4^{3-} levels did not increase in E. coli following an osmotic upshock [60]. By limiting the available anions to Cl^-, Kepes et al. [116] observed a requirement for Cl^- for net K^+ uptake that showed an apparent K_m of perhaps 30 mM Cl^-. Direct Cl^- uptake measurements were not made, but under osmotic upshock conditions with Cl^- as sole available anion, it was "safely assumed" that approximately 120 mM Cl^- uptake accompanied the 120 mM K^+ increase. This very recent report [116] provides for the first time conditions with which it should be possible

TABLE 6-8 Chloride Content of Bacterial Cells[a]

Species	Cl⁻ in Cells (mmol/kg cell water)	Cl⁻ in Medium (mM)	Ratio Cell:Medium
S. aureus	8 ± 3	150	0.053
S. oranienburg	< 5	150	< 0.03
Micrococcus halodenitrificans	55 ± 6	1,000	0.055
Vibrio costicolus	139 ± 25	1,000	0.14
Sarcina morrhuae	$3,660 \pm 250$	4,000	0.91
Halobacterium salinarium	$3,610 \pm 70$	4,000	0.90
Halobacterium sp.	4,900	3,500	1.4

[a]The first six lines are the data from [35]; the last line contains data from [75] with the added assumption (from Ref. 35) that the Halobacterium contains 1 ml water per gram dry weight.

to directly study bacterial chloride movements. Either Cl⁻ or perhaps in some cases substantial pools of dicarboxylic amino acids [136] may provide electroneutrality for the primary cation K^+ under osmotic equilibrium conditions. One can imagine Cl⁻ movement during rapid deplasmolysis followed by slower displacement of Cl⁻ by newly synthesized organic anions.

The chloride transport system of the filamentous ascomycete N. vasinfecta has recently been studied in some detail [139-142]. We discuss this eucaryotic microbe here because of the lack of comparable procaryotic studies. N. vasinfecta concentrated Cl⁻ some 890-fold above the external concentration to about 55 mM Cl⁻ (internal) [140]. Cl⁻ uptake was against its electrochemical gradient [142] and was energy dependent, temperature dependent, and inhibited by dinitrophenol, cyanide, anaerobiosis, and DCCD [140]. Cl⁻ uptake was saturable and kinetic analysis indicated two systems operating with about equal V_{max} values of 2.4 μmol min^{-1} g^{-1} (dry weight) and with K_m values of 6.4 and 100 μM Cl⁻, respectively. The transport system was highly specific: only Br⁻ inhibited Cl⁻ uptake; I⁻, F⁻, NO_3^-, SO_4^{2-}, HCO_3^-, and $H_2PO_4^-$ at 1 mM did not significantly inhibit the rate of uptake of 0.1 mM Cl⁻ [140]. Concomitant cation uptake was not required. The Cl⁻ accumulated by N. vasinfecta was retained during incubation in non-chloride containing medium but was readily released by the membrane active antibiotic nystatin under conditions where cellular macromolecules were not released. The rate of uptake of Cl⁻ by N. vasinfecta was under regulatory control and increased some four to five times during 4-hr incubation in buffer or distilled water [139]. The increase in rate required RNA and protein synthesis but did not seem a usual "derepression" phenomenon as found with bacterial transport systems, since the increase in rate was not inhibited by 0.1 mM Cl^- but rather required the <u>absence</u> of both glucose and K^+ from the suspension medium [139]. A regulatory decrease in rate of Cl^- uptake also occurred, and preincubation for 1 hr in the presence of 1 mM Cl^- or Br^- decreased the rate of $^{36}Cl^-$ uptake by 75% [140]. Surprisingly, since Cl^- uptake was energy dependent, the addition of glucose inhibited chloride uptake by about 50% [141]. Glucose also stimulated Cl^- efflux [141], which in the absence of glucose occurred at only 5% the rate of influx [140]. No suggestion of saturation kinetics for the Cl^- efflux process were seen over the range of internal Cl^- concentrations from 8 to 80 mM [141]. This is consistent with failure to find saturation of most other microbial efflux systems.

The most thoroughly studied chloride transport system is that of the mammalian red blood cell, but I will consider this system only briefly in this review of bacterial transport. The kinetics to equilibrium with this anion exchange system are so rapid that special "stop flow" filtration procedures had to be devised [212] in order to measure efflux with rate constants of $1/0.32$ sec for $^{38}Cl^-$, $1/1.7$ sec for $^{82}Br^-$, $1/3.3$ sec for $^{18}F^-$, and $1/17$ sec for $^{131}I^-$ at 23°C. The specificity of the red blood cell chloride carrier is very broad indeed with rates for "poorer" substrates ranging down to $1/10^4$ that for chloride [80]. The K_m for Cl^- was 35 mM internal Cl^- and the V_{max} 270 μmol/liter of cells per minute at 0°C [80]. Br^-, F^-, and I^- exchanged, but more slowly than Cl^-[212]; SO_4^{2-} and HCO_3^- ($K_i = $ 6 mM HCO_3^-) utilized the same transport system [80], as does also $H_2PO_4^-$ with an uptake K_m of 80 mM and V_{max} of 2.8 mmol/liter of cells per minute at 23°C [106]. Cl^- flux from the red blood cells occurred at a rate approximately 10^6 times that of K^+ efflux from the same cells. In each case, the anion carrier functioned in non-energy–dependent carrier-mediated facilitated diffusion, and thus differs from the Cl^- active transport systems that were studied or hypothesized for microbial systems. This is not surprising since the microbial cells experience variable external chloride conditions and exposure to variable osmotic strength, while the red blood cells exist in stable high Cl^- conditions. The intestinal HCl excretion system is a more likely mammalian source for active Cl^- transport.

The chloride transport protein of human red blood cell membranes has been identified by experiments involving radioactive labeling with highly specific inhibitors [32,106]. This protein differs from the anion binding proteins of bacteria in that it is an intrinsic membrane protein that spans the membrane and is considered to contain the anion channel [179]. Approximately 3×10^5 anion transport sites occur per red blood cell, and these apparently consist of dimers of a protein of about 100,000 molecular weight [32,106,179] which constitutes about 30% of the total red blood cell membrane protein. With estimated exchange rates of 10^{11} Cl^- ions per cell per second [80], one calculates 3×10^5 Cl^- ions per carrier per second, about the same order as turnover numbers for some enzyme reactions.

E. Molybdate Transport

Although molybdenum has long been recognized as an obligatory part of the nitrogenase enzyme responsible for nitrogen fixation in many bacterial types (see Refs. 33,239), it is only recently that the first report appeared [54] of a highly specific MoO_4^{2-} transport system with Clostridium pasteur-ianum. $^{99}MoO_4^{2-}$ uptake was energy dependent, absolutely requiring sucrose, and inhibited by inhibitors of glycolysis. O_2 also completely inhibited MoO_4^{2-} uptake. Michaelis-Menten saturation kinetics were found with a K_m of 48 μM MoO_4^{2-} and a V_{max} of 55 nmol min^{-1} g^{-1} (dry weight) cells. Since SO_4^{2-} was a competitive inhibitor of comparable affinity ($K_i = 30$ μM SO_4^{2-}), the question of whether the system is basically a molybdate or a sulfate system naturally arose. Thiosulfate, a potent competitive inhibitor of sulfate transport in S. typhimurium [46] (Sec. III.B), was without effect on MoO_4^{2-} uptake. VO_3^{2-} was also without effect. WO_4^{2-} was accumulated by this transport system (K_m of 24 μM WO_4^{2-}), and could replace MoO_4^{2-} in the formation of a nonfunctional nitrogenase enzyme in some bacteria (but not in C. pasteurianum) [54]. Once within the cells, very little of the MoO_4^{2-} was exchangeable [54].

High MoO_4^{2-} or high NH_4^+ in the growth medium repressed the level of MoO_4^{2-} transport activity in C. pasteurianum [55], the former presumedly in order to regulate the internal level of MoO_4^{2-} and the latter since NH_4^+ is a more efficient nitrogen source than N_2. As long as NH_4^+ was present in growth medium, no detectable MoO_4^{2-} uptake (nor nitrogenase activity) could be detected [55]. MoO_4^{2-} transport appeared, i.e., was derepressed, over a period of hours following NH_4^+ starvation by a process dependent on protein synthesis. This derepression was independently repressible by high (near 1 mM) MoO_4^{2-} [55]. Following the addition of NH_4^+ to a culture of C. pasteurianum fixing N_2, MoO_4^{2-} transport activity persisted and was not subject to feedback inhibition. However, no additional transport function was synthesized, and the rate of MoO_4^{2-} uptake per cell gradually decreased due to cell growth [55]. Since C. pasteurianum can grow under conditions of high SO_4^{2-} and low NH_4^+, one might assume that separate transport mechanisms exist for SO_4^{2-} and MoO_4^{2-} [54]. Similarly, MoO_4^{2-} repressed formation of the MoO_4^{2-} transport system but SO_4^{2-} did not [55].

Although no one (to my knowledge) has looked for a MoO_4^{2-} transport system in E. coli, molybdenum is an essential constituent of the formate dehydrogenase and nitrate reductase enzymes (e.g., Ref. 57), and pleotropic mutants defective in MoO_4^{2-} assimilation were found at the chlD locus determining chlorate resistance [7,76]. These mutants lack both enzyme activities unless grown on high (10^{-4} to 10^{-3} M) MoO_4^{2-}. ChlD mutants, however, appeared to be defective in some posttransport step of insertion of MoO_4^{2-} into the enzyme [202], and both chlD$^+$ and chlD$^-$ cells had comparable intracellular MoO_4^{2-} levels [76].

There was no indication of the involvement of an extracellular chelate in MoO_4^{2-} transport in C. pasteurianum. However, Bacillus thuringiensis produces a highly specific MoO_4^{2-} coordinating compound under iron starvation conditions [117] that may play a transport function. This peptide chelate containing threonine, glycine, and alanine (molecular weight 550 without MoO_4^{2-}) did not reconstitute nitrogenase activity in vitro under conditions where bacterial and fungal MoO_4^{2-} peptides isolated from nitrate reductase were interchangeable [117] and differed from the MoO_4^{2-} peptide found in nitrogenase as well [117,239].

Until there are further studies of bacterial MoO_4^{2-} transport, we need to pay particular attention to the related systems in eucaryotic microbes. MoO_4^{2-} transport has been studied in the filamentous fungi, Penicillium and Aspergillus [216], but the specificity of this transport system appeared to include $^{35}SO_4^{2-}$, $^{35}S_2O_3^{2-}$, $^{75}SeO_4^{2-}$, and $^{99}MoO_4^{2-}$. S_2^{2-} and SO_3^{2-} appeared to be transported by different systems. That MoO_4^{2-} was being transported on the sulfate system was argued for by the existence of a mutant equally defective in uptake of sulfate, selenate, and molybdate [216].

F. Other Anion Systems

Transport systems for additional inorganic anions have not been demonstrated for bacterial cells, although there are several "trace" elements forming anions known to be required or to stimulate growth of different bacteria or to be involved in particular enzymes. Furthermore we might postulate a HCO_3^- system for bacteria utilizing CO_2 for growth. Among the additional trace systems that we can consider are those for BO_3^{3-}, SeO_4^{2-}, CrO_4^{2-} and WO_4^{2-} in different bacteria. Borate is commonly part of the growth medium of photoheterotrophic bacteria, but no boron-dependent enzymes are known. Chromium is a component of electron transport chains as reviewed by Mertz [138]. Whether the chromium enters the cells as CrO_4^{2-} is not known, nor is it known whether a special CrO_4^{2-} transport system might exist under some growth conditions in addition to the CrO_4^{2-} uptake via the SO_4^{2-} transport system. Selenium is a required component of formate dehydrogenase, the first enzyme of CO_2 fixation [5,205]. ^{75}Se incorporation paralleled the synthesis of formate dehydrogenase in E. coli [187] and Clostridium formicoaceticum [5]. ^{75}Se was associated with a particular 110,000 molecular weight polypeptide from the purified E. coli enzyme [57]. However, no direct studies of $^{75}SeO_4^{2-}$ transport have been performed to my knowledge. It is possible that sufficient SeO_4^{2-} can enter the cells via a relatively nonspecific SO_4^{2-} transport system, but with our central thesis of separate systems for essential cellular ions, I would postulate that a specific trace SeO_4^{2-} transport system could be demonstrated. E. coli "adapted" to high-selenium low-sulfur medium (10 μM Se:23 μM S) accumulated 40% Se:60% S. The selenium was incorporated into proteins as selenomethionine [107].

Silicon has no known required role in procaryotic microbes, although it can be incorporated into organosilicate compounds by some bacteria. Heinen [102] made preliminary characterization of a silicate uptake system in subcellular respiratory fragments of Proteus mirabilis. There is a single report of a $^{31}Si(OH)_4$ transport system in heterotrophic diatoms [6]. The energy-dependent concentrative uptake of silicon by diatoms showed high substrate affinity with a K_m of 4.5 μM $Si(OH)_4$ [6]. Energy inhibitors prevented silicic acid uptake, and germanic acid was a high affinity competitive inhibitor with a K_i of 2 μM. Once taken within the cells as monomeric silicic acid, the silicic acid was polymerized into the silica of the diatom's intracellular "shell".

VO_3^- was growth stimulatory and was incorporated into iron-deficient green algae. VO_3^- appeared involved in chlorophyll biosynthesis [137]. Vanadium may also be involved as a functional analog of molybdenum in the nitrogenase of Azotobacter [9]. If some of these trace anions are essential under special conditions of carbon, nitrogen, and energy production, then we must postulate that specialized high affinity transport systems will assure efficient uptake under limiting conditions.

CO_2 can be fixed as an ancillary or even primary source of cellular carbon. In such cases, the question of passive diffusion of CO_2 across the cellular membrane or carrier-mediated transport of HCO_3^- is rather parallel with the problem of NH_3 vs NH_4^+ movements discussed in Sec. II.G. With a pK_a of 3.8, essentially all of the CO_2 will exist as HCO_3^- at physiological pH values. I know of no studies of HCO_3^- transport as opposed to "assimilation". The use of formate dehydrogenase mutants (chlF in E. coli; Table 6-1) should permit analysis of HCO_3^- transport separate from subsequent metabolism.

IV. COMPARATIVE ASPECTS

A. Uptake vs Extrusion Pumps

Most of this chapter has been concerned with the mechanisms by which bacterial cells accumulate needed ionic nutrients. In a few cases such as Ca^{2+} and Na^+, we have discussed energy-dependent processes that reduce the internal level of cations below that in the external medium or below the Nernst equation equilibrium. Even in the case of active uptake systems, there is generally a mechanism of regulated efflux from the cells. Efflux of most intracellular cations and anions is rather slow and often "coupled" with uptake (e.g., Refs. 101,190,192,195), making it likely that egress is carrier mediated rather than occurring through nonspecific diffusion across the membrane.

Two questions arise: (1) Does egress occur by means of the same carriers as utilized in uptake? and (2) What are the kinetic parameters and energy coupling of egress of ions normally concentrated against an electrochemical gradient? Answers to these questions are most unsatisfactory and incomplete at this time. Proton gradient coupled transport systems such as the inducible lacY gene permease in E. coli may utilize the same "carrier" or channel for uptake and for egress [120,234], while there is evidence that directly ATP-coupled systems such as those for galactose [232] and glutamate [84] utilize separate pathways for uptake and for exit. There is extensive data that the lacY permease can function in both directions, and that under partially energy-inhibited conditions, this system functions to carry out "facilitated transport" of β-galactosides to concentration equilibrium [120,234]. The effect of energy poisons was to reduce the apparent K_m of the exit process to a value of 0.7 mM lactose in energy-poisoned cells from 16 mM lactose with energized cells [234]. Wilson [232] could not find saturation levels for exit of galactose in his studies of egress. No such detailed studies of egress have been performed with cations, and there are some hints that separate mechanisms may be utilized for uptake and for egress. First, the energy poisoning conditions that "uncouple" lactose transport allowing rapid efflux via the lacY carrier have a very different effect with the cations K^+, Mg^{2+} and Mn^{2+}, where in each case movement in both directions is frozen [190,192,195]. Downhill egress as well as uphill concentration requires energy. With the K^+ mutants, two genes (trkB and trkC) primarily affect egress rates without affecting uptake, suggesting that different systems are functioning in the two directions. The egress defect of the trkB gene was measurable in membrane vesicles that required valinomycin for uptake [16]. And finally, in studies of Mn^{2+} regulation in B. subtilis [66], we found 10-fold changes in rate of uptake without discernible changes in the rate of exit, again most readily interpretable as resulting from different systems functioning in different directions. Studies of regulation of constant cation levels under conditions of varying external concentrations should prove very informative.

Winkler [233,234] has introduced a means of estimating the K_m for egress in systems where one cannot obtain intracellular concentrations high enough to saturate the egress system. Assuming that both processes follow Michaelis-Menten kinetics and have the same maximum rate, then the highest concentration ratio (intracellular:extracellular) that is achieved by the cells should equal the ratio of the K_m values for uptake and egress:

$$V_{uptake} = \frac{V_{max-u}[S_{out}]}{K_{m-u} + [S_{out}]} = V_{egress} = \frac{V_{max-o}[S_{in}]}{K_{m-o} + [S_{in}]}$$

at equilibrium when uptake equals egress. Since the maximum "uptake ratio" occurs when the external substrate is low, below the K_m level for uptake, and since the internal substrate never reaches saturating levels (near or above K_m for egress), this equation can be simplified to:

$$\frac{[S_{in}]}{[S_{out}]} = \frac{K_{m-out}}{K_{m-in}}$$

This calculation was introduced in studies of the lacY permease system where there is extensive evidence for changes in K_m of uptake and exit during energization [234]. It is not clear whether such calculations are appropriate to cation systems in the absence of such data.

B. Microrequirements vs Macrorequirements and the Possibility of Passive Diffusion

We have now discussed each of the cation and anion systems one by one in terms of what is known and/or anticipated. Here we compare the properties of the systems in terms of kinetic parameters and substrate specificity. Generally the properties of the transport system will be governed by (1) the intracellular requirements of enzymes and metabolic processes; (2) the availability of the inorganic nutrient in normally experienced growth milieu, and perhaps (3) the abundance of the ion in the primeval seawater in which it is thought that bacteria first arose [48, 68]. It is possible that the apparently arbitrary selection of intracellular cation species (K^+-in, Na^+-out; Mg^{2+}-in, Ca^{2+}-out) reflected the relatively greater abundance of K^+ and Mg^{2+} in ancient seawater [48].

The division into "macro" and "micro" ions is clearly arbitrary. K^+, Na^+, Mg^{2+}, Ca^{2+}, Cl^-, SO_4^{2-}, and PO_4^{3-} are generally available in higher concentrations than other ionic species. The kinetic parameters of the transport systems for these ions generally reflect this with high K_m values (low affinities) and high V_{max} values. This is not always the case, since scavenging transport systems with very high affinities (Kdp for K^+, Pst for P_i) have evolved, undoubtedly for competitive scavenging under nutrient limiting growth conditions. Macronutrient transport systems need relatively little substrate specificity against more rare ionic species, since the alternative substrates would not occur in sufficient concentrations to limit entrance of the primary substrate. For example, the Mg^{2+} transport systems have to be highly substrate specific against Ca^{2+}, a naturally abundant divalent cation, but are not so specific as to exclude Co^{2+}, Ni^{2+}, and Mn^{2+} [109, 161]. Competitive uptake of the more rare alternative substrate provides the cells with low levels of these (sometimes) essential micronutrients without the need for specific micronutrient transport systems.

However, under conditions where growth is limited by the availability of essential micronutrients such as Fe^{3+}, Zn^{2+}, Mn^{2+}, and MoO_3^{2-}, highly substrate-specific transport systems provide a growth advantage. These micronutrient systems tend to have very high affinities (low K_m values), moderate to low V_{max} values, and require high substrate specificity to exclude the related macronutrients. For example, the Mn^{2+} and Fe^{3+} systems totally exclude Mg^{2+} and Ca^{2+}. While K^+ and Mg^{2+} usually serve charge neutralizing roles governed by considerations of ionic size and charge density, the micronutrient cations sometimes play similar stabilization or substrate-enzyme fitting roles but sometimes serve as electron donors or recipients in the enzymatic processes. Micronutrient systems are, in general, subject to repression-induction and/or feedback inhibition-type regulatory control which (from a teleonomic point of view) protects the cells from accumulating high toxic levels under those rare conditions when the external abundance is high.

Bacteria differ from the cells of higher organisms, and especially mammals, in that there is no constancy of extracellular environment. Given widely fluctuating growth conditions, one would not expect passive diffusion to survive the tests of natural selection in highly competitive growth settings. Cells with high affinity transport systems would have growth advantages under conditions of a limiting nutrient. Similarly, cells would find it difficult to retain ionic nutrients if these could diffuse into the very dilute aqueous environment. Therefore, I would be very surprised to find any examples of truly passive diffusion in microbial inorganic ion systems. The basic impermeability of the cell membrane also makes this mode of entrance improbable. The exception here may, of course, be those ions that are in equilibrium with a nonionic relatively lipid-soluble form such as the NH_4^+-NH_3 pair and HCO_3^--CO_2. In the stable milieu of animal cells, these substances appear to cross cell membranes by passive diffusion of the un-ionized form, and the ambient fluid maintains high concentrations of the needed material. With bacterial cells, one would expect that passive diffusion of NH_3 and CO_2 across the membranes could still occur, but that this would only dominate cellular uptake under those occasional conditions of high substrate repression and inhibition of normal trace scavenging active transport systems.

V. CONCLUSION

Studies of bacterial cation and anion transport mechanisms are in their infancy. Analogous studies of amino acid and sugar transport systems (Chaps. 2 to 5) have drawn much more attention, and the mechanism(s) of

amino acid and sugar transport are much better understood. Our starting
point was the hypothesis that for each and every inorganic cation and anion
species that bacterial cells require for growth, one or more highly specialized active transport systems will have been evolved. These systems will
have a degree of specificity for the transport substrate commensurate with
the requirements for growth and the abundances of different ions in natural
environments. Systems carrying out transport of more than one species of
inorganic cation or anion will be no more common than has been found for
amino acid and sugar systems. The very limited range of each such system
will also be analogous to that found with amino acids and sugars, and with
what would be expected with recognition sites hypothetically determined by
amino acid sequences within the carrier membrane proteins. In addition
to the many cations and anions that must be concentrated for growth, there
are also cations and anions whose intracellular concentrations are below
that in common growth media. For these, the process of evolution will have
provided highly specific energy-dependent egress systems with properties
otherwise similar to those of uptake systems. Both uptake and egress
systems will consist of membrane embedded carriers, undoubtedly proteins,
determined by genes, and in general tightly regulated. Sometimes, transport facilitating coordination complexes will be dissociable from the transport proteins. The iron transport chelates and ionophores provide models
for these components. Given this basic thesis, I have tried to organize
the limited experimental data that exist. Where there are no bacterial
data, I have moved over to eucaryotic microbes and sometimes even to
mammalian cells, on the hypothesis that the basic mechanisms of ion
transport will be common to all cellular forms. The further I moved
from sound experimental data, undoubtedly the more mistakes in interpretation and in hypothesis that I will have made. I hope that the distinctions
between results and hypotheses have been made sufficiently clear so as not
to entrap the casual reader, or myself. I would welcome information about
the errors that I have made and the gaps that I have left unfilled. With
such help, this first attempt at a broad coverage review of bacterial cation
and anion transport will have served its primary goal of bringing forth such
questions to experimental attack.

RECENT DEVELOPMENTS

For such a wide ranging and speculative review, space considerations
preclude more than a listing of pertinent papers that have appeared or have
come to my attention during the last year.

Cations

Potassium

Lanyi, J. K. and M. P. Silverman, The state of intracellular K$^+$ in Halo-
bacterium cutirubrum, Canad. J. Microbiol., 18: 993-995 (1972).
Masui, M. and S. Wada, Intracellular concentrations Na$^+$, K$^+$, and Cl$^-$ of
a moderately halophilic bacterium, Canad. J. Microbiol., 19: 1181-
1186 (1973).
Pressman, B. C., Biological applications of ionophores, Ann. Rev.
Biochem., 45: 501-530 (1976).
Ryabova, I. D., G. A. Gorneva, and Yu. A, Ovchinnikov, Effect of
valinomycin on ion transport in bacterial cells and on bacterial growth,
Biochim. Biophys. Acta, 401: 109-118 (1975).

Iron

Dailey, Jr., H. A., Purification and characterization of the membrane-bound
ferrochelatase from Spirillum itersonii, J. Bacteriol., 132: 302-307.
(1977).
Pugsley, A. P. and P. Reeves, The role of colicin receptors in the uptake
of ferrienterochelin by Escherichia coli K-12, Biochem. Biophys.
Res. Commun., 74: 903-911 (1977).
Woodrow, G. C., L. Langman, I. G. Young, and F. Gibson, Mutations
affecting the citrate-dependent iron uptake system in Escherichia coli,
J. Bacteriol., 133 (3) March issue (1978).

Manganese

Haavik, H. I., Possible functions of peptide antibiotics during growth of
producer organisms: Bacitracin and metal(II) ion transport, Acta
Path. Microbiol. Scand., 84B, 117-124 (1976).
Ose, D. E. and I. Fridovich. Superoxide dimutase: Reversible removal of
manganese and its substitution by cobalt, nickel, or zinc, J. Biol.
Chem., 251, 1217-1218 (1976).
Villafranca, J. J., D. E. Ash, and F. C. Wedler, Manganese (II) and
substrate interaction with unadenylated glutamine synthetase (Escherichia coli W). I. Temperature and frequency dependent nuclear magnetic resonance studies, Biochemistry, 15, 536-543 (1976).

Zinc

Chlebowski, J. F. and J. E. Coleman, Zinc and its role in enzymes. In
Metal Ions in Biological Systems, Volume 6 (H. Sigel, ed.) Marcel Dekker,
New York 1976, pp. 1-140.

Hayman, S, J. S. Gatmaitan, and E. K. Patterson, The relationship of extrinsic and intrinsic metal ions to the specificity of a dipeptidase from Escherichia coli B., Biochemistry, 13, 4486-4494 (1974).

Sodium

Lanyi, J. K. and R. E. MacDonald, Existence of electrogenic hydrogen/ sodium ion antiport in Halobacterium halobium cell envelope vesicles, Biochemistry, 15, 4608-4613 (1976).
Lanyi, J. K., V. Yearwood-Drayton, and R. E. MacDonald, Light-induced glutamate transport in Halobacterium halobium envelope vesicles. I. Kinetics of light-dependent and soidum-gradient dependent uptake, Biochemistry, 15, 1595-1603 (1976).
Rodriquez-Navarro, A. and J. Asensio, An efflux mechanism determines the low entry rate of lithium in yeasts, FEBS Letters, 75, 169-172 (1977).

Protons

Harold, F. M. and J. Van Brunt, Circulation of H^+ and K^+ across the plasma membrane is not obligatory for bacterial growth, Science, 197 372-373 (1977).
Stoeckenius, W., The purple membrane of salt-loving bacteria, Scientific Amer. 234, 38-46 (1976).

Ammonium

O'Gara, F. and K. T. Shanmugam, Regulation of nitrogen fixation by Rhizobia. Export of fixed N_2 as NH_4^+, Biochim. Biophys. Acta, 437, 313-321 (1976).
Roon, R. J., J. S. Levy, and F. Larimore, Negative interactions between amino acid and methylamine/ammonia transport systems of Saccharomyces cerevisiae, J. Biol. Chem., 252, 3599-3604 (1977).
Stevenson, R. and S. Silver, Methyammonium uptake by Escherichia coli: Evidence for a bacterial NH_4^+ transport system, Biochem. Biophys. Res. Commun., 75, 1133-1139 (1977).

Inorganic Anions

Phosphate

Burns, D. J. W., and R. E. Beever, Kinetic characterization of the two phosphate uptake systems in the fungus Neurospora crassa, J. Bacteriol., 132: 511-519 (1977).

Willsky, G. R., Inorganic phosphate transport and alkaline phosphatase
    regulation in Escherichia coli. Ph.D. Thesis, Tufts University
    School of Medicine (1977).

Sulfate

Breton, A. and Y. Surdin-Kerjan, Sulfate uptake in Saccharomyces
    cerevisiae: Biochemical and genetic study, J. Bacteriol., 132:
    224-232 (1977).

ACKNOWLEDGMENTS

The research in our laboratory has been supported by grants from the
National Science Foundation (BMS71-01456) and the National Institutes of
Health (AI08062). The ideas presented have been developed with the
active participation of many people (who are often listed as authors of the
papers in the references). I would like especially to thank Eric Eisenstadt
and Paula Jasper, whose PhD theses on cation transport span my time of
learning about this topic. Different sections of the manuscript have had
benefit of suggestions and corrections by experts on specific cations or
anions including M. Failla, F. Harold, R. G. Gerdes, F. Gibson, J. E.
Lusk, M. H. Malamy, B. P. Rosen, H. Rosenberg, and R. Stevenson.
Without Kathelen Farrelly's efforts in keeping track of the bibliography
and in reducing my sketches to the current figures, the task of writing
this manuscript would not have been completed.

REFERENCES

1. L. W. Adler and B. P. Rosen, Functional mosaicism of membrane
    proteins in vesicles of Escherichia coli, J. Bacteriol., 129: 959-966
    (1977).
2. K. Altendorf, F. M. Harold, and R. D. Simoni, Impairment and
    restoration of the energized state in membrane vesicles of a mutant
    of Escherichia coli lacking adenosine triphosphatase, J. Biol. Chem.,
    249:4587-4593 (1974).
3. K. Altendorf, H. Hirata, and F. M. Harold, Accumulation of lipid-
    soluble ions and of rubidium as indicators of the electrical potential
    in membrane vesicles of Escherichia coli, J. Biol. Chem., 250:
    1405-1412 (1975).
4. K. Altendorf and W. Zitzmann, Identification of the DCCD-reactive
    protein of the energy transducing adenosinetriphosphatase complex
    from Escherichia coli, FEBS Lett., 59:268-272 (1975).

5. J. R. Andreesen, E. El Ghazzawi, and G. Gottschalk, The effect of ferrous ions, tungstate, and selenite on the level of formate dehydrogenase in Clostridium formicoaceticum and formate synthesis from CO_2 during pyruvate fermentation, Arch. Microbiol., 96: 103-118 (1974).

6. F. Azam, B. B. Hemmingsen, and B. E. Volcani, Role of silicon in diatom metabolism. V. Silicic acid transport and metabolism in the heterotrophic diatom Nitzschia alba, Arch. Microbiol., 97: 103-114 (1974).

7. B. J. Bachmann, K. B. Low, and A. L. Taylor, Recalibrated linkage map of Escherichia coli K-12, Bacteriol. Rev., 40:116-167 (1976).

8. R. S. Beauchamp, S. Silver, and J. W. Hopkins, Uptake of Mg^{2+} by KB cells, Biochim. Biophys. Acta, 225:71-76 (1971).

9. J. R. Benemann, C. E. McKenna, R. F. Lie, T. G. Traylor, and M. D. Kamen, The vanadium effect in nitrogen fixation by Azotobacter, Biochim. Biophys. Acta, 264:25-38 (1972).

10. R. L. Bennett, Studies on the utilization of orthophosphate by wild type and arsenate resistant strains of Escherichia coli, Ph.D. thesis, Tufts University (1973).

11. R. L. Bennett and M. H. Malamy, Arsenate resistant mutants of Escherichia coli and phosphate transport, Biochem. Biophys. Res. Commun., 40:496-503 (1970).

12. E. A. Berger and L. A. Heppel, Different mechanisms of energy coupling for the shock-sensitive and shock-resistant amino acid permeases of Escherichia coli, J. Biol. Chem., 249:7747-7755 (1974).

13. P. Bhattacharyya, Active transport of manganese in isolated membranes of Escherichia coli, J. Bacteriol., 104:1307-1311 (1970).

14. P. Bhattacharyya, Active transport of manganese in isolated membrane vesicles of Bacillus subtilis, J. Bacteriol., 123:123-127 (1975).

15. P. Bhattacharyya and E. M. Barnes, Jr., ATP-Dependent calcium transport in isolated membrane vesicles from Azotobacter vinelandii, J. Biol. Chem., 251:5614-5619 (1976).

16. P. Bhattacharyya, W. Epstein, and S. Silver, Valinomycin induced uptake of potassium in membrane vesicles from Escherichia coli, Proc. Natl. Acad. Sci. USA, 68:1488-1492 (1971).

17. A. E. Blaurock, Bacteriorhodopsin: a trans-membrane pump containing α-helix, J. Mol. Biol., 93:139-158 (1975).

18. L. I. Boguslavsky, A. A. Kondrashin, I. A. Kozlov, S. T. Metelsky, V. P. Skulachev, and A. G. Volkov, Charge transfer between water and octane phases by soluble mitochondrial ATPase (Fl), bacteriorhodopsin and respiratory chain enzymes, FEBS Lett., 50:223-226 (1975).

19. D. H. Boxer and R. A. Clegg, A transmembrane location for the proton-translocating reduced ubiquinone →nitrate reductase segment of the respiratory chain of Escherichia coli, FEBS Lett., 60:54-57 (1975).

20. G. Bradfield, P. Somerfield, T. Meyn, M. Holby, D. Babcock, D. Bradley, and I. H. Segel, Regulation of sulfate transport in filamentous fungi, Plant Physiol., 46:720-727 (1970).

21. V. Braun, R. E. W. Hancock, K. Hantke, and A. Hartmann, Functional organization of the outer membrane of Escherichia coli: phage and colicin receptors as components of iron uptake systems. J. Supramol. Struct., 5:37-58 (1976).

22. J. Broach, C. Neumann, and S. Kustu, Mutant strains (nit) of Salmonella typhimurium with a pleiotropic defect in nitrogen metabolism, J. Bacteriol., 128:86-98 (1976).

23. F. Bronner, and T. S. Freund, Calcium accumulation during sporulation of Bacillus megaterium, in Spores V (H. O. Halvorson, R. Hanson, and L. L. Campbell, eds.), American Society for Microbiology, Washington, D.C., 1972, pp. 187-190.

24. F. Bronner, W. C. Nash and E. E. Golub, Calcium transport in Bacillus megaterium, in Spores VI (P. Gerhardt, R. N. Costilow, and H. L. Sadoff, eds.), American Society for Microbiology, Washington, D.C., 1975, pp. 356-361.

25. C. M. Brown, D. S. Macdonald-Brown, and S. O. Stanley, Inorganic nitrogen metabolism in marine bacteria: nitrate uptake and reduction in a marine pseudomonad, Marine Biol., 31:7-13 (1975).

26. K. A. Brown and C. Ratledge, Iron transport in Mycobacterium smegmatis: ferrimycobactin reductase (NAD(P)H:ferrimycobactin oxidoreductase), the enzyme releasing iron from its carrier, FEBS Lett., 53:262-266 (1975).

27. G. F. Bryce and N. Brot, Studies on the enzymatic synthesis of the cyclic trimer of 2,3-dihydroxy-N-benzoyl-L-serine in Escherichia coli, Biochemistry, 11:1708-1715 (1972).

28. F. Bucheder and E. Broda, Energy-dependent zinc transport by Escherichia coli, Eur. J. Biochem., 45:555-559 (1974).

29. J. N. Burnell, P. John, and F. R. Whatley, Phosphate transport in membrane vesicles of Paracoccus denitrificans, FEBS Lett., 58:215-218 (1975).

30. J. N. Burnell, P. John, and F. R. Whatley, The reversibility of active sulfate transport in membrane vesicles of Paracoccus denitrificans, Biochem. J., 150:527-536 (1975).

31. B. R. Byers, Iron transport in gram-positive and acid-fast bacilli, in Microbial Iron Metabolism (J. B. Neilands, ed.), Academic, New York, 1974, pp. 83-105.

32. Z. I. Cabantchik and A. Rothstein, Membrane proteins related to anion permeability of human red blood cells. I. Localization of disulfonic stilbene binding sites in proteins involved in permeation. J. Membr. Biol., 15:207-226 (1974).

33. J. Cardenas and L. E. Mortenson, Role of molybdenum in dinitrogen fixation by Clostridium pasteurianum, J. Bacteriol., 123:978-984 (1975).

34. J. R. Chipley and H. M. Edwards, Jr., Cationic uptake and exchange in Salmonella enteritidis, Can. J. Microbiol., 18:509-513 (1972).

35. J. H. B. Christian and J. A. Waltho, Solute concentrations within cells of halophilic and non-halophilic bacteria, Biochim. Biophys. Acta, 65:506-508 (1962).

36. V. P. Cirillo, Membrane potentials and permeability, Bacteriol. Rev., 30:68-79 (1966).

37. J. G. Cobley and B. A. Haddock, The respiratory chain of Thiobacillus ferrooxidans: the reduction of cytochromes by Fe^{2+} and the preliminary characterization of rusticyanin a novel "blue" copper protein, FEBS Lett., 60:29-33 (1975).

38. A. Coddington, Biochemical studies on the nit mutants of Neurospora crassa, Mol. Gen. Genet., 145:195-206 (1976).

39. G. N. Cohen and J. Monod, Bacterial permeases, Bacteriol. Rev., 21:169-194 (1957).

40. G. B. Cox, F. Gibson, and L. McCann, Reconstitution of oxidative phosphorylation and the adenosine triphosphate-dependent transhydrogenase activity by a combination of membrane fractions from uncA⁻ and uncB⁻ mutant strains of Escherichia coli K12, Biochem. J., 134:1015-1021 (1973).

41. J. Cuppoletti and I. H. Segel, Kinetics of sulfate transport by Penicillium notatum. Interactions of sulfate, protons, and calcium, Biochemistry, 14:4712-4718 (1975).

42. R. Damadian, Biological ion exchanger resins. I. Quantitative electrostatic correspondence of fixed charge and mobile counter ion, Biophys. J., 11:739-760 (1971).

43. G. W. Dietz, Dehydrogenase activity involved in the uptake of glucose 6-phosphate by a bacterial membrane system, J. Biol. Chem., 247:4561-4565 (1972).

44. D. R. Di Masi, J. C. White, C. A. Schnaitman, and C. Bradbeer, Transport of vitamin B12 in Escherichia coli: common receptor sites for vitamin B12 and the E colicins on the outer membrane of the cell envelope, J. Bacteriol., 115:506-513 (1973).

45. L. A. Drachev, A. A. Jasaitis, A. D. Kaulen, A. A. Kondrashin, E. A. Liberman, I. B. Nemecek, S. A. Ostroumov, A. Y. Semenov, and V. P. Skulachev, Direct measurement of electric current generation by cytochrome oxidase, H^+-ATPase and bacteriorhodopsin, Nature, 249:321-324 (1974).

46. J. Dreyfuss, Characterization of a sulfate- and thiosulfate-transporting system in Salmonella typhimurium, J. Biol. Chem., 239:2292-2297 (1964).

47. J. Dreyfuss and A. B. Pardee, Regulation of sulfate transport in Salmonella typhimurium, J. Bacteriol., 91:2275-2280 (1966).

48. F. Egami, Origin and early evolution of transition element enzymes. J. Biochem. (Tokyo), 77:1165-1169 (1975).

49. E. Eisenstadt, Potassium content during growth and sporulation in Bacillus subtilis, J. Bacteriol., 112:264-267 (1972).

50. E. Eisenstadt, S. Fisher, C.-L. Der, and S. Silver, Manganese transport in Bacillus subtilis W23 during growth and sporulation, J. Bacteriol., 113:1363-1372 (1973).

51. E. Eisenstadt and S. Silver, Calcium transport during sporulation in Bacillus subtilis, in Spores V (H. O. Halvorson, R. Hanson, and L. L. Campbell, eds.), American Society for Microbiology, Washington, D.C., 1972, pp. 180-186.

52. E. Eisenstadt and S. Silver, Restoration of cation transport during germination, in Spores V (H. O. Halvorson, R. Hanson, and L. L. Campbell, eds), American Society for Microbiology, Washington, D. C., 1972, pp. 443-448.

53. D. J. Ellar, M. W. Eaton, C. Hogarth, B. J. Wilkinson, J. Deans, and J. La Nauze, Comparative biochemistry and function of forespore and mother-cell compartments during sporulation of Bacillus megaterium cells, in Spores VI (P. Gerhardt, R. N. Costilow, and H. L. Sadoff, eds.), American Society for Microbiology, Washington, D.C., 1975, pp. 425-433.

54. B. B. Elliott and L. E. Mortenson, Transport of molybdate by Clostridium pasteurianum, J. Bacteriol., 124:1295-1301 (1975).

55. B. B. Elliott and L. E. Mortenson, Regulation of molybdate transport by Clostridium pasteurianum, J. Bacteriol., 127:770-779 (1976).

56. T. Emery, Biosynthesis and mechanism of action of hydroxamate-type siderochromes, in Microbial Iron Metabolism (J. B. Neilands, ed.), Academic, New York, 1974, pp. 107-123.

57. H. G. Enoch and R. L. Lester, The purification and properties of formate dehydrogenase and nitrate reductase from Escherichia coli, J. Biol. Chem., 250:6693-6705 (1975).

58. W. Epstein and M. Davies, Potassium-dependent mutants of Escherichia coli K-12, J. Bacteriol., 101:836-843 (1970).

59. W. Epstein and B. S. Kim, Potassium transport loci in Escherichia coli K-12, J. Bacteriol., 108:639-644 (1971).

60. W. Epstein and S. G. Schultz, Cation transport in Escherichia coli. V. Regulation of cation content. J. Gen. Physiol., 49:221-234 (1965).

61. W. Epstein and S. G. Schultz, Cation transport in Escherichia coli. VI. K^+ exchange, J. Gen. Physiol., 49:469-481 (1966).

62. M. L. Failla, Zinc: Functions and transport in microorganisms, in Microorganisms and Minerals (E. D. Weinberg, ed.), Dekker, New York, 1977, pp. 151-214.

63. M. L. Failla, C. D. Benedict, and E. D. Weinberg, Accumulation and storage of Zn^{2+} by Candida utilis, J. Gen. Microbiol., 94:23-36 (1976).

64. M. L. Failla and E. D. Weinberg, Cyclic accumulation of zinc by Candida utilis during growth in batch culture, J. Gen. Microbiol., 99: 85-97 (1977).

65. R. H. Fillingame, Identification of the dicyclohexylcarboiimide-reactive protein component of the adenosine 5'-triphosphate energy-transducing system of Escherichia coli, J. Bacteriol., 124:870-883 (1975).

66. S. Fisher, L. Buxbaum, K. Toth, E. Eisenstadt, and S. Silver, Regulation of manganese accumulation and exchange in Bacillus subtilis W23, J. Bacteriol., 113: 1373-1380 (1973).

67. L. Frank and I. Hopkins, Sodium-stimulated transport of glutamate in Escherichia coli, J. Bacteriol., 100:329-336 (1969).

68. E. Frieden, The chemical elements of life, Sci. Am., 227:52-60 (1972).

69. G. E. Frost, Iron transport in Escherichia coli, Ph.D. thesis, Australian National University, Canberra, Australia (1974).

70. G. E. Frost and H. Rosenberg, The inducible citrate-dependent iron transport system in Escherichia coli, Biochim. Biophys. Acta, 330: 90-101 (1973).

71. G. E. Frost and H. Rosenberg, Relationship between the tonB locus and iron transport in Escherichia coli, J. Bacteriol., 124:704-712 (1975).

72. P. B. Garland, J. A. Downie, and B. A. Haddock, Proton translocation and the respiratory nitrate reductase of Escherichia coli, Biochem. J., 152:547-559 (1975).

73. R. G. Gerdes and H. Rosenberg, The relationship between the phosphate-binding protein and a regulator gene product from Escherichia coli, Biochim. Biophys. Acta, 351:77-86 (1974).

74. R. G. Gerdes, K. P. Strickland, and H. Rosenberg, Restoration of phosphate transport by the phosphate binding protein in spheroplasts of Escherichia coli, J. Bacteriol., 131: 512-518 (1977).

75. M. Ginzburg, B. Z. Ginzburg, and D. C. Tosteson, The effect of anions on K^+-binding in a Halobacterium species, J. Membr. Biol., 6:259-268 (1971).

76. J. H. Glaser and J. A. DeMoss, Phenotypic restoration by molybdate of nitrate reductase activity in chlD mutants of Escherichia coli, J. Bacteriol., 108:854-860 (1971).

77. J. Goldsmith, J. P. Livoni, C. L. Norberg, and I. H. Segel, Regulation of nitrate uptake in Penicillium chrysogenum by ammonium ion, Plant Physiol., 52:362-367 (1973).

78. E. E. Golub and F. Bronner, Bacterial calcium transport: energy dependent calcium uptake by membrane vesicles from Bacillus megaterium, J. Bacteriol., 119:840-843 (1974).

79. D. E. Green, G. Blondin, R. Kessler, and J. H. Southard, Paired moving charge model of energy coupling. III. Intrinsic ionophores in energy coupling systems, Proc. Natl. Acad. Sci. USA, 72:896-900 (1975).

80. R. B. Gunn, M. Dalmark, D. C. Tosteson, and J. O. Wieth, Characteristics of chloride transport in human red blood cells, J. Gen. Physiol., 61:185-206 (1973).

81. Th. Günther and F. Dorn, Über den K-Transport bei der K-Mangelmutante E. coli B525, Z. Naturforsch., 21B:1082-1088 (1966).

82. S. K. Guterman, Colicin B: mode of action and inhibition by enterochelin, J. Bacteriol., 114:1217-1224 (1973).

83. S. L. Hackette, G. E. Skye, C. Burton, and I. H. Segel, Characterization of an ammonium transport system in filamentous fungi with methylammonium-^{14}C as the substrate, J. Biol. Chem., 245:4241-4250 (1970).

84. Y. S. Halpern, H. Barash, and K. Druck, Glutamate transport in Escherichia coli K-12: nonidentity of carriers mediating entry and exit, J. Bacteriol., 113:51-57 (1973).

85. R. E. W. Hancock, K. Hantke, and V. Braun, Iron transport in Escherichia coli K-12: involvement of the colicin B receptor and of a citrate-inducible protein, J. Bacteriol., 127:1370-1375 (1976).

86. K. Hantke and V. Braun, Membrane receptor dependent iron transport in Escherichia coli, FEBS Lett., 49:301-305 (1975).

87. K. Hantke and V. Braun, A function common to iron-enterochelin transport and action of colicins B, I, V in Escherichia coli, FEBS Lett., 59:277-281 (1975).

88. F. M. Harold, Conservation and transformation of energy by bacterial membranes, Bacteriol. Rev., 36:172-230 (1972).

89. F. M. Harold, Membranes and energy transduction in bacteria. Curr. Topics in Bioenergetics, 6:83-149 (1977).

90. F. M. Harold and K. Altendorf, Cation transport in bacteria: K^+, Na^+, and H^+. Curr. Topics in Membranes and Transport, 5:1-50 (1974).

91. F. M. Harold and J. R. Baarda, Inhibition of potassium transport by sodium in a mutant of Streptococcus faecalis, Biochemistry, 6:3107-3110 (1967).

92. F. M. Harold and J. R. Baarda, Interaction of arsenate with phosphate-transport systems in wild-type and mutant Streptococcus faecalis, J. Bacteriol., 91:2257-2262 (1966).

93. F. M. Harold and J. R. Baarda, Gramicidin, valinomycin and cation permeability of Streptococcus faecalis, J. Bacteriol., 94:53-60 (1967).

94. F. M. Harold and J. R. Baarda, Inhibition of membrane transport in Streptococcus faecalis by uncouplers of oxidative phosphorylation and its relationship to proton conduction, J. Bacteriol., 96:2025-2034 (1968).

95. F. M. Harold and J. R. Baarda, Inhibition of membrane-bound adenosine triphosphatase and of cation transport in Streptococcus faecalis by N,N'-dicyclohexylcarbodiimide, J. Biol. Chem., 244: 2261-2268 (1969).

96. F. M. Harold, J. R. Baarda, and E. Pavlasova, Extrusion of sodium and hydrogen ions as the primary process in potassium ion accumulation by Streptococcus faecalis, J. Bacteriol., 101:152-159 (1970).

97. F. M. Harold, R. L. Harold, and A. Abrams, A mutant of Streptococcus faecalis defective in phosphate uptake, J. Biol. Chem., 240:3145-3153 (1965).

98. F. M. Harold, R. L. Harold, J. R. Baarda, and A. Abrams, A genetic defect in retention of potassium by Streptococcus faecalis, Biochemistry, 6:1777-1784 (1967).

99. F. M. Harold and E. Levin, Lactic acid translocation: terminal step in glycolysis by Streptococcus faecalis, J. Bacteriol., 117:1141-1148 (1974).

100. F. M. Harold and D. Papineau, Cation transport and electrogenesis by Streptococcus faecalis. II. Proton and sodium extrusion, J. Membr. Biol., 8:45-62 (1972).

101. F. M. Harold and E. Spitz, Accumulation of arsenate, phosphate, and aspartate by Streptococcus faecalis, J. Bacteriol., 122:266-277 (1975).

102. W. Heinen, Ion accumulation in bacterial systems. III. Respiration-dependent accumulation of silicate by a particulate fraction from Proteus mirabilis cell-free extracts, Arch. Biochem. Biophys., 120:101-107 (1967).

103. R. Henderson, The structure of the purple membrane from Halobacterium halobium: analysis of the X-ray diffraction pattern, J. Mol. Biol., 93:123-138 (1975).

104. R. Henderson and P. N. T. Unwin, Three-dimensional model of purple membrane obtained by electron microscopy, Nature, 257:28-32 (1975).

105. H. Hirata, K. Altendorf, and F. M. Harold, Role of an electrical potential in the coupling of metabolic energy to active transport by membrane vesicles of Escherichia coli, Proc. Natl. Acad. Sci. USA, 70:1804-1808 (1973).

106. M. K. Ho and G. Guidotti, A membrane protein from human erythrocytes involved in anion exchange, J. Biol. Chem., 250:675-683 (1975).

107. R. E. Huber, I. H. Segel, and R. S. Criddle, Growth of Escherichia coli on selenate, Biochim. Biophys. Acta, 141:573-586 (1967).

108. P. L. Jasper, The potassium transport system of Rhodopseudomonas
     capsulata, J. Bacteriol., 133 (3) March issue (1978).
109. P. Jasper and S. Silver, Magnesium transport in microorganisms,
     in Microorganisms and Minerals (E. D. Weinberg, ed.), Dekker,
     New York, 1977, pp. 7-47.
110. P. L. Jasper and S. Silver, Divalent cation transport in Rhodo-
     pseudomonas capsulata, J. Bacteriol., 133 (3) March issue (1978).
111. H. R. Kaback, Transport studies in bacterial membrane vesicles,
     Science, 186:882-892 (1974).
112. R. J. Kadner and G. L. Liggins, Transport of vitamin B$_{12}$ in
     Escherichia coli: genetic studies, J. Bacteriol., 115:514-521 (1973).
113. S. Kahane, M. Marcus, H. Barash, Y. S. Halpern, and H. R. Kaback,
     Sodium-dependent glutamate transport in membrane vesicles of
     Escherichia coli K-12, FEBS Lett., 56:235-239 (1975).
114. L. P. Kayushin and V. P. Skulachev, Bacteriorhodopsin as an electro-
     genic proton pump: reconstitution of bacteriorhodopsin proteolipo-
     somes generating $\Delta\Psi$ and ΔpH, FEBS Lett., 39:39-42 (1974).
115. J. D. Kemp and D. E. Atkinson, Nitrite reductase of Escherichia
     coli specific for reduced nicotinamide adenine dinucleotide, J.
     Bacteriol., 92:628-634 (1966).
116. A. Kepes, J. Meury, A. Robin, and J. Jimeno-Abendano, Some ion
     transport systems in E. coli (transport of potassium and anionic sugars).
     In Biochemistry of Membrane Transport (G. Semenza and E. Carafoli,
     eds.), Springer-Verlag, Berlin, 1977, pp. 633-647.
117. P. A. Ketchum and M. S. Owens, Production of a molybdenum-
     coordinating compound by Bacillus thuringiensis, J. Bacteriol.,
     122:412-417 (1975).
118. S. Kida, The biological function of the R2a regulatory gene for
     alkaline phosphatase in Escherichia coli, Arch. Biochem. Biophys.,
     163:231-237 (1974).
119. D. Kleiner, Ammonium uptake by nitrogen fixing bacteria. I.
     Azotobacter vinelandii, Arch. Microbiol., 104:163-169 (1975).
120. A. L. Koch, The role of permease in transport, Biochim. Biophys.
     Acta, 79:177-200 (1964).
121. G. B. Kolata, Water structure and ion binding: a role in cell
     physiology? Science, 192:1220-1222 (1976).
122. H. E. Kubitschek, M. L. Freedman, and S. Silver, Potassium
     uptake in synchronous and synchronized cultures of Escherichia coli,
     Biophys. J., 11:787-797 (1971).
123. F.-C. Kung, J. Raymond, and D. A. Glaser, Metal ion content of
     Escherichia coli versus cell age, J. Bacteriol., 126:1089-1095 (1976).
124. J. Leong and J. B. Neilands, Mechanisms of siderophore iron trans-
     port in enteric bacteria, J. Bacteriol., 126:823-830 (1976).
125. A. Lewis, J. Spoonhower, R. A. Bogomolni, R. H. Lozier, and
     W. Stoeckenius, Tunable laser resonance Raman spectroscopy of
     bacteriorhodopsin, Proc. Natl. Acad. USA, 71:4462-4466 (1974).

126. F. J. Lombardi, J. P. Reeves, and H. R. Kaback, Mechanisms of active transport in isolated bacterial membrane vesicles. XIII. Valinomycin-induced rubidium transport, J. Biol. Chem., 248:3351-3365 (1973).

127. R. H. Lozier, R. A. Bogomolni, and W. Stoeckenius, Bacteriorhodopsin: a light-driven proton pump in Halobacterium halobium, Biophys. J., 15:955-962 (1975).

128. M. Luckey, J. R. Pollack, R. Wayne, B. N. Ames, and J. B. Neilands, Iron uptake in Salmonella typhimurium: utilization of exogenous siderochromes as iron carriers, J. Bacteriol., 111:731-738 (1972).

129. R. K. J. Luke and F. Gibson, Location of three genes concerned with the conversion of 2,3-dihydroxybenzoate into enterochelin in Escherichia coli K-12, J. Bacteriol., 107:557-562 (1971).

130. J. E. Lusk and E. P. Kennedy, Magnesium transport in Escherichia coli, J. Biol. Chem., 244:1653-1655 (1969).

131. J. E. Lusk, R. J. P. Williams, and E. P. Kennedy, Magnesium and the growth of Escherichia coli, J. Biol. Chem., 243:2618-2624 (1968).

132. R. E. Marquis and E. L. Carstensen, Electric conductivity and internal osmolality of intact bacterial cells, J. Bacteriol., 113:1198-1206 (1973).

133. G. A. Marzluf, Genetic and metabolic controls for sulfate metabolism in Neurospora crassa: isolation and study of chromate-resistant and sulfate transport-negative mutants, J. Bacteriol., 102:716-721 (1970).

134. G. A. Marzluf, Regulation of sulfate transport in Neurospora by transinhibition and by inositol depletion, Arch. Biochem. Biophys., 156:244-254 (1973).

135. G. A. Marzluf, Uptake and efflux of sulfate in Neurospora crassa, Biochim. Biophys. Acta, 339:374-381 (1974).

136. J. C. Measures, Role of amino acids in osmoregulation of non-halophilic bacteria, Nature, 257:398-400 (1975).

137. H.-U. Meisch and H.-J. Bielig, Effect of vanadium on growth, chlorophyll formation and iron metabolism in unicellular green algae, Arch. Microbiol., 105:77-82 (1975).

138. W. Mertz, Chromium occurrence and function in biological systems, Physiol. Rev., 49:163-239 (1969).

139. A. G. Miller and K. Budd, The development of an increased rate of Cl^- uptake in the ascomycete Neocosmospora vasinfecta, Can. J. Microbiol., 21:1211-1216 (1975).

140. A. G. Miller and K. Budd, Halide uptake by the filamentous ascomycete Neocosmospora vasinfecta, J. Bacteriol., 121:91-98 (1975).

141. A. G. Miller and K. Budd, Influence of exogenous sugars and polyols on Cl^- influx and efflux by the ascomycete Neocosmospora vasinfecta, J. Bacteriol., 126:690-698 (1976).

142. A. G. Miller and K. Budd, Evidence for a negative membrane potential and for movement of Cl⁻ against its electrochemical gradient in the ascomycete Neocosmospora vasinfecta, J. Bacteriol., 128:741-748 (1976).

143. P. Mitchell, Transport of phosphate across the osmotic barrier of Micrococcus pyogenes: specificity and kinetics, J. Gen. Microbiol., 11:73-82 (1954).

144. P. Mitchell, Chemiosmotic Coupling and Energy Transduction, Glynn Research Ltd., Bodmin, Cornwall, 1968.

145. P. Mitchell, A chemiosmotic molecular mechanism for proton-translocating adenosine triphosphatases, FEBS Lett., 43:189-194 (1974).

146. J. B. Neilands (ed.), Microbial Iron Metabolism, Academic, New York, 1974.

147. D. L. Nelson and E. P. Kennedy, Magnesium transport in Escherichia coli. Inhibition by cobaltous ion, J. Biol. Chem., 246:3042-3049 (1971).

148. D. L. Nelson and E. P. Kennedy, Transport of magnesium by a repressible and a nonrepressible system in Escherichia coli, Proc. Natl. Acad. Sci. USA, 69:1091-1093 (1972).

149. I. G. O'Brien, G. B. Cox, and F. Gibson, Enterochelin hydrolysis and iron metabolism in Escherichia coli, Biochim. Biophys. Acta, 237:537-549 (1971).

150. P. Oehr and K. Willecke, Citrate-Mg^{2+} transport in Bacillus subtilis. Studies with 2-fluoro-L-erythro-citrate as a substrate, J. Biol. Chem., 249:2037-2042 (1974).

151. D. Oesterhelt, M. Meentzen, and L. Schuhmann, Reversible dissociation of the purple complex in bacteriorhodopsin and identification of 13-cis and all-trans-retinal as its chromaphores, Eur. J. Biochem., 40: 453-463 (1973).

152. D. Oesterhelt and W. Stoeckenius, Functions of a new photoreceptor membrane, Proc. Natl. Acad. Sci. USA, 70:2853-2857 (1973).

153. D. Oesterhelt and W. Stoeckenius, Rhodopsin-like protein from the purple membrane of Halobacterium halobium, Nature New Biol., 233:149-152 (1971).

154. N. Ohta, P. R. Galsworthy, and A. B. Pardee, Genetics of sulfate transport by Salmonella typhimurium, J. Bacteriol., 105:1053-1062 (1971).

155. A. B. Pardee, Purification and properties of a sulfate-binding protein from Salmonella typhimurium, J. Biol. Chem., 241:5886-5892 (1966).

156. A. B. Pardee, Crystallization of a sulfate-binding protein (permease) from Salmonella typhimurium, Science, 156:1627-1628 (1967).

157. A. B. Pardee, Membrane transport proteins, Science, 162:632-637 (1968).

158. A. B. Pardee and L. S. Prestidge, Cell-free activity of a sulfate binding site involved in active transport, Proc. Natl. Acad. Sci. USA, 55:189-191 (1966).

159. A. B. Pardee, L. S. Prestidge, M. B. Whipple, and J. Dreyfuss, A binding site for sulfate and its relation to sulfate transport into Salmonella typhimurium, J. Biol. Chem., 241:3962-3969 (1966).

160. A. B. Pardee and K. Watanabe, Location of sulfate-binding protein in Salmonella typhimurium, J. Bacteriol., 96:1049-1054 (1968).

161. M. H. Park, B. B. Wong, and J. E. Lusk, Mutants in three genes affecting transport of magnesium in Escherichia coli: genetics and physiology, J. Bacteriol., 126:1096-1103 (1976).

162. J. A. Pateman, E. Dunn, J. R. Kinghorn, and E. C. Forbes, The transport of ammonium and methylammonium in wild type and mutant cells of Aspergillus nidulans, Mol. Gen. Genet., 133:225-236 (1974).

163. W. H. N. Paton and K. Budd, Zinc uptake by Neocosmospora vasinfecta, J. Gen. Microbiol., 72:173-184 (1972).

164. W. J. Payne, Reduction of nitrogenous oxides by microorganisms, Bacteriol. Rev., 37:409-452 (1973).

165. G. Pilwat and U. Zimmermann, Untersuchungen uber den Kalium-transport bei Escherichia coli B. II. Die Abhangigkeit der intra-zellularen Kaliumkonzentration von der extrazellularen Kalium- und Natriumkonzentration im stationaren Zustand bei E. coli B525, Z. Naturforsch., 27B:62-67 (1972).

166. A. P. Pugsley and P. Reeves, Iron uptake in colicin B-resistant mutants of Escherichia coli K-12, J. Bacteriol., 126:1052-1062 (1976).

167. E. Racker and P. C. Hinkle, Effect of temperature on the function of a proton pump, J. Membr. Biol., 17:181-188 (1974).

168. E. Racker and W. Stoeckenius, Reconstitution of purple membrane vesicles catalyzing light-driven proton uptake and adenosine tri-phosphate formation, J. Biol. Chem., 249:662-663 (1974).

169. A. S. Rae and K. P. Strickland, Uncoupler and anaerobic resistant transport of phosphate in Escherichia coli, Biochem. Biophys. Res. Commun., 62:568-576 (1975).

170. D. B. Rhoads and W. Epstein, Energy coupling to net K transport in Escherichia coli K-12, J. Biol. Chem., 252: 794-1401 (1977).

171. D. B. Rhoads, F. B. Waters, and W. Epstein, Cation transport in Escherichia coli. VIII. Potassium transport mutants, J. Gen. Physiol., 67:325-341 (1976).

172. R. N. Robertson and N. K. Boardman, The link between charge separation, proton movement and ATPase reactions, FEBS Lett., 60:1-6 (1975).

173. R. J. Roon, H. L. Even, P. Dunlop, and F. L. Larimore, Methyl-amine and ammonia transport in Saccharomyces cerevisiae, J. Bacteriol., 122:502-509 (1975).

174. B. P. Rosen, Restoration of active transport in a Mg^{2+}-adenosine triphosphatase-deficient mutant of Escherichia coli, J. Bacteriol., 116:1124-1129 (1973).

175. B. P. Rosen and L. W. Adler, The maintenance of the energized membrane state and its relation to active transport in Escherichia coli, Biochim. Biophys. Acta, 387:23-36 (1975).

176. B. P. Rosen and J. S. McClees, Active transport of calcium in inverted membrane vesicles of Escherichia coli, Proc. Natl. Acad. Sci. USA, 71:5042-5046 (1974).

177. H. Rosenberg, R. G. Gerdes, and K. Chegwidden, Two systems for the uptake of phosphate in Escherichia coli, J. Bacteriol., 131:505-511 (1977).

178. H. Rosenberg and I. G. Young, Iron transport in the enteric bacteria, in Microbial Iron Metabolism (J. B. Neilands, ed.), Academic, New York, 1974, pp. 67-82.

179. A. Rothstein, Z. I. Cabantchik, and P. Knauf, Mechanism of anion transport in red blood cells: role of membrane proteins, Fed. Proc., 35:3-10 (1976).

180. R. H. Schloemer and R. H. Garrett, Nitrate transport system in Neurospora crassa, J. Bacteriol., 118:259-269 (1974).

181. R. H. Schloemer and R. H. Garrett, Uptake of nitrite by Neurospora crassa, J. Bacteriol., 118:270-274 (1974).

182. P. Scholes, P. Mitchell, and J. Moyle, The polarity of proton translocation in some photosynthetic microorganisms, Eur. J. Biochem., 8:450-454 (1969).

183. S. G. Schultz, N. L. Wilson, and W. Epstein, Cation transport in Escherichia coli. II. Intracellular chloride concentration, J. Gen. Physiol., 46:159-166 (1962).

184. H. Scribner, E. Eisenstadt, and S. Silver, Magnesium transport in Bacillus subtilis W23 during growth and sporulation, J. Bacteriol., 117: 1224-1230 (1974).

185. H. E. Scribner, J. Mogelson, E. Eisenstadt, and S. Silver, Regulation of cation transport during bacterial sporulation, in Spores VI (P. Gerhardt, R. N. Costilow, and H. L. Sadoff, eds.), American Society for Microbiology, Washington, D.C., 1975, pp. 346-355.

186. K. T. Shanmugam and R. C. Valentine, Microbial production of ammonium ion from nitrogen, Proc. Natl. Acad. Sci. USA, 72:136-139 (1975).

187. A. C. Shum and J. C. Murphy, Effects of selenium compounds on formate metabolism and coincidence of selenium-75 incorporation and formic dehydrogenase activity in cell-free preparations of Escherichia coli, J. Bacteriol., 110:447-449 (1972).

188. S. Silver, Active transport of magnesium in Escherichia coli, Proc. Natl. Acad. Sci. USA, 62:764-771 (1969).

189. S. Silver, Calcium transport in microorganisms, in <u>Microorganisms and Minerals</u> (E. D. Weinberg, ed.), Dekker, New York, 1977, 49-103.

190. S. Silver and D. Clark, Magnesium transport in <u>Escherichia coli</u>. Interference by manganese with magnesium metabolism, <u>J. Biol. Chem.</u>, <u>246</u>:569-576 (1971).

191. S. Silver and P. Jasper, Manganese transport in microorganisms, in <u>Microorganisms and Minerals</u> (E. D. Weinberg, ed.), Dekker, New York, 1977, pp. 105-149.

192. S. Silver, P. Johnseine, and K. King, Manganese active transport in <u>Escherichia coli</u>, <u>J. Bacteriol.</u>, <u>104</u>:1299-1306 (1970).

193. S. Silver, P. Johnseine, E. Whitney, and D. Clark, Manganese-resistant mutants of <u>Escherichia coli</u>: physiological and genetic studies, <u>J. Bacteriol.</u>, <u>110</u>:186-195 (1972).

194. S. Silver and M. L. Kralovic, Manganese accumulation by <u>Escherichia coli</u>: evidence for a specific transport system, <u>Biochem. Biophys. Res. Commun.</u>, <u>34</u>:640-645 (1969).

195. S. Silver, E. Levine, and P. M. Spielman, Cation fluxes and permeability changes accompanying bacteriophage infection of <u>Escherichia coli</u>, <u>J. Virol.</u>, <u>2</u>:763-771 (1968).

196. S. Silver, K. Toth, P. Bhattacharyya, E. Eisenstadt, and H. Scribner, Changes and regulation of cation transport during bacterial sporulation, in <u>Comparative Biochemistry and Physiology of Transport</u> (L. Bolis, K. Bloch, S. E. Luria, and F. Lynen, eds.), North-Holland, Amsterdam, 1974, pp. 393-408.

197. S. Silver, K. Toth, and H. Scribner, Facilitated transport of calcium by cells and subcellular membranes of <u>Bacillus subtilis</u> and <u>Escherichia coli</u>, <u>J. Bacteriol.</u>, <u>122</u>:880-885 (1975).

198. R. D. Simoni and A. Shandell, Energy transduction in <u>Escherichia coli</u>. Genetic alteration of a membrane polypeptide of the (Ca^{2+}, Mg^{2+})·ATPase complex, <u>J. Biol. Chem.</u>, <u>250</u>:9421-9427 (1975).

199. C. L. Slayman, W. S. Long, and C.Y.-H. Lu, The relationship between ATP and an electrogenic pump in the plasma membrane of <u>Neurospora crassa</u>, <u>J. Membr. Biol.</u>, <u>14</u>:305-338 (1973).

200. C. W. Slayman, The genetic control of membrane transport, in <u>Current Topics in Membranes and Transport</u>, Vol. 4 (F. Bronner and A. Kleinzeller, eds.), Academic, New York, 1973, pp. 1-174.

201. G. A. Snow, Mycobactins: iron-chelating growth factors from mycobacteria, <u>Bacteriol. Rev.</u>, <u>34</u>:99-125 (1970).

202. G. T. Sperl and J. A. DeMoss, <u>chlD</u> Gene function in molybdate activation of nitrate reductase, <u>J. Bacteriol.</u>, <u>122</u>:1230-1238 (1975).

203. G. F. Sprague, Jr., R. M. Bell, and J. E. Cronan, Jr., A mutant of <u>Escherichia coli</u> auxotrophic for organic phosphates: evidence for two defects in inorganic phosphate transport, <u>Mol. Gen. Genet.</u>, <u>143</u>:71-77 (1975).

204. S. E. Springer and R. E. Huber, Evidence for a sulfate transport system in Escherichia coli K-12, FEBS Lett., 27:13-15 (1972).

205. T. C. Stadtman, Selenium biochemistry, Science, 183:915-922 (1974).

206. L. F. Strenkoski and B. T. DeCicco, pH-conditional, ammonia assimilation-deficient mutants of Hydrogenomonas eutropha: evidence for the nature of the mutation, J. Bacteriol., 105:296-302 (1971).

207. I. Suzuki, U. Dular, and S. C. Kwok, Ammonia or ammonium ion as substrate for oxidation by Nitrosomonas europaea cells and extracts, J. Bacteriol., 120:556-558 (1974).

208. R. L. Switzer, The inactivation of microbial enzymes in vivo. Ann. Rev. Microbiol., 31:135-157 (1977).

209. P. Thipayathasana, Isolation and properties of Escherichia coli ATPase mutants with altered divalent metal specificity for ATP hydrolysis, Biochim. Biophys. Acta, 408:47-57 (1975).

210. J. Thompson and R. A. MacLeod, Na^+ and K^+ gradients and α-aminoisobutyric acid transport in a marine pseudomonad, J. Biol. Chem., 248:7106-7111 (1973).

211. A. Torriani, Alkaline phosphatase subunits and their dimerization in vivo, J. Bacteriol., 96:1200-1207 (1968).

212. D. C. Tosteson, Halide transport in red blood cells, Acta Physiol. Scand., 46:19-41 (1959).

213. T. Tsuchiya and B. P. Rosen, Restoration of active calcium transport in vesicles of an Mg^{2+}-ATPase mutant of Escherichia coli by wild-type Mg^{2+}-ATPase, Biochem. Biophys. Res. Commun., 63:832-838 (1975).

214. T. Tsuchiya and B. P. Rosen, Characterization of an active transport system for calcium in inverted membrane vesicles of Escherichia coli, J. Biol. Chem., 250:7687-7692 (1975).

215. T. Tsuchiya and B. P. Rosen, Calcium transport driven by a proton gradient in inverted membrane vesicles of Escherichia coli, J. Biol. Chem., 251:962-967 (1976).

216. J. W. Tweedie and I. H. Segel, Specificity of transport processes for sulfur, selenium, and molybdenum anions by filamentous fungi, Biochim. Biophys. Acta, 196:95-106 (1970).

217. D. W. Urry, A molecular theory of ion-conducting channels: a field-dependent transition between conducting and nonconducting conformations, Proc. Natl. Acad. Sci. USA, 69:1610-1614 (1972).

218. D. W. Urry, M. M. Long, M. Jacobs, and R. D. Harris, Conformation and molecular mechanisms of carriers and channels, Ann. N.Y. Acad. Sci., 264:203-220 (1975).

219. A. S. Visser and P. W. Postma, Permeability of Azotobacter vinelandii to cations and anions, Biochim. Biophys. Acta, 298:333-340 (1973).

220. G. Vogel and R. Steinhart, ATPase of Escherichia coli: purification, dissociation, and reconstitution of the active complex from isolated subunits, Biochemistry, 15:208-216 (1976).

221. C. C. Wang and A. Newton, An additional step in the transport of iron defined by the tonB locus of Escherichia coli, J. Biol. Chem., 246:2147-2151 (1971).

222. R. Wayne and J. B. Neilands, Evidence for common binding sites for ferrichrome compounds and bacteriophage Ø80 in the cell envelope of Escherichia coli, J. Bacteriol., 121:497-503 (1975).

223. M. Webb, Interrelationships between the utilization of magnesium and the uptake of other bivalent cations by bacteria, Biochim. Biophys. Acta, 222:428-439 (1970).

224. P. L. Weiden, W. Epstein, and S. G. Schultz, Cation transport in Escherichia coli. VII. Potassium requirement for phosphate uptake, J. Gen. Physiol., 50:1641-1661 (1967).

225. A. A. Weiss, S. Silver, and T. G. Kinscherf, Cation transport alteration association with plasmid-determined resistance to cadmium, J. Bacteriol., submitted (1978).

226. I. C. West and P. Mitchell, Proton/sodium ion antiport in Escherichia coli, Biochem. J., 144:87-90 (1974).

227. K. Willecke, E.-M. Gries, and P. Oehr, Coupled transport of citrate and magnesium in Bacillus subtilis, J. Biol. Chem., 248:807-814 (1973).

228. D. B. Willis and H. L. Ennis, Ribonucleic acid and protein synthesis in a mutant of Bacillus subtilis defective in potassium retention, J. Bacteriol., 96:2035-2042 (1968).

229. G. R. Willsky, R. L. Bennett, and M. H. Malamy, Inorganic phosphate transport in Escherichia coli: involvement of two genes which play a role in alkaline phosphatase regulation, J. Bacteriol., 113: 529-539 (1973).

230. G. R. Willsky and M. H. Malamy, The loss of the phoS periplasmic protein leads to a change in the specificity of a constitutive inorganic phosphate transport system in Escherichia coli, Biochem. Biophys. Res. Commun., 60:226-233 (1974).

231. G. R. Willsky and M. H. Malamy, Control of the synthesis of alkaline phosphatase and the phosphate-binding protein in Escherichia coli, J. Bacteriol., 127:595-609 (1976).

232. D. B. Wilson, Properties of the entry and exit reactions of the betamethyl galactoside transport system in Escherichia coli, J. Bacteriol., 126:1156-1165 (1976).

233. H. H. Winkler, Energy coupling of the hexose phosphate transport system in Escherichia coli, J. Bacteriol., 116:203-209 (1973).

234. H. H. Winkler and T. H. Wilson, The role of energy coupling in the transport of β-galactosides by Escherichia coli, J. Biol. Chem., 241:2200-2211 (1966).

235. G. C. Woodrow, Ph.D. thesis, Australian National University, Canberra, Australia (1976).

236. G. C. Woodrow, I. G. Young, and F. Gibson, Mu-induced polarity in the Escherichia coli K-12 ent gene cluster: evidence for a gene (entG) involved in the biosynthesis of enterochelin, J. Bacteriol., 124:1-6 (1975).

237. I. G. Young, L. Langman, R. K. J. Luke, and F. Gibson, Biosynthesis of the iron-transport compound enterochelin: mutants of Escherichia coli unable to synthesize 2,3-dihydroxybenzoate, J. Bacteriol., 106:51-57 (1971).

238. M. Zarlengo and A. Abrams, Selective penetration of ammonia and alkylamines into Streptococcus faecalis and their effect on glycolysis, Biochim. Biophys. Acta, 71:65-77 (1963).

239. W. G. Zumft and L. E. Mortenson, The nitrogen-fixing complex of bacteria, Biochim. Biophys. Acta, 416:1-52 (1975).

Chapter 7

TRANSPORT OF PEPTIDES IN BACTERIA

John W. Payne

Department of Botany
University of Durham
Durham, DH1 3LE England

and

Charles Gilvarg

Department of Biochemical Sciences
Frick Chemical Laboratory
Princeton University
Princeton, New Jersey

  I. INTRODUCTION 326
 II. ABBREVIATIONS AND TERMINOLOGY 327
III. GENERAL OBSERVATIONS ON PEPTIDE TRANSPORT
     AND UTILIZATION 327
     A. Theoretical Models for Peptide Transport 327
     B. Nutritional Aspects 330
     C. Distinction Between Amino Acid and Peptide Transport . . . 331
     D. Role of Proteases and Peptidases in Peptide Utilization . . . 333
 IV. METHODS OF STUDYING PEPTIDE TRANSPORT 337
     A. Genetic and Culture Considerations 337
     B. Direct Methods 338
     C. Indirect Methods 339
  V. PROPERTIES OF PEPTIDE TRANSPORT SYSTEMS 340
     A. Distinction Between Dipeptide and Oligopeptide Transport . . 341
     B. Energetics of Peptide Transport 342
     C. Role of Binding Proteins in Peptide Transport 345
     D. Kinetic Parameters of Peptide Transport 345
     E. Regulation of Peptide Transport Systems 346

VI. STRUCTURAL REQUIREMENTS FOR PEPTIDE TRANSPORT
        IN MICROORGANISMS 347
    A. N-Terminal α-Amino Group 347
    B. C-Terminal Carboxyl Group 354
    C. Stereospecificity 358
    D. α-Peptide Bond 363
    E. Peptide Bond Nitrogen Atom 364
    F. α-Hydrogen Atom 365
    G. Amino Acid Side Chains 365
    H. Other Parameters: Future Studies 368
        REFERENCES 369

I. INTRODUCTION

It has been recognized since the beginning of the century that peptides are
especially valuable in the nutrition of many microbial species, but only in
recent years have the details of peptide utilization begun to be understood.
It is now clear that to be utilized nutritionally peptides must first be hydrolyzed to their free amino acid constitutents. This hydrolysis may conceivably occur before, during, or after uptake into the microbial cytoplasm:
this review will concern itself mainly with the third of these alternatives,
namely the transport of intact peptides across the cytoplasmic membrane
and their subsequent intracellular hydrolysis. Most of the information on
microbial peptide transport has been obtained using bacteria, especially
Escherichia coli. However, in the last several years extensive information
has been obtained for additional bacterial species and for other microorganisms, e.g., yeast and fungi, and this now allows us to present
broader based conclusions than was possible in previous reviews on
microbial peptide transport and metabolism [7, 41, 99, 131, 138-140, 143,
169, 170, 185]. We shall in addition discuss the pertinence of these peptide
transport studies to an understanding of the many and varied biological
activities that are being ascribed to peptides in nature, and will consider
the likely direction and application of future studies in this area.
    Although recent studies have shown that peptide transport in micro-
organisms and in mammals (especially intestinal absorption) share many
common features, and much fruitful collaboration has taken place between
workers in the two areas, we shall not discuss mammalian transport here;
this aspect has been the subject of several recent reviews [96-98, 113-115,
188-191, 202].

II. ABBREVIATIONS AND TERMINOLOGY

Standard three-letter abbreviations have been used for free and peptide
bound amino acids (see Ref. 65). Amino acids are in the L-configuration
unless otherwise indicated. An oligopeptide is considered to comprise
three or more amino acid residues.

Confusion over the use of the term permease is widespread [23], not
least in respect to peptide transport. We shall here continue to apply the
criteria most recently discussed [7], namely that the terms (di) oligopeptide
permease(s) and (di)oligopeptide transport system(s) are interchangeable.
As it is presently not known whether the oligopeptide transport system
comprises one or several specific, genetically determined components,
we shall accordingly continue (see Ref. 7) to use the symbols OPT and opt
(oligopeptide transport system) for the phenotype and genotype, respectively,
rather than use OPP and opp (oligopeptide permease) as was used in earlier
publications [5, 6, 29]. Thus, an oligopeptide transport deficient mutant
might be altered in some specific component which might not permanently
form part of a membrane bound permease. This situation could arise for
example, if the mutation were in an oligopeptide binding protein that occurred not as a "peripheral" permease protein (as conceived in the fluid mosaic
membrane model of Singer [177]), perhaps intermittently dissociating from
the membrane into the periplasm, but existed in a permanently free form
in the periplasm, serving to take peptides across the cell envelope and
deliver them in the vicinity of the permease, as has been speculated by
others [125].

III. GENERAL OBSERVATIONS ON
PEPTIDE TRANSPORT AND
UTILIZATION

A. Theoretical Models for Peptide
Transport

There are a number of possible routes by which exogenous peptides, such
as are derived from partial hydrolysis of proteins, might be utilized by
microorganisms. Several possible models are shown in Fig. 7-1; related
schemes have been discussed earlier for bacteria [7, 140] and for intestinal
peptide transport [97, 98]. It is important to have these various possibilities clear from the outset, although experimental evidence relating to the
operation of the various models will only be presented in later sections.

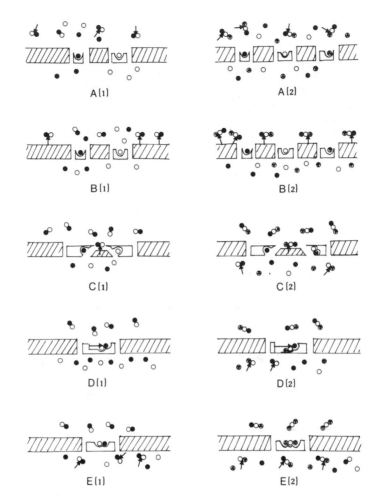

FIG. 7-1 Theoretical models for peptide utilization. The cytoplasmic
membrane is represented by crosshatched section, and contains a number
of specific carriers. Periplasm, cell wall, and cell exterior are above,
and cell interior is below the membrane. Peptidase activity is represented
by an arrow. Hydrolysis and transport of dipeptides is on the left (A1, etc.),
and of oligopeptides is on the right (A2, etc.). (A1, A2) Hydrolysis by free
peptidases either extracellularly or within the cell wall or periplasm,
followed by amino acid transport. (B1, B2) Hydrolysis by membrane bound
peptidases followed by amino acid transport. (C1, C2) Hydrolysis by membrane bound peptidases followed by transport of cleavage products through
associated transport sites. (D1, D2) Group translocation in which membrane bound hydrolase carrier serves both to hydrolyze and to transport
peptides. (E1, E2) Transport of intact peptides followed by intracellular
hydrolysis. (See text for discussion.)

In Models A1 and A2 exogeneous peptides are hydrolyzed by free extra-cellular, cell wall, or periplasmic enzymes, and the liberated amino acids are transported by specific amino acid permeases. Good experimental evidence exists for the occurrence of this system in several microbial species (see Sec. III.D1a). In this case, competition for uptake by a permease could be observed between a free exogenous amino acid and a related amino acid residue supplied in peptide linkage, as well as between residues from di- and oligopeptides.

The second model B has much in common with A, but it is also related to schemes C and D in that all three envisage that peptide transport and hydrolysis are both functions associated with the cell wall/membrane. In B1 and B2, peptidases bound to the cell wall or externally to the cytoplasmic membrane liberate free amino acids (and initially smaller peptides in B2) that can diffuse outwards or be absorbed through amino acid permeases. The possibilities for competition are as for A above. However, the following observations all indicate the absence of this mechanism is most microorganisms: (1) clear distinctions between amino acid and peptide transport mutants; (2) lack of competition between, and different rates of uptake for, free and peptide-bound amino acids; and (3) general failure to demonstrate the presence of such hydrolases (see Sec. III.D1a).

In the third scheme, variations of which have been considered in relation to intestinal peptide transport [98,188], it is envisaged that a membrane bound hydrolase generates cleavage products (without any out-ward release) that are taken up by associated carriers available only to products released by the hydrolase and inaccessible to identical substances in free solution. The hydrolysis and translocation are considered to act in sequential fashion, thus amounting to a group translocation process; but the peptidase and the carrier are separate functional entities. One could envisage one or two carriers associated with a dipeptidase (C1). For oligopeptides, an aminopeptidase of broad specificity might sequentially pass amino acids to a carrier of equally broad specificity (not shown), or alternatively an oligopeptidase might pass on a mixture of an amino acid and dipeptide (C2) or a mixture of smaller peptide cleavage products (not shown). Specificity would reside in the peptidases, and there might con-ceivably be one or more for dipeptides and one or more different ones for oligopeptides. This scheme can accommodate most of the available infor-mation on microbial peptide transport, e.g., the distinction between amino acid and peptide transport mutants (see Sec. III.C), or the general failure to demonstrate membrane-associated peptidases (see Sec. III.D1) if the liberated amino acids cannot be released into free solution. However, it fails to explain the fact that certain peptides naturally resistant to intra-cellular peptidase activity can be accumulated intact (see Sec. III.C), and that mutations in intracellular peptidases and environmental manipulations that inactivate intracellular peptidases can lead to an intracellular build-up of uncleaved peptides (see Sec. III.D2).

The fourth scheme (D) is a simple group translocation mechanism [24] in which a peptide binds from the outside and undergoes an obligatory coupled process of hydrolysis and expulsion of cleavage products on the inside. Here, peptidase activity and translocation activity are both properties of a single carrier entity. It could be speculated that the relatively high-standard free energy of hydrolysis of peptide bonds might be used to help fuel this translocation process. Similar comments and qualifications apply to the occurrence of this model as were made for (C) above.

Model E, although simple, readily accounts for all the important characteristics of peptide transport observed in E. coli and many other bacterial species (see below). Here specific di- and oligopeptide carriers, distinct from amino acid carriers, deliver intact peptides to the cell interior where they are hydrolyzed to free amino acids. This is "true" peptide transport, and this review will consider the considerable biochemical and genetic evidence for its occurrence. However, it is important to note that in many instances the available evidence does not permit us to exclude involvement of additional uptake models such as are indicated by (C) and (D).

B. Nutritional Aspects

The broad nutritional aspects of peptide transport have been discussed elsewhere [99,140], and only a few points will be made here. Bacteriological handbooks that list recommended culture media make it abundantly clear that peptides (in the form of varied commercial peptones) are nutritionally important for a vast array of microorganisms. However, it is likely that the peptones are probably acting only as (preferred) sources of amino acids. If peptones are recommended components of the culture media for a particular species, it may reasonably be assumed that the species can transport peptides. More detailed knowledge of the nature of the recommended peptone provides preliminary information on the most likely mode of peptide utilization, e.g., by model A/B or E (see Fig. 7-1). Thus, "proteoses" contain a much larger percentage of polypeptides than do "peptones" and will therefore be utilized efficiently only by those species that possess extracellular hydrolytic activity able to cleave the polypeptides to fragments that can traverse the microbial envelope. Information on the size distribution of peptides in commercial peptones has been obtained from gel filtration studies, and related to efficiency of microbial utilization [52,55,120,142,149,153,204,217,218].

The nutritional fate of different peptides varies among species, and additionally, for any particular species can be markedly changed by varying the constituents of the culture medium. With E. coli auxotrophs and related species, it is observed that the growth rate and growth yield are almost always identical upon equivalent amounts of free and peptide bound amino acid. However, many species exist for which this is not the case. For example, with amino acid auxotrophs of Pseudomonas, the growth rate on certain peptides was slower than with the free amino acids, while in contrast, the growth yield of a histidine auxotroph was 5 to 10 times greater on histidine peptides than on the free amino acid [20]. It appears that with E. coli, peptide transport and hydrolysis are not rate limiting for protein synthesis and cell growth [133], whereas in Pseudomonas peptide cleavage and/or uptake may limit the rate of supply of a required amino acid for protein synthesis [21]. Increased cell yield on a peptide compared with that on the free, required amino acid is frequently observed when the peptide protects against the intracellular degradation of the free residue within the amino acid pool [20,40,79,182]. Such observations clearly have significance for the general cultivation of microorganisms, and in particular to present-day commercial production of bacterial (single-cell) protein.

Peptide stimulation of growth is most commonly observed with fastidious microorganisms. The general explanation for the effect, provided by studies of "strepogenin" (reviewed in Refs. 49 and 99), is that under many culture conditions, peptides act as more efficient sources of required amino acids and serve to overcome imbalances in complex media such as competition for uptake between a limiting and nonlimiting free amino acid. Interactions such as these between constituents of complex growth media frequently give rise to an apparent growth requirement for peptides, and the strepogenin story is testimony to this [75-77,155,208,209].

There is at present considerable interest in the role of peptides in the nutrition of microorganisms and mammals in nature [100]. A recent report [152] indicates that peptides may also play a role in the nutrition of invertebrate metazoa. In the laboratory, peptides have long been used in culture media for microorganisms, and increasingly, defined peptides are being used in media for tissue culture [45,69,150]. Although the known biological actions of simple peptides are vast [101], there are very few microbiological instances in which it is the peptide per se that mediates the effect; when it does occur it is usually inhibitory, and occasional evidence for stimulatory effects are not conclusive [60].

C. Distinction Between Amino Acid
   and Peptide Transport

For those microorgansims unable to cleave exogenously supplied peptides outside the cytoplasmic membrane (i.e., Models A and B of Fig. 7-1 do not apply) there is extensive, unambiguous evidence for the operation of distinct amino acid and peptide transport systems. The two classes show distinct kinetic parameters and structural requirements, the substrates do not compete with one another for uptake, and mutants can be isolated that are deficient in one or the other process. We shall not discuss amino acid

transport here; it has been considered in relation to peptide transport
previously [138,143], and is discussed at length in Chap. 5.

Early studies with a number of bacterial species indicated that peptides
are frequently nutritionally superior to equivalent concentrations of their
constituent amino acids [40,75-77,79,105,109,112,122,148,166,187,208,
209], and these results contributed to the idea of separate uptake systems
for the two types. More recent studies have provided further examples
of the above effect (see Refs. 20,99).

Competition studies indicate that there are distinct amino acid and
peptide transport systems. Thus, competition for uptake between amino
acids that share a common transport system has frequently been demonstrated [50,123,124], but competition occurs only between free amino acids
and is not observed when one of the amino acids is in peptide linkage. In
growth studies, we have observed this effect extensively in our studies with
E. coli [7,138,140,143,185], as have others [29,59,88]; it has similarly
been reported for Leuconostoc mesenteroides [103,166,214], Streptococci
[15], Pseudomonas [20], Salmonella [5], Neurospora [203], and Saccharomyces [119]. Many of these studies reported the reciprocal case in which
peptide uptake was found not be to inhibited by amino acids that compete
for uptake with the constituent amino acids.

More definitive evidence is provided by studies using isotopically
labeled substrates. Cohen and Rickenberg [26] first showed that uptake
of labeled valine into E. coli was competitively inhibited by unlabeled
valine, leucine, and isoleucine, but not by peptides containing these amino
acids, although the peptides could be utilized nutritionally. The evidence
was even more persuasive when radioactive peptides became available.
Thus, Leach and Snell [85] reported that the uptake of glycine by Lactobacillus casei was about 10 times faster, and more extensive, from
glycine peptides than with the free amino acid. In further studies [86]
it was shown that none of the free protein amino acids inhibited uptake of
radiolabeled glycine peptides. Related conclusions using radiolabeled
substrates have been obtained with Streptococci [15], Pseudomonas [20],
E. coli [27,29,74,87,88,179], L. mesenteroides [103,214], and Bacteroides ruminicola [154].

Mutants that are deficient in a transport system for a free amino acid
but readily take up the amino acid in peptide linkage have been described
for strains of E. coli [29,46-48,73,74,87], Salmonella typhimurium [81],
and Lactobacillus delbrueckii [148]. Examples of analogous "natural
mutants" have been described. Thus, studies with B. ruminicola have
shown its inability to absorb free [^{14}C]proline and [^{14}C]glutamate, although
these amino acids can apparently be accumulated when present in polypeptides with molecular weights approaching 2000 [153,154]. Bacteroides
melaninogenicus also showed limited ability to ferment amino acids
although peptides could be utilized [197]. A similar situation apparently
exists with Fusiformis necrophorus, which seems not to be able to synthe-

size proline or to accumulate it in free form. Its apparent requirement
for exogenous proline was met by peptone, and the growth so obtained was
further enhanced by addition of polyproline (mean molecular weight 2000),
but not by several defined, small oligopeptides of proline. Conclusions on
peptide utilization with these species are somewhat confused. Certainly
in the last instance unambiguous evidence for (poly)peptide utilization was
not presented, and other strains of Bacteroides can utilize an extensive
range of free amino acids including proline [63]. Preliminary observations
with Mycoplasma indicate that they may have deficiencies in the absorption
of certain required amino acids ([146]; Cirillo personal communication); an
implication of this observation is that peptides may satisfy the amino acid
requirements [180]. However, evidence that these species possess membrane bound peptidase activity (see Sec. III.D1) [22,146] makes it possible
that in this case peptide uptake could occur by Models C and D (Fig. 7-1).

The alternative class of mutants deficient in peptide transport but
retaining normal amino acid transport capability has been described in
E. coli, using radioactive amino acids (Barak and Gilvarg, unpublished
observations) and growth tests [6,11,29,38,43,118,128-130,132,136,137,
141], and in S. typhimurium [5,66]. These same growth characteristics
(i.e., failure to utilize peptides but growth on free amino acids) are found
with peptidase mutants [74,104,111,184], and it is therefore essential to
eliminate this possible explanation for any selected mutants.

Finally, the observation that first, impermeant substances can be taken
into bacteria by coupling them to peptides [5,37,38], and second, that peptides per se may have biological properties not shown by their constituent
amino acids [9,17,43,61,173,194], both clearly distinguish between amino
acid and peptide transport systems.

D. Role of Proteases and Peptidases in Peptide Utilization

Microorganisms differ in their complement of proteases and peptidases and
also in their cellular locations. These features dictate which model(s) of
uptake (see Fig. 7-1) occur for any species. Earlier reviews dealing with
this topic have appeared [139,185].

1. Locations of Protease and Peptidase Activities

In considering this aspect, it is useful to keep in mind the various modes
of peptide utilization depicted in Fig. 7-1.

(a) Extracellular Hydrolases. Extracellular peptidases are not produced by those species, e.g., Escherichia, Salmonella, in which peptide

transport has been studied intensively and shown to occur by Model E, Fig. 7-1. However, extensive evidence does exist for the secretion of extracellular hydrolases by many microorganisms. Although various roles have been suggested for these enzymes, it is clear for at least some species [198] that they function to hydrolyze large molecular weight substrates to peptides that can either be transported directly, or that can act as substrates for extracellular peptidases (see review, Refs. 20, 203, 204, 219). Extracellular peptidase activity has been reported for many species, e.g., strains of Arthrobacter [12, 160], Bacillus licheniformis [49a], Bacillus subtilis [157, 195], Aeromonas proteolytica [90, 91, 236, 276, 281], and Neurospora crassa [56, 203, 204]. In several instances it appears that the secretion of extracellular peptidases can be regulated by the levels of exogenous amino acids and peptides [91, 102, 195], and that exogenous proteins can act similarly for protease secretion [25, 30, 31]. Only in the above case of N. crassa has the presence of both extracellular peptidase activity and peptide transport systems been confirmed, although several authors have pointed out the advantages of this coupled process for microorganisms that grow in environments normally rich in protein materials [20, 21, 195, 201, 203, 204]. Sarid et al. [161, 162] made the puzzling observation that an E. coli auxotroph could apparently utilize polyproline (n = 100), although it was an intracellular enzyme that cleaved the substrate, and other studies indicate that the polyproline would be too large to be absorbed. As growth was observed only after a lag of 24 hr, it is possible that reversion to Pro$^+$ had occurred.

(b) Periplasmic and Cell Wall Hydrolases. Periplasm is the name given to a region of the gram-negative cell envelope between the cytoplasmic membrane and the outer cell membrane. It is the location of a variety of degradative enzymes that are selectively released by a specific osmotic shock procedure [57, 58]. However, repeated attempts to detect among these periplasmic enzymes hydrolytic activity towards any of a variety of peptide substrates have been generally unsuccessful [21, 93, 176, 192]. These same studies indicated that peptido-hydrolases were not released during E. coli spheroplast formation, and therefore were presumably not associated with the cell envelope. However, recent studies [82-84, 117] are not altogether in agreement with these conclusions. Thus, a hydrolytic enzyme activity, termed "aminoendopeptidase", was reported [82] to be released from cells of E. coli (and other gram-negative bacteria) by osmotic shock and by EDTA-lysozyme spheroplast treatment; furthermore, 50% of the total activity (i.e., relative to that found in toluenized cells) was directly detectable with suspensions of intact cells (substrate, L-alanine-p-nitroanilide). Additionally, the release of activity by osmotic shock or by spheroplasting was not as efficient (about 30% of that for the periplasmic alkaline phosphatase). The periplasmic aminoendopeptidase is constitutively produced, but its differential rate of synthesis is increased fourfold when cell growth is limited by inorganic phosphate [82, 83, 117]. Derepression of aminoendopeptidase is simultaneous with that of alkaline phosphatase. Increasing the concentration of inorganic phosphate in the medium to levels that prevent derepressed synthesis has no effect on the constitutive amino-endopeptidase synthesis. Of the growth limitations tested, phosphate was specific; limitation for neither carbon nor nitrogen caused derepressed aminoendopeptidase synthesis. It is clearly of considerable importance that the substrate specificity and absolute activity of this recently described enzyme be taken into consideration when interpreting results of peptide utilization in E. coli. However, the fact that we are here considering only true peptide (i.e., protein degradation products) utilization allows us to express reservations on applying the name aminoendopeptidase based on the published substrate specificty of the enzyme. Endopeptidase activity was determined by the hydrolysis of ^{125}I-labeled proteins; casein and serum albumin were "good" substrates, whereas the Fc fragment from purified rabbit immunoglobulin C was poorly hydrolyzed. Based on this evidence, use of the term endopeptidase seems satisfactory. In contrast, the demonstrated aminopeptidase substrate specificity is much less satis-factory when one may be tempted from its name, as we are here, to extra-polate to its possible involvement in true peptide hydrolysis. Thus, the relative rates of hydrolysis (L-ala-p-nitroanilide = 100) of (1) the p-nitroanilide derivatives of ala-, leu-, and phe-, were 100, 32, and 0, respectively; (2) the β-naphthylamide derivatives of ala-, leu-, pro-, lys-, and asp-, were 16, 3, 2, 5, and 0.8, respectively; and 0 activity was detected against Ala-Gly-Gly, Ala-Val, Ala-Pro, and Ala-Ser. No further peptide substrates were tested. Thus, we are not at this time convinced that, when interpreting the studies of true peptide transport and utilization in E. coli described here, we need be over-concerned with this enzyme. It seems probable that true peptides are not sufficiently good substrates that their "external" hydrolysis by the constitutive enzyme (under test conditions for peptide transport/utilization, derepressed enzyme synthesis should not occur) will significantly contribute to total peptide uptake/utilization. Its true function has yet to be described.

It has been reported [145,186] that with Streptococcus lactis most of the cell protease activity is localized within the cell envelope and is released during spheroplast formation.

(c) Membrane Bound Hydolases. With intestinal peptide absorption the involvement of membrane bound peptidases is well documented [97,98]. However, numerous reports testify to the general absence of membrane bound peptidases in microorganisms, which if true would rule out peptide transport by Models C and D, Fig. 7-1. Some reservation has been placed on this conclusion, however, because of the common difficulty of detecting membrane-associated enzyme activity using broken cell preparations in vitro [139]. It has been reported that peptidases are associated with the cytoplasmic membrane of Mycoplasma laidlawii [22,146] and Bacillus stearothermophilus [158].

(d) Intracellular Hydrolases. In contrast to the above findings, a wide range of peptidase and protease activities has been demonstrated in cytoplasmic extracts from many species of microorganisms. Reviews have appeared on the general characteristics of these enzymes [13,62,147] and their probable functions [44,139,140,151,185]. Intracellular peptidase action coupled to true peptide transport (model E, Fig. 7-1), is clearly implicated in peptide utilization in several species, and partial characterization of some of the enzymes involved has been reported for the following species: E. coli [16,54,61,67,94,126,127,133,134,169,170,174,183-185, 193,212,213], S. typhimurium [66,104,110,111], Pseudomonas putida and Pseudomonas maltophilia [20,21], Streptococcus pneumoniae [68], Saccharomyces cerevisiae [10,119], and N. crassa [167,203-205]. These studies show that in any one strain different peptidases exist that have widely overlapping specificities. (However, some of the earlier studies indicating the existence of multiple enzymes should perhaps be reconsidered in the light of recent work showing how specificity for peptide bonds can be dramatically changed by addition of different cations [54,95,126,127,174], or by different associations between enzyme subunits [181]). These enzymes clearly perform many functions in addition to their nutritional role in cleaving transported peptides; in spite of this, a strain can carry mutations in at least four distinct peptidases and still be fully viable [104,111] and show only slight changes in growth response to peptides [66,174].

In E. coli many different studies indicate that the intracellular peptidases are constitutive (reviewed Ref. 132). They are not induced by adding nutritional peptides to the growth medium, and it seems clear that these enzymes perform many functions in addition to this nutritional one. The peptidase activity measured in vitro against various nutritional peptides has been calculated [133,134] to be significantly in excess of that needed to meet the amino acid requirements (from peptide linkage) of an auxotroph growing at maximum rate. However, extensive studies have shown that this activity may actually occur in a partially latent form within the cell (reviewed by Simmonds [169,170]), and that additionally, activity may vary throughout the growth cycle [121].

2. Independence of Peptidase Activity
    and Peptide Transport

In most studies on microbial peptide transport, it has proved impossible to detect intact peptides within the cell; instead only the free amino acids are found. This feature makes it difficult to determine whether uptake involves an obligatory hydrolytic step (Fig. 7-1, Models C and D) or true peptide transport followed by rapid and complete intracellular peptidase action (Fig. 7-1, Model E). However, there are some biochemical studies (with E. coli) indicating that a few peptides can be accumulated intact.

Based on these results and other circumstantial evidence, it has become customary to extrapolate to the conclusion that in all species for which it can be demonstrated that extracellular peptidase activity is absent, the transport of all peptides is by Model E, Fig. 7-1. Although we feel no pressing need to challenge this conclusion in any particular case, it is important to recognize that the available evidence does not usually rule out the operation (perhaps additionally) of coupled hydrolytic uptake such as shown in Fig. 7-1, Models C and D.

The following evidence supports the view that peptide transport is independent of cleavage in E. coli. First, membrane bound peptidases are absent (see Sec. III.D1). Second, dipeptides can be accumulated intact under environmental conditions that depress intracellular peptidase activity [106,169,170,173]; in several instances the accumulated dipeptides can be bacteriostatic, although their amino acid components are not [43,135,141, 194,199]. Certain oligopeptides of ornithine and lysine are also bactericidal, and there is good evidence that this occurs because they are accumulated intact by E. coli (and by S. typhimurium). Certain chemically modified peptides, or peptides containing nonprotein amino acids, may be resistant to intracellular peptidase activity and yet satisfy the requirements of the transport systems, thereby being accumulated intact. Examples are peptides containing β-alanyl [136], sarcosyl [132], and ε-acetyl lysine residues [131]. Finally, peptide transport mutants can be isolated that have unchanged intracellular peptidase activity; in addition, mutations in intracellular peptidases can cause the intracellular build up of intact, transported peptides.

IV. METHODS OF STUDYING PEPTIDE TRANSPORT

A. Genetic and Culture Considerations

When carrying out experiments on peptide transport, or when comparing the results from different laboratories, it is important to keep in mind the way in which differences in growth conditions and use of different strains may affect the results [72,88,123,124]. The possible complicated regulatory controls (see Sec. V.E) of amino acid and peptide transport make this particularly important. The fact that cells grown in batch culture will not be homogeneous, especially "starved" cells grown into a state of nutrient limitation [58a,185a], should also be kept clearly in mind when assessing kinetic evidence for different transport systems that apparently display different affinities and/or efficiencies, and corroborative genetic evidence should be obtained when possible. It goes without saying that the purity of commercial peptides must always be checked (e.g., by electrophoresis, thin layer chromatography) for the presence of free amino acids, peptide homologs, etc., and to ensure no modifications occur on autoclaving.

Peptide transport in microorganisms can be studied either directly, i.e., by measuring uptake of the peptide, or indirectly by observing a specific biological response resulting from the uptake of the peptide.

B. Direct Methods

1. Radioactive Peptides

Defined peptides containing isotopically labeled amino aicd residues are the ideal substrates for peptide transport studies; unfortunately, only a few species are commercially available. The difficulty and expense in synthesizing a variety of these substances has meant that a limited range of peptides has been used by only a few workers in studies with E. coli [27,29,74, 87,88,131,179], S. typhimurium [66], L. casei [85,86,216], Streptococcus faecalis [15,78], L. mesenteroides [103,166,214], P. putida [20], B. ruminicola [154,211], Bacteroides amylophilus [63], Clostridium perfringens [51,53], and N. crassa [204,205]. In most studies., labeled materials were recovered either in protein (TCA-insoluble material) or as free amino acid, rather than as intact peptide. In general, the results indicated that uptake of amino acids was faster and more extensive when these were in peptide form. Kinetic parameters for peptide transport systems were reported by De Felice et al. [29], Cowell [27], Yoder et al. [214], Mayshak et al. [103], and Leach and Snell [86].

Use of radioactive peptides allows the study of peptide transport in osmotically shocked cells [27], spheroplasts, and membrane vesicles that are not susceptible to study by conventional indirect methods.

2. Chemical Determinations

Although common analytical methods, e.g., automatic amino acid analyses, are used in studies on mammalian peptide transport [97,98], their limited sensitivity makes them generally unsuitable for microbial studies. However, in studies with E. coli on the relative rates of transport and utilization of individual peptides present within complex mixtures of peptides, the disappearance of the peptides from the growth medium has been successfully monitored using a peptide mapping procedure with fluorescamine as detecting agent (G. Bell and J. W. Payne, unpublished observations).

3. Fluorescent Derivatives

It has been suggested [7] that fluorescent peptide derivatives could be useful in studies of peptide uptake by individual mammalian cells, using a fluorescence microscope. Substitution of the terminal carboxyl group (see

Sec. VI.C) or amino acid side chain (See Sec. VI.B) might not affect the transport characteristics of the peptide, but the substituent should be made as small as possible. The principle may have application to some larger microorganisms but probably not to most bacteria.

C. Indirect Methods

1. Growth Measurements

Most of the results on peptide transport described in this review have been obtained by measuring the growth response of a microorganism to peptides in the growth medium. Detailed information has been obtained using auxotrophs in which growth is dependent upon an external source of amino acid. For success, the procedure requires that the microorganism lack extracellular peptidase activity (Fig. 7-1, A, B) so that peptides per se may be transported. However, in some cases, it is possible to apply the method to organisms that produce extracellular peptidases [203].

Although clearly recognized as an indirect way of measuring a permeation process (see Ref. 138), the method has provided results that form the basis for present understanding of peptide transport, and allows critical appraisal of the particular isotopically labeled peptides that should be synthesized and the experiments in which they should be employed.

Disadvantages of the method (other than its indirect nature) are the relative insensitivity of the absorbance measurements used to measure growth, the high concentrations and relatively large amounts of peptides required, and the inability to quickly monitor the biological response over a short time.

2. Competition Experiments

The ability of different peptides to share a common transport system has been established in many of the studies described later by showing that they compete with one another during transport, as judged by growth response changes. Thus, growth of an auxotroph is inhibited if a peptide containing an essential amino acid residue is competitively excluded from the cell by another peptide devoid of the essential residue. Alternatively, growth may be enhanced if such competition prevents uptake of a toxic peptide.

3. Use of Toxic Peptides

With toxic peptides it is the inhibition of growth that indicates peptide transport. They are particularly useful when studying peptide transport in microorganisms that are not auxotrophic, and when growth responses to peptides cannot easily be monitored. They are also useful in selecting peptide transport deficient mutants.

4. Enzyme Synthesis

In recent experiments (G. Bell, G. M. Payne, and J. W. Payne, manuscript in preparation), the changes in concentration of several inducible and derepressible enzymes have been used to study peptide transport in whole cells of E. coli amino acid auxotrophs. The method has many considerable advantages over the conventional growth and competition methods (Sec. IV.C1, C2). For example, (1) experiments can be carried out using peptide concentrations around their K_m for uptake, allowing facile determination of these parameters; (2) there are economic advantages in that for any test only about 1% of the amount of peptide employed in the growth method is required; (3) competition experiments, transport mutants, and toxic peptides can be employed as in growth tests; and (4) sensitive colorimetric/fluorimetric enzyme assays can be made within minutes of adding peptides to the bacterial culture.

A related method using amino acid–starved bacteria has been reported [19], and numerous reports have appeared on the effects of peptides on intracellular enzyme synthesis (see Refs. 18,64].

5. Radioisotope Incorporation

This method (G. Bell and J. W. Payne, manuscript in preparation) employs a double amino acid auxotroph of E. coli, say, Lys⁻, Met⁻. Incorporation of radioactivity from free lysine into trichloracetic acid precipitated protein is studied as a function of added methionine peptides. The procedure is related to Sec. IV.C4, and has many of its advantages (1-3, above). It is particularly useful for studying competition between amino acids, dipeptides, and oligopeptides.

V. PROPERTIES OF PEPTIDE
    TRANSPORT SYSTEMS

In this section we describe some of the molecular aspects of peptide transport established for several types of microorganisms. Although most of the properties discussed apply to all peptides, it is nevertheless desirable at an early stage to consider dipeptides and oligopeptides separately.

A. Distinction Between Dipeptide and
   Oligopeptide Transport

The existence of separate transport systems for dipeptides and oligopeptides
is indicated by (1) the results of competition studies, (2) differences in
their structural requirements for transport, and (3) the isolation of appropriate mutants, and genetic mapping.

In competition studies (see Sec. IV.C,D) with strains of E. coli [128],
dipeptides were found to compete with tripeptide uptake when measured by
their effect on the utilization of a nutritional tripeptide or the toxicity produced by an inhibitory tripeptide [129]. However, the converse does not
occur, i.e., oligopeptides compete for uptake with each other but only to a
limited extent with dipeptides [128-130,132,137]. These findings with
E. coli strain W (reviewed in Ref. 143) have been extended to other strains
of E. coli. With strain K12, growth of a proline auxotroph on Pro-Gly and
Pro-Gly-Gly was competitively inhibited by di- and trilysine, respectively,
but trilysine did not inhibit uptake of Pro-Gly [130]. Similarly, the toxicity
of Gly-Val-Gly was counteracted by diglycine and triglycine, respectively
[129]. Analogous results have been reported using a range of nutritional
and toxic peptides [6,8,29,74,118,178]. Recently, competition studies
carried out with the following species have provided evidence for separate
dipeptide and oligopeptide transport systems: S. typhimurium [5], P. putida
and P. maltophilia [20], and N. crassa [203-205].

The situation regarding dipeptide uptake in N. crassa is complicated
and uncertain [203,205]. The existence of extracellular peptidase activity
ensures that their amino acid residues can be utilized following transport
by amino acid permeases. In general it appears that dipeptides per se are
not taken up. Exceptions were found however, and certain of these dipeptides appeared to be accumulated by an uptake system which was sensitive
to inhibition by free amino acids. Confusion arose in cases where a single
dipeptide, e.g., His-Leu, would support growth of a histidine auxotroph
but not of a leucine mutant. It is possible that if a specific dipeptide transport
system exists in N. crassa, it has a poor K_m, and that most dipeptides
were tested at too low a concentration.

Several examples can be found in the earlier literature to indicate that
distinct systems may occur in the following species, although this interpretation was not generally placed on the evidence at the time. In L.
mesenteroides [166] and Lactobacillus arabinosus [33], tripeptides failed
to compete with dipeptide uptake, although dipeptides did compete with one
another. Shankman et al. [163] mentioned that several dipeptides could
partially reverse the inhibitory effect of Val-Val-D-Val on L. arabinosus.

In E. coli the transport systems for the two classes of peptides differ in their structural specificites; a dipeptide must possess a free C-terminal carboxyl group but no such requirement exists for oligopeptides [141]. However, dipeptides without this group, or with it substituted, can enter via the oligopeptide transport system. Similar structural requirements have been reported for S. typhimurium [5] and P. putida and P. maltophilia [20]. This feature is considered in Sec. VI. C.

Mutants that are deficient in oligopeptide transport but retain dipeptide transport ability have been described in strains of E. coli [6, 8, 29, 118, 128, 130, 135, 136] and S. typhimurium [5]. An oligopeptide transport-deficient mutant of N. crassa has been described [205]; however, in this case, the parent appeared unable to utilize dipeptides. Dipeptide transport mutants were isolated by Kessel and Lubin [74], De Felice et al. [29], and Umbarger and coworkers [194, 199].

B. Energetics of Peptide Transport

This important feature of peptide transport has received little attention to date, but is surely an area demanding study (see Chap. 12).

It is useful to consider this feature by reference to the ways in which peptides may enter microbial cells (Fig. 7-2). At physiological concentrations passive diffusion (Fig. 7-2A) is not important for peptides, as indicated by all competition, genetic studies, etc. In addition, the pronounced temperature dependence of peptide uptake is incompatible with simple diffusion [86, 154, 216]. However, it remains a possibility that with high concentrations of rather small lipophilic peptides, uptake by this process may occur. A little circumstantial evidence exists for this speculation. Thus, Kessel and Lubin [74] reported that a dipeptide transport-deficient strain of E. coli could still accumulate [^{14}C]diglycine when the peptide was present at high concentration. However, the transport defect was uncharacterized, and it remains a possibility that facilitated diffusion occurred, or even active transport through a separate system with a poor K_m. Cowell [27] calculated that not more than 10% of the total uptake of gly-gly into E. coli was by a diffusion-controlled process. Likewise, Payne [128, 129] showed that at high concentrations certain lipophilic tripeptides could still enter an oligopeptide transport mutant of E. coli. Although it was shown that this uptake was not susceptible to competition by other oligopeptides known to use the deficient oligopeptide transport system, recent evidence [8, 118] makes it likely that the lipophilic peptides were being taken up through an alternative oligopeptide permease(s). Competition studies with other peptides, known to use the alternative permease, would indicate the validity of this speculation.

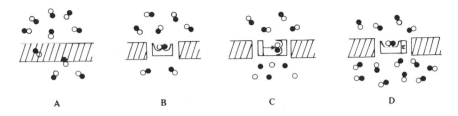

FIG. 7-2 Mechanisms of peptide transport in relation to energy coupling. The microbial cytoplasmic membrane is represented by crosshatched sections, peptide carriers by open sections. Exterior of cell is above, the interior below. (A) Passive diffusion; (B) facilitated diffusion, (C) group translocation: peptidase activity is shown by an arrow, (D) active transport: energy coupling component is represented by E. (See text for discussion.)

Group translocation could take a variety of forms, but the most likely for peptides appears to be one coupled to hydrolysis (Fig. 7-2C, and see Fig. 7-1). It could require an external energy source and/or the energy released by peptide bond cleavage might be used [15,24,159]. The difficulties in identifying this mode of peptide uptake were considered in Sec. III.A.

Facilitated diffusion (Fig. 7-2B) is a carrier-mediated process that is not directly coupled to an energy supply (although energy may be required indirectly to maintain the carrier in its operational conformation). This process cannot therefore transport peptides against a concentration gradient. No evidence has been cited for its involvement in peptide transport, although it is possible that in energy-starved E. coli (where active transport may be curtailed), the uptake of radioactivity from isotopically labeled dipeptides occurred by facilitated diffusion (or conceivably Model C, Fig. 7-2) [27, 168]. The rapid intracellular hydrolysis of peptides would allow their continuous accumulation by facilitated diffusion.

There is extensive indirect and direct evidence that peptides are actively transported (Fig. 7-2D) by many microbial species. When the uptake of isotope from a radioactive peptide is shown to require addition of an energy source, this provides indirect evidence for active transport. This requirement has been shown for dipeptides with E. coli [27,29,74, 106,168], P. putida [20], L. casei [85,86], and L. mesenteroides [103, 214], and for oligopeptides with E. coli [27,29,131,179], P. putida [20], S. typhimurium [66], B. ruminicola [154], P. cerevisiae [163], N. crassa [205], and L. casei [216]. Many of these workers also reported that common energy poisons, e.g., dinitrophenol, inhibited peptide uptake. To provide direct evidence for active transport, it is necessary to show the

intracellular accumulation of peptide in an unmodified form against a concentration gradient. The intense activity of intracellular peptidases makes
it difficult to demonstrate the accumulation of intact peptide. However,
this difficulty has been overcome in several ways: (1) by manipulating
culture conditions to produce cells with low peptidase activity, (2) by use
of peptidase-deficient mutants, and (3) by use of peptides that are resistant
to peptidase action. In the first case, there is evidence that peptidase
activity in E. coli varies with culture conditions and growth phase [121,
134,169,170], and advantage was taken of the fact that low activity occurs
in stationary-phase cells (fully-aged) to demonstrate their active accumulation of intact, radioactive Gly-Leu [89,168]. As examples of the second
approach, Kessel and Lubin [74] reported the active transport of Gly-Gly
by a diglycine peptidase-deficient mutant of E. coli, and Jackson et al. [66]
described similar findings for oligopeptide transport in S. typhimurium
peptidase mutants. Examples of peptides that are transported but are
resistant to intracellular hydrolysis are ε-substituted lysine peptides [131]
and one with N-methylpeptide bonds [132].

Cowell [27] has reported the most detailed studies in this area, investi-
gating Gly-Gly transport in E. coli. Using strains ML308-225, he reported
that, "glycylglycine transport in ML308-225 is mediated primarily by a
single, saturable transport...," and that uptake by other means "...which
may describe a diffusion-controlled process, can be calculated to contribute only about 10% to the total uptake." Unfortunately, the oligopeptide
transport ability of the strain was not studied, so it is not known whether
the 10% represented (partial) uptake by the oligopeptide transport system
or if it was indeed diffusion. Nevertheless, the Gly-Gly transport system
behaved like a shock-sensitive transport system (see Sec. V.C). The
initial rate of Gly-Gly transport was decreased 85% after osmotic shock
treatment. Gly-Gly was not transported by membrane vesicles. (Triglycine
was also not transported by membrane vesicles, but no evidence was presented that the tripeptide could actually enter whole cells either.) The
experiments indicated that the active transport of Gly-Gly, like other shock-
sensitive transport systems (e.g., glutamine), has an obligatory requirement for phosphate bond energy, but not for respiration or the energized
state of the membrane (as, e.g., proline uptake). These conclusions rested
on the following evidence: (1) Gly-Gly transport was strongly inhibited by
arsenate (which causes a drastic depletion of intracellular ATP levels);
(2) oxidizable energy sources such as D-lactate, succinate, and ascorbate
could not serve as energy sources for Gly-Gly transport in strain DL-54,
which lacks oxidative phosphorylation (strain DL-54 is a mutant of strain
ML308-225 that is deficient in Ca^{2+} and Mg^{2+}-stimulated adenosine 5'-
triphosphatase activity); (3) substantial Gly-Gly transport was obtained
when energy was supplied only from adenosine-5'-triphosphate produced
by glycolysis (i.e., anaerobic transport assays with glucose as the energy
source in strain DL-54); and (4) when the Ca^{2+}-Mg^{2+}-adenosine triphos-

phatase activity was absent, but substrate-level phosphorylations and
electron transport were functioning (mutant DL-54 using glucose as energy
source), Gly-Gly transport showed significant resistance to the uncouplers
dinitrophenol (DNP) and carbonylcyanide-p-trifluoromethoxyphenylhydrazone. The relation of these results to oligopeptide transport remains to be
established.

C. Role of Binding Proteins in Peptide Transport

The involvement of periplasmic binding proteins released by osmotic shock
[57,58] in amino acid and sugar transport (see Chap. 4, 5, and 11) is well
established. However, although it has been found by various workers that
dipeptide ([27]; B. P. Rosen, personal communication 1973; Barak and
Gilvarg, unpublished observations) and oligopeptide transport (Barak and
Gilvarg, unpublished observations) into E. coli is significantly decreased
by osmotic shock, in no case has a binding protein been detected. Cowell
[27] suggested that failure to detect binding activity for [14C]Gly-Gly in the
shock fluid might be caused by enzymic cleavage of the dipeptide, although
other studies (see Sec. III.D1b) have indicated that peptidase activity is not
released by the osmotic shock procedure. Furthermore, Cowell [27]
reported that membrane vesicles of E. coli [70,71] failed to transport
[14C]Gly-Gly, and a similar result was found for [14C]Gly-Gly-Gly (Barak
and Gilvarg, unpublished observations). However, Kaback (personal
communication) found [14C]Gly-Gly uptake in membrane vesicles.

It is clear that further insight into the molecular basis of peptide transport requires that the possible involvement of binding proteins be clarified.
In this context, it will be of interest to isolate and to characterize the products of the known di- and oligopeptide transport genes. If binding proteins
do function in peptide transport, it will be of some interest to compare their
specificities for binding with the established structural requirements for
peptide transport (see Sec. VI).

D. Kinetic Parameters of Peptide Transport

Limited data are available on the kinetic constants for peptide transport.
Using isotopically labeled peptides, De Felice et al. [29] reported the
following for uptake by E. coli K12: Gly-Gly: K_m, 1 x 10^{-5} M and V_{max},
10 μmol min^{-1} g^{-1} of cells (dry weight); Gly-Gly-Gly: K_m, 1.1 x 10^{-6} M
and V_{max}, 5 μmol min^{-1} g^{-1} of cells (dry weight). Cowell [27] reported
the following figures for Gly-Gly uptake by E. coli: K_m, 3.17 ± 0.79 x
10^{-6} M and V_{max}, 4.68 ± 0.40 μmol min^{-1} g^{-1} of cells (dry weight). Wolfinbarger and Marzluf [205] reported the following for uptake of Gly-Leu-Tyr

into $\underline{\text{N. crassa}}$: K_m, 3.4×10^{-5} M and V_{max}, 2.07 μmol min^{-1} g^{-1} of cells (dry weight). The following studies with radioactive peptides have provided information on rates of uptake and saturating concentrations for transport, but for various reasons kinetic constants were not calculated: for $\underline{\text{E. coli}}$ [74, 87, 89, 131, 179], $\underline{\text{P. putida}}$ [20], and $\underline{\text{L. casei}}$ [85, 86].

Kinetic constants similar to those above have been obtained using the enzyme synthesis method (see Sec. IV.C.4); (G. Bell, G. M. Payne, and J. W. Payne, unpublished results; [19]).

E. Regulation of Peptide Transport
     Systems

All the available evidence indicates that peptide transport systems are constitutive (early studies reviewed in Refs. 143, 169, 170). It is also clearly shown by the rapid uptake of radiolabeled peptides into micro-organisms grown in media devoid of peptides (see references in Sec. V.D). Apart from this fact, it is surprising that no studies have been made of the possible regulation of peptide transport (Fig. 7-3). The topic was recently discussed [138], but the following points can be made here.

Although separate regulation of peptide transport may occur, it seems useful to consider the topic as an extension of amino acid uptake, for trans-ported peptides are rapidly cleaved and their constituents join the amino acid pool.

Accordingly, considering first the effect of peptides on amino acid uptake, it can be speculated that peptides could stimulate or inhibit amino acid uptake. The former might occur by a process of "counterflow" in which an internal amino acid generated by peptide hydrolysis can be exchanged

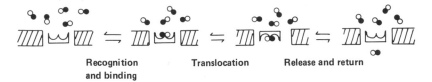

Recognition   **Translocation**   **Release and return**
and binding

FIG. 7-3 Hypothetical stages in active transport of a peptide. The micro-bial cytoplasmic membrane is represented by crosshatched sections and a peptide carrier by open sections. Exterior of cell is above and interior below. Initial step involves specific binding of peptide to some "recognition component" of transport system. Subsequent steps (some or all of which must be energy-linked for active transport) involve transfer across the membrane (translocation), intracellular release, and return of the system to a form able to carry out further transport.

for an exogenous amino acid [14,15]. Alternatively, amino acid permeases might be induced by peptides. Inhibition of amino acid transport might occur by feedback control resulting from peptide uptake and hydrolysis; the feedback regulator could be an amino acid residue itself or its metabolite. Alternatively, an effective inhibition of amino acid uptake might occur if efflux were stimulated by peptide-derived amino acids that built up in the pool. Evidence relating to these various speculations is given in Payne [138].

It is more difficult to assess the possible effects of the converse situation in which amino acids affect peptide transport. The possibility that amino acid and peptide transport may be coordinately controlled in certain microorganisms, and may vary depending upon whether they are acting as nitrogen sources or for protein synthesis alone, is a likely area for future study.

VI. STRUCTURAL REQUIREMENTS FOR PEPTIDE TRANSPORT IN MICRO-ORGANISMS

In recent years much effort has gone into delineating the structural specificities of the peptide transport systems. The initial studies, carried out almost exclusively with E. coli, have been reviewed [7,138,139,143]. More recent studies have extended the conclusions to other species. All these results are summarized below, in which di- and oligopeptides are considered separately for each of the organisms investigated to more clearly indicate recorded similarities and differences. Structural requirements for mammalian peptide transport have been reviewed [7,97,98].

A. N-Terminal α-Amino Group

1. Dipeptides

(a) Escherichia. Early experiments indicated that substitution of the α-N-terminal group of dipeptides destroyed their nutritional activity (Table 7-1, items 1-6) [171,175]. However, at this time it was not established whether this resulted from a failure in transport or hydrolysis. Dunn and Dittmer [32] reported that the chloro-, acetyl-, and benzyloxycarbonyl derivatives of β-2-thienyl-DL-Ala-Gly were up to several thousand times less inhibitory than the free peptide. In competitive tests, benzyloxycarbonyl derivatives of Phe-Gly and Gly-Phe failed to nullify the inhibitory effect of β-2-thienyl-Ala-Gly, although the free peptides were effective at low concentrations. Gilvarg and Katchalski [42] were the first to perform systematic studies that showed α-N-substitution can cause a transport

TABLE 7-1 The Effect of Various α-N-Terminal Substituents on the Utilization, Transport, and Hydrolysis of Peptides by Microorganisms[a]

Organism	Auxotroph	Amino Acid or Peptide	Name	N-Terminal Substituent Structure	Charge	Fate of Peptide Utilization	Transport	Hydrolysis	Ref.	Comments
1. E. coli	Pro	Phe–Pro	unsub.	NH_2-	+ve	+	+[a]	+[a]	171	
2. E. coli	Pro	Phe–Pro	acetyl	$CH_3-CO-NH-$	neut.	-	n.d.	n.d.	171	
3. E. coli	Phe, Tyr	Gly–Phe; Gly–Tyr	unsub.	NH_2-	+ve	+	+	+	175	
4. E. coli	Phe–Tyr	Gly–Phe; Gly–Tyr	cbz	$Ph-CH_2-O-CO-NH-$	neut.	-	n.d.	n.d.	175	
5. E. coli	Pro	Pro_2	unsub.	NH_2-	+ve	+	+[a]	+[a]	171	
6. E. coli	Pro	Pro_2	cbz	$Ph-CH_2-O-CO-NH-$	neut.	-	n.d.	n.d.	171	
7. E. coli	Lys, Arg	Lys_2, Arg_2	unsub.	NH_2-	+ve	+	+	+	42	
8. E. coli	Lys, Arg	Lys_2, Arg_2	acetyl	$CH_3-CO-NH-$	neut.	-	-	+	42, 92	
9. E. coli	Gly/Ser	Gly, Gly_2	unsub.	NH_2-	+ve	+	-	+	128,129	
10. E. coli	Gly/Ser	Gly_2	acetyl	$CH_3-CO-NH-$	neut.	-	n.d.[c]	+	129	
11. E. coli	Gly/Ser	Gly, Gly_2	propionyl	$CH_3-CH_2-CO-NH-$	neut.	-	n.d.[c]	+	129	
12. E. coli	Gly/Ser	Gly, Gly_2	succinyl	$COOH-(CH_2)_2-CO-NH-$	-ve	-	n.d.[c]	+	129	
13. E. coli	Gly/Ser	Gly, Gly_2	glutaryl	$COOH-(CH_2)_3-CO-NH-$	-ve	-	n.d.[c]	+	129	
14. E. coli	Gly/Ser	Asp–Gly, Glu–Gly	unsub.	NH_2-	+ve	+	+	+	129	
15. E. coli	Gly/Ser	Gly_2	methyl	CH_3-NH-	+ve	±	+[b]	+	129,137	
16. E. coli	Gly/Ser	Gly–Ser	unsub.	NH_2-	+ve	+	+	+	129	
17. E. coli	Gly/Ser	Gly–Ser	methyl	CH_3-NH-	+ve	+	+	+	129	
18. E. coli	Gly/Ser	Gly_2	ethyl	CH_3-CH_2-NH-	+ve	+	n.d.[b]	±	137	
19. E. coli	Gly/Ser	Gly_2	propyl	$CH_3-(CH_2)_2-NH-$	+ve	-	n.d.[b]	±	137	
20. E. coli	Gly/Ser	Gly_2	butyl	$CH_3-(CH_2)_3-NH-$	+ve	-	n.d.[b]	±	137	
21. E. coli	Gly/Ser	Gly_2	isopropyl	$(CH_3)_2-CH_2-NH-$	+ve	-	n.d.[b]	±	137	
22. E. coli	Gly/Ser	Gly_2	isobutyl	$(CH_3)_2-(CH_2)_2-NH-$	+ve	-	n.d.[b]	±	137	
23. E. coli	Gly/Ser	Gly_2	dimethyl	$(CH_3)_2-N-$	+ve	-	n.d.[c]	±	137	
24. E. coli	Leu	Gly–Leu	methyl	CH_3-NH-	+ve	+	+[a]	+[a]	17	
25. E. coli	Met	Met_2	acetyl	$CH_3-CO-NH-$	neut.	-	n.d.	+	11	t-boc Derivatives also not utilized, quoted in [119]
26. S. typhimurium	Pro	Pro_2	unsub.	NH_2-	+ve	+	+	+	66	Auxotroph used was peptidase deficient
27. S. typhimurium	Pro	Pro_2	t-boc	$(CH_3)_3-C-O-CO-NH-$	neut.	-	n.d.	+	66	

#	Organism	Amino acid	Peptide	Substituent	Formula	Charge				Ref.	Notes
28.	S. typhimurium	-	Gly-Phe	cbz	Ph-CH$_2$-O-CO-NH-	neut.	n.d.	n.d.	–	111	
29.	S. typhimurium		Gly-Phe	benzoyl	Ph-CO-NH-	neut.	n.d.	n.d.	–	111	
30.	P. putida	Met	Gly-Met	unsub.	NH$_2$-	+ve	+	+	+	20,21	
31.	P. putida	Met	Gly-Met	acetyl	CH$_3$-CO-NH-	neut.	–	n.d.,b	–	20,21	
32.	P. putida	Met	Gly$_2$ Leu$_2$	unsub.	NH$_2$-	+ve	–d	n.d.,b	n.d.	20,21	
33.	P. putida	Met	Gly$_2$, Leu$_2$	cbz	Ph-CH$_2$-O-CO-NH-	neut.	–d	n.d.,c	+a	20,21	
34.	Lactobacilli	e	Leu$_2$, Val$_2$	unsub.	NH$_2$-	+ve	+	+a	+a	164	Utilizability was 20-50% of unsub. with diff. strains
35.	Lactobacilli	e	Leu$_2$, Val$_2$	phthalyl	Ph-COOH-CO-NH-	-ve	±	±	±	164	
36.	L. casei	e	Ala-Glu	unsub.	NH$_2$-	+ve	+	+a	+a	207	
37.	L. casei	e	Ala-Glu	chloroacetyl	CH$_2$Cl-CO-NH-	neut.	–	n.d.	n.d.	207	
38.	S. cerevisiae	Met	Met$_2$, Gly-Met	unsub.	NH$_2$-	+ve	+	+a	+	119,10	Growth rate & yield equal to unsub. peptides
39.	S. cerevisiae	Met	Met$_2$, Gly-Met	acetyl	CH$_3$-CO-NH-	neut.	+	+a	+f	119	
40.	S. cerevisiae	Met	Met-Gly	unsub.	NH$_2$-	+ve	+	+a	+	119	Hydrolysis checked by peptidase assay
41.	S. cerevisiae	Met	Met-Gly	acetyl	CH$_3$-CO-NH-	neut.	–	–	–	119	
42.	S. cerevisiae	Met	Gly-Met	t-boc	(CH$_3$)$_3$-C-O-CO-NH-	neut.	+	+a	+f	119	Growth approx. equal to the acetyl peptide
43.	S. cerevisiae	Met	Met$_2$	t-boc	(CH$_3$)$_3$-C-O-CO-NH-	neut.	±	±a	±f	119	Growth very inferior to acetyl peptide
44.	E. coli	Lys	Lys$_3$, Lys$_4$, Lys$_3$,	unsub.	NH$_2$-	+ve	+	+a	+a	42	Acetyl trilysine gave linear growth
45.	E. coli	Lys	Lys$_3$, Lys$_4$, Lys$_4$,	acetyl	CH$_3$-CO-NH-	neut.	–	n.d.,c	+	42,92 / 128	
46.	E. coli	Arg	Arg$_3$, Arg$_4$	unsub.	NH$_2$-	+ve	+	+a	+a	42,92	
47.	E. coli	Arg	Arg$_3$, Arg$_4$	acetyl	CH$_3$-CO-NH-	neut.	–	n.d.,c	+	42,92	
48.	E. coli	Gly/Ser	Gly$_3$, Gly$_4$	unsub.	NH$_2$-	+ve	+	+a	+	128,129	
49.	E. coli	Gly/Ser	Gly$_3$	acetyl	CH$_3$-CO-NH-	neut.	–	n.d.,c	+	129	
50.	E. coli	Gly/Ser	Gly$_3$	propionyl	CH$_3$-CH$_2$-CO-NH-	neut.	–	n.d.,c	+	129	
51.	E. coli	Gly/Ser	Gly$_3$	succinyl	COOH-(CH$_2$)$_2$-CO-NH-	-ve	–	n.d.,c	+	129	
52.	E. coli	Gly/Ser	Gly$_3$	glutaryl	COOH-(CH$_2$)$_3$-CO-NH-	-ve	–	n.d.,c	+	129	
53.	E. coli	Gly/Ser	Gly$_3$	methyl	CH$_3$-NH-	+ve	+	+a,b	+	129,137	
54.	E. coli	Gly/Ser	Gly$_3$	ethyl	CH$_3$-CH$_2$-NH-	+ve	±	+a,b	±	137	
55.	E. coli	Gly/Ser	Gly$_3$	propyl	CH$_3$-(CH$_2$)$_2$-NH-	+ve	+	+a,b	±	137	

TABLE 7-1 (Cont.)

Organism	Auxotroph	Amino Acid or Peptide	Name	N-Terminal Substituent Structure	Fate of Peptide				Ref.	Comments
					Charge	Utilization	Transport	Hydrolysis		
56. E. coli	Gly/Ser	Gly_3	butyl	$CH_3-(CH_2)_3-NH-$	+ve	+	$+^{a,b}$	±	137	
57. E. coli	Gly/Ser	Gly_3	isopropyl	$(CH_3)_2-CH_2-NH-$	+ve	+	$+^{a,b}$	±	137	
58. E. coli	Gly/Ser	Gly_3	isobutyl	$(CH_3)_2-(CH_2)_2-NH-$	+ve	+	$+^{a,b}$	+	137	
59. S. typhimurium	Pro	Pro_3	unsub.	NH_2-	+ve	+	$+^a$	+	66	
60. S. typhimurium	Pro	Pro_3	t-boc	$(CH_3)_3-C-O-CO-NH-$	neut.	-	n.d.	+	66	
61. Lactobacilli	e	Leu_3, Leu_4	unsub.	NH_2-	+ve	+	$+^a$	$+^a$	164	
62. Lactobacilli	e	Leu_3, Leu_4	phthalyl	$Ph-COOH-CO-NH-$	-ve	±	$±^a$	+	164	
63. S. cerevisiae	Met	Met_3, Met-Gly-Met	unsub.	NH_2-	+ve	+	$+^a$	±	119	Growth equal to the unsub. peptides
64. S. cerevisiae	Met	Met_3, Met-Gly-Met	acetyl	$CH_3-CO-NH-$	neut.	+	$+^a$	$+^f$	119	Hydrolysis checked in direct assays
65. S. cerevisiae	Met	Met-Gly-Gly	unsub.	NH_2-	+ve	±	$±^a$	+	119	
66. S. cerevisiae	Met	Met-Gly-Gly,	acetyl	$CH_3-CO-NH-$	neut.	-	n.d.	±	119	
67. S. cerevisiae	Met	Gly-Gly-Met, Met_3	unsub.	NH_2-	+ve	+	$+^a$	$+^a$	119	
68. S. cerevisiae	Met	Gly-Gly-Met, Gly-Met-Met-Met_3	t-boc	$(CH_3)_3-C-O-CO-NH-$	neut.	±	$±^a$	$±^a$	119	
69. N. crassa	Leu	Gly-Leu-Tyr	unsub.	NH_2-	+ve	+	$+^a$	$+^a$	205	
70. N. crassa	Leu	Gly-Leu-Tyr	acetyl	$CH_3-CO-NH-$	neut.	n.d.	-	n.d.	205	

[a]Abbreviations used are: Gly_2, Gly_3, are Gly-Gly and Gly-Gly-Gly, respectively, and similar notation is used for other homopeptides; neut = neutral; unsub. = unsubstituted; cbz = benzyloxycarbonyl; t-boc = tertiarybutyloxycarbonyl; n.d. = not determined; (+), (-), and (±) indicate that transport, hydrolysis, or growth are extensive, not detectable, or slight, respectively. (a) when peptide is utilized (+), transport (+) and hydrolysis (+) are assumed; (b) peptide is good competitive inhibitor of transport although its own uptake may or may not be extensive; (c) no competition demonstrated against unsubstituted peptides; (d) peptide does not contain required amino acid, used only for competition; (e) microorganism has multiple amino acid requirements including those in the peptide; (f) hydrolysis not detected in direct assays although peptide is utilized.

defect. In this and a later study [92] it was shown that α-acetyl dilysine and α-acetyl diarginine were not utilized by appropriate auxotrophs although the mutants possessed intracellular peptidase activity able to release the required free amino acids (Table 7-1, items 7, 8). Acetylation of an ε-amino group did not prevent dilysine uptake. It was concluded, therefore, that a free α-amino group is essential for dipeptide transport. Seeking an explanation for the effect, Payne [129] argued that α-acetylation might prevent a dipeptide from binding to the transport system in various ways, e.g., (1) by steric hindrance, (2) by converting an essential primary amino group to a secondary one, or (3) by neutralization of the positive charge on the amino group. In an attempt to decide between the last two possibilities, he prepared a series of N-substituted peptides that had different charges on the amino terminus (Table 7-1, items 10-17). The structures of certain of these N-substituted products mean that they can alternatively be regarded as peptides that lack α-amino groups. Thus, α-N-acetyl-, α-N-propionyl-, α-N-succinyl-, and α-N-glutaryl-glycine (Table 7-1, items 10-14), respectively, are α-amino group deficient analogs of glycyl-, alanyl-, aspartyl-, and glutamyl-glycine, respectively, and similarly for the substituted homologs shown in Table 7-1. Considered from the more usual standpoint, the acyl derivatives are substituted diglycine peptides in which the net charge on the α-amino group is neutralized (Table 7-1, items 10, 11), or made negative (Table 7-1, items 12, 13). None of these acyl derivatives was nutritionally active for a glycine auxotroph, although enzymic activity was present that liberated free glycine. The additional observation that they lacked competitive ability towards other dipeptides indicated that they were transport defective. In contrast, several α-N-methyl dipeptides (Table 7-1, items 15, 17, 24) were nutritionally active and did compete with a variety of other peptides. It was therefore concluded from this study that the dipeptide transport system can in fact tolerate a substituted α-amino group [possibility (2) above]. However, acceptance of an N-alkyl group (positively charged) but not an N-acyl group (neutral or negative) did not provide unambiguous evidence regarding a charge requirement because the large acyl groups might be sterically hindered, whereas no such effect might occur with the small methyl substituent. This possibility was eliminated by using various N-alkyl peptides (Table 7-1, items 15, 18-22) in which the net positive charge was maintained, but the substituents were varied in size, some being larger or equal in size to at least some of the acyl groups [137]. All these N-alkyl derivatives were utilized and all were active in competition tests. It should also be noted that the analogous dipeptides with N-terminal prolyl residues are transported [130]. It appeared, therefore, that substitution of the N-terminus with retention of positive charge allowed the derivatives to be handled by the dipeptide transport system. However, this conclusion was modified when it was found that α-N-dimethyl diglycine which has a positive N-terminus was not utilized nor was it competitively active (Table 7-1, item 23) [137].

It appeared, therefore, that the dipeptide transport system has specificity for a protonated primary or secondary N-terminal amino group.

(b) Salmonella. Although several studies have been made with this organism, little attention has been given to N-terminal specificity. Jackson et al. [66] reported that N-acylation prevented utilization of Pro-Pro (Table 7-1, items 26, 27), although free proline could be liberated by intracellular peptidase action. However, other studies [111] show that certain N-acyl peptides are not cleaved (Table 7-1, items 28, 29).

(c) Pseudomonas. Recent studies [20, 21] indicate that N-acyl derivatives are not utilized. Although this may be attributed to lack of enzymatic machinery for their hydrolysis in certain cases, they also lack competitive ability, indicating a failure at the transport level (Table 7-1, items 30-34).

(d) Lactobacilli. Shankman et al. [164] reported that various α-N-acyl substituents prevented utilization of leucine and valine dipeptides by a number of species of lactobacilli, and provided evidence that the defect was in transport not hydrolysis (Table 7-1, items 34, 35). Chloroacetyl-Ala-Glu was not utilized by L. casei although the free peptide stimulated growth [207] (Table 7-1, items 36, 37).

(e) Other Bacteria. We have been unable to find in the literature relevant information for other bacterial species.

(f) Yeast. Dunn and Dittmer [32] reported that substituting the N-terminal group of Gly-β-2-thienylalanine and Gly-Phe with the benzyloxycarbonyl group destroyed their respective inhibitory and competitive properties in a strain of S. cerevisiae. In recent studies [10, 119] using S. cerevisiae G1333, conflicting evidence was presented regarding the role of the α-amino group. Several N-acetyl (Ac) and t-butyloxycarbonyl (t-boc) dipeptides of Gly and Met (Ac-Met-Met, Ac-Gly-Met, t-boc-Gly-Met) were utilized by a methionine auxotroph, whereas others (t-boc-Met-Met, Ac-Met-Gly) were not. In the absence of extracellular peptidases it would appear that certain N-acylated dipeptides were transported into yeast cells. However, more extensive studies are required to establish the generality of this conclusion, for at this time the results presented do not exclude alternative explanations, e.g., the presence of membrane bound or extracellular deacylase activity with variable activity towards the above substrates. It is possible that the failure of several peptide derivatives to support growth might result from their resistance to hydrolysis rather than inability to enter the cells. Unfortunately, no peptidase activities were detected against any of the tested derivatives (the utilizable or nonutilizable), and growth response could not be correlated to the presence or absence of the appropriate peptidase activity [119]. It remains a possibility, therefore, that the nonutilizable peptides were actually absorbed but were not hydrolyzed. It appeared to be the case that peptides (free or substituted) carrying a C-terminal glycine residue were utilized least efficiently. It was pointed out [10], however, that peptidase activity sometimes appears to be localized in lysosomal-like vacuoles, and that a peptide may have to cross two membrane barriers (the periplasmic and the vacuolar) before being hydrolyzed. This possibility makes the utilization of peptides in yeast potentially more complex than in bacteria, and the interpretation of results more difficult.

(g) Fungi. Appropriate studies have not been carried out for dipeptides because in general they appear not to be transported by N. crassa (see Sec. V.A).

2. Oligopeptides

(a) Escherichia. A variety of scattered reports in the literature [11, 32], together with systematic studies [42,92,129,137], have revealed the same specificity requirements as for dipeptides (see Sec. VI.B1a). Certain of these results are summarized in Table 7-1, items 44 to 58.

The structural requirement of the additional oligopeptide transport system recently described [8,118] is considered separately in Sec. VI.J.

(b) Salmonella. Jackson et al. [66] reported that N-acetylation prevented utilization (but not hydrolysis) of triproline by a proline auxotroph (Table 7-1, items 59, 60). No other evidence is available.

(c) Pseudomonas. Cascieri and Mallette [20] stated that cbz-Gly_3 (0.1 mM) did not compete with Gly-Met-Gly (0.1 mM) for growth of a Met auxotroph of P. putida, although free Gly_3 was shown to compete.

(d) Lactobacilli. Shankman et al. [164] provided analogous results for oligopeptides as for dipeptides, indicating N-acylation prevented transport but not hydrolysis (see Sec. VI.B1d). The growth stimulating ability of a number of oligopeptides was lost after N-acylation [108,206,207].

(e) Other Bacteria. No relevant information has been found for other bacteria.

(f) Yeast. The variable effect of N-acylation reported for dipeptides in a methionine auxotroph of S. cerevisiae [10,119] is also found for oligopeptides. N-acetylation of trimethionine and Met-Gly-Met did not diminish their utilizability, but the original poor utilizability of Met-Gly-Gly was further diminished by N-acetylation. Neither Gly-Met-Gly nor its N-acetyl derivative supported growth. Substituting the N-terminus of Gly-Met-Met,

Gly-Gly-Met, and Met-Met-Met with a t-butyloxycarbonyl group diminished but did not destroy utilizability. Similar difficulties attended efforts to assay peptidase activity as for peptides (see Sec. VI.B1f), and the discussion there is relevant to the case of oligopeptides also.

(g) Fungi. In N. crassa it has been reported that -N-acetyl-Gly-Leu-Tyr has no competitive ability in preventing transport of the free peptide [205].

B. C-Terminal Carboxyl Group

1. Dipeptides

(a) Escherichia. The early literature contains few results concerning the role of the C-terminus in dipeptide transport. Simmonds et al. [175] and Simmonds and Griffith [172] reported that several dipeptide amides were nutritionally inferior to the free peptides for auxotrophic strains. Furthermore, quite different growth responses were seen with, e.g., Phe-Gly-NH_2, depending on whether it was being tested with a glycine or phenylalanine auxotroph. However, they showed that the amides could be hydrolyzed by intracellular enzymes and concluded that uptake was rate limiting for growth. Kessel and Lubin [74] showed that Gly-Gly-methyl ester was a poor competitive inhibitor of [^{14}C]Gly-Gly accumulation. Payne and Gilvarg [141,143] reported that Leu-Gly-amide and Gly-Leu-amide were much less efficient than the free dipeptides in supporting growth of a glycine auxotroph. However, more importantly they showed that lysylcadaverine, i.e., dilysine lacking its C-terminal carboxyl group, was utilized by a lysine auxotroph. Additionally, by use of appropriate mutants it was shown that this dipeptide analog was transported by the oligopeptide transport system and not by the dipeptide permease. This result led to the speculation [143] that the dipeptide transport system has a requirement for a free C-terminal carboxyl group. Although dipeptides with a free C-terminal carboxyl group can to some extent use the oligopeptide transport system (as judged by competition studies), it would be predicted [138] from the C-terminal requirement of the dipeptide system that competition should be observed between oligopeptides and C-terminal substituted dipeptides, e.g., dipeptide amides, that can enter only by the OPT. In a recent study of dipeptide amide utilization [61] using an extensive range of leucine peptide amides, it was shown that substrates with the structure Leu-x-amide (where x is any amino acid) were utilized by a variety of leucine auxotrophs as well as the free peptide, but that compounds with the structure x-Leu-amide (where x is not leucine) were utilized much less effectively than the corresponding dipeptides. Enzyme tests indicate that this probably arose not because of the poor transport of the latter substrates, but rather the inability to rapidly liberate leucine from the amide when it was supplied in peptide form (but not in free form) (Table 7-2, items 6-14). Similar conclusions were reached for a phenylalanine auxotroph utilizing phenylalanine peptide amides (Table 7-2, items 12-14). Analogous enzymic specificity has been reported in a fungus [200].

(b) Salmonella. There is a single report [111] that a leucine auxotroph could utilize Leu-Ala-amide and Leu-Trp-amide. In view of the indicated uptake of dipeptide amides by the oligopeptide transport system in E. coli, and the high frequency at which oligopeptide transport-deficient mutants occur, it is surprising that Miller and MacKinnon [111] isolated peptidase/ amidase-deficient mutants of S. typhimurium rather than oligopeptide transport mutants when selecting for a strain unable to utilize Leu-Ala-amide.

(c) Pseudomonas. A methionine auxotroph of P. putida had a generation time of 1 hr on Gly-Met and 6.7 hr on Gly-Met-ethyl ester, each at 0.1 mM [20]. Linear growth was observed on the latter compound. Furthermore, Gly-Gly-ethyl ester did not compete for uptake with Gly-Met at an 8:1 molar ratio, although Gly-Gly at the same ratio was a good competitor. The authors concluded that the dipeptide ethyl esters were not transported by this dipeptide transport system, and they suggested that the ethyl derivatives probably entered the cells by slow, passive diffusion, followed by intracellular hydrolysis. However, they did not investigate whether the derivatives competed with oligopeptides for a presumptive oligopeptide permease.

(d) Lactobacilli, Other Bacteria, Yeast, and Fungi. We have found no relevant information in the literature.

2. Oligopeptides

(a) Escherichia. It is clear that the oligopeptide permease differs from the dipeptide system in having no requirement for a free C-terminal carboxyl group. Thus, decarboxylated trilysine, i.e., dilysine cadaverine, supports growth of a lysine auxotroph and was shown to enter by the oligopeptide permease [141]. Triornithine amide inhibits cells after entering via the OPT [141].

Becker and Naider [11] showed that a number of methyl esters of methionine oligopeptides were effective sources of the required amino acid for a Met auxotroph and confirmed that these substituted compounds were transported by the oligopeptide transport system. Payne [136] showed that the tripeptide Gly-Gly-β-Ala could be utilized by a glycine auxotroph but not by a derived opt⁻ mutant, and pointed out that the tripeptide could be

TABLE 7-2 The Effect of Various C-Terminal Carboxyl Group Modifications on the Utilization, Transport, and Hydrolysis of Peptides by Microorganisms[a]

Organism	Auxotroph	Peptide	C-Terminal Modification	Fate of Peptide			Ref.	Comments
				Utilization	Transport	Hydrolysis		
1. E. coli	Phe, Gly/Ser	Gly-Phe; Phe-Gly	none	+	+[a]	+	172	Rate and extent of growth less than with Phe, or dipeptides.
2. E. coli	Phe	Phe-Gly	amide	±	±[a]	+	172	Growth is inferior to that on Phe-Gly-NH$_2$.
3. E. coli	Phe	Gly-Phe	amide	±	±[a]	+	172	Cleavage of peptide bond precedes and is more rapid than cleavage of amide bond.
4. E. coli	Gly/Ser	Phe-Gly	amide	-	n.d.	+	172	
5. E. coli	Gly	Gly-Gly	methyl ester	n.d.	±	-	74	Competitive inhibition of (^{14}C) gly-gly transport was measured
6. E. coli	Leu	Leu-Ala	none	+	+[a]	+[a]	61	Study carried out with five separate strains and equivalent growth found for amide and free peptide.
7. E. coli	Leu	Leu-Ala	amide	+	+[a]	+	61	
8. E. coli	Leu	Leu$_2$, Leu-Ile, Leu-Gly, Leu-Phe	none	+	+[a]	+	61	Direct assays show rapid release of free leucine in all cases
9. E. coli	Leu	Leu$_2$, Leu-Ile, Leu-Gly, Leu-Phe	amide	+	+[a]	+	61	Same as 8 above.
10. E. coli	Leu	Gly-Leu, Phe-Leu, Ile-Leu	none	+	+[a]	+	61	Same as 8 above.
11. E. coli	Leu	Gly-Leu, Phe-Leu, Ile-Leu	amide	±	n.d.	±	61	Direct assays show slow release of free leu from amide bond, presumed to be rate limiting for growth.
12. E. coli	Phe	Phe-Leu, Leu-Phe	none	+	+[a]	+	61	
13. E. coli	Phe	Phe-Leu	amide	+	+[a]	+	61	
14. E. coli	Phe	Leu-Phe	amide	±	n.d.	±	61	Direct assays show slow release of free phe from amide bond, presumed to be rate limiting for growth.
15. E. coli	Lys	Lys$_2$	none	+	+[a]	+	141	Transport shown to be by oligopeptide, but not dipeptide transport system.
16. E. coli	Lys	Lys$_2$	decarboxylated	+	+[a]	+	141	
17. S. typhimurium	Leu	Leu-Ala, Leu-Trp	none	+	+[a]	+[a]	111	
18. S. typhimurium	Leu	Leu-Ala, Leu-Trp	amide	+	+[a]	+[a]	111	
19. P. putida	Met	Gly-Met	none	+	+[a]	+	20	
20. P. putida	Met	Gly-Met	ethyl ester	±	±[a]	+	20, 21	Slow linear growth; direct assays showed good hydrolysis.

21.	P. putida	Met	Gly-Gly	none	n.d.[b]	+	20,21	Dipeptide, but not dipeptide ester competed for transport with gly-met.
22.	P. putida	Met	Gly-Gly	ethyl ester	n.d.[b]	+	20,21	Similar growth with unsubstituted peptides.
23.	E. coli	Met	Met$_3$, Met$_4$, Met$_5$	methyl ester	+	±[a]	11	
24.	E. coli	Lys	Lys$_3$	none	+	+[a]	141	
25.	E. coli	Lys	Orn$_3$	decarboxylated	n.d.[b]	+[a]	141	
26.	E. coli	Lys	Lys$_2$	amide	+	+	141	Peptide amide inhibits variety of strains.
27.	E. coli	Lys + Thr	Lys$_2$	homoserine	+	+[a]	38	Peptide meets Lys requirement, Thr in medium. Thr mutation is in homoserine kinase ∴ homoserine cannot satisfy the requirement. Does not enter opt$^-$ mutant.
28.	E. coli	Lys + Thr	Lys$_2$	homoserine phosphate	+	+[a]	38	Peptide meets Thr requirement, Lys in medium. Same Thr mutant as above. Homoserine phosphate itself is impermeant and cannot meet Thr requirement. Does not enter opt$^-$ mutant.
29.	E. coli	Lys + Thr	Lys$_2$	homoserine	+	+[a]	38	Thr mutation is in aspartic semialdehyde dehydrogenase. Peptide meets Lys or Thr requirement. Does not enter opt$^-$ mutant.
30.	E. coli	Lys + Thr	Lys$_2$	homoserine phosphate	+	+[a]	38	Peptide meets Thr or Lys requirement even when no periplasmic alkaline phosphatase produced. Homoserine phosophate meets Thr requirement only when alk. phos. present. Same Thr mutant as in 29. Does not enter opt$^-$ mutant.
31.	S. typhimurium	His	Gly$_2$	Histidinol	+	+[a]	5	Peptide meets His requirement in mutant blocked in first enzyme of His pathway but not one blocked in histidinol phosphate phosphatase. Does not enter opt$^-$ mutant. Histidinol phosphate is impermeant.
32.	P. putida	Met	Gly-Met-Gly	none	+	+	20,21	
33.	P. putida	Met	Gly-Met-Gly	ethyl ester	±	+	20,21	
34.	S. cerevisiae	Met	Met$_3$, Met$_4$, Met$_5$	none	+	+[a]	119	Slow linear growth, direct assays showed good hydrolysis.
35.	S. cerevisiae	Met	Met$_3$, Met$_4$, Met$_5$	methyl ester	±	±[a]	119	After initial lag, growth rates similar to those found for the free peptides.

[a]Abbreviations used are: n.d., not determined; orn, ornithine; (+), (−), and (±) indicate that utilization, transport, or hydrolysis are extensive, not detectable, or slight, respectively; Lys$_2$, Lys$_3$ are Lys–Lys and Lys–Lys–Lys, respectively, and similar notation is used for other homopeptides. (a) when peptide is utilized (+), transport (+) and hydrolysis (+) are assumed; (b) peptide does not contain required amino acid, used only for competition.

regarded as Gly-Gly-Asp, devoide of its C-terminal α-carboxyl group. The nonessential nature of this group has been used to bring impermeant compounds into the cell after attachment to the carboxyl group, and these provide further examples of allowable modifications (see Table 7-2) [38].

(b) <u>Salmonella.</u> Although systematic studies have not been carried out it is likely that this organism also shows no requirement for a free C-terminal carboxyl group. Thus, Ames et al. [5] showed that the impermeant residue histidinol phosphate could be attached to the C-terminus and the resultant substituted tripeptide transported by the oligopeptide permease.

(c) <u>Pseudomonas.</u> Conversion of the carboxyl group in Gly-Met-Gly to an ethyl ester caused linear growth of a Met auxotroph of <u>P. putida</u> with a generation time of 3.1 hr compared with 1 hr for the free tripeptide; peptidase activity on the peptide ester appeared not to be rate limiting [20, 21].

(d) <u>Lactobacilli, Other Bacteria, and Fungi.</u> Several synthetic oligopeptide amides and esters were reported to show equal strepogenin activity (i.e., growth stimulation) as the free peptides [108,210]. Shankman et al. [164] showed that the growth promoting activity of Val-Val-Val-methyl ester was about 20% to 50% that obtained with the free tripeptide with seven strains. In later studies [163], the tripeptide methyl ester was shown to competitively inhibit the uptake of unsubstituted oligopeptides.

(e) <u>Yeast.</u> With a methionine auxotroph of <u>S. cerevisiae</u> Naider et al. [119] reported that the methyl esters of tri-, tetra-, and pentamethionine all supported growth. However, long lags were observed, compared with the free peptides, although once started, growth rates were similar for all compounds. Although it is likely that these results reflect similar requirements as in bacteria, competition studies were not performed, and it was not confirmed that the esters shared an uptake system used by oligopeptides, neither was the possibility ruled out that the methyl group was cleaved by an extracellular esterase.

C. Stereospecificity

Microbial peptidases are usually specific for peptide bonds formed from L-amino acid residues. In addition, microorganisms are commonly unable to convert the D-stereoisomers into utilizable L-forms. For these reasons, the use of auxotrophic growth tests to assess transport of peptides containing D-residues is of limited value. It is necessary therefore to use radiolabeled D-peptides and/or to measure their competitive abilities.

1. Dipeptides

(a) Escherichia. Simmonds et al. [175] reported that D-Leu-Gly failed to support growth of a leucine auxotroph although the L-isomer was active. More significantly, it was found [173] that D-Leu-Gly had no competitive effect on the utilization of Leu-Gly (Table 7-3, items 1-2). Unambiguous results indicating a stereospecific requirement for dipeptide transport have been provided by studies using radioactive dipeptides. Levine and Simmonds [89] showed that unlabeled D-Leu-Gly was without effect on the transport of isotopically labeled Leu-Gly (Table 7-3, item 3). Kessel and Lubin [74] showed that the presence of D-residues dramatically decreased the ability of a range of dipeptides to competitively inhibit the transport of [^{14}C]Gly-Gly (Table 7-3, items 4, 5).

(b) Salmonella. We have found no relevant information in the literature.

(c) Pseudomonas. The observation [20] that growth of a methionine auxotroph of P. putida on Gly-Met was inhibited by Leu-Leu but not by Leu-D-Leu or D-Leu-Leu, indicates that dipeptide transport is probably a stereospecific process in this species also. However, the result of the control to show that Leu-Leu did not inhibit growth on Met was not reported. The transport system and intracellular peptidases have the same stereochemical requirements [21].

(d) Lactobacilli. Early growth studies [80,164,215] indicated that the presence of D-residues in a number of dipeptides prevented their utilization although the L-stereoisomers supported growth. However, it was not unambiguously demonstrated whether the transport or hydrolysis were primarily affected. Peters et al. [148] showed that D-Ala-His served as a good histidine source for L. delbrueckii. In other studies, Kihara et al. [78] showed that dipeptides containing D-Ala could be absorbed by S. faecalis. Thus, D-Ala-Gly and D-Ala-Ala could satisfy the cells' requirement for D-Ala, but Ala-D-Ala and D-Ala-D-Ala would not. The tritiated compounds were absorbed in the order D-Ala > Ala-D-Ala = D-Ala-Ala > D-Ala-Gly >> D-Ala-D-Ala. These results are difficult to interpret, however, because of the presence of extracellular peptidases, and the possible presence of cell wall enzymes involved in incorporation of D-Ala-D-Ala into peptidoglycan.

Use of radioactive dipeptides has provided unambiguous evidence for the stereochemical specificity of dipeptide transport. Thus, Gly-D-Ala has no competitive effect on the uptake of [^{14}C]Gly-Ala by L. casei [86].

TABLE 7-3　The Stereospecificity of Peptide Utilization in Microorganisms[a]

Organism	Auxotroph	Peptide Substrate	Peptide Competitor	Utilization	Transport	Competition	Hydrolysis	Ref.	Comments
					Fate of Peptide Substrate				
1. E. coli	Leu	D-leu-gly	none	-	b	n.a.	-	175	The L-stereoisomer is hydrolysed and utilized
2. E. coli	Leu	Leu-gly	D-leu-gly	+	+[a]	-	+	175,89	
3. E. coli	Leu	(^{14}C)-leu-gly	D-leu-gly	n.d.	+	-	±	89	
4. E. coli	Gly	(^{14}C)-gly-gly	D-ala-D-ala, D-leu-gly, D-ala-gly	n.d.	+	-	-	74	Strain was diglycyl peptidase deficient. L stereoisomers had significant competitive ability.
5. E. coli	Gly	(^{14}C)-gly-gly	ala-D-ala, D-ala-ala	n.d.	+	±	-	74	The competitors were significantly inferior to the L-stereoisomers.
6. P. putida	Met	gly-met	D-leu-leu, leu-D-leu	+	+[a]	-	+	20	leu-leu competitively inhibited growth.
7. L. casei	c	(^{14}C)-gly-ala	gly-D-ala	n.d.	+	-	n.d.	86	
8. E. coli	Lys, W.T.	orn$_3$	D-ala-D-ala-D-ala	n.a.	+	-	-	131	Competitor failed to overcome orn$_3$ and val$_3$ toxicity, although LLL isomer was effective.
9. E. coli	several, K-12	val$_3$	D-ala-D-ala-D-ala	n.a.	+	-	+	131	
10. E. coli	gly/ser	gly$_3$	D-ala-D-ala-D-ala	+	+	-	+	131	Competitor did not prevent utilization of gly$_3$ LLL isomer did compete.
11. E. coli	Lys, W.T.	(^{14}C)-acetyl lys$_3$	D-ala-D-ala-D-ala	n.a.	+	-	-	131	Competitor did not inhibit uptake of radioactivity although LLL isomer did compete.
12. E. coli	Met	met-met-D-met, met-met-D-met-ome	none	+	+	n.a.	+	11	Growth equal to that on LLL isomers.
13. E. coli	Met	D-met-met-met, D-met-met-met-ome	none	+	+	n.a.	+	11	Strain was mutant selected for growth on DLL. Peptide shown to be transported by Opt.

Organism								Ref.	Comments
14. P. putida	Met	gly-met-gly	D-leu-gly-gly	+	+[a]	−	+	20	Leu-gly-gly competitively inhibited growth on the peptide substrate.
15. Lactobacilli	c	(^{14}C)-val-val-D-val	none	+	n.a.	n.a.	n.a.	163	
16. Lactobacilli	c	(^{14}C)-val-val-D-val	val_3	±	+	n.a.	n.a.	163	All other stereoisomers of val_3 failed to compete for transport of radioactive substrate.
17. Lactobacilli	c	(^{14}C)-val-val-D-val	val-leu-D-val	±	+	n.a.	n.a.	163	

[a] All amino acid residues are of the L-form unless indicated otherwise. The abbreviations used are: n.d., not determined; n.a., not applicable; W.T., wild-type; Leu_2, Leu_3 are Leu-Leu and Leu-Leu-Leu, and similar notation is used for other homopeptides; (+), (−), (±), indicate extensive, not detectable, slight, respectively, for utilization, transport, competition, or hydrolysis. (a) when peptide is utilized (+), transport (+) and hydrolysis (+) are assumed; (b) when hydrolysis is not detected (−) and peptide is not utilized (−) transport may or may not occur; (c) species requires variety of amino acids.

2. Oligopeptides

(a) Escherichia. Payne [131] provided evidence for the stereospecificity of oligopeptide transport. Thus, D-Ala-D-Ala-D-Ala was unable to competitively overcome the inhibitory effects of triornithine or trivaline with K12 strains, although the tri-L-alanine at similar concentrations reversed the toxicity. The D-isomer (but not the L-) also failed to compete against utilization of triglycine by a glycine auxotroph, or the uptake of radioactive lysyl oligopeptides. It had earlier been postulated [143] that the presence of a D-isomer at the C-terminus of a tripeptide should have little effect on the transport of the peptide, and Becker and Naider [11] provided evidence for this view. In a detailed investigation, they showed that of eight diastereomers of trimethionine, only LLL and LLD (and their corresponding methyl esters) supported growth of a Met auxotroph of strain K12. Unfortunately, absence of peptidase activity towards most of the other stereisomers prevented any conclusions regarding their transport. They also found that a mutant could be isolated that grew upon the DLL form and showed that it was transported by the oligopeptide transport system. Although stereospecific requirements appear to be established for oligopeptide transport, there remains uncertainty as to the tolerance towards D-residues at specific positions in the sequence.

(b) Salmonella. We have found no relevant information in the literature.

(c) Pseudomonas. Cascieri and Mallette [20] stated that in a growth test, using a methionine auxotroph of P. putida, Leu-Gly-Gly (1.0 mM) but not D-Leu-Gly-Gly (1.0 mM) competed with Gly-Met-Gly (0.1 mM), but they presented no data.

(d) Lactobacilli. Similar results have been found as for E. coli above. Shankman et al. [165] found that of the eight stereoisomers of L- and D-Val$_3$, LLD was inhibitory to Pediococcus cerevisiae, and the LLL isomer was the only one able to reverse the inhibition. Later it was shown [163] that radioactive LLD-Val$_3$ was actively transported into this strain. The peptides LLL-Val$_3$ and Val-Leu-D-Val inhibited this uptake, but none of the other stereoisomers of Val$_3$ were effective. LLL- and LLD-Val$_3$ were the only stereoisomers to be readily utilized by a variety of strains [164] although L. casei and S. faecalis showed slight growth response to DDL-Val$_3$.

(e) Fungi. Five tripeptides with the sequence Arg-D-X-Phe (X = Ala, Tyr, Val, Phe, Leu) showed antibiotic activity on several fungi and pathogenic molds. When X = Asn, Glu, Ser, the peptides were inactive. Addition of the corresponding L-amino acid relieved toxicity, suggesting that the peptides were accumulated and hydrolyzed, but actual uptake studies were not performed [34,35] (See also Sec. VI.F).

D. α-Peptide Bond

1. Dipeptides

(a) Escherichia. A variety of β-, γ-, and ε-linked dipeptides failed to support growth of appropriate auxotrophs [131,136]. However, in general, the strains lacked peptidase activity able to release the required amino acids from these unusual peptides, so the growth tests failed to provide unambiguous results relating to transport specificity. The further observations that these peptides failed to compete with nutritionally effective α-linked dipeptides does, however, indicate a transport requirement for the α-peptide linkage. If substantiated, this conclusion accords with the requirement for both the N-terminal α-amino group and the C-terminal carboxyl group for dipeptide transport (see Sec. VI.B1a and VI.C1a), since only the presence of an α-peptide bond can preserve the fixed spatial arrangement of these two groups in the dipeptide molecule. However, as pointed out by Barak and Gilvarg [7], these observations do not prove a need for a peptide bond for uptake; they might alternatively reflect an overall topographical requirement for a certain distance between the N- and C-termini of the transported molecule. No studies have yet been performed to check this speculation.

It should be mentioned that certain unusual peptides were utilized. Thus, γ-Glu- ε-Lys satisfied the amino acid requirements of appropriate auxotrophs [131], and β-Ala-His (carnosine) supported growth of a histidine auxotroph [136]. In the latter case, it was speculated that a separate uptake system may exist. Kessel and Lubin [74] showed that carnosine had little effect on the uptake of [^{14}C]Gly-Gly.

(b) Salmonella. We have found no relevant information in the literature.

(c) Pseudomonas. Cascieri and Mallette [20] reported that Gly- β-Ala and β-Ala-His (each at up to 10-fold molar excess) did not inhibit growth of a P. Putida Met auxotroph on Gly-Met.

(d) Lactobacilli. L. delbrueckii was shown to utilize β-Ala-His [148]. In contrast, P. cerevisiae failed to use the substance [39], although α-linked histidyl peptides supported normal growth.

(e) Other Bacteria. Several strains of Corynebacteriae utilized   -Ala-His [116]. Rowlands et al. [159a] reported that γ-glutamyl peptides

were not taken up by Staphylococcus aureus, and Dalen [28] showed carnosine and homocarnosine (α-aminobutyryl-His) were also not utilized by
S. aureus.

2. Oligopeptides

The only studies reported are for E. coli. Payne [136] showed that Gly-Gly-
β-Ala entered by the oligopeptide transport system during utilization by a
glycine auxotroph. However, the peptide may be regarded as Gly-Gly-Asp
devoid of its C-terminal α-carboxyl group, making the result not unexpected
in view of the absence of C-terminal carboxyl requirement (see Sec. VI.C2a).
In contrast, β-Ala-Gly-Gly was not nutritionally active although the cell
contained enzymes able to liberate glycine. This single example indicates that
the OPT probably has specificity for the first α-linkage, and although not
tested, one might anticipate also for the second.

E. Peptide Bond Nitrogen Atom

Study of this feature in E. coli [132] led to the most important finding that
peptides in which the peptide bond nitrogen is substituted (methylated) are
transported via the di- and oligopeptide permeases and may be accumulated
intact, for the substituted bonds are resistant to intracellular peptidase
activity. These substrates therefore provide good evidence for the separate
nature of transport and hydrolysis (see Sec. III.A). They are also ideal for
determining whether peptide transport is an active process and for measuring its kinetic parameters. As such they have proved invaluable in
studies on intestinal peptide absorption [1-4].
    Resistance to hydrolysis generally prevents utilization of these peptides,
but they can be effective competitors of transport. Thus, glycylsarcosine
(Gly-N-methyl-Gly) competitively inhibits utilization of a variety of unsubstituted dipeptides. The N-terminal residue in Gly-Gly-Sar was in fact
utilized by a glycine-requiring auxotroph of E. coli. Competition studies
with utilizable and toxic oligopeptides, and lack of growth of an opt mutant
of the glycine auxotroph all indicated that this substituted oligopeptide used
the oligopeptide transport system. In a Met auxotroph of P. putida, Gly-
Sar was reported not to compete with the utilization of Gly-Met [20]. Six
tripeptides with the sequence Arg-N-α-methyl-X-Phe (X = Ala, Gly, Val,
Leu, Phe, Ile) showed antibiotic activity in fungi and pathogenic molds, but
not towards B. subtilis or E. coli; when X = Glu, Ser, Thr, the peptides
were inactive. Addition of the corresponding nonmethylated amino acid
relieved toxicity, except for N-methyl gly for which no neutralizing amino
acid was found. It seemed likely that the peptides were accumulated and
hydrolyzed, but this was not shown directly [36].

F. α-Hydrogen Atom

Several reports indicate that the hydrogen atom on an α-carbon of an amino acid residue is not involved in the mechanism of oligopeptide transport (in Lactobacilli). Thus, Gly-[^{14}C]-AIB-Ala (AIB is α-aminoisobutyrate in which the α-hydrogen atom of alanine is replaced by a methyl group) was readily absorbed by L. casei [216], and its uptake was competitively inhibited by triglycine [179]. These represent a further group of peptides, analogous to those with N-methylated peptide bonds (see Sec. VI.F), that could be useful in demonstrating active transport, etc., for it is established that peptides containing AIB residues are resistant to the usual α-proteases [144]. Analogous tripeptides with cycloleucine (cyl, 1-amino-cyclopentane-1-carboxylic acid, which is also devoid of an α-hydrogen) as the central residue, (Gly-[^{14}C]-cyl-Ala; Gly-[^{14}C]-cyl-Val), were rapidly accumulated by L. casei [216].

G. Amino Acid Side Chains

Di- and oligopeptides formed from the protein amino acids can form an enormous variety of sequences, and there is no reason not to expect that all theoretical possibilities can occur in partial protein digests. Most such peptides will be of mixed character, e.g., Lys-Ala-Glu-Tyr rather than specifically basic, or neutral or aromatic, e.g., Lys-Arg-Gly-Ala-Leu, or Phe-Phe-Trp. It might be anticipated therefore that a general peptide transport system would not show specificity with respect to side chains. The evidence from a variety of microorganisms supports this view. Nevertheless, only a small fraction of the total possible peptides has been studied, and the possibility of secondary systems with particular specificities must remain open.

Conclusions on side chain specificity have been obtained from competition studies and from use of mutants. When different peptides compete for transport, it may be assumed they share a permease. However, inability to demonstrate competition in any instance may simply reflect the differing affinities of the peptides for the permease rather than any absolute side-chain requirements of the system. The use of mutants is seemingly more clear-cut. When a single mutation results in simultaneous inability to transport two (or more) peptides, it is generally taken as evidence that the peptides share a single permease. However, the mutations are generally uncharacterized, and it remains a possibility that the defect has occurred in some component (e.g., energy coupling) that is common to several peptide permeases that may have other distinguishable characteristics (e.g., side-chain specificity!). With these caveats, evidence is presented for the lack of side-chain specificity found with various microbial species.

1. Dipeptides

The evidence already presented that dipeptides can use two distinct transport systems, the dipeptide and the oligopeptide system, makes it particularly difficult to interpret competition data. For example, the partial inhibition of uptake or growth that is frequently obtained in competition experiments might arise from complete inhibition of nonspecific uptake through the oligopeptide transport system with no effect on the specific dipeptide system. It therefore is clearly desirable in studies on dipeptide uptake to eliminate the oligopeptide permease component by use of oligopeptide transport-deficient mutants. Unfortunately, experiments of this type have not been carried out.

(a) E. coli. Some representative studies will be cited. Levine and Simmonds [88, 89] showed that competition for entry was shown by Leu-Gly and Gly-Leu. Kessel and Lubin [74] reported (variable) inhibition of [^{14}C]Gly-Gly uptake by all 17 dipeptides tested. In growth tests, the following competitive interactions were demonstrated: Gly$_2$ with both Lys$_2$ and Gly-Sar, Gly-Sar with Gly-Pro, Lys$_2$ with Pro-Gly, and Gly-Val with both Gly$_2$ and Sar-Gly [128,130,132].
A dipeptide transport-deficient mutant (dpt$^-$) that failed to accumulate [^{14}C]Gly-Gly was also unable to utilize Leu-Gly or Gly-Leu as glycine source [74]. De Felice et al. [29] recently reported another dpt$^-$ mutant that was isolated from an opt$^-$ strain of K12. This mutant (although still sensitive to valine) was resistant to the toxic dipeptides, Val-Val, Gly-Val, Val-Leu, and Leu-Val, and failed to accumulate [^{14}C]Gly-Gly. Several strains of K12 isolated as resistant to Gly-Leu [194] also proved to be transport-defective and cross-resistant, therefore, to Gly-Val.

(b) Salmonella and Other Bacteria. We have found no relevant information in the literature.

(c) Pseudomonas. In growth tests using a Met auxotroph of P. putida, the following dipeptides inhibited growth on Gly-Met: Gly$_2$, Ala$_2$, Leu$_2$, Ala-Gly, Gly-Ala, Ala-His, Pro-Lys, Phe-Gly, Lys-Lys, Gly-Pro [20].

(d) Lactobacilli. Leach and Snell [86] showed, using radioactive peptides, that Gly-Ala and Ala-Gly used a common transport system in L. casei. In L. mesenteroides, Gly-Val, Gly-Leu, Gly-Gly, Gly-Ser, and Ala-Phe also share a common system. However, uptake of [^{14}C]Gly-Ala by L. casei was not inhibited by Pro-Phe or His-His [86].

(e) Yeast. Too few peptides have been studied to reach any general conclusions, although the results obtained already indicate differences from bacteria [10,19]. A Met auxotroph of S. cerevisiae grew on all nine meth-

ionine dipeptides tested, although the growth curves varied somewhat. In
contrast, three different lysine auxotrophs all failed to grow on Lys_2, Lys-
Gly, and Gly-Lys, although free lysine was released when the peptides
were incubated with cell extracts. Unfortunately, neither competition
studies nor isolation of peptide transport-deficient mutants has been
reported, and there is also no evidence available concerning a possible
distinction between di- and oligopeptide transport.

(f) Fungi. No meaningful summary can be made at this time, for it
appears that, in general, dipeptides are not transported [203-205].

2. Oligopeptides

(a) Escherichia. Discussion is limited to the general oligopeptide
transport system, specificities of the secondary system(s) are not considered [8,118]. The peptides that have been tested and shown to share the
oligopeptide transport system in strain W by their failure to be transported
in opt⁻ mutants, or by competition experiments, include some with basic
amino acid residues (Lys, Arg), neutral residues (Gly, Ala, Leu, Val),
aromatic residues (Tyr, Trp, Phe), hydroxy residues (Ser, Thr), and sulfur
residues (Met). Oligopeptides containing nonprotein amino acids are also
transported by this system, e.g., ornithine [6,43,128], ε-acetylated lysine
[42,131], and norleucine and norvaline [129]. However, many of the competition experiments indicated that oligopeptides could have widely different
affinities for the transport system (see, e.g., Ref. 128).

Similar conclusions apply to strain K12 although the evidence is less
extensive. Several opt⁻ mutants, isolated as resistant to triornithine,
showed cross-resistance to other toxic tripeptides like trivaline [6,29]
and Lys-Lys-p-F-Phe [6]. Some were also shown unable to utilize nutritional oligopeptides. [^{14}C]Triglycine was not accumulated by an opt⁻
mutant [29].

(b) Salmonella. Proline auxotrophs of S. typhimurium have been shown
to utilize a variety of proline oligopeptides [66] and to lose this ability in
an opt⁻ mutant. Using a growth assay, competition was demonstrated
between nutritional proline peptides and peptides lacking the required
amino acid. This study also provided evidence for additional oligopeptide
permease. Ames et al. [5] reported that opt⁻ mutants of a His auxotroph were isolated as resistant to either; Lys_3, Norleu-Gly-Gly, Gly-
Gly-Norleu, or Gly-Gly-Norval were all cross-resistant, and each was
unable to utilize Gly-His-Gly as histidine source, although the parent
strain could.

(c) Pseudomonas. Using a Met auxotroph of P. putida, Cascieri and
Mallette [20] reported that growth on Gly-Met-Gly (0.1 mM) was inhibited

by the following tripeptides (0.5 mM): trilysine, Ser-His-Asp, Thr-Gly-Gly, Ala-Gly-Gly, and Gly-Gly-Gly. Additionally, growth of a histidine auxotroph on Ser-His-Asp or Gly-His-Gly was inhibited by Gly-Gly-Gly and Gly-Met-Gly, respectively.

(d) Lactobacilli. Only a little circumstantial evidence is available to suggest broad specificity. Merrifield [107] noted that the inhibitory effect of Thr-His-Leu-Val-Glu and Thr-Thr-Asn-Glu-Ala-Lys on L. casei was competitively relieved by Ser-His-Leu-Val-Glu. Dunn et al. [33] noted that Leu-Phe-Gly and Leu-β-2-Thienylalanyl-Gly competed for uptake in L. arabinosus. Young et al. [216] showed, using radioactive peptides, that peptides containing "unnatural" amino acid residues shared a system with "normal" oligopeptides in L. casei.

(e) Other Bacteria. We have found no relevant information.

(f) Yeast. Similar reservations apply here as mentioned for dipeptide uptake (see Sec. VI.H1e). A methionine auxotroph of S. cerevisiae utilized a variety of methionine oligopeptides, but apparent specificity towards peptides containing C-terminal Met was noted [119]. Homooligopeptides of lysine failed to support growth of several lysine auxotrophs, although they could be hydrolyzed by cell extracts.

(g) Fungi. An extensive range of defined oligopeptides, and mixed peptides present in Neopeptone, have been shown to support growth of leucine, methionine, lysine, and histidine auxotrophs of N. crassa [203, 204]. A mutant isolated as resistant to Gly-Leu-Tyr (but still sensitive to the toxic element Tyr) was characterized as transport deficient because of its inability to utilize a variety of previously nutritionally effective oligopeptides [204] and its failure to accumulate Gly-Leu-[^{3}H]-Tyr [205]. Competitive inhibition of uptake of the radioactive tripeptide was shown by Met-Ala-Ser, Gly-Leu-Gly-Leu, Gly-Ala-Leu, Gly-Leu-Tyr, Leu-Leu-Leu, and a mixture of peptides, of chain length about three to five, present in Neopeptone fractions [205]. It appears that a general oligopeptide permease is present analogous to that found in, e.g., E. coli.

H. Other Parameters: Future Studies

It is clear from this survey that much work is needed with various microorganisms to decide how general are the structural specificities established for E. coli. Even for E. coli more data are required in relation to stereochemistry, α-hydrogen atoms, and peptide bond nitrogen atoms. However, it may also be fruitful, rather than treating each element of peptide structure separately, to consider the overall topography of di- and oligopeptides.

Model building should prove useful here. Such considerations may generate structures compatible with the structural and spatial specificites of peptide transport but in fact, chemically rather dissimilar from peptides. Future studies should attempt to explain the reasons for the relative affinities of different peptides for the transport system.

REFERENCES

1. J. M. Addison, D. Burston, D. M. Matthews, J. W. Payne, and S. Wilkinson, Evidence for active transport of the tripeptide glycylsarcosylsarcosine by hamster jejunum in vitro, Clin. Sci. Mol. Med., 46:39P (1974).

2. J. M. Addison, D. Burston, D. M. Matthews, J. W. Payne, and S. Wilkinson, Competition between the tripeptide glycylsarcosylsarcosine and other di- and tripeptides for uptake by hamster jejunum in vitro, Clin. Sci. Mol. Med., 48:5P-6P (1975).

3. J. M. Addison, D. Burston, J. W. Payne, S. Wilkinson, and D. M. Matthews, Evidence for active transport of tripeptides by hamster jejunum in vitro Clin. Sci. Mol. Med., 49:305-312 (1975).

4. J. M. Addison, D. Burston, J. A. Dalrymple, D. M. Matthews, J. W. Payne, M. H. Sleisenger, and S. Wilkinson, A common mechanism for transport of di- and tripeptides by hamster jejunum in vitro, Clin. Sci. Mol. Med., 49:313-322 (1975).

5. B. N. Ames, G. F. Ames, J. D. Young, D. Isuchiya, and J. Lecocq, Illicit transport, the oligopeptide permease, Proc. Natl. Acad. Sci. USA, 70:456-458 (1973).

6. Z. Barak and C. Gilvarg, Triornithine-resistant strain of Escherichia coli: isolation, definition and genetic studies, J. Biol. Chem., 249:143-148 (1974).

7. Z. Barak and C. Gilvarg, Peptide Transport, In Biomemebranes Vol. 7, (H. Eisenberg, E. Katchalski-Katzir, and L. A. Manson, eds.), Plenum, New York, 1975, pp. 167-218.

8. Z. Barak and C. Gilvarg, Specialized peptide transport system in Escherichia coli, J. Bacteriol., 122:1200-1207 (1975).

9. Z. Barak, S. Sarid, and E. Katchalski, Inhibition of protein synthesis in Escherichia coli B by tri-L-ornithine, Eur. J. Biochem., 34:317-324 (1973).

10. J. M. Becker, F. Naider, and E. Katchalski, Peptide utilization in yeast: studies on methionine and lysine auxotrophs of Saccharomyces cerevisiae, Biochim. Biophys. Acta, 291:388-397 (1973).

11. J. M. Becker and F. Naider, Stereospecificity of tripeptide utilization in a methionine auxotroph of Escherichia coli K-12, J. Bacteriol., 120:191-196 (1974).

12. U. Bjare, B. Hofsten, and A. C. Ryden, Cell bound proteolytic enzymes of Arthrobacter, Biochim. Biophys. Acta., 20:134-136 (1970).
13. P. Boyer, The Enzymes: Hydrolysis: Peptide Bonds, Vol. 3 (3rd ed.) Academic, New York, 1971.
14. T. D. Brock and G. Moo-Penn, An amino acid transport system in Streptococcus faecium, Arch. Biochem. Biophys., 98:183-190 (1962).
15. T. D. Brock and S. O. Wooley, Glycylglycine uptake in Streptococci and a possible role of peptides in amino acid transport, Arch. Biochem. Biophys., 105:51-57 (1964).
16. J. L. Brown, Purification and properties of dipeptidase M from Escherichia coli B, J. Biol. Chem., 248:409-416 (1973).
17. N. F. Bukland, D. Sanborn, and I. N. Hirshfield, Particular influence of leucine peptides on lysyl-transfer ribonucleic acid ligase formation in a mutant of Escherichia coli K-12, J. Bacteriol., 116:1477-1478 (1973).
18. T. Cascieri and M. F. Mallette, Stimulation of lysine decarboxylase production in Escherichia coli by amino acids and peptides, Appl. Microbiol., 26:975-981 (1973).
19. T. Cascieri and M. F. Mallette, New method for study of peptide transport in bacteria, Appl. Microbiol., 27:457-463 (1974).
20. T. Cascieri and M. F. Mallette, Peptide utilization by Pseudomonas putida and Pseudomonas maltophilia, J. Gen. Microbiol., 92:283-285 (1976).
21. T. Cascieri and M. F. Mallette, Intracellular peptide hydrolysis by Pseudomonas putida and Pseudomonas maltophilia, J. Gen. Microbiol., 92:296-303 (1976).
22. G. L. Choules and W. R. Gray, Peptidase activity in the membranes of Mycoplasma laidlawii, Biochem. Biophys. Res. Commun., 45:849-855 (1971).
23. H. N. Christensen, Biological Transport (2nd ed.), Benjamin, London, 1975, pp. 450-452.
24. H. N. Christensen, Biological Transport (2nd ed.), Benjamin, London, 1975, pp. 247-271.
25. B. L. Cohen, J. E. Morris, and H. Brucker, Regulation of two extracellular proteases of Neurospora crassa by induction and by carbon-nitrogen and sulfur-metabolite repression, Arch. Biochem. Biophys., 169:324-330 (1975).
26. G. N. Cohen and H. V. Rickenberg, Concentration specifique reversible des aminoacides chez Escherichia coli, Ann. Inst. Pasteur, 91:693-720 (1956).
27. J. L. Cowell, Energetics of glycylglycine transport in Escherichia coli, J. Bacteriol., 120:139-146 (1974).
28. A. B. Dalen, The influence of pH and histidine dipeptides on the production of Staphylococcal α-toxin, J. Gen. Microbiol., 79:265-274 (1973).

29. M. DeFelice, J. Guardiola, A. Lamberti, and M. Iaccarino, Escherichia coli K-12 mutants altered in the transport systems for oligo- and dipeptides, J. Bacteriol., 116:751-756 (1973).

30. H. Drucker, Regulation of exocellular proteases in Neurospora crassa: Role of Neurospora proteases in induction, J.Bacteriol., 116:593-599 (1973).

31. H. Drucker, Regulation of exocellular proteases in Neurospora crassa: Metabolic requirements of the process, J. Bacteriol., 122:1117-1125 (1975).

32. F. W. Dunn and K. Dittmer, The synthesis and microbiological properties of some peptide analogues, J. Biol. Chem., 188:263-272 (1951).

33. F. W. Dunn, J. Humphreys, and W. Shive, Utilization of tripeptides, Arch. Biochem. Biophys. 71:475-476 (1957).

34. K. Eisele, Sequence variation on an antibiotically active tripeptide, Z. Naturforsch, 30:541-543 (1975).

35. K. Eisele, Antibiotically active tripeptides of the sequence L-arg-D-X-L-phe, Experientia, 31:764 (1975).

36. A. Eisele and K. Eisele, Antibiotically active tripeptides with the sequence L-Arginine-N-α-methyl-L-X-L-Phenylalanine, FEBS. Lett., 55:153-155 (1975).

37. T. E. Fickel, The oligopeptide permease of E. coli as a vehicle for the transport of impermeant substances and its accessibility to large oligopeptides, PhD. thesis, Princeton University, 1973.

38. T. E. Fickel and C. Gilvarg, Transport of impermeant substances in E. coli by way of oligopeptide permease, Nature New Biol., 241:161-163 (1973).

39. H. A. Florsheim, S. Makineni, and S. Shankman, The isolation, identification and synthesis of a peptide growth factor for P. cerevisiae, Arch. Biochem. Biophys., 97:243-249 (1962).

40. E. F. Gale, The arginine, ornithine and carbon dioxide requirements of Streptococci (Lancefield group D) and their relation to arginine dehydrolase activity, Brit. J. Exp. Path., 26:225-233 (1945).

41. C. Gilvarg, Peptide transport in bacteria. In Peptide Transport in Bacteria and Mammalian Gut, CIBA Foundation Symposium, Associated Scientific Publishers, Amsterdam, 1972, pp. 11-16.

42. C. Gilvarg and E. Katchalski, Peptide utilization in Escherichia coli, J. Biol. Chem., 240:3093-3098 (1965).

43. C. Gilvarg and Y. Levin, Response to Escherichia coli to ornithyl peptides, J. Biol. Chem., 247:543-549 (1972).

44. A. L. Goldberg and J. F. Dice, Intracellular protein degradation in mammalian and bacterial cells, Ann. Rev. Biochem., 43:835-869 (1974).

45. O. Grahl-Nielsen, P. Ødegaard, and G. L. Tritsch, Oligopeptides as sources of indispensible amino acids for mammalian cells in culture, In Vitro, 9:414-420 (1974).

46. J. Guardiola, M. DeFelice, T. Klopotowski, and M. Iaccarino, Multiplicity of isoleucine, leucine, and valine transport systems in Escherichia coli K12, J. Bacteriol., 117:382-392 (1974).

47. J. Guardiola, M. DeFelice, T. Klopotowski, and M. Iaccarino, Mutations affecting the different transport systems for isoleucine, leucine, and valine in Escherichia coli K-12, J. Bacteriol., 117:393-405 (1974).

48. J. Guardiola, and M. Iaccarino, Escherichia coli K12 mutants altered in the transport of branched-chain amino acids, J. Bacteriol., 108: 1034-1044 (1971).

49. B. M. Guirard and E. E. Snell, Nutritional requirements of microorganisms, in, The Bacteria, Vol. 4. (I. C. Gunnsalus and R. Y. Stanier, eds.), Academic, New York, 1962, pp. 33-93.

49a. F. F. Hall, H. O. Kunkel, and J. M. Prescott, Multiple proteolytic enzymes of Bacillus licheniformis, Arch. Biochem. Biophys., 114: 145-153 (1966).

50. Y. S. Halpern, Genetics of amino acid transport in bacteria, Ann. Rev. Genet., 8:103-133 (1974).

51. A. H. W. Hauschild, Incorporation of C^{14} from amino acids and peptides into protein by Clostridium perfringens Type D, J. Bacteriol., 90:1569-1574 (1965).

52. A. H. W. Hauschild, Peptides for toxinogenesis of Clostridium perfringens Type D, J. Bacteriol., 90:1793-1794 (1965).

53. A. H. W. Hauschild, Selective effect of pH on the production of exocellular protein by Clostridium perfringens Type D, J. Bacteriol., 92:800-801 (1966).

54. S. Hayman, J. S. Gatmaitan, and E. K. Patterson, The relationship of extrinsic and intrinsic metal ions to the specificity of a dipeptidase from Escherichia coli B, Biochemistry, 13:4486-4494 (1974).

55. D. A. H. Hearfield and A. W. Phillips, Fractions of trypsinized casein giving different patterns of bacterial growth, Nature, 190:266-267 (1961).

56. V. Heiniger and Ph. Matill, Protease secretion in Neurospora crassa, Biochem. Biophys. Res. Commun., 60:1425-1432 (1974).

57. L. A. Heppel, The concept of periplasmic enzymes, in Structure and Function of Biological Membranes (L. I. Rothfield, ed.), Academic, New York, 1971, pp. 223-247.

58. L. A. Heppel, B. P. Rosen, I. Friedberg, E. A. Berger, and J. H. Weiner, Studies on binding proteins, periplasmic enzymes and active transport in Escherichia coli, in The Molecular Basis of Biological Transport, Miami Winter Symposia, Vol. 3 (F. J. Woessner and F. Huising eds.), Academic, New York, 1972, pp. 133-156.

58a. D. Herbert, The chemical composition of microorganisms as a function of their environment, Symp. Soc. Gen. Microbiol., 11:391-416 (1961).

59. M. L. Hirsch and G. N. Cohen, Peptide utilization by a leucine-requiring mutant of Escherichia coli, Biochem. J., 53:25-30 (1953).

60. I. N. Hirshfield, F. M. Yeh, and L. E. Sawyer, Metabolites influence control of lysine transfer ribonucleic acid synthetase formation in Escherichia coli K-12, Proc. Natl. Acad. Sci. USA., 72:1364-1367 (1975).

61. I. N. Hirshfield and M. B. Price, Utilization of selected leucine amides by Escherichia coli, J. Bacteriol., 122:966-975 (1975).

62. H. Holzer, H. Betz, and E. Ebner, Intracellular proteinases in microorganisms, in Current Topics in Cellular Regulation, Vol. 9 (B. L. Horecker, ed.), Academic, New York, 1975 pp. 103-156.

63. W. A. Hullah and T. H. Blackburn, Uptake and incorporation of amino acids and peptides by Bacteroides amylophilus, Appl. Microbiol., 21:187-191 (1971).

64. P. W. Ifland, E. Ball, F. W. Dunn, and W. Shive, Peptide and keto acid utilization in replacing phenylalanine in adaptive enzyme synthesis, J. Biol., Chem., 230:897-904 (1958).

65. IUPAC-IUB Commission on Biochemical Nomencalutre, Symbols for amino acid derivatives and peptides, Recommendations 1971, J. Biol. Chem., 247:977-983 (1972).

66. M. B. Jackson, J. M. Becker, A. S. Steinfeld, and F. Naider, Oligopeptide transport in proline peptidase mutants of Salmonella typhimurium, J. Biol. Chem., 251:5300-5309 (1976).

67. G. L. Johnson and J. L. Brown, Partial purification and characterization of two peptidases from Neurospora crassa, Biochim. Biophys. Acta. 370:530-540 (1974).

68. M. K. Johnson, Physiological roles of Pneumococcal peptidases, J. Bacteriol. 119:844-847 (1974).

69. J. E. Jones, F. Naider, and J. M. Becker, Hydrolysis of oligopeptides by sera used in cell and tissue culture, In Vitro, 11:41-45 (1975).

70. H. R. Kaback, Transport, Ann. Rev. Biochem., 39:561-598 (1970).

71. H. R. Kaback, Transport across isolated bacterial cytoplasmic membranes, Biochim. Biophys. Acta, 265:367-416 (1972).

72. A. Kepes and G. N. Cohen, Permeation, in The Bacteria, Vol. 5 (I. C. Gunsalus and R. Y. Stanier, eds.), Academic, New York, 1962, pp. 179-221.

73. D. Kessel and M. Lubin, Transport of proline in Escherichia coli, Biochim. Biophys. Acta, 57:32-43 (1962).

74. D. Kessel and M. Lubin, On the distinction between peptidase activity and peptide transport, Biochim. Biophys. Acta. 71:656-663 (1963).

75. H. Kihara and E. E. Snell, Peptides and bacterial growth. II. L-alanine peptides and growth of Lactobacillus casei, J. Biol. Chem., 197:791-800 (1952).

76. H. Kihara and E. E. Snell, Peptides and bacterial growth. VIII.
    The nature of Strepogenin, J. Biol. Chem., 235:1409-1414 (1960).
77. H. Kihara and E. E. Snell, Peptides and bacterial growth IX.
    Release of double inhibitions with single peptides, J. Biol. Chem.,
    235:1415-1419 (1960).
78. H. Kihara, M. Ikawa, and E. E. Snell, Peptides and bacterial
    growth. X. Relation of uptake and hydrolysis to utilization of D-
    alanine peptides for growth of Streptococcus faecalis, J. Biol. Chem.,
    236:172-176 (1961).
79. H. Kihara, O. A. Klatt, and E. E. Snell, Peptides and bacterial
    growth. III. Utilization of tyrosine peptides by Streptococcus
    faecalis, J. Biol. Chem., 197:801-807 (1952).
80. W. A. Krehl and J. S. Fruton, The utilization of peptides by lactic
    acid bacteria, J. Biol. Chem., 173:479-485 (1948).
81. S. G. Kustu and G. F. Ames, The his P protein, a known histidine
    transport component in Salmonella typhimurium, is also an arginine
    transport component, J. Bacteriol., 116:107-113 (1973).
82. A. Lazdunski, M. Murgier, and C. Lazdunski, Evidence for an amino-
    endopeptidase localized near the cell surface of Escherichia coli,
    Eur. J. Biochem., 60:349-355 (1975).
83. A. Lazdunski, C. Pellissier, and C. Lazdunski, Regulation of
    Escherichia coli K10 aminoendopeptidase activity: Effects of
    mutations involved in the regulation of alkaline phosphatase, Eur. J.
    Biochem., 60:357-362 (1975).
84. C. Lazdunski, J. Busuttil, and A. Lazdunski, Purification and
    properties of a periplasmic aminoendopeptidase from Escherichia
    coli, Eur. J. Biochem., 60:363-369 (1975).
85. F. R. Leach and E. E. Snell, Occurrence of independent uptake
    mechanisms for glycine and glycine peptides in Lactobacillus casei,
    Biochim. Biophys. Acta, 34:292-293 (1959).
86. F. R. Leach and E. E. Snell, The absorption of glycine and alanine
    and their peptides by Lactobacillus casei, J. Biol. Chem. 235:3523-
    3531 (1960).
87. E. M. Levine and S. Simmonds, Metabolite uptake by serine-glycine
    auxotrophs of Escherichia coli, J. Biol. Chem., 235:2902-2909 (1960).
88. E. M. Levine and S. Simmonds, Effect of cultural conditions on
    growth and metabolite uptake by serine-glycine auxotrophs of Escher-
    ichia coli, J. Bacteriol., 84:683-693 (1962).
89. E. M. Levine and S. Simmonds, Further studies on metabolite uptake
    by serine-glycine auxotrophs of Escherichia coli, J. Biol. Chem.,
    237:3718-3724 (1962).
90. C. D. Litchfield and J. M. Prescott, Regulation of proteolytic
    enzyme production by Aeromonas proteolytica. I. Extracellular
    endopeptidase, Can. J. Microbiol., 16:17-22 (1970).

91. C. D. Litchfield and J. M. Prescott, Regulation of proteolytic enzyme production by Aeromonas proteolytica II. Extracellular aminopeptidase, Can. J. Microbiol., 16:23-27 (1970).

92. R. Losick and C. Gilvarg, Effect of α-acetylation on utilization of lysine oligopeptides in Escherichia coli, J. Biol. Chem., 241:2340-2346 (1966).

93. A. T. Matheson and T. Murayama, The limited release of ribosomal peptidase during formation of Escherichia coli spheroplasts, Can. J. Biochem., 44:1407-1415 (1966).

94. A. T. Matheson, L. P. Visentin, A. Boutet, and C. F. Rollin, Localization of the basic aminopeptidases in Escherichia coli, Cana. J. Biochem., 49:1340-1346 (1971).

95. Y. Matsumura, N. Minamiura, J. Kukumoto, and T. Yamamoto, Intracellular peptidase of Bacillus subtilis. III. Effects of metal ions on activity and specificity of Aminopeptidase of Bacillus subtilis, Agr. Biol. Chem., 35:975-982 (1971).

96. D. M. Matthews, Rates of peptide uptake by small intestine, in Peptide Transport in Bacteria and Mammalian Gut, CIBA Foundatio Symposium, Associated Scientific Publishers, Amsterdam, 1972, pp. 71-88.

97. D. M. Matthews, Absorption of peptides by mammalian intestine, in Peptide Transport in Protein Nutrition (D. M. Matthews and J. W. Payne, eds.), North Holland, Amsterdam, and American Elsevier, New York, 1975, pp. 61-146.

98. D. M. Matthews, Intestinal absorption of Peptides, Physiol. Rev. 55:537-608 (1975).

99. D. M. Matthews and J. W. Payne, Peptides in the nutrtion of microorganisms and peptides in relation to animal nutrition, in Peptide Transport in Protein Nutrition (D. M. Matthews and J. W. Payne, eds.), North Holland, Amsterdam, and American Elsevier, New York, 1975, pp. 1-60.

100. D. M. Matthews and J. W. Payne, Peptide Transport in Protein Nutrition, Frontiers of Biology, Vol. 37, North Holland, Amsterdam, and American Elsevier, New York, 1975.

101. D. M. Matthews and J. W. Payne, Occurrence and biological activities of peptides, in Peptide Transport in Protein Nutrition (D. M. Matthews and J. W. Payne, eds.), North Holland, Amsterdam, and American Elsevier, New York, 1975, pp. 392-463.

102. B. K. May and W. H. Elliott, Characteristics of extracellular protease formation by Bacillus subtilis and its control by amino acid repression, Biochim. Biophys. Acta, 157:607-615 (1968).

103. J. Mayshak, O. C. Yoder, K. C. Beamer, and D. C. Shelton, Inhibition and transport kinetic studies involving L-leucine, L-valine, and their dipeptides in Leuconostoc mesenteroides, Arch. Biochem. Biophys., 113:189-194 (1966).

104. G. L. McHugh, and C. G. Miller, Isolation and characterization of
      proline peptidase mutants of Salmonella typhimurium, J. Bacteriol.,
      120:364-371 (1974).
105. J. O. Meinhart and S. Simmonds, Metabolism of serine and glycine
      peptides by mutants of Escherichia coli strain K-12, J. Biol. Chem.,
      216:51-65 (1955).
106. N. Meisler and S. Simmonds, The metabolism of glycyl-L-leucine by
      Escherichia coli, J. Gen. Microbiol., 31:109-123 (1963).
107. R. B. Merrifield, Competitive inhibition of a strepogenin active
      peptide by related peptides, J. Biol. Chem., 232:43-54 (1958).
108. R. B. Merrifield and D. W. Woolley, The synthesis of L-seryl-L-
      histidyl-L-leucyl -L-valyl-L-glutamic acid, a peptide with strepogenin
      activity, J. Americ. Chem. Soc., 78:4646-4649 (1956).
109. A. Miller, A. Neidle, and H. Waelsch, Chemical stability and meta-
      bolic utilization of asparagine peptides, Arch. Biochem. Biophys.,
      56:11-21 (1955).
110. C. G. Miller, Genetic mapping of Salmonella typhimurium peptidase
      mutations, J. Bacteriol., 122:171-176 (1975).
111. C. G. Miller and K. MacKinnon, Peptidase mutants of Salmonella
      typhimurium, J. Bacteriol., 120:355-363 (1974).
112. H. K. Miller and H. Waelsch, The utilization of glutamine and
      asparagine peptides by microorganisms, Arch. Biochem. Biophys.,
      35:184-194 (1952).
113. M. D. Milne, Transport of amino acids and peptides in the gut and
      the kidney, in Scientific Basis of Medicine Annual Reviews (I.
      Guilliland and E. Francis, eds.), Athlone, London, 1971, pp. 161-
      177.
114. M. D. Milne, Peptides in genetic errors of amino acid transport, in
      Peptide Transport in Bacteria and Mammalain Gut, CIBA Foundation
      Symposium, Associated Scientific Publishers, Amsterdam, 1972,
      pp. 93-102.
115. M. D. Milne and A. M. Asatoor, Peptide absorption in disorders of
      amino acid transport, in Peptide Transport in Protein Nutrition (D.
      M. Matthews and J. W. Payne, eds.), North Holland, Amsterdam,
      and American Elsevier, New York, 1975, pp. 167-182.
116. J. H. Mueller, The utilization of carnosine by Diphtheria bacillus,
      J. Biol. Chem., 123:421-432 (1938).
117. M. Murgier, A. Lazdunski, and C. Lazdunski, Control of amino-
      peptidase and alkaline phosphatase synthesis in Escherichia coli:
      Evidence for a regulation by inorganic phosphate independent of pho
      R and pho B phosphatase regulatory genes, FEBS Lett., 64:130-134
      (1976).
118. F. Naider and J. M. Becker, Multuplicity of oligopeptide transport
      systems in Escherichia coli, J. Bacteriol., 122:1208-1215 (1975).

119. F. Naider, J. M. Becker, and E. Katzir-Katchalski, Utilization of methionine-containing peptides and their derivatives by a methionine-requiring auxotroph of Saccharomyces cerevisiae, J. Biol. Chem., 249:9-20 (1974).

120. K. Nekvasolova, J. Sidlo, and J. Haza, Effect of peptide groups isolated from enzymic casein hydrolysate on growth and toxinogenesis of Clostidium perfringens (welchii), J. Gen. Microbiol., 62:3-16 (1970).

121. A. Nishi and S. Hirose, Further observations on the rhythmic variation in peptidase activity during the cell cycle of various strains of Escherichia coli, J. Gen. Appl. Microbiol., 12:293-297 (1966).

122. T. P. O'Barr, H. Levin, and H. Reynolds, Some interrelationships of amino acids in the nutrition of Leuconostoc mesenteroides, J. Bacteriol. 75:429-435 (1958).

123. D. L. Oxender, Amino acid transport in microorganisms, in Metabolic Transport, Vol. VI of Metabolic Pathways (3rd ed.), (L. E. Hokin, ed.), Academic, New York, 1972, pp. 133-185.

124. D. L. Oxender, Membrane transport, Ann. Rev. Biochem., 41:777-814 (1972).

125. D. L. Oxender, Binding proteins and transport, in Biological Transport (2nd ed.), (H. N. Christensen, ed.), Benjamin, London, 1975, pp. 232-246.

126. E. K. Patterson, J. S. Gatmaitan, and S. Hayman, Substrate specificity and pH dependence of dipeptidases purified from Escherichia coli B and mouse ascites tumor cells, Biochem., 12:3701-3709 (1973).

127. E. K. Patterson, J. S. Gatmaitan, and S. Hayman, The effect of Mn^{2+} on the activities of a zinc metallodipeptidase from a mouse ascites tumor, Biochem., 14:4261-4266 (1975).

128. J. W. Payne, Oligopeptide transport in Escherichia coli: specificity with respect to side chain and distinction from dipeptide transport, J. Biol. Chem., 243:3395-3403 (1968).

129. J. W. Payne, The requirement for the protonated α-amino group for the transport of peptides in Escherichia coli, Biochem. J., 123:245-253 (1971).

130. J. W. Payne, The utilization of prolyl peptides by Escherichia coli, Biochem. J., 123:255-260 (1971).

131. J. W. Payne, Mechanisms of bacterial peptide transport, in Peptide Transport in Bacteria and Mammalian Gut, CIBA Foundation Symposium, Associated Scientific Publishers, Amsterdam, 1972, pp. 17-32.

132. J. W. Payne, Effects of N-methyl peptide bonds on peptide utilization by Escherichia coli, J. Gen. Microbiol., 71:259-265 (1972).

133. J. W. Payne, The characterization of dipeptidase from Escherichia coli, J. Gen. Microbiol., 71:267-279 (1972).

134. J. W. Payne, Variations in the peptidase activities of Escherichia coli in response to environmental changes, J. Gen. Microbiol., 71:281-289 (1972).
135. J. W. Payne, Discussion to mechanisms of bacterial peptide transport, in Peptide Transport in Bacteria and Mammalian Gut, CIBA Foundation Symposium, Associated Scientific Publishers, Amsterdam, 1972, p. 38.
136. J. W. Payne, Peptide utilization in Escherichia coli: Studies with peptides containing β-alanyl residues, Biochim. Biophys. Acta, 298:469-478 (1973).
137. J. W. Payne, Peptide transport in Escherichia coli: Permease specificity towards terminal amino group substituents, J. Gen. Microbiol., 80:269-276 (1974).
138. J. W. Payne, Transport of Peptides in Microorganisms, in Peptide Transport in Protein Nutrition (D. M. Matthews and J. W. Payne, eds.), North-Holland, Amsterdam, and American Elsevier, New York, 1975, pp. 283-364.
139. J. W. Payne, Microbial Peptidohydrolases, in Peptide Transport in Protein Nutrition (D. M. Matthews and J. W. Payne, eds.), North-Holland, Amsterdam, and American Elsevier, New York, 1975, pp. 365-391.
140. J. W. Payne, Peptides and Microorganisms, Adv. Microbial Physiol., 13:55-113 (1976).
141. J. W. Payne and C. Gilvarg, The role of the terminal carboxyl group in peptide transport in Escherichia coli, J. Biol. Chem. 243:335-340
142. J. W. Payne and C. Gilvarg, Size restriction on peptide utilization in Escherichia coli, J. Biol. Chem., 243:6291-6299 (1968).
143. J. W. Payne and C. Gilvarg, Peptide transport, Adv. Enzymol. 35: 187-244 (1971).
144. J. W. Payne, R. Jakes, The primary structure of alamethicin, Biochem. J., 117:757-766 (1970).
145. L. E. Pearce, N. A. Skipper, and B. D. W. Jarvis, Proteinase activity in slow lactic acid-producing variants of Streptococcus lactis, Appl. Microbiol., 27:933-937 (1974).
146. M. Pecht, E. Giberman, A. Keysary, J. Yariv, and E. Katchalski, Hydrolysis of alanine oligopeptides by an enzyme located in the membrane of Mycoplasma laidlawii, Biochim. Biophys. Acta, 290:267-273 (1972).
147. G. E. Perlman and L. Lorand, eds.), Methods in Enzymology: Proteolytic Enzymes, Vol. 19, Academic, New York, 1970.
148. V. J. Peters, J. M. Prescott, and E. E. Snell, Peptides and bacterial growth. IV. Histidine peptides as growth factors for Lactobacillus delbruecki 9649, J. Biol. Chem., 202:521-532 (1953).

149. A. W. Phillips and P. A. Gibbs, Techniques for the fractionation of microbiologically active peptides derived from casein, Biochem. J., 81:551-556 (1961).

150. L. Pickart and M. M. Thaler, Tripeptide in human serum which prolongs survival of normal liver cells and stimulates growth in neoplastic liver, Nature New Biol. 243:85-87 (1973).

151. M. J. Pine, Turnover in intracellular proteins, Ann. Rev. Microbiol., 26:103-126 (1972).

152. C. Pinnock, B. Shane, and E. L. R. Stokstad, Stimulatory effects of peptides on growth of the free-living nematode Caenorhabditis briggsae (38615), Proc. Soc. Exp. Biol. Med., 148:710-713 (1975).

153. K. A. Pittman and M. P. Bryant, Peptides and other nitrogen sources for growth of Bacteroides ruminicola, J. Bacteriol, 88:401-410 (1964).

154. K. A. Pittman, S. Lakshmanan, and M. P. Bryant, Oligopeptide uptake by Bacteroides ruminicola, J. Bacteriol., 93:1499-1508 (1967).

155. J. M. Prescott, V. J. Peters, and E. E. Snell, Peptides and bacterial growth, V. Serine peptides and growth of Lactobacillus delbrueckii 9649, J. Biol. Chem., 203:533-540 (1953).

156. J. M. Prescott, S. H. Wilkes, F. W. Wagner, and K. J. Wilson, Aeromonas aminopeptidase: improved isolation and some physical properties, J. Biol. Chem., 246:1756-1764 (1971).

157. L. E. Ray and F. W. Wagner, Characteristics of an aminopeptidase activity from the cultural fluid of Bacillus subtilis, Can. J. Microbiol., 18:853-859 (1972).

158. G. Roncari and H. Zuber, Thermophilic aminopeptidases from Bacillus stereothermophilus I. Isolation, specificity and general properties of the thermostable aminopeptidase, Int. J. Prot. Res., 1:45-61 (1969).

159. S. Roseman, A bacterial phosphotransferase system and its role in sugar transport, in The Molecular Basis of Biological Transport, Miami Winter Symposia, Vol. 3 (J. F. Woessner and F. Huijing, eds.), Academic, New York, 1972, pp. 181-219.

159a. D. A. Rowlands, E. F. Gale, A. T. Folkes, and D. H. Marrian, Accumulation of three glutamic acids within Staphylococcus aureus incubated with derivatives of glutamic acid, Biochem. J., 65:519-526 (1957).

160. A. C. Ryden, Separation and characterization of three proline peptidases from a strain of Arthrobacter, Acta Chem. Scand., 25:847-858 (1971).

161. S. Sarid, A. Berger, and E. Katchalski, Proline iminopeptidase, J. Biol. Chem., 234:1740-1746 (1959).

162. S. Sarid, A. Berger, and E. Katchalski, Proline Iminopeptidase. II. Purification and comparison with iminodipeptidase (prolinase), J. Biol. Chem., 237:2207-2212 (1962).

163. S. Shankman, V. Gold, S. Higa, and R. Squires, On the mode of action of a peptide inhibitor of growth of P. cerevisiae, Biochem. Biophys. Res. Commun., 9:25-31 (1962).

164. S. Shankman, S. Higa, H. A. Florsheim, Y. Schvo, and V. Gold, Peptide Studies. II. Growth promoting activity of peptides of L-leucine and L- and D-valine for lactic acid bacteria, Arch. Biochem. Biophys., 86:204-209 (1960).

165. S. Shankman, S. Higa, and V. Gold, Peptide Studies: Inhibition of bacterial growth by di- and tripeptides, Texas Rep. Biol. Med., 19: 358-369 (1961).

166. D. C. Shelton and W. E. Nutter, Uptake of valine and glycylvaline by Leuconostoc moesenteroides, J. Bacteriol., 88:1175-1181 (1964).

167. D. Siepen, P. H. Yu, and M. R. Kula, Proteolytic enzymes of Neurospora crassa: Purification and some properties of five intracellular proteinases, Eur. J. Biochem., 56:271-281 (1975).

168. S. Simmonds, The role of dipeptidases in cells of Escherichia coli K12, J. Biol. Chem., 241:2502-2508 (1966).

169. S. Simmonds, Peptidase activity and peptide metabolism in Escherichia coli K12, Biochem., 9:1-9 (1970).

170. S. Simmonds, Peptidase activity and peptide metabolism, in Peptide Transport in Bacteria and Mammalian Gut, CIBA Foundation Symposium, Associated Scientific Publishers, Amsterdam, 1972, pp. 43-53.

171. S. Simmonds, and J. S. Fruton, The utilization of proline derivatives by mutant strains of Escherichia coli, J. Biol. Chem., 174:705-715 (1948).

172. S. Simmonds and D. D. Griffith, Metabolism of phenylalanine-containing peptide amides in Escherichia coli, J. Bacteriol., 83: 256-263 (1962).

173. S. Simmonds, J. I. Harris, and J. S. Fruton, Inhibition of bacterial growth by leucine peptides, J. Biol. Chem., 188:251-262 (1951).

174. S. Simmonds, K. W. Szeto, and C. G. Fletterick, Soluble tri- and dipeptidases in Escherichia coli K12, Biochem. 15:261-270 (1976).

175. S. Simmonds, E. L. Tatum, and J. S. Fruton, The utilization of phenylalanine and tyrosine derivatives by mutant strains of Escherichia coli, J. Biol. Chem., 169:91-101 (1947).

176. S. Simmonds and N. O. Toye, Peptidases in spheroplasts of Escherichia coli K-12, J. Biol. Chem., 241:3852-3860 (1966).

177. S. J. Singer, The molecular organization of membranes, Ann. Rev. Biochem., 43:805-833 (1974).

178. R. L. Smith and F. W. Dunn, Bacterial utilization of oligopeptides containing β-2-thienylalanine and phenylalanine, J. Biol. Chem., 245:2962-2966 (1970).

179. R. L. Smith, E. G. Archer, and F. W. Dunn, Uptake of [^{14}C]-labeled tri-, tetra-, and pentapeptides of phenylalanine and glycine by Escherichia coli, J. Biol. Chem., 245:2967-2971 (1970).

180. P. F. Smith, T. A. Langworthy, and M. R. Smith, Polypeptide nature of growth requirement in yeast extract for Thermoplasma acidophilum, J. Bacteriol., 124:884-892 (1975).

181. E. Stoll, L. H. Ericsson, and H. Zuber, The function of the two subunits of thermophilic aminopeptidase I, Proc. Natl. Acad. Sci. USA, 70:3781-3784 (1973).

182. D. Stone and H. D. Hoberman, Utilization of proline peptides by a prolineless mutant of Escherichia coli, J. Biol. Chem., 202:203-208 (1953).

183. A. J. Sussman, Peptidase activity in E. coli K12, PhD Thesis, Princeton University (1970).

184. A. J. Sussman and C. Gilvarg, Peptidases in Escherichia coli K12 capable of cleaving lysine homopeptides, J. Biol. Chem., 245:6518-6524 (1970).

185. A. J. Sussman and C. Gilvarg, Peptide transport and metabolism in bacteria, Ann. Rev. Biochem., 40:397-408 (1971).

185a. D. W. Tempest, Continuous culture in microbiological research, Adv. Microbiol. Physiol., 4:223-250 (1970).

186. T. D. Thomas, D. B. W. Jarvis, and N. A. Skipper, Localization of proteinase(s) near the cell surface of Streptococcus lactis, J. Bacteriol., 118:329-333 (1974).

187. F. Tokita, Peptides as amino acid sources for growth of lactic acid bacteria, Milchwissenschaft, 21:220-222 (1966).

188. A. M. Ugolev, Membrane Digestion and Peptide Transport, in Peptide Transport in Bacteria and Mammalian Gut, CIBA Foundation Symposium, Associated Scientific Publishers, Amsterdam, 1972, pp. 127-137.

189. A. M. Ugolev, Membrane (contact) digestion, in Biomembranes: Intestinal Absorption, Vol. 4A, (D. H. Smyth, ed.), Plenum, London, 1974, pp. 285-362.

190. A. M. Ugolev and P. DeLaey, Membrane digestion. A concept of enzymic hydrolysis on cell membranes, Biochim. Biophys. Acta, 300:105-128 (1973).

191. A. M. Ugolev, A. A. Gruzdkov, P. DeLaey, V. V. Egorova, N. N. Iezuitova, G. G. Koltushkina, N. M. Timofeeva, E. Ch. Tulyaganova, V. A. Tsvetkova, M. Yu. Chernyakhovskaya, and G. G. Shcherbakov, Characterization of multisubstrate processes during membrane digestion, Nahrung, 19:299-318 (1975).

192. E. J. Van Lenten and S. Simmonds, Dipeptidases in spheroplasts and osmotically shocked cells prepared from Escherichia coli K12, J. Biol. Chem., 242:1439-1444 (1967).

193. V. M. Vogt, Purification and properties of an aminopeptidase from
      Escherichia coli, J. Biol. Chem., 245:4760-4769 (1970).
194. R. A. Von der Haar and H. E. Umbarger, Isoleucine and valine
      metabolism in Escherichia coli. XIX. Inhibition of isoleucine bio-
      synthesis by glycylleucine, J. Bacteriol. 112:142-147 (1972).
195. F. W. Wagner, A. Chung, and L. E. Roy, Characterization of the
      aminopeptidase from Bacillus subtilis as an extracellular enzyme,
      Can. J. Microbiol., 18:1883-1891 (1972).
196. F. W. Wagner, S. H. Wilkes, and J. M. Prescott, Specificity of
      Aeromonas amino peptidase towards amino acid amides and dipep-
      tides, J. Biol. Chem., 247:1208-1210 (1972).
197. A. Wahren and R. J. Gibbons, Amino acid fermentation by Bacter-
      oides melaninogenicus, Antonie van Leeuwenhoek J. Microbiol.
      Serol. 36:149-159 (1970).
198. A. Wahren and T. Holme, Amino acid and peptide requirement of
      Fusiformis necrophorus, J. Bacteriol., 116:279-284 (1973).
199. J. J. Wasmuth and H. E. Umbarger, Role for free isoleucine or
      glycylleucine in the repression of threonine deaminase in Escherichia
      coli, J. Bacteriol., 117:29-39 (1974).
200. J. R. Whittaker and P. V. Caldwell, Unusual kinetic behavior of
      Endothia parasitica protease in hydrolysis of small peptides, Arch.
      Biochem. Biophys., 159:188-200 (1973).
201. S. H. Wilkes, M. E. Bayliss, and J. M. Prescott, Specificity of
      Aeromonas aminopeptidase towards oligopeptides and polypeptides,
      Eur. J. Biochem., 34:459-466 (1973).
202. G. Wiseman, Absorption of protein digestion products, in Biomem-
      branes: Intestinal Absorption, Vol. 4A (D. H. Smuth, ed.), Plenum,
      London, 1974, pp. 363-481.
203. L. Wolfinbarger and G. A. Marzluf, Peptide utilization by amino
      acid auxotrophs of Neurospora crassa, J. Bacteriol., 119:371-378
      (1974).
204. L. Wolfinbarger and G. Marzluf, Size restriction on utilization of
      peptides by amino acid auxotrophs by Neurospora crassa, J. Bacteriol.,
      122:949-956 (1975).
205. L. Wolfinbarger and G. A. Marzluf, Specificity and regulation of
      peptide transport in Neurospora crassa, Arch. Biochem. Biophys.,
      171:637-644 (1976).
206. D. W. Woolley, The position of strepogenin in the protein molecules,
      J. Biol. Chem., 171:443-445 (1947).
207. D. W. Woolley, Strepogenin activity of derivatives of glutamic acid,
      J. Biol. Chem., 172:71-81 (1948).
208. D. W. Woolley and R. B. Merrifield, Specificities of peptides,
      Science, 128:238-240 (1958).
209. D. W. Woolley and R. B. Merrifield, Anomalies of the structural
      specificity of peptides, Ann. N.Y. Acad. Sci., 104:161-171 (1963).

210. D. W. Woolley, R. B. Merrifield, C. Ressler, and V. du Vigneaud, Strepogenin activity of synthetic peptides related to oxytocin, Proc. Soc. Exp. Biol. Med., 89:669-670 (1955).

211. D. W. Wright, Metabolism of peptides by rumen microorganisms, Appl. Microbiol., 15:547-550 (1967).

212. A. Yaron and D. Mlynar, Aminopeptidase P, Biochem. Biophys. Res. Commun., 32:658-663 (1968).

213. A. Yaron, D. Mylnar, aand A. Berger, A dipeptidocarboxypeptidase from E. coli, Biochem. Biophys. Res. Commun., 47:897-902 (1972).

214. O. C. Yoder, K. C. Beamer, P. B. Cipolloni, and D. C. Shelton, Kinetic study of L-valine and glycyl-L-valine uptake by Leuconostoc mesenteroides, Arch. Biochem. Biophys., 110:336-342 (1965).

215. O. C. Yoder, K. C. Beamer, and D. C. Shelton, Structural and stereochemical specificity of transport for glycine, valine and their dipeptides in L. mesenteroides, Fed. Proc., 24:352 (1965).

216. E. A. Young, D. O. Bowen, and J. F. Diehl, Transport studies with peptides containing unnatural amino acids, Biochem. Biophys. Res. Commun., 14:250-255 (1964).

217. P. Ziska, Untersuchungen zur molekularen Verteilung von Polypeptiden und Peptiden in handelsublichen Kasein - und Fleischpeptonen mittels Gelfiltration, Arch. Hyg., 151:370-376 (1967).

218. P. Ziska, Untersuchungen zur molekularen Verteilung von Polypeptiden und Peptiden inhandelsublichen. Spezialpeptonen mittles Gelfiltration, Arch. Hyg., 152:73-76 (1968).

Chapter 8

TRANSPORT OF CARBOXYLIC ACIDS

William W. Kay*

Department of Biochemistry
University of Saskatchewan
Saskatoon, Saskatchewan, Canada

I. INTRODUCTION 386
II. MONOCARBOXYLATE TRANSPORT 386
   A. Hexuronates 386
   B. Hexonates 387
   C. Glycolytic Intermediates 390
   D. Acetate and Propionate 392
   E. Short-Chain Fatty Acids (C_4 and C_5) 393
   F. Medium and Long-Chain Fatty Acids (C_6-C_8) 394
III. DICARBOXYLATE TRANSPORT. 395
   A. Escherichia coli .395
   B. Bacillus subtilis .398
   C. Pseudomonas species. 401
   D. Azotobacter vinelandii . . . . , . .401
   E. Rhodopseudomonas spheroides 401
IV. TRICARBOXYLATE TRANSPORT 401
   A. Aerobacter aerogenes 404
   B. Azotobacter vinelandii 404
   C. Pseudomonas fluorescens 405
   D. Bacillus subtilis .405
   E. Salmonella typhimurium 407
V. SUMMARY 407
   RECENT DEVELOPMENTS 408
   REFERENCES. 408

*Present address: Department of Biochemistry, University of
Victoria, Victoria, B.C., Canada

I. INTRODUCTION

Among the plethora of intensively studied bacterial transport systems, the transport systems for carboxylic acids have received relatively modest attention. This is indeed surprising, since the existence of permeation systems for carboxylic acids was recognized very early in metabolic studies with Aerobacter species and Pseudomonads [1-3]. As well, many of these organic acids are important metabolites not only as sources of carbon and energy but also as vital intermediates for gluconeogenesis and anaplerosis [4,5].

The fundamental advances in our current understanding of bacterial transport mechanisms have been made principally with carbohydrates and amino acids and not particularly the carboxylic acids, since inherent problems of rapid catabolism, multiple biochemical pathways, and electrogenic complications make this area somewhat more complex and difficult to study. In spite of these obstacles, recent progress in this area, following principles established with other bacterial systems, has been dynamic and rapid, and has resulted in the detailed characterization of at least one system which may well prove to be most amenable to molecular analysis and reassembly.

It is the purpose of this article to outline comparatively the various carboxylic acid transport systems of bacteria and describe their regulation, genetics, and molecular characterization where possible.

II. MONOCARBOXYLATE TRANSPORT
(TABLE 8-1)

A. Hexuronates

The hexuronic acids D-glucuronate and D-galacturonate are metabolized via the hexuronic acid pathway in Escherichia coli [6,7]. These substrates share a common transport system and are actively transported in mutants defective in glucuronic acid isomerase, the first enzyme of the hexuronic acid pathway. Glucuronic acid appears to be the inducer since such mutants are constitutive for hexuronic acid transport [8]. The existence of a mobile hexuronate carrier which undergoes substrate and energy-dependent conformational changes has been postulated, since the transport inactivation by N-ethylmaleimide (NEM) and 1-fluoro-2,4-dinitrobenzene (FDNB) can be accelerated either by these substrates or by respiratory poisons [8].

B. Hexonates

1. 2-Keto-3-deoxygluconate

Another intermediate of this pathway, 2-keto-3-deoxygluconate (KDG) has been shown to enter E. coli by an active transport system in KDG kinase mutants [9,10]. This system has high affinity for KDG and low affinity for D-glucuronate [12]. The transport of KDG is normally cryptic in wild-type strains but could be studied in constitutive strains selected on the basis of their ability to grow on KDG as a sole carbon source. It can be readily differentiated from the specific glucuronate system [8] with respect to sub- trate specificity and inducibility [13]. For example, KDG is not taken up in the wild-type strain, while D-glucuronate accumulation is significantly and competitively inhibited by D-galacturonate [14] but not by KDG [13]. KDG is the true inducer of its own transport system since KDG kinase mutants are endogenously induced for KDG transport. Structural (kdgT), operator (kdgP^C) and regulator (kdgR) mutants have been mapped (Fig. 8-1), and these genes constitute a regulon with KDG kinase and KDG aldolase, the other enzyme of this catabolic pathway [9,11,13,14]. Figure 8-3 des- cribes the relative transport of KDG in these various mutants. More is now known of the genetics and regulation of this carboxylate transport sys- tem than any other.

2. Gluconate

Gluconate is actively transported both into gluconokinase mutants [15] and membrane vesicles of E. coli [17]. Although specific kinetic parameters have not been described, there appear to be both high- and low-affinity systems for gluconate. Mutants in each system have been located near the malA region of the chromosome (Fig. 8-1) along with several genes involved in gluconate catabolism [15], including a regulatory gene which is cotrans- ducible with the low-affinity system [18]. Gluconate transport mutants have also been isolated from strains with genetic lesions at 6-phosphoglu- conate dehydrogenase and 6-phosphogluconate dehydrase [19], as a conse- quence of the resistance to the toxic effects of accumulated 6-phosphoglu- conate. Cyclic AMP appears to be required for induction of gluconate uptake, since induction is absent in adenylcyclase mutants [20].

   Active transport of gluconate has also been demonstrated using mem- brane vesicle preparations of Pseudomonas aeruginosa [21,22]. The vesicle system is saturable (K_m, 20 μM) comparable to whole cell uptake (40 μM)

TABLE 8-1 Characteristics of Bacterial Monocarboxylate Permeability Systems

Organism	System[a]	Substrates	K_m^c (µM)	V_{max} (nmol min^{-1} mg^{-1})	Inducer	Catabolite Repression	Mutants	Ref.
E. coli	A.T.	D-glucuronate D-galacturonate	--	--	D-glucuronate			8,14
E. coli	A.T.	KDG	250	20	KDG	+	kdgT	9-11, 13
E. coli	U	KDG D-glucuronate	200 1500	10 --	KDG	+	kdgP^c	16,18, 20
E. coli	A.T.	D-gluconate	--	--	D-gluconate	+	gntR	
P. aeruginosa							gntA	
P. aeruginosa	A.T.	D-gluconate	40	--	--		gntB	21,22
E. coli	U	D L-lactate D-, L-lactate	-- --	-- --	D- or L-lactate	+		23,24
E. coli	A.T.	Pyruvate	--	--	--			26
E. coli	U	Pyruvate	--	--	--		usp	27
R. spheroides	U	Pyruvate D- and L-lactate Propionate Butyrate Acetate	13	5.6	-- -- --			28
S. typhimurium	U	3-PG, PEP, and 3-PG	100	--	3-PG, 2-PG, or PEP	+	pgtA pgtC	30

Organism	System[a]	Substrate[b]	K_m[c]		Acylation	Product of acylation	Gene	Ref.
E. coli K12	U.	Acetate	36	25				31
		Propionate	220					
E. coli W	U.	Acetate	6	1.6				32
E. coli K12	U.	Acetate	740 / 30–40	10–20	+		--	23
		Propionate	--	--				
		Glycolate	--	--				
S. typhimurium	U.	Acetate			+		--	33
		Propionate						
E. coli	V.A.	Butyrate	22(125)	0.11(10)	-	Butyrate	--	40
		3- and 2-Butenoate	--	--				
		4-Pentenoate	--	--				
		Valerate	5000	60				
E. coli	V.A.	Oleate	15(34)		+	Oleyl-,	oldD,	44,46
		Decanoate	71		+	Decanoyl-, or	dec	
		Octanoate	90			Octanoyl-CoA		

[a]A. T. refers to active transport, U. to uptake, V.A., to vectorial acylation.

[b]Substrates for the transport system are arranged in decreasing orders of affinity.

[c]K_m refers to the apparent K_m of uptake or transport as derived from saturation kinetics with whole cells or vesicles.

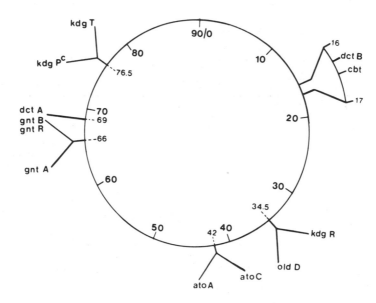

FIG. 8-1 Location of genes affecting carboxylate transport or uptake systems on the E. coli chromosome. Genes are arranged according to Taylor
and Trotter [105], and the genotypic interpretations are: dctB, dicarboxylate membrane binding protein 1; cbt, carboxylate periplasmic binding protein; kdgR, KDG transport regulatory gene; oldD, acyl-CoA synthetase;
atoC, butyrate uptake regulatory gene; atoA, butyrate uptake system (butyryl-
CoA:acetate-CoA-transferase); gntA, high-affinity gluconate transport system; gntB, low-affinity gluconate regulatory gene; gntR, high-affinity gluconate transport regulatory gene (genetic order with gntB is undetermined);
dctA, dicarboxylate membrane binding protein (SBP2, possibly structural
or regulatory); kdgP^c, KDG transport operator mutant; kdgT, KDG transport system.

and is induced by gluconate. Membrane vesicles have been shown to couple
gluconate transport to electron flow via a FAD-linked L-malate dehydrogenase or D-glucose dehydrogenase.

C. Glycolytic Intermediates

1. D- and L-Lactate

DL-lactate was shown to be actively transported into a L-lactate oxidase
mutant of E. coli [23], but an unusually low accumulation was observed.

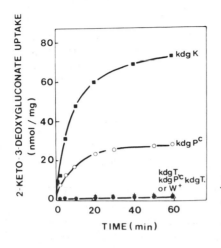

FIG. 8-2 Time course of 2-keto-3-deoxygluconate uptake in different E. coli strains [10]. The genotype symbols are: (■-■) kdgK (KDG kinase mutant), (○-○) kdgP^C (KDG transport operator mutant (+ - +), kdgT (KDG transport mutant), (△-△) kdgP^C, kdgT (KDG operator and transport double mutant), (●-●) wild-type strain.

L-Lactate and glycolate were weak competitive inhibitors, indicating a specificity which extended to the D-isomer. This system was strongly repressed in adenylcyclase mutants, and specific escape from catabolite repression for L- but not D-lactate was interpreted as evidence for a specific L-lactate permeation system. However, a common transport system of equal affinity for both stereoisomers has been found in membrane vesicles of E. coli as indicated by their mutual competitive inhibition of transport, and by similar competitive inhibition patterns with L-aspartate, L-glutamate, glycolate, glyoxylate, glycerate, and pyruvate. This monocarboxylate transport system could be differentiated from the membrane bound L- and D-lactate dehydrogenases, since vesicles prepared from cells not induced for these enzymes were found to transport normally [24]. Furthermore, 2-hydroxy-3-butynoate, and inhibitor of the D-lactate dehydrogenase, had no effect on D-lactate transport [25]. Membrane vesicles from D-lactate dehydrogenase mutants transported both L- and D-lactate at a high rate [24]. Both glyoxylate and glycolate were found to be competitive inhibitors of D- and L-lactate transport in membrane vesicles of E. coli, Bacillus subtilis and a Pseudomonas species [26].

2. Pyruvate

Pyruvate has been shown to be actively transported into membrane vesicles of E. coli energized with reduced phenazine methosulfate but not into vesicles of B. subtilis or of a Pseudomonas species [26]. In vesicles from all these species, pyruvate inhibited the transport of both D- and L-lactate, suggesting the existence of a system with overlapping substrate specificities for these acids. However, a mutant specifically defective in the uptake of

pyruvate (usp) and not D- or L-lactate has been reported [27], suggesting
the existence of a separate specific pyruvate transport system. Pyruvate
has also been shown to be taken up by a system with high affinity in Rhodopseudomonas spheroides [28]. This system appears to have a wide substrate
specificity, as it is competitively inhibited by a variety of short-chain fatty
acids as well as lactate.

3. Glycolate

Glycolate is actively transported in E. coli glycolate oxidase mutants by a
system related to glyoxylate catabolism. A transport mutant has been
reported which is impermeable to glycolate, but apparently not to glyoxylate,
since the latter could still induce the synthesis of malate synthase-G [29].
These results suggest but do not prove the existence of separate glycolate
and glyoxylate transport systems. Reversion studies of a pleiotropic mutant
suggests a common regulation of malate synthase-G and the glycolate transport protein(s).

4. Phosphoenolpyruvate, 2-Phosphoglycerate, and 3-Phosphoglycerate

The phosphorylated glycolytic intermediates phosphoenolpyruvate (PEP),
2-phosphoglycerate (2-PG) and 3-phosphoglycerate (3-PG) have been shown
to be taken up into whole cells of Salmonella typhimurium by a common system [30] different from the pyruvate system or the α-glycerophosphate
(glpT) system. Structural gene mutants (pgtA) which were unable to take up
these three substrates were mapped at 74 min on the S. typhimurium chromosome, and constitutive mutants (pgtC) able to use these substrates as a
phosphate source at low concentration were also characterized. This uptake
system could be induced by growth on either of these three substrates. The
K_m for 3-PG was approximately 100 μM and the uptake rate was strongly
dependent on K^+ ions.

D. Acetate and Propionate

The mechanism of acetate and propionate transport in bacteria has tenuously
resisted elucidation because of the inability to isolate mutants either specifically devoid of transport activity or effectively blocked in an early metabolic step. As a result, all studies have been concerned with the overall
uptake mechanism. In E. coli K12, acetate and propionate appear to share
a common uptake system with K_m values of 36 and 220 μM, respectively
[31], and an even lower affinity system has been measured with E. coli B
[32]. When measured in acetokinase or phosphotransacetylase mutants

totally devoid of enzyme activity, acetate uptake continues normally [23], suggesting these enzymes are not involved in the uptake process, which is likely since the K_m for acetokinase is several orders of magnitude higher than that for uptake. However, citrate synthase mutants were largely defective in acetate accumulation, suggesting that a facilitated diffusion mechanism may in fact be operative. Nearly identical results have been obtained with whole cells of S. typhimurium [33]. High affinity kinetics may have been due to a new acetothiokinase activity found in an acetokinase-deficient strains [23], thereby suggesting a vectorial acylation [34] mechanism. Unlike the acetokinase and phosphotransacetylase reactions, the K_m values for thiokinase and the uptake system are of a similar order of magnitude. This enzyme is, however, unfortunately soluble and may activate acetate only after permeation, perhaps by free diffusion as suggested for butyrate and valerate [35]. It would thus be unlikely to be directly involved in acetate translocation; however, it is always possible that there occurs in vivo a closer association of the enzyme with the membrane. It may well be that acetate is also taken up by the short-chain fatty acid system since it has been reported to competitively inhibit butyrate uptake [36], although other studies [31] do not support this observation. Acetate has been suggested to permeate Azotobacter vinelandii as the undissociated acid or by an OH^- exchange mechanism [37].

E. Short-Chain Fatty Acids (C_4 and C_5)

Similar to the acetate uptake studies, early investigations of the mechanism of butyrate and valerate transport were unhappily complicated by metabolism. E. coli will not normally use these acids as sole carbon sources, but can gain this ability after derepression for both the fad (fatty acid degradation) regulon and ato (acetoacetate) operons [38,39]. Specific translocation first appeared not to be involved, since the uptake process could be divisible into two elements, a pH-dependent diffusion and subsequent derepressed acetyl-CoA:butyrate-CoA transferase [41,42], which will act to translocate butyrate as the CoA ester even in membrane vesicles [42]. The kinetics for both butyrate uptake and the transferase are nearly identical. The CoA- and ATP-dependent vectorial acylation and accumulation of butyrate in E. coli membrane vesicles are shown in Fig. 8-3. The enzyme appears also to be localized on the inside of the membrane. This enzyme must be loosely associated with the membrane since it has been reported as largely soluble by other workers [39]. Butyryl-CoA synthetase, however, does not suffice as a translocator even in constitutive strains [39]. Both butyrate and valerate are mutually competitively inhibitory during whole cell uptakes as were various monounsaturated short-chain fatty acids [40]. The system appears to be inducible by butyrate.

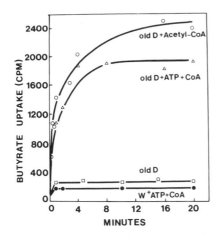

FIG. 8-3 Time course of butyrate
uptake by E. coli [42]. Membrane
vesicles were incubated with labeled
butyrate for the time intervals shown
and collected by filtration. Butyrate
was recovered from the vesicles as
butyryl-CoA.

F. Medium and Long-Chain Fatty Acids (C_6-C_8)

The transport of medium and long-chain fatty acids appears to be inextricably
dependent upon metabolism. The uptake of long-chain fatty acids is inducible by their CoA esters, principally oleate, since a mutant lacking the acyl-
CoA synthetase remains uninducible for uptake as well as three other enzymes of fatty acid degradation (thiolase, crotonase, and hydroxyacyl-CoA
dehydrogenase) even in the presence of oleate [43]. Furthermore, derepression is mediated by a mutation in the dec gene which governs the regulation of this regulon (Fig. 8-1). Apparently, substrate recognition decreases with decreasing chain length down to C_8 fatty acids. From studies on
the specificity of fatty acid uptake, respiration, and β-oxidation enzymes,
it was concluded that the acyl-CoA synthetase reaction was the rate-limiting
step in the overall uptake process [44]. These authors suggested that permeation and acylation were connected and coined the term "vectorial acylation" to describe this system in analogy to "vectorial phosphorylation" [45],
as applied to the bacterial phosphotransferase system. Vectorial acylation
was suggested for the following reasons: the substrate specificities for
uptake and activation were the same, there was no apparent efflux and no
fatty acid appeared intracellularly, acyl-CoA synthetase mutants were
incapable of fatty acid uptake, and the acyl-CoA synthetase was partially
membrane bound. Also, competition for intracellular coenzyme A inhibits
oleate uptake [46]. In opposition to this view is the disturbing fact that the
apparent affinity for fatty acid uptake is an order of magnitude higher than
for activation [41], but other workers have reported activation affinities
similar to the uptake affinity, using a separate enzyme assay [47]. Other
investigators have reported that decanoate is actively transported in similar

derepressed mutants [48], although this has been criticized as adsorption and desorption phenomena [44].   Both the expression of the fatty acid uptake mechanism and degradation enzymes require cyclic AMP and the CR protein [49].   The uptake of palmitate by Bacillus megaterium also appears to be closely linked to metabolism [50], but has not been investigated in detail.

III.   DICARBOXYLATE TRANSPORT
   (TABLE 8-2)

Of all the carboxylate transport systems of bacteria studied so far, the TCA cycle dicarboxylates have probably received the most attention.   The result has been the elucidation of a bacterial transport system, the C_4-dicarboxylate transport system of E. coli, which is currently becoming one of the most amenable systems for an in-depth analysis of the molecular mechanisms in metabolite transport.

A.   Escherichia coli

The uptake of C_4-dicarboxylates by E. coli was first shown to be less complicated than first thought by the discovery of a gene dct which is required for the uptake of succinate, fumarate, L-malate, and L-aspartate [51]. The transport of these four substrates is mutually competitive and simultaneously controlled by induction and catabolite repression [52].   These results were confirmed in other strains [53-55], as well as S. typhimurium [56].   The dicarboxylates were also shown to be actively transported in membrane vesicles from induced cells [57,26].   This system is important as a permeation route for these organic acids as sole carbon sources.   L-aspartate, however, can also enter the cell on another system, ast [58], which permits enough substrate to enter the cell and serve either as a nitrogen source, or for anaplerosis, or both.
    The dicarboxylate transport system has recently been described in greater detail and resolved into several components, all of which are required for active transport of the C_4-dicarboxylates in intact cells [59]. Genetically, two new genes were discovered, dctB and cbt [59,60].   The two dctA and B genes were suggested to code for two membrane proteins, each of which is required to effect active transport in membrane vesicles [61].   The third gene, cbt [59], which is closely linked to dctB (Fig. 8-1), was shown to code for a low molecular weight periplasmic binding protein which, once purified by affinity chromatography, avidly binds succinate, fumarate, and L-malate as well as D-lactate, which agrees with the phenotype of the cbt mutants [62].   Since no carboxylate binding protein could be recovered from cbt⁻ mutants, it is likely that this gene either regulates the synthesis or codes for the structure of this dicarboxylate binding protein.

TABLE 8-2 Properties of Bacterial Dicarboxylate Permeability Systems

Organism	System	Succ		Fum		L-Mal		OAA		Asp		Inducers	Mutants	Ref.
		K_m[a]	V_{max}[b]	K_m	V_{max}	K_m	V_{max}	K_m	V_{max}	K_m	V_{max}			
E. coli	U.	30	25	30	25	+	+	−	−	30	25	C_4-Dicarboxylates	*dct*	51, 52
	A.T.		nr[c]				nr			3.5	1.5		*ast*	58
	A.T.	20	3	20–30	+		+		+	+		C_4-Dicarboxylates	*dct*	53–55
	A.T.	14	20	20–30	nr	20–30		nr		nr		C_4-Dicarboxylates	*dctA, dctB,*	59
	A.T. (vesicles)	11	nr	12	nr	22	nr	nr		nr			*cbt dctA dctB*	61
	B.P.	40 (K_d)		55 (K_d)		34 (K_d)		nr		1300 (K_d)			*cbt*	62
	SBP I	23 (K_d)		47 (K_d)		47 (K_d)		nr		nr			*dctB?*	63
	SBP II	2.3 (K_d)		7 (K_d)		7 (K_d)		nr		nr			*dctA*	
S. typhimurium	U.	14	6.7	21	9.8	19	6.6	+	+	+		C_4-Dicarboxylates	*dct*	56
	A.T.	180	17.4	50	33.2	100	20.1	+	−	+			*dct*	33
B. subtilis	A.T.	−	−	700	50	400	400	nr	+	nr		L-Malate	−−	69
	U.	−	−											
	A.T.	100	20	140	18	400		+	+	+		L-Malate	−−	70–73
	A.T. (fluoro-malate)	nr			nr	55	nr	nr		nr			−−	74
	A.T. (vesicles)	4.3	0.6	7.5	3.7	13.5	1.6	−		+			−−	71
P. putida	U.	12.5	5.8	52	nr	27		nr		nr		C_4-Dicarboxylates	−−	78
	A.T.	10	1.0	66	nr	20		nr		nr		C_4-Dicarboxylates	−−	
R. spheroides	U.	2.7	50	0.8	67	2.3	52	nr		−		C_4-Dicarboxylates		28
A. vinelandii	U.	+	+	+	+	+	+	+		nr		C_4-Dicarboxylates		79

[a] Apparent K_m (μM).
[b] V_{max} (nmol min^{-1} mg^{-1} dry weight of cells).
[c] Not reported.

This protein appears to be different in substrate specificity from the D-and L-lactate transport system in membrane vesicles [24].

Two additional succinate binding proteins (SBP1 and 2) have been detergent solubilized from membrane vesicles of E. coli, separated, and purified by affinity chromatography in the presence of detergent [63]. These results correlate rather well with the demonstration of the two genes (dctA and B) coding for elements of dicarboxylate transport in membrane vesicles [61]. Mutants harboring the dctA gene were found to produce a reduced amount of the SBP2 protein. These two proteins differ both in molecular weight (14,262 and 20,054 for SBP1 and SBP2, respectively) and in amino acid composition, in a way that suggests the SBP1 is more deeply embedded in the membrane due to increased hydrophobicity [64]. The immense importance of this system is that while many periplasmic binding proteins implicated in transport have been purified, no other intrinsic membrane proteins involved in bacterial transport have been purified reproducibly in an active form. A tentative model relating the possible transport roles and protein-protein interactions of these components has been formulated [65] and is schematically represented in Fig. 8-4. It is important to emphasize that in the absence of any one of the three known dicarboxylate binding proteins, in dctA, B, or cbt mutants, no effective transport occurs in whole cells, although membrane vesicles from cbt mutants transport dicarboxylates well. These results suggest that the periplasmic protein sequesters dicarboxylates in the periplasmic space, a function puzzlingly not required in vesicles, but vital to whole cells since osmotically shocked cells are also unable to transport the dicarboxylates [59]. The requirement for both dct genes suggests that SBP1 and 2 act in a concerted manner, not autonomously, although this conclusion must be taken with some caution as yet until rigorous proof of the correlation between these genes and SBP1 and 2 is available. Also, since biphasic kinetics have not been observed with vesicles, it is likely but not mandatory that only one component has access to the outside of the membrane, that is, SBP2. The high affinity of SBP2 for the substrate may aid in the capture or the transmittance of the dicarboxylate from the PBP to SBP2. Recent results suggest that both SBP1 and 2 are exposed to the external and internal face of the membrane, since they could be crosslinked with a nonpenetrating cleavable crosslinking reagent in right-side-out and inside-out vesicles. The SPB1-SBP2 could be purified on affinity columns, cleaved, and the individual components subsequently recovered [65]. These results strongly favor the channel-type of transport model (Fig. 8-2).

The energization of dicarboxylate transport has been investigated in membrane vesicles from E. coli [66]. From studies with various metabolic inhibitors, D-lactate oxidation inhibitors, and ATPase mutants, it was concluded that energization is mediated by the passage of electrons and not necessarily via the oxidation-reduction of a carrier. Efflux studies indicated that the K_m for efflux was 1.4 to 2.2 mM. However, studies with

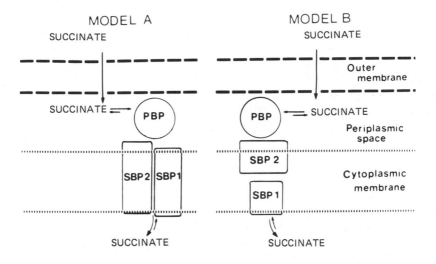

FIG. 8-4 Two potential models describing the transport of C_4-dicarboxylates in E. coli [65]. In both models, succinate diffuses in through the porous outer membrane and is perhaps sequestered by PRP, the periplasmic carboxylate binding protein. In Model A, the two membrane localized succinate binding proteins (SBP1 and 2) interact to form a hydrophilic channel to facilitate the transport across the cytoplasmic membrane. In Model B SBP1 and 2 are arranged vertically such that these proteins act as sequential membrane carriers for succinate translocation.

whole cells strongly suggest that C_4-dicarboxylate transport is obligatorily coupled to proton uptake in a way in which the complex is electroneutral, with the exception of aspartate which was cationic [67]. These results emphasize that E. coli in its evolutionary wisdom brilliantly solved the electrogenic problem of transporting an organic divalent anion while simultaneously providing an efficient means to tap the electrochemical proton gradient in order to affect active transport. The kinetics of succinate and succinate induced proton uptake are shown in Fig. 8-5.

B. Bacillus subtilis

The transport of C_4-dicarboxylates has also been investigated both in intact cells of B. subtilis appropriately blocked in various TCA cycle enzymes [68-70], as well as in membrane vesicles [26,71]. Like the C_4-dicarboxylate system of E. coli, the B. subtilis system has broad specificity for the C_4-dicarboxylates. L-Malate appears to be the physiological inducer of the system [69,72], and the system could be gratuitously induced by bromo-

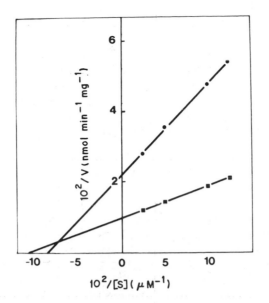

FIG. 8-5 K_m and V_{max} determinations of succinate uptake and succinate-induced proton uptake in a lightly buffered medium [67]. Succinate uptake (●-●) and succinate-induced proton uptake (■-■) were calculated and used in a double reciprocal plot.

succinate [72]. Some confusion currently exists over the kinetic para-meters for transport. In whole cells, dicarboxylate transport is charac-terized by biphasic kinetics of both high and low affinities [69,70,72] with DL-tartarate having specificity for the low-affinity system [69]. Studies with the nonmetabolizable malate analog 3-fluoro-L-erythromalate indicate that two systems coexist; one of which is inducible by L-malate with high and low substrate affinities, but one of these was yet an order of magnitude higher than previously reported with blocked mutants [74]. In membrane vesicles, however, only an extremely high-affinity system constitutive for these dicarboxylates exists [71], and comparatively, the membrane system has a wider substrate specificity than exhibited by whole cells in that L-glutamate and L-aspartate are more strongly competitive. This is in con-trast to the C_4-dicarboxylate transport system of E. coli where the kinetic parameters for transport are similar in whole cells and membrane vesicles. There is no apparent explanation for such variation in kinetic parameters unless multiple systems do exist.

α-Ketoglutarate is transported by a separate low-affinity transport system (K_m, 6.7 mM) which is induced by the addition of α-ketoglutarate or gratuitously in a α-ketoglutarate-deficient mutant [69].

In all these studies, dicarboxylate transport was either inducible by C_4-dicarboxylates, specifically by L-malate, or catabolite-repressed by glucose. No clear structural gene mutants have been reported as yet, as might be expected from multiple systems. Two mutants have been isolated which are defective in the utilization of the dicarboxylates [72]. One of these, an apparent transport-defective mutant, exhibited this phenotype only in glutamate minimal medium. High concentrations of L-malate were required to induce uptake. Another mutant appeared to be constitutive for dicarboxylate uptake. The regulation appears to be even more complex, since α-ketoglutarate or glutamate prevent induction of dicarboxylate uptake especially in α-ketoglutarate dehydrogenase-deficient mutants [70,73]. Prevention of inducer formation as suggested for aconitase mutants [75] was also ruled out.

Commensurate with the above evidence of transport multiplicity in B. subtilis was the demonstration that membrane vesicles of this organism, when induced for transport, contain three proteins as demonstrated by SDS polyacrylamide gels after double labeling with N-ethylmaleimide [76]. High concentrations of L-malate were found to protect against reactions with NEM. meso-Tartarate, a powerful competitive inhibitor of L-malate transport, protected transport inhibition by NEM and also labeling of two (44,000 and 33,000 dalton) membrane proteins. No transport components have yet been isolated from gram-positive bacteria.

In membrane vesicles of B. subtilis, the C_4-dicarboxylates are actively transported by a system with broad substrate specificity, and similar to intact cells, transport was energized by electron donors [26]. In intact cells of succinate dehydrogenase mutants, transport was strongly stimulated by polyvalent inorganic anions. The absence of an anion permeability mechanism in arsenate-resistant mutants precludes succinate transport, but partial escape from this restriction was mediated by derepression of a phosphate transport system [73]. A decline in both the initial rate of succinate transport and the accumulation of succinate in the presence of divalent cations in a phosphate transport mutant underscores the required coordination between phosphate transport and divalent cation transport. Double mutants for both phosphate and divalent cation transport were severely defective in succinate transport [77]. Succinate was not transported by an anion exchange diffusion mechanism since phosphate efflux was absent during succinate transport. Divalent cations are preferred over monovalent cations presumably to preserve electroneutrality during dicarboxylate transport. The affinity for succinate transport was drastically lowered in K^+-containing media (K_m, 20 to 35 mM) as compared to Mg^{2+} (K_m, 0.3 to 1 mM). However, K^+ was still found to be required intracellularly, since K^+ retention mutants were defective in succinate uptake, and the addition of valinomycin enhanced this defect. Thus, the complex ion requirements for dicarboxylate transport in B. subtilis largely reflect the coordination of ion transport systems to compensate for organic anion accumulation [77].

C. Pseudomonas Species

A C_4-dicarboxylate active transport system has been demonstrated for Pseudomonas putida [78]. The kinetics and regulation appear to be similar to other gram-negative and gram-positive organisms described above. The main difference seems to be a different order of dicarboxylate substrate specificity and a relatively low velocity of transport.

Vesicles of another Pseudomonad transport succinate on a system which is also competitively inhibited by L-aspartate, however not actively in spite of the fact that other mono- and dicarboxylates were actively transported [26].

D. Azotobacter vinelandii

No TCA cycle mutants or transport mutants have been described for A. vinelandii. As a consequence only a qualitative estimation of the number and function of TCA intermediate transport systems currently exists. From an analysis primarily of oxygen consumption data it has been concluded that separate translocation systems operate for the C_4-dicarboxylates, α-keto-glutarate, and also one for malonate [79]. The induction of the C_4-dicarboxylate transport system requires either of these C_4 acids, however, various analogs are gratuitous inducers, and α-ketoglutarate also induces its own translocator [80, 81].

E. Rhodopseudomonas spheroides

C_4-dicarboxylate uptake systems have been quantitated in R. spheroides [28]. As with many of the organisms already cited, the substrate range is broad, each one competitively inhibiting the uptake of the other with the exception of aspartate. Uptake was induced by succinate and repressed by fructose.

IV. TRICARBOXYLATE TRANSPORT
   (TABLE 8-3)

Citrate permeation systems have been known to exist in bacteria for over 25 years. A variety of early studies compiled evidence for their existence based primarily on induction, oxygen uptake, growth, and enzymological data [1-3, 82-85]. None of these early studies attempted to define rigorously

TABLE 8-3 Properties of Bacterial Tricarboxylate Transport System

Organism	System[a]	Substrate[b]	K_m^c	V_{max}	Inducer	Cat. Rep.	Ref.
S. diacetilactis	U.	citrate	48	114	citrate	nr[d]	86
A. aerogenes	U.	citrate	250	–	citrate	nr	87
	U.	citrate	143	nr	nr	nr	88
	A.T. (vesicles)	citrate	240	nr	citrate	nr	89
A. vinelandii	U. (oxidation)	isocitrate	1000	nr	citrate	nr	
		citrate	2500	nr	citrate	nr	93
P. fluorescens I	A.T.	citrate	30	20	citrate	nr	94
		threo-D-isocitrate	nr	nr	citrate	nr	
II	A.T.	citrate	nr	nr	tricarballylate	nr	94
		cis-aconitate	nr				
		tricarballylate	nr				

B. subtilis						
	A.T.	citrate	2300	150	citrate	+	98
	A.T.	citrate	800	-	citrate	+	96
	A.T.	citrate - Mg^{2+}	450	145	citrate	nr	99
	A.T.	fluorocitrate	300	105	citrate	nr	100
S. typhimurium	A.T.	citrate	2.3	27	citrate	+	33
		dl-fluorocitrate	29	nr	citrate		

[a] A.T. refers to active transport and U to uptake.

[b] Substrates for the transport system are arranged in decreasing orders of affinity.

[c] K_m refers to the apparent K_m of uptake or transport.

[d] Not reported.

either the number or specific properties of the transport systems. Twenty years regrettably elapsed before the first definitive experiments were reported, although in this time one study did describe the properties of a citrate transport mechanism in Streptococcus diacetilactis by measuring the rate of disappearance of labeled citrate from the incubation medium [86]. In this case citrate utilization was deemed to be inducible and energy dependent, and to display saturation kinetics.

A. Aerobacter aerogenes

Citrate-induced cells of A. aerogenes rapidly catabolize labeled citrate, making uptake measurements tenuous. However, the system has been studied under conditions whereby most of the label resides in the glutamate pool [87, 88]. Under these conditions it was found that citrate was taken up by a high-affinity temperature-dependent system which was largely inactivated by osmotic shock. The shock effect was due to the loss of aconitase required for metabolism [88]. Membrane vesicles of cells anaerobically grown on citrate were found to accumulate citrate when energized by D-lactate [89]. The vesicles were characterized by kinetics similar to whole cell uptakes, and citrate transport was strongly competitively inhibited by fluorocitrate and hydroxycitrate isomers. Sodium ions were specifically required for citrate transport (K_m 250 μM for Na^+). Oxaloacetate decarboxylase had previously been suggested as a candidate for the citrate transport protein in these cells, based on the evidence [90] that: citrate is a competitive inhibitor of this membrane bound sodium activated enzyme; the K_i for citrate inhibition and the K_m for uptake and enzymic functions were both induced by citrate. Although interesting, this model appears to be unlikely since oxaloacetate is not a competitive inhibitor of citrate transport, and the activity of the enzyme and the transport system differ by two orders of magnitude [89]. Other strains of A. aerogenes possess potassium-dependent citrate uptake systems [91], but the difference in cation requirement is clearly a strain-dependent phenomenon [92].

B. Azotobacter vinelandii

Citrate and isocitrate were suggested to be translocated separately from other TCA cycle intermediates, from evidence based primarily on oxygen uptake data [93]. Magnesium or calcium ions are required in a 1:1 stoichiometry for citrate or isocitrate induction, suggesting that these tricarboxylates are translocated as the cation complex [79]. It was felt that citrate and isocitrate must be transported by separate translocators, since

EDTA specifically inhibited citrate oxidation but not that of isocitrate [93]. This interpretation must be considered cautiously because of the possibility of differential inhibition of metabolism rather than specifically of transport.

C. Pseudomonas fluorescens

The formation of an inducible citrate transport system in Pseudomonads has long been suspected [2,3,84], however, rapid metabolism made the detailed characterization of this system difficult. This problem was first circumvented by inhibiting the aconitase of P. fluorescens with fluorocitrate [94]. Under these conditions citrate was partially accumulated. The affinity of this system for citrate was high and reasonably close to that determined by polarographic measurements. Only citrate and isocitrate were transported, as deduced from exchange studies. Another strain in P. fluorescens harbored two tricarboxylate systems: one system similar to that already described and another of broader substrate specificity was induced by tricarballylate. This system recognized not only citrate but also cis-aconitate and tricarballylate, as well as trans-aconitate, albeit poorly [94]. These results confirm from the transport point of view the variability of different strains of P. fluorescens to metabolize various tricarboxylates [95]. Although a valuable technique, fluorocitrate poisoning has the inherent disadvantage of being a likely competitive inhibitor of citrate transport, making the interpretation of rate data difficult. In the above studies it was a relatively poor competitive inhibitor of citrate transport (50% inhibition at 2 mM) making this inhibitor particularly useful with Pseudomonas.

D. Bacillus subtilis

The first unequivocal demonstration of an active transport system for citrate in bacteria was that of B. subtilis aconitase mutants [96]. With these mutants, citrate was accumulated over 100-fold, but by a system of depressingly low affinity. Citrate transport was apparently constitutive in these mutants, likely due to endogenous induction by intracellularly accumulated citrate; indeed, these strains excrete citrate into the growth medium [97]. In the wild-type strain, citrate uptake was induced by citrate, cis-aconitate, or isocitrate [98], in accord with their ability to competitively inhibit citrate uptake.

The active transport of a polyanion such as citrate of course raises an electrogenic problem which must be circumvented either by the cotransport of neutralizing equivalents of cations on separate carriers, by the simultaneous efflux of equivalent intracellular anions, or in such a way that citrate and cations are cotransported on the same carrier, either at separate sites or as a complex. In B. subtilis, citrate was shown to be transported as a citrate-Mg^{2+} complex. This conclusion was based on evidence that: there is a divalent cation requirement for citrate transport, Mg^{2+} transport was dependent on citrate, both citrate transport and citrate-dependent Mg^{2+} transport were simultaneously inhibited, the kinetic parameters for citrate-dependent Mg^{2+} and Mg^{2+}-dependent citrate transport were similar, and citrate-Mg^{2+} transport remained unaltered in divalent cation transport mutants [99]. The kinetics of Mg^{2+}-citrate dependent transport are illustrated in Fig. 8-6. The transport as a divalent cation complex still leaves one negative charge to be compensated for, but a strong requirement for intracellular K^+ suggests that other ions are involved. It is also possible that in excess, Mg^{2+} may be transported by yet another system to preserve electroneutrality [99].

The citrate analog 2-fluoro-L-erythro-citrate was shown to be a more convenient and useful probe of the citrate transport mechanism [100]. This analog was actively transported by B. subtilis cells and could induce the

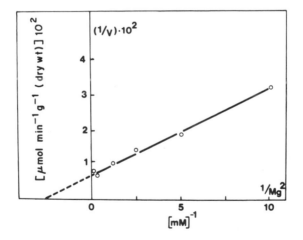

FIG. 8-6 Citrate-dependent $^{28}Mg^{2+}$ transport as a function of Mg^{2+} concentration in whole cells of B. subtilis [99]. Studies were carried out in the presence of 7.7 mM citrate with cells induced for citrate transport. The difference of data between the total transport and that inhibited by m-chlorophenyl carbonylcyanidehydrazone (CCCP) (noncitrate-dependent $^{28}Mg^{2+}$ transport) are plotted as a double reciprocal plot.

system, likely via the inhibition of aconitase. Mutants resistant to this toxic analog were defective in citrate-Mg^{2+} transport. Membrane vesicles from induced cells accumulated citrate when energized by ascorbate-PMS. This accumulation is particularly Mg^{2+}-dependent, and Na^+ was required as well but not as stringently [100].

The induction of citrate transport in B. subtilis was not dependent upon lipid synthesis, neither in glycerol auxotrophs deprived of glycerol [101] nor in cerulenin (an antibiotic which quantitively inhibits de novo lipid synthesis) treated cells [102].

E. Salmonella typhimurium

Citrate was actively transported in both aconitase mutants and fluorocitrate poisoned cells [33]. The system was of high affinity and was strongly competitively inhibited by dl-fluorocitrate and weakly inhibited by cis-aconitate, tricarballylate, and trans-aconitate, whereas other TCA intermediates were ineffective. Mutants devoid of citrate transport activity were isolated and were specific for citrate. Two mutants with high levels of citrate transport were isolated. Citrate induced the transport system, and aconitase mutants were particularly dependent on an exogenous energy source.

The uptake of tricarboxylates in S. typhimurium wild-type cells has been studied polarographically [103]. At least two transport systems have been inferred from induction data. The first was induced by citrate, isocitrate, and/or cis-aconitate and took up citrate and isocitrate. A second system similarly induced required Mg^{2+} ions and high pH. A third system induced by tricarballylate was also postulated to transport this substrate, as well as citrate and cis-aconitate. Tricarballylate dehydrogenase is induced in S. typhimurium [104] and makes this conclusion tenuous since the product of the reaction is cis-aconitate.

V. SUMMARY

The transport of carboxylic acids into bacteria occurs by a wide variety of systems which recognize a host of substrates. These systems can be described by mediated transport, active transport, and vectorial acylation mechanisms and have various forms of metabolic control exerted over them.

The bacterial systems which have yielded the most information have been those principally described in strains with a well-developed genetic system, such as E. coli, B. subtilis, and S. typhimurium. It is to be expected that further major advances in this somewhat complicated area of transport will be the result of the advantages of the biochemical genetics afforded by these strains. It is worth emphasizing that as a result of these approaches there has evolved what is probably the most finely resolved system, the C_4-dicarboxylate transport system of E. coli.

RECENT DEVELOPMENTS

Dicarboxylate Transport in E. coli

Recent results from the laboratory of T.C.Y. Lo have further elucidated
the organization of the dicarboxylate transport components in the inner
membrane of E. coli. The components appear to be organized according
to model A depicted in Fig. 8-4. Substrate binding studies with inside-out
and right side-out residues place the high affinity component SBP2 with the
substrate recognition site external and the lower affinity component SBP1
with its substrate binding site internal. Both proteins transcend the membrane [65].

Recent results from H. R. Kaback's group have shown that succinate
transport in membrane vesicles is driven by the electrochemical proton
gradient ($\Delta \widetilde{\mu}_{H^+}$) at pH 5.5, but that at pH 7.5, transport is driven by the
electrical component ($\Delta \psi$) [17]. This is in contrast to D-lactate, gluconate,
and glucuronate, which are driven primarily by ΔpH at pH 5.5, but also by
$\Delta \psi$ at pH 7.5.

REFERENCES

1. J. J. R. Campbell and F. N. Stokes, J. Biol. Chem., 190:853 (1951).
2. J. T. Barrett and R. F. Kallio, J. Bacteriol., 66:187 (1953).
3. M. Kogut and E. P. Podoski, Biochem. J., 55:800 (1953).
4. H. L. Kornberg, FEBS Symp., 19:5 (1969).
5. B. D. Sanwal, Bacteriol Rev., 34:20 (1970).
6. G. Ashwell, in Methods in Enzymology (S. P. Colowick and N. O.
   Kaplan, eds.), Academic, New York, 1962.
7. J. Robert-Baudouy, M. L. Didier-Fichet, J. Jimeno-Abendano, G.
   Novel, R. Potalies, and F. Stoeber, C.R. Acad. Sc. Paris, 271D:
   255 (1970).
8. J. Jimeno-Abendano and A. Kepes, Biochem. Biophys. Res. Commun.,
   54:1342 (1973).
9. J. Pouyssegur and A. Lagarde, Mol. Gen. Genet., 121:163 (1970).
10. A. E. Lagarde and J. M. Pouyssegur, Eur. J. Biochem., 36:328
    (1973).
11. J. Pouyssegur and F. Stoeber, C.R. Acad. Sci. Paris, 274:2249
    (1972).
12. J. M. Pouyssegur and F. R. Stoeber, Eur. J. Biochem., 21:363
    (1971).
13. J. M. Pouyssegur and F. R. Stocber, Eur. J. Biochem., 30:479
    (1972).
14. F. Autissier and A. Kepes, Biochimie 54:93 (1972).
15. J. M. Pouyssegur, Mol. Gen. Genet., 113:31-42 (1971).

16. N. Zwaig, R. N. deZwaig, T. Isturiz, and M. Wecksler, J. Bacteriol., 114:469 (1973).
17. S. Ramos and H. R. Kaback, Biochemistry 16:854 (1977).
18. R. N. de Zwaig, N. Zwaig, T. Isturiz, and R. S. Sanchez, J. Bacteriol., 114:463 (1973).
19. B. Bachi and H. L. Kornberg, Biochem. J., 150:123 (1975).
20. J. M. Pouyssegur, P. Faik, and H. L. Kornberg, Biochem. J., 140:193 (1974).
21. J. D. Stinnett, L. F. Guymon, and R. G. Eagon, Biochem. Biophys. Res. Commun., 52:284 (1973).
22. L. F. Guymon and R. G. Eagon, J. Bacteriol., 117:1261 (1974).
23. T. D. K. Brown, PhD Thesis, University of Leicester, U.K. (1972).
24. L. de Jong and W. N. Konings, Adv. Microbiol. Physiol., 15:175 (1977).
25. S. A. Short, H. R. Kaback, and L. D. Kohn, Proc. Natl. Acad. Sci. USA, 71:5032 (1974).
26. A. Matin and W. Konings, Eur. J. Biochem., 34:58 (1973).
27. H. L. Kornberg and J. Smith, Biochim. Biophys. Acta, 148:591 (1967).
28. J. Gibson, J. Bacteriol., 123:471 (1975).
29. L. N. Ornston and M. K. Ornston, J. Bacteriol., 101:1088 (1970).
30. M. H. Saier, Jr., D. L. Wenzel, B. U. Feucht, and J. J. Judice, J. Biol. Chem., 250:5089 (1975).
31. W. W. Kay, Biochim. Biophys. Acta, 264:508 (1972).
32. C. Wagner, R. Odom, and T. Briggs, Biochem. Biophys. Res. Commun., 47:1036 (1972).
33. W. W. Kay and M. Cameron, Arch. Biochem. Biophys. (In press).
34. K. Klein, R. Steinberg, B. Fiethen, and P. Overath, Eur. J. Biochem., 19:442 (1971).
35. J. P. Salanitro and W. S. Wegener, J. Bacteriol., 108:885 (1971).
36. J. P. Salanitro and W. S. Wegener, J. Bacteriol., 108:893 (1971).
37. A. S. Visser and P. W. Postma, Biochim. Biophys. Acta, 298:333 (1973).
38. E. Vanderwinkel, M. DeVlieghere, and J. V. Meersche, Eur. J. Biochem., 22:115 (1971).
39. G. Pauli and P. Overath, Eur. J. Biochem., 29:553 (1972).
40. J. P. Salanitro and W. S. Wegener, J. Bacteriol., 108:893 (1971).
41. E. Vanderwinkel, P. Furmanski, H. C. Reeves, and S. J. Ajl, Biochem. Biophys. Res. Commun., 33:902 (1968).
42. F. E. Frerman, Arch. Biochem. Biophys., 159:444 (1973).
43. P. Overath, G. Pauli, and H. U. Schairer, Eur. J. Biochem., 7:559 (1969).
44. K. Klein, R. Steinberg, B. Fiethen, and P. Overath, Eur. J. Biochem., 19:442 (1971).
45. H. R. Kaback, in The Molecular Basis of Membrane Function (D. C. Tosteson, ed.), Prentice-Hall, Englewood Cliffs, New Jersey, 1969.
46. F. E. Frerman and W. Bennett, Arch. Biochem. Biophys., 159:434 (1973).

47. D. Samuel, J. Estroumza, and G. Ailhaud, Eur. J. Biochem., 30: 275 (1975).
48. G. Weeks, M. Shapiro, R. O. Burns,and S. J. Wakil, J. Bacteriol., 97:827 (1969).
49. G. Pauli, R. Ehring, and P. Overath, J. Bacteriol., 117:1178 (1974).
50. A. J. Fulco, J. Biol. Chem., 247:3503 (1972).
51. W. W. Kay and H. L. Kornberg, FEBS Lett., 3:93 (1969).
52. W. W. Kay and H. L. Kornberg, Eur. J. Biochem., 18:274 (1971).
53. A. A. Herbert and J. R. Guest, J. Gen. Microbiol., 63:151 (1971).
54. S. Murakawa, K. Izaki, and H. Takahashi, Agr. Biol. Chem., 36: 2397 (1972).
55. S. Murakawa, K. Izaki, and H. Takahashi, Agr. Biol. Chem., 36: 2487 (1972).
56. J. L. Parada, M. V. Ortega, and G. Carillo-Castaneda, Arch. Microbiol., 94:65 (1973).
57. S. Murakawa, K. Izaki, and H. Takahashi, Agr. Biol. Chem., 35: 1992 (1971).
58. W. Kay, J. Biol. Chem., 246:7373 (1971).
59. T. C. Y. Lo, M. K. Rayman, and B. D. Sanwal, J. Biol. Chem., 247:6323 (1971).
60. T. C. Y. Lo and B. D. Sanwal, Mol. Gen. Genet., 140:303 (1975).
61. M. Khalil, T. C. Y. Lo, and B. D. Sanwal, J. Biol. Chem., 247: 6332 (1972).
62. T. C. Y. Lo and B. D. Sanwal, J. Biol. Chem., 250:1600 (1975).
63. T. C. Y. Lo and B. D. Sanwal, Biochem. Biophys. Res. Commun., 63:278 (1975).
64. T. C. Y. Lo, PhD Thesis, University of Toronto, Toronto, Canada (1973).
65. T. C. Y. Lo, J. Supramolecular Struct., 4:463 (1977).
66. T. C. Y. Lo, M. K. Rayman, and B. D. Sanwal, Can. J. Biochem., 52:854 (1974).
67. S. J. Gutowski and H. Rosenberg, Biochem. J., 152:647 (1975).
68. O. K. Ghei and W. W. Kay, FEBS Lett., 20:137 (1972).
69. R. E. Fournier, M. N. McKillen, and A. B. Pardee, J. Biol. Chem., 247:5587 (1972).
70. O. K. Ghei and W. W. Kay, J. Bacteriol., 114:65 (1973).
71. A. Bisschop, H. Doddema, and W. N. Konings, J. Bacteriol., 124: 613 (1975).
72. O. K. Ghei and W. W. Kay, Can. J. Microbiol., 21:527 (1975).
73. O. K. Ghei and W. W. Kay, Biochim. Biophys. Acta, 401:440 (1975).
74. K. Willecke and R. Lange, J. Bacteriol., 117:373 (1974).
75. F. Ramos, J. M. Wiame, J. Wynants, and J. Bechet, Nature 793: 70 (1962).
76. R. E. Fournier and A. B. Pardee, J. Biol. Chem., 249:5948 (1974).
77. O. K. Ghei and W. W. Kay, unpublished observations.

78. R. E. Dubler, W. A. Toscano, Jr., and R. A. Hartline, Arch. Biochem. Biophys., 160:422 (1974).

79. P. W. Postma and K. Van Dam, Biochim. Biophys. Acta, 249:515 (1971).

80. A. J. J. Reuser and P. W. Postma, FEBS Lett., 21:145 (1972).

81. A. J. J. Reuser and P. W. Postma, Eur. J. Biochem., 33:584 (1973).

82. S. Dagley and W. A. Dawes, J. Bacteriol., 66:259 (1953).

83. S. Dagley and W. A. Dawes, Nature 172:345 (1953).

84. P. H. Clarke and P. M. Meadow, J. Biol. Chem., 190:853 (1959).

85. A. M. Williams and P. W. Wilson, J. Bacteriol., 67:353 (1954).

86. R. J. Harvey and E. B. Collins, J. Bacteriol., 83:1065 (1962).

87. E. I. Villarreal-Moguel and J. Ruiz-Herrera, J. Bacteriol., 98: 552 (1969).

88. L. S. Wilkerson and R. G. Eagon, Arch. Biochem. Biophys., 149: 209 (1972).

89. C. L. Johnson, Y. A. Cha, and J. R. Stern, J. Bacteriol., 121: 682 (1975).

90. D. S. Sachan and J. R. Stern, Biochem. Biophys. Res. Commun., 45:402 (1971).

91. R. G. Eagon and L. S. Wilkerson, Biochem. Biophys. Res. Commun., 46:1944 (1972).

92. L. S. Wilkerson and R. G. Eagon, J. Bacteriol., 120:121 (1974).

93. P. W. Postma, PhD Thesis, University of Amsterdam, Amsterdam, The Netherlands (1973).

94. H. G. Lawford and G. R. Williams, Biochem. J., 123:571 (1971).

95. W. W. Altekar and M. R. R. Rao, J. Bacteriol., 85:604 (1963).

96. K. Willecke and A. B. Pardee, J. Biol. Chem., 246:1032 (1971).

97. R. A. Carls and R. S. Hanson, J. Bacteriol., 106:848 (1971).

98. M. N. McKillen, K. Willecke, and A. B. Pardee, in The Molecular Basis of Biological Transport (J. F. Woessner and F. Huijung, eds.), Academic, New York (1972).

99. K. Willecke, E. M. Gries, and P. Oehr, J. Biol. Chem., 248:807 (1973).

100. P. Oehr and K. Willecke, J. Biol. Chem., 249:2037 (1974).

101. K. Willecke and L. Mindrich, J. Bacteriol., 106:514 (1971).

102. W. Wille, E. Eisenstadt, and K. Willecke, Antimicrob. Ag. Chemother., 8:231 (1975).

103. I. Imai, T. Iyima, and T. Hasegawa, J. Bacteriol., 114:961 (1975).

104. J. R. Stern, Abst. Am. Soc. Microbiol., 212 (1973).

105. A. L. Taylor and C. D. Trotter, Bacteriol. Rev., 36:504 (1972).

Chapter 9

THE ROLE OF THE OUTER MEMBRANE IN ACTIVE TRANSPORT

Robert J. Kadner and Philip J. Bassford, Jr.*

Department of Microbiology
University of Virginia School of Medicine
Charlottesville, Virginia

I. INTRODUCTION 414
II. BARRIER PROPERTIES OF THE OUTER MEMBRANE 415
   A. Structure of the Outer Membrane 415
   B. Barrier Properties of the Outer Membrane 419
   C. Specific Transport Systems in the Outer Membrane 424
III. ROLE OF THE LAMBDA RECEPTOR IN MALTOSE
    TRANSPORT 426
IV. ROLE OF THE OUTER MEMBRANE IN IRON TRANSPORT . . . 430
   A. Iron Transport Systems in the Gram-negative Bacteria . . . 430
   B. Individual Iron Chelate Transport Systems 431
   C. Interactions Among Siderophore Uptake Systems 440
V. VITAMIN B$_{12}$ UPTAKE 442
   A. General Characteristics of Uptake 442
   B. Role of the Outer Membrane 443
   C. Role of the tonB Product 445

_____

*Present address: Department of Microbiology and Molecular Genetics,
Harvard Medical School, Boston Massachusetts

VI. FUNCTIONAL PROPERTIES OF THE tonB PRODUCT AND THE
     POSSIBLE ROLE OF MEMBRANE ADHESION SITES 447
     A. Functional Properties of the tonB and bfe Products 447
     B. Membrane Adhesion Sites 451
     REFERENCES 455

I. INTRODUCTION

The characteristic cell envelope of gram-negative bacteria consists of a
multilayered complex of two distinct membranes separated by a rigid peptidoglycan layer. Both membranes appear identical in gross morphology
and both contain protein and phospholipid. The cytoplasmic or inner membrane is the subject of all of the other chapters in this book, since this
membrane is the site of essentially all energy coupling and nutrient transport processes. In addition, it serves as the matrix for the synthesis of
lipid, peptidoglycan, and lipopolysaccharide (LPS), as well as being intimately involved in the synthesis and segregation of the bacterial chromosome, and possibly in the synthesis of certain proteins. Befitting its multiplicity of functions, the cytoplasmic membrane has attracted considerable
attention, while most investigators considered the outer membrane to be a
rather inert structure. However, some recent studies have been directed
toward the properties of the outer membrane, and they have been rewarded
with most interesting results. Although this review is primarily concerned
with the role of the outer membrane in transport, it is especially timely to
present some of the more general findings relating to the structure and
function of the outer membrane.

A large part of the attraction of the outer membrane for the investigator
stems from its relatively simple composition and from the ease with which
the presence of its constituents can be manipulated genetically. It has
proved relatively simple to obtain mutants missing specific proteins from
this membrane. Mutants producing altered lipopolysaccharides have long
been known, and the phospholipid composition can be manipulated by use of
the various well-known mutants blocked in phospholipid or unsaturated fatty
acid biosynthesis. In addition, this membrane exhibits a very marked
asymmetric distribution of its constituents, correlating well with its functional properties.

Owing to the necessary brevity of the first portion of this chapter on
general aspects of the structure and barrier functions of the outer membrane, the reader is urged to consult other review articles dealing with

these topics. Especially recommended are two recent reviews by Nikaido [1,2] on the structure and function of the outer membrane, one by Braun [3] on the murein lipoprotein, one by Inouye [4] on the assembly of the outer membrane, and several general reviews on the bacterial cell surface [5-7].

II. BARRIER PROPERTIES OF THE
OUTER MEMBRANE

A. Structure of the Outer Membrane

1. Composition and Isolation

The arrangement of the multiple layers of the envelope of gram-negative bacteria seems to be well documented. Under normal conditions, turgor pressure causes the cytoplasmic membrane to be tightly pressed against the peptidoglycan, or rigid layer. The peptidoglycan can be recovered as an intact, cell shaped sacculus following disruption of the cell and solubilization of proteins by detergents [8]. Its chemical structure is well defined [9], and most notable is the fact that less than one-half of the potential cross-bridges between peptide chains are formed [8]. Covalently attached to 10 to 12% of the diaminopimelate residues is the unique murein lipoprotein [10], whose structure has been clearly elucidated by Braun and his colleagues (reviewed in Ref. 3). The bulk of this lipoprotein is embedded in the outer membrane and is retained there, in part, by three fatty acyl residues attached to the amino terminal cysteine. This protein anchors the outer membrane to the peptidoglycan. This protein can also be recovered from the outer membrane in a free state, and the ratio of bound:free molecules is dependent on the strain and growth conditions [11,12].

Discussion of the properties of the outer membrane must start with mention of the methods for its isolation and possible resultant artifacts. The two most common methods for disrupting cells involve either the passing of cells through a French pressure chamber or the lysing of spheroplasts either osmotically or by sonic oscillation, the spheroplasts having been prepared by the action of lysozyme plus EDTA [13-15]. A recent procedure [16] employs the lysis of spheroplasts prepared by lysozyme treatment of osmotically shocked cells, thereby obviating the requirement for EDTA treatment to allow the lysozyme access to the peptidoglycan. Membranes released by either procedure of cell disruption are then separated from each other by isopycnic sucrose gradient centrifugation, which usually separates cleanly the denser outer membrane fractions from the cytoplasmic membrane fractions. Both linear and discontinuous sucrose gradients have been employed [13,15]. In a few cases [17,18], particle electrophoresis has been used for this separation. Another method which permits the isolation solely of outer membrane components takes advantage of the selective release of portions of the outer membrane during the formation of spheroplasts and their aggregation upon exposure to pH 5 [19]. Many of these methods suffer from the likelihood that the material obtained may differ substantially from the outer membrane of the intact cell. For example, treatment with EDTA during formation of spheroplasts does result in the release of a considerable portion of the cellular LPS [20], and hydrolysis of the peptidoglycan can be followed by the redistribution of the LPS to both sides of the bilayer [21].

There is a relatively reliable method for solubilizing selectively the proteins of the outer and cytoplasmic membranes from a preparation of total cellular membranes. This method takes advantage of the finding that essentially all the proteins of the cytoplasmic membrane are soluble in 2% Triton X-100 (or other nonionic detergent) in the presence of Mg^{2+}, whereas those of the outer membrane are soluble in these detergents only when divalent cations have been chelated by EDTA [22,23].

The outer membrane is composed of proteins, phospholipids, and LPS, along with cations such as Mg^{2+} and polyamines. The relative amounts of these constituents are dependent on the growth conditions and on the method of isolation. Proteins comprise about half the weight of the outer membrane, and these include 50 to 70% of the total membrane protein [14,15]. The protein composition of the outer membrane is relatively simple, befitting the paucity of its functions; depending on the strain and growth conditions, 15 to 25 protein bands are usually resolved by SDS-polyacrylamide gel electrophoresis [15,24]. An idealized profile of an electropherogram of the outer membrane proteins of E. coli K12 is shown in Fig. 9-1, which was provided by Schnaitman. Methods allowing higher resolution than that provided by the tube gel methods, such as slab gel or two-dimensional techniques, have revealed the presence of considerably more minor species of proteins, some of which may be released by osmotic shock ([25]; C. Schnaitman, personal communication).

Two to five species of proteins are present in greatest amount. These include the free form of the murein lipoprotein (7000 molecular weight) and three or four proteins with molecular weights in the range of 32,000 to 36,000. These latter proteins are relatively resistant to solubilization in SDS and, depending on the temperature of their solubilization, exhibit different mobilities during electrophoresis [24,26,27]. There is no evidence for the presence of any of these proteins in both membranes.

The phospholipid composition of the two membrane fractions from several species has been determined. Relative to the cytoplasmic membrane, the outer membrane contains less phospholipid [28] and a slightly different distribution of the classes, a higher proportion of phosphatidyl ethanolamine being present in the outer membrane [14,17]. The distribution of the fatty acids on the phosphatides is similar for the two membranes [17].

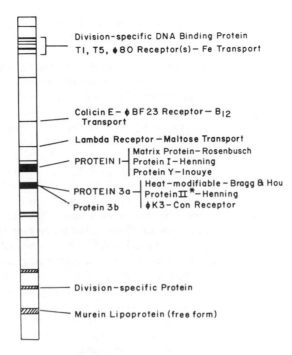

FIG. 9-1 An idealized profile of a SDS-polyacrylamide gel electrophero-gram of the outer membrane proteins from E. coli K12. Probable functions and designations of the major proteins are presented. (Courtesy of C. A. Schnaitman.)

The other lipid moiety present in, and unique to, the outer membrane is LPS, whose detailed molecular structure is known for many of the sero-types of Salmonella [1] and recently for several strains of Escherichia coli [29].

2. Asymmetry of Structure

Mühlradt and coworkers [30] have shown that LPS reactive with specific antibody covers 15 to 25% of the surface of the cell, and that this density is independent of the length of the polysaccharide chain. As long as the under-lying peptidoglycan layer is intact, all of the LPS reactive with antibody is on the outer face of the outer membrane [21]. Removal of this layer with trypsin or lysozyme at 25 to 37°C results in a very rapid randomization of the LPS onto both faces of the outer membrane. This reorientation may be very important for consideration of studies of the reconstitution of this

membrane from its constituents. Studies by Smit et al. [28] confirmed that
the number of LPS molecules per unit surface area was independent of the
length of the saccharide chain and comprised 22% of the total surface area
of the outer membrane. If the LPS in the intact cell is all in one leaflet of
the outer membrane, this means that LPS represents almost half (44%) the
external surface area. The remainder of the outer face could contain protein
and phospholipid, or either alone.

Evidence suggests that protein is present in the outer face, and in
rather high density. First, it would seem that phage and colicin receptors,
many of which are protein in nature, must be accessible to the exterior.
Electron microscopy provides more direct information. The general
pattern of freeze-fracture profiles of E. coli [31] and Salmonella typhimurium [28] are similar. The inner face of the cytoplasmic membrane
contains many particles, whereas the outer face of this membrane is relatively devoid of particles. Fracture profiles of the outer membrane, which
are only rarely obtained with strains producing LPS with normal length
saccharide chains, were also asymmetrical. The outer, or concave, face
appears densely studded with particles having an apparent diameter of 80 Å,
whereas the inner, or convex, face is rather smooth. In the deep-rough
mutants producing a very truncated LPS, the ease of fracture through the
outer membrane greatly increased. In addition, the density of particles
is decreased in proportion to the decrease in protein content of the outer
membrane in these mutants, with the concomitant appearance of smooth
regions in the outer face [28, 32]. Most of the outer membrane proteins
can be cross-linked with bifunctional imidoester reagents [33], consistent
with the existence of extensive protein-protein interactions over the entire
surface of the membrane. Thus, in addition to all of the LPS, most of the
protein of the outer membrane appears to reside in the outer face.

The location of the phospholipids in the outer membrane is less definite,
but they are probably localized to the inner leaflet. The phospholipids of
the intact cells of a strain with normal LPS are resistant to phospholipase
action [34]. The amount of phospholipid in the membrane is sufficient to
cover almost one leaflet. In the deep-rough mutants with very short LPS
chains, the amount of phospholipid is increased [28], and a portion of it
can be acted upon by phospholipases [34]. These results suggested to
Nikaido [2] that all the phospholipid resides in the inner face of the outer
membrane and appears in the outer face to compensate for the loss of protein which occurs in mutants producing greatly truncated LPS. This would
allow formation of regions of phospholipid-phospholipid bilayer, which
could account for the increased ease of fracture through these mutant
membranes. In addition, the permeability properties relative to hydrophobic permeants may be different for this type of bilayer than those in the
usual case where protein is present or where LPS is on one leaflet. In this
context, regions containing LPS and protein are considerably less fluid than
regions of phospholipid [35], although the presence of LPS does not alter the

permeability properties of phospholipid vesicles [36]. Overath and his coworkers [37] demonstrated that a much smaller fraction of the phospholipids of the outer membrane, relative to those of the cytoplasmic membrane, were able to participate in lipid phase transitions. They suggested that a large part of the lipids of this membrane form fluid domains, possibly as monolayers. These permeability properties will be discussed subsequently; the important point is that many functional properties of the outer membrane are consistent with Nikaido's model of its asymmetric structure.

B. Barrier Properties of the Outer Membrane

The most likely function of the outer membrane is to serve as a penetration barrier both to protect the peptidoglycan from hydrolysis by the muramidases and to protect the cytoplasmic membrane from disruption by the detergents present in, for example, the habitat of the enteric bacteria. Evidence has accumulated that there is a barrier function against both hydrophilic and hydrophobic compounds, owing to different features of the structure of the cell envelope.

1. Permeability of Hydrophobic Compounds

Study of the properties of mutants producing LPS molecules with shorter than normal saccharide chains has implicated LPS as affecting the permeation of hydrophobic compounds and has shown that the outer membrane of normal cells is less permeable to hydrophobic compounds than expected. These mutants, and especially the deep-rough mutants, were found to be more sensitive to various antibiotics, dyes, and detergents [38,39], and to allow a much more rapid penetration of the basic dye, crystal (gentian) violet, into the cell [40]. Conversely, many mutants isolated on the basis of their increased sensitivity to various agents produced altered LPS structures. Three classes of mutants which were somewhat more resistant to penicillin but somewhat more sensitive to cholate produced LPS with less than the normal amount of core saccharides [41]. Some of these mutants showed the increased rate of uptake of gentian violet [40]. Mutants selected for sensitivity to novobiocin produced greatly truncated LPS molecules lacking the outer core, and some were lacking even heptose [42]. These mutants were resistant to phage T4; mutants selected for T4 resistance included strains with deranged outer membrane permeability. Even penetrability of DNA was increased by the genetic truncation of LPS [43].

Exposure of normal cells to EDTA resulted in loss of approximately one-half the cellular LPS. Such treatment rendered the cells sensitive to such normally impermeable agents as actinomycin D [21,44,45], or a wide variety of antibiotics, dyes, and detergents [46]. Loss of LPS was not solely responsible for the alterations in permeability, since mutants resistant to the EDTA-induced changes in sensitivity still released LPS, although not quite as much, following EDTA treatment [47]. The half of the cellular LPS that is refractory to removal was shown to be identical chemically to the removable pool [45]. These two pools are in equilibrium [48], and it has been proposed that they differ only with respect to the LPS being associated solely with other like molecules or in a complex domain with membrane proteins [1]. The presumption was that one of these associations was stabilized by Mg^{2+} bridges.

Some evidence for LPS-protein complexes came from the finding that the deep-rough mutants had a lower protein content in their outer membrane [49,50]. Most of the protein species, including certain functional proteins [51], were affected. It was mentioned previously that these mutants have altered permeability properties and appear to leak periplasmic proteins [52]. The decrease in protein content is compensated for by increased amounts of phospholipid, some of which are accessible to external phospholipases. It is thought most likely that the abnormally short LPS does not provide sufficient stabilization to retain the usual complement of protein, with the excess proteins being shed into the medium.

Crucial to this discussion is the recent finding of Nikaido [53] that the increased permeability observed in the deep-rough mutants affected primarily hydrophobic molecules, those with finite solubility in octanol. He also showed that the outer membrane was the significant permeability barrier to one such permeant, nafcillin. The permeation of octanol-insoluble probes was much less affected by alteration of the LPS, if at all.

Finally, the fact that many, but not all [59], mutants lacking specific outer membrane proteins are altered in their permeability properties [54-58] strengthens the contention that the presence of only LPS-protein domains on the outer face is primarily responsible for the outer membrane's abnormally low permeability to hydrophobic molecules.

2. Permeability of Hydrophilic
    Compounds

Several different lines of investigation had pointed to the existence of a permeability barrier external to the cytoplasmic membrane. Accessibility of penicillins and cephalosporins to their target enzymes and to β-lactamases, both located in the periplasm, was clearly restricted for certain of these antibiotics [60]. Detailed kinetic studies of transmembrane effects on the influx and efflux of β-galactosides led Robbie and Wilson [61] to postulate the existence of a partial diffusion barrier external to the cytoplasmic membrane. This unstirred layer effect limiting the diffusion of substrates from their transport systems required specially designed experiments for its detection, although similar effects can be seen with many other transport systems. It is generally thought that the effect of the outer membrane in the usual assays for the uptake of nutrients or transport substrates is minimal.

There are two important series of experiments directed toward the permeability properties of the cell envelope. Studies by Payne and Gilvarg [62], reviewed in Chap. 7, demonstrated a size limitation for the utilization of peptides by amino acid auxotrophs of E. coli. For several series of peptides of known composition, there was a clear relationship between the cellular utilization of a peptide and its apparent molecular dimensions (Stokes radius), estimated from its partition coefficient on gel filtration. Peptides could be utilized if they had a free amino terminus and if their partition coefficient K_D on a Sephadex G-15 column was less than 0.19. Peptides eluting earlier were not utilized, even though they had the same composition as the smaller peptides and could be hydrolyzed completely by peptidases extracted from the cell. The lack of competition between peptides of the utilized and excluded sizes indicated that this discrimination was not at the level of the transport system. There was competition between peptides of proper size but different in composition. It was concluded that peptides with molecular weights greater than 560 (Stokes radii in the range of 4 to 6 Å) are excluded from access to the peptide transport system. The true exclusion limit may be somewhat larger, since technical difficulties prevented the use of the wide ranges of external concentrations of the probe peptides necessary to determine accurately the size of the diffusion pore.

In a complementary study with a series of oligosaccharides, Nakae and Nikaido [63] avoided the possible discrimination at the cytoplasmic membrane by measuring permeation into the periplasmic space of intact plasmolyzed cells. There was rapid and complete equilibration of sucrose and the trisaccharide raffinose into the periplasm. The tetrasaccharide stachyose (molecular weight, 666) was partially excluded, and oligosaccharides of molecular weight greater than 900 to 1000 were fully excluded. These results agree well with those from Gilvarg's group and indicate the presence of nonspecific pores or channels for hydrophilic compounds. The radius of this pore penetrating the cell envelope is proposed to be responsible for the apparent exclusion threshold for permeants as a function of their molecular dimensions.

This diffusion pore could span the outer membrane, the peptidoglycan, or both layers. Early results had shown that treatment of cells with ampicillin or lysozyme increased the ease of access of ampicillin to the periplasmic penicillinases, suggesting that peptidoglycan constituted at least part of the penetration barrier [64]. A similar conclusion drawn by Barak and Gilvarg [65] was based on their findings that certain phosphorylated peptides were better substrates for the periplasmic alkaline phosphatase than for the peptide transport system in the cytoplasmic membrane. Their results suggested that alkaline phosphatase resides external to a barrier for peptides, presumably the peptidoglycan.

Other evidence favors the view that the primary penetrability barrier of the cell envelope is the outer membrane. The permeability of the peptidoglycan cell wall of Bacillus megaterium was extensively studied by Scherrer and Gerhardt [66]. Their data indicated that the wall does not totally exclude permeants with molecular weights under 10^5 (Stokes radii in the range of 100 Å). The authors concluded the exclusion limit for the wall was as low as 11 Å (molecular weight 1200), although permeants with molecular weight around 5000 could enter half the space of the wall. Even their lowest value for the exclusion limit is larger than the exclusion limit of the cell envelope of gram-negative cells. In addition, the exclusion threshold for the peptidoglycan of gram-negative cells is probably larger than this value. This layer in E. coli is much thinner and looser in comparison to the thick, three-dimensional, highly cross-linked mesh present in gram-positive cells.

Nakae and Nikaido [63] showed that the penetration barrier toward oligosaccharides remained intact in cells whose peptidoglycan layer had been disrupted, either by treatment with lysozyme (without EDTA) or by growth in the presence of penicillin. They showed that isolated outer membrane vesicles, free of peptidoglycan, had permeability properties similar to those of the intact cell, showing that peptidoglycan is not required for at least a major portion of the size discrimination behavior of the cell envelope.

3. Constituents of the Outer Membrane
    Diffusion Pore

The nature and composition of the postulated nonspecific pore for hydrophilic compounds across the outer membrane is of considerable current interest. Since bilayers of phospholipid and LPS are impermeable to these permeants [36], the pore is probably composed, at least in part, of protein. Two candidates have been proposed.

Inouye [67] has presented a model for the formation of complexes of the murein lipoprotein which could generate the pore. Braun [68] had shown that this protein could exist in a conformation with considerable α-helicity. This structure could be arranged such that the hydrophobic side chains are predominantly oriented in one direction along the long axis of the helical molecule, with the hydrophilic residues facing other directions. Inouye pointed out that the aggregation of the lipoprotein, perhaps into a hexamer, could form a pore of proper internal dimensions and of sufficient length to span the outer membrane. Several findings strengthened this proposal. It would have been difficult to formulate complex formation if all of this protein were covalently attached to the peptidoglycan. Inouye et al. [69] found that only one-third of this protein is bound, and that the bound pool is formed from the free pool; the chemical structure of the two forms were essentially identical [70,71]. Thus, there are sufficient molecules free to interact to form these proposed complexes. In support of this view, Henning and Haller [59] isolated mutants lacking all the major outer membrane proteins except the murein lipoprotein. These mutants showed no gross derangement in morphology or metabolism as might have been expected for a cell lacking the hydrophilic pore. Recently, Wu and Lin [72] have described mutants which are deficient in the amount of bound lipoprotein and in which the free form appears to be altered. The growth of these strains was aberrant, but this was not correlated with the changes in the lipoprotein. Von Meyenburg [73] described some mutants of E. coli B/r which exhibited decreased affinities for possibly all of their transport substrates, including carbohydrates, amino acids, and inorganic ions. It is possible that the envelope defects in these strains affected the function of the pores.

The other candidates proposed as constituents of the pore are the major outer membrane proteins with molecular weights in the range of 32,000 to 36,000. Like the lipoprotein, these are also present in about 10^5 copies per cell [22,23]. Some of these proteins form complexes associated with the peptidoglycan which are resistant to mild detergent extraction [74]. These complexes exhibit hexagonal symmetry. Support for their role in pore formation comes from studies of the reconstitution of outer membrane vesicles. Removal of detergent from the separated constituents allowed formation of membrane vesicles; all three major constituents were necessary for proper morphology [75,76]. Vesicles formed from phospholipids and LPS were impermeable to sucrose [36]. Addition of solubilized proteins from the outer membrane, but not those from the cytoplasmic membrane [77], resulted in formation of topologically closed sucrose-permeable structures [78]. The outer membrane proteins were separated into four fractions by gel chromatography on Sepharose 4-B in the presence of 0.1% SDS, conditions insufficient to disrupt all protein-protein interactions. Two of the fractions, one eluting at the void volume and one just thereafter, were effective at generating sucrose-permeable vesicles upon combination with phospholipid and LPS. The most active fraction contained only three of the major outer membrane proteins (36,000, 35,000, and 34,000), which appear to be the analogs in S. typhimurium of the E. coli outer membrane proteins 1A and 1B, in the terminology of Schnaitman [27]. The active fractions contained none of the murein lipoprotein, and purified lipoprotein was completely inactive in this pore formation assay.

In the spirit of compromise, it can be proposed that both proteins are involved in pore formation in the intact cell. In accord with the results of Nakae [78], the pore could be lined only with the major outer membrane proteins. However, in the cell, these proteins are maintained in the proper structural orientation by their interaction with the murein lipoprotein. Genetic loss of the major outer membrane proteins, while not being lethal to

the cell [59], is nonetheless not without marked effect, such as colicin
tolerance [79-81] and possibly altered permeability properties of the outer
membrane [82,83].

C. Specific Transport Systems in the
    Outer Membrane

1. Specific Functions in the Outer
    Membrane

Specific functions for the outer membrane are hard to find. A phospholipase A activity is the only known enzymatic activity shown in this membrane [84]. The permeation of hydrophilic compounds through the presumptive pore appears to be nonspecific, responding only to the molecular
dimensions of the permeant.
      However, it has long been recognized that phage and colicin receptors
must be located on the surface of the cell exposed to the environment. In
a few cases, referenced later, some phage receptors have been shown by
direct determination to be located in the outer membrane. Since sensitivity
to various lethal agents would seem to place the cell at a severe disadvantage, the presence of phage receptors on the cell surface raises the question of the nature of the selective pressures maintaining their presence.
Some previously described phage receptors have been shown to include
flagella, pili, capsules, and numerous of the saccharide residues on LPS
(reviewed in Ref. 85). All these structures are advantageous to the cell,
especially for survival under natural conditions. Recent findings have
shown that many of the outer membrane proteins that are phage receptors
also have functions of obvious advantage to the cell. Some phages and
colicins use as receptors the major outer membrane proteins, which are
involved in maintenance of the proper structure of the outer membrane
[80,81,86,88]. Several other outer membrane receptors are essential
components of specific nutrient transport systems. The remainder of this
chapter is concerned with the characterization of these proteins and of the
transport systems in which they participate. The one characteristic of
these transport systems that accounts for the need for participation of
outer membrane components is that the nutrients they transport are too
large to permeate efficiently, if at all, through the diffusion pores in the
outer membrane. Clearly, specific proteins are needed for the translocation of these large molecules across that barrier.

2. Phage and Colicin Receptors

The entry of phages and colicins is intimately related to these transport
systems, and much of the following discussion will assume some familiar-

ity with the properties of colicins and phage entry. There have been several recent reviews on this subject [89-91]. Several general points should be emphasized here. Different colicins can share the same receptor but have different modes of action; conversely, colicins can use different receptors but have the same mode of action. To effect lethality, some colicins must enter the cytoplasm whereas others need not [92]. A large number of the bacterial mutants surviving exposure to colicins have nonfunctional or absent receptors; these are termed resistant, and the colicin is not even able to bind to that cell. Many mutants retain functional receptors but are not killed by the bound colicin; these are termed tolerant. There are many mechanisms for tolerance, but none have been clearly defined. Tolerance can result from alteration of the target of that colicin, but in most cases, tolerance is probably due to failure of the colicin to be transmitted from the receptor to its target. Many of the examples of this latter class exhibit, in some manner, alterations in the normal surface structure. For example, some colicin-tolerant mutants are sensitive to dyes and detergents [57, 58]; in some cases, this was accompanied by changes in the LPS [93]. Those mutants in which the presence of receptor has not been determined will be referred to as insensitive.

From studies of patterns of cross-resistance, Davies and Reeves [57, 58, 94] found that all of the colicins could be specifically assigned into one of two exclusive groups (A and B). For our considerations, this grouping operates as follows. The Group A colicins (including A, E1, E2, E3, K, and L) are dependent on the function of the tolA gene. Mutants in this locus, as well as other mutants tolerant to these colicins, exhibit the altered surface properties exemplified by sensitivity to dyes and detergents. The Group B colicins (including B, Ia, Ib, M, and V) are independent of tolA function, but are dependent on the function of another locus, tonB. Mutants tolerant to these colicins do not exhibit the altered surface properties. This fundamental distinction proves to be extremely important, as will be shown later.

3. The Molecular Masquerade

The results presented in this chapter will show that the same receptor molecule for phages and/or colicins also functions in nutrient transport. Clearly, the latter is the physiological role of the receptor. An important topic concerns the relation of the binding sites of the receptor for its various ligands. Does the recognition site on the phage or colicin resemble the steric configuration of the specific nutrient? Is the nutrient a constituent of the phage? For example, it is known that tetrahydrofolate is an essential component of phage T4 and T6, and is necessary for phage DNA ejection [95, 96]. None of the evidence on this question with the systems described later supports the presence of the nutrient on the lethal agent.

Thus, the possibility still exists, especially in light of the strictly competitive interactions between the nutrient and the lethal agent, that selective
pressures have allowed portions of a phage tail or colicin molecule to masquerade as a nutrient, such as a ferric chelate.

III. ROLE OF THE LAMBDA RECEPTOR
    IN MALTOSE TRANSPORT

A relationship between the adsorption of phage lambda (λ) and the metabolism of maltose in E. coli has long been recognized. Lederberg [97]
first reported that nearly 80% of spontaneous λ-resistant isolates had
specifically lost the ability to utilize maltose. Most of these mutations
were in the malA region (at 74 min on the revised genetic map [98]), with
the remainder in the malB region (at 90 min). The mutations in the remaining λ^r Mal$^+$ isolates were also in the malB region, closely linked to
a gene involved in maltose transport (malK) [99]. These latter mutations
defined the lamB cistron, which has been shown to be the structural gene
for the lambda receptor. Since the other mutations in the malB region
resulted in the loss of the ability to utilize maltose, owing to inactivation
of at least the malK locus by either polar mutations in malK or deletions
of malK and lamB, it was proposed that lamB was part of an operon for
maltose utilization [100]. Efficient induction of lambda receptor activity
is dependent on growth in the presence of maltose [101] and subject to
catabolite repression [102,103].

The malB region includes three genes involved in the transport of
maltose and maltodextrins. These three genes and lamB are organized
into two divergent operons which may overlap in their promoter regions
[104]. The leftward operon includes malE and malF, and the other is
comprised of malK and lamB. The malE product is a periplasmic maltose
binding protein of 44,000 molecular weight [105]. The malK product is
thought to be the maltose permease associated with the cytoplasmic membrane. The product of malF is unknown, but has been suggested to play a
role in energy coupling for transport [104]. The products of these three
genes comprise one transport system, and all three are necessary for
growth on, and transport of, maltose. Maltose uptake is energy dependent
and allows accumulation against a concentration gradient; transported
maltose is, however, subject to considerable metabolism. This transport
system is induced by growth in the presence of maltose [101]. The induction of this system, and of the other enzymes of catabolism, is under
positive control by the product of the malT gene, located in the malA
region [106,107]. Those mutations in malA that conferred resistance to
phage λ caused alterations in the malT-coded activator, preventing expression of all maltose operons. This demonstrated that expression of the lamB
gene is regulated coordinately with the enzymes for maltose transport.

The lamB product has been identified as the phage receptor. Randall-Hazelbauer and Schwartz [108] demonstrated receptor activity in cell extracts, as assayed by phage inactivation. This phage-receptor interaction was reversible, unless either the complex was treated with chloroform or ethanol or the phage or receptor was derived from variant strains demonstrating an irreversible interaction. Alternatively, DNA ejection occurred following phage adsorption to outer membrane complexed with cytoplasmic membrane [110]. The reversible interaction exhibited a dissociation constant in the range of 5×10^{-12} M [109]. The irreversible inactivation of phage on its interaction with receptor appears to require a conformation different from that present in solubilized wild-type receptor. The receptor activity was localized to the outer membrane by several criteria [108]. It was solubilized in the presence of detergents and partially purified, allowing its identification as a protease-sensitive protein of 55,000 molecular weight. Strains deleted for lamB specifically lack both receptor activity and this protein band on electropherograms of the outer membrane proteins, identifying this lamB product as the λ receptor.

Despite the fact that loss of the lambda receptor did not prevent growth on maltose, there was some evidence that it might be involved in maltose transport. Hazelbauer [111] showed that mutants in lamB were depressed in their chemotactic response to maltose, particularly at low external concentrations. Szmelcman and Hofnung [112] examined the effect of lamB mutations on the utilization and transport of maltose. Mixed cultures of a lamB mutant and its parent were grown in a chemostat with limiting carbon source. With maltose as carbon source, the lamB strain was at a severe disadvantage, and nearly 100% of the population was λ sensitive within four generations. In contrast, there was no discrimination with lactose or glycerol as carbon source. Growth of the lam+ strain was slightly favored with glucose as carbon source; it was suggested that this might be due to the failure of the lamB strain to utilize efficiently the maltose or maltodextrins which could be formed from glucose and released into the medium.

Although the results suggested that lamB strains might be defective in maltose transport, no defect was observed in the usual uptake assay with substrate concentrations in the range of 1 mM. Using fully induced wild-type and lamB strains, Szmelcman and Hofnung [112] measured [^{14}C]maltose uptake as a function of substrate concentration. At low external levels (3.5 μM), a number of strains with lamB nonsense mutations never exhibited more than 5% of the parental rate of uptake. Strains with lamB missense mutations exhibited from 6 to 91% of the wild-type rate. In all cases, the uptake of thiomethyl-β-D-galactoside was unaffected. Raising the external maltose concentration from 3.5 μM to 1 mM progressively eliminated the differences in uptake rates between the parent and mutant strains. Incubation of lambda-sensitive cells for 1 hr at 25°C with antiserum prepared against the partially purified receptor reduced the initial rate of uptake of 3.5 μM maltose to less than 20% of that observed in the presence of control serum. Other transport systems were unaffected by this treatment. These authors concluded that the lamB product is a component of the specific maltose transport system but is required only at limiting substrate concentrations. This conclusion was supported by Hazelbauer's [113] finding that the lamB receptor was not directly involved in maltose chemotaxis, but was only required at low maltose concentrations to facilitate access of maltose to its chemoreceptor in the periplasm. He had shown earlier that the periplasmic maltose binding protein served as the maltose chemoreceptor [111].

The role of the λ receptor in maltose transport and its relationship to phage adsorption are still unsettled. There is no detectable binding of maltose or any other maltodextrin to this receptor, either in vivo or in vitro [114]. These transport substrates provide no protection for sensitive cells against the lethal adsorption of λ phage.

Maltose transport differs from most other sugar transport systems in its absolute requirement for the periplasmic maltose binding protein [105]. Szmelcman et al. [114] determined a K_m of 1 μM for maltose in a wild-type strain. From fluorescence quenching experiments, a K_D of 1 μM was determined for the binding of maltose to the purified binding protein. The rate of maltose uptake conformed to a hyperbolic relationship with respect to the external maltose concentration. It was concluded from these results that the maltose binding protein is the sole recognition site of the maltose transport system and that the outer membrane in a lam^+ strain does not represent a significant permeability barrier to the access of maltose to its binding protein. However, in lamB mutants, the apparent K_m for maltose was more than 100-fold higher, although the V_{max} was unaltered. Thus, diffusion across the outer membrane is rate limiting at low substrate concentrations, and the lamB product allows free passage through this barrier.

An even more striking result was seen with maltotriose, the next higher maltodextrin. The K_m for maltotriose transport in a wild-type strain (2 μM) was 13-fold greater than the K_D for its binding to the maltose binding protein (0.16 μM) [114], suggesting that, even in the presence of functional lamB product, this trisaccharide is not freely accessible to the periplasm. Mutants in lamB showed almost no transport of maltotriose, and such mutants utilized even elevated concentrations of this sugar poorly. This result indicates that a functional lamB product is much more critical for the utilization of maltotriose and possibly higher maltodextrins than it is for that of maltose. These are carbon sources enteric bacteria are likely to encounter frequently in their usual environment. It was suggested that the maltose transport system has evolved to accommodate these higher maltodextrins, since they bind to the maltose binding protein with higher affinity than does maltose itself.

Why should the lambda receptor be required for the diffusion of low concentrations of maltose across the outer membrane, inasmuch as the disaccharide should easily diffuse through the hydrophilic pores? One possible hypothesis is that, since the recognition site for maltose transport and chemotaxis is on a protein located not at the cytoplasmic membrane but rather somewhere in the periplasm, the role of the receptor might be to position the binding protein so that its recognition site is oriented for the most efficient binding of incoming substrate [112]. Perhaps maltose diffusing through the outer membrane pores would have to equilibrate over the entire volume of the periplasm before interacting with the binding protein, whereas substrate translocated across the outer membrane by the receptor could be transmitted directly to the binding protein with little dilution.

The distribution of the lambda receptor among the enteric bacteria was studied by Schwartz and LeMinor [115]. All wild-type strains of E. coli, all of which were Mal$^+$, possessed receptor activity as determined by assays of detergent-solubilized extracts, yet most of these strains failed to support growth of the phage. In some cases, receptor activity was masked in the intact cell and could not even be detected with antiserum unless the outer membrane had been solubilized. In other cases, adsorption to intact cells occurred, but phage development was blocked at some subsequent stage. Among strains of Shigella, most Mal$^+$ strains possessed receptor activity; many of the strains adsorbed the phage and some even yielded plaques. Most of the Mal$^-$ isolates lacked receptor activity. No receptor activity was detected in extracts prepared from several strains of Levinea (Citrobacter) or Salmonella, although all of these strains were Mal$^+$. It would be interesting to see if these strains possessed a maltose-inducible outer membrane protein which could cross-react immunologically with the lambda receptor from E. coli. The masking of phage receptors, usually by LPS, is advantageous to the cell since the receptor is no longer accessible to the lethal phage but presumably can still function in substrate transport.

In summary, the lambda receptor appears to be involved in the translocation of maltose and maltodextrins across the outer membrane to the periplasmic maltose binding protein. The importance of this function increases with increasing length of the substrate. There appears to be no overlap between the substrate binding site and the phage binding site on the receptor. Perhaps now these possibly distinct binding sites can be detected by genetic means. Along this line, a mutant of E. coli in which the adsorption of lambda phage is temperature sensitive has been isolated [116]; it would be of interest to determine whether transport of low concentrations of maltose is also temperature sensitive. Studies on the number of these receptors and on their site of export to the outer membrane will be presented in Sec. VI.

IV. ROLE OF THE OUTER
    MEMBRANE IN IRON TRANSPORT

A. Iron Transport Systems in the
   Gram-negative Bacteria

Iron is an essential trace element for almost all living organisms, but the availability of iron in natural conditions is extremely low. At pH 7, the solubility of ferric ion in aqueous solution is in the range of 10^{-16} M. To satisfy the requirement for iron in their essential metabolic processes, microorganisms have developed means for the solubilization of iron from polynuclear complexes and for the efficient transport and accumulation of this iron. These systems utilize low molecular weight ferric ion chelators, generally termed siderophores, whose structures, biosynthesis, and general properties have been described elsewhere [117,118]. Catechols and secondary hydroxamates are the usual chemical moieties represented among these chelators.

Under conditions of iron limitation, the enteric bacteria (E. coli, S. typhimurium and Enterobacter aerogenes) secrete into their culture medium a chelator known as enterochelin [119] or enterobactin [120]. The provision of iron for cell metabolism clearly involves this compound, and it serves to solubilize ferric iron, to compete with other chelators for iron, and to mediate, in part, its transport into the cell. As a fascinating example of evolutionary interactions, these bacteria are also able to utilize iron complexed to other ligands, including chelators of different chemical nature produced by other species of bacteria and fungi. For example, both E. coli and S. typhimurium possess specific transport systems for a variety of fungal hydroxamate siderophores, and some strains of E. coli have an inducible transport system for ferric citrate. These strains also exhibit a low-affinity iron uptake system , probably independent of chelators, whose existence is based on the growth responses of mutant strains to ferric and ferrous salts. Details of these systems are provided in several recent reviews on the topic of iron transport in bacteria [121-123].

The molecular weights of several siderophores approach the exclusion threshold of the outer membrane of the gram-negative bacteria [63]. This portion of the chapter is concerned with the current evidence for the involvement of outer membrane components in iron transport. The first evidence for this involvement came from observations that mutants selected for resistance or tolerance to certain phages or colicins were defective in iron transport. However, only very recently has this analysis allowed the identification of specific gene products as outer membrane components and the assignment to them of function in iron metabolism. The following section is concerned with the relevant properties of the individual iron chelate transport systems, and subsequent sections will describe inter-

actions among these components. As an aid in following the genetic terminology, Table 9-1 provides a listing of loci implicated in iron transport
in E. coli K12.

B. Individual Iron Chelate Transport Systems

1. Ferrichrome System

The fact that tonA mutants of E. coli are resistant to phages T1, T5, and
φ80 and to colicin M [94] led Fredericq and Smarda [124] to conclude that
these four entities share a common surface receptor. This was verified
by the purification by Braun et al. [125] of an 85,000 molecular weight
polypeptide which could inactivate both phage T5 and colicin M. Hancock
and Braun [171] demonstrated that this protein could reversibly adsorb
phages T1 and φ80. That this receptor protein is a component of the outer
membrane was shown by the separation of the two membranes by the method
of Osborn et al. [14,125].
     The involvement of this tonA product in iron transport was first sug-
gested by Wayne and Neilands [126], and was based on the resistance of
tonA mutants to the siderophore-like antibiotic, albomycin [163]. Luckey
and coworkers [127] had isolated a number of albomycin-resistant mutants
of S. typhimurium (termed sid) which were defective in iron transport
mediated by a number of different hydroxamate-type siderophores, including the fungal products, ferrichrome and rhodotorulic acid. Many of these
sid mutations were mapped at a location on the S. typhimurium genetic map
corresponding to the tonA region on the E. coli map (near panC). Wayne
and Neilands [128] then demonstrated that ferrichrome specifically protected E. coli ton$^+$ cells from killing by phages φ80, φ80vir, and λh80, but
did not protect against λ-clear or a hybrid of φ80 with the host range of λ.
This protection resulted from competition between φ80 phage particles and
ferrichrome for a site on the cell surface. No protection occurred with
EDTA, citrate, or a number of other siderophores, including enterochelin
and rhodotorulic acid. These authors suggested that the tonA product
functions as the outer membrane component of the specific ferric-ferrichrome uptake system.
     Hantke and Braun [129] also noted the albomycin resistance of tonA
mutants and demonstrated that ferrichrome specifically protected sensitive
cells against the lethal adsorption of colicin M. They showed that ferrichrome-mediated iron uptake was totally lacking in tonA mutants, whereas
ferric-citrate and ferric-enterochelin uptake was unaffected. Further
support for the common receptor site hypothesis came from the demonstration by Luckey et al. [130] that ferrichrome and phage T5 compete for
binding to a partially purified receptor complex. However, they did not

TABLE 9-1   Genetic Loci in E. coli K12 Linked to Iron Transport

Locus	Map[a] Position	Ref.	Comments
entA, B, C	13	141,144,149	Enterochelin biosynthesis; enzymes which convert chorismate to DHBA; will respond to exogenously supplied enterochelin or DHBA in the culture medium
entD, E, F, G	13	140,143,144, 149	Enterochelin biosynthesis; enzymes which convert DHBA to enterochelin; will respond to exogenously supplied entero-chelin in the culture medium
fesB	13	147,148	B component of enzyme enterochelin esterase; accumulate enterochelin but cannot hydrolyze it to DHBS and thus cannot use ferric-enterochelin as an iron source; specifically blocked in utilization of ferric-enterochelin
fep	13	149	Ferric-enterochelin premease; hyperexcrete enterochelin; specifically blocked in transport of enterochelin
tonA	3	94,126,128-130	Resistance to bacteriophages T1, T5, and φ80, colicin M and albomycin; specifically blocked in binding and transport of iron-ferrichrome
tonB(exbA)	27	94,129,134,149-149-151,153-158	Resistance to bacteriophages T1 and φ80, and albomycin; tolerance to colicins B, D, G, H, Ia, Ib, M, Q, S1, and V; chromium sensitive; hyperexcrete enterochelin; blocked in transport of all iron chelates
exbB	64	94,153-156	Tolerance to colicins B, D, G, H, Ia, Ib, M, Q, S1, and V; hyperexcrete enterochelin; methionine auxotropy; partially defective in transport of ferric-enterochelin

Gene	Map position	References	Properties
exbC	?	94, 156	Tolerance to colicins B, D, G, H, and M; hyperexcrete enterochelin; partially defective in transport of ferric-enterochelin
cbt	13	94, 156	Tolerance to colicins B and D; chromium sensitive; hyper-excrete enterochelin; specifically defective in binding and transport of ferric-enterochelin; identical to cbr
cbr	?	159	Resistance to colicins B and D; specifically defective in ferric-enterochelin binding and transport
cir	43	94, 136, 148, 164, 165, 167	Resistance to colicins Ia, Ib, and S1; tolerance to colicins Q and V; implied as having a role in iron transport (see text); missing 74K protein
feuA	?	136, 160–162	Resistance to colicins Ia, and Ib, specifically defective in ferric-enterochelin transport; missing 74K protein
feuB	73	136, 160, 161	Resistance to colicin B; specifically defective in ferric-enterochelin transport; missing 81K protein

[a] According to the recalibrated linkage map of Bachmann et al. [98].

measure ferrichrome binding to the receptor and could not rule out the involvement of more than one protein in this receptor activity. The presence of an analogous activity in S. typhimurium is indicated by the observation by Luckey and Neilands [131,132] that ferrichrome specifically protected sensitive cells against phage ES18, and that phage ES18-resistant mutants are also albomycin resistant and utilize ferrichrome poorly.

2. Citrate

The growth properties of strains unable to synthesize enterochelin indicated that a citrate-mediated iron transport system is present in E. coli K12 [133] and B/r [134] and in E. aerogenes [135], but not in E. coli W or S. typhimurium [133]. Iron uptake in Fep⁻ or Ent⁻ mutants of E. coli K12 was markedly stimulated by citrate. This system was not present unless the culture had been grown, at least briefly, in the presence of citrate (1 mM); both induction and uptake were quite specific for citrate [133]. The initial rates of iron uptake mediated by a mixture of citrate and enterochelin were additive relative to the rates with either ligand alone.

    This citrate-mediated uptake system is not present in tonB mutants [134] and is also specifically lacking in certain as yet uncharacterized mutants [133].

    A protein has been detected in the outer membrane of cells grown under conditions of iron limitation in the presence of 1 mM citrate [136]. This protein migrates on electrophoresis as if 81,000 in molecular weight and is very near the position of another 81,000 protein to be described later. It is most easily detected in feuB mutants lacking this other protein. However, the identification of this protein as a component of the inducible ferric citrate transport system must still be considered tentative.

3. Rhodotorulic Acid and Other
    Hydroxamate Siderophores

By using strains unable to synthesize enterochelin, Luckey et al. [127] first showed that S. typhimurium was able to utilize iron complexed to a wide variety of hydroxamate siderophores. Many phenotypic classes of albomycin-resistant derivatives lacked the ability to respond to certain of these chelators. The lesion in many of these mutants was linked by transduction to panC, in the region analogous to tonA in E. coli. Frost and Rosenberg [137] showed that E. coli also could respond to one of these siderophores, rhodotorulic acid. In accordance with the growth responses, this chelator stimulated iron uptake. This activity was present in tonA, fep, exb, and cbt mutants, but was missing in tonB mutants [158].

Although E. coli and S. typhimurium produce only enterochelin, E. aerogenes can produce both enterochelin and a dihydroxamate, aerobactin [138]. The presence in all of these strains of hydroxamate-siderophore uptake mechanisms might indicate that such strains formerly possessed the ability to synthesize this type of compound. The retention of this transport activity would provide quite a marked selective advantage for such strains in competition with other species under conditions of iron limitation. The relationships of the binding sites for the various hydroxamate siderophores and the genetic basis for the various phenotypes among albomycin-resistant mutants are so far unknown but of great interest.

4. Enterochelin System

(a) Enterochelin Synthesis and Metabolism. The ferric iron chelator of the enteric bacteria is a catechol compound called enterochelin [119] or enterobactin [120]. It is a cyclic trimer of 2,3-dihydroxy-N-benzoyl-L-serine units in ester linkage [119] and forms a 1:1 complex with ferric ion with a binding constant in the range of 10^{30} [139]. The iron can be released from this ligand either by low pH or by hydrolytic cleavage of the ester bonds.

Six steps in the biosynthesis of enterochelin from chorismate and serine have been identified by genetic and biochemical means [140-143]. Mutants unable to synthesize enterochelin grow poorly on low-iron media with glucose as carbon source, and imperceptibly on that with succinate [146]. Normal growth properties are restored by supplementation with enterochelin (1 μM) or its linear trimer or dimer or, in strains blocked before its synthesis, with 2,3-dihydroxybenzoate [135]. As will be shown later, these strains also respond to other chelators. Growth is normal in media with excess iron (50 to 200 μM). Enterochelin synthesis in normal cells is repressed by growth in excess iron, with repression occurring at intracellular iron levels above 2×10^{-18} mol/cell [145].

There is clear evidence for the uptake of ferric-enterochelin. Iron uptake is markedly stimulated by enterochelin; it is even dependent on enterochelin when assayed in the presence of nontransported iron chelators (such as nitrilotriacetate) to eliminate low-affinity uptake [133]. Frost and Rosenberg [133] demonstrated the uptake of label from both [^{55}Fe]- and [^{14}C]enterochelin. With a strain blocked in enterochelin degradation, there was comparable uptake of both moieties. The parental strain capable of enterochelin degradation showed minimal accumulation of the enterochelin label, although the iron was accumulated normally. In this case, there was the rapid appearance of the degradation product, dihydroxybenzoylserine, in the medium. The mutant strain was lacking the activity of a ferric-

enterochelin esterase (fesB), which thus appears to be necessary for the
release of transported iron [147,148]. A number of mutants (termed fep)
which are specifically defective in the uptake of ferric-enterochelin have
been isolated.

The fep locus is closely linked to the six ent genes and to the fesB
locus [147,149].

(b) The Involvement of Outer Membrane Receptors. The existence of
other mutants defective in ferric-enterochelin uptake led to the implication
of outer membrane components in this transport process. Some of the
mutants resistant to phages T1 and ϕ80 and insensitive to several of the
Group B colicins carry lesions in tonB [94]. Wang and Newton [150,151]
first reported that E. coli strains carrying point mutations or deletions at
tonB required fivefold higher concentrations of iron for normal growth.
These strains were sensitive to chromium, apparently because its presence
decreases the solubility of free iron. Synthesis of enterochelin was elevated and was not subject to repression by iron supplementation [153].
This, combined with the inability of these mutants to respond to citrate,
suggested a role for the tonB product in the transport of iron chelates.
Further characterization of the role and nature of the tonB product will be
presented later. The important point here is that mutations affecting phage
adsorption also altered iron utilization.

Guterman and Luria [152] observed that a number of mutants of E. coli
selected for insensitivity to the Group B colicin B excreted a colicin inhibitor into the medium. Guterman [153] separated these mutants into two
classes, exbA and exbB. The ExbA⁻ phenotype was identical to that of
TonB⁻, in resistance to phages T1 and ϕ80, sensitivity to chromium,
increased iron requirement, excretion of enterochelin, and genetic linkage
to trp. In contrast, the exbB mutants were partial methionine auxotrophs,
sensitive to phages T1 and ϕ80, resistant to chromium, and altered in a
region of the chromosome near metC, distant from tonB [154]. Mutants in
this class also hyperexcreted enterochelin. No defect in iron transport was
detected, but subsequent workers have noted a partial loss of ferric-enterochelin uptake [156]. The inhibitor of colicin B which is excreted by these
strains was identified by a number of criteria as enterochelin [155]. This
inhibition of colicin did not result from destruction of the colicin, but
rather from interference with its adsorption to sensitive cells.

This protective effect of enterochelin against colicin B led Guterman
[153] to suggest that these two compounds competed for a common site on
the cell surface. Utilizing an entA strain of E. coli blocked in enterochelin
synthesis, Wayne et al. [157] documented the specific competition between
enterochelin and colicin B for binding to a surface receptor. Davies and
Reeves [94] and Pugsley and Reeves [156] observed that enterochelin protected sensitive cells against both colicins B and D. They isolated two
classes of colicin B-insensitive mutants in addition to the previously

described tonB and exbB classes. These new mutants, termed exbC and cbt (for colicin B tolerance), also hyperexcreted enterochelin, further supporting the existence of a site on the surface responsible both for the lethal adsorption of colicins B and D and for the binding of enterochelin as part of its transport.

The assumption that resistance to these colicins resulted directly from the hyperexcretion of enterochelin was dispelled by the demonstration by Frost and Rosenberg [158] that tonB mutants, which were also blocked in enterochelin synthesis, were still insensitive to killing by colicin B. Subsequently, Pugsley and Reeves [156] showed that aroE derivatives (unable to synthesize enterochelin) of tonB, exbB, exbC, and cbt strains were still tolerant to colicins B and D and sensitive to chromium. All four of these mutant strains were defective in enterochelin-mediated iron transport but, except for the tonB strain, had normal or near-normal iron uptake mediated by citrate, ferrichrome, or rhodotorulic acid, as well as normal activity of the low-affinity system. Thus, the hyperexcretion of enterochelin by the parental (aroB$^+$) colicin B-insensitive strains is most likely the consequence of the derepression of the enterochelin biosynthetic pathway by the low intracellular iron concentration resulting from the inability to utilize the extracellular enterochelin and the absence of other chelators.

(c) Identification of the Ferric-Enterochelin Receptor. Pugsley and Reeves [156] determined the cellular binding of [^{55}Fe^{3+}]enterochelin by the aroE colicin B-tolerant strains and an aroE colicin-sensitive strain. Whole cells of all strains grown in iron-rich medium bound equal amounts of ferric-enterochelin. Growth with limiting iron produced a sevenfold increase in binding activity in the colicin-sensitive strain, but no increase in whole cell binding activity in any of the tolerant strains. In contrast to these results with whole cells, the binding activity in isolated outer membrane preparations from all strains except the aroE, cbt double mutant was higher when the membranes were obtained from cells grown with limiting iron. This evidence indicated that the basal level of ferric-enterochelin binding observed in all these strains represented nonspecific binding to cell wall components. Resolution of the outer membrane proteins from the cells grown with limiting or excess iron revealed the presence of two high molecular weight proteins which were present in all iron-starved aroE strains but which were not present in aroE strains grown in iron-rich medium, nor were they present in the aro$^+$ colicin-sensitive strain grown in either limiting or excess iron. It was suggested that these proteins were specifically involved in iron transport.

It was later found that the outer membranes of all of the colicin B-tolerant strains, including the cbt strain, when grown under iron-limiting conditions, had more colicin B and colicin D neutralizing activity than did outer membrane fractions from the same strains grown in iron-rich medium [159]. Again, this same result was obtained with the aroE derivatives of these strains, ruling out the possibility that the colicin neutralizing activity was due to enterochelin in the preparations. Pugsley and Reeves suggested that one of the high molecular weight proteins appearing in the outer membranes of iron-starved colicin B-tolerant strains was a component of the receptor complex for colicins B and D and, in addition, for ferric-enterochelin; the available evidence strongly suggested that these colicins and ferric-enterochelin share a common receptor.

More recently, Pugsley and Reeves [159] have reported the partial purification of these high molecular weight protein species, which they have termed proteins P and Q, in order of decreasing molecular weight. From the ability of these proteins to neutralize colicin killing in vitro, they have identified protein P as being the receptor for colicins B and D. The ability of this isolated protein to bind ferric-enterochelin was not determined. [These authors also identified protein Q as being the receptor for colicin M and phages T1, T5, and ⌀80. However, this result does not agree with the results of Braun et al. [160] or Luckey et al. [130] (See below).] Pugsley and Reeves also described the isolation of a new class of mutants which they called cbr (for colicin B resistance). These mutants were phenotypically similar to the cbt mutants they had previously isolated, except that they totally lacked receptor activity for colicins B and D. The cbr mutants were specifically defective in enterochelin-mediated iron transport, while retaining functional transport activity for other iron chelates. However, none of 27 independently isolated cbr mutants was missing protein P.

Braun and coworkers [160] noted that a tonB strain grown under iron limitation overproduced three high molecular weight membrane proteins, designated 74K, 81K, and 83K according to their apparent molecular weights from polyacrylamide gel electropherograms. In contrast to the previous results, the level of these three proteins was depressed in cells grown in iron-rich media. The tonB strain, when grown in limiting iron, exhibited elevated binding capacity for the colicins B, Ia, and Ib. From a strain blocked in enterochelin synthesis (aroB), Hantke and Braun [161] had previously isolated a number of mutants insensitive to colicins B, Ia, Ib, or V. Several of these isolates fell into two classes which had not previously been described. Since these mutants were also specifically defective in ferric-enterochelin uptake, the mutant classes were termed feuA (ferric-enterochelin uptake) and feuB. Even when grown in limiting iron, strains mutant at feuA were resistant to colicins Ia and Ib and were missing the 74K outer membrane protein [162]. The possible involvement of the colicin I receptor in iron transport is discussed later.

Mutants in the second class, feuB, did not adsorb colicin B and were specifically missing the 81K outer membrane protein, suggesting its role as the colicin B receptor. This protein is thus a leading candidate for the role of receptor for colicin B and ferric-enterochelin [136]; however, the binding of enterochelin to the mutant cells and the binding of colicin B to isolated outer membranes was not measured.

As yet, no function has been assigned to the 83K protein, which is present in the outer membrane of all strains examined following growth on limiting iron. According to Braun et al. [160], it is not the tonA product, which has been identified as a 78,000 protein. Unlike the 83K protein, the level of this 78,000 protein does not vary with the level of iron supplementation, supporting the earlier observation of Luckey et al. [130] that the level of phage T5 receptor activity was unaffected by the level of iron supplementation.

(d) Possible Involvement of the Colicin I Receptor. The two Group B colicins Ia and Ib share a common receptor in the outer membrane of E. coli [164]. Receptor activity is lost in strains mutant in the cir (colicin I receptor) locus, and such strains are resistant to colicins Ia, Ib, and S1 and are tolerant to colicins Q and V [94,165]. The level of this receptor, assayed by the binding of [^{125}I]colicin Ia, is affected by the cellular availability of iron. For example, mutants conditionally blocked in heme biosynthesis (hemA) suffer an 80% reduction in colicin I receptor activity; this level returns to normal with restoration of heme biosynthesis by supplementation with α-aminolevulinic acid [166]. Furthermore, Konisky et al. [167] found an inverse relationship between the number of colicin receptors and the cellular utilization of exogenous iron. Mutants blocked in ferric-enterochelin transport or utilization (ent, fesB, or fep) had a greater number of receptors than the parental strain. The addition of citrate, ferrichrome, or 100 μM ferrous salts to the growth medium of each of these mutants, or of enterochelin to that of the ent mutant, significantly reduced the number of receptors and thereby the sensitivity of these strains to colicin I. Only the addition of ferrous salts reduced the receptor level in a tonB strain. It thus appears that colicin I receptor level is derepressed under conditions of iron limitation.

As mentioned previously, Braun and coworkers [160-162] isolated feuA mutants specifically resistant to colicins Ia and Ib. These mutants lacked the 74K protein which was normally induced under iron-limiting conditions, indicating that this protein is a component of the colicin I receptor. These feuA mutants are also defective in ferric-enterochelin uptake, suggesting that the colicin I receptor, as well as the colicin B receptor, might be involved in enterochelin uptake. However, the genetic location of feuA appears to be different from that of cir, which also codes for resistance to the I colicins. Nonetheless, a cir mutant also lacks the 74K protein [136]. Furthermore, Wayne and Neilands [168] found that strains of E. coli lacking colicin I receptor activity had apparently normal iron transport properties and that enterochelin did not protect sensitive cells against the lethal adsorption of the I colicins. Thus it could be, as Konisky et al. [167] suggested, that the colicin I receptor is involved in some other aspect of iron metabolism or heme synthesis and that the defect in iron metabolism manifested in feuA mutants might represent pleiotropic alterations.

It is impossible at this time to identify the genetic locus coding for the colicin B-ferric enterochelin receptor protein in the outer membrane. Part of the confusion stems from the use of different strains and experimental conditions in the various labs. However, three separate groups have reported the isolation of mutants lacking colicin B- and enterochelin-receptor activity, and perhaps shortly a gene assignment can be made. Several points can be concluded at this time. There are a number of high molecular weight outer membrane proteins whose level is responsive to the state of iron sufficiency of the cell. Controlled in like manner is a receptor site responsible for the binding of the colicins B and D and of ferric-enterochelin. Binding to this receptor appears to be the essential first step in the uptake of these molecules.

C. Interactions Among Siderophore Uptake Systems

These occasionally contradictory conclusions and the apparent surplus of protein candidates for specific iron transport systems, coupled with the existence of quite pleiotropic mutations, such as tonB, point to the operation of complex interactions of the outer membrane receptors among each other and with the steps of uptake subsequent to binding. The nature of the intimate relationship between the group B colicins and iron metabolism is only beginning to surface. The level of most of these receptors, with the apparent exception of the 78,000 tonA product, responds to the level of iron supplementation. Of greater relevance is the finding by Wayne et al. [157] that all siderophores capable of supporting growth of a strain on iron-limiting medium protected that strain against killing by the Group B colicins B, Ia and V. For example, ferrichrome protected sensitive cells against colicins B, Ia, V, and M; SidA$^-$ strains unable to utilize ferrichrome exhibited only the protection by ferrichrome against colicin M owing to the specific competition for receptor. Identical responses were seen with enterochelin, which provided protection against all of the Group B colicins with the exception that, for strains defective in enterochelin utilization (fesB), only the specific protection against colicin B was obtained. This general protection was not a direct result of the failure of the colicins to adsorb to the protected cell, although this could be mediated through repression of receptor synthesis.

This type of response might be analogous to the effect of tonB mutations. As mentioned, such mutants are found among those resistant to phages T1 and ϕ80 and are defective in enterochelin-mediated iron uptake [150]. Hantke and Braun [129] showed such strains to be albomycin resistant and defective in ferric-ferrichrome uptake. These mutants are, in fact, missing all chelate-mediated iron uptake systems [156,158]. Measurement of the kinetics of $^{59}FeCl_3$ accumulation revealed only a 10-fold increase in the

K_m for iron (from 0.36 to 3.6 μM) in a tonB mutant relative to the parental strain [134]. The V_{max} remained the same. The implications of these results will be mentioned later.

The resistance of tonB mutants to phages T1 and ϕ80 is not the result of loss of the receptor, the tonA product. Both phages do adsorb to tonB cells, but do so reversibly [169,170]. Furthermore, phage T5, which shares the same tonA-coded receptor, is independent of tonB function. The tonB$^+$ product is necessary for the irreversible adsorption of phages T1 and ϕ80, an energy-dependent process utilizing the proton motive gradient [171]. The loss of the capacity for irreversible adsorption of these phages following cell disruption has prevented the localization of the tonB function [172]. These mutants are also tolerant to all of the Group B colicins [94]. Colicin receptor activity is still present and is usually in elevated amounts as a result of the iron limitation. The tonB mutation could result in a major alteration in membrane organization, thereby disrupting the normal communication between the receptors and the subsequent step of uptake. However, this mutation has no effect on amino acid or phosphate transport [158] and does not result in the obvious surface defects (sensitivity to dyes, drugs, and detergents) often associated with tolerance to the Group A colicins (such as tolA mutants). Some recent evidence concerning tonB function will be presented in Sec. VI.A.

One final point for consideration here concerns the metabolism of iron chelates subsequent to their passage through the outer membrane. The identification of the one gene proposed to code for the ferric-enterochelin permease in the cytoplasmic membrane (fep) is by no means conclusive [149]. Frost and Rosenberg [158] stated that all of the iron chelate transport systems involve energy-dependent, membrane-associated permeases, but these may be quite complex. From studies of the uptake of labeled iron and labeled ferrichrome, Leong and Neilands [173] proposed the operation of at least three mechanisms for ferrichrome-mediated transport. In the more rapid processes, either the iron is reduced on the cytoplasmic membrane without the chelator entering the cell, or the siderophore-iron complex is transported, with the siderophore being immediately expelled. In the third mechanism, the iron-siderophore complex is accumulated within the cell, with subsequent release and reduction of the iron. The first two mechanisms are probably not operative with enterochelin, which requires esterase cleavage for reduction [147] and release [148] of the iron.

In conclusion, the process of iron uptake in these bacteria is of awesome complexity, involving many constituents of the outer membrane whose precise roles and relationships are as yet undeciphered. The necessity of iron for the cell is unquestioned. The need for mediated transport across the outer membrane is less clear. The usual siderophores are of a size such that they are subject to at least partial exclusion by the diffusion pores through the outer membrane, but the complete lack of transport in many of these mutants would imply that there is essentially no nonmediated diffusion of iron chelates across the outer membrane. Perhaps the components in the outer membrane play a further role in the uptake of iron beyond that of translocating the iron-chelate complex across the outer membrane.

V. VITAMIN B_{12} UPTAKE

A. General Characteristics of Uptake

The utilization of cyanocobalamin (B_{12}) by enteric bacteria represents another interesting evolutionary situation in that this vitamin is neither synthesized nor required by these strains. Two B_{12}-dependent reactions have been recognized in E. coli and S. typhimurium. One reaction is the B_{12}-dependent homocysteine transmethylase (coded by metH) which catalyzes the terminal step of methionine biosynthesis. This reaction can be bypassed by the B_{12}-independent transmethylase (coded by metE); this enzyme is specifically repressed by growth in the presence of B_{12} [174]. Mutants in metE require either methionine or B_{12} [175] and can respond to B_{12} at concentrations as low as 10^{-12} M; only about 20 molecules per cell are sufficient to support normal growth [176]. The other B_{12}-dependent reaction is an ethanolamine deaminase, whose activity is required for the utilization of ethanolamine as nitrogen source; half-maximal growth with ethanolamine requires 0.75 nM B_{12} [177].

The uptake of B_{12} in E. coli was first described by Oginsky [178] and has recently been studied in detail by Bradbeer and his colleagues and by Taylor and coworkers. Many of their results will be presented in Chap. 10 since the major consideration here is the role of the outer membrane in B_{12} uptake. Both groups found that B_{12} uptake was specific and capable of accumulation of substrate with high affinity and against a concentration gradient [179-181]. Osmotic shock both reduced the rate of uptake and caused the release of B_{12} binding material; however, the significance of this released material in the transport process has not been established [182,183].

The time course of uptake revealed a biphasic process, with a very rapid, energy-independent initial phase (occurring at 0°C or in the presence of energy poisons) [179,180]. This was followed by a slower, energy-dependent phase which could proceed for 30 min or more. It is apparently the proton motive force that provides the energy for B_{12} accumulation [180]. Taylor et al. [181] determined a K_m of 10 nM for this concentrative phase, but disregarded the initial phase as nonspecific binding. However, present evidence supports the contention of DiGirolamo and Bradbeer [179] that this initial phase is a specific aspect of the transport process. At the time, this was based on the high affinity of the initial phase (K_D of 4 nM), the similar substrate specificities of binding and uptake, and the finding that no mutant lacking the initial phase exhibited normal secondary uptake [176].

Mutants defective in B_{12} uptake were isolated in a metE strain by selection for the inability to utilize B_{12} efficiently as a methionine source [176]. Some of the mutants were altered in B_{12} uptake, and these fell into two classes. Some (termed BtuA⁻) had normal levels of binding, but low or negligible secondary uptake activity. The second class (termed BtuB⁻) showed essentially no initial phase and very small levels of secondary uptake. Both classes of mutants were unable to respond to B_{12} at concentrations below 10^{-6} M. It was subsequently shown that the BtuB⁻ phenotype was the product of two mutations [184], such that these strains were both altered in transport and deficient in the B_{12}-dependent homocysteine transmethylase (metH). Recombinants that were metH⁺ btuB were still totally deficient in B_{12} uptake but were able to utilize it for growth, although with a 5- to 10-fold decrease in efficiency. This points out an unusual feature of this system: because growth responses can be obtained at concentrations far below that which can be detected with labeled B_{12}, the utilization of B_{12} as methionine source for growth is a far more sensitive assay of its uptake than is the accumulation of isotopically labeled B_{12}.

B. Role of the Outer Membrane

The involvement of outer membrane components in this transport process was first detected by biochemical techniques. The majority of the B_{12} binding activity released by osmotic shock was excluded from Sephadex columns and was associated with membranous material [183]. Separation of the outer and cytoplasmic membranes by sucrose gradient centrifugation or by differential solubilization revealed that almost all of the cellular B_{12} binding activity was associated with the outer membrane [179, 180, 183]. There were about 200 binding sites per cell, with K_D in the range of 0.8 nM. There was no detectable B_{12} binding activity associated with the cytoplasmic membrane, and membrane vesicles do not accumulate B_{12} [179, 180].

Experiments by DiMasi et al. on the effect of the energy-dissipating Group A colicins E1 and K showed that, although both colicins inhibited the secondary phase of B_{12} uptake, colicin E1 also inhibited the initial binding phase [185]. Colicin E3, which uses the same receptor as E1 but is not an energy poison, also inhibited B_{12} binding to cells and, at high enough concentrations, inhibited secondary uptake. Conversely, B_{12} protected sensitive cells against both the lethality and the specific biochemical inhibitions produced by colicins E1 and E3, but not K. Half-maximal protection was obtained with concentrations of B_{12} (1 to 6 nM) in the range of its binding constant for the outer membrane receptor. The BtuA⁻ transport mutants were sensitive to the E colicins, whereas BtuB⁻ mutants lacking the initial phase were resistant (in fact, these mutants are sensitive to elevated levels of these lethal agents, as will be explained below). Mutants selected for resistance to the E colicins or phage BF23, which also shares the receptor, were unable to utilize, transport, or bind B_{12}. The btu loci were very close to bfe, the previously described locus for the receptor for the E colicins and phage BF23 [186,187]. Finally, the colicin E3 receptor had previously been purified from the outer membrane by Sabet and Schnaitman [188]; its molecular weight of 60,000 and presence in about 220 copies per cell agreed well with the values for the B_{12} receptor. Thus, B_{12}, the E colicins, and phage BF23 share the bfe-coded polypeptide as the initial receptor site for their binding to the cell surface.

The relationship of the bfe locus to the BtuA and B phenotypes has recently been clarified [189]. Among mutants selected for resistance to the E colicins or phage BF23 were found mutants with the BtuB⁻ phenotype (very low binding and transport activity, somewhat depressed utilization of B_{12}, and moderate sensitivity to high multiplicities of phage BF23), in addition to those fully resistant to the phage and having severely depressed B_{12} utilization. By direct assay, it was shown that the BtuB⁻-type mutants still retain 0.5 to 2 functional receptors per cell [190]. It is thought that the BtuA⁻ phenotype is also the result of mutation at the bfe locus, based on two lines of evidence. There was no complementation for B_{12} transport or utilization between BtuA⁻, BtuB⁻, or bfe mutant alleles [189], indicating that mutants of all three phenotypes have lesions in the same cistron. Second, it was possible to obtain second-site B_{12}-utilizing revertants of bfe mutants which still lacked receptor activity. These were obtained either directly by selection or indirectly among phage T4-resistant isolates. Many of these had grossly deranged outer membrane function, as judged from their increased sensitivity to drugs and detergents. Presumably, their ability to utilize B_{12} resulted from increased nonspecific permeability of the outer membrane. If BtuA⁻ mutants were altered in the cytoplasmic membrane-associated B_{12} permease, then utilization of B_{12} in these revertants should still be dependent on this BtuA⁺ function. Such was not the case, as these revertants still responded to B_{12} following introduction of BtuA⁻ by transduction. The same class of B_{12} responders with outer membrane defects was found among revertants of the BtuA⁻ phenotype. Thus, it is possible that the BtuA⁻ phenotype results from an alteration in the bfe locus which merely confers an unusual response on the receptor, such that its binding functions are normal, but that it is specifically defective in the coupling to the subsequent step for B_{12} uptake, but not for that of colicin or phage.

Although the appellation bfe has historical precedence, we feel that the term btu should apply to this locus, since it more accurately reflects the physiological role of this gene's product in the transport of B_{12}.

The question of the relationship of the binding sites on this receptor for these disparate ligands is still uncertain. Binding activity for all five ligands (B_{12}, phage BF23, and the three E colicins) can be lost by a single mutation. There are, however, some indications that some binding sites are separable. The receptor purified by Sabet and Schnaitman [188] had full receptor activity against colicins E2 and E3, which have closely related structures, but had little activity towards E1 (approximately 6% of that of the intact cell). Hill and Holland [191] described mutants resistant to colicins E2 and E3 (owing to inability to adsorb these colicins) which were still sensitive to colicin E1. Similarly, mutants have been isolated which are sensitive to the E colicins and phage BF23 but which exhibit no detectable B_{12} binding; the lesion in two of these mutants is cotransducible with argH, as is bfe [189]. Other mutants with different combinations of sensitivity or resistance and B_{12} binding have been obtained, suggesting the presence of multiple binding sites.

On the other hand, the binding of B_{12} to whole cells, outer membrane fragments, and solubilized receptor was competitively inhibited by the E colicins and by phage BF23 [185,190]. Phage binding was subject to half-maximal inhibition by B_{12} concentrations in the range of 0.5 to 2 nM, which is close to the dissociation constant for binding to the transport receptor [190]. Also, tolerant mutants allowed the demonstration of protection by colicin against the phage, although competitive kinetics could not be demonstrated for technical reasons [192].

In conclusion, a receptor in the outer membrane is necessary for the efficient accumulation of B_{12} by normal cells. This requirement for receptor is not obligate, but can be bypassed either by elevated concentrations of substrate or by nonspecific genetic or physiological derangements of the barrier functions of the outer membrane.

C. Role of the tonB Product

Prompted by the loss of a potential candidate for the energy-dependent step of B_{12} uptake (BtuA) and also by the previously described results with the iron uptake systems, the effect of tonB mutations on B_{12} utilization was measured [193]. Strains carrying either point or deletion mutations in tonB were devoid of the secondary phase of B_{12} uptake; binding to the outer membrane receptors was unaltered (Fig. 9-2). In addition, vitamin B_{12} auxotrophs that were also tonB mutants required greatly elevated levels of B_{12} (as much as 10^{-6} M) for normal growth. Other point mutants in tonB were less severely restricted in this function, suggesting that altered activity was still present. The specificity of this mutation was shown by the fact that the uptake of amino acids and maltose was unaltered.

This observation allowed two important conclusions. First, the tonB product is not just a specific component of iron-chelate transport systems, but is also intimately involved in this other transport system which is dependent on outer membrane components. This might suggest the location of this product in association with the outer membrane. Second, the bfe

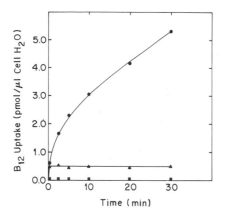

FIG. 9-2   Time course of vitamin B_{12} uptake in a wild-type strain ($\bullet$), and in bfe ($\blacksquare$) and tonB ($\blacktriangle$) mutants derived from it. The external B_{12} concentration was 10 nM. A minor contribution of B_{12} binding to filters in the absence of cells was subtracted.

product can interact independently with the products of both the tolA locus, necessary for the entry of the E colicins [57, 58, 94], and the tonB locus, necessary for B_{12} uptake. Mutations at tolA or any of a number of other loci conferring tolerance to the E colicins and other Group A colicins had no specific effect on B_{12} uptake [194]. Conversely, tonB mutants were fully sensitive to the Group A colicins. The structural and functional characterization of the tolA and tonB products is very rudimentary. It may be that these products are involved in maintenance of complex assemblies of outer membrane proteins so as to allow their proper function as transport systems. Heller and coworkers [195] have suggested that most of the outer membrane proteins are arranged in complex asymmetrical arrays. More information on this topic will be presented in a subsequent section.

Another genetic locus involved in B_{12} uptake has been recently identified [189], but its relationship to the outer membrane is not clear. Vitamin B_{12} auxotrophs carrying mutations in this locus, termed btuC, were defective in their response to B_{12}, but this was noticeable only at concentrations below 10^{-9} M. In transport assays, B_{12} uptake was somewhat depressed relative to that in the parental strain. The most marked characteristic of this mutations was that double mutants in both btuC and bfe or in btuC and tonB were totally unable to respond to B_{12} at any concentration. It has also proved impossible to obtain from these double mutants any responders able to utilize B_{12}, except for bfe$^+$ or ton$^+$ revertants. The genetic location of the btuC locus was near 38 min on the revised linkage map [98], 180° from bfe and clearly separated from tonB. The function of the product of this gene is unknown; it may be involved in a route of B_{12} uptake separate from the normal bfe-tonB system.

VI. FUNCTIONAL PROPERTIES OF THE tonB PRODUCT AND THE POSSIBLE ROLE OF MEMBRANE ADHESION SITES

Some characteristics of the outer membrane associated transport systems described herein, as well as other systems involved in the uptake of bacteriocins and phage nucleic acids, are suggestive of the requirement for a precise orientation of certain membrane components and for the interaction of many of these components. It is possible that the tonB product (and also that of tolA, although even less is known about it) plays a role in the maintenance of this organization. This section presents some recent information on the properties and functions of the tonB product and its interactions, and concludes with a very speculative model for a mechanism of communication between surface receptors and the interior of the cell. Membrane adhesion sites play a prominent role in this model. Their involvement has already been invoked by numerous authors and has recently been reviewed [203].

A. Functional Properties of the tonB and bfe Products

Some insight into the nature of the bfe and tonB products was sought from measurements of their functional stability after the specific inactivation of the expression of their genes. In the first series of experiments, a mutant bfe allele was introduced by conjugation into a bfe$^+$ recipient, and the kinetics of appearance of the mutant phenotype was determined [192]. In the reverse experiment in which the bfe$^+$ allele was introduced into receptorless recipients, it was shown that functional receptor (assayed by the appearance of sensitivity to colicin E3 or phage BF23) appeared soon after mating in essentially all the exconjugants. On the other hand, it was expected that resistance, the mutant phenotype, would appear only after a considerable lag after the entry of the mutant bfe allele. This lag could be as long as eight generation times, in order to allow the approximately 200 resident wild-type receptors to be diluted out through cell division. In fact, colicin E3-resistant recombinants began to appear as soon as the recombinant population initiated exponential growth. Similar results had previously been reported for the receptors for phages T6 [196,197] and T1 [198]. An important finding from this experiment was that the appearance of recombinants insensitive to colicins E2 and E3 always preceded the appearance of insensitivity to phage BF23, which also adsorbs to the bfe receptor. The further appearance of colicin E3-insensitive recombinants was halted immediately upon the inhibition of protein synthesis with spectinomycin, despite the fact that the total number of recombinants increased by about 50%. These results suggested that the bfe-coded receptor could exist on the cell surface in a state nonfunctional for the productive adsorption of colicins.

In order to obtain independent verification that the receptor was still present after gene inactivation, another experimental system was developed. Oeschger and Woods [199] have described a temperature-sensitive suppressor (SupD^{ts}) whose presence allows normal transcription of genes with amber mutations when at the permissive temperature. Shifting to the restrictive temperature (41°C) results in the nonsuppressed, premature termination of transcription of that gene. Expression of other genes is unaffected. Consistent with the results from the previous experiment, shifting a Bfe^{am} SupD^{ts} strain to the nonpermissive temperature resulted in the rapid appearance of insensitivity to colicin E3, followed somewhat later by the acquisition of insensitivity to phage BF23 [201]. Essentially all the cells in the population became resistant to colicin E3 within 60 min after the temperature shift. Strikingly, the total B$_{12}$ transport capacity of the culture remained constant for more than three cell doublings after the shift (Fig. 9-3A). In this and subsequent experiments, controls with strains carrying temperature-independent suppressor alleles showed that the suppressed product itself was not temperature sensitive.

These results showed that, even after further cessation of its synthesis, there was no loss of receptor activity assayed by its role in B$_{12}$ uptake activity. Yet that receptor rapidly lost its ability to transmit bound colicin or phage BF23 to their proper sites for their lethal action. This was interpreted to mean that only newly synthesized bfe product is functional for interaction with the tolA system to allow E colicin uptake. Furthermore, the transition of the receptor from the state functional for colicin uptake to that tolerant to colicin requires protein synthesis.

Even more striking results were obtained when a TonB^{am} SupD^{ts} strain was investigated by this protocol. In this case, the shift to the nonpermissive temperature was followed by the rapid and complete loss of B$_{12}$ uptake capacity (Fig. 9-3B). Vitamin B$_{12}$ binding was unaffected, and the amount of receptor increased after the shift in synchrony with total cell protein. Another tonB-dependent function, sensitivity to colicin D, was also rapidly lost following the temperature shift with this strain.

As a corollary to these experiments, the effect of the inhibition of general protein synthesis was measured. Growth inhibition by any of a large number of antibiotics or antimetabolites led to the rapid and extensive loss of B$_{12}$ uptake activity, with a good correlation between the extent of growth inhibition and the rate of decline of B$_{12}$ uptake capacity [194]. Uptake activity was never completely lost, and B$_{12}$ receptor activity was unaffected. Other tonB-dependent functions which also declined markedly upon inhibition of protein synthesis included sensitivity to the Group B colicins D [200], Ia, and B [202]. Sensitivity to the Group A colicins E1 and K was the same in

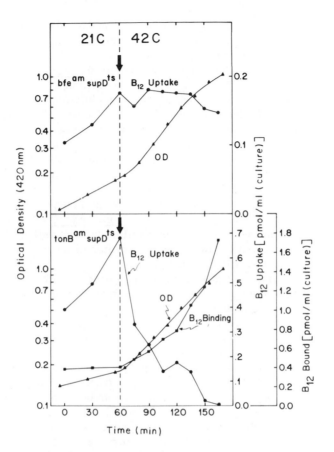

FIG. 9-3 This figure presents the effect of the cessation of synthesis of the bfe and tonB products on the uptake of vitamin B_{12} in E. coli. The strains carry the allele of supD such that suppression is temperature sensitive in these strains. The strain represented in the top panel carries an amber mutation in bfe; that in the lower panel has an amber mutation in tonB. The strains were grown initially at 21°C, and at the time indicated by the arrow were shifted to 42°C. Represented on the graph are the responses of the optical density of the culture (▲—▲), showing the expected increased growth rate at the higher temperature; B_{12} uptake activity (●—●); and B_{12} binding activity (■—■). The suppressed level of binding activity in the bfe mutant is too low to be measured reliably. (Data redrawn from Ref. 201.)

growing and spectinomycin-treated cells. All of these results indicate that the tonB product is functionally unstable; whether it is also structurally unstable is unknown at this time.

These results strengthen the contention that the bfe product can inter-act independently with both the tolA system for colicin E uptake and the tonB system for B_{12} uptake. It has proved possible to interfere with the interaction of the receptor with each of these systems by both genetic and physiological means. The various tolerant mutants are altered in their response to the E colicins or B_{12}, but not to both. Cells infected with phage f1 are considerably less sensitive to colicins K and E1 [219], but are not specifically altered in B_{12} uptake [194]. It must be noted that the receptor may enter into yet a third interaction so as to allow the effective adsorption of phage BF23. Infection by this phage is dependent on bfe function, but is unaffected by the function of tolA or tonB [22]. The kinetics of appearance of phage-insensitive cells following inactivation of bfe^+ expression suggests that the presence of receptor alone may not be sufficient for effective phage DNA ejection or uptake.

Evidence as to the location of the tonB product is indirect and ambigu-ous. The effect of the tonB mutation on iron transport led Wang and Newton [134] to implicate its product in the transport process. Since the mutation apparently affected the K_m component of uptake whereas energy poisons depressed the V_{max} component, they removed tonB involvement from the energy-dependent step. Instead, they implied that this product facilitates access of iron chelates to the energy-dependent permease, similar to the proposed role of the lamB product in maltose transport. It is, however, unclear how this role as facilitator of uptake can be reconciled with the absolute dependence of uptake on tonB function. It must also be noted that their experiments were carried out in the presence of both citrate and, for the tonB mutants, derepressed levels of enterochelin. It is apparent from subsequent work that tonB mutants totally lack chelate-mediated iron uptake, suggesting that the original kinetic analysis of the effect of the tonB mutation may be misleading. However, it is clear that the tonB product is required only for those transport systems employing outer membrane components. This, and the involvement of this product in the energy-dependent stage of phage adsorption, is consistent with some manner of association of this product with the outer membrane.

Further evidence for the location of the tonB product external to the periplasm came from studies of Frost and Rosenberg [158] on the growth characteristics of a tonB strain also blocked in enterochelin biosynthesis (aroB). Although the iron requirements of this double mutant were not satisfied by the addition of enterochelin, citrate, or a number of hydroxa-mate siderophores, the strain did respond upon addition of the enterochelin precursors, shikimate or 2,3-dihydroxybenzoate (DHBA). This growth response did not result from the transport of iron chelated to DHBA or any other enterochelin precursor, since the tonB aroB entF triple mutant (also

blocked in the formation of enterochelin from DHBA) did not respond.
These authors suggested that, in the tonB aroB double mutant growing with
shikimate or DHBA, enterochelin was synthesized slowly from these precursors and secreted into the periplasm, where it chelated the iron that
had diffused into that space and mediated its transport into the cell. They
correctly predicted that increased supplementation with these precursors
would decrease the growth response on low-iron medium, since the greater
amounts of enterochelin secreted into the medium would chelate most of the
free iron and prevent its entry into the periplasm. These results suggest
that high-affinity iron transport across the cytoplasmic membrane is also
chelate-mediated and that ferric chelates formed in the periplasm bypass
the tonB-dependent step. A very important question is whether the uptake
of ferric chelates from the periplasm is as efficient as uptake of chelates
from the medium by ton$^+$ cells.

B. Membrane Adhesion Sites

Membrane adhesion sites were first demonstrated by Cota-Robles [204] in
electron micrographs of thin sections of plasmolyzed cells of E. coli. They
appear as regions of apparent contact between the cytoplasmic membrane
and the cell envelope. Their morphology is usually poorly defined, and
their numbers and dimensions may be quite different in the intact cell from
what is observed under the hypertonic conditions required for their visual-
ization. Bayer [205] enumerated 200 to 400 such sites per cell and estimated that they underlie about 5% of the cell surface area. The first evidence
suggestive of their function was provided by the demonstration by Bayer
[206] that the majority of T phages adsorbed to cells were in close proximity to these adhesion sites. This result was obtained whether the cells
were subjected to plasmolysis either before or after phage adsorption,
although in many cases phage adsorption was severely depressed by the
presence of plasmolyzing concentrations of sucrose. There have been
other reports of decreased uptake of phages or colicins in plasmolyzed cells
[220-222]. In many, but not all cases, the maximum number of phages
which can be adsorbed is similar to the number of adhesion sites. These
results led to the proposal that functional phage receptors are located
(primarily or exclusively) at adhesion sites; this model can provide a
mechanism for the appearance of phage DNA in the cytoplasm without its
obligate passage through the periplasm.
　　Direct evidence for a role of adhesion sites came from the demonstration by Mühlradt and his coworkers [207] that these regions represent
the areas in which LPS first appears on the outer membrane following its
synthesis on the cytoplasmic membrane. In these experiments, newly
synthesized LPS was identified by ferritin-labeled antibodies specific for
the O-specific side chains of S. typhimurium LPS. The strains employed

were galE mutants and dependent on supplementation with galactose for the addition to newly synthesized LPS of the O-specific portion. Electron microscopy of freeze-etched samples revealed that, within 30 sec after the addition of galactose, the newly synthesized LPS appeared at about 220 discrete sites on the cell surface (corresponding to 8% of the surface area). Further incubation at 20 to 37°C resulted in the migration of this LPS from these sites to cover the surface completely within 2 to 3 min. This mobility of the LPS was greatly restricted at 0°C. Electron micrographs of thin sections revealed that 65% of the detectable adhesion sites were labeled with new LPS after a 33 sec pulse of galactose, or that 86% of the export sites were located over detectable adhesion sites.

By utilizing the marked change in density of the LPS made in a galE mutant after galactose supplementation, Kulpa and Leive [208] demonstrated that the newly synthesized LPS in E. coli also appears in the outer membrane in at least 20 discrete regions. They proposed that this change in density of the new LPS, and the adhesion sites that contain it, could allow the isolation of the adhesion sites. Unfortunately, this method is not applicable to K12 strains of E. coli, whose LPS does not change sufficiently in density in this type of experiment. Perhaps some related procedure can be developed for the isolation of adhesion sites. Clearly, this isolation is necessary for verification of many of the roles ascribed to these regions; their low amount and the presence of other structures containing both cytoplasmic and outer membrane fractions makes this difficult.

There is considerable evidence that a portion of the bacterial chromosome is associated with the membrane [209-212], as are the assembly and DNA replication systems for several phages [213-215]. Complexes of DNA and membrane have been isolated and contain biochemical and morphological markers of both the cytoplasmic and the outer membrane [18].

Given their role in the export of LPS, adhesion sites could also be involved in the export of outer membrane phospholipids and proteins. Furthermore, if their postulated role as the channels for penetration of many phages and colicins is correct, then they may also be instrumental in those nutrient transport systems which employ the same receptors as the phage or colicins. These three possible functions of adhesion sites (export of outer membrane constituents, entry of bacteriocins and phage nucleic acids, and a direct route for entry of B_{12} and iron chelates) can be combined in a model. Assume that phage receptors must be located over adhesion sites for the proper uptake of bound colicins or ejected phage DNA. The preferential localization of receptors at adhesion sites (since they only represent 5 to 10% of the surface area) could be the result of their having just been exported there. Just like the LPS, the newly exported receptors could then diffuse away from the adhesion sites, either through nonspecific lateral migration or by being driven away by the continued export of outer membrane constituents. This would generate two populations of receptors differing in their proximity to adhesion sites. These two classes may also differ in the efficacy with which bound ligands are transmitted to the interior of the cell. Evidence in support of this model is indirect, but some salient findings will be summarized here.

One feature of this model is that it proposes the existence of functionally distinct classes of receptors. Following a proposal of Nomura and Maeda [216], Shannon and Hedges [217] have suggested such a heterogeneity among colicin E2 receptors. Colicins usually exhibit single-hit killing kinetics despite the fact that a "lethal unit" usually contains many colicin molecules. These authors showed that the release of a "lethal dose" of colicin from a cell did not rescue that cell from the lethal consequences of the initial adsorption. They proposed that only a minority (less than 10%) of the total receptors were lethal receptors capable of irreversible colicin adsorption. Thus, the lethal dose of colicin released from the first cell had been reversibly attached to the majority of ineffective receptors, whereas a few colicin molecules remained irreversibly attached to the original cell at one of the few receptors in proper orientation or location for effective uptake.

The experiments on the inactivation of the bfe locus [192] support and extend this contention. Only newly synthesized receptors appeared to be effective for the lethal adsorption of colicins E2 and E3 and phage BF23. "Old" receptor was fully functional for uptake of B_{12} but not of colicin, providing evidence of this functional heterogeneity. Uptake of colicin E3 and B_{12} differ in their separate dependencies on tolA and tonB function, respectively. With all the assumptions made so far, the placement of tolA function at adhesion sites seems justified. According to the proposed model, receptor would first appear on the surface in the vicinity of an adhesion site, within the realm specified by the tolA-dependent structure. As protein synthesis or membrane growth continues, this receptor would be forced away from this adhesion site, perhaps by the emergence of new receptor, whether functional or not. In accordance with the observed effect of spectinomycin, cessation of protein synthesis would prevent the movement of receptor away from this region. The migrated receptor would still be fully capable of binding colicin, but such binding would be reversible and ineffectual in terms of colicin-produced lethality.

The migrating receptor could then enter into a state of interaction with the tonB product or a structural region specified by this product. This is proposed from the ability of the bfe product to function for B_{12} uptake even though it is no longer functional for colicin E3 uptake. This proposal does not necessarily exclude the tonB product from adhesion sites since it certainly would be convenient to find that B_{12} enters the cell directly, without passage across the periplasm. There is no evidence with respect to the periplasmic residence of B_{12} or iron chelates. There is indirect evidence that B_{12} or iron chelates in the periplasm can be utilized, but the efficiency of this process is unknown [158, 189].

Finally, mention should be made of the lambda receptor, whose properties appear to contradict all of the predictions for an adhesion site-dependent system. This receptor can be synthesized in very large quantities, such that the entire surface of the cell can be covered with specifically adsorbed lambda phage [108]. The number of receptor molecules can approach the number of major outer membrane proteins, about 10^5 per cell (D. Diedrich, personal communication). The distribution of receptors on the cell surface following induction of receptor synthesis was most consistent with the appearance of receptor only in the region of the division septum with insertion occurring only during the last quarter of the division cycle [218]. However, the role of the receptor in maltose or maltotriose uptake seems to be strictly that of facilitating the translocation of the substrate across the outer membrane and enhancing its access to the obligately required periplasmic maltose binding protein [112]. The status of tonB has no effect on maltose uptake [193]. This system clearly differs from the B_{12} and iron chelate uptake systems in that the role of the outer membrane component in maltose transport appears primarily to be a means to bypass the exclusion limits of the diffusion pores in the outer membrane. The outer membrane components for the other two systems may play a more direct role in the entry of substrate into the cell.

In conclusion, recent discoveries have demonstrated the existence and some of the properties of transport systems whose normal functioning is dependent on constituents of the outer membrane. These systems are further related in that they share a requirement for the product of the tonB gene. The properties of these novel uptake systems have been only rudimentarily described, and future research will be directed toward the description of the involvement and interaction of membrane components in these processes. Key to the understanding of these nutrient transport systems is the uptake of colicins and phage DNA, which is mediated, in part, by the same outer membrane components as the nutrients employ. Hence, any explanations of these transport systems must account for, and may provide insight into, the uptake of the colicins and mechanisms of colicin insensitivity. Thus, in addition to their intrinsic interest, these nutrient transport systems promise to reveal valuable information on the synthesis and organization of the outer membrane and its relationship to the rest of the cell.

ACKNOWLEDGMENTS

The work in the authors' laboratory was supported by a research grant from the U.S. Public Health Service (GM19078). R. J. Kadner is the recipient of a Research Career Development Award (GM00019). We are grateful for the information contributed by J. Konisky, J. B. Neilands, W. Boos, and V. Braun. We especially appreciate the ideas and advice of C. Schnaitman, C. MacGregor, C. Bradbeer, and D. Diedrich.

REFERENCES

1. H. Nikaido, Bacterial Membranes and Walls (L. Leive, ed.), Dekker, New York, 1973, p. 131-208.
2. H. Nikaido, In press.
3. V. Braun, Biochim. Biophys. Acta, 415:335-377 (1975).
4. M. Inouye, Membrane Biogenesis (A. Tzagoloff, ed.), Plenum, New York, 1975, pp. 351-392.
5. J. W. Costerton, J. M. Ingram, and K.-J. Cheng, Bacteriol. Rev., 38:87-110 (1974).
6. M. E. Bayer, Ann. N.Y. Acad. Sci., 235:6-28 (1974).
7. V. Braun, and K. Hantke, Ann. Rev. Biochem., 43:89-121 (1974).
8. V. Schwarz and W. Leutgeb, J. Bacteriol., 106:588-595 (1971).
9. V. Braun, H. Gnirke, U. Henning, and K. Rehn, J. Bacteriol., 114:1264-1270 (1973).
10. V. Braun and K. Rehn, Eur. J. Biochem., 10:426-438 (1969).
11. S. Halegoua, A. Hirashima, and M. Inouye, J. Bacteriol., 120:1204-1208 (1974).
12. V. Braun and H. Wolff, J. Bacteriol., 123:888-897 (1975).
13. T. Miura and S. Mizushima, Biochim. Biophys. Acta, 193:268-276 (1969).
14. M. J. Osborn, J. E. Gander, E. Parisi, and J. Carson, J. Biol. Chem., 247:3962-3972 (1972).
15. C. A. Schnaitman, J. Bacteriol., 104:890-901 (1970).
16. B. Withold, M. Boekhout, M. Brock, J. Kingman, and H. van Heerikhuizen, Anal. Biochem., 74:160-170 (1976).
17. D. A. White, W. J. Lennarz, and C. A. Schnaitman, J. Bacteriol., 109:686-690 (1972).
18. W. L. Olsen, H.-G. Heidrich, K. Hannig, and P. H. Hofschneider, J. Bacteriol., 118:646-653 (1974).
19. H. Wolf-Watz, S. Normark, and G. D. Bloom, J. Bacteriol., 115:1191-1197 (1973).
20. L. Leive, Proc. Natl. Acad. Sci. USA, 53:745-750 (1965).
21. P. F. Mühlradt and J. R. Golecki, Eur. J. Biochem., 51:343-352 (1975).
22. C. A. Schnaitman, J. Bacteriol., 108:545-552 (1971).
23. C. A. Schnaitman, J. Bacteriol., 108:553-563 (1971).
24. G. F.-L. Ames, J. Biol. Chem., 249:634-644 (1974).
25. G. F.-L. Ames and H. Nikaido, Biochemistry, 15:616-623 (1976).
26. J. Uemura and S. Mizushima, Biochim. Biophys. Acta, 413:163-176 (1975).
27. C. A. Schnaitman, J. Bacteriol., 118:442-453 (1974).
28. J. Smit, Y. Kamio, and H. Nikaido, J. Bacteriol., 124:942-958 (1975) (1975).

29. J.-P. Benedetto, M. Bruneteau, and G. Michel, Eur. J. Biochem.,
    63:313-320 (1976).
30. P. F. Mühlradt, J. Menzel, J. R. Golecki, and V. Speth, Eur. J.
    Biochem., 43:533-539 (1974).
31. A. P. Van Gool and N. Nanninga, J. Bacteriol., 108:474-481 (1971).
32. A. J. Verkleij, E. J. J. Lugtenberg, and P. H. J. Th. Ververgaert,
    Biochim. Biophys. Acta, 426:581-586 (1976).
33. I. Haller and U. Henning, Proc. Natl. Acad. Sci. USA, 71:2018-
    2021 (1974).
34. Y. Kamio and H. Nikaido, Biochemistry, 15:2561-2570 (1976).
35. S. Rottem, M. Hasin, and S. Razin, Biochim. Biophys. Acta, 375:
    395-405 (1975).
36. H. Nikaido and T. Nakae, J. Infec. Dis., 128:S30-S34 (1973).
37. P. Overath, M. Brenner, T. Gulik-Krzywicki, E. Schechter, and
    L. Letellier, Biochim. Biophys. Acta, 389:358-369 (1975).
38. R. J. Roantree, T. Kuo, D. G. MacPhee, and B. A. D. Stocker,
    Clin. Res., 17:157 (1969).
39. S. Schlecht and G. Schmidt, Zentralbl. Bakteriol. Abt. I. Orig.,
    212:505-511 (1969).
40. P. Gustafsson, K. Nordstrom, and S. Normark, J. Bacteriol., 116:
    893-900 (1973).
41. K. G. Eriksson-Grennberg, K. Nordstrom, and P. Englund, J.
    Bacteriol., 108:1210-1223 (1971).
42. S. Tamaki, T. Sato, and M. Matsuhashi, J. Bacteriol., 105:968-975
    (1971).
43. H. Bursztyn, V. Sgaramella, O. Ciferri, and J. Lederberg, J.
    Bacteriol., 124:1630-1634 (1975).
44. L. Leive, J. Biol. Chem., 243:2373-2380 (1968).
45. L. Leive, V. K. Shovlin, and S. E. Mergenhagen, J. Biol. Chem.,
    243:6384-6391 (1968).
46. L. Leive, Ann. N.Y. Acad. Sci., 235:109-127 (1974).
47. M. J. Voll and L. Leive, J. Biol. Chem., 245:1108-1114 (1970).
48. S. R. Levy and L. Leive, Proc. Natl. Acad. Sci. USA, 61:1435-
    1439 (1968).
49. G. F.-L. Ames, E.N. Spudich, and H. Nikaido, J. Bacteriol.,
    117:406-416 (1974).
50. J. Koplow and H. Goldine, J. Bacteriol., 117:527-543 (1974).
51. L. L. Randall, J. Bacteriol., 123:41-46 (1975).
52. R. T. Irvin, A. K. Chatterjee, K. E. Sanderson, and J. W. Coster-
    ton, J. Bacteriol., 124:930-941 (1975).
53. H. Nikaido, Biochim. Biophys. Acta, 433:118-132 (1976).
54. S. Normark, J. Bacteriol., 108:51-58 (1971).
55. H. C. Wu, Biochim. Biophys. Acta, 290:274-289 (1972).
56. R. A. Weigand and L. I. Rothfield, J. Bacteriol., 125:340-345 (1976).
57. J. K. Davies and P. Reeves, J. Bacteriol., 123:102-117 (1975).

58. R. E. W. Hancock and P. Reeves, J. Bacteriol., 127:98–108 (1976).
59. U. Henning and I. Haller, FEBS Lett., 55:161–164 (1975).
60. M. II. Richmond and R. B. Sykes, Adv. Microb. Physiol., 9:31–88 (1973).
61. J. P. Robbie and T. H. Wilson, Biochim. Biophys. Acta, 173:234–244 (1969).
62. J. W. Payne and C. Gilvarg, J. Biol. Chem., 243:6291–6299 (1968).
63. T. Nakae and H. Nikaido, J. Biol. Chem., 250:7359–7365 (1975).
64. L. G. Burman, K. Nordstrom, and G. D. Bloom, J. Bacteriol., 112:1364–1374 (1972).
65. Z. Barak and C. Gilvarg, Biomembranes, 7:167–218 (1975).
66. R. Scherrer and P. Gerhardt, J. Bacteriol., 107:718–735 (1971).
67. M. Inouye, Proc. Natl. Acad. Sci. USA, 71:2396–2400 (1974).
68. V. Braun, J. Infec. Dis., 128:59–515 (1973).
69. M. Inouye, J. Shaw, and C. Shen, J. Biol. Chem., 247:8154–8159 (1972).
70. A. Hirashima, H. C. Wu, P. S. Venkateswaran, and M. Inouye, J. Biol. Chem., 248:5654–5659 (1973).
71. V. Braun, K. Hantke, and U. Henning, FEBS Lett., 60:26–28 (1975).
72. H. C. Wu and J. J.-C. Lin, J. Bacteriol., 126:147–156 (1976).
73. K. Von Meyenburg, J. Bacteriol., 107:878–888 (1971).
74. J. P. Rosenbusch, J. Biol. Chem., 249:8019–8029 (1974).
75. J. Sekizawa and S. Fukui, Biochim. Biophys. Acta, 307:104–117 (1973).
76. P. D. Bragg and C. Hou, Biochim. Biophys. Acta, 274:478–488 (1972).
77. K. Nakamura and S. Mizushima, Biochim. Biophys. Acta, 413:371–393 (1975).
78. T. Nakae, J. Biol. Chem., 251:2176–2178 (1976).
79. J. Foulds and C. Barrett, J. Bacteriol., 116:885–892 (1973).
80. B. Rolfe and K. Ondera, Biochem. Biophys. Res. Commun., 44:767–773 (1971).
81. J. K. Davies and P. Reeves, J. Bacteriol., 123:372–373 (1975).
82. I. R. Beacham, R. Kahana, L. Levy, and E. Yagil, J. Bacteriol., 116:957–964 (1973).
83. I. R. Beacham, D. Haas, and E. Yagil, J. Bacteriol., 129:1034–1044 (1977).
84. F. R. Albright, D. A. White, and W. J. Lennarz, J. Biol. Chem., 248:3968–3977 (1973).
85. A. A. Lindberg, Ann. Rev. Microb., 27:205–241 (1973).
86. C. J. Schmitges and U. Henning, Eur. J. Biochem., 63:47–52 (1976).
87. C. Schnaitman, D. Smith, and M. Forn de Salas, J. Virol., 15:1121–1130 (1975).
88. P. J. Bassford, Jr., D. L. Diedrich, C. A. Schnaitman, and P. Reeves, J. Bacteriol., 131:608–622 (1977).

89. K. G. Hardy, Bacteriol. Rev., 39:464-515 (1975).
90. S. E. Luria, Bacterial Membranes and Walls (L. Leive, ed.), Dekker, New York, 1973, pp. 293-320.
91. I. B. Holland, Adv. Microb. Physiol., 12:56-139 (1975).
92. C. Lau and F. M. Richards, Biochemistry, 15:666-671 (1976).
93. K. G. Eriksson-Grennberg and K. Nordstrom, J. Bacteriol., 115: 1219-1222 (1973).
94. J. K. Davies and P. Reeves, J. Bacteriol., 123:96-101 (1975).
95. L. M. Kozloff, M. Lute, L. K. Crosby, N. Rao, V. A. Chapman, and S. S. DeLong, J. Virol., 5:726-739 (1970).
96. L. M. Kozloff, C. Verses, M. Lute, and L. K. Crosby, J. Virol., 5:740-753 (1970).
97. E. M. Lederberg, Genetics, 40:580-581 (1955).
98. B. F. Bachmann, K. B. Low, and A. L. Taylor, Bact. Rev., 40: 116-167 (1976).
99. J. P. Thirion and M. Hofnung, Genetics, 71:207-216 (1972).
100. M. Schwartz, Ann. Inst. Pasteur, 113:685-704 (1967).
101. H. Wiesmeyer and M. Cohn, Biochim. Biophys. Acta, 39:440-447 (1960).
102. W. V. Howes, J. Bact., 90:1188-1193 (1965).
103. T. Yokota and T. Kasuga, J. Bact., 109:1304-1306 (1972).
104. M. Hofnung, Genetics, 76:169-184 (1974).
105. O. Kellermann, and S. Szmelcman, Eur. J. Biochem., 47:139-149 (1974).
106. M. Hofnung, D. Hatfield, and M. Schwartz, J. Bact., 117:40-47 (1974).
107. M. Hofnung and M. Schwartz, Mol. Gen. Genet., 112:117-132 (1971).
108. L. Randall-Hazelbauer and M. Schwartz, J. Bact., 116:1436-1446 (1973).
109. M. Schwartz, J. Mol. Biol., 99:185-201 (1975).
110. V. Zgaga, M. Medić, E. Salaj-Šmic, D. Novak, and M. Wrischer, J. Mol. Biol., 79:697-708 (1973).
111. G. Hazelbauer, J. Bact., 122:206-214 (1975).
112. S. Szmelcman and M. Hofnung, J. Bact., 124:112-118 (1975).
113. G. L. Hazelbauer, J. Bact., 124:119-126 (1975).
114. S. Szmelcman, M. Schwartz, T. J. Silhavy, and W. Boos, Eur. J. Biochem., 65:13-19 (1976).
115. M. Schwartz and L. LeMinor, J. Virol., 15:679-685 (1975).
116. T. Shinozawa and H. Shida, J. Bacteriol., 126:1025-1029 (1976).
117. G. C. Rodgers and J. B. Neilands, Handbook of Microbiology, Vol. II (Microbial composition) (A. L. Laskin and H. A. Lecheuslier, eds.), CRC Press, Cleveland, Ohio, 1973, pp. 823-830.
118. J. B. Neilands (ed.), Microbial Iron Metabolism, Academic, New York, 1974.

119. I. G. O'Brien and F. Gibson, BBA, 215:393-402 (1970).
120. J. R. Pollack and J. B. Neilands, BBRC, 38:989-992 (1970).
121. C. E. Lankford, CRC Crit. Rev. Microbiol., 2:273-332 (1973).
122. J. B. Neilands, Inorganic Biochemistry (G. Fichhorn, ed.), Elsevier, Amsterdam, 1973, pp. 167-202.
123. II. Rosenberg and I. G. Young, Microbial Iron Metabolism (J. B. Neilands, ed.), Academic, New York, 1974, pp. 67-81.
124. P. Fredericq and J. Smarda, Ann. Inst. Pasteur, 118:767-774 (1970).
125. V. Braun, K. Schaller, and H. Wolff, Biochim. Biophys. Acta, 323:87-97 (1973).
126. R. Wayne and J. B. Neilands, Abstracts 168th Meeting of the American Chemical Society, Volume 168, 1974.
127. M. Luckey, J. R. Pollack, R. Wayne, B. N. Ames, and J. B. Neilands, J. Bacteriol., 111:731-738 (1972).
128. R. Wayne and J. B. Neilands, J.Bact., 121:497-503 (1975).
129. K. Hantke and V. Braun, FEBS Lett., 49:301-305 (1975).
130. M. Luckey, R. Wayne and J. B. Neilands, BBRC, 64:687-693 (1975).
131. M. Luckey and J. B. Neilands, Pacific Slope Biochemical Conference Abstracts, June, 1975.
132. M. Luckey and J. B. Neilands, J. Bact., 127:1036-1037 (1976).
133. G. E. Frost and H. Rosenberg, Biochim. Biophys. Acta, 330:90-101 (1973).
134. C. C. Wang and A. Newton, J. Biol. Chem., 246:2147-2151 (1971).
135. I. G. Young, G. B. Cox, and F. Gibson, Biochim. Biophys. Acta, 177:401 (1967).
136. R. E. W. Hancock, K.Hantke, and V. Braun, J. Bact., 127:1370-1375 (1976).
137. G. E. Frost and H. Rosenberg, Proc. Austral. Biochem. Soc., 6:49 (1973).
138. F. Gibson and D. I. Magrath, Biochim. Biophys. Acta, 192:175-184 (1969).
139. I. G. O'Brien, G. B. Cox, and F. Gibson, Biochim. Biophys. Acta, 237:537-549 (1971).
140. R. K. J. Lucke and F. Gibson, J. Bact., 107:557-562 (1971).
141. I. G. Young, L. Langman, R. K. J. Luke, and F. Gibson, J. Bact., 106:515-7 (1971).
142. J. R. Pollack, B. N. Ames, and J. B. Neilands, J. Bact., 104:635-639 (1970).
143. G. C. Woodrow, I. G. Young, and F. Gibson, J. Bacteriol., 124:1-6 (1975).
144. G. F. Bryce, R. Weller, and N. Brot, BBRC, 42:871-879 (1971).
145. G. F. Bryce and N. Brot, Arch. Biochem. Biophys., 142:399-406 (1971).

146. I. G. O'Brien, G. B. Cox, and F. Gibson, *Biochim. Biophys. Acta*, 201:453-460 (1970).
147. L. Langman, I. G. Young, G. E. Frost, H. Rosenberg, and F. Gibson, *J. Bacteriol.*, 112:1142-1149 (1972).
148. R. J. Parra, L. Langman, I. G. Young, and F. Gibson, *Arch. Biochem. Biophys.*, 153:74-78 (1972).
149. G. B. Cox, F. Gibson, R. K. J. Luke, N. A. Newton, I. G. O'Brien, and H. Rosenberg, *J. Bacteriol.*, 104:219-226 (1970).
150. C. C. Wang and A. Newton, *J. Bacteriol.*, 98:1135-1141 (1969).
151. C. C. Wang and A. Newton, *J. Bacteriol.*, 98:1142-1150 (1969).
152. S. K. Guterman and S. E. Luria, *Science*, 164:1414 (1969).
153. S. K. Guterman, *BBRC*, 44:1149-1155 (1971).
154. S. K. Guterman and L. Dann, *J. Bacteriol.*, 114:1225-1230 (1973).
155. S. K. Guterman, *J. Bacteriol.*, 114:1217-1224 (1973).
156. A. P. Pugsley and P. Reeves, *J. Bacteriol.*, 126:1052-1062 (1976).
157. R. Wayne, K. Frick, and J. B. Neilands, *J. Bact.*, 126:7-12 (1976).
158. G. E. Frost and H. Rosenberg, *J. Bact.*, 124:704-712 (1975).
159. A. P. Pugsley and P. Reeves, *Biochem. Biophys. Res. Commun.*, 70:846-853 (1976).
160. V. Braun, R. E. W. Hancock, K. Hantke, and A. Hartmann, *J. Supramol. Struct.*, 5:37-58 (1976).
161. K. Hantke and V. Braun, *FEBS Lett.*, 59:277-281 (1975).
162. R. E. W. Hancock and V. Braun, *FEBS Lett.*, 65:208-210 (1976).
163. H. Zähner, E. Hütter, and R. Bachmann, *Arch. Microbiol.*, 36:325-349 (1960).
164. J. Konisky, B. S. Cowell, and M. J. R. Gilchrist, *J. Supramol. Struct.*, 1:208-219 (1973).
165. C. Cardelli and J. Konisky, *J. Bact.*, 119:379-385 (1974).
166. M. J. R. Gilchrist and J. Konisky, *J. Bact.*, 125:1223-1225 (1976).
167. J. Konisky, S. Soncek, K. Frick, J. K. Davies, and C. Hammond, *J. Bacteriol.*, 127:249-257 (1976).
168. R. Wayne and J. B. Neilands, Annual Meeting, *Am. Soc. Biol. Chem. Abst.*, June, 1976.
169. A. Garen and T. T. Puck, *J. Exp. Med.*, 94:177-189 (1951).
170. A. Garen, *Biochim. Biophys. Acta*, 14:163-172 (1954).
171. R. E. W. Hancock and V. Braun, *J. Bact.*, 125:409-415 (1976).
172. W. Weidel, *Ann. Rev. Microbiol.*, 12:27-48 (1958).
173. J. Leong and J. B. Neilands, *J. Bacteriol.*, 126:823-830 (1976).
174. L. Milner, C. Whitfield, and H. Weissbach, *Arch. Biochem. Biophys.*, 133:413-419 (1969).
175. B. D. Davis, and E. S. Mingioli, *J. Bacteriol.*, 60:17-28 (1950).
176. P. M. DiGirolamo, R. J. Kadner, and C. Bradbeer, *J. Bacteriol.*, 106:751-757 (1971).
177. G. W. Chang and J. T. Chang, *Nature*, 254:150-151 (1975).
178. E. L. Oginsky, *Arch. Biochem. Biophys.*, 36:71-79 (1952).

179. P. M. DiGirolamo and C. Bradbeer, J. Bacteriol., 106:745-750 (1971).
180. C. Bradbeer and M. L. Woodrow, J. Bacteriol., 128:99-104 (1976).
181. R. T. Taylor, S. A. Norrell, and M. L. Hanna, Arch. Biochem. Biophys., 148:366-381 (1972).
182. R. T. Taylor, M. P. Nevins, and M. L. Hanna, Arch. Biochem. Biophys., 149:232-243 (1972).
183. J. C. White, P. M. DiGirolamo, M. L. Fu, Y. A. Preston, and C. Bradbeer, J. Biol. Chem., 248:3978-3986 (1973).
184. R. J. Kadner and G. L. Liggins, J. Bacteriol., 115:514-521 (1973).
185. D. R. DiMasi, J. C. White, C. A. Schnaitman, and C. Bradbeer, J. Bacteriol., 115:506-513 (1973).
186. R. S. Buxton, Mol. Gen. Genet., 113:154-156 (1971).
187. P. E. Jasper, E. Whitney, and S. Silver, Genet. Res., 19:305-312 (1972).
188. S. F. Sabet and C. A. Schnaitman, J. Biol. Chem., 248:1797-1806 (1973).
189. P. J. Bassford, Jr., and R. J. Kadner, J. Bacteriol., 132:796-805 (1977).
190. C. Bradbeer, M. L. Woodrow, and L. I. Khalifah, J. Bacteriol., 125:1032-1039 (1976).
191. C. Hill and I. B. Holland, J. Bacteriol., 94:677-686 (1967).
192. P. J. Bassford, Jr., R. J. Kadner, and C. A. Schnaitman, J. Bacteriol., 129:265-275 (1977).
193. P. J. Bassford, Jr., C. Bradbeer, R. J. Kadner, and C. A. Schnaitman, J. Bacteriol., 128:242-247 (1976).
194. R. J. Kadner and P. J. Bassford, Jr., J. Bacteriol., 129:254-264 (1977).
195. I. Heller, B. Hoehn, and U. Henning, Biochemistry, 14:478-484 (1975).
196. J. Leal, and H. Marcovich, Mol. Gen. Genet., 130:345-359 (1974).
197. J. Leal and H. Marcovich, Ann. Inst. Pasteur, 120:467-474 (1971).
198. W. Hayes, J. Gen. Microb., 16:97-119 (1957).
199. M. P. Oeschger and S. L. Woods, Cell, 7:205-212 (1976).
200. K. Timmis and A. J. Hedges, Biochim. Biophys. Acta, 262:200-207 (1972).
201. P. J. Bassford, Jr., C. A. Schnaitman, and R. J. Kadner, J. Bacteriol., 130:750-758 (1977).
202. R. J. Kadner and P. J. Bassford, Jr., J. Bacteriol., in press (1978).
203. M. E. Bayer, in, Membrane Biogenesis (A. Tzagoloff, ed.), Plenum, New York, 1975, pp. 393-427.
204. E. H. Cota-Robles, J. Bacteriol., 85:499-503 (1963).
205. M. E. Bayer, J. Gen. Microbiol., 53:395-404 (1968).
206. M. E. Bayer, J. Virol., 2:346-356 (1968).

207. P. F. Mühlradt, J. Menzel, J. R. Golecki, and V. Speth, Eur. J. Biochem., 35:471-481 (1973).
208. C. F. Kulpa, Jr., and L. Leive, J. Bacteriol., 126:467-477 (1976).
209. P. H. Van Knippenberg, G. A. H. Duijts, and M. S. T. Euwe, Mol. Gen. Genet., 112:197-207 (1971).
210. P. H. Van Knippenberg and G. A. H. Duijts, Mol. Gen. Genet., Genet., 112:208-220 (1971).
211. M. A. McIntosh and C. F. Earhart, J. Bacteriol., 122:592-598 (1975).
212. P. Dworsky, J. Bacteriol., 126:64-71 (1976).
213. S. Klaus, R. Geuther, and D. Noack, Mol. Gen. Genet., 115:93-96 (1972).
214. R. C. Miller, Jr., J. Virol., 10:920-924 (1972).
215. M. S. Center, J. Virol., 12:847-854 (1973).
216. M. Nomura and A. Maeda, Zentrbl. Bakteriol. Orig. A, 196:216-239 (1965).
217. R. Shannon and A. J. Hedges, J. Bacteriol., 116:1136-1144 (1973).
218. A. Ryter, H. Shuman, and M. Schwartz, J. Bacteriol., 122:295-301 (1975).
219. H. Smilowitz, J. Virol., 13:100-106 (1974).
220. M. M. Alemohammad and C. J. Knowles, J. Gen. Microbiol., 82:125-142 (1974).
221. E. M. Holland and I. B. Holland, Biochim. Biophys. Acta, 281:179-191 (1972).
222. R. E. W. Hancock and P. Reeves, J. Bacteriol., 121:983-993 (1975).

Chapter 10

TRANSPORT OF VITAMINS AND ANTIBIOTICS

ROBERT J. KADNER

Department of Microbiology
University of Virginia School of Medicine
Charlottesville, Virginia

I. INTRODUCTION 463
II. VITAMIN UPTAKE 465
   A. Cyanocobalamin (Vitamin B_{12}) 466
   B. Thiamine (Vitamin B_1) 470
   C. Folates 472
   D. Biotin 475
   E. Pyridoxine (Vitamin B_6) 477
   F. Other Vitamins 478
   G. Summary 479
III. ANTIBIOTIC TRANSPORT 481
   A. General Features 481
   B. Antimetabolites 482
   C. Antibiotics 483
   REFERENCES 487

I. INTRODUCTION

Other chapters in this book examined the properties of the transport systems for the major classes of bacterial nutrients, describing the primary current interest in the mechanisms of energy coupling for solute accumulation. This chapter is addressed to the uptake of two classes of permeants which have not received nearly so much attention and, in fact, are often ignored. Vitamins are usually transported and required in much lower amounts than other nutrients but, as will be shown, the systems for vitamin transport span the entire spectrum of proposed transport mechanisms. The uptake of vitamins is a very important topic and one which is only beginning to receive the critical analysis it deserves.

This chapter is directed at pointing out the unusual and interesting aspects of the transport systems included under these headings. Hence, it will make no attempt at being comprehensive in its coverage. Rather, a limited number of fairly well-studied systems will be described to point out the current state of the literature in this area and to allow the individual facets of each transport system to surface.

To be sure, a considerable portion of the literature in these areas is concerned with such currently mundane topics as whether there are specific transport systems for these substrates and whether active transport really occurs. An affirmative answer to these two questions cannot be assumed and is questionable in many cases. It is necessary to review some of the criteria that must be met in order to state that a particular process employs a transport system and whether the substrate is actively transported. The presence of a carrier in the membrane requires some degree of substrate specificity as indicated by the lack of uptake of substrate analogs and by the ability of other substrate analogs to compete for uptake with the normal substrate. This specificity must be demonstrated at the level of transport and not just by interference with the effect of the normal substrate on growth of the cell. The dependence of the rate of uptake on the external substrate concentration should show some degree of saturability. Often there is some low-affinity or nonsaturable diffusion process detectable at high substrate concentrations. In all cases examined in adequate detail so far, the uptake of the water-soluble vitamins and antibiotics is carrier mediated. Evidence for the uptake of fat-soluble compounds is less extensive, and this uptake can employ passive diffusion through the membrane and accumulation governed by the difference in pH on the two sides of the membrane.

Evidence for the operation of active transport is more difficult to obtain. Many systems exhibit the accumulation of labeled substrates at concentrations far greater than the external concentration. This accumulation can be reduced or abolished by the presence of the classical energy poisons. It is unfortunate that many studies have used iodoacetate, which can inhibit glycolysis and can serve as a general sulfhydryl reagent. Neither the accumulation of substrate nor the sensitivity to energy poisons justify the conclusion of the involvement of an active transport process according to the classical definition. Without exception, all of the vitamins undergo some intracellular metabolic alteration during their conversion to the coenzyme form. In many cases, this metabolically altered form is trapped within the cell, thereby accounting for the apparent energy dependence of accumulation. Measurement of the intracellular fate of transported substrates is nowhere more essential than in the study of vitamin transport. The nature and extent of efflux or chase of substrates can indicate whether the accumulated species are free to leave the cell by reversal of the influx process. Uptake into membrane vesicles would be very useful for the characterization of the transport process. This would allow a better definition of the nature of the energy coupling, and even more important, would reduce the contribution of the binding of substrate to intracellular macromolecules. In light of the relatively small amount of vitamin accumulated, this binding may be quite significant, thereby abolishing the free pool of substrate capable of efflux.

There are certain obvious disadvantages to the study of vitamin transport. Many of the earlier studies used organisms, such as the lactic acid bacteria, which are naturally auxotrophic for many vitamins. These strains require complex growth media and are not particularly amenable to genetic analysis. Furthermore, the rates of uptake and the steady-state levels of accumulation of many of these substrates are considerably lower than for amino acids or sugars. There can be extensive macromolecular binding and alteration of the substrate. Starvation of an auxotroph for a vitamin in order to study the regulation of the vitamin transport system can have profound effects on cellular metabolism, including membrane function. Finally, some of the substrates are unstable or reactive with cell constituents.

There are, however, certain compensating advantages to this study. Characterization of antibiotic transport systems should aid in the design of more effective (i.e., more permeable) agents. Also, since vitamin metabolism is a major target for antimicrobial chemotherapy, a thorough knowledge of vitamin transport and metabolism must profit the designer of drugs. Finally, the student of transport must take note that although vitamin uptake systems are usually present in relatively low amount, they do display quite high substrate affinities, often several orders of magnitude greater than those of systems for amino acids or sugars. This tight binding might make feasible the detection and isolation of the long-elusive membrane carriers. Finally, growth responses can be obtained in appropriate strains at quite low concentrations. Although serious errors can result from minor levels of contamination, especially in complex media, this can allow a very sensitive assay of vitamin utilization, often more sensitive than the usual uptake assays with radioisotopes.

II. VITAMIN UPTAKE

This description of vitamin uptake will focus on several well-defined systems which portray some important feature of general interest as well as some possibly unique aspect. When useful, each section will begin with a brief summary of the structure, synthesis, and metabolism of the vitamin and of the role of the coenzyme. The primary focus will be on the uptake

into the enteric bacteria and then into other bacteria (primarily the Lacto-
bacteriaceae). In a few cases, mention will be made of the analogous
transport processes in yeast, especially as this compares or contrasts
with the processes in the bacteria. It must be reiterated that this is not
meant to be a comprehensive review. Some of the comparative aspects of
these transport systems are summarized in Table 10-1. General aspects
of microbial utilization of vitamins have been published [1].

A. Cyanocobalamin (Vitamin B_{12})

1. Enteric Bacteria

Current research on B_{12} uptake has focused on the involvement of a receptor protein located in the outer membrane of the cell envelope of Escherichia coli; this has been reviewed in Chapter 9.
     The enteric bacteria do not synthesize B_{12} and do not normally require
it. Two B_{12}-dependent reactions are known: a homocysteine transmethylase providing an alternate route of methionine biosynthesis [2], and an
ethanolamine deaminase required for the utilization of ethanolamine as
nitrogen source [3]. A B_{12} concentration as low as 0.75 nM allowed a
half-maximal growth rate with ethanolamine as nitrogen source [3]; meth-
ionine-B_{12} auxotrophs, blocked in the B_{12}-independent methylase, achieved
half-maximal growth rates with 0.2 nM B_{12} [4]. The structure and synthesis of B_{12} is too complicated for mention here. The primary metabolic
alterations of the transported vitamin are the replacement of the cyano
group on the sixth coordination position of the cobalt atom with either a
methyl group to form the cofactor for the methyl transferase or a 5'-deoxy
adenosyl group to form the cofactor for the ethanolamine deaminase.
     The uptake of B_{12} in E. coli was first described by Oginsky [5,6] and
has recently been studied in detail in groups headed by Bradbeer and by
Taylor. Both groups demonstrated that uptake appeared to be a carrier-
mediated process that was energy, pH, and temperature dependent [7,8].
Whether it is an active transport process allowing accumulation of unaltered
substrate in the cytoplasm is not clear. In experiments of DiGirolamo and
Bradbeer [7], there was considerable conversion of transported cyanocobalamin to derivative forms, although the authors claimed transport
could occur without this alteration. On the other hand, Taylor et al. [8]
found most of the accumulated label as the original CN-B_{12}, but argued
that the majority of this was bound to the methyl transferase.
     As described in Chapter 9, the time course of uptake was biphasic,
with a very rapid, energy-independent initial phase followed by a slower
energy-dependent phase. It is now clear that the initial phase represents
binding to a receptor protein embedded in the outer membrane and is an
essential first step of the uptake process. The overall uptake process

exhibits a very high substrate affinity [K_m of 4 to 10 nM] but a not overly exacting substrate specificity [7,8], consistent with earlier reports of the ability of a variety of cobamides and cobalamines to satisfy the methionine auxotrophy [9,10].

Both groups of investigators showed that uptake was severely depressed following osmotic shock [7,8]. Proteins with B_{12} binding activity were isolated both from the osmotic shock fluid and from the first wash of the cells with Tris-EDTA by Taylor et al. [8]; this was confirmed by White et al. [11]. This binding activity was distributed into two fractions following Sephadex gel filtration. One fraction was in the excluded volume and represented the outer membrane receptor in membranous fragments. The second fraction eluted at the position corresponding to globular proteins of 22,000 molecular weight. Dissociation constants for B_{12} binding by both fractions were similar, 6 nM [8], or 0.3 to 0.8 nM [11]. The amount of the lower molecular weight, presumably periplasmic B_{12} binding protein, corresponded to the release of two to five molecules per cell. The role of this protein in transport is not known. It appears to have similar specificity and regulation as the overall transport process [12]; none of the B_{12} transport mutants isolated so far have altered levels of this protein [13].

Rather little is known about the secondary phase of B_{12} uptake. It is dependent on the function of the tonB gene, mutations in which also result in loss of iron chelate transport employing outer membrane receptor proteins ([14]; cf, Chap. 9). Based on the effect of uncouplers and anaerobiosis on B_{12} uptake in wild-type strains and in mutants lacking the Ca^{2+}, Mg^{2+}-ATPase, Bradbeer and Woodrow [15] proposed that the driving force for B_{12} accumulation is the proton motive gradient rather than a phosphorylated intermediate.

2. Lactobacilli

Lactobacillus leichmanii requires B_{12} for growth, in part because B_{12} is an essential cofactor for the ribonucleotide reductase of this organism [17]. Cells of this and related strains bind quite large amounts of B_{12}, up to 0.5 μg/mg dry weight [18]. The majority of this was bound to a protein (near 22,000 molecular weight) located in the cell wall fraction [14]. Material bound to the wall apparently could subsequently enter the cell and be utilized to support growth. The remainder of the B_{12} appeared to be bound to a similarly sized protein associated with the ribosomes [19,20]. This ribosomal B_{12} binding activity could be released by salt washing; this activity was also found on ribosomes of E. coli, and the reassociation of the protein with ribosomes did not demonstrate species-specificity [21].

The uptake of B_{12} must pose a particular difficulty for cells, probably owing to its size. In both gram-negative and gram-positive bacteria, the involvement of a receptor protein external to the cytoplasmic membrane is

TABLE 10-1 Properties of Some Vitamin Transport Systems

Strain	Conditions	K_m (nM)	V_{max}^a	pH_{opt}	$Temp_{opt}$	Intracellular Species	Ref.
Thiamine							
L. fermenti		480		6.8	50	TPP; Unaltered in P_i-starved	25
E. coli	Uptake:	830	110	7.2		TPP	29,31,32
	Binding to binding prot:	20					
S. cerevisiae		180		4.5		Unaltered	44
Folates							
L. casei	PteGlu	45	24			Gradually converted to polyglutamates and other derivatives	52
		350					53
		28	14				58
	5-methyl-H4-PteGlu	900	23.7				58
		27	15.6				
	PteGlu3	320	13.5				
	PteGlu4	1,900	6.0				
	PteGlu5	3,700					
P. cerevisiae	5-formyl-H4PteGlu	400	100	6.0			62
	5-methyl-H4PteGlu	400	160				

								Ref.
Folate-responder	PteGlu	6,600	400	5.6				
Biotin								
L. plantarum	− glucose	7.7	3.3	None	None		Free d-biotin	71,72
	+ glucose	31.5	9.7	5.6, 7.4	37			
S. cerevisiae		330	39	4.0	30		Both free and bound	73
E. coli K12		140	6.6	6.6	37		Free and unaltered	78
Pyridoxine								
S. typhimurium	Pyridoxine:	200		8.1			All as the phosphate	82
	Pyridoxal	120					or pyridoxal	
	pH:	3.5	6.0	3.5	6.0			
S. cerevisiae	Pyridoxine	460	790	40	30	30	Unaltered, at least	84
	Pyridoxal	1740	8510	34	--	&	initially	
	Pyridoxamine:	20600	2200	--	11.6	50		

[a]Picomoles of substrate accumulated per minute per milligram dry weight.

obvious. Whether this receptor is necessary for the complete transport process is not known, but not impossible. Certainly, this appears to be the situation in eucaryotic cells in which B_{12} uptake requires its prior binding to specific serum proteins, primarily transcobalamin II [22,23]; this entire complex may then be taken up by the cell.

B. Thiamine (Vitamin B_1)

Just as B_{12} uptake is of special interest owing to the involvement of the extracytoplasmic membrane receptor, thiamine uptake has received considerable attention focusing on both the role of periplasmic binding proteins in transport and the nature of the uptake observed in the absence of the metabolic trap which normally functions.

Studies of thiamine uptake in Lactobacillus fermenti by Neujahr [24] in 1963 had shown that thiamine-deficient cells had higher levels of energy-dependent (glucose or ATP) thiamine uptake activity than did thiamine-sufficient cells. The specificity of uptake was shown by the inhibition by the analog pyrithiamine and by the phosphorylated derivatives of thiamine [25]. Under normal conditions, most of the transported thiamine was rapidly converted to thiamine pyrophosphate [TPP], the normal coenzyme form of this vitamin [26]. However, the rate of phosphorylation could be markedly reduced by starvation for phosphate without a concomitant decrease in the rate of uptake. Under these conditions, up to 70% of the total pool could be composed of free thiamine, representing a concentration gradient of about 250-fold. She showed that thiamine uptake was also present in a number of other bacterial species, although L. fermenti exhibited by far the highest level of uptake and the most dramatic response to thiamine starvation and glucose supplementation [27].

As is often the case, a more detailed analysis of thiamine uptake was possible in E. coli, owing in large part to the availability of mutants blocked in thiamine synthesis and transport. In this organism, thiamine is synthesized by the condensation of a hydroxymethyl pyrimidine-P moiety and the thiazole-P moiety to form thiamine phosphate [TP], and its phosphorylation to TPP. Free exogenous thiamine is converted to TPP by a membrane bound thiamine kinase activity [28] which appears to catalyze two sequential phosphorylation steps separable in mutants. Exogenous thiamine accumulated as TPP even at the earliest time point measured, with essentially no detectable thiamine in the cells [29]. The K_m of the kinase for thiamine was in the range of the K_m for transport (800 nM), so a considerable degree of phosphorylation was not unexpected. The important question was whether transport required thiamine kinase action, possibly as a group translocation process.

Evidence for a transport component separate from the kinase was provided by the finding that the analogs pyrithiamine and oxythiamine competitively inhibited thiamine uptake at concentrations below those which inhibited the kinase in vitro [29]. Furthermore, starting from a mutant blocked in the synthesis of the thiazole moiety, a double mutant was obtained which required greatly increased levels of thiamine supplementation [30]. Although the mutant required 150-times higher levels of thiamine than the parental strain, it was unaltered in its response to thiazole, which does not compete with thiamine for uptake. Direct assay revealed markedly reduced rates of thiamine uptake, but normal rates of efflux. The levels of four of the thiamine biosynthetic enzymes and of the membrane bound thiamine kinase were near normal. The properties of this mutant showed that at least a portion of the transport system is independent of the thiamine kinase.

Several laboratories have reported that thiamine uptake in several strains of E. coli is sensitive to osmotic shock [31–35]. A thiamine binding protein of 32,000 to 36,000 molecular weight was released into the shock fluid. Thiamine binding to this protein exhibited a dissociation constant of 20 to 29 nM and a similar substrate specificity as the uptake system [31,32]. Both uptake activity and the level of this binding protein responded similarly to repression or derepression by supplementation with thiamine or adenine in the growth medium, respectively [33,34,36]. There is little further information on the role of this binding protein in transport in E. coli.

There are some interesting results concerning the role of phosphorylation in transport. Both TP and TPP are very effective inhibitors of thiamine uptake and its binding to the purified binding protein. They are also fully capable of supporting the growth of a thiamine auxotroph; this utilization was greatly depressed in the thiamine transport mutant [37]. This finding prompted the isolation of TP- and TPP-requiring mutants, defective in the separate steps of phosphorylation of exogenous thiamine [38–40]. Growth of the TPP$^-$ auxotroph was inhibited by TP or free thiamine, in parallel with their inhibition of TPP uptake [41]. This indicates that thiamine and its phosphorylated derivatives share a common uptake system, of which the periplasmic binding protein is most likely a constituent. Uptake of thiamine into the TPP$^-$ auxotroph was identical to that into the parent strain, except that all of the intracellular material accumulated as TP [39,40]. The mutant blocked in the first kinase step (responding to TP or TPP, but not to thiamine) also transported thiamine. The rate of thiamine uptake into this strain in the absence of glucose was comparable to that into the other strains under identical conditions. Free thiamine was accumulated into these cells at concentrations 25-fold higher than the external. However, the addition of glucose, which stimulated uptake in the other strains, caused a pronounced efflux or expulsion of intracellular free thiamine from the TP auxotroph [39,40]. Thus, phosphorylation appears to be necessary for the effective accumulation of thiamine in the cell even though free thiamine can be transported and perhaps even accumulated without obligatory phosphorylation.

These results raise questions concerning the nature of energy coupling in this system. It could appear that thiamine enters the cell by facilitated diffusion and is then trapped within the cell by its phosphorylation, as has been proposed for some sugar transport systems. This view loses credence with the realization that the phosphorylated forms are substrates for the transport system for uptake, and presumably also for efflux. The apparent accumulation of free thiamine in the absence of glucose and its expulsion upon addition of glucose is curious. It is crucial to know whether the phosphorylated forms are able to leave the cell, either in the presence or absence of metabolic energy. The mechanism of energy coupling in this system remains an interesting problem for future study.

There are separate and distinct uptake systems for the two constituent moieties of thiamine in E. coli. Hydroxymethyl pyrimidine, but not its mono- or diphosphate esters, could be utilized by the appropriate auxotroph [42]. Uptake of hydroxyethylthiazole was dramatically stimulated by., if not dependent on, the presence and metabolism of the other moiety, hydroxymethyl pyrimidine [43].

Thiamine transport in the yeast Saccharomyces cerevisiae is similar in general outline to other vitamin transport systems in this organism, as will be shown later. There was extensive accumulation of thiamine, primarily in the free form [44]. Mutants specifically defective in thiamine uptake were obtained by selection for resistance to the analog pyrithiamine [45]. Finally, uptake in the parental strain appears to be subject to regulation by the repression controlled by growth in exogenous thiamine [46]. A thiamine binding activity was found in these cells, but its location and nature is unresolved [46].

C. Folates

Folate transport is of primary interest owing to its relationship to the pathways of one-carbon metabolism which are primary targets for chemotherapy directed against both numerous infectious agents and several oncogenic states. The action of such folate antagonists as the sulfonamides, amethopterin (Methotrexate), and trimethoprim has focused attention on the synthesis and interconversions of the folate derivatives (especially on the role of the dihydrofolate reductase reaction), and on their transport. Discussions of folate metabolism and transport become quite complicated owing to the almost bewildering number of folate derivatives differing in oxidation state, number of glutamyl residues, and the nature of the substituents at N^5, N^{10}, or elsewhere on the pteroyl moiety, each of which is necessary for some different reaction in these pathways.

It may come almost as a relief to the reader to find no mention of folate transport in E. coli. The capacity to transport folates is rather rare. This is to be expected from the susceptibility of many bacterial species to sulfonamides even in the presence of exogenous folate. Although gram-negative bacteria do not utilize exogenous folate, many do release variable and often rather large amounts of folates and related compounds into the culture medium [47]. The last mention of E. coli in this section will be to point out the presence of dihydropteroyl hexaglutamate [H_2PteGlu$_6$, in the terminology to be employed here] and of dihydrofolate reductase as structural components of coliphages T4 and T6 [48,49]. Apparently, the reduction by NADPH of the dihydrofolate in the phage tail, catalyzed by the dihydrofolate reductase also present in the tail plate, causes some conformational change necessary for irreversible phage attachment and DNA ejection [50].

Most of the studies of folate transport have employed the naturally occurring folate auxotrophs Lactobacillus casei, which has a rather high but nonspecific requirement for folate, or Pediococcus cerevisiae. It was first shown by Wood and Hitchings [51] that these two strains and Streptococcus faecalis, but not E. coli, could assimilate folate [PteGlu] and/or 10-formyl tetrahydrofolate [10-formyl-H_4PteGlu]. Cooper [52] documented the uptake of labeled folate by L. casei and suggested that there might be multiple transport systems for different classes of folate derivatives. Recent studies in L. casei by Henderson and Huennekens [53] verified the existence of a carrier-mediated system for the uptake of PteGlu, 5-methyl-H_4PteGlu, and amethopterin. The K_m values for uptake for the latter two compounds were very similar to the K_i values for their competitive inhibition of PteGlu uptake. PteGlu uptake was also inhibited by a variety of other folate derivatives and analogs. PteGlu and amethopterin effectively accelerated the net efflux of preloaded 5-methyl-H_4PteGlu. Uptake of these three substrates were equally inhibited by exposure of the cells to iodoacetate. Finally, this activity was under apparent repression control by the level of folate supplementation in the growth medium.

Folates are usually present in cells as polyglutamate derivatives. The predominant chain length is species dependent, being the triglutamate in Bacillus subtilis [54], the tetraglutamate in S. faecalis, the hexaglutamate in E. coli, and the octaglutamate in L. casei and Lactobacillus plantarum [55]. Formation of the higher polyglutamates in L. casei was inhibited by the presence of amethopterin [56], consistent with the proposal that the tetrahydro derivatives are the substrates for glutamate addition [55]. Although most folate auxotrophs respond only to the mono- or diglutamate, L. casei can grow on oxidized pteroyl polyglutamates with as many as seven glutamyl residues [PteGlu$_7$] [57]. Accordingly, Shane and Stokstad [58] investigated the transport and metabolism of several polyglutamate derivatives. They confirmed that PteGlu, 5-methyl-H_4PteGlu, and amethopterin share an iodoacetate-sensitive active transport system; their K_m for PteGlu (30 nM) was an order of magnitude lower than that reported by Henderson and Huennekens [53], but close to that found by Cooper [52]. The transport activity of different species agreed well with the ability of these strains to respond to the various folate derivatives. For example, although the polyglutamates of 5-methyl-H_4PteGlu, which do not support growth, did appear to enter L. casei, they were not accumulated above their exogenous concentration. On the other hand, the polyglutamates of PteGlu were accumulated without prior hydrolysis. Although the K_m for uptake increased with increasing glutamyl chain length [from 28 nM for n = 1 to 3,700 nM for n = 5], the V_{max} for uptake remained relatively constant. The polygluta-mates were competitive inhbitiors of the uptake of both PteGlu and 5-methyl-H_4PteGlu; the similarity of the K_i values for uptake was consistent with the involvement of a single transport system.

Shane and Stokstad [58] showed further that transported PteGlu was converted to higher polyglutamates, the addition of the third glutamyl residue being the rate-limiting step. Exit of labeled folate was first-order, stimulated by glucose, and enhanced by the presence of external unlabeled folate. When cells were loaded in the presence of 500 nM PteGlu, the bulk of the internal folate was as the diglutamate. In this case, the exit rate was relatively independent of the time of incubation with labeled substrate prior to dilution. On the other hand, cells loaded with 7 nM PteGlu formed appreciable amounts of the higher polyglutamates; in this case, the rate of efflux decreased dramatically with the time of loading. The folate that was released comprised various derivatives, but only of the mono- or digluta-mate. Hence, the progressive decline in rate of exit presumably resulted from the conversion to polyglutamates which were unable to leave the cell. This is another example of the peculiar form of metabolic trap seen with thiamine transport in which a metabolic product which appears to be unable to leave the cell is nonetheless a substrate for the transport system. A possible answer to this enigma could be that the coenzyme forms (thiamine pyrophosphate or folate polyglutamates) bind so tightly to intracellular enzymes that their level in the cytoplasm is never high enough to allow significant efflux. The polyglutamates were more effective at supporting growth [58], and this may be related to their more efficient binding to many enzymes, including thymidylate synthetase [59] and dihydrofolate reductase [60].

S. faecalis requires and transports oxidized folates [1]. It does not respond to 5-methyl-H_4PteGlu; this compound could be accumulated, al-though without further metabolism [61].

The other folate transport system studied in detail is that of P. cere-visiae, which normally utilizes folinate [5-formyl-H_4PteGlu] for growth, but not PteGlu or 5-methyl-H_4PteGlu. Wild-type strains could transport both reduced folates [62]. Although the methyl derivative was not meta-bolized or utilized, it, along with H_4PteGlu, did share the transport system with folinate, suggesting a specificity for reduced folates. Several labor-atories have reported the isolation of mutants which respond to folate; this was associated with the acquisition of the capacity to transport PteGlu [63, 64]. Folate uptake into these mutants was carrier-mediated, energy-dependent, and concentrative. It was competitively inhibited by the reduced folates and amethopterin, suggesting that the mutation altered the existing carrier to broaden its substrate specificity and allow the binding of oxidized folates. Transport in this folate-utilizing strain was also altered in the response to 5-methyl-H_4PteGlu (sevenfold increase in V_{max}), along with the appearance of sensitivity to amethopterin [65]. This altered uptake system was still subject to repression. From this folate-utilizing strain, amethopterin-resistant mutants were selected. One such mutant, which was 1000-fold more resistant to amethopterin, had 60-times higher activity of a somewhat altered dihydrofolate reductase [66]. In addition, its transport system was altered, such that it exhibited a 10-fold lower affinity for amethopterin [67].

One of the main attractions of the study of vitamin transport is the possibility that the quite high substrate affinity of these transport systems might enable the detection and isolation of the putative membrane carriers. Along this line, some very intriguing results have recently been presented by Henderson and Huennekens [68]. They showed that cells of L. casei, grown in limiting folate and assayed at 4°C to reduce transport activity, exhibited folate binding. This binding was rapid, saturable, and, in contrast to the complete uptake process, insensitive to sulfhydryl reagents. The binding isotherm revealed only a single component whose dissociation constant [36 nM] was very close to the K_m for folate transport reported by Shane and Stokstad [58]. Folate binding activity, representing at least 80% of the folate binding activity of the cell, was solubilized from lysozyme-treated cells by sonication in the presence of 5% Triton X-100 and 5 μM folate. A protein was purified to homogeneity and had a subunit molecular weight of 28,000. The levels of folate transport, folate binding by whole cells, and this folate binding protein responded identically to the repressive effects of growth with increasing levels of folate supplementation. Substrate had to be present during solubilization for recovery of binding activity; substrate thus bound was not available for exchange with added folate and was released only by denaturation of the protein.

This technique of solubilization in the presence of substrate has recently allowed the detection and isolation of analogous membrane-associated binding proteins specific for thiamine, nicotinate, and biotin [69]. The involvement of these proteins in the transport process has not yet been proven, but this is clearly a most interesting approach.

D. Biotin

The transport of biotin in several species has been extensively investigated. One interesting aspect for consideration here concerns the regulation of uptake activity, especially the relationship of the control of the transport system to the control of the biotin biosynthetic enzymes.

Following preliminary indication of a carrier-mediated process [70], the accumulation of biotin was first studied in the biotin auxotroph L. plantarum by Waller and Lichstein [71]. The temperature-, pH-, and energy-dependent uptake was measured by bioassay, and transported biotin was found to be either free and unaltered (extracted by boiling water) or bound by covalent attachment to enzymes (released only by acid hydrolysis). The only major metabolic alteration undergone by biotin in its conversion to the coenzyme is its covalent attachment to apoenzymes, such as various carboxylases. In the absence of glucose, there was a low level of uptake which was independent of pH and temperature, but which was competitively inhibited by the analog homobiotin. Addition of glucose allowed extensive accumulation of free biotin and also increased the rate of its attachment to protein. At low external biotin concentrations, the glucose stimulation was entirely represented by the increased conversion to the bound form. Energization increased the apparent K_m for biotin from 7.7 to 31.5 nM [72].

Uptake into the biotin-requiring yeast, S. cerevisiae, was measured with isotopically labeled biotin by Rogers and Lichstein [73]. As in the previous case, transported material was bound or free and unaltered, and uptake exhibited all the features expected for a carrier-mediated process. Concentration gradients of over 1000-fold could be generated upon addition of glucose. Two regulatory phenomena were observed with this system. During uptake, an overshoot occurred such that, after a period of continual uptake, a partial net efflux of accumulated biotin occurred, usually down to a final steady-state level. This overshoot is not unique to this system and will be mentioned later. Second, the activity of the system, measured by initial rates, was lower in cells grown in the presence of excess biotin [74]. The increase in activity following removal of exogenous biotin by the addition of avidin required protein synthesis. Further evidence for the operation of a repression control was that activity was not inhibited by high intracellular biotin pools per se, and that this regulation affected only the V_{max} component of uptake [75]. It had previously been shown that the energy-dependent portion of biotin uptake in L. plantarum was subject to similar regulation [76].

The accumulation of biotin by prototrophic strains of E. coli was demonstrated first by Pai [77] and then by Prakash and Eisenberg [78]. Transported vitamin was primarily in the free form in biotin prototrophs; considerable portions were protein bound in biotin auxotrophs. The properties of this system resembled those in yeast, with a temperature and pH optimum for transport close to the optimal growth conditions for that organism. The system was stimulated by glucose to maintain a maximal concentration gradient of almost 40-fold. Substrate specificity was also similar to that in yeast, being fairly broad and requiring only an intact ureido ring. Some substrate was bound at 0°C, but this was discounted as nonspecific adsorption. Biotin uptake was not affected by osmotic shock, and no biotin binding activity was released by this procedure [35].

A mutant with depressed biotin uptake (10% of wild-type initial rate and steady-state level) was found among strains resistant to the growth inhibitory analog α-dehydrobiotin [79]. Based on the normalcy of proline uptake, this strain was considered defective in a specific component of biotin uptake. The affected genetic locus, termed bioP, was placed at 83 min, very near metE. Similar mutants were described by Pai [80], who showed that the residual biotin uptake activity was less sensitive to competitive inhibition by the analog. Another class of dehydrobiotin-resistant mutants, termed P, had decreased uptake of both biotin and proline, and presumably carried some generalized defect of membrane function [79].

As in other species, biotin uptake in E. coli was subject to repression by growth in the presence of biotin [78]. Mutants in bioR (dhbB), obtained by selection for dehydrobiotin resistance [79, 80], exhibit constitutive and elevated production of the biotin biosynthetic enzymes. These strains accumulate and excrete large amounts of biotin and its precursors, and these strains had reduced levels of biotin uptake [77]. However, a bioR strain that was also blocked in biotin synthesis exhibited normal regulation of the biotin uptake activity, as controlled by the level of biotin supplementation [81]. Thus, the depressed uptake in the bioR mutant was the result of the repression of the uptake system by the high intracellular levels of biotin. This further indicated that the specific regulatory system for the biotin permease is distinct from the regulatory system governing the biotin biosynthetic enzymes. This points out the hazards in trying to define the regulatory controls of transport systems; careful and extensive genetic analysis is requisite.

E. Pyridoxine (Vitamin B$_6$)

Pyridoxine transport is worthy of mention for two reasons. First, it is extremely complicated, with numerous possible substrates, considerable intracellular metabolism, and the availability of various mutants. Second, it places in sharp focus the often radical difference in transport properties between bacteria and yeasts or fungi.

Cells of Salmonella typhimurium exhibit a single system for the uptake of pyridoxine and pyridoxal, but not pyridoxamine [82]. Although these substrates could be concentrated up to 25-fold, they were present primarily as phosphorylated derivatives. This fact, combined with the lack of stimulation by glucose and the lack of effect of the usual energy poisons, led to the proposal that pyridoxine enters via a facilitated diffusion mechanism and is then trapped through the action of pyridoxal kinase. There was no indication of a pyridoxine binding protein released by osmotic shock when precautions were taken to eliminate the nonspecific binding by contaminating amounts of pyridoxal.

Pyridoxine uptake in E. coli, although stimulated by glucose and inhibited by DNP, was also associated with the almost complete conversion of the vitamin to the phosphorylated forms [83]. Here too, there was no concentration gradient of free vitamin. So, until direct evidence is presented, it is reasonable to presume that E. coli also exhibits only facilitated diffusion-mediated uptake coupled with the metabolic trap of a kinase, which may represent the step of energy dependence.

Shane and Snell [84] have provided a most extensive analysis of pyridoxine uptake in the yeast Saccharomyces carlsbergensis. Kinetic analysis revealed the operation of two separate uptake systems. Under conditions in which both systems are functional, the initial intracellular product of pyridoxine uptake was free pyridoxine (90% of the intracellular vitamin after 10 min). This was subsequently metabolized, primarily to pyridoxamine-P and pyridoxal-P. The two transport systems differed in their substrate specificities, such that one (optimal pH, 3.5) transported pyridoxal effectively, but pyridoxamine poorly; the other (optimal pH, 6.0) transported pyridoxamine well, but not pyridoxal. Pyridoxine was transported by both, and its phosphate by neither. This substrate specificity was reflected in the ability of these substrates to inhibit pyridoxine uptake at these appropriate pH values. The 5'-deoxy analogs were very effective inhibitors of uptake, consistent with the conclusion that phosphorylation was not required for transport. Both systems exhibited dual temperature optima (30° and 50°C). Substrate accumulation was energy dependent and afforded considerable concentration gradients. Under certain conditions, this system exhibited the overshoot phenomenon.

This is but another example of the usual situation in the fungi, in which substrate is accumulated in unaltered form against a considerable concentration gradient by an energy-dependent process. In most cases, this contrasts with the uptake of the same substrate in bacteria in which extensive metabolic alterations are observed. This is fully consistent with recent demonstrations that other transported substrates in fungi are sequestered into a relatively nonaccessible pool, perhaps in membrane bound vesicles [85-87]. This makes it very difficult to determine the site of energy coupling for fungal transport, whether it is at the cytoplasmic membrane or at the step of entry into the storage vesicle.

F. Other Vitamins

Some information is available concerning the transport of several other vitamins. They are placed in this section either because the characterization is incomplete or because they do not portray unique features of note.

1. Niacin

Little information was found concerning the transport of nicotinate (niacin). Mutants blocked in the synthesis of nicotinamide adenine dinucleotide (NAD) are known [88, 89], and these will respond to nicotinate, nicotinamide, or NAD [90]. Mutants selected for resistance to 6-amino-nicotinamide [91] lacked activity of nicotinamide deamidase and had lost the ability to respond to nicotinamide or NAD. This was taken to indicate that nicotinamide is not an intermediate in the synthesis of NAD and that NAD cannot be utilized directly but must be broken down to nicotinamide prior to entry into the cell [90, 92]. Cells did take up nicotinamide much more efficiently than nicotinate [93].

2. Pantothenate

Pantothenate is a constituent of coenzyme A and of acyl carrier proteins in bacteria. Pantothenate transport has been demonstrated in both Pseudomonas fluorescens and E. coli [94, 95]. In both species, uptake was energy, pH, and temperature dependent, with a K_m near 30 μM. Transport occurred without concomitant phosphorylation. In the Pseudomonas strain, transport activity was induced by the presence of pantothenate or its analogs or by various utilizable carbohydrates [96]. Transport activity was reduced by osmotic shock treatment, concomitant with the release of pantothenate binding material [94, 95]. The level of this inducible binding protein correlated with the level of pantothenate uptake in P. fluorescens [97].

3. Lipoic Acid

The uptake of lipoic acid by a variety of bacterial species was demonstrated [98]. It is uncertain whether this uptake is carrier mediated, since this substrate could dissolve in the membrane and appear to be concentrated as a function of the pH values on the two sides of the membrane. It is now known that one result of the energization of a cell is the establishment of a proton motive force, which includes both a proton diffusion potential (interior negative) and a pH gradient (interior alkaline) [99,100]. This pH difference can be used to trap the nondiffusible anion of a lipid soluble acid.

G. Summary

The general impression of vitamin transport in bacteria is that most of these systems do afford accumulation of their substrate with high efficiency from the medium, and that these systems do employ specific membrane bound carriers. In a few cases, there are mutants lacking some component of the specific vitamin uptake process. However, with a few exceptions, most of the vitamins are accumulated by the sequential process of the carrier-mediated translocation of substrate across the cytoplasmic membrane, followed by the metabolic alteration of the substrate by an energy-requiring step, such as kinase action. Note that this situation is not that simple because, in some cases, the intracellular product (thiamine pyrophosphate or folate polyglutamates) is itself a substrate for the transport system.

Although the existence of any cellular system is its own justification, the evolutionary or selective advantage of vitamin transport and the nature of their regulation could be questioned. Obviously, a strain auxotrophic for a vitamin would be expected to have a highly efficient uptake system for that substrate. The need is less obvious for prototrophic strains. Apparently, the ability to acquire preformed vitamins from the environment provides a metabolic savings in excess of the cost of formation and operation of the transport system. Most of the vitamin transport systems were shown to be subject to repression by the presence in the growth medium of sufficient levels of the vitamin. This means that, as the external concentration of the vitamin increases, the cell economically reduces the number of carriers to a level such that the cell's vitamin requirements can still be met by the reduced number of carriers. However, this must pose a problem for cells in an environment very low in that vitamin. Not only is the transport system, which is derepressed under these conditions, a means of acquisition, but it also represents a potential leak of endogenously synthesized vitamins from the cell. This problem is easily solved if the intracellular form of the vitamin is not a substrate for the transport system. The other systems may have circumvented this problem by the operation of a gated efflux mechanism whereby the rate of influx exhibits its usual hyperbolic dependence on substrate concentration, but efflux does not proceed until a set intracellular concentration of that vitamin has been reached.

It is possible that this type of regulation explains the observation of the overshoot phenomenon observed in several vitamin transport systems in yeast, such as those for biotin and folate. Uptake proceeds at a linear rate for up to 20 to 30 min, after which an efflux process appears to be activated, such that there then ensues the net loss of material from the cell. This does not appear to represent a sluggish form of counterflow or the trapping of transported material by intracellular pools of unlabeled vitamin [84]. Rather, this phenomenon probably is an expression of a process frequently observed with fungal transport systems, termed transinhibition, in which the influx process is responsive to the level of substrate already in the cell. Becker and Lichstein [101] proposed for biotin uptake by S. cerevisiae that influx proceeded at constant rate until the internal pool of the vitamin reached high enough levels to bring about this transinhibition; this would inhibit the

rate of influx while allowing efflux to proceed at rates determined solely by
the internal concentration of substrate. The situation is probably more
complicated than this owing to the probability that transported vitamins, in
fungi, can also be sequestered into vesicles where they are inaccessible to
the transport systems [85-87], perhaps until the vesicles are loaded. The
application of these explanations to bacterial transport systems is doubtful,
since vesicles and transinhibition processes are rare, if not nonexistent.

In conclusion, it may be worthwhile to repeat one of the major attractions of the study of vitamin transport: the isolation of membrane carriers.
The high binding affinity of carrier for substrate should improve the chance
of its detection. The apparent lack of energy coupling for the actual transport process may overcome the possible impediment raised by Kaback and
his colleagues [102,103], that the "unenergized carrier" is either inaccessible or has such a low substrate affinity as to be undetectable. By no means
has this area of research been fully explored.

III. ANTIBIOTIC TRANSPORT

A. General Features

Nowhere is the evolutionary widsom for, or even the possession of, the
existence of a transport system less obvious than in the case of the antibiotics. This subject, which has been reviewed by Franklin [104], is still
rather neglected and detailed pictures of the mechanisms of antibiotic entry
are very rare. This is perhaps rather surprising, since in most cases the
major determinant of the susceptibility of an organism to any agent is that
agent's ability to achieve a sufficient concentration at the target site.
Three general mechanisms of uptake could be proposed: passive diffusion
based on the dissolution of a hydrophobic antibiotic in the membrane, and
its distribution in response to pH gradients or macromolecular binding;
entry mediated by a carrier specific for that antibiotic or its structural
analogs; or entry mediated by a transport system for a normal nutrient.
Diffusion through water-filled pores may explain translocation across the
outer membrane of the gram-negative bacteria, but is unsatisfactory as an
explanation for passage across the cytoplasmic membrane. The third
mechanism (use of a nutrient transport system) is certainly employed for
the entry of various antimetabolites, as will be shown in Sec. III.B. Evidence for the discrimination between the first two mechanisms is often not
available, but some salient findings will be presented in later sections. In
addition to the question of the existence of specific transport systems, the
existence and mechanism of energy coupling processes is even more open
to question, especially in light of the extensive macromolecular binding
undergone by these substrates. Perhaps in no other field is the design
and interpretation of transport assays more challenging.

Antibiotic transport is a field in which genetic approaches are quite important to the identification of the existence of specific transport systems. The literature on antibiotic resistance is enormous, but primarily defines either mutational alteration in the target site, such as the ribosome, or the plasmid-coded resistance associated with the production of drug-inactivating enzymes. These will not be discussed here. There is another large literature concerning the intrinsic resistance of gram-negative cells to a wide variety of antibiotics. This resistance is predominantly a result of the unusual permeability properties of the outer membrane, and also is beyond the scope of this chapter. This therefore excludes from consideration the uptake of the penicillins, cephalosporins, and polymyxins, and other antibiotics which act exterior to the cytoplasmic membrane and for which transport systems are not necessary. It is important to remember, though, that the outer membrane does pose a quite significant barrier to the entry of a large number of antibiotics and, in fact, is the primary barrier to the uptake of a large number of antibiotics. This is clearly demonstrated by the acquisition of sensitivity by cells in which the integrity of the outer membrane has been compromised, either by removal of the cell wall during spheroplast formation, partial removal of lipopolysaccharide by treatment with chelators, or by mutational events leading to the formation of lipopolysaccharide with severely truncated polysaccharide chains. This chapter is concerned only with the existence of specific transport systems in the cytoplasmic membrane.

B. Antimetabolites

The term antimetabolites will be applied to those potential chemotherapeutic agents that are analogs of normal cell constituents and which inhibit growth by interfering with the metabolism of that compound. Such compounds would be expected to enter cells by means of the transport system for the normal compound, just as amino acid analogs enter on the amino acid uptake system.

One example which is well documented is the uptake of D-cycloserine, which is a competitive antagonist of D-alanine in reactions of cell wall biosynthesis [105]. Mora and Snell [106] showed that it competed with D-alanine for uptake into S. faecalis, although D-, but not L-, cycloserine also competed with L-alanine. Neuhaus and his colleagues [107] showed that there are probably transport systems for L- and D-alanine, and that glycine and D-cycloserine share the D-alanine transport system in E. coli. There are two D-alanine uptake systems, both of which carry glycine and D-cycloserine. This was verified by the isolation of D-cycloserine-resistant mutants, and the demonstration that they had sequentially lost the D-alanine transport activities in parallel with their acquisition of resistance to D-cycloserine. They remained unchanged in their sensitivity to L-cycloserine [108].

An analogous situation was seen for fosfomycin, which is an analog of phosphoenolpyruvate and is lethal to cells by virtue of its blockage of the phosphoenolpyruvate-dependent reaction in the formation of muramic acid for cell wall biosynthesis [109]. Fosfomycin can enter bacterial cells on either of two inducible transport systems, that for α-glycerol phosphate or that for hexose phosphates [110,111]. Mutations at many sites give rise to resistance to this compound, either by producing an alteration of the target enzyme [112], or by loss of one or both of the transport systems [113].

Another example is the case of the sideromycins, which are complex molecules structurally related to the iron-chelating siderophores [114]. They enter via the siderophore transport systems and then kill by means of molecular groupings attached to the siderophore-like moiety. Their uptake is blocked by the presence of the natural siderophore or by mutational loss of that siderophore's uptake system [115,116].

A final interesting example is provided by naturally occurring antimicrobial agents which contain an inhibitory moiety linked to a peptide containing normal amino acids [117]. This is taken into the cell by the very nonspecific peptide transport system, free from competition by the endogenous level of free amino acids.

C. Antibiotics

There is no evidence at present that antibiotics (naturally occurring inhibitors of macromolecular synthesis) enter the cells by means of usual nutrient transport systems. There is even considerable doubt whether they enter via transport systems at all. At the present time, genetic approaches have not provided definitive information. Among mutants selected for resistance to any antibiotic are those which provide only a limited degree of resistance and in which the target reaction is still sensitive to the antibiotic.

It is presumed in most cases that these represent decreased uptake of the drug. For example, streptomycin-resistant mutants of S. typhimurium fall into two classes: those mapping at strA produce altered ribosomes which do not bind streptomycin, whereas those mapping distantly at strB are resistant to several aminoglycoside antibiotics and have sensitive ribosomes [118]. Other examples of these two mechanisms of resistance include nalidixic acid in E. coli [119] and aminoglycoside in Mycobacterium tuberculosis [120] and B. subtilis [121]. Does this mean that the mutations to low-level resistance result in loss of a specific antibiotic transport mechanism? In contrast to the high-level resistance, the low-level resistance is frequently associated with considerable cross-resistance to similar or unrelated antibiotics. For example, in E. coli, some of the mutants selected for resistance to chloramphenicol (cmlB) have acquired partial resistance to tetracycline [122]. This example is presented not to imply

that these different antibiotics share a transport system, but to point out
the hazards which especially prey on investigators studying gram-negative
bacteria; these mutations quite likely are alterations of the outer membrane.

There are two lines of evidence which may be applicable to the problem
of antibiotic uptake. These come from studies of the uptake of the tetracyclines and of the aminoglycosides.

1. Tetracyclines

The tetracyclines are a class of effective broad-spectrum antibiotic inhibitors of protein synthesis whose basic structures are based on four substituted six-membered rings and at least one basic dimethylamino moiety.
Taking advantage of their high coloration, it was found that they are accumulated by bacteria and that this accumulation required glucose and Mg^{2+}
[123]. They appear not to be transported by mammalian cells, although
their ribosomes are sensitive to these drugs [124].

Isotopically labeled tetracyclines are accumulated by an energy-
dependent process from rather low external concentrations [125,126].
Loaded tetracycline was free to exit and this efflux was stimulated by
energy poisons [127], arguing that the observed accumulation was not
solely the result of binding to cellular macromolecules. There was rather
appreciable energy-independent binding in both Staphylococcus aureus [127]
and E. coli [128]. This binding is probably reflected in the energy-independent and nonconcentrative binding of tetracycline (K_m = 30 μM) to membrane vesicles derived from E. coli [129]. Since the preparation of these
vesicles involved treatment with the detergent Triton X-100, which effectively solubilizes the cytoplasmic membrane [130], it is most likely that
they represent vesicles from the outer membrane alone and would not be
expected to exhibit energy-dependent processes. The tetracyclines have a
high affinity for divalent cations [131], and these appear to stimulate uptake
[123].

The evidence for a tetracycline carrier in the membrane is still somewhat inconclusive. The energy-dependent uptake of tetracycline, measured
after exposure of cells to drug for 10 min, was a linear function of the
external concentration from 6 to 600 μM. Over this range of concentrations,
there was no evidence of saturability nor of the inhibition of tetracycline
uptake by chlorotetracycline [132]. A similar result was seen in studies of
the transport of tetracycline in relation to the closely related minocycline,
which differs in its possession of an extra dimethylamino moiety. In this
study, azide had little effect on tetracycline uptake, but completely inhibited that of minocycline [133]. The sulfhydryl reagent, p-hydroxymercuri-
benzoate, inhibited tetracycline uptake, but augmented that of minocycline.
There was no inhibition by either drug of the uptake of the other. These
results argue against the presence of a specific carrier. As will be men-

tioned later, the plasmid R64 conferred moderate resistance to tetracycline, associated with decreased uptake of the drug. The uptake of minocycline was little affected by the presence of the plasmid.

The fluorescence of chlorotetracycline (ClTet) is enhanced by chelation with Mg^{2+} or Ca^{2+} and further enhanced by transfer to a nonpolar environment. The addition of ClTet to cells of S. aureus resulted in a time-dependent enhancement of fluorescence, which was thought to represent the transfer of the drug-cation complex into the membrane to a nonpolar region inaccessible to exogenous EDTA or the quenching Mn^{2+} ion [134,135]. This apparent uptake was energy dependent and saturable (K_m = 107 μM). It is possible that the saturation resulted from either the concentration-dependent self-quenching of the drug or the saturation of the membrane with the drug such that larger amounts were transferred to the aqueous cytoplasm. The temperature dependence of this process could be explained in terms of the movement or dissolution of the drug complex in the fluid regions of the membrane.

These results are consistent with the absence of a specific carrier, and with the dissolution of the substrate, possibly as a magnesium-tetracycline complex, in the hydrophobic regions of the membrane. The apparent energy-dependent accumulation can represent simply the distribution of a lipid-soluble ion across the membrane in response to a potential or pH gradient.

Many transferrable plasmids (R-factors) in E. coli confer resistance to tetracycline. This resistance is induced by growth in the presence of sublethal concentrations of tetracycline, but not other antibiotics to which the R-factor confers resistance [136]. The resistance appears to result from the decreased uptake of the drug [137-139]. Induction requires a period of protein synthesis following exposure to the inducer [138,139], but does not, at least in S. aureus, require concurrent lipid synthesis [140]. However, the evidence for decreased permeation, in contrast to depressed intracellular binding, is still not conclusive, especially in light of the uncertainty concerning the mechanism of this drug's uptake. R-factor-mediated tetracycline resistance appears to be transiently overcome by osmotic shock treatment [141]. There has been a preliminary report describing an R-factor-specific tetracycline-inducible protein inserted into the membranes of minicells derived from R^+ cells [142]. Furthermore, a recent report described the presence of specific antigens present in the envelopes of cells of E. coli or S. aureus resistant to tetracycline [143]. There was no immunological cross-reaction between these two antigens, which had molecular weights in the range of 50,000 and 32,000, respectively. It is difficult to see how the addition of a protein to the membrane could specifically retard the passive permeation of this drug, and future investigations with this system should be addressed to the detailed characterization of the mechanism of tetracycline uptake and resistance.

2. Aminoglycosides

The aminoglycosides present a similar, although more complex, situation. These compounds, typified by streptomycin, are large, water-soluble, and strongly cationic. They are bactericidal inhibitors of protein synthesis. Despite numerous studies, the mechanism of their entry is still not well understood. The uptake of streptomycin into Bacillus megaterium was linear with time and was reduced by chloramphenicol, anaerobiosis, or low temperature [144]. In E. coli, most investigators have found a very rapid initial phase which is followed, after a variable lag period, by extensive secondary uptake [145-149]. The secondary uptake coincided with the release of intracellular materials and increased permeability toward β-galactosides [145-147]. Rapid uptake of streptomycin was also seen with cells whose membranes had been disrupted with toluene or polymyxin. Davis and his coworkers [145-147] interpreted these results as showing that the initial effect of the antibiotic was to disrupt the integrity of the membrane, thereby circumventing the normal exclusion barrier and allowing the extensive binding of the drug to ribosomes, which is seen as the rapid uptake phase.

The onset of the rapid uptake phase appears to require protein synthesis [148, 149], as was also the case with tetracycline uptake. Treatment of cells with chloramphenicol or lethal levels of streptomycin prevented the appearance of the secondary phase of uptake, while not affecting the initial phase. After the onset of the secondary phase, further uptake is refractory to the presence of chloramphenicol. It was proposed that the requirement for protein synthesis represents the formation of an inducible streptomycin permease [148]. The possession of such a transport system seems very unlikely from a genetic standpoint. Another explanation would be that some step of protein synthesis is needed to allow the disruption of the membrane, which then allows the entry of large amounts of the drug.

Another interesting observation is the striking dependence of aminoglycoside action on the aerobic metabolism of the cell. These drugs are ineffective against anaerobes and even against sensitive strains of E. coli when grown under anaerobic conditions [150]. Sensitivity to these drugs is also reduced under catabolite-repressing conditions (presence of glucose) [151], which also depresses the level of the respiratory system. Also, among mutants selected for resistance to various aminoglycosides are many which are defective in oxidative metabolism. Neomycin is the most frequently used drug for the isolation of these mutants, and these can be blocked in any of a wide number of steps of aerobic metabolism [152, 153]. In addition to these results in E. coli, similar mutants have been obtained in the gram-positive bacteria Bacillus licheniformis [154] and B. subtilis [155]. Oxidative metabolism is not necessary for the the conversion of these drugs to an active form, so it is most likely that the requirement for aerobic metabolism is at the level of the transport process [156-158].

A recent study [156-158] has examined these requirements for protein synthesis and for aerobic metabolism for streptomycin uptake. The most important factor for investigators in this field to note is the marked dependence of streptomycin uptake on the composition of the growth and uptake medium. The striking dependence on protein synthesis at the time of addition of the antibiotic was again demonstrated. One important finding was that streptomycin uptake remained energy dependent even at a time at which the cells were losing intracellular adenine. This is clearly not in accord with the "membrane damage" theory of streptomycin uptake. Second, accumulation in sensitive strains of both E. coli and Pseudomonas aeruginosa was not qualitatively different from that in strains resistant to these drugs owing to alterations of their ribosomal drug binding sites. All energy-dependent binding phases were present in the resistant mutants, although reduced. This showed that a considerable portion of the observed uptake was the result of ribosomal binding, and that membrane disruption was not one of the phases of uptake. Under certain conditions, these resistant mutants could concentrate streptomycin above its concentration in the medium.

Two major questions remain. Can the existence of a streptomycin carrier be eliminated, perhaps through studies of uptake specificity and kinetics, or what is the requirement for protein synthesis? Second, what is the nature of the requirement for aerobic metabolism? It is not the requirement for a proton motive gradient, since one can be generated anaerobically by reversal of the ATPase. Also, the requirement is not just for aerobic metabolism, since many of the mutants selected for loss of some aerobic function remain sensitive to these drugs and differ from the mutants selected for drug resistance in that the latter are often defective in normal substrate transport activities [153].

There are still many gaps in our understanding of the uptake of these antimicrobial agents. The filling of these gaps may add not only to the better design of chemotherapeutic agents but also to a better understanding of the structure and function of the bacterial membrane.

REFERENCES

1. S. A. Koser, Vitamin Requirements of Bacteria and Yeasts, C. C. Thomas, Springfield, Ill., 1968.
2. B. D. Davis and E. S. Mingioli, J. Bacteriol., 60:17-28 (1950).
3. G. W. Chang and J. T. Chang, Nature, 254:150-151 (1975).
4. R. J. Kadner and G. L. Liggins, J. Bact., 115:514-521 (1973).
5. E. L. Oginsky, Arch. Biochem. Biophys., 36:71-79 (1952).
6. E. L. Oginsky and P. H. Smith, J. Bact., 65:183-186 (1953).
7. P. M. DiGirolamo and C. Bradbeer, J. Bact., 106:745-750 (1971).
8. R. T. Taylor, S. A. Norrell, and M. L. Hanna, Arch. Biochem. Biophys., 148:366-381 (1972).

9. J. E.Ford, E. S. Holdsworth, and S. K. Kon, Biochem. J., 59:86-93 (1955).
10. E. L. Smith, Vitamin B$_{12}$, Wiley, New York, 1960.
11. J. C. White, P. M. DiGirolamo, M. L. Fu, Y. A. Preston, and C. Bradbeer, J. Biol. Chem., 248:3978-3986 (1973).
12. R. T. Taylor, M. P. Nevins, and M. L. Hanna, Arch. Biochem. Biophys., 149:232-243 (1972).
13. P. M. DiGirolamo, R. J. Kadner, and C. Bradbeer, J. Bact., 106:751-757 (1971).
14. P. J. Bassford, Jr., C. Bradbeer, R. J. Kadner, and C. A. Schnaitman, J. Bact., 128:242-247 (1976).
15. C. Bradbeer and M. L. Woodrow, J. Bact., 128:99-104 (1976).
16. R. L. Blakley and H. A. Barker, Biochem. Biophys. Res. Commun., 16:391-397 (1964).
17. R. L. Blakeley, J. Biol. Chem., 240:2173-2179 (1965).
18. T. Sasaki and K. Kitihara, Biochim. Biophys. Acta, 74:170-172 (1963).
19. T. Sasaki, J. Bact., 109:169-178 (1972).
20. S. Kashket, J. T. Kaufman, and W. S. Beck, Biochim. Biophys. Acta, 64:447-457,458-469 (1962).
21. S. Kashket and W. S. Beck, Biochim. Biophys. Acta, 129:350-358 (1966).
22. L. Ellenbogen, in Cobalamin (B. M. Babior, ed.), Wiley, New York, 1975, pp. 111-140.
23. F. M. Huennekens, P. M. DiGirolamo, K. Fujii, G. B. Henderson, D. W. Jacobson, V. G. Neef, and J. I. Rader, in Advances in Enzymol. Regulation (G. Weber, ed.), Pergamon, New York, 1975, pp. 131-153.
24. H. Y. Neujahr, Acta. Chem. Scand., 17:1902-1906 (1963).
25. H. Y. Neujahr, Acta. Chem. Scand., 20:771-785 (1966).
26. H. Y. Neujahr, Acta. Chem. Scand., 20:786-798 (1966).
27. H. Y. Neujahr, Acta. Chem. Scand., 20:1513-1517 (1966).
28. I. Miyata, T. Kawasaki, and Y. Nose, Biochem. Biophys. Res. Commun., 27:601-606 (1967).
29. T. Kawasaki, I. Miyata, K. Esaki, and Y. Nose, Arch. Biochem. Biophys., 131:223-230 (1969).
30. T. Kawasaki, I. Miyata, and Y. Nose, Arch. Biochem. Biophys., 131:231-237 (1969).
31. T. Nishimune and R. Hayashi, Biochim. Biophys. Acta, 244:573-583 (1971).
32. T. Nishimune and R. Hayashi, Biochim. Biophys. Acta, 328:124-132 (1973).
33. A. Iwashima, A. Matsuura, and Y. Nose, J. Bact., 108:1419-1421 (1971).

34. Y. Nose, A. Iwashima, and A. Nishino, in Thiamine (eds., C. J. Gubler, M. Fujiwara, and P. M. Dreyfus, eds.), Wiley, New York, 1976, pp. 105-120.

35. T. W. Griffith and F. R. Leach, Arch. Biochem. Biophys., 150: 658-663 (1973).

36. T. Kawasaki and K. Esaki, Arch. Biohcem. Biophys., 142:163-169 (1971).

37. A. Matsuura, A. Iwashima, and Y. Nose, J. Vitaminol., 18:29-33 (1972).

38. H. Nakayama and R. Hayashi, J. Bact., 112:1118-1126 (1972).

39. T. Kawasaki and K. Yamada, Biochem. Biophys. Res. Commun., 47:465-471 (1972).

40. T. Kawasaki and K. Yamada in Thiamine (C. J. Gubler, M. Fujiwara, and P. M. Dreyfus, eds.), Wiley, New York, 1976, pp. 83-94.

41. H. Nakayama and R. Hayashi, J. Bact., 118:32-40 (1974).

42. H. Nakayama and R. Hayashi, J. Vitaminol., 17:64-72 (1971).

43. H. Yamasaki, H. Sanemori, K. Yamada, and T. Kawasaki, J. Bact., 116:1280-1286 (1973).

44. A. Iwashima, H. Nishino, and Y. Nose, Biochim. Biophys. Acta, 330:222-234 (1973).

45. A. Iwashima, Y. Wakabayashi, and Y. Nose, Biochim. Biophys. Acta, 413:243-247 (1975).

46. A. Iwashima and Y. Nose, J. Bact., 128:855-857 (1976).

47. K. Iwai, M. Kobashi, and H. Fujisawa, J. Bact., 104:197-201 (1970).

48. L. M. Kozloff, M. Lute, and L. K. Crosby, J. Virol., 6:754-759 (1970).

49. L. M. Kozloff, M. Lute, and C. M. Baugh, J. Virol., 11:637-641 (1973).

50. C. J. Male and L. M. Kozloff, J. Virol., 11:840-847 (1973).

51. R. C. Wood and G. H. Hitchings, J. Biol. Chem., 234:2381-2385 (1959).

52. B. A. Cooper, Biochim. Biophys. Acta, 208:99-109 (1970).

53. G. B. Henderson and F. M. Huennekens, Arch. Biochem. Biophys., 164:722-728 (1974).

54. D. N. Hintze and J. L. Farmer, J. Bact., 124:1236-1239 (1975).

55. J. P. Brown, F. Dobbs, G. E. Davidson, and J. M. Scott, J. Gen. Microb., 84:163-172 (1974).

56. Y. S. Shin, K. U. Buehring, and E. L. R. Stokstad, J. Biol. Chem., 249:5772-5777 (1974).

57. T. Tamura, Y. S. Shin, M. A. Williams, and E. L. R. Stokstad, Anal. Biochem., 49:517-521 (1972).

58. B. Shane and E. L. R. Stokstad, J. Biol. Chem., 250:2243-2253 (1975) (1975).

59. R. L. Kisluik, Y. Gaumont, and C. M. Baugh, J. Biol. Chem., 249:4100-4103 (1974).

60. M. Friedkin, L. T.Plante, E. J. Crawford, and M. Crumm, J. Biol. Chem., 250:5614-5621 (1975).
61. P. G. McElwee and J. M. Scott, Biochem. J., 127:901-905 (1972).
62. F. Mandelbaum-Shavit and N. Grossowicz, J. Bact., 104:1-7 (1970).
63. F. Mandelbaum-Shavit and N. Grossowicz, J. Bact., 114:485-490 (1973).
64. S. F. Zakrzewski and B. Grzelakowska-Sztabert, J. Biol. Chem., 248:2684-2690 (1973).
65. F. Mandelbaum-Shavit and N. Grossowicz, J. Bact., 123:400-406 (1975).
66. F. Mandelbaum-Shavit, Biochim. Biophys. Acta, 428:664-673 (1976).
67. F. Mandelbaum-Shavit, Biochim. Biophys. Acta, 428:674-682 (1976).
68. G. B. Henderson, E. M. Zevely, and F. M. Huennekens, Biochem. Biophys. Res. Commun., 68:712-717 (1976).
69. G. B. Henderson and E. M. Zevely, Fed. Proc., 35:30 (1976).
70. H. C. Lichstein and R. B. Ferguson, J. Biol. Chem., 233:243-244 (1958).
71. J. R. Waller and H. C. Lichstein, J. Bact., 90:843-852 (1965).
72. J. R. Waller and H. C. Lichstein, J. Bact., 90:853-856 (1965).
73. T. O. Rogers and H. C. Lichstein, J. Bact., 100:565-572 (1969).
74. T. O. Rogers and H. C. Lichstein, J. Bact., 100:565-572 (1969).
75. J. F. Cicmanec and H. C. Lichstein, J. Bact., 119:718-725 (1974).
76. H. C. Lichstein and J. R. Waller, J. Bact., 81:65-69 (1961).
77. C. H. Pai, J. Bact., 112:1280-1287 (1972).
78. O. Prakash and M. A. Eisenberg, J. Bact., 120:785-791 (1974).
79. M. A. Eisenberg, B. Mee, O. Prakash, and M. R. Eisenberg, J. Bact., 122:66-72 (1975).
80. C. H. Pai, Mol. Gen. Genet., 134:345-357 (1975).
81. C. H. Pai, J. Bact., 116:494-496 (1973).
82. J. H. Mulligan and E. E. Snell, J. Biol. Chem., 251:1052-1056 (1976).
83. N. Oya, Vitamins, 41:222-229 (1970) [Japanese].
84. B. Shane and E. E. Snell, J. Biol. Chem., 251:1042-1051 (1976).
85. K. N. Subramanian, R. L. Weiss, and R. H. Davis, J. Bact., 115:284-290 (1973).
86. R. L. Weiss, J. Biol. Chem., 248:5409-5413 (1973).
87. R. L. Weiss, J. Bact., 126:1173-1179 (1976).
88. G. J. Tritz, T. S. Matney, and R. K. Gholson, J. Bact., 102:377-381 (1970).
89. G. J. Tritz, T. S. Matney, and J. L. R. Chandler, J. Bact., 104:45-49 (1970).
90. R. K. Gholson, G. J. Tritz, T. S. Matney, and A. J. Andreoli, J. Bact., 99:895-896 (1969).
91. B. J. White, S. J. Hochhauser, N. M. Cintron, and B. Weiss, J. Bact., 126:1082-1088 (1976).

92. E. S. Dickinson and T. K. Sundaram, J. Bact., 101:1090-1091 (1970).
93. J. McLaren, D. T. C. Ngo, and B. M. Olivera, J. Biol. Chem., 248:5144-5159 (1973).
94. P. Mäntsälä, Acta. Chem. Scand., 26:127-135 (1972).
95. P. Mäntsälä, Acta. Chem. Scand., 27:445-452 (1973).
96. P. Mäntsälä, H. Kokkonen, and V. Nurmikko, Acta. Chem. Scand., 26:136-142 (1972).
97. P. Mäntsälä, Acta. Chem. Scand., 28:78-84 (1974).
98. D. C. Sander and F. R. Leach, Biochim. Biophys. Acta, 82:41-49 (1974).
99. S. H. Collins and W. A. Hamilton, J. Bact., 126:1224-1231 (1976).
100. S. Ramos, S. Schuldiner, and H. R. Kaback, Proc. Natl. Acad. Sci. USA, 73:1892-1896 (1976).
101. J. M. Becker and H. C. Lichstein, Biochim. Biophys. Acta, 282:409-420 (1972).
102. S. Schuldiner, G. K. Kerwar, R. Weil, and H. R. Kaback, J. Biol. Chem., 250:1361-1370 (1975).
103. G. Rudnick, H. R. Kaback, and R. Weil, J. Biol. Chem., 250:6847-6851 (1975).
104. T. J. Franklin, CRC Crit. Rev. Microb., 1:253-272 (1973).
105. J. L. Strominger, Fed. Proc., 21:134 (1962).
106. J. Mora and E. E. Snell, Biochemistry, 2:136-141 (1963).
107. R. J. Wargel, C. A. Shadur, and F. C. Neuhaus, J. Bact., 103:778-788 (1970).
108. R. J. Wargel, C. A. Shadur, and F. C. Neuhaus, J. Bact., 105:1028-1035 (1971).
109. F. M. Kahan, J. S. Kahan, P. J. Cassidy, and J. Kropp, Ann. N.Y. Acad. Sci., 235:364-368 (1974).
110. H. B. Woodruff, F. M. Kahan, H. Wallick, A. K. Miller, and E. O. Stapley, in Microbial Drug Resistance (S. Mitsuhashi and H. Hashimoto, eds.), University Park Press, Tokyo, 1974, pp. 539-559.
111. R. J. Kadner and H. H. Winkler, J. Bact., 113:895-900 (1973).
112. H. C. Wu and P. S. Venkateswaran, Ann. N.Y. Acad. Sci., 235:587-592 (1974).
113. J. C. Cordaro, T. Melton, J. P. Stratis, M. Atagün, C. Gladding, P. E. Hartman, and S. Roseman, J. Bact., 128:785-793 (1976).
114. G. A. Snow, Bact. Rev., 34:99-125 (1970).
115. H. Zähner, E. Bachman, R. Hutter, and J. Neusch, Pathol. Mikrobiol., 25:708 (1962).
116. M. Luckey, J. R. Pollack, R. Wayne, B. N. Ames, and J. B. Neilands, J. Bact., 111:731-738 (1972).
117. H. Diddens, H. Zähner, E. Kraas, W. Göhring, and G. Jung, Eur. J. Biochem., 66:11-23 (1976).
118. T. Yamada and J. Davies, Mol. Gen. Genet., 110:197-210 (1971).
119. M. W. Hane and T. H. Wood, J. Bact., 99:238-241 (1969).

120. M. Tsukamura and S. Mizuno, J. Gen. Microb., 88:269-274 (1975).

121. C. Goldthwaite and I. Smith, Mol. Gen. Genet., 114:190-204 (1972).

122. E. C. R. Reeve, Genet. Res. (Camb.), 11:303-309 (1968).

123. K. Arima and K. Izaki, Nature (Lond.), 200:192-193 (1963).

124. T. J. Franklin, Biochem. J., 87:449-453 (1963).

125. K. Izaki and K. Arima, J. Bact., 89:1335-1339 (1965).

126. T. J. Franklin and B. Higginson, Biochem. J., 116:287-297 (1970).

127. B. L. Hutchings, Biochim. Biophys. Acta, 174:734-748 (1969).

128. J. R. DeZeeuw, J. Bact., 95:498-506 (1968).

129. T. J. Franklin, Biochem. J., 123:267-273 (1971).

130. C. A. Schnaitman, J. Bact., 108:545-552 (1971).

131. A. Albert, Nature (Lond.), 172:201 (1953).

132. A. M. Reynard and L. F. Nellis, Biochem. Biophys. Res. Commun., 48:1129-1132 (1972).

133. V. E. Del Bene and M. Rogers, Antimicrob. Ag. Chemother., 7:801-806 (1975).

134. M. E. Dockter and J. A. Magnuson, J. Supramolec. Struct., 2:32-44 (1974).

135. M. E. Dockter and J. A. Magnuson, Arch. Biochem. Biophys., 168:81-88 (1975).

136. K. Izaki, K. Kiuchi, and K. Arima, J. Bact., 91:628-633 (1966).

137. K. Izaki and K. Arima, Nature (Lond.), 200:384-385 (1963).

138. T. J. Franklin, Biochem. J., 105:371-378 (1967).

139. D. Sompolinsky, T. Krawitz, Y. Zaidenzaig, and N. Abramova, J. Gen. Microb., 62:341-349, 351-362 (1970).

140. I. Chopra, J. Gen. Microb., 91:433-436 (1975).

141. T. J. Franklin and S. J. Foster, Biochem. J., 121:287-292 (1971).

142. S. B. Levy and L. McMurry, Biochem. Biophys. Res. Commun., 56:1060-1068 (1974).

143. A. Wojdani, R. R. Avtalion, and D. Sompolinsky, Antimicrob. Ag. Chemother., 9:526-534 (1976).

144. R. Hancock, J. Gen. Microbiol., 28:493-501, 503-516 (1962).

145. N. Anand, B. D. Davis, and A. K. Armitage, Nature (Lond.), 185:22-24 (1960).

146. D. T. Dubin and B. D. Davis, Biochim. Biophys. Acta, 52:400-402 (1961).

147. D. T. Dubin, R. Hancock, and B. D. Davis, Biochim. Biophys. Acta, 74:476-489 (1963).

148. C. Hurwitz and C. L. Rosano, J. Bact., 82:1193-1201 (1962).

149. K. Andry and R. C. Bockrath, Nature, 251:534-536 (1974).

150. M. Kogut, J. W. Lightbrown, and P. Isaacson, J. Gen. Microbiol., 39:155-164, 165-183 (1965).

151. M. Artman and S. Wethamer, J. Bact., 120:542-544 (1974).

152. B. I. Kanner and D. L. Gutnick, J. Bact., 111:287-289 (1972).

153. B. P. Rosen, J. Bact., 116:1124-1129 (1973).

154. S. R. Goodman, B. L. Marrs, R. J. Narconis, and R. E. Olson, J. Bact., 125:282–289 (1976).
155. H. Taber and G. M. Halfenger, Antimicrob. Ag. Chemother., 9:251–259 (1976).
156. L. E. Bryan, H. M. Van Den Elzen, and M. S. Shahrabdi, in Microbial Drug Resistance (S. Mitsuhashi and H. Hashimoto, eds.), University Park Press, Tokyo, 1974, pp. 475–490.
157. L. E. Bryan and H. M. Van Den Elzen, J. Antibiot., 28:696–703 (1975).
158. L. E. Bryan and H. M. Van Den Elzen, Antimicrob. Ag. Chemother., 9:928–938 (1976).

Chapter 11

BACTERIAL TRANSPORT PROTEINS

David B. Wilson and Jeffrey B. Smith[*]

Department of Biochemistry, Molecular and Cell Biology
Cornell University
Ithaca, New York

I. INTRODUCTION 495
II. THE BACTERIAL PROTON-TRANSLOCATING ATPase 498
   A. Purification of the F_1 Portion of the Enzyme 500
   B. Purification of the Complete F_0F_1 Complex 503
   C. The F_1 Subunits 506
   D. The F_0 Subunits and Bacteriorhodopsin 515
   E. Kinetic Properties and Bound Nucleotides 517
   F. Mutants with Lesions in the ATPase Complex 519
   G. Role of the Membrane ATPase in Nutrient Transport
      and Chemotaxis 520
III. RESPIRATORY CHAIN LINKED DEHYDROGENASES 521
IV. NUTRIENT TRANSPORT PROTEINS 525
   A. Membrane Bound Transport Systems 525
   B. Binding Proteins 528
   C. Outer Membrane Proteins 545
V. CONCLUDING REMARKS 546
   RECENT DEVELOPMENTS 548
   REFERENCES 547

I. INTRODUCTION

There has been considerable progress in the identification and isolation
of bacterial transport proteins during the last 5 years. This is especially
true for membrane proteins where the techniques for extracting and
purifying them have been greatly improved.

---

[*]Present address: Imperial Cancer Research Fund, London, WC2A,
3PX, England

In this period, a whole new class of proton translocating proteins has been recognized. These proteins include the proton translocating ATPase, bacteriorhodopsin, and some of the proteins in the electron transport chain. All of these proteins appear to function to convert energy from specific sources, i.e., ATP, light, and respiratory substrates into a form of energy called the proton motive force. This energy is produced by transporting protons through a membrane so as to create a pH gradient and a membrane potential, and the total energy in these two components is called the proton motive force. Because of their role in energy conversion and because of the very special properties of protons, it seems likely that the mechanism or mechanisms of transport used by these proteins may differ from those used by proteins transporting most other molecules.

Also in this period there has been much controversy about the role of membrane bound dehydrogenases in driving transport by those systems which are retained in membrane vesicles, the class of membrane bound transport systems. One group postulated that membrane dehydrogenases interacted directly with the membrane carriers of these transport systems [1,2]. In this model, substrate oxidation by a dehydrogenase would reduce the associated membrane carrier, and this reaction would provide the driving force for active transport by the carrier. These studies also led to the proposal that the membrane bound D-lactate dehydrogenase played a major role in energizing active transport by Escherichia coli membrane bound transport systems, although other dehydrogenases could also function in this process. However, it is now generally accepted that these proposals are not correct and that dehydrogenases function as part of a proton translocating respiratory chain. In this model the membrane bound dehydrogenases feed electrons into a common electron transport chain, and the movement of electrons through this chain is coupled to the transport of protons through the membrane to generate a proton motive force. The carriers of the membrane bound class of transport systems derive the energy to drive active transport from the proton motive force. The evidence for this chemiosmotic model is discussed in Refs. 3 to 6.

The early studies on the role of membrane dehydrogenases in active transport demonstrate a serious problem that can arise when membrane vesicles are used to study active transport. These vesicles have the advantage that almost all the cytoplasm is removed during vesicle formation, so there is little or no metabolism of transport substrates, and transport is dependent on the addition of an energy source. However, the structure of the membrane may be altered during vesicle formation in a complex way that could lead to the presence in the vesicle population of many different classes of vesicles. The evidence for this statement is presented in Chap. 1. Because of the complexity of the vesicle population there is no predictable relationship between the rate of respiration of a given substrate and its ability to drive active transport. This is illustrated in the case of NADH which is respired actively by membrane vesicles when it is added externally but usually causes only a small stimulation of nutrient transport [7]. However, when NADH is generated inside the vesicles its respiration drives transport very well [8].

The ability of a compound to drive transport in a given vesicle will depend on the orientation of its dehydrogenase, the permeability of the vesicle to the compound, the orientation of the transport system being studied, and the integrity of the electron transport chain within the vesicle. At present we only have information about these properties for the entire population and do not know what individual classes of vesicles make up this population.

Because of the differences in the function of various transport proteins, this chapter is divided into sections; Section II contains a discussion of the structure and function of the large multimeric proton translocating ATPase and the light-driven proton pump, bacteriolrhodopsin; Sec. III contains a discussion of the isolation and properties of those dehydrogenases which feed electrons directly into the respiratory chain; Section IV contains a discussion of the structure and function of proteins implicated in the transport of molecules other than protons (nutrient transport systems).

It is now well established that there are at least three distinct classes of nutrient transport systems in bacteria [3-5]. One class of systems couples the transport of molecules across the cell membrane with a chemical modification of the substrate, a process called group translocation. A second class of systems transports molecules against a concentration gradient with no change in the transported molecule. All of the component proteins of this class are firmly bound to the cell membrane and are retained in isolated membrane vesicles, so that members of this class are called membrane bound transport systems.

The third class also concentrates molecules without modification, but transport systems in this class require at least one component which is lost during the formation of membrane vesicles. This component is a soluble protein which specifically binds the substrates of the transport system. Therefore, members of this class are called binding protein transport systems.

There is strong evidence that each class utilizes a different mechanism to couple energy to the accumulation of substrates. The group translocation systems utilize the chemical energy of the modification reaction to accumulate the substrate. The membrane bound systems utilize the proton motive force across the membrane to drive active transport, while the binding protein transport systems utilize ATP or a related compound to drive active transport [3-5].

Because of the major differences between the mechanisms of transport used by each class, information about members of one class cannot be generalized to members of either of the other classes.

The proteins involved in group translocation systems are discussed in Chap. 2, while the proteins involved in membrane bound and binding protein transport system are discussed in detail in Sec. IV of this chapter.

Most of the proteins discussed in Sec. IV are binding proteins because they are selectively removed from whole cells as stable soluble proteins in the absence of detergent by the osmotic shock procedure [9]. However, the lactose M protein is no longer the only membrane bound transport protein to be recognized, as potential carriers for succinate, malate, and alanine have now been identified.

II. THE BACTERIAL PROTON-
   TRANSLOCATING ATPase

Most, if not all, bacteria have an ATPase which is localized in the membrane of the cell and has a paramount role in the interconversion of chemical and osmotic forms of energy. The membrane ATPase in procaryotes is now believed to specifically translocate protons like the ATPase in mitochondria and chloroplasts. The proton-translocating ATPase is a central feature of the chemiosmotic hypothesis which was formulated by the British biochemist Peter Mitchell in 1961 [10]. Chemiosmotic ideas produced a revolution in bioenergetics and have now become the predominant paradigm.

A considerable amount of support for the chemiosmotic hypothesis has been provided by its successful application in recent years to bacterial bioenergetics. Exceptionally lucid discussions of bacterial bioenergetics from a chemiosmotic perspective have been provided by Harold [3,11]. To briefly sketch the chemiosmotic theory as it applies to procaryotes, the export of protons by the bacterial cell generates what is referred to as a proton motive force (pmf) across the cell membrane, which is relatively impermeable to protons. The pmf is the composite of the transmembrane pH gradient (ΔpH) and electrical potential ($\Delta\psi$). Some bacteria, which lack a respiratory chain and a photosynthetic system, for example S. faecalis, use the membrane ATPase to export protons as the only available means for producing a pmf. The pmf is the form of energy which bacteria use to drive the transport of a variety of nutrients (but not all), as well as for the reduction of $NADP^+$ by NADH and motility. Bacteria which have a respiratory chain or a photosynthetic system, for example, E. coli or Halobacterium halobium, use these electron transporting systems as an alternative to the ATPase to eject protons from the cell. In addition, the pmf which is produced during the transfer of electrons may be used to drive the synthesis of ATP from ADP and inorganic phosphate. The synthesis of ATP in response to the pmf generated by substrate oxidation or photosynthesis occurs via the reversal of the proton translocation by the ATPase. The various processes in procaryotes which produce and consume a pmf are summarized in Fig. 11-1.

As far as we know, the proton-translocating ATPase is the only ATPase present in bacterial membranes. Bacteria do contain additional enzymes with ATPase activity, but these enzymes are mainly involved in nucleic acid

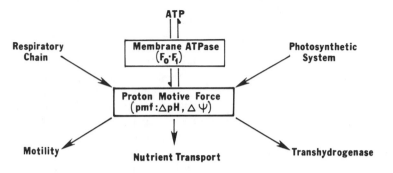

FIG. 11-1 The reactions in procaryotes responsible for the formation and utilization of the proton motive force. The transhydrogenase catalyzes the transfer of reducing equivalents from NADH to $NADP^+$ (see also Chap. 1 and reviews by Harold [3,11]).

metabolism and do not appear to be membrane bound [12-14]. One of these enzymes is the transcription termination factor (called rho) which is an RNA-dependent ATPase [12]. The possibility that the membrane ATPase might share a common subunit with rho has recently been raised, and work is in progress to test this possibility (A. Das and D. Court, personal communication).

In contrast to the situation in procaryotes, there are at least two membrane ATPases in eucaryotes besides the mitochondrial proton pump. These are the Na^+/K^+ transport ATPase, which is widely distributed in the plasma membrane of mammalian cells, and the Ca^{2+} transport ATPase of the sarcoplasmic reticulum. All of the proton-translocating ATPases, whether of bacterial, mitochondrial, or chloroplast origin, are remarkably similar to one another, with respect to a number of catalytic and structural properties [15]. By contrast, the proton-pump ATPase seems to bear little obvious resemblance to either the Na^+/K^+- or Ca^{2+}-transport ATPase. Since the ATPase in membranes from mitochondria, chloroplasts, and bacteria appear to be so similar to one another, it has been suggested that they evolved from a common ancestral protein [16-18], as would be predicted by the endosymbiont theory of the origin of mitochondria and chloroplasts [19].

Although many of the concepts which have been exceedingly beneficial to the rapid progress in bacterial bioenergetics in recent years originated from studies of mitochondria and chloroplasts, this discussion will be limited to the membrane ATPase of bacteria with emphasis on progress since a previous review of the subject [18]. Those who are interested in the membrane ATPase in mitochondria and chloroplasts are referred to the excellent reviews that have recently appeared [15,20-23].

A highly distinctive feature of the proton-translocating ATPase is that it consists of subunits which differ greatly with respect to their level of integration into the phospholipid bilayer of the membrane. One class of subunits composes a lipoprotein complex, called F_0,* which is intrinsic to the membrane. The other class of subunits composes a portion of the enzyme, referred to as F_1, which is rather peripheral to the membrane proper. The F_1 portion becomes soluble in water once it has been detached from F_0, whereas the solubilization of F_0 requires the use of detergents. In electron micrographs of negatively stained membranes from mitochondria, chloroplasts, and bacteria, F_1 has the appearance of knobs about 100 Å in diameter which project from one side of the membrane via a stalk as originally shown by Kagawa and Racker [24]. These rather unusual organizational features are depicted in Fig. 11-2. Table 11-1 lists some of the activities of the complete complex and its two major components.

A. Purification of the F_1 Portion of
     the Enzyme

The purification of the multisubunit F_1 portion of the proton-translocating ATPase became possible after methods were developed for separating F_1 from F_0. F_1 is apparently connected to F_0 through hydrophobic and ionic interactions with Mg^{2+} possibly providing a cationic bridge between anionic sites on F_1 and F_0. The removal of divalent cations and exposure to a low ionic strength seem to be the necessary conditions for dissociating F_1 from F_0 in bacteria [18]. The first procedure used to detach a bacterial F_1 from F_0 was the aqueous release method of Abrams [25]. This procedure consists of several washes with Tris buffers at low ionic strength and neutral pH in the absence of mutivalent cations and is quite selective in its release of F_1 from membranes of S. faecalis [18]. A simpler method, which has been found to work for the enzyme from E. coli as well as several other bacteria, is to extract isolated membranes with a dilute EDTA solution [26]. Both procedures usually solubilize more than 70% of the total F_1 and have the advantage of leaving the membranes and F_0 functionally intact. Furthermore, the release of F_1 from E. coli membranes with EDTA becomes quite selective for F_1 if the membranes are first washed with a relatively high

---

*Although F_1 was originally applied by Racker to the coupling factor ATPase from mitochondria, it is now used to designate that portion of the proton-translocating ATPase from mitochondria, chloroplasts, or bacteria which is peripheral to the membrane. BF_1 refers to the F_1 from bacteria TF_1, the F_1 from the thermophilic bacterium called PS3; CF_1, the F_1 from chloroplasts. F_0 refers to the portion of the proton-translocating ATPase which is an integral component of the membrane; F_0F_1 refers to the complete ATPase complex. F_1 contains five separate polypeptides which are called α, β, γ, δ, and ε, in order of decreasing size.

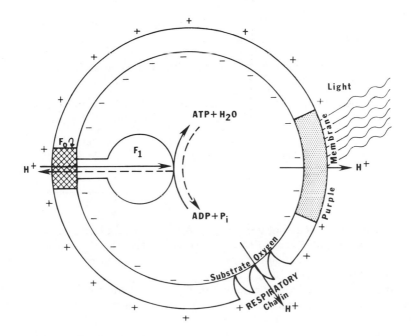

FIG. 11-2 Schematic of the reversible proton-translocating ATPase (F_0F_1).
According to the chemiosmotic hypothesis of Mitchell [10], which has
become the predominant paradigm in bioenergetics, the function of electron
transport during respiration and photosynthesis is to eject protons from the
cell, thereby creating a pH difference and a membrane potential. This
proton motive force (pmf) is used by the proton-translocating ATPase,
operating in reverse, to generate ATP. Alternatively, in the absence of a
functioning respiratory chain or photosynthetic system, the ATPase in
bacteria uses glycolytic ATP to eject protons from the cell creating a pmf
which is the driving force for motility, reduction of $NADP^+$ by NADH, and
solute transport.

concentration of EDTA (10 mM) which does not detach F_1 from F_0, and
then treated with a lower concentration of EDTA (0.5 mM) to solubilize F_1
[27]. The prior treatment of the membranes with 10 mM EDTA removes
some proteins, including dehydrogenases, which would otherwise be extracted by 0.5 mM EDTA along with the ATPase, and improves the specific
activity of the solubilized ATPase about fourfold [27]. Salton and Nachbar
[28] first described the differential response of the membrane ATPase and
NADH dehydrogenase of Micrococcus lysodeiktieus to solubilization by EDTA.
Since a low ionic strength as well as the removal of divalent cations is a
requirement for solubilizing the bacterial F_1, treatment of the membranes

TABLE 11-1   Activities of the Proton-Translocating ATPase (F_0F_1) and Its Two Major Components (F_0 and F_1)

F_1:	ATPase; structural role of blocking the proton-conducting pathway in F_0
F_0:	Proton ionophore; site that reacts with DCCD which inactivates proton ionophore and inhibits F_0-F_1 activities; receptor for F_1
F_0F_1:	Synthesis of ATP from ADP and P_i; ATP-dependent H^+ translocation; ATPase; exchange reactions, e.g., ATP-P_i and ATP-ADP

with 10 mM EDTA does not detach F_1. An additional advantage of washing the membranes with 10 mM EDTA before extracting the F_1 is that the volume of the extraction buffer can be reduced to one-third to one-fourth the level used without a 10 mM EDTA wash.

An even more selective removal of F_1 from isolated E. coli membranes was achieved with chloroform and EDTA [27], following the procedure that Beechey and coworkers [29] devised for extracting the F_1 from mitochondria. However, the exposure to chloroform removes one of the subunits from the F_1 molecule, making it unable to recombine with F_1-depleted membranes unless this subunit, which is referred to as delta, is replaced [27]. Detergents have been used to solubilize the F_1 from E. coli [30,31], but one of these preparations also was shown to be deficient in the delta subunit.

Conventional chromatographic techniques have been used to purify the F_1 portion of the proton-translocating ATPase, since the behavior of F_1, once it has been detached from the membrane, is not different from that of nonmembrane enzymes.

Apparently homogeneous preparations of the F_1 from several different bacteria which are widely separated phylogenetically have been described (Table 11-2). In less than a week, it is now possible to obtain about 100 mg of EC-F_1 from 400 g (wet weight) of E. coli. The preparation is better than 95% pure as judged by gel electrophoresis [27]. Since F_1 from bacteria is inactivated by exposure to low temperature once it has been detached from the membrane, as is the F_1 from mitochondria [32] and chloroplasts [33], it is necessary to protect the E. coli F_1 by adding 10% glycerol to all buffers along with a reducing agent, EDTA, and ATP. In the presence of these reagents the enzyme can be frozen and stored at -80°C for several months without losing either ATPase activity or the ability to restore various energy transducing reactions to F_1-depleted membrane vesicles [27].

A new procedure was recently devised for purifying F_1 from E. coli [34]. In the new method, the enzyme, which has been released from membranes washed with 10 mM EDTA, is first precipitated at -10°C with 20%

methanol in the presence of 50 mM $MgCl_2$. The F_1 in the precipitate is selectively dissolved in a small volume of buffer containing 0.5 mM EDTA, 0.5 mM DTT, 10% glycerol, and 1 mM TrisCl, pH 7.3, which increases its specific activity about fourfold. Then the enzyme is cold inactivated by freezing it in high salt following a modification of the method reported by Vogel and Steinhart [35]. Cold inactivation dissociates $EC-F_1$ into subunits which in the dissociated state are more readily resolved from certain contaminants by molecular sieve chromatography. Finally, the subunits are reassociated by dialysis against MgATP to give an apparently homogeneous F_1 with a high specific ATPase activity and the capability of restoring varoius energy transducing reactions to F_1-depleted membrane vesicles [34].

B. Purification of the Complete F_0F_1 Complex

Kagawa and coworkers [36] have obtained the complete proton-translocating ATPase in a very highly purified form from a thermophilic aerobe called PS3. This organism, which was isolated from a hot spring, provided a source of an exceedingly stable ATPase complex [37]. Prior to extracting isolated thermophile membranes with Triton X-100, they were washed with cholate which removed some contaminating proteins and improved the solubilization of F_0F_1 by the Triton [36]. The F_0F_1 complex was purified in the presence of Triton X-100 by chromatography on a molecular sieve column and a DEAE-cellulose column, which removed the bulk of the impurities. After being reconstituted into liposomes the purified F_0-F_1 catalyzed the various energy transducing reactions which are characteristic of the native complex [36, 38].

The F_0F_1 from E. coli has been partially purified after extraction from isolated membranes with deoxycholate and centrifugation in a sucrose gradient [39, 40]. Bragg used Triton X-100 to remove components of the respiratory chain from BF_0F_1 [41]. The ATPase activity of these F_0F_1 preparations was stimulated by phospholipids and was more sensitive to inhibition by DCCD than the activity of F_1 itself [39-41]. However, since the preparations have not been shown to be capable of catalyzing any of the energy transformations associated with the native complex, it remains to be clarified whether the preparations from E. coli contain all the F_0F_1 subunits or only the one(s) needed for a DCCD-sensitive ATPase.

Baron and Thompson [42] have reported that the nonionic detergent, octylglucoside, solubilizes a substantial portion of the DCCD-sensitive ATPase from S. faecalis without loss of activity [42]. This detergent is especially attractive for use in the purification of the enzyme because it can be removed by simple dialysis.

TABLE 11-2   Comparison of the F_1 Portion of the Proton-Translocating ATPase from Bacteria and Organelles[a]

Source	Molecular Weight (x 10^{-3})	Subunit Composition and Size[b] (x 10^{-3})					Sedimentation Coefficient	Latent[c]	Cold Labile[d]	Ref.
		α	β	γ	δ	ε				
E. coli	380	58	52	31	18	16	11-13	No	Yes	26, 44, 45
S. typhimurium	284	57	52	31	22	13	--	No	Yes	41
S. faecalis	385	60	55	37	20	12	13.4	No	Yes	18, 47, 48
M. lysodeikticus	345	52.5	47	41.5	28.5	--	13.6	Yes	Yes	49, 50
Bacillus megaterium KM	379	68	65	--	--	--	13.6	No	Yes	51, 52
Bacillus stearothermophilus	280	--	--	--	--	--	11.9	No	No	53
Alcaligenes faecalis	--	59	54	43	--	12	13	Yes	Yes	54, 55
Rhodospirillum rubrum	350	54	50	32	13	7, 5	13.1	No	Yes	56, 57
Mycobacterium phlei	250	--	--	--	--	--	--	Yes	No	58

PS3	380	56	53	32	15.5	11	--	No	No	37
Proteus L-forms	360	64	58	--	--	--	12.5	No	Yes	59
Beef heart mitochondria	1384	54	50	33	17.3	11	12.2	Yes	Yes	15
Chloroplast	325	59	56	37	17.5	13	13.8	Yes	Yes	15

[a] After Harold [3].

[b] Molecular weights are from gel electrophoresis in the presence of SDS, with the exception of the values for δ and ε from E. coli which were determined by equilibrium centrifugation [64].

[c] The hydrolytic activity of the F_1 from some bacteria is absent unless it is treated with trypsin which apparently inactivates the inhibitory subunit (ε).

[d] The ATPase activity of most of the F_1 molecules is inactivated by exposure to low temperatures (0 to 4°C) which destabilizes subunit interactions.

C. The F_1 Subunits

The F_1 portion of the bacterial membrane ATPase is a large multimeric
protein having a molecular weight close to 350,000. Some properties of the
F_1 molecules from a variety of different bacteria are summarized in Table
11-2. Each F_1 molecule contains at least five different kinds of polypeptides
commonly referred to as α, β, γ, δ, and ε in order of decreasing size
(Fig. 11-3), following the nomenclature adopted for the chloroplast F_1 [43].
The α and β polypeptides, which have very similar electrophoretic mobilities on SDS gels, compose the bulk of the F_1 molecule, possibly as much
as 85%. The molecular weights of the five F_1 polypeptides from a wide
variety of bacteria are given in Table 11-2. The values for the five different F_1 polypeptides range from 60,000 for α to about 10,000 for ε and are
quite uniform from a wide variety of bacteria (Table 11-2). Even more

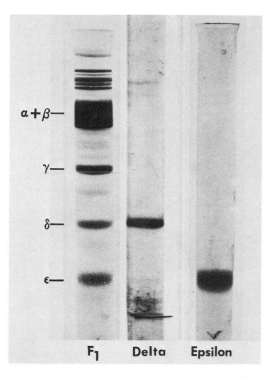

FIG. 11-3 SDS-gel electrophoresis of the purified F_1 portion of the E. coli
membrane ATPase and active delta and epsilon subunits [64]. The α and β
subunits are smeared together due to overloading of the gel so that the delta
and epsilon bands would stand out.

remarkable is the similarity in size of the subunits from bacteria, mito-chondria, and chloroplasts (Table 11-2). Since in the case of the F_1 from E. coli there is now convincing functional evidence that each of the five different polypeptides is an authentic subunit of BF_1 (see below), it is likely that all bacterial F_1 molecules are composed of five separate subunits. However, since the smaller δ and ε polypeptides represent such a minor proportion of the total protein of the F_1 molecule, they can easily escape detection unless rather large amounts of F_1 (30 to 40 μg) are electrophoresed on an SDS gel.

The stoichiometry of the five subunits of the F_1 from E. coli and PS3 have been estimated by isolating the F_1 from bacteria grown in a mixture of radioactive amino acids [46,60]. After purification of labeled F_1 it was electrophoresed on acrylamide gels in the presence of SDS, and the radio-activity was measured in each subunit after slicing it from the gel. To obtain the subunit ratios from these data, the amount of radioactivity in each subunit was divided by the molecular weight of the subunit. This approach yielded a subunit formula of $\alpha_3 \beta_3 \gamma$, δ, ε for both the E. coli and PS3 F_1^* molecules [46,60]. This subunit formula is rather provisional since the molecular weights of the α, β, and γ subunits have as yet only been esti-mated by gel electrophoresis in the presence of SDS. Until they are veri-fied by other methods, the molecular weight values of the F_1 subunits are tentative, since some proteins are known to behave anomalously on SDS gels. Furthermore, the fragmentation of BF_1 and its reconstruction from isolated subfragments has led Vogel and Steinhart [35] to suggest a subunit stoichiometry of $\alpha_2 \beta_2 \gamma_2 \delta_{1-2} \varepsilon$. This proposal is consistent with data presented for the cysteine content and the distribution of sulfhydryl groups and disulfide bonds in the subunits of the beef heart mitochondrial F_1 [60a], but similar data are not yet available for BF_1. Electron micrographs of F_1 show it to have a characteristic appearance suggesting a planar hexa-gonal array of subunits [61,62]. Hexagonal models for F_1 have been pro-posed [15,46] and may be possible with either of the above stoichiometries. Some laboratories favor the $\alpha_3 \beta_3 \gamma$, δ, ε, stoichiometry based on the tenta-tive subunit molecular weights and the relative staining intensities of the subunits on acrylamide gels, which is especially unsound, since the staining of proteins in gels with hydrophobic dyes is known to be unreliable for quantitative purposes. New approaches to determining subunit stoichio-metry will have to be developed to ascertain the correct answer.

---

*Although Kagawa et al. [60] reported 3:3:1 for the ratio of $\alpha : \beta : \gamma$ of PS3 F_1, there is a discrepency in their report which was apparently over-looked by the authors. From the data in the report [60] we calculate a ratio of only 1.7:1.7:1.0 $\alpha : \beta : \gamma$ for TF_1 (Table I in Ref. 60) whereas for puri-fied $TF_0 F_1$ (Table II in Ref. 60) we calculate a ratio of 2.7:2.7:1.

Bragg and Hou [46] used a bifunctional reagent with a cleavable disulfide, dithiobis(succinimidyl propionate), to investigate subunit interactions in BF_1. Treatment of purified BF_1 with this reagent produced the following aggregates: $\alpha\beta$, $\alpha\delta$, $\beta\gamma$, $\beta\varepsilon$ and $\gamma\varepsilon$, indicating that the two subunits of each aggregate are in proximity to one another [46]. The absence of $\alpha\alpha$ and $\beta\beta$ aggregates (among others) was taken as evidence for a model of BF_1 in which three α and three β subunits are arranged alternately at the vertices of a hexagon [46]. However, the absence of $\alpha\alpha$ and $\beta\beta$ aggregates may have been due to the limited reactivity of the crosslinking reagent used or the difficulty of analyzing the composition of complex subunit aggregates by gel electrophoresis.

Cross-linking studies of the chloroplast F_1 molecule suggested that two β subunits are close together and led to a model in which the α and β subunits are in positions to cross-link with every type of subunit [63]. This model places γ and ε at one side of a tetramer composed of two α and two β subunits and δ on the other side [63]. The minimal subunit stoichiometry of the chloroplast F_1 which is consistent with the cross-linking results and interactions with nucleotides and other ligands is $\alpha_2\beta_2\gamma\delta\varepsilon_2$ [63].

A considerable amount of new information about the bacterial F_1 molecule is being provided by the recent progress in the isolation and characterization of the individual subunits of the molecule. Some of the functions of the individual subunits of the F_1 portion of the membrane ATPase from E. coli are summarized in Table 11-3. Active delta and epsilon subunits have been purified to apparent homogeneity from E. coli [64]. Delta (δ) and epsilon (ε) are the smallest of the five BF_1 subunits, and each represents only about 5% of the total F_1 protein. The purification of δ and ε was facilitated by a simple and efficient method for separating them from the larger F_1 subunits [65]. Treating the purified BF_1 with 50% pyridine selectively precipitates only the three larger subunits, leaving δ and ε in solution. After removing the pyridine both δ and ε are active, and can be separated from one another by chromatography on a single molecular sieve column from which both subunits elute in a good yield and in an essentially homogeneous form (Fig. 11-3).

Epsilon strongly inhibits the hydrolytic activity of the purified ATPase, indicating that ε has a role in the regulation of the enzyme in vivo [64,66]. Epsilon maximally inhibits the purified ATPase by about 80%, and the inhibition shows purely noncompetitive kinetics with respect to substrate. Epsilon is very potent, since half-maximal inhibition of the purified ATPase was observed when the enzyme and added inhibitor were present in approximately stoichiometric amounts [64]. Epsilon has a molecular weight of about 16,000 by molecular sieve chromatography, which is somewhat larger than earlier estimates by gel electrophoresis in the presence of SDS [26,44]. The larger value is apparently correct since it was confirmed by sedimentation equilibrium centrifugation and gel electrophoresis in the presence of SDS and urea [64], which indicates that gel electrophoresis in SDS without urea probably underestimates the size of epsilon.

TABLE 11-3 Functions of the Subunits of the F_1 Portion of the Membrane ATPase from E. coli

α and β:	Major components of the F_1 headpiece; appear to be sufficient for ATPase activity
γ:	Required for assembly of α and β after cold inactivation; may be required for inhibition of F_1 ATPase by ϵ
δ:	Required for attachment of F_1 to F_0; may be stalk
ε:	Inhibits the ATPase activity of purified BF_1 implying a regulatory role for ε in vivo

It is somewhat curious that the ATPase inhibitor was isolated from purified BF_1 which seems to have full hydrolytic activity. Thus it would appear that most of the endogenous in the purified BF_1 is active but bound to a noninhibitory site on BF_1. It is known that the chloroplast ATPase inhibitor can be reversibly displaced from its inhibitory site to another site on the molecule where it is not inhibitory [67]. Assuming that there is only one ε present endogenously per molecule of purified $EC-F_1$, it may be that the inhibition of the ATPase requires two or three molecules of ε per BF_1. According to this view, the endogenous epsilon would be expected to be relatively nonexchangeable, but not in itself sufficient for inhibition until an exchangeable site(s) becomes occupied by ε. Studies of the interaction of epsilon with BF_1 from which all of the inhibitor subunit has been removed would be expected to be quite illuminating; however, a method has not yet been devised for removing the endogeneous ε from the purified enzyme without destroying its activity.

The importance of the epsilon subunit as a regulator of the membrane ATPase in vivo has not yet been assessed in bacteria, although the functionally homologous protein inhibitor in mitochondria [68] seems to have a dynamic influence upon whether the mitochondrial ATPase is operating primarily in the direction of ATP synthesis or utilization [69]. Some recent evidence suggests that the inhibitory subunit of the ATPase in yeast mitochondria is subject to repression when glucose is used as a carbon source for growth [70]. Since antibodies to the purified ϵ of BF_1 have been obtained [71], it is now possible to develop a radioimmunoassay for ε which could be used to quantitate ε in cell extracts and find out whether there is any ε present in the cytoplasm of E. coli, and if the total ε content of the cell varies with different growth conditions.

The presence of an active inhibitory subunit probably accounts for the fact that the hydrolytic activity of the membrane ATPase in some bacteria is latent (Table 11-2). The ATPase in membranes from the strict aerobe, M. lysodeikticus, for example, is almost completely devoid of any hydrolytic activity until the membranes are treated with trypsin [72]. The ATPase from

E. coli is activated somewhat by trypsin [73], and trypsin completely reverses the   inhibition of the BF_1 in vitro [71]. It is well known that the F_1
from both mitochondria and chloroplasts contains an inhibitory subunit that
is responsible for the latency of the organelle ATPases [67, 68]. Thus, although the only ATPase inhibitor protein that is known to exist in bacteria is
the one from E. coli, it seems likely that the   polypeptide of other bacterial
F_1 molecules, both latent and nonlatent, will be found to be an inhibitory subunit.

A recent result has suggested a possible similarity between a protein
associated with ribosomes and the inhibitory subunit of the proton-translocating ATPase. A protein about the same size as the   subunit of $EC-F_1$
has been isolated from E. coli ribosomes and shown to inhibit elongation
factor G_1-dependent GTPase activity [74]. It would be quite interesting to
compare these two proteins, which inhibit the different purine triphosphatase
activities, to see how similar they are to one another.

The delta subunit of BF_1, like ε, has a distinct functional role in the
overall operation of the proton-translocating ATPase. The earliest clues
about the function of delta were provided by observations suggesting that it
was essential for the role of the enzyme in the various energy transformations in membrane vesicles, but not for the simple hydrolysis of ATP by the
purified enzyme [75]. Furthermore, preparations of purified BF_1 which lack
δ, in contrast to preparations containing δ, were found to be unable to recombine with inverted membrane vesicles depleted of BF_1 [44]. The inability
of F_1 to reattach to F_0 of course explains the failure of the δ-deficient enzyme preparations to restore energy coupling reactions to the BF_1-depleted
vesicles [64, 65]. Convincing evidence that δ has a role in the attachment of
BF_1 to F_0 has been obtained with purified δ. The apparently homogenous
δ completely restored the capacity of BF_1 missing δ to recombine with
F_0 in inverted membrane vesicles, confirming previous results with impure
δ [65, 66]. Moreover, the attachment of BF_1 to F_0 exhibited an absolute
dependence on δ, and one or two delta molecules were sufficient to reattach one F_1 molecule to F_0 [64]. Thus the purified δ is highly active and is
probably located at the junction between F_1 and F_0. Delta may be the
stalk seen in electron micrographs connecting the F_1 headpiece to the F_0
basepiece which is embedded in the membrane.

The purified δ elutes from a calibrated molecule sieve column at an
apparent molecular weight of 35,000, which is nearly twice the size obtained
by gel electrophoresis in the presence of SDS [64]. Moreover, values
between 17,000 and 18,000 have been obtained by sedimentation equilibrium
centrifugation both in the presence and absence of 6 M guanidine hydrochloride. The latter result implies that the active δ is a monomer and not a
dimer. Therefore, the higher value obtained by molecular sieve chromatography is probably due to δ being rather elongated in shape as would be
expected of the stalk. The circular dichroism spectrum of δ suggests that
it contains more than 50% α-helical structure [64].

From the obvious functional homology between δ of BF_1 and nectin,
which is required for attaching the BF_1 in S. faecalis to the membrane [76],

it seems likely that nectin is the recently discovered delta subunit of the streptococcal enzyme [48]. Nectin has not yet been purified sufficiently, however, to permit its identification on acrylamide gels, so its electrophoretic mobility has not yet been compared to that of the streptococcal δ. One apparent difference between nectin and δ of BF_1 involves the interaction with divalent cations.

The in vitro reassembly of the E. coli F_1F_0 complex occurs via the following two steps:

$$\left(F_1 \text{ missing } \delta + \delta\right) \xrightleftharpoons[\text{pH 9}]{\text{pH 7}} \left(F_1 - \delta\right) \text{(soluble)} \tag{1}$$

$$\left(F_1 - \delta\right) + F_0 \text{ (particulate)} \xrightleftharpoons[\text{EDTA}]{\text{Mg}^{2+}} F_1F_0 \text{ (particulate)} \tag{2}$$

which is nearly the same as the scheme obtained for the reassembly of the streptococcal ATPase-membrane complex [18]. The homogeneous δ from E. coli can bind to F_1 missing δ even in the absence of membranes [Eq. (1)]. In fact, the complex was so stable it was isolated by molecular sieve chromatography [64]. Furthermore, the $F_1 - \delta$ complex formed in vitro was functional because it reattached to F_1-depleted membrane vesicles [Eq. (2)] and restored the coupling of metabolic energy to the transhydrogenase [64]. The second step in the above scheme requires the addition of a divalent cation, whereas the first one does not, at least when the components are from E. coli. Moreover, both components of the soluble $F_1 - \delta$ complex from E. coli were purified in the presence of EDTA, and the complex between them was formed and isolated in the presence of 1 mM EDTA. However, Mg^{2+} was apparently required for the formation of a complex between nectin (thought to be the δ subunit of the streptococcal enzyme) and the rest of the F_1 ATPase [18] and may have a role in stabilizing the interactions between the F_1 subunits in the streptococcal enzyme [48].

An alternative scheme for the reassembly of F_1F_0 complex would involve the formation of a particulate $\delta - F_0$ complex which would then serve as a receptor for F_1 missing δ. However, this reassembly sequence could not be demonstrated in vitro. If δ is the site on F_1 which interacts with a component of F_0 as the present data implies, then δ by itself might be expected to form a complex with F_0, but no such complex has as yet been detected [27]. The $F_0 - \delta$ complex would be expected to be less stable than the $F_1 - \delta$ complex, since essentially all the δ goes with F_1 when it is detached from the membrane.

Certainly the δ subunit of BF_1 has a role in attaching the F_1 headpiece to the membrane, but δ is probably not a static structural component of enzyme. If δ is sandwiched betweeen the F_1 and F_0 proteins, then the proton motive force which is developed across the membrane would have to be transmitted through δ to F_1 where the catalytic site(s) for ATP synthesis and hydrolysis is located. Accordingly, it may be possible to modify δ in a way that would not impair its capacity to connect F_1 to F_0, but that would modify or block the transfer of energy into and out of the F_1 headpiece.

Some new insights about the structural organization and functional properties of the α, β, and γ subunits, which compose the major portion of the F_1 molecule, are currently being obtained by reversibly dissociating it into subfragments. This was accomplished by Vogel and Steinhart [35] by freezing BF_1 in salt solutions. Many multimeric enzymes besides the solubilized F_1 are susceptible to inactivation by freezing in salt solution, which apparently weakens hydrophobic and ionic interactions responsible for holding subunits together. The subfragments of BF_1 themselves were devoid of any activity but reassociated to form the active ATPase when the temperature was raised and the ionic strength lowered. The specific activity of the reconstituted enzyme maximally reached about 70% of the value observed before the enzyme was inactivated. Maximal reconstitution required the addition of the substrate MgATP, but Mg^{2+} alone was partially effective, suggesting that Mg^{2+} may be a structural constituent of the enzyme [35].

Vogel and Steinhart [35] separated the dissociated BF_1 into three fragments by ion exchange chromatography. A fragment called I_A contained the α, γ, and ε polypeptides, and had a molecular weight of about 100,000. Another called I_B was about the same size as I_A but contained the δ polypeptide in addition to α, γ, and ε. The third fragment (II) was apparently homogeneous β in a monomeric form since it was about half the size of the other two fragments. A titration of I_A vs II showed that maximal reconstitution of enzyme activity occurred at a protein ratio of about 0.5 μg II per μg of I_A. Since this corresponds to an equimolar mixture of the fragments, a stoichiometry of $(\alpha\gamma\varepsilon)_2\beta_2$ was suggested for the native enzyme [35]. As would be predicted from studies of the purified δ subunit [64-66], the capacity of the reconstituted ATPase to transduce energy in F_1-depleted membrane vesicles was only restored after incubating I_A and II with I_B, which contained δ.

Some recent studies of the reassociation of subunits after fragmentation by freezing BF_1 in salt solution have implicated the   polypeptide in the assembly of a catalytically active unit in vitro [77, 80]. A chaotrophic anion was included in the salt solution used to fragment the enzyme, and the α and β subunits were apparently converted to monomeric forms [77]. Chromatography on a hydroxy apatite column yielded two major peaks of protein. Both peaks contained α and β in roughly equal amounts and a much smaller amount of ε. Gamma was present in the second peak only and was considerably more abundant by comparison to α and β, than in the native enzyme. There was no δ in either peak since BF_1 missing δ was used for these experiments. Lowering the ionic strength by dialysis against MgATP at room temperature resulted in the virtually complete restoration of hydrolytic activity to the fractions containing γ, but the γ-free fractions remained devoid of any activity. These results implicate the γ subunit in the reconstitution of ATPase activity after cold inactivation, whereas δ is dispensible. The reconstituted enzyme of course was unable to reattach to F_1-depleted membranes unless purified δ was added, which made the enzyme capable of transducing energy for various reactions in depleted membrane vesicles. Moreover, a titration of an excess γ fraction with one missing γ revealed that several-fold more ATPase activity was reconstituted than obtained with the excess γ fraction alone. This synergistic reconstitution of the ATPase indicates that the excess γ restored activity to the fraction missing γ. However, the α and β subunits appear to be sufficient for ATPase activity, since it is retained after treatment of the purified enzyme with trypsin, which removes the γ, δ, and ε polypeptides [45]. Furthermore, McCarty and Fagan [78] found that when chloroplasts are illuminated in the presence of NEM the γ subunit of CF_1 becomes specifically modified. The reaction of the γ subunit with a single molécule of NEM virtually abolished photophosphorylation, but has only a slight effect on ATPase activity [78, 79]. Therefore, since γ itself does not appear to be involved in catalyzing ATP hydrolysis, it may have an organizational role in directing the association of α and β to reform the catalytic unit after cold inactivation, besides being required for ATP synthesis. Active δ and ε subunits added either separately or together did not restore activity to the $\alpha\beta$ fraction, nor did δ and/or ε prevent the restoration of activity achieved with the excess γ fraction [80].

     BF_1 from which γ, δ, and ε have been removed by digestion with trypsin has full ATPase activity but is no longer sensitive to inhibition by the purified ε subunit [64, 66]. Since the δ subunit is known not to be required for inhibition by ε, it appeared that the removal of γ from BF_1 caused it to become insensitive to ε. The argument would be considerably strengthened, however, if adding γ back to trypsin-treated BF_1 could be shown to restore sensitivity to ε. Recently we found that the $\gamma\varepsilon$-rich fraction restored full ε sensitivity to the BF_1 from which γ, δ, and ε had been removed by digestion with trypsin [80]. Since the sensitivity to inhibition by ε was regained when γ was added back, it may be concluded that the insensitivity of BF_1 containing only the α and β subunits is due to the absence of γ and not to a defect in α or β. Structural evidence that the γ and ε interact with one another was recently reported by Bragg and Hou [46] who showed that γ and ε become covalently connected when BF_1 is reacted with a cross-linking reagent. The γ and ε subunits of the chloroplast F_1 also become connected when CF_1 is reacted with bifunctional reagents [63].

     Recently a novel property of the BF_1 subunits was revealed when the two larger subunits were separated from the others after cold inactivation [80]. The fraction containing only α and β in approximately equal amounts becomes enormously viscous after it is dialyzed against MgATP. The $\alpha\beta$

fraction has an intrinsic viscosity of about 10,000 (g/ml) after overnight
dialysis against MgATP, which is about 50 times greater than the viscosity
of purified BF_1 [80]. The high viscosity of the $\alpha\beta$ fraction apparently
reflects an intrinsic property of one or both of the two larger subunits
which is normally prevented by the other subunits. The $\alpha\beta$ fraction
exhibits the highest viscosity values, and the addition of the smallest of
the BF_1 subunits, ε, to the $\alpha\beta$ fraction decreases its viscosity [80].
The highest viscosity of the $\alpha\beta$ fraction is independent of ATPase activity
since the $\alpha\beta$ fraction is devoid of catalytic activity unless it is combined
with the excess γ fraction before dialysis against MgATP, which decreases
the viscosity. The excess γ fraction by itself becomes fully active hydro-
lytically after dialysis against MgATP, but it does not become highly viscous. The high viscosity of the $\alpha\beta$ fraction may reflect the formation of
fibers.

Munoz and coworkers have characterized denatured α and β subunits
of the F_1 from M. lysodeikticus [81]. The two subunits were purified to
apparent homogeneity by preparative gel electrophoresis in the presence
of urea. The purified subunits reacted positively towards a glycoprotein
stain, and chemical analysis confirmed the presence of carbohydrate suggesting that both subunits from M. lysodeikticus are glycoproteins [81].

Kagawa and coworkers have recently succeeded in what may become a
landmark in bioenergetics by reconstituting the complete proton-translocating ATPase from component polypeptides which were isolated from the
thermophilic aerobe PS3 [36,37,60]. The thermophile F_1 (TF_1) is by far
the most stable F_1 known [37]. TF_1 is not cold-labile and shows maximal
activity at 70°C. In general, the concentrations of a variety of protein
denaturants required to inactivate TF_1 were 5- to 10-fold higher than those
necessary to inactivate the F_1 from beef heart mitochondria [37]. The fact
that TF_1 has an amino acid composition which is remarkably similar to that
of F_1 from beef heart mitochondria makes it unlikely that TF_1 is stabilized
by a higher percentage of proline, disulfide bonds, or hydrophobic amino
acids, which have been suggested to account for the thermophily of other
enzymes [82]. The five TF_1 subunits, which are quite similar in size to
those of other F_1 molecules, were dissociated from one another by
treatment with SDS and urea [37], and each was purified to apparent homogeneity [60]. Then TF_1 was reconstructed by combining the isolated subunits and removing SDS and urea [60]. The reconstitution of hydrolytic
activity required only the three larger polypeptides, but all five were
needed to reconstitute a TF_1 capable of transducing energy in TF_1-depleted
vesicles containing TF_0. These results are consistent with some of the
conclusions about the functions of the BF_1 subunits (Table 11-3).

D. The F_0 Subunits and Bacteriorhodopsin

The purified TF_0F_1 from the thermophile contained three polypeptides in addition to the five TF_1 subunits [36,37]. In contrast to TF_1 alone the complete TF_0F_1 complex required phospholipids for ATPase activity and was inhibited by DCCD and tributyltin. The purified TF_0F_1 was inserted into liposomes made by sonicating purified phospholipids, using the cholate dialysis procedure of Kagawa and Racker [83]. Addition of ATP to the vesicles containing TF_0F_1 caused them to accumulate protons and develop a membrane potential (positive inside), as indicated by the quenching of 9-aminoacridine fluorescence and enhancement of 1-anilinonaphthalene-8-sulfonate fluorescence, respectively [36,38]. The purified TF_1 portion alone, of course, was incapable of transducing energy in liposomes. These results indicate that the functional TF_0F_1 in the reconstituted vesicles has the opposite orientation to the enzyme in vivo. Furthermore, when purple membranes containing bacteriorhodopsin were included in the liposomes along with the purified TF_0F_1, illumination of the vesicles produced a net synthesis of ATP [38]. The light-dependent ATP formation was abolished by an uncoupler of oxidative phosphorylation or by an inhibitor of TF_0F_1. Previously, Racker and Stoeckenius [84] described similar experiments with phospholipid vesicles containing bacteriorhodopsin and a partially purified F_0F_1 preparation from mitochondrial.

Bacteriorhodopsin is a novel photosynthetic system which Stoeckenius and coworkers [85,86] discovered in the salt-loving bacterium, H. halobium. This system utilizes a rhodopsin-like pigment instead of chlorophyll to capture energy from the sun.

Bacteriorhodopsin is localized in separate regions or islands in the cytoplasmic membrane, referred to as purple patches. These contain exclusively bacteriorhodopsin, which composes 75% of the total mass of the patch, with lipid making up the remainder. The purple color is due to the cofactor retinal which is covalently linked to the protein in a 1:1 stoichiometry.

Oesterhelt and Stoeckenius [87] obtained evidence that bacteriorhodopsin functions in vivo as a proton pump which expels hydrogen ions from the bacterium in response to activation by light. Racker and Stoeckenius [84] showed that vesicles, which were reconstituted with phospholipids and bacteriorhodopsin as the only polypeptide, catalyzed the translocation of protons in response to illumination. The protons were transported from the outside to the inside, in the opposite direction to that of the intact bacteria, indicating that the bacteriorhodopsin in the phospholipid vesicles has an orientation which is the opposite of what it is in the intact bacterial membrane [84]. The proticity generated by the light-driven proton pump may be used to drive ATP synthesis via the F_1F_0 complex as well as other endergonic processes such as active solute transport.

The purple membrane components form an extremely regular two-dimensional array [88] which provided the basis for a most remarkable achievement. Henderson and Unwin [89] used electron microscopy of tilted, unstained specimens to obtain a 7 Å resolution map of the purple membrane. Each bacteriorhodopsin molecule of molecular weight 26,000 has dimensions of 25 x 35 x 45 Å and is composed of 70 to 80% α-helices [89]. The protein in the membrane was shown to contain seven closely packed α-helical segments which extend roughly perpendicular to the plane of the phospholipid bilayer for most of its thickness. The longest dimension of the molecule is perpendicular to the plane of the membrane and parallel to the helices. The protein is thus an example of an "intrinsic" protein which has a globular shape, and is almost certainly exposed on both sides of the membrane and surrounded by lipids arranged as a bilayer in areas between the helices and between separate protein molecules.

The experiments with bacteriorhodopsin and the ATPase complex are some of the best demonstrations of chemiosmotic coupling and show that the purified TF_0F_1 contains all the components necessary for coupling ATP synthesis and hydrolysis to proton translocation. Moreover, each of the eight different polypeptides which are present in the purified TF_0-F_1 has now been purified to apparent homogeneity and used to reconstitute a functional TF_0F_1 complex [60]. This achievement may be expected to open the way for a definitive assessment of the structural and functional relationships of all of the subunits of a proton-translocating ATPase.

One of the requisite functions of the proton-translocating ATPase is the transfer of protons through the relatively impermeable phospholipid bilayer of the membrane. Evidence that the F_0 in E. coli contains a specific proton conducting pathway has been obtained from studies of mutants lacking ATPase activity due to a defect in the F_1 portion of the enzyme [90-92]. The genetic lesion in F_1 greatly enhances the permeability of the membrane to protons which is apparently due to an exposed proton conducting pathway in F_0 [90-92]. The high permeability of the mutants to protons impairs the energization of certain nutrient transport systems as would be expected if they depended on a pH gradient or membrane potential. Furthermore, treatment of the mutants with DCCD, which seals the proton leak, restores the coupling of energy for nutrient transport [90-92]. At low concentration, DCCD inhibits the hydrolytic activity of the E. coli F_0F_1 complex by reacting with a component of F_0 [93,94]. It is now apparent that blockage of the proton conducting pathway in F_0 may itself be sufficient to cause the inhibition of the F_1 ATPase by DCCD [95-97]. Previously, the reaction of F_0 with DCCD was envisaged as initiating conformational changes which were transmitted to the F_1 portion of the molecule [98].

A DCCD-reactive component of the F_0 in E. coli has been identified after labeling isolated membranes with radioactive carbodiimide [93, 94]. The labeled component is a protein which migrates with an apparent molecular weight of about 10,000 on SDS gels. The fact that its labeling was greatly diminished in membranes from mutant strains containing a DCCD-resistant ATPase provided strong evidence that the labeled protein is a component of F_0. The DCCD-reactive protein from E. coli is apparently a proteolipid with similar properties to the analogous protein which was previously identified in mitochondria [99]. Since there is a large discrepancy in the apparent molecular weight of the DCCD-reactive proteolipid from mitochondria estimated on SDS gels in the presence and absence of urea (B. I. Kanner, personal communication), the size mentioned above for the E. coli proteolipid may be regarded as tentative. A further characterization of the proteolipid component of F_0 would be expected to shed light on the mechanism of proton translocation via the membrane ATPase as well as its inhibition by DCCD.

E. Kinetic Properties and Bound
   Nucleotides

The enzyme is relatively specific for purine nucleoside triphosphates and requires a divalent metal ion for activity. Optimal activity is observed with Ca^{2+} or Mg^{2+} at a ratio of metal ion to ATP of about one-half. Unliganded Ca^{2+} or Mg^{2+} is inhibitory [31, 45, 100, 101]. ADP is a potent product inhibitor of the ATPase, and the inhibition is competitive with ATP [101]. Some other compounds which inhibit the enzyme are listed in Table 11-4. The undissociated neutral azide, diphenylphosphorazidate, is a more selective inhibitor of the ATPase than azide itself [102]. Bathophenanthroline, which is a lipophilic metal chelating agent, inhibits the ATPase activity of purified BF_1 and the membrane bound enzyme, but the inhibition may not be due to chelation of metal ions [103]. NBD-Cl(7-chloro-4-nitrobenzene-2-oxa-1, 3-diazole) is a potent inhibitor of the BF_1 ATPase which appears to react with a tryosyl residue in the β subunit of the enzyme [45, 107]. The α and β subunits seem to be sufficient for hydrolytic activity since activity is unaffected by treatment with a protease which apparently removes the three smaller F_1 subunits (γ, δ, and ϵ) from the enzyme [45]. Phospholipids are required for the ATPase activity of the BF_1 complex but not for the activity of the F_1 portion alone [39-41, 106].

The membrane ATPase from S. faecalis [47] and E. coli [18, 108-110] contains firmly bound nucleotides like the enzyme in mitochondria and chloroplasts [13, 110]. Abrams and coworkers [47] reported that the nucleotides in the complex formed in vivo with the S. faecalis ATPase are in a relatively nonexchangeable state. The complex formed by incubating intact cells with ^{32}P-labeled inorganic phosphate contained ^{32}P-labeled ADP and

TABLE 11-4   Kinetic Properties of the Membrane ATPase from E. coli[a]

Specificity for nucleoside triphosphates: [31,100,101]	BF_1: ATP > GTP > ITP >> UTP > CTP
Activators: [31,45,100,101]	BF_1: Mg^{2+} or Ca^{2+} and to a lesser degree Zn^{2+}, Co^{2+}, Ni^{2+}, and Mn^{2+}. $Mg^{2+}/ATP = 0.5$ for optimal activity.
K_m for ATP:	BF_1: 0.2 to 0.4 mM [44]; 0.6 [101].
K_i for ADP:	BF_1: 0.3 mM [101].
Inhibitors: [31,44,45,64, 66,100-105]	BF_1: NaN_3, diphenylphosphorazidate; bathophenathroline, pyrophosphate, P_i, quercetin, Dio-9, DCCD, NBD-Cl, AMP-PNP, divalent metal ions, antibodies to purified BF_1, ε subunit of BF_1. The complete F_0F_1 complex is more sensitive to DCCD than the F_1 portion alone. At low concentrations Dio-9 stimulates the ATPase activity of the F_1 portion but inhibits the activity of the complete complex.
Phospholipids: [39-41,106]	Required for ATPase activity of partially purified F_0F_1 complex but not for activity of the F_1 portion alone.
Subunits involved: [45]	The α and β polypeptides of BF_1 appear to be sufficient for the expression of maximal hydrolytic activity.
Amino acids involved: [45,107]	A tyrosyl residue appears to be involved in ATP hydrolysis.

[a]See Pedersen [15] for a similar compilation for the mitochondrial ATPase.

ATP which were not displaced by unlabeled nucleotides [47]. The purified streptococcal enzyme was also shown to bind nucleotides reversibly in vitro [47]. Thus the bacterial enzyme, like the one from mitochondria [110], contains two kinds of tightly bound nucleotides which differ with respect to exchangeability with added nucleotides.

The nucleotides in purified BF_1 are released when it is inactivated by freezing at a high ionic strength [105], as was previously shown to be true for the mitochondrial ATPase [110]. When BF_1 is reactivated in the presence of [^{3}H]ATP, both ADP and ATP become firmly bound to the enzyme [105].

Boyer [111] and Slater [110] and coworkers have presented evidence supporting the novel view that during oxidative phosphorylation in mitochondria, part of the energy from electron transport is used to release preformed ATP from the catalytic site on F_1. Other conceivable roles for some of the bound nucleotides are in subunit assembly and interactions and the regulation of the enzyme.

F. Mutants with Lesions in the ATPase
    Complex

The resurgence of interest in the membrane ATPase from bacteria has stemmed in part from the isolation and characterization of ATPase mutants. The use of mutants with alterations in the ATPase complex or respiratory chain has considerably simplified the investigation of the relationships between various energy transductions in bacteria. Mutants lacking ATPase activity have been instrumental in ascertaining the source of energy for various nutrient transport systems as well as other membrane processes [3]. Information about the genetics and physiological properties of various mutants can be found in some excellent reviews devoted entirely to the subject [112, 113].

Investigations are now underway to identify the components of the ATPase complex which carry the different genetic lesions. Some of the mutants which have been isolated have alterations in the F_0 portion of the enzyme. Many of the mutants belonging to this class have an F_0F_1 complex whose ATPase activity has become resistant to inhibition by DCCD [93, 112, 113, 114, 115]. Comparisons of the polypeptide composition of membranes from the F_0 mutants with the parental strains has led to the tentative identification of two polypeptides as components of F_0 [93, 94, 115].

Other mutants are known to have defects in the F_1 portion of the enzyme. The defective F_1 has now been purified from some mutants lacking ATPase activity, and work is in progress to identify which of the five F_1 subunits contain the lesion. One mutant which has ATPase activity is deficient in coupling energy to various membrane processes and may have an alteration in the   subunit of BF_1 [75]. A new class of F_1 mutants has an ATPase with an altered response to divalent metal ions [116]. In the future the biochemical definition of the ATPase mutants would be expected to have an increasingly important role in unraveling the functions of the individual subunits of the complex.

G. Role of the Membrane ATPase in
Nutrient Transport and Chemotaxis

The membrane ATPase in E. coli supplies the cell with energy in a form
which is utilized by various transport systems. Some nutrient transport
systems are known to require ATP or a closely related metabolite as an
energy donor. This is apparently the case with the systems in E. coli
which have a periplasmic binding protein as one of their components as
first shown by Berger [117]. Certain transport systems in gram-positive
bacteria also appear to rely directly on phosphate bond energy, even though
these organisms do not have periplasmic binding proteins [3]. Also it has
been known for many years that the uptake of some sugars by bacteria occurs
by the transfer of phosphate from phosphoenolpyruvate to the sugar as it
passes through the membrane. This vectorial phosphorylation is catalyzed
by the phosphoenolpyruvate:sugar phosphotransferase system (see Chap.
2). While ATP may be supplied by oxidative or photosynthetic phosphorylation via the reversal of the membrane ATPase, the ATPase does not participate in the utilization of phosphate bond energy for any of these transport
systems, since mutant strains lacking the ATPase have normal transport
providing a fermentative source of ATP is available [117].

Other nutrient transport systems which have been thoroughly studied
in isolated membrane vesicles are coupled to the transmembrane pH gradient or electrical potential [3,6]. These chemiosmotically coupled systems enjoy a wide distribution in bacteria [2] and may be energized by
a respiratory substrate or, under anaerobic conditions, by ATP produced
fermentatively. The utilization of ATP by these transport systems is of
course mediated by the proton-translocating ATPase. Both types of energy
donors generate an electrochemical gradient of protons which is the
immediate driving force for nutrient translocation. The model proposed
by Kaback and Barnes [118], which depicted the transport carriers as
intermediates in the electron transfer chain as the basis for coupling
between respiration and nutrient uptake, has been abandoned.

The membrane ATPase has the same role in chemotaxis as in nutrient
transport, namely supplying energy in a utilizable form. The immediate
source of energy for motility, which results from the rotation of the fla-
gellar bundle, appears to be the electrochemical gradient of protons [119].
To swim towards an attractant or away from a repellant (positive or negative chemotaxis), bacteria seem to require an additional energy donor,
presumably ATP [119]. The same periplasmic binding proteins which are
involved in nutrient transport (see Sec. IV) are also the chemoreceptors
that reversibly bind the sensed chemicals and initiate a signal to the
flagella. The signal apparently determines the direction of rotation of the
flagella; counterclockwise rotation produces forward swimming, whereas
clockwise rotation causes tumbling [120]. Recent observations suggest that
the chemoreceptors respond to attractants and repellants by influencing the

degree of methylation of a membrane protein called MCP [121]. Methylation and demethylation of MCP may be the mechanism for transmitting the stimulus from the chemoreceptor to the flagella [121,122]. Thus the ATP which is required for chemotaxis is presumably needed for the synthesis of S-adenosylmethionine. Since the same binding proteins which are the specific chemoreceptors are also involved in specific nutrient transport systems which can only be energized by ATP (or a closely related compound), it seems possible that methylation of a membrane protein might be the basis of energy input to these transport systems.

Besides the indirect role of the ATPase in supplying energy for chemotaxis, a new role of the enzyme in chemotaxis has come to light recently [123]. Taxis towards Ca^{2+} and Mg^{2+} was found to be absent in a mutant lacking F_1 ATPase activity (AN120) and one having a lesion in the F_0 portion of the enzyme (SW46) [123]. Taxis towards other attractants was found to be normal in the mutants, as was Ca^{2+} and Mg^{2+} taxis in the strains from which the F_0-F_1 mutants were derived [123]. Additional evidence was obtained indicating that the cation responses are mediated by a receptor with properties similar to those known to be displayed by F_0F_1 [123]. Thus the membrane ATPase appears to serve as the chemoreceptor for divalent metal ions.

III. RESPIRATORY CHAIN LINKED
     DEHYDROGENASES

Most bacterial dehydrogenases are cytoplasmic enzymes which transfer electrons to pyridine nucleotide coenzymes. However, there are a number of dehydrogenases which are directly linked to the respiratory (electron transport) chain present in the plasma membrane. Many examples of this class were recently reviewed [124], however in most cases, while the activities have been demonstrated, the enzymes have not been purified and characterized. In this section I will discuss only those enzymes which have been purified or characterized. These are mainly from E. coli, where there are seven enzymes which have been shown to belong to this class. All but one of these, the anaerobic α-glycerol phosphate dehydrogenase, are tightly bound to the plasma membrane.

The most thoroughly studied of these enzymes is D-lactate dehydrogenase from E. coli. This enzyme has been solubilized and purified to homogeneity by two groups [125,126]. Although they used different procedures for solubilizing and purifying the enzyme, they found identical properties for their pure preparations. This enzyme makes up about 1% of the protein in washed membranes and is a single polypeptide chain with a molecular weight of about 73,000. The enzyme contains a single molecule of tightly bound FAD which is not attached by covalent bonds. The enzyme has a rather sharp pH optimum at pH 8. The K_m for D-lactate of the purified enzyme is 6×10^{-4} M, while for the membrane bound enzyme, the K_m is 2×10^{-3} M.

Both groups have succeeded in binding the purified enzyme to membrane vesicles prepared from a strain lacking this enzyme [127, 128]. The reconstituted vesicles show good D-lactate oxidase activity. Respiration of D-lactate generates sufficient proton motive force to drive active transport by membrane bound transport systems at nearly the same rate as native vesicles. The enzyme is present on the outside surface of reconstituted vesicles, while it is on the inside surface of the membrane in both whole cells and normal vesicles. The ability of the reconstituted enzyme to feed electrons into the respiratory chain in a functional way suggests that it may be directly coupled to a mobile component such as a quinone. However it is possible that it is coupled to a protein component which is accessible from both sides of the membrane in vesicles. At present there is no strong direct evidence which identifies the respiratory component to which D-lactate dehydrogenase transfers electrons.

The aerobic α-glycerol phosphate dehydrogenase of E. coli, which can transfer electrons to either oxygen or nitrate, has been solubilized from membranes by extraction with sodium chloride and deoxycholate [129]. The solubilized enzyme has been purified to homogeneity and also makes up about 1% of the protein in washed membranes. The molecular weight of the native enzyme is about 80,000, while when the enzyme is dissociated by SDS there is a single subunit with a molecular weight around 35,000. The enzyme contains tightly bound but not covalently attached flavin which appears to be FAD. The enzyme bound flavin is reduced by α-glycerol phosphate and is not readily autooxidized. The enzyme shows nearly full activity over the range of pH from 6 to 9. The purified enzyme has a K_m for glycerol phosphate of 8×10^{-4} M, very different from the K_m of the membrane bound enzyme, which is 3×10^{-2} M. Phosphate is a noncompetitive inhibitor of the purified enzyme, causing a maximum inhibition of 50%.

The purified enzyme can bind to vesicles not induced for glycerol metabolism and reconstitute α-glycerol phosphate oxidation [127]. α-Glycerol phosphate respiration in reconstituted vesicles drives active transport as well as it does in vesicles from glycerol-induced cells. The reconstituted enzyme is on the outside of the membrane vesicles, while in spheroplasts the enzyme appears to be on the inside surface of the plasma membrane. Thus, like D-lactate dehydrogenase, α-glycerol phosphate dehydrogenase probably feeds electrons into a mobile component of the electron transport chain; however, there is no strong direct evidence which identifies this component. A similar membrane bound respiratory α-glycerol phosphate dehydrogenase has been studied in Propionibacterium arabionosium, although the enzyme was not solubilized [130].

E. coli also contains an anaerobic α-glycerol phosphate dehydrogenase which can be coupled to either fumarate reductase or nitrate reductase but is inhibited by oxygen. This enzyme is not tightly bound to the membrane and has been purified 40-fold [131]. It has a molecular weight of about 80,000 and requires flavin. It functions with either FAD or FMN but needs both to show maximal activity. Its K_m for FAD is 3×10^{-7} M, for FMN is 3×10^{-4} M, and for α-glycerol phosphate is 2×10^{-4} M, but the kinetics are complex with this substrate. The pH optimum is 7.7.

There is evidence that this dehydrogenase is bound to fumarate reductase in vivo and that this complex is present in the membrane [132]. Surprisingly, the coupled activity present in the particulate fraction from cell extracts does not require added flavin, even though the α-glycerol phosphate dehydrogenase activity of this particulate fraction shows the same flavin requirement as the partially purified enzyme.

Formate dehydrogenase is another anaerobic dehydrogenase which has been studied in E. coli. This enzyme appears to make up 2% of the protein in washed membranes prepared from cells grown under anaerobic conditions and was solubilized by extraction with deoxycholate [133]. The solubilized enzyme has been purified to a nearly homogeneous state. The enzyme tends to aggregate and is a more complex protein than the other dehydrogenases discussed in this chapter, as it dissociated into three polypeptides in SDS. The native enzyme has a molecular weight of about 600,000 while the subunits have molecular weights of around 110,000, 32,000, and 20,000. The two largest subunits are present in equimolar amounts, and there are probably four of each in the native enzyme. The amount of the smallest polypeptide varies from 0.2 to 1 times the molar value of the other subunits. It is not yet known whether it is a weakly bound subunit or an impurity.

The pure enzyme contains selenium, heme, molybdenum, nonheme iron, and acid-labile sulfide. The selenium is present only in the largest subunit while the other compounds have not yet been localized. Quantitative measurements give a stoichiometry of about one molecule each of heme, selenium, and molybdenum, 14 molecules of nonheme iron, and 13 molecules of acid-labile sulfide per molecule of the largest subunit.

The K_m for formate is 1 to 2×10^{-4} M. There is some evidence that this enzyme can react directly with nitrate reductase, but further work is needed to identify the component or components to which it transfers electrons.

There appear to be a number of DPNH dehydrogenases present in E. coli. At least 80% of the DPNH dehydrogenase activity in crude extracts is in the soluble fraction. This is probably a different enzyme from the membrane bound enzyme which is the only activity coupled to the respiratory chain. The membrane bound enzyme has been solubilized with Triton and purified to near homogeneity [134]. It represents about 3% of the protein in washed membranes from cells grown in rich medium or glucose-minimal medium, but is about 1.5% of the protein in cells grown on succinate or lactate.

The molecular weight of the native enzyme was not reported, but the molecular weight in SDS is 38,000. The pure enzyme requires FAD and its K_m is 4×10^{-6} M [135]. The K_m for NADH (3×10^{-4} M) is nearly the same for both the pure enzyme and the membrane bound enzyme. A number of adenine nucleotides inhibit the activity of both the pure and membrane bound activities. NAD^+ has a K_i of 2×10^{-2} M for the pure enzyme and 7×10^{-5} M for the membrane bound enzyme, while AMP and ADP have similar K_i values of 5×10^{-4} M and 8×10^{-4} M, respectively, for both forms of the enzyme. The pH activity profile of both forms of the enzyme are similar, with maximum activity at pH 6.5 and a steady drop in activity as the pH is increased to 9.

Two forms of NADH dehydrogenase are present when Triton extracts of membranes are separated by polyacrylamide gel electrophoresis. One band has the same mobility as the pure enzyme while the nature of the other band has not been determined. It is not yet known which component in the respiratory chain accepts electrons from this enzyme.

In a study of NADH dehydrogenase in Propionibacterium shermanii, it was reported that 80% of the activity is in the soluble fraction while, 20% is membrane bound [136]. These workers purified the soluble activity 30-fold and characterized its enzymatic activity. However, the enzyme was not pure. This preparation gave two peaks of activity after electrophoresis on polyacrylamide gels, and since the membrane fraction gave the same two peaks, these workers proposed that the membrane and soluble enzymes were the same. However, nothing was done to solubilize the membrane bound enzyme before electrophoresis, so it seems probable that the activity bands which were seen after electrophoresis of the membranes came from contaminating soluble enzyme.

A soluble NADH dehydrogenase has been purified to homogeneity from Peptostreptococcus elsdensi [137] and characterized as a protein of molecular weight 63,000 containing a modified FAD cofactor. However there was no evidence presented to indicate that this enzyme feeds into the electron transport chain.

The membrane bound NADPH dehydrogenase activity from Acholeplasma laidlawii has been solubilized by Triton extraction but has not been purified or characterized [138].

There are two reports of the partial purification of succinate dehydrogenase from E. coli. In the one the enzyme was extracted from an acetone powder and purified about twofold and was still very impure. The molecular weight of the enzyme was estimated to be 100,000 [139]. In the other report the enzyme was solubilized by deoxycholate and partially purified [140]. In this case the extracted enzyme appeared to be complexed with cytochrome b_1 and was not characterized further. Succinate dehydrogenase from M. lysodeikitcus has been extracted from washed membranes with DOC [141]. Most of the extracted activity was still soluble after centrifugation for 2 hr at 130,000 x $\underline{g}$. However this activity was not purified further. The K_m for succinate of the solubilized enzyme was 2.6×10^{-4} M.

There is a report that the membrane succinate dehydrogenase from Azotobacter vinelandii can be stimulated sixfold by a factor present in the soluble fraction [142]. This protein was purified to homogeneity, had a molecular weight of about 200,000, and appeared to contain a single class of subunit of molecular weight 46,000. This factor stimulates the succinate oxidase and succinate dehydrogenase activity of the membrane to the same extent without altering the K_m for succinate or the activity of the other membrane bound dehydrogenases. The protein has no succinate dehydrogenase activity by itself and does not contain any light absorbing cofactors.

The final respiratory dehydrogenase present in E. coli is specific for D-amino acids. This enzyme has not been solubilized but has been studied in membranes [143]. It has a pH optimum of 8.3, and the K_m for D-alanine is 6.6 mM. A similar activity has also been reported in Pseudomonas aeruginosa [144].

IV. NUTRIENT TRANSPORT PROTEINS

A. Membrane Bound Transport Systems

The classical example of a membrane bound transport system is the E. coli lactose permease. This is one of the few systems where the membrane carrier has been identified. However, it has only been isolated in an inactive form, so it has not been possible to study the function of the isolated protein. The carrier is called the M protein and it was identified by the fact that it contains a sulfhydryl group in which it can be protected from reaction with N-ethylmaleimide by certain transport substrates. When E. coli membranes were reacted with unlabeled N-ethylmaleimide in the presence of the lactose analog, thiodigalactoside, washed free of N-ethylmaleimide, and then reacted with [^{14}C]-N-ethylmaleimide, a large part of the label was bound to the M protein [145]. The labelled M protein was solubilized in 2% SDS and purified. This protein represents 0.5 to 1% of the total protein of induced cells and has a subunit molecular weight of 29,000. The protein as isolated does not bind substrates of the transport system, but the binding properties of the M protein have been studied in intact E. coli membranes [146].

The evidence that the M protein is the carrier comes mainly from the results of genetic studies of lactose permease. Every mutant in which lactose permease is missing or altered has been shown to map in the Y gene of the lactose operon. Since the M protein is missing in strains carrying certain mutations in the Y gene, and the M protein and lactose permease are coregulated, the Y gene is clearly the structural gene for the M protein [147]. There is a recent preliminary report on the successful reconstitution of lactose transport using a chaotropic extract of induced membranes as the source of the M protein and vesicles prepared from a Y⁻ strain [148].

The first strong evidence for the isolation of a functional bacterial membrane carrier has been reported recently [149]. The thermophilic bacterium PS3 was used in this work in the hope that it would have evolved more stable proteins than other bacteria. The carrier for alanine was studied and assayed by reconstituting the partially purified protein with liposomes. The lipid-protein mixture was sonicated in the presence of 2% sodium cholate and 1% sodium deoxycholate and then dialyzed for 20 hr at 40°C to remove the detergent and form phospholipid vesicles containing the carrier. These vesicles were loaded with potassium chloride. Valino-mycin and alanine were added, and the uptake of alanine into the vesicles was measured by filtration. The movement of potassium, catalyzed by valinomycin down its concentration gradient, builds up a membrane poten-tial which presumably drives the uptake of alanine. Alanine uptake was linear with added protein and inhibited by uncouplers of oxidative phosphory-lation. The alanine carrier was purified 13-fold but was still impure and was not characterized further.

Three proteins that appear to be involved in the transport of succinate in E. coli have been purified using affinity chromatography on a column of aspartate-coupled sepharose. One of these was isolated from osmotic shock fluid [150] and had a subunit molecular weight of 15,000. This protein bound succinate with a K_d of 4 x 10^{-5} M and the K_i values for the inhibition of succinate binding by fumarate, malate, and lactate were 5.5 x 10^{-5} M, 3.4 x 10^{-5} M, and 1.3 x 10^{-3} M, respectively. The relationship of this protein to succinate transport is discussed in Chap. 8, and the author's interpre-tation of these data is that this protein functions to transport acids through the outer membrane rather than being a shock releasable binding protein of the type discussed in Sec. IV.B.

The other two succinate related proteins were extracted from induced E. coli membranes with 4% lubrol 17A-10 [151]. After high speed centri-fugation, the supernatant was chromatographed on an aspartate-coupled Sepharose column equilibrated with 0.5% lubrol. The column was washed with 0.2 M succinate and two peaks of protein were eluted. Both proteins bound succinate at a level of one molecule per 10,000 molecular weight of protein. The K_d of succinate binding to the first peak eluted was 2.3 x 10^{-5} M, and the K_d of the second peak was 2.3 x 10^{-6} M. The binding was inhi-bited when the proteins were treated with N-ethylmaleimide or p-chloro-mercuribenzoate. Succinate binding to both peaks was inhibited by fumarate or malate but not by any other compound tested. These proteins were not characterized further but seem to be promising candidates for membrane carriers of the succinate transport system.

Besides detecting the M protein, the technique of specifically labeling membrane carriers with radioactive N-ethylmaleimide has also been used to identify three proteins in the membranes of Bacillus subtilis which appear to function in the transport of C_4-dicarboxylic acids [152]. In these experiments, cells induced for C_4-dicarboxylic acid transport were reacted with unlabeled N-ethylmaleimide in the presence of L-malate, washed, and then reacted with [^{3}H]-N-ethylmaleimide. Uninduced cells were treated identically except that [^{14}C]-N-ethylmaleimide was used in the last reaction. The two lots of cells were mixed and membranes prepared. The membranes were solubilized in SDS, and the proteins separated by electrophoresis in SDS polyacrylamide gels. There were three distinct peaks which had ratios of [^{3}H:^{14}C] counts higher than the original extract. The mobilities of these peaks correspond to molecular weight values of 33,000, 44,000, and 62,000. The 33,000 peak had much less label than the other two peaks. In addition, when the above experiment was repeated using the competitive inhibitor of dicarboxylic transport, mesotartaric acid, as a protective agent, the 33,000 peak was not present but the other two were. The 62,000 and 44,000 peaks were not removed from the membrane by either sonication or osmotic shock. These two proteins appear to be the carriers for one or both of the two malate transport systems present in B. subtilis [153].

There are two reports of experiments where substrates were used to protect transport systems against reaction with affinity labels [154,155]. The preliminary results suggest that it should be possible to use this technique to identify membrane carriers; however this has not yet been done for a bacterial system.

There is a report of the solubilization of amino acid binding activities for five different amino acids from purified E. coli membranes [156]. The membranes were extracted with 2% Brij 36-T at room temperature. This procedure solubilized 30% of the membrane protein, and this soluble fraction was chromatographed on a column of Sephadex G-100 and the fractions assayed for proline binding. Three peaks of proline binding were found, and the workers concentrated on the last peak to be eluted. The binding activity in this peak was inhibited by several sulfhydryl reagents, by boiling for 5 min, or by treatment with 2.7 M urea. Proline binding was also inhibited by a 100fold excess of proline but not inhibited by a 100-fold excess of any other amino acid. This peak also bound lysine, serine, tyrosine, and glycine. There are two puzzling aspects to this report. The first is that the peak being studied eluted at a volume of buffer three times larger than the volume of the G-100 column, and few proteins would be expected to elute in such a position. The second is that the sum of the binding activities present in this peak gives a value of one amino acid binding site per 20,000 daltons of protein. This calculation implies that a single gel filtration step has given a fraction containing nearly pure amino acid carriers. There have been no further studies reported on these binding activities.

Some preliminary work on the solubilization of potential membrane carriers for lysine and cystine is discussed in the section on amino acid binding proteins.

B. Binding Proteins

The best studied bacterial transport proteins are the binding proteins. These are soluble proteins consisting of a single polypeptide chain with molecular weights in the range of 26,000 to 43,000. These proteins appear to be present in the region between the outer wall and plasma membrane of some gram-negative bacteria, the so-called periplasmic space. The proteins in the periplasmic space can be specifically released from certain gram-negative bacteria by a gentle procedure called osmotic shock [9].

Binding proteins do not have any known enzymatic activity but do bind specific substrates with K_d values in the range of 10^{-8} to 10^{-5} M. The high affinity and specificity of substrate binding to these proteins has been utilized to develop a microassay for certain amino acids [157]. This procedure is applicable to most other compounds for which there is a binding protein. Specific binding proteins have been identified which can bind ions, amino acids, sugars, and vitamins. The properties of the purified binding proteins are summarized in Table 11-5. In every case studied, the properties of each binding protein closely resemble the kinetic properties of a specific transport system. In addition to this circumstantial evidence, there is conclusive evidence for three different binding proteins that they are essential components of specific transport system. The evidence is the isolation of mutants which have lost the specific transport system as the result of a mutation in the structural gene for the binding protein [158-160]. Furthermore, in each of these cases, genetic experiments show that at lease one other gene product besides the binding protein is required for a functional transport system [161].

1. Binding Proteins for Ions

The first binding protein to be isolated was the sulfate binding protein from Salmonella typhimurium [162]. This protein was isolated in a pure state from the shock fluid of cells derepressed for sulfate transport. About 20% of the protein released by osmotic shock from derepressed cells was sulfate binding protein, which means that the sulfate binding protein represents almost 1% of the protein in the cell. This comes to about 20,000 molecules of sulfate binding protein per cell. The sulfate binding protein is a single polypeptide chain of molecular weight 31,000 [163]. Its amino acid composition was determined, and it contains only one methionine residue and no residues of one-half cystine. The purified preparation of sulfate binding protein contained less than 1% lipid or carbohydrate, and no tightly bound cations. The protein was crystallized, and a preliminary report of its crystallographic properties has been published [164]. These studies as well as hydrodynamic studies indicate that the protein is asymmetric, with an axial ratio for an ellipsoid structure of 4. The hydrodynamic and crystallographic data can be fitted by a cylinder 112 Å long and 27 Å in diameter, although other models are possible. The ORD and CD spectra of this protein indicate an α-helix content of about 36%.

The purified protein has one sulfate binding site with a K_d of 2 x 10^{-5} M, when binding is measured under the conditions used to measure sulfate uptake. The binding activity is not very sensitive to pH in the range from 5.3 to 8.5 and is quite stable to heat. The binding constant is decreased with increasing ionic strength. This decrease is identical to the decrease seen for the binding of sulfate to Dowex 1, which suggests that ionic forces function in the binding of sulfate to the protein. The sedimentation coefficient of the binding protein is not changed by sulfate binding, nor is there a measurable change in the tryptophan fluorescence of the protein upon addition of sulfate. There is a very small change in the ORD spectrum of the protein when sulfate is bound, but overall the evidence indicates that sulfate binding does not cause a major conformation change in the protein.

The properties mentioned above and the coregulation of the sulfate binding protein with sulfate transport provide strong, though indirect, evidence that it is an essential component of the sulfate transport system. Osmotic shock, which releases 80% of the binding protein from the cell, inhibits sulfate uptake by 80%. The K_m for sulfate uptake of 2 x 10^{-5} M is the same as the K_d of the sulfate binding protein. The specificity of ion binding to the protein is the same as the specificity of transport. Finally, mutations in the cysA and cysB gene produce cells with greatly reduced levels of sulfate transport and sulfate binding protein [165]. Since the residual binding protein in these cells is identical to the binding protein from wild-type cells, neither of these genes is the structural gene for the binding protein, but transport activity and binding protein are coregulated. Despite an extensive search no mutants in the gene for the sulfate binding protein have been found. All attempts at restoring transport to shocked cells or to mutants lacking sulfate uptake by the addition of purified sulfate binding protein or crude shock fluid have been unsuccessful.

A phosphate binding protein has been purified from E. coli [166]. This protein represents 4% of the protein released by osmotic shock. It is a single polypeptide chain of molecular weight 41,000. The protein contains a single binding site for phosphate with a K_d of 0.8 x 10^{-7} M and appears to function in a high-affinity phosphate transport system [167]. Mutants lacking the binding protein were isolated from a strain which is supposed to contain only the one phosphate transport system. These mutants have an altered K_m for phsophate, a reduced V_{max}, and are no longer sensitive to arsenate. It seems likely that this strain actually contains two phosphate transport systems, one of which requires the phosphate binding protein, however this has not been proven. It was reported that the addition of purified binding

TABLE 11-5 Bacterial Binding Proteins

Organisms	Substrates	Molecular Weight	Dissociation Constant (μM)	K_m of Transport (μM)	Ref.
S. typhimurium	sulfate	31,000	20	20	163
E. coli	phosphate	41,000	0.08		166
E. coli	leucine		0.6	0.5	173,174
	isoleucine	36,000	0.6	0.4	
	valine		10	--	
E. coli	leucine	36,000	0.6	0.5	181
S. typhimurium	histidine	26,000	0.15 – 1	0.3 – 1	186,187
E. coli	glutamine	25,000	0.3	0.1	190
E. coli	glutamate		0.7		192
	asparate		1.2		
E. coli	cystine	27,000	0.01	0.02	199
	DAP		17	14	
E. coli	lysine		3	0.5	
E. coli	arginine	27,000	1.5		203
	ornithine		5.0	1.4	

Organism	Substrate	Molecular weight	K_d	K_d	Reference
E. coli	arginine	28,000	0.026	0.03	205
Comamonas sp.	phenylalanine	25,000		0.1	208
	tyrosine			0.05	
	tryptophan			0.8	
E. coli	galactose	35,000	0.5	1	210, 211
	glucose			0.5	
S. enteritidis	galactose	35,000			228
	glucose				
E. coli	arabinose	38,000	4	2	230, 231
S. typhimurium	ribose	31,000		0.3	234
E. coli	ribose	30,000	0.3	0.2	235
E. coli	maltose	37,000		160	
E. coli	Thiamine	40,000	0.3	0.05	237
E. coli	vitamin B12	22,000	0.01	0.006	239

protein could reconstitute phosphate uptake into shocked cells and spheroplasts. However these results could not be repeated [168]. The phosphate
binding protein is the product of the phoS gene [169] which was first identified as the site of mutations allowing the constitutive synthesis of alkaline
phosphatase [170]. The synthesis of this protein is regulated by phosphate
in a complex fashion [171].

There is evidence that the high affinity potassium transport system in
E. coli involves a periplasmic binding protein [172], but the protein has not
been characterized at the present time.

2. Amino acid binding proteins

The first amino acid binding protein to be studied was a leucine binding
protein from E. coli [173,174]. In cells derepressed for leucine transport,
this protein represents about 1% of the protein released by osmotic shock.
This protein (LIV binding protein) is a single polypeptide chain of molecular
weight 36,000. The purified protein has been crystallized and does not
contain any carbohydrate or phosphate. The amino acid compositions
reported by the two groups [175,176] are similar except for the values of
tryptophan. Both groups find one half cystine residue per molecule.
The two groups report different frictional coefficients for this protein:
$f/f_0 = 1.03$ and 1.28. The last value seems the best since the molecular
weight determined in the ultracentrifugation experiments, which gave a
value of f/f_0 of 1.03, was $26,000$, which is well below the values found in
all other experiments. The α-helix content of the LIV binding protein from
its ORD spectrum was reported to be in the range from 20 to 40% by two
groups [177,178]. This protein can be reversibly denatured by heat or
urea but does not appear to change conformation significantly upon binding
leucine.

There is a single binding site on the LIV protein which binds leucine,
isoleucine, and valine with K_d values of 6×10^{-7} M, 6×10^{-7} M, and $1 \times$
10^{-5} M, respectively. This protein also binds threonine, alanine, and
possibly serine [179]. Binding activity is relatively constant in the pH
range from 4 to 9 and is not altered by changes in ionic strength. Binding
is not inhibited by metal ions or N-ethylmaleimide.

Studies of leucine transport in E. coli are complicated by the presence
of at least three distinct transport systems [179,180]. In the strain studied
in Oxender's laboratory the LIV transport system with which the LIV binding protein is associated appears to be responsible for most of the leucine
transport, and these workers were able to show a good correlation between
leucine transport and the LIV binding protein. Thus, osmotic shock greatly
inhibited leucine transport and released most of the LIV binding protein
from the cell. Growth in the presence of increasing concentrations of leucine caused a parallel repression of leucine transport and the amount of the

leucine binding protein. The K_m values for leucine, isoleucine, and valine transport are very similar to the K_d values for the LIV protein. Finally, a mutant was isolated which had a threefold increase in the LIV binding protein and in leucine transport compared to the wild-type strain.

Another leucine binding protein (leucine specific) has been purified from E. coli and crystallized [181, 182]. It has a molecular weight of 36,000 and contains a single binding site for leucine with a K_d of 6 x 10^{-7} M. This protein is very specific in its binding, and only the leucine analog, trifluoroleucine, inhibits leucine binding to this protein. This protein is probably associated with a second leucine transport system called the leucine-specific system. Mutants have been isolated which have increased levels of the leucine-specific transport system and leucine-specific binding protein [179]. In addition, a mutant was isolated which has lost the leucine-specific transport system and has an altered leucine-specific binding protein. The LIV and leucine-specific binding proteins are structurally related as they show immunological cross-reaction and even appear to share some common sequences, as their trypsin fingerprints have some peptides in common [183].

There is evidence of a third leucine binding protein in E. coli which probably is associated with an isoleucine transport system, as it binds isoleucine more tightly than leucine [179, 184].

It was reported that the addition of purified LIV binding protein or leucine-specific binding protein to osmotically shocked E. coli restored leucine transport, and that the LIV protein restored isoleucine transport while the leucine-specific protein did not [185]. However in these experiments the amount of binding protein added was capable of binding about 10-times more substrate than the observed transport stimulation, and binding proteins are known to bind to the membrane filters used in the transport assay. Furthermore, the stimulation was complete within the first minute, which is consistent with a binding reaction. Finally, Oxender's laboratory has never been able to reconstitute transport in shocked cells despite many attempts.

Osmotic shock releases several histidine binding proteins from S. typhimurium. However, more than 90% of the total histidine bound at a histidine concentration of 10^{-7} M is present in the major peak called the hisJ protein [46]. The major peak of histidine binding activity from S. typhimurium has bee purified by two groups [158, 187], and although there are some differences in the reported binding properties, the bulk of the evidence indicates that both groups isolated the hisJ protein. The protein has a molecular weight of 26,000, and amino acid analysis showed the presence of two residues of one-half cystine [158]. The protein has a single binding site for histidine.

The K_d values reported by the two groups (0.15 μM and 1 μM) differ somewhat. The binding is very specific for histidine, as the only amino acid that causes any inhibition of histidine binding is arginine. The two papers also reported different effects of pH on binding. One group reported no changes in binding when the pH was increased from 5 to 7 with sharp drop in binding as the pH was increased from 9 to 10.

The other paper reported a peak of binding at about pH 6. The binding dropped steadily as the pH increased from pH 6, reaching 0 at pH 9. This last result seems the most probable since histidine has a pK around pH 6. Both groups showed that the binding activity was very stable to heating.

There is a strong correlation between the level of the high-affinity histidine transport system and the amount of the hisJ protein in Salmonella mutants. Mutants in the hisJ gene have very low levels of transport and hisJ protein. Mutants in the dhuA gene, which appears to be a regulatory locus, have about a fivefold increase in both histidine transport and the hisJ protein. Osmotic shock releases the hisJ protein and strongly inhibits transport. A revertant of a hisJ mutant has been isolated in which histidine transport is temperature sensitive [186]. In this strain the hisJ protein has different chromatographic properties and is more readily inactivated by heating. This indicates that the hisJ locus is the structural gene for the histidine binding protein. Since most hisJ mutants are missing the binding protein and lack the high-affinity transport system this protein must be an essential component of the high-affinity transport system. Mutants in a second locus hisP also lack histidine transport but have normal levels of the binding protein. Therefore the high affinity histidine transport system requires at least one protein in addition to the hisJ protein. It is probable, but not yet demonstrated, that the hisP product is an integral membrane protein.

Recent genetic experiments indicate that there is a direct interaction between the hisJ protein and the hisP protein [188]. Mutants in the hisJ gene which showed normal binding but defective histidine transport were plated on D-histidine. This selects for strains which have regained the histidine transport system. A strain was isolated which still contained the original hisJ mutation and also had a mutation in the hisP gene. This strain had increased levels of histidine transport. When the new hisP gene was transferred to a strain with the wild-type hisJ gene, histidine transport still occurred, so that the mutant hisP protein retains the ability to interact with the normal histidine binding protein. It was also possible to start with a hisP mutant and isolate a strain with improved histidine transport which was shown to have a mutation in the hisJ gene. These experiments provide strong evidence that the products of the hisJ gene and the hisP gene interact directly but do not answer the question: What is the role of the hisP gene product?

HisP mutations not only inactivate histidine transport but also prevent the utilization of arginine as a nitrogen source [189]. The high-affinity arginine transport system is normal in hisP mutants, and the specific transport system involved in growth on arginine has not been identified.

The glutamine binding protein of E. coli is associated with the only major transport system for glutamine in E. coli. This makes the study of the relationship between the transport system and the binding protein simpler than for most substrates which are transported by several systems. This protein was purified from osmotic shock fluid and represents about 3% of the protein in shock fluid [190]. Its molecular weight is 25,000, and the protein was shown to contain no half cystine residues by amino acid analysis. The ORD spectrum indicates an α-helix content of 45% which is not altered by the addition of glutamine. The protein has a single binding site for glutamine with a K_d of 3 x 10^{-7} M, which is fairly close to the K_m value for glutamine transport of 0.8 x 10^{-7} M. No other natural amino acid inhibits glutamine binding, and the only glutamine analogs which inhibit are γ-glutamyl hydroxamate and γ-glutamyl hydrazide. These compounds are also the only ones that inhibit glutamine transport in whole cells. The binding activity is constant in the pH range from 3 to 9 and is not altered by changing the KCl concentration from 0.01 to 0.25 M. There is a decrease in the λ_{max} of tryptophan fluorescence as well as a decrease in its intensity upon the binding of glutamine to the binding protein. The K_d for glutamine measured by the fluorescence shift is 3 x 10^{-7} M; identical to the value measured by equilibrium dialysis. The two tryptophan residues in the protein can be oxidized by N-bromosuccinimide without inhibiting binding, and this oxidation is prevented when glutamine is bound. The change in fluorescence upon binding was used to measure the rates of glutamine binding and dissociation from the binding protein. The value of k_{on} determined for the reaction gln + BP $\rightleftarrows$ gln BP was 9.8 x 10^{-7} M^{-1} while the value of k_{off} was 16 sec^{-1}. This gives a K_d of 1.6 x 10^{-7} M which is in good agreement with the values determined above. The change in conformation upon glutamine binding has also been detected by NMR spectroscopy [191].

A mutant was isolated which has a rate of glutamine transport three times higher than the wild type and also contains about three times more binding protein than the wild-type strain. A second mutant has 10% of the wild-type level of glutamine transport, and this strain contains only 1/10 the amount of glutamine binding protein. In addition, osmotic shock inhibits glutamine transport 90% and releases most of the binding protein. In summary, all the properties of the binding protein are consistent with it being an essential component of the glutamine transport system.

A protein which binds both glutamate and aspartate has been purified from E. coli K12 [192]. This protein represents only 0.3% of the protein in osmotic shock fluid and is a single polypeptide chain with a molecular weight of 30,000. Amino acid analysis showed that the protein contains two residues of half cystine. The protein is reversibly denatured by 6 M urea or 6 M guanidine-HCl. When the protein is denatured in the presence of a reducing agent, no activity can be recovered. Since the reducing agent by itself has no effect on binding, the inactivation by the combination of denaturing and reducing agents suggest that the protein contains a buried disulfide linkage, which is important either in determining the overall structure of the protein or in the binding reaction. The protein contains a single binding site which binds glutamate with a K_d of 7×10^{-7} M and asparate with a K_d of 1.2×10^{-6} M. The fluorescence spectrum of the protein is shifted to lower wavelengths and decreased slightly when glutamate is bound. Glutamate binding is completely inactivated when the tryptophan residues of the protein are oxidized by N-bromosuccinimide, and the oxidation is not prevented by the presence of glutamate. Glutamate binding increases sharply as the pH increases from 4 to 5. Binding is constant in the range from pH 5 to 9 and then drops somewhat at pH 10.

There are at least three systems in E. coli which can transport glutamate [193]. The system which has been most studied is stimulated by sodium ions and is retained in membrane vesicles [194, 195]. The glutamate-aspartate binding protein appears to be associated with a minor glutamate-aspartate transport system which can be inhibited by cysteic acid. Cysteic acid also inhibits the binding of glutamate and aspartate to the purified binding protein.

It was reported that spheroplasts had lost 80% of their glutamate transport. This transport could be completely restored by the addition of a fraction containing the proteins released during spheroplast formation, which contained the glutamate-aspartate binding protein, along with many other proteins [196]. Since it is now clear that the binding protein is associated with a minor glutamate transport system, it could not be responsible for the restoration of 80% of the cells' glutamate transport. Furthermore, the reconstituted spheroplasts were not tested for other transport systems, so that the spheroplast fluid may have had a general effect on transport such as stimulating energy metabolism rather than causing a specific reconstitution of glutamate transport. Although all the discrepancies in the literature of glutamate transport are not yet fully explained, the results do suggest that the glutamate binding protein is probably an essential component of a single glutamate transport system rather than playing the more indirect role proposed in earlier reports [197, 198].

A cystine binding protein has been purified from E. coli cells. This protein represents about 3% of the shock fluid protein and is a single polypeptide chain of molecular weight 27,000 [199]. When crude shock fluid was chromatographed on DEAE cellulose, there were two distinct peaks of cystine binding activity. Both peaks were purified to homogeneity, and they were identical in every property tested. The protein was shown to contain three residues of half cystine by amino acid analysis. The protein has a single binding site with a K_d for L-cystine of 1×10^{-8} M and an K_d for α, ε-diaminopimelic acid (DAP) of 1.7×10^{-5} M. The cystine binding activity does not change in the pH range from 3 to 10, and binding is also unaffected by raising the sodium chloride concentration from 0.05 M to 0.5 M. In wild-type E. coli cells there are two systems which transport cystine [200]. One of these (cystine-general) transports DAP while the other does not. In a mutant strain which contains only the cystine-general transport system, osmotic shock inhibits cystine transport 83% while releasing most of the cystine binding protein. Several mutants were isolated which are missing cystine-general transport and have less than 10% of the cystine binding protein of wild-type cells. The final evidence indicating that the cystine binding protein functions in the cystine general transport system is that most of the large number of cystine analogs tested for their ability to inhibit cystine uptake by the cystine transport system and cystine binding to the binding protein inhibit the two reactions to the same extent.

The other cystine transport system (cystine-specific) is not inhibited by osmotic shock carried out on cells grown in minimal medium. However, when the cells are grown in a medium containing yeast extract and bacto-tryptone, the cystine-specific transport system is induced and is inhibited 80% by osmotic shock. Furthermore, a cystine binding activity is released in the shock fluid which is different from the cystine binding protein, as it does not bind DAP. This activity is somewhat unstable and could not be purified. The exact relationship between this cystine binding activity and the cystine-specific transport system has not been determined, but it seems likely that the protein is a component of the transport system.

Lysine is transported in E. coli by at least two systems. One of these (lysine-specific) resembles the cystine-specific transport system in that it is induced in cells grown in rich medium and is sensitive to osmotic shock only in cells grown in rich medium. A lysine binding activity is released by osmotic shock of rich medium-grown cells which is unstable and has not been purified [201]. A lysine binding activity which could be the same as the one referred to above has been extracted from E. coli membranes using 1 mM Tris-HCl, pH 7.2 [202]. This activity was chromatographed on Sephadex G-100, and it eluted in the low molecular weight region but was not purified further. The K_d for lysine of this activity is 8×10^{-6} M, which is close to the K_m value of 1×10^{-5} M for the lysine-specific transport system. However, this protein has not been studied further.

The other lysine transport system (LAO system) is shock sensitive in cells grown in minimal medium. A lysine binding protein has been purified from such cells, and this protein represents about 2% of the protein in the shock fluid [203]. The LAO protein is a single polypeptide chain of molecular weight 27,000. Its amino acid composition was determined, and it contains two residues of half cystine. This protein has a single binding site with K_d values of 3×10^{-6} M, 0.5×10^{-6} M, and 1.5×10^{-6} M for lysine, ornithine, and arginine, respectively. These values are close to the K_m values of 5×10^{-7} M and 1.4×10^{-6} M, respectively, for lysine and ornithine uptake by the LAO transport system. Furthermore, osmotic shock inhibits transport by the LAO system by 90% and releases most of the LAO binding protein.

Although the LAO binding protein binds arginine and arginine inhibits the transport of lysine and ornithine by the LAO system in whole cells, arginine does not appear to be transported by the LAO system. Arginine appears to be transported by a single system in E. coli which is associated with an arginine binding protein [204,205]. This protein represents 3% of the protein in shock fluid and has a molecular weight of 28,000 [205]. Its amino acid composition is quite different from that of the LAO protein, although it also contains two residues of half cystine. These residues appear to form an interchain disulfide linkage, as no sulfhydryl groups could be detected even in the presence of 8 M urea. Antisera prepared against the LAO protein and against the arginine binding protein showed no cross-reaction against the other protein when tested by Ouchterlony double diffusion precipitation. The arginine binding protein has a single binding site for arginine with a K_d of 3×10^{-8} M. This site is extremely specific, as arginine binding is not inhibited by any natural amino acid or by any arginine analog tested. The K_m for arginine uptake is 2.6×10^{-8} M, which is very close to the K_d of binding, and arginine transport is also very specific. Osmotic shock inhibits arginine transport by 95% and releases most of the arginine binding protein. These results indicate that the arginine binding protein is probably an essential component of the E. coli arginine transport system.

It was reported [204] that partially purified arginine binding protein could partially restore arginine transport in osmotically shocked cells. In these experiments arginine transport was only inhibited 30% by osmotic shock. However, in later work using pure arginine binding protein there was no evidence for reconstitution [205].

Even though the LAO protein and arginine binding protein do not appear to be structurally similar, there is evidence that they may both interact with a common membrane component. This evidence is the isolation of mutants that have lost the LOA transport system and arginine transport as the result of a single mutation [206]. Thsee mutants have normal levels of unaltered LAO and arginine binding proteins, so that the mutation must affect some other component of each transport system [207].

A phenylalanine binding protein has been purified from the gram-negative organism Comamonas SP [208]. This protein is a single polypeptide chain with a molecular weight of about 25,000. Its amino acid composition was determined, and it does not contain histidine and may not contain cystine. The protein has an isolelectric point of 9.0. Reaction of the protein with either 1-fluoro-2,4-dinitrobenzene or maleic anhydride, both of which react with amino groups, completely inhibits phenylalanine binding. The protein has a single binding site with K_d values of 1×10^{-7} M, 5×10^{-8} M, and 3×10^{-7} M for phenylalanine, tryosine, and tryptophan, respectively. Binding decreases somewhat as the pH is increased from 5 to 9, and there is no binding at pH 4 or 10. Binding is reversibly inhibited by 2.0 M urea and is slightly stimulated by 0.2 M sodium chloride. Growth on phenylalanine causes a larger increase in phenylalanine uptake than in the level of the binding protein. However, in these experiments the level of the binding protein was measured in osmotic shock fluid, and there is only a partial release of this binding protein by osmotic shock. There is also a difference between the value of the K_m of phenylalanine transport of 1×10^{-5} M and the K_d of binding; however, there may well be several transport systems for phenylalanine, and the K_m was determined at high concentrations of phenylalanine [209]. Although the data on the relationship of this protein to aromatic amino acid uptake is very incomplete, the protein closely resembles the other binding proteins and thus probably plays a role in an aromatic transport system.

3. Sugar Binding Proteins

In an early report the E. coli galactose binding protein was purified [210] and shown to have a molecular weight of 35,000. Its amino acid composition is similar to that of the E. coli leucine binding protein. This protein has no enzymatic activity but binds both galactose and glucose with K_d values of 1×10^{-6} M and 0.5×10^{-6} M, respectively. The binding activity is stable to heating and is not changed in the pH range from 5 to 9.

Later work on the protein was recently reviewed [211], and showed that the protein is a single polypeptide chain. The ORD, CD, and infrared spectra of the protein [212] were determined, and they indicate that there is a low content of α-helix (less than 10%) and a high content of β-structure. The addition of galactose to the protein caused no change in any of the above spectra, which indicates that the binding of substrate does not result in a large conformational change. However, there was a substantial change in tryptophan fluorescence when galactose was added. Glucose also caused a change in fluorescence although it was not as large as when galactose was added. Further evidence that substrate binding causes a conformational change in the galactose binding protein is a change in the rate of migration of the protein during polyacrylamide gel electrophoresis at pH 8.4 when galactose is present. This change appears to result from a change in the effective charge of the protein since the addition of galactose causes no change in its molecular weight or sedimentation coefficient. In addition, the isoelectric point as determined by isoelectric focusing changes from 5.4 to 5.3 when galactose is present.

The K_d values for galactose and glucose calculated from the change in fluorescence are both about 1×10^{-6} M. These values are in good agreement with the values originally reported [210]. However, these values do not agree with the results of the more recent binding studies. These studies showed two different binding constants with the pure binding protein preparation. In one study binding was measured by ultrafiltration and about 0.2 molecules of galactose were bound per molecule of protein with a K_d of 5×10^{-7} M, and there was evidence for further binding with a much higher K_d [213]. In more extensive studies where binding was measured by equilibrium dialysis, a high affinity site was again detected, but this time it had a K_d of 1×10^{-7} M and only 0.1 molecule of galactose was bound per molecule of protein [212]; there was also binding with a K_d of 1×10^{-5} M. The data has to be extrapolated to give the number of binding sites, which was two molecules of galactose per molecule of protein. A model has been proposed [211] based on the observed conformational change upon the binding of galactose to the protein and the presence of two different binding constants. In this model, the galactose binding protein in the absence of substrate is predominantly in conformation I, with a binding site A for galactose having a K_d of 10^{-5} M. The binding of substrate to this site favors the conversion of the protein into form II, in which the K_d of site A is now 1×10^{-7} M, and which has a second site with a K_d of 1×10^{-5} M. This model is clearly incorrect since it predicts that there will be one molecule of galactose bound to each protein molecule with a K_d of 1×10^{-7} M, while only 1/10 that much galactose was actually bound with this K_d. In terms of binding one could try to salvage the model by proposing that site A initially had a K_d of 1×10^{-7} M and substrate binding converted it to a site with a K_d of 10^{-5} M. However, this model is not possible for thermodynamic reasons.* It is therefore not possible to explain the recent binding data by proposing interconvertible forms of the protein. A more reasonable explanation comes from a recent report [214] which showed that purified galactose binding protein contains tightly bound glucose which will interfere with binding measurements at low substrate concentrations if small volumes are used in the assay. In fact, the original binding studies were carried out in a 10 ml volume while the more recent measurement utilized a volume of only 0.3

---

*This model gives the following set of partial reactions:

$$K_2 = \frac{[II]}{[I]} \qquad \begin{array}{ccc} I + gal \xrightleftharpoons{\hspace{1cm}} I\text{-}gal & & \\ \updownarrow & & \updownarrow \\ II + gal \xrightleftharpoons{\hspace{1cm}} II\text{-}gal & & \end{array} \qquad K_1 = \frac{[II\text{-}gal]}{[I\text{-}gal]}$$

$$K_a = 10^7$$
$$K_a = 10^5$$

K_1 must be greater than 1 since all the observed binding in the presence of a high galactose concentration occurs with a K_a of 10^5. $K_2 = K_1 \, 10^7/10^5 = 100 \, K_1$. This means that less than 1% of the protein would be in form I in the absence of galactose so that this model would never predict any significant amount of binding with a K_a of 10^7.

to 0.2 ml. However, experiments have been reported [215] showing that the purified preparation of galactose binding protein used in the later studies does not contain any bound ligand and still shows the anomalous binding. It seems that the only explanation for these results is either that the binding measurements are incorrect or that the binding protein preparation contains only 10% of the normal activity and the rest is partially inactivated. In any case it seems highly probable that the galactose binding protein, like every other binding protein studied, contains a single substrate binding site and that the K_d value for galactose binding to this site is 1×10^{-6} M.

It has been proposed that the complex formed between the galactose binding protein and its substrates does not dissociate but can exchange with external substrate [214]. This is not a reasonable model thermodynamically, and when experiments are correctly designed to take into account the high affinity of the binding protein, dissociation can be shown directly [216].

There have been several genetic studies of the β-methylgalactoside transport system with which the galactose binding protein is associated [217-221]. The results of these studies show that the β-methylgalactoside transport system requires at least two proteins. One is the galactose binding protein which is encoded in the mg1B cistron, and another (or, less likely, others) is encoded in the mg1A and mg1C cistrons. The mg1 cistrons form a negatively controlled operon which is induced by D-fucose or D-galactose and strongly repressed by growth on glucose [222].

Besides a role in transport, the galactose binding protein also appears to function in the chemotactic response of E. coli to galactose [223]. The role of the binding protein in chemotaxis differs in part from its role in transport, since one binding protein mutant has been isolated which has normal galactose chemotaxis and very little transport activity. In addition, mutants having normal transport and no galactose chemotaxis have also been found [224]. Most mutants which change the K_m of transport show a similar change in the K_m of chemotaxis. These results suggest that the binding site for galactose functions in both transport and chemotaxis but that there is a region on the protein which functions only in transport and another region which functions only in chemotaxis. Presumably these specific regions are involved in the interaction with the other component or components which function only in transport (mglA, mglC gene products) or only in chemotaxis (trg product).

There have been several reports [185,225] of the reconstitution of galactose transport in shocked cells by the addition of crude or purified galactose binding protein. In the first report there was only a small increase in transport, and the transport stimulation was studied in cells lacking an energy source. In the next report the absolute value of the transport stimulation was significantly lower than the amount of galactose binding protein added, so that the reported increase in transport was probably just due to binding to the added binding protein. Other workers have not been able to obtain reproducible reconstitution of galactose transport by the addition of purified binding protein to shocked cells.

There has been one report of the reconstitution of chemotaxis in shocked cells [226]. In these experiments purified galactose binding protein gave a fourfold increase in the number of cells responding to the galactose gradient. The amount of chemotaxis in reconstituted cells was close to the value for unshocked cells. Furthermore, when an altered binding protein was used to reconstitute wild-type shocked cells, the properties of the chemotactic response were characteristic of the strain from which the binding protein was isolated, not the original strain. Transport was not studied in this report.

A protein that binds glucose and galactose has also been purified from osmotic shock fluid from Salmonella enteritidis [227]. This protein represented 5% of the protein in shock fluid and has a molecular weight of 35,000. Its amino acid composition is quite different from that of the E. coli galactose binding protein. It has a single binding site for galactose and glucose, but the properties of this binding were not studied. The protein binds fatty acids, and there is a 50% increase in the amount of galactose bound in the presence of certain fatty acids [228]. The significance of the fatty acid binding and the role of the binding protein in sugar transport have not been studied in any detail.

A glucose binding protein has been extracted from P. aeruginosa by treatment with 0.2 M $MgCl_2$ followed by osmotic shock [229]. The protein has not yet been purified or characterized, but a number of preliminary observations suggest that this protein is part of a binding protein transport system. The same paper also describes a protein released by the same treatment which binds C_4-carboxylic acids, but it was not purified and there is not sufficient data reported to tell whether this protein is an outer membrane carrier or a true periplasmic binding protein.

A binding protein for arabinose has been isolated from E. coli B[230] and from E. coli K12 [231]. This protein represents about 2% of the total protein in induced E. coli B and about 1% of the protein in E. coli K12. The protein appears to be a single polypeptide chain of molecular weight 38,000. Recently the partial amino acid sequence [232] and an X-ray structure at 3.5 Å resolution [233] have been reported for the protein from E. coli B. This protein appears to be related to the galactose binding protein, as antisera prepared against either protein cross-react with the other protein when they are both prepared from E. coli B. It has a single binding site for arabinose with a K_d of 2 x 10^{-6} M. This protein also binds D-galactose, D-xylose, and D-fucose. It appears to function in the high affinity arabinose transport system ($K_m = 4 \times 10^{-6}$ M), since the specificity of this system and the binding protein are very similar. Finally, mutants which have lost the high affinity transport system appear to have lost the binding protein or contain an altered binding protein.

A ribose binding protein has been isolated from S. typhimurium [234] and E. coli [235]. The ribose binding protein from S. typhimurium is a single polypeptide chain of molecular weight 31,000. Amino acid analysis showed that the protein contains no half cystine residues or tryptophan residues, although the composition was not reported. The protein has a single binding site which is extremely specific for ribose and has a K_d of 3×10^{-7} M. The E. coli protein is a single polypeptide chain of molecular weight 30,000. This protein has an isoelectric point of 6.6 and a single binding site for ribose with a K_d of 2×10^{-7} M. The binding is very specific as only ribulose showed any inhibition of ribose binding. The K_m for ribose uptake of 3×10^{-7} M is very similar to the K_d of ribose binding, and the specificity of transport and binding are the same. Chemotaxis for ribose also shows a similar specificity [236]. There is a strong cross-reaction between the E. coli protein and an antiserum prepared against the S. typhimurium ribose binding protein.

A maltose binding protein has been purified from E. coli and represents 0.3% of the total protein of induced cells [160]. This protein, which is released by osmotic shock, is a single polypeptide chain of molecular weight 37,000. The protein was homogeneous by a number of tests, but the binding of maltose was heterogeneous. One type of binding had a K_d of 1.5×10^{-6} M and there were 0.5 molecules of maltose bound per molecule binding protein. The other type of binding had a K_d of 1×10^{-5} M and also gave 0.5 molecules bound per molecule of protein. This result may indicate a partial inactivation during the isolation, the presence of two conformational states of the protein, or the presence of tightly bound maltose. The maltose binding protein and maltose transport are both induced 10-fold by growth in the presence of maltose and are both repressed by growth on glucose or by a malT mutation. The malE gene was shown to be the structural gene for the maltose binding protein, and malE mutants have lost the maltose transport system. This shows that the maltose binding protein is an essential component of the maltose transport system. The binding protein also is required for maltose chemotaxis [236]. The malJ mutants inactivate maltose transport but have normal levels of the maltose binding protein. These results indicate that the maltose binding protein is an essential component of the maltose transport system and that at least one additional protein is also required for the function of the maltose transport system.

4. Vitamin Binding Proteins

A binding protein for thiamine has been purified from E. coli [237]. This protein represents about 0.4% of the protein in shock fluid and is a single polypeptide chain with a molecular weight of about 40,000. The protein has a single binding site per molecule with a K_d of 5×10^{-8} M for thiamine and 5×10^{-7} M for thiamine pyrophosphate. The thiamine transport system has a K_m for thiamine of 3×10^{-7} M and is inhibited 85% by osmotic shock [238]. All the properties of the thiamine binding protein and thiamine transport system suggest that the thiamine binding protein is a component of the thiamine transport system.

A binding protein for vitamin B_{12} has been reported in the shock fluid of E. coli [239] but the protein was only partially purified. It has a molecular weight of about 22,000 and a K_d for vitamin B_{12} of 6×10^{-9} M. The K_m for vitamin B_{12} uptake is 1×10^{-8} M and uptake is inhibited 90% by osmotic shock. These experiments measured uptake into whole cells and not the initial binding phase of B_{12} transport [240]. The size difference and ease of release clearly distinguish this binding protein from the outer membrane B_{12} receptor discussed in Chap. 9.

5. Mechanisms of Binding Protein Transport Systems

Recent work on the β-methylgalactoside transport system has provided strong evidence that binding proteins do not function as the actual membrane carriers [241,242]. These experiments showed that mutants in which the galactose binding protein was inactivated still retained the ability to transport β-methylgalactoside with the same V_{max} as the wild-type strain, although the K_m was 1000-fold higher. Mutants in the mglA or mglC cistrons had totally lost the ability to transport β-methylgalactoside. The only uncertainty in these experiments is that none of the mglB mutants studied were deletions or chain termination mutants, and so they contained a mutated binding protein. It is therefore possible that the transport remaining in these mutants utilized the mutant binding protein. This seems unlikely, since the properties of this transport were the same in a number of different mglB mutants which contained binding proteins with different properties. This work suggests that the mglA and/or mglC gene products are the carriers for this system.

It should be emphasized that even if some transport does occur in the absence of the binding protein, the tremendous increase in the K_m when the binding protein is inactivated completely abolishes the physiological role of these high-affinity transport systems, which is to utilize compounds present in the medium in low amounts. In the case of the β-methylgalactoside transport system, wild-type strains have a K_m for β-methylgalactoside of 2×10^{-5} M and can build an internal pool 10^4 times higher than the external concentration, while the mutant strains have a K_m of 10^{-2} M and can achieve at most a 10-fold concentration of β-methylgalactoside.

A recent study on the exit of β-methylgalactoside showed that exit was normal in mglA and mglC mutants and that exit was normal even when the synthesis of the mgl operon was completely repressed by growth on glucose [243]. These results indicate that if the mglA and/or mglC products are the carriers for the β-methylgalactoside entry reaction, then exit must occur by a different carrier. The fact that most of the properties of the exit reaction are completely different from those of the entry reaction is consistent with this conclusion. However, the exit reaction did show one similarity to the entry reaction: it was greatly stimulated by ATP or a related compound [243].

Since it is difficult to imagine why bacteria would contain two membrane carriers for a compound, each of which would function in only one direction, it seems possible that the mglA and mglC products are not the carriers but rather function to couple energy to the transport system. In this model, the carrier would have the properties seen in the exit reaction. The binding protein would bind substrate and then this complex would bind to the carrier. Energy would be used to dissociate the substrate from the binding protein and the substrate would enter the cell via the carrier. The binding protein would then dissociate and revert to its high-affinity form. The above model postulates the existence of another gene product for the β-methylgalactoside transport system besides the products of the mgl operon, and at present there is no genetic evidence for this product. However, the isolation of transport mutants is not easy, so it is possible that the β-methylgalactoside transport system requires the product of yet another locus. The determination of the mechanism of binding protein transport systems will require the identification and isolation of the other protein or proteins which functions in these systems.

C. Outer Membrane Proteins

There is one class of outer membrane proteins that appear to transfer large substrates through the outer membrane. The known examples of this class are discussed in Chap. 9. There is also some evidence that there may be a second class of outer membrane proteins which function to transport small highly charged anions through the outer membrane. The best studied protein of this class is the shock releasable succinate binding protein which is discussed on p. 526 .

Another protein that may belong to this class has been associated with α-glycerol phosphate transport [244]. This protein is released by osmotic shock of an E. coli K12 strain and was identified by two-dimensional polyacrylamide gel electrophoresis. The protein was coregulated with α-glycerol phosphate transport and was missing in certain mutants lacking the α-glycerol phosphate transport system. The protein has not been purified or shown to bind α-glycerol phosphate. Since α-glycerol phosphate is transported in membrane vesicles, this protein cannot be a classical binding protein [202] . Rather, if it functions in transport, it seems likely to function to transport α-glycerol phosphate through the outer membrane.

The finding that glucose-6-phosphate transport is inhibited by osmotic shock [245] even though glucose-6-phosphate transport is retained in membrane vesicles [246,247] suggests that there may be an outer membrane receptor for glucose-6-phosphate, although this has not been shown directly. It is curious that the three potential outer membrane proteins discussed above, which all function to transport small anions, are released by osmotic shock, while the receptors for maltose, vitamin B_{12}, and iron ferrichrome are all hydrophobic proteins which are firmly bound to the outer membrane.

V. CONCLUDING REMARKS

The work discussed in this review indicates that the techniques for extracting and purifying membrane proteins have developed to the point where it is now possible to investigate their structure and function in solute transport, even though it is still a difficult task. For example, much more is known about the water-soluble F_1 portion of the protein – translocating ATPase than the hydrophabic F_0 portion which cannot be solubilized without using detergents. However, considerable progress is being made in the purification and reconstitution of the complete F_0-F_1 complex in phospholipid vesicles. Two other areas of active research on the bacterial ATPase enzyme are the isolation, characterization, and determination of the function of its individual subunits and the role of the many nucleotide binding sites present on the enzyme.

Now that a number of membrane bound dehydrogenases have been purified, an important next step will be to identify the component or components to which these enzymes transfer electrons, and to determine the topology of the flow of protons and electrons across the membrane so as to delineate the role of these enzymes in proton translocation.

Although the water-soluble binding proteins are the best characterized components of nutrient transport systems, some membrane carriers have been isolated in an inactive form, and one has now been partially purified and used to reconstitute transport in phospholipid vesicles. It seems likely that the membrane bound transport systems present in vesicles will be the first to be completely understood, since members of this class appear to require only a single protein and a membrane potential or proton gradient. In contrast, the nutrient transport systems which have a binding protein as one of their components will probably be more difficult to understand, since they require at least two proteins and little is known about the mechanism of the coupling of energy to these transport systems. However, the additional effort that will be required to understand these systems should be worthwhile, since it seems almost certain that they have a different energy coupling mechanism and they might even use a different mechanism for moving substrates through the phospholipid bilayer of the membrane. In any case, it seems likely that during the next few years a great deal of progress will be made in both the isolation of individual transport proteins and in determining the detailed mechanism of the different classes of transport systems.

RECENT DEVELOPMENTS

Further studies on the E. coli succinate system have been reported which show that the two-membrane proteins are adjacent and span the membrane, with one having its binding site on the inside surface and the other on the outside surface of the membrane [248]. Two papers have reported very preliminary work on the isolation and reconstitution of proline transport proteins [249,250]. A glutamate binding protein has been isolated from Halobacterium halobium membranes and reconstituted into phospholipid vesicles. The reconstituted system catalyzes facilitative diffusion but not transport [251]. Surprisingly, the reconstituted system does not require sodium ions even though the transport system present in the membranes does.

More complete studies on glutamate transport have proven that the glutamate binding protein is an essential component of a low activity glutamate transport system in E. coli [252]. The conflicting reports on the binding properties of the E. coli galactose binding protein have been resolved by the demonstration that the binding experiments showing two binding affinities resulted from impurities in the galactose and that there is only a single binding site [253]. The unusual binding curve reported for the maltose binding protein has also been shown to result from an artifact [254]. As a result, all work on all binding proteins indicates that they have a single ligand binding site. The complete sequence of the arabinose binding protein has been published as has the results of an x-ray study to 2.5 Å resolution [255,256].

The ability of the phosphate binding protein to restore phosphate transport to spheroplasts has been demonstrated. These experiments appear to contain the controls necessary to rule out some of the artifacts present in previous reconstitution studies and ought to convince any remaining holdouts that binding proteins do function in the transport process [257].

ACKNOWLEDGMENTS

J. B. Smith is a postdoctoral fellow in Professor Leon Heppel's laboratory and holds a fellowship (1 F32 GM02419-01) from the National Institutes of Health. The work of D. B. Wilson was supported by Public Health Service Grant 501 AM 11923 from the National Institute of Arthritis, Metabolism, and Digestive Diseases. We thank Dr. F. M. Harold for making his review available to us before publication.

REFERENCES

1. H. R. Kaback, Biochim. Biophys. Acta, 265:367 (1972).
2. H. R. Kaback, Science, 186:882 (1974).
3. F. M. Harold, Curr. Top. Bioenerg., 6:83 (1976).
4. R. D. Simoni and P. W. Postma, Ann. Rev. Biochem., 44:523 (1975).
5. W. Boos, Ann. Rev. Biochem., 43:123 (1974).
6. S. Ramos, S. Schuldiner, and H. R. Kaback, Proc. Natl. Acad. Sci. USA, 73:1892 (1976).
7. F. J. Lombardi and H. R. Kaback, J. Biol. Chem., 247:7844 (1972).
8. M. Futai, J. Bacteriol., 121:861 (1975).
9. L. A. Heppel, in, Structure and Function of Biological Membranes, (L. I. Rothfield, ed.), Academic, New York, 1964, p. 224.
10. P. Mitchell, Nature (Lond.), 191:144 (1961).
11. F. M. Harold, Bacteriol. Rev., 36:172 (1972).
12. C. Lowery-Goldhammer and J. P. Richardson, Proc. Natl. Acad. Sci. USA, 71:2003 (1974).
13. S. Wickner and J. Hurwitz, Proc. Natl. Acad. Sci. USA, 72:3342 (1975).
14. F. G. Nobrega, F. H. Rola, M. Pasetto-Nobrega, and M. Oishi, Proc. Natl. Acad. Sci. USA, 69:15 (1972).
15. P. L. Pedersen, Bioenergetics, 6:243 (1975).
16. D. O. Lambeth, H. A. Lardy, A. E. Senior, and J. C. Brooks, FEBS Lett., 17:330 (1971).
17. J. K. Weltman and R. M. Dowben, Proc. Natl. Acad. Sci. USA, 70:3230 (1973).
18. A. Abrams and J. B. Smith, in, The Enzymes (P. D. Boyer, ed.), Academic, New York, 1974.
19. R. Y. Stainer, in, Organization and Control in Prokaryotic and Eukaryotic Cells (H. P. Charles and B. C. J. G. Knight, eds.), Cambridge Univ. Press, Cambridge, 1970.
20. A. E. Senior, Biochim. Biophys. Acta, 301:249 (1973).
21. H. S. Penefsky, in, The Enzymes (P. D. Boyer, ed.), Academic, New York, 1974.
22. P. Mitchell and J. Moyle, Biochem. Soc. Spec. Publ., 4:91 (1974).
23. E. Racker, Biochem. Soc. Transact., 3:27 (1975).
24. Y. Kagawa and E. Racker, J. Biol. Chem., 241:2475 (1966).
25. A. Abrams, J. Biol. Chem., 240:3675 (1965).
26. P. D. Bragg and C. Hou, FEBS Lett., 28:309 (1972).
27. J. B. Smith and P. C. Sternweis, Biochemistry 16:306 (1977).
28. M. R. J. Salton and M. S. Nachbar, in, Autonomy and Biogenesis of Mitochondria and Chloroplasts (N. K. Boardman, A. W. Linnane, and R. M. Smillie, eds.), North-Holland, Amsterdam, 1971.

29. R. B. Beechey, S. A. Hubbard, P. E. Linnett, A. D. Mitchell, and E. A. Munn, Biochem. J., 148:533 (1975).
30. D. J. Evans, Jr., J. Bacteriol., 104:1203 (1970).
31. R. L. Hanson and E. P. Kennedy, J. Bacteriol., 114:772 (1973).
32. M. E. Pullman, H. S. Penefsky, A. Datta, and E. Racker, J. Biol, Chem., 235:3322 (1960).
33. R. E. McCarty and E. Racker, Brookhaven Symp. Biol., 19:202 (1966).
34. L. A. Heppel, R. J. Larson, J. B. Smith, and P. C. Sternweis, unpublished observations (1976).
35. G. Vogel and R. Steinhart, Biochemistry, 15:208 (1976).
36. N. Sone, M. Yoshida, H. Hirata, and Y. Kagawa, J. Biol. Chem., 250:7917 (1975).
37. M. Yoshida, N. Sone, H. Hirata, and Y. Kagawa, J. Biol. Chem., 250:7910 (1975).
38. M. Yoshida, N. Sone, H. Hirata, Y. Kagawa, Y. Takeuchi, and K. Ohno, Biochem. Biophys. Res. Commun., 67:1295 (1975).
39. J. F. Hare, Biochem. Biophys. Res. Commun., 66:1329 (1975).
40. S. M. Hasan, Fed. Proc., 35:1032 (1976).
41. P. D. Bragg and C. Hou, Arch. Biochem. Biophys., 174:553 (1976).
42. C. Baron and T. E. Thompson, Biochim. Biophys. Acta, 382:276 (1975).
43. N. Nelson, D. W. Deters, H. Nelson, and E. Racker, J. Biol. Chem., 248:2049 (1973).
44. M. Futai, P. C. Sternweis, and L. A. Heppel, Proc. Natl. Acad. Sci. USA, 71:2725 (1974).
45. N. Nelson, B. I. Kanner and D. L. Gutnick, Proc. Natl. Acad. Sci. USA, 71:2720 (1974).
46. P. D. Bragg and C. Hou, Arch. Biochem. Biophys., 167:311 (1975); Biochem. Biophys. Res.Commun., 72:1042 (1976). 1042 (1976).
47. A. Abrams, C. Jensen, and D. Morris, J. Supramol. Struc., 3:261 (1975). J. B. Smith and A. Abrams, Fed. Proc., 33:192 (1974).
48. A. Abrams, C. Jensen, and D. Morris, Biochem. Biophys. Res. Commun., 69:804 (1976).
49. J. M. Andreu, J. A. Albendea, and E. Munoz, Eur. J. Biochem., 37:505 (1973).
50. S. Ishikawa, J. Biochem. (Tokyo), 60:598 (1966).
51. R. Mirsky and V. Barlow, Biochim. Biophys. Acta, 274:556 (1972).
52. R. Mirsky and V. Barlow, Biochim. Biophys. Acta, 291:480 (1973).
53. A. Hachimori, N. Muramatsu, and Y. Nosoh, Biochim. Biophys. Acta, 206:426 (1970).
54. R. Adolfsen and E. N. Moudrianakis, Biochemistry, 10:2247 (1971).
55. R. Adolfsen, J. A. McClung, and E. N. Moudrianakis, Biochemistry, 14:1727 (1975).

56. B. C. Johansson, M. Baltscheffsky, and H. Baltscheffsky, Eur. J. Biochem., 40:109 (1973).
57. B. C. Johansson and M. Baltscheffsky, FEBS Lett., 53:221 (1975).
58. T. Higashi, V. K. Kalra, S.-H. Lee, E. Bogin, and A. F. Brodie, J. Biol. Chem., 250:6541 (1975).
59. H. Monteil, G. Roussel, and D. Boulouis, Biochim. Biophys. Acta, 382:465 (1975).
60. Y. Kagawa, N. Sone, M. Yoshida, H. Hirata, and H. Okamoto, J. Biochem. (Tokyo), 80:141 (1976); M. Yoshida, N. Sone, H. Hirata, H. Okamoto, and Y. Kagawa, Tenth International Congress of Bio-Chemistry, Abstracts (1976), p. 340.
61. E. Munoz, J. H. Freer, D. J. Ellar, and M. R. J. Salton, Biochim. Biophys. Acta, 150:531 (1968).
62. H. P. Schnebli, A. E. Vatter, and A. Abrams, J. Biol. Chem., 245:1122 (1970).
63. B. A. Baird and G. G. Hammes, J. Biol. Chem., 251:6953 (1976).
64. J. B. Smith and P. C. Sternweis, Fed. Proc., 35:1029 (1976); P. C. Sternweis and J. B. Smith, Biochemistry 16:4020 (1977).
65. J. B. Smith and P. C. Sternweis, Biochem. Biophys. Res. Commun., 62:764 (1975).
66. J. B. Smith and P. C. Sternweis, J. Supramol. Struct., 3:248 (1975).
67. N. Nelson, H. Nelson, and E. Racker, J. Biol. Chem., 247:7657 (1972).
68. M. E. Pullman and G. C. Monroy, J. Biol. Chem., 238:3762 (1963).
69. R. J. Van de Stadt, B. L. De Boer, and K. Van Dam, Biochim. Biophys. Acta, 292:338 (1973).
70. M. Satre, M.-B. De Jerphanion, J. Huet, and P. V. Vignais, Biochim. Biophys. Acta, 387:241 (1975).
71. J. B. Smith, P. C. Sternweis, and R. J. Larson, J. Supramolec. Struc., in press.
72. E. Munoz, M. R. J. Salton, M. H. Ng, and M. T. Schor, Eur. J. Biochem., 7:490 (1969).
73. F. J. R. M. Neiuwenhuis, J. A. M. v. d. Drift, A. B. Voet, and K. Van Dam, Biochim. Biophys. Acta, 368:461 (1974).
74. Y. Kuriki and F. Yoshimura, J. Biol. Chem., 249:7166 (1974).
75. P. D. Bragg, P. L. Davies, and C. Hou, Arch. Biochem. Biophys., 159:664 (1973).
76. C. Baron and A. Abrams, J. Biol. Chem., 246:1542 (1971).
77. J. B. Smith, P. C. Sternweis, R. J. Larson, and L. A. Heppel, J. Cell. Physiol., 89:567 (1976).
78. R. E. McCarty and J. Fagan, Biochemistry, 12:1503 (1973).
79. R. P. Magnusson and R. E. McCarty, J. Biol. Chem., 250:2593 (1975).
80. R. J. Larson and J. B. Smith, Biochemistry, 16:4266 (1977).

81. J. M. Andreu, J. Carreira, and E. Munoz, FEBS Lett., 65:198 (1976).
82. R. Singleton and R. E. Amelunxen, Bacteriol. Rev., 37:320 (1973).
83. Y. Kagawa and E. Racker, J. Biol. Chem., 246:5477 (1971).
84. E. Racker and W. Stoeckenius, J. Biol. Chem., 249:662 (1974).
85. W. Stoeckenius, in, The Structural Basis of Membrane Function: Proceedings of the International Symposium, Tehran, May 5-7, 1976 (Y. Hatefi and L. Djaradi-Ohaniance, eds.), Academic, New York, 1976.
86. D. Oesterhelt and W. Stoeckenius, Nature New Biol., 233:149 (1971).
87. D. Oesterhelt and W. Stoeckenius, Proc. Natl. Acad. Sci. USA, 70: 2853 (1973).
88. A. E. Blaurock and W. Stoeckenius, Nature New Biol., 233:152 (1971).
89. R. Henderson and P. N. T. Unwin, Nature, 257:28 (1975).
90. B. P. Rosen, J. Bacteriol., 116:1124 (1973).
91. B. P. Rosen, Biochem. Biophys. Res. Commun., 53:1289 (1973).
92. K. Altendorf, F. M. Harold, and R. D. Simoni, J. Biol. Chem., 249:4587 (1974).
93. R. H. Fillingame, J. Bacteriol., 124:870 (1975).
94. K. Altendorf and W. Zitzmann, FEBS Lett., 59:268 (1975).
95. P. Mitchell, FEBS Lett., 33:267 (1973).
96. P. Mitchell and J. Moyle, Biochem. Soc. Spec. Publ., 4:91 (1974).
97. P. Mitchell, FEBS Lett., 43:189 (1974).
98. F. M. Harold, J. R. Baarda, C. Baron, and A. Abrams, J. Biol. Chem., 244:2261 (1969).
99. K. J. Cattell, C. R. Lindop, I. G. Knight, and R. B. Beechey, Biochem. J., 125:169 (1971).
100. P. L. Davies, and P. D. Bragg, Biochim. Biophys. Acta, 266:273 (1972).
101. H. Kobayashi and Y. Anraku, J. Biochem. (Tokyo), 71:387 (1972).
102. E. Kin and Y. Anraku, J. Biochem. (Tokyo), 76:667 (1974).
103. I. L. Sun, D. C. Phelps, and F. L. Crane, FEBS Lett., 54:253 (1975).
104. M. P. Roisin and A. Kepes, Biochim. Biophys. Acta, 305:249 (1973).
105. J. B. Smith, unpublished observations (1977).
106. F. R. J. M. Nieuwenhuis, A. A. M. Thomas, and K. Van Dam, Biochem. Soc. Trans., 2:512 (1974).
107. D. W. Deters, E. Racker, N. Nelson, and H. Nelson, J. Biol. Chem., 250:1041 (1975).
108. M. Maeda, H. Kobayashi, M. Futai, and Y. Anraku, Biochem. Biophys. Res. Commun., 70:228 (1976).
109. T. Tsuchiya and B. P. Rosen, J. Bacteriol., 127:154 (1976).

110. E. C. Slater, J. Rosing, D. A. Harris, R. J. Van de Stadt, and
      A. Kemp, Jr., In Membrane Proteins in Transport and Phosphoryl-
      ation (G. F. Azzone, M. E. Klingerberg, E. Quagliariello, and N.
      Siliprandi, eds.), North Holland, Amsterdam, 1974,, p. 137; D. A.
      Harris, J. Rosing, R. J. Van de Stadt, and E. C. Slater, Biochim.
      Biophys. Acta, 314:149 (1973).
111. P. D. Boyer, R. L. Cross and W. Momsen, Proc. Natl. Acad. Sci.
      USA, 70:2837 (1973).
112. B. I. Kanner, N. Nelson, and D. L. Glutnick, Biochim. Biophys.
      Acta, 396:347 (1975).
113. G. B. Cox and F. Gibson, Biochim. Biophys. Acta, 346:1 (1974);
      F. Gibson and G. B. Cox, Ess. Biochem., 9:1 (1973).
114. A. Abrams, J. B. Smith, C. Baron, J. Biol. Chem., 247:1484 (1972).
115. R. D. Simoni and A. Schandell, J. Biol. Chem., 250:9421 (1975).
116. P. Thipayathasana, Biochim. Biophys. Acta, 408:47 (1975).
117. E. Berger, Proc. Nat. Acad. Sci. USA, 70:1514 (1973).
118. H. R. Kaback and E. M. Barnes, Jr., J. Biol. Chem., 246:5523
      (1971).
119. S. H. Larsen, J. Adler, J. Gargus and R. W. Hogg, Proc. Natl.
      Acad. Sci. USA, 71:1239 (1974).
120. J. Adler, Tenth International Congress of Biochemistry, Abstracts,
      p. 13 (1976).
121. E. N. Kort, M. F. Goy, S. H. Larsen, and J. Adler, Proc. Natl.
      Acad. Sci. USA, 72:3939 (1975).
122. M. S. Springer, E. N. Kort, S. H. Larsen, G. W. Ordal, R. W.
      Reader, and J. Adler, Proc. Natl. Acad. Sci. USA, 72:4640 (1975).
123. R. S. Zukin and D. E. Koshland, Jr., Science, 193:405 (1976).
124. N. S. Gel'man, M. A. Lukoyanova, and D. N. Ostrovskii, Bio-
      membranes, Vol. 6, Plenum, New York, 1975, p. 131.
125. M. Futai, Biochemistry, 12:2468 (1973).
126. L. D. Kohn and H. R. Kaback, J. Biol. Chem., 248:7012 (1973).
127. M. Futai, Biochemistry, 13:2327 (1974).
128. S. A. Short, H. R. Kaback, and L. K. Kohn, J. Biol. Chem., 250:
      4291 (1975).
129. J. H. Weiner and L. A. Heppel, Biochem. Biophys. Res. Commun.,
      47:1360 (1972).
130. N. Sone and S. Kitsutani, J. Biochem., 72:291(1972).
131. W. S. Kistler and E. C. C. Lin, J. Bacteriol., 112:539 (1972).
132. K. Miki and E. C. C. Lin, J. Bacteriol., 114:767 (1973).
133. H. G. Enoch and R. L. Lester, J. Biol. Chem., 250:6693 (1975).
134. G. F. Dancey, A. E. Levine, and B. M. Shapiro, J. Biol. Chem.,
      251:5911 (1976).
135. G. F. Dancey and B. M. Shapiro, J. Biol. Chem., 251:5921 (1976).
136. A. C. Schwartz and A. E. Krause, Zeit. Allg. Mikrobiol., 15:99
      (1975).

137. S. G. Mayhew and V. Massey, Biochim. Biophys. Acta., 351:364 (1974).
138. Z. N'Eman, I. Kahane, and S. Razin, Biochim. Biophys. Acta, 249:169 (1971).
139. I. C. Kim and P. D. Bragg, Can. J. Biochem., 49:1098 (1971).
140. R. Schaff, R. W. Hendler, N. Nanninga, and A. H. Burgess, J. Cell Biol., 53:1 (1972).
141. J. J. Pollock, R. Linder, and M. R. J. Salton, J. Bacteriol., 107: 230 (1971).
142. K. Shimada, I. Mabuchi, and T. Yamada, Biochim. Biophys. Acta, 351:364 (1974).
143. R. P. Ravnio, L. D. Straus, and W. T. Jenkins, J. Bacteriol., 115:567 (1973).
144. J. E. Norton, G. S. Bulmer, and I. R. Sokatch, Biochim. Biophys. Acta, 78:136 (1963).
145. C. F. Fox and E. P. Kennedy, Proc. Natl. Acad. Sci. USA, 54:891 (1965).
146. J. R. Carter, C. F. Fox, and E. P. Kennedy, Proc. Natl. Acad. Sci. USA, 80:725 (1968).
147. C. F. Fox, J. R. Carter, and E. P. Kennedy, Proc. Natl. Acad. Sci. USA, 57:698 (1967).
148. C. R. Muller, K. Altendorf, B. Kholand, and H. Sanderman, FEBS Symp. Abstr., p. 137 (1976).
149. H. Hirata, N. Sone, M.Yoshida, and Y. Kagawa, Biochem. Biophys. Res. Commun., 69:665 (1976).
150. T. C. Y. Lo and B. D. Sanwal, J. Biol. Chem., 250:1600 (1975).
151. T. C. Y. Lo and B. D. Sanwal, Biochem. Biophys. Res. Commun., 63:278 (1975).
152. R. E. Fournier and A. B. Pardee, J. Biol. Chem., 249:5948 (1974).
153. K. Willicke and R. Lange, J. Bacteriol., 117:373 (1974).
154. S. I. Chavin, FEBS Lett., 14:269 (1971).
155. E. A. Bayer, T. Viswanatha, and M. Wilchek, FEBS Lett., 60:309 (1975).
156. A. Gordon, F. J. Lombardi, and H. R. Kaback, Proc. Natl. Acad. Sci. USA, 69:358 (1972).
157. R. Oshima, R. C. Willis, and C. E. Furlong, J. Biol. Chem., 249:6033 (1974).
158. G. F. Ames and J. E. Lever, J. Biol. Chem., 247:4309 (1972).
159. W. Boos, J. Biol. Chem., 247:5414 (1972).
160. O. Kellermann and S. Szmelcman, Eur. J. Biochem., 47:139 (1974).
161. G. W. Ordal and J. Adler, J. Bacteriol., 117:517 (1974).
162. A. B. Pardee, L. S. Prestidge, M. B. Whipple, and J. Dreyfuss, J. Biol. Chem., 241:3962 (1966).
163. A. B. Pardee, J. Biol. Chem., 241:5886 (1968).
164. R. Langridge, H. Shinagawa, and A. B. Pardee, Science, 169:59 (1970).

165. N. Ohta, P. R. Galsworthy, and A. B. Pardee, J. Bacteriol., 105: 1053 (1971).

166. N. Medvecky and H. Rosenberg, Biochim. Biophys. Acta, 211:158 (1970).

167. G. R. Willsky, R. L. Bennett, and M. H. Malamy, J. Bacteriol., 113:529 (1973).

168. A. S. Rae, K. P. Strickland, N. Medvecky, and H. Rosenberg, Biochim. Biophys. Acta, 433:555 (1976).

169. R. G. Gerdes and H. Rosenberg, Biochim. Biophys. Acta, 351:77 (1974).

170. H. Aono and N. Otsuji, J. Bacteriol., 95:1182 (1968).

171. G. R. Willsky and M. H. Malamy, J. Bacteriol., 127:595 (1976).

172. D. B. Rhoads, F. B. Waters, and W. Epstein, J. Gen. Physiol., 67:325 (1976).

173. J. R. Piperno and D. L. Oxender, J. Biol. Chem., 241:5732 (1966).

174. Y. Anraku, J. Biol. Chem., 243:3116 (1968).

175. Y. Anraku, J. Biol. Chem., 243:3123 (1968).

176. W. R. Penrose, G. E. Michoalds, J. R. Piperno, and D. L. Oxender, J. Biol. Chem., 243:5921 (1968).

177. W. R. Penrose, R. Zand, and D. L. Oxender, J. Biol. Chem., 245: 1432 (1970).

178. K. Berman and P. D. Boyer, Biochemistry, 11:4650 (1972).

179. M. Rahmanian, D. R. Claus, and D. L. Oxender, J. Bacteriol., 116:1258 (1973).

180. J. Wood, J. Biol. Chem., 250:4477 (1975).

181. C. E. Furlong and J. H. Weiner, Biochem. Biophys. Res. Commun., 38:1076 (1970).

182. H. Amanuma and Y. Anraku, J. Biochem. (Tokyo), 76:1165 (1974).

183. C. E. Furlong, R. C. Willis, R. G. Morris, G. D. Schellenberg, and N. H. Gerber, Fed. Proc., 32:517 (1973).

184. C. E. Furlong, C. Cirakoglu, R. C. Willis, and P. A. Santy, Anal. Biochem., 51:297 (1973).

185. Y. Anraku, Kobayashi, H. Amanuma, and A. Yamaguchi, J. Bio-Chem. (Tokyo), 74:1249 (1973).

186. G. F. Ames and J. Lever, Proc. Natl. Acad. Sci. USA, 66:1096 (1970).

187. B. P. Rosen and F. D. Vasington, J. Biol. Chem., 246:5351 (1971).

188. G. F. Ames and E. N. Spudich, Proc. Natl. Acad. Sci. USA, 73: 1877 (1976).

189. S. G. Kistu and G. F. Ames, J. Bacteriol., 116:107 (1973).

190. J. H. Weiner and L. A. Heppel, J. Biol. Chem., 246:6933 (1971).

191. G. P. Kreishman, D. E. Robertson, and C. Ho, Biochem. Biophys. Res. Commun., 53:18 (1973).

192. R. C. Willis and C. E. Furlong, J. Biol. Chem., 250:2574 (1975).

193. G. D. Schellenberg, Fed. Proc., 35:1617 (1976).

194. L. Frank and I. Hopkins, J. Bacteriol., 100:329 (1969).
195. Y. S. Halpern, H. Barash, S. Dover, and K. Druck, J. Bacteriol., 114:53 (1973).
196. H. Barash and Y. S. Halpern, Biochem. Biophys. Res. Commun., 45:681 (1971).
197. S. Kahane, M. Marcus, E. Metzer, and Y. S. Halpern, J. Bacteriol., 125:770 (1976).
198. R. C. Willis and C. E. Furlong, J. Biochem., 250:2581 (1975).
199. E. A. Berger and L. A. Heppel, J. Biol. Chem., 247:7684 (1972).
200. L. Leive and B. Davis, J. Biol. Chem., 240:4362 (1965).
201. B. P. Rosen, Fed. Proc., 31:1061 (1971).
202. L. A. Heppel, B. P. Rosen, I. Friedberg, E. A. Berger, and J. H. Weiner, in, The Molecular Basis of Biological Transport, Vol. 3 (J. F. Woessner, Jr. and F. Huijing, eds.), Academic, New York, 1973, p. 1.
203. B. P. Rosen, J. Biol. Chem., 246:3653 (1971).
204. Q. H. Wilson and J. T. Holden, J. Biol. Chem., 244:2743 (1969).
205. B. P. Rosen, J. Biol. Chem., 248:1211 (1973).
206. J. Schwartz, W. K. Maas, and E. J. Simon, Biochim. Biophys. Acta, 32:582 (1959).
207. B. P. Rosen, J. Bacteriol., 116:627 (1973).
208. H. Kuzuya, K. Bromwell, and G. Guroff, J. Biol. Chem., 246:6371 (1971).
209. G. Guroff and K. Bromwell, Arch. Biochem. Biophys., 137:379 (1970).
210. Y. Anraku, J. Biol. Chem., 243:3116 (1968).
211. W. Boos, in, Current Topics in Membranes and Transport, Vol. 5 (F. Bronner and A. Kleinzeler, eds.), Academic, New York 1974, p. 51.
212. W. Boos, A. S. Gordon, R. E. Hall, and H. D. Price, J. Biol. Chem., 247:917 (1972).
213. W. Boos and A. S. Gordon, J. Biol. Chem., 246:621 (1971).
214. G. Richarme and A. Kepes, Eur. J. Biochem., 45:127 (1974).
215. T. J. Silhavy and W. Boos, Eur. J. Biochem., 54:163 (1975).
216. T. J. Silhavy, S. Szmelcman, W. Boos, and M. Schwartz, Proc. Natl. Acad. Sci. USA, 72:2120 (1975).
217. A. K. Ganesan and B. Rotman, J. Mol. Biol., 16:42 (1968).
218. W. Boos and M. O. Sarvas, Eur. J. Biochem., 13:526 (1970).
219. W. Boos, J. Biol. Chem., 247:5414 (1972).
220. G. W. Ordal and J. Adler, J. Bacteriol., 117:309 (1974).
221. J. Lengeler, K. O. Hermann, H. J. Unsold, and W. Boos, Eur. J. Biochem., 19:457 (1971).
222. A. R. Robbins, J. Bacteriol., 123:69 (1975).
223. J. Adler, Science, 166:1588 (1969).
224. G. W. Ordal and J. Adler, J. Bacteriol., 117:517 (1974).

225. Y. Anraku, J. Biol. Chem., 242:793 (1967).
226. G. L. Hazelbauer and J. Adler, Nature New Biol., 230:101 (1971).
227. J. R. Chipley, Microbios., 9:35 (1974).
228. J. R. Chipley and B. A. Lessely, Microbios., 9:179 (1974).
229. M. W. Stinson, M. A. Cohen, and J. M. Merrick, J. Bacteriol., 128:423 (1976).
230. R. W. Hogg and E. Englesberg, J. Bacteriol., 100:423 (1969).
231. R. Schleif, J. Mol. Biol., 46:185 (1969).
232. R. W. Hogg and M. A. Hermodson, Fed. Proc., 35:1587 (1976).
233. G. N. Phillips, Jr., V. K. Mahajan, and F. A. Quiocho, Fed. Proc., 35:1367 (1976).
234. R. Aksamit and B. Koshland, Jr., Biochem. Biophys. Res. Commun., 48:1348 (1972).
235. R. C. Willis and C. E. Furlong, J. Biol. Chem., 249:6926 (1974).
236. J. Adler, G. L. Hazelbauer, and M. M. Dahl, J. Bacteriol., 115: 824 (1973).
237. T. W. Griffith and F. R. Leach, Arch. Biochem. Biophys., 159:658 (1973).
238. T. Kawasaki, I. Miyata, K. Esaki, and Y. Nose, Arch. Biochem. Biophys., 131:223 (1969).
239. R. T. Taylor, S. A. Norrell, and M. L. Hanna, Arch. Biochem. Biophys., 148:366 (1972).
240. P. M. DiGirolamo and C. Bradbeer, J. Bacteriol., 106:745 (1971).
241. A. R. Robbins and B. Rotman, Proc. Natl. Acad. Sci. USA, 72:423 (1975).
242. A. R. Robbins, R. Guzman, and B. Rotman, J. Biol. Chem., 251: 3112 (1976).
243. D. B. Wilson, J. Bacteriol., 126:1156 (1976).
244. T. J. Silhavy, I. Hartig-Beecken, and W. Boos, J. Bacteriol., 126: 951 (1976).
245. L. A. Heppel, J. Gen. Physiol., 54:95 (1969).
246. G. W. Dietz, J. Biol. Chem., 247:4561 (1972).
247. H. R. Kaback and E. M. Barnes, Jr., J. Biol. Chem., 246:5523 (1971).
248. T. C. Y. Lo, J. Supramol. Struc., 7 (in press).
249. H. Amanuma, K. Motojima, A. Yamaguchi, and Y. Anraku, Biochem. Biophys. Res. Commun., 74:366-373 (1977).
250. Soon-Ho Lee, N. S. Cohen, A. J. Jacobs, and A. F. Brodie, J. Supramol. Struc., 7 (in press).
251. J. K. Lanyi, J. Supramolec. Struc., 6:169-177 (1977).
252. G. D. Schellenberg and C. E. Furlong, J. Biol. Chem., 252:9055 (1977).
253. R. S. Zukin, P. G. Strange, L. R. Heavey, and D. E. Koshland, Jr. Biochemistry, 16:381-386 (1977).

254. M. Schwartz, A. Kellerman, S. Szmelcman, and G. L. Hazelbauer, Eur. J. Biochem., 71:167–170 (1976).

255. R. W. Hogg and M. A. Hermodson, J. Biol. Chem., 252:5135–5141 (1977).

256. F. A. Quiocho, G. L. Gilliland, and G. N. Phillips, Jr., J. Biol. Chem., 252:5142–5149 (1977).

257. R. G. Gerdes, K. P. Strickland, and H. Rosenberg, J. Bacteriol., 131:512–518 (1977).

Chapter 12

ENERGETICS OF ACTIVE TRANSPORT

Barry P. Rosen

Department of Biological Chemistry
University of Maryland School of Medicine
Baltimore, Maryland

and

Eva R. Kashket

Department of Microbiology
Boston University School of Medicine
Boston, Massachusetts

   I. INTRODUCTION 560
  II. DEVELOPMENT OF EARLY CONCEPTS OF ENERGY
       COUPLING. 561
      A. Studies with Intact Cells 561
      B. Studies with Membrane Vesicles 562
 III. COUPLING MECHANISMS 563
      A. The Chemiosmotic Hypothesis 563
      B. Other Modes of Energy Coupling 574
      C. Measurements of the Components of the Proton Motive
         Force 578
  IV. COUPLING OF PRIMARY SYSTEMS TO SECONDARY ACTIVE
       TRANSPORT. 582
      A. Proton/Solute Symports 582
      B. Other Cations Associated with Solute Transport 591
   V. QUANTITATIVE ASPECTS OF PROTON GRADIENT COUPLED
       SOLUTE TRANSPORT. 594

VI.   ENERGY COUPLING OF TRANSPORT SYSTEMS SENSITIVE
          TO OSMOTIC SHOCK.   599
VII. BACTERIOCINS AND ENERGY TRANSDUCTION   603
      A. Mechanisms of Bacteriocin Action   603
      B. Active Transport and Bacteriocins   606
VIII. AVENUES OF FUTURE STUDY.   608
      RECENT DEVELOPMENTS   611
      REFERENCES   612

I. INTRODUCTION

Bacterial cells have the capacity to translocate against a concentration
gradient a considerable number of small molecules, including such classes
of compounds as sugars, amino acids, and ions. To accomplish such work,
energy must be derived from catabolism and applied to transport work.
The mechanisms by which metabolic energy is coupled to transport are still
only partially understood and are of much current interest. The basic
mechanisms for the conservation and transduction of energy in bacterial
cytoplasmic membranes are most likely analogous to those in other membranes, such as the inner mitochondrial membrane and the thylakoid membrane of chloroplasts.

The purpose of this review is to provide a summary of current concepts in the area of bacterial bioenergetics as they relate to active transport, and to indicate where, in our opinion, the future of the field lies.
There have been a number of reviews published lately on this topic [28,58,
60,61,88,150,207]. The recent reviews by Harold [60,61] and Hamilton
[58] are especially recommended. With such an abundance of comprehensive articles, we feel that the most effective approach for this chapter
is the reporting of a few areas in detail.

The question of energy coupling of bacterial transport systems has
been pondered for nearly as long as bacterial permeation itself has been
studied, with possible models proposed several decades ago. But it is
only within the past few years that evidence has accumulated favoring
particular schemes, and, in fact, it is only within the last year or so that
nearly general agreement has been reached on one of the major models,
the chemiosmotic proposal of Peter Mitchell. Since it is rare that areas
as controversial as this converge on a single or a small number of compatible ideas within the lifetime of those who proposed the concepts, it
becomes of interest to trace (in an admittedly incomplete way) the development of the field. With the perspective gained from a historical review,
the present state of knowledge and the direction in which future research
will go can be understood.

II. DEVELOPMENT OF EARLY CONCEPTS OF ENERGY COUPLING

A. Studies with Intact Cells

The pioneering work of Gale [39-46] established that the passage of amino acids across the bacterial membrane was an energy requiring process. Various mechanisms had been proposed for the accumulation of substances within cells, such as binding to intracellular sites. For example, an early suggestion was that potassium uptake might be the result of electrostatic charges built up within the cell during glycolysis [185]. In the first careful study of the energy requirement for active transport of glutamate by Staphylococcus aureus cells, Gale and Paine in 1951 [45] demonstrated that glutamate accumulated as the free amino acid, and that accumulation was independent of glutamate metabolism or protein synthesis. The energy for accumulation was not necessarily derived from glycolysis, since accumulation was inhibited by compounds such as sodium azide and 2,4-dinitrophenol (DNP). Thus Gale early recognized that the source of energy was in some way related to oxidative phosphorylation [40,42,45].

Much of the experimental effort on bacterial transport in the 1950s was descriptive: categorizing transport systems by determining the kinetic parameters and specificities of solute uptake systems. There were fundamental difficulties which prevented an intensive investigation of energetics at that time. One problem was that different transport systems showed different sensitivities to the same inhibitor. For example, most amino acid transport systems were inhibited by sodium azide, yet the transport of α-methylglucoside was stimulated by this compound [105]. On the other hand, α-methylglucoside uptake was inhibited by iodoacetate, an inhibitor of glycolysis, while, as mentioned above, amino acid transport did not depend on glycolysis. With results such as those, it was difficult to formulate universal mechanisms by which metabolic energy is funneled into transport. A second major difficulty was the fact that physiological studies relied on the use of intact cells. Since intact cells of most bacterial species have sufficient endogenous reserves of metabolizable compounds to energize transport, a biochemical investigation of the source of energy could not be undertaken until there was a better understanding of the physiology of bacteria.

The past decade has seen remarkable progress due principally to three advances: (1) the technical achievement of developing cell-free vesicle preparations that are capable of active transport, (2) the conceptual advance of a chemiosmotic mechanism for energy conservation and transduction within membranes, and (3) the recognition that there is no single mechanisms of energy coupling, but rather a small number of different mechanisms applicable to the various transport systems.

B. Studies with Membrane Vesicles

In 1960 Kaback [85] described the preparation of closed membrane vesicles obtained from cells of Escherichia coli by lysis of penicillin spheroplasts. These vesicles were capable of concentrating proline by a process stimulated by an added energy source [93]. The first vesicular systems were little better than whole cells for studying energetics because these preparations contained significant amounts of soluble enzymes and endogenous substrates.

Kaback later changed from penicillin to lysozyme for the production of spheroplasts [86]. Vesicles prepared from lysozyme–derived spheroplasts did not utilize glucose for driving transport reactions, but the addition of oxidizable energy substrates, particularly D–lactate, resulted in a marked enhancement of active transport of a number of solutes [9,90,91]. The authors concluded that the energy derived from electron transport was used to drive active transport directly without the involvement of oxidative phosphorylation. This conclusion was based on several premises which were later shown to be incorrect, namely, that ATP has no effect on transport [192,219,220] and that the vesicles were incapable of coupling electron transport to ATP synthesis [218]. Nonetheless, ATP synthesis did not occur during the transport reactions, and the general conclusion that "... the effect of D-(-)-lactate and other electron transport chain substrates on amino acid transport is apparently not exerted through the production of stable high energy phosphate compounds such as ATP or phosphoenolpyruvate" [91] was undoubtedly warranted. The importance of these observations should not be underestimated, since they conclusively demonstrated that some active transport systems derive their energy from the electron transport chain (as does oxidative phosphorylation) and are not obligatorily coupled to the hydrolysis of high-energy phosphate compounds.

The interpretation of those results led Kaback and coworkers in a direction that generated a great deal of controversy in the following years. First, it was stated that energy from ATP hydrolysis not only was not obligatorily coupled to active transport, but was indeed incapable of such coupling [16,88,110]. Second, it was found that not all respiratory substrates were equally effective in energizing transport reactions; in every bacterial species examined, one electron donor was usually found to be more effective than others [88,90,91]. In the case of E. coli, D-(-)-lactate was more effective for energizing uptake than other electron donors, which in some cases were actually oxidized more rapidly than D-lactate [90]. This observation led to the hypothesis that transport carrier proteins were obligatorily coupled to the electron transport chain, and, in fact, were components of the chain, specifically located between the flavoprotein of the D-lactate dehydrogenase and cytochrome b_1 [10,90]. In that model, the transport carrier functioned by alternate oxidation and reduction of sulfhydryl groups within the protein. The oxidized form of the carrier was thought to be exposed to the external medium, where it would bind its substrate. Reduction of the sulfhydryl group by passage of electrons from D-lactate dehydrogenase would allow the binding site of the carrier to face the internal medium, where the carrier would release its substrate because of a decrease in affinity for the transport substrate. The carrier would then donate electrons to cytochrome b_1 and become reoxidized. The model did not specify how the reoxidized carrier reoriented itself to face the external medium without at the same time returning the substrate out of the vesicles.

A number of experimental results was consistent with the model [90]. Inhibition of the electron transport at sites after cytochrome b_1 caused efflux of accumulated substrates. According to the model, such inhibition would result in the carrier remaining in a reduced, low-affinity form on both sides of the membrane, allowing efflux. Oxamate, on the other hand, is an inhibitor of D-lactate dehydrogenase, but did not cause efflux. It was reasoned that if the carrier were located in the electron transport chain after the dehydrogenase, it would remain in the oxidized, high-affinity form, which was assumed to be inaccessible to the internal substrate. However, other experimental observations were inconsistent with the hypothesis. The most important discrepancy was that uncouplers of oxidative phosphorylation, which inhibit transport in intact cells [166], completely inhibited uptake and caused efflux of any substrate accumulated by vesicles [9,10]; yet uncouplers were without effect on electron transport. Moreover, oxamate was later shown to be a much less effective inhibitor of D-lactate dehydrogenase after D-lactate oxidation had begun than had been first assumed [5]. In addition, this model did not hold for anaerobic organisms, which lack oxidative metabolism.

For these and other reasons, it became apparent that the model of Kaback was not valid, and the model has since been abandoned by Kaback himself [180,200]. However, that working hypothesis was important in that it generated new ideas and approaches and spurred much work utilizing both vesicle preparations and intact cells.

III. COUPLING MECHANISMS

A. The Chemiosmotic Hypothesis

1. Introduction of the Hypothesis

Some of the earliest investigations into the phenomenon of active transport in bacteria were performed by Mitchell, working both in parallel and in collaboration with Gale [44,46,139-143,152,153]. Studies on the transport of phosphate in bacteria led Mitchell to consider mechanisms by which phosphate could pass through a biological lipid membrane and be released

in a free form inside of the cell [139-143, 152, 153]. It is interesting that
one of the early models on which Mitchell speculated entailed direct coupling
of transport to oxidation-reduction reactions through the formation and
cleavage of disulfide bonds, a model remarkably similar to the later ideas
of Kaback [10, 90].

In his early attempts to define the mechanism of active transport,
Mitchell proposed that phosphate transport may be the result of covalent
attachment and detachment from a transmembrane enzyme [143]. Expand-
ing on Lipmann's ideas on group transfer [125], Mitchell later proposed
the concept of vectorial metabolism [146]. Enzymes in solution have been
presumed to act as point sources, with substrates and products diffusing
into and out of the active site of the enzyme from all directions. In effect,
such reactions are scalar, having magnitude only. However, it is more
likely that enzyme systems in cells act in a vectorial manner, having the
property of direction as well as magnitude. Distinctions such as these are
of little consequence under many circumstances, such as with soluble
enzymes catalyzing reactions in vitro. But if proteins are embedded within
a membrane, substrate may be able to approach the active site from one
direction only, and product may be released, only in a different direction.
In that case, it is necessary to consider the vectorial nature of the reaction:
substrate approaches from one side of the membrane and product is released
on the other side. It is a simple conceptual step from transmembrane
enzymes to transmembrane proteins which have no apparent catalytic
function in a chemical reaction, but bind substrate on one side of the membrane and release it on the other: indeed, this is the function of carrier
proteins, or porters, in Mitchell's designation. Such porters include the
membrane-linked proteins required for soluble transport and the members
of the electron transport chain.

Three forms of vectorial metabolism were proposed by Mitchell [146].
The first type was termed "group translocation" and involved the spatial
transfer of chemical groups by enzymes. One example of a group translocation reaction, which will receive more attention below, is the flavin-linked
dehydrogenase, which in effect transfers hydrogen atoms across membranes. Another group translocation system which has received wide
attention is the phosphoenolpyruvate:sugar phosphotransferase system
studied by Roseman and his coworkers [112], in which glycosyl units are
transferred across the membrane (see Sec. III.B and Chaps. 2 and 3).

Electron transfer systems are examples of a second type of mechanism
for vectorial metabolism. That electron transfer systems span the membrane of mitochondria and chloroplasts has been well documented [61].
Group translocation systems such as flavin-linked dehydrogenases and
electron transfer systems such as cytochromes can together form single
systems for the net translocation of protons, as, for example, the respiratory chain of bacteria, mitochondria, and chloroplasts. We shall define
such an overall complex for the translocation of protons or other substrates

as a "primary active transport system," meaning an enzymatic complex which converts chemical energy into electrochemical energy.

The third form of vectorial metabolism is solute transport, which can be either primary or secondary. "Secondary" refers to a reliance on primary transport systems for the immediate source of energy. Thus, secondary transport systems transfer energy from one electroosmotic gradient into another.

2. The Mitchell Hypothesis and Oxidative Phosphorylation

At about the same time as Mitchell was formulating his ideas, an intensive search was in progress for the elusive "high energy intermediate" of oxidative phosphorylation. Since Mitchell considered the electron transport chain and the ATP synthetase complex as primary active transport systems for protons, a most logical application of his hypothesis was an explanation of the mechanism of the coupling of oxidation to phosphorylation [144].

A detailed description of oxidative phosphorylation is beyond the scope of this chapter (for some excellent reviews, see Refs. 60, 61, and 150). It is of value to describe some of the most important features, however, for two reasons. First, the process of oxidative phosphorylation can be considered as the coupling of two primary active transport systems, and so provides an excellent example of the universality of the chemiosmotic hypothesis. Second, much of the controversy over the validity of the hypothesis and, in turn, much of the evidence in favor of it, derives from that literature.

The main tenet of Mitchell's hypothesis is that metabolic energy is conserved at the level of the membrane not in the form of a chemical high-energy intermediate, but as an electrochemical gradient of hydrogen ions. An essential requirement for the maintenance of such a gradient is a topologically intact membrane which is also essentially impermeable to protons. According to the hypothesis, the primary event resulting from the transfer of electrons down the respiratory chain is the separation of protons and hydroxyl ions on opposite sides of the membrane. Such a separation results in a chemical gradient of hydrogen ions, detectable as a difference in pH values in the aqueous phases on either side of the membrane, as well as an electrical membrane potential difference across the lipid membrane. The sum of these potentials can be represented as an electrochemical potential of hydrogen ions, $\Delta\tilde{\mu}_{H^+}$, across the membrane, given by the expression

$$\Delta\tilde{\mu}_{H^+} = F\Delta\psi - 2.3\ RT\Delta pH \tag{1}$$

where F is the Faraday, $\Delta\psi$ is the electrical gradient across the membrane, R is the gas constant, T is temperature in degrees Kelvin, and ΔpH is the chemical gradient of protons. By analogy with the electromotive force, emf, of electrochemical cells, Mitchell has rearranged Eq. (1) to

$$\Delta p = \frac{\Delta \widetilde{\mu}_{H^+}}{F} = \frac{F \Delta \psi - 2.3 \, RT \Delta pH}{F} = \Delta \psi - Z \Delta pH \qquad (2)$$

where Z is a combination of constants equal to about 59 mV/pH unit at 37°C, Δp or pmf is the proton motive force, analogous to the emf. Thus both the emf and pmf are forces which are derived from the potential energy of charged species separated by a membrane.

It should be pointed out here that all the experimental evidence is equally well explicable in terms of an electrochemical gradient of hydroxyl ions. It is convenient, however, to regard these gradients in terms of protons in analogy with other membrane systems that involve cations, such as the eucaryotic Na^+/K^+-ATPase [205] and the Ca^{2+}-ATPase of sarcoplasmic reticulum [136]. It should also be pointed out that, while the actual numbers of protons present inside organelles or bacterial cells are very small (for example, at an internal pH of 7, an E. coli cell apparently would contain approximately 10 to 30 protons), the availability of protons is very large, since association and dissociation reactions of protons with other ionic species are very much more rapid than the transport fluxes considered here [148].

The second component of the process of oxidative phosphorylation is the formation of ATP by the enzyme complex called the ATPase. In both bacterial cells and organelles of eucaryotes, the ATPase is a complex of several functional units, each composed of a number of different polypeptide chains [165] (see Chap. 11). For the purposes of this discussion, we can consider the complex simply to be composed of two units, the F_0 or intrinsic membrane portion, and the F_1 which is the extrinsic portion. The F_1 (often designated with a prefix to indicate its source, e.g., BF_1 for bacterial F_1, and CF_1 for the chloroplast protein) exhibits the enzymatic activity of the complex: it hydrolyzes ATP to give ADP and P_i (inorganic phosphate) in the absence of energy and even when removed from the membrane. It synthesizes ATP in the presence of energy, but only when bound to the F_0. The F_0 portion is considered to form a channel through which protons can cross the membrane, either from the external medium to the active site of the F_1 or in the reverse direction. The F_0 and F_1 form the unit (sometimes termed the F_0F_1, the dicyclohexylcarbodiimide (DCC)-sensitive ATPase, or the ATP synthetase complex) which cleaves ATP molecules in an anisotropic manner by means of the active site of the F_1, while protons are directed through the F_0 into the external medium. This electrogenic extrusion of protons can establish a pmf (Fig. 12-1). Indeed, this alternate mechanism for establishing a proton gradient across the membrane (by means of the ATPase activity in the direction of hydrolysis) is the only means by which anaerobes which lack respiratory activity and

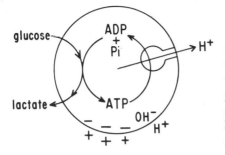

FIG. 12-1 Establishment of an electrochemical gradient of protons under anaerobic conditions. ATP produced during the fermentation of glucose can be hydrolyzed by the BF_0F_1, which catalyzes electrogenic extrusion of protons.

facultative anaerobes in the absence of a terminal electron acceptor (such as oxygen or nitrate) can generate a pmf [66].

The coupling of phosphorylation to oxidation thus occurs by means of proton influx through the ATPase. The reaction occurs in the direction of synthesis when the pmf established by the electron transport chain exceeds that which could be generated by ATP hydrolysis. Assuming a stoichiometry of two protons translocated per molecule of ATP synthesized, a pmf of 210 mV would allow for ATP:ADP = 1 inside of the cell [161]. The ratio of ATP to ADP would increase 10-fold for each rise of 30 mV in the pmf. When the pmf derived from electron transport falls below 210 mV, the ratio of ATP to ADP becomes less than unity; the lower the pmf, the more the direction of the reaction is toward ATP hydrolysis. A high pmf thus causes the synthesis of ATP, and, at the same time, completes a proton circuit. The sum of these two processes, oxidation and phosphorylation, is shown in Fig. 12-2.

Mitchell's hypothesis contains an explanation for the action of uncouplers of oxidative phosphorylation. A number of compounds had been found to prevent the phosphorylation of ADP. In mitochondria and a few bacterial species these compounds also stimulate respiration. The mechanisms by which these uncouplers, which are apparently dissimilar structurally, can produce the same effect was unknown until Mitchell pointed out that each is a weak organic acid [145,147]. If an uncoupler is lipid soluble in both the protonated and unprotonated form, then it is capable of crossing the membrane in either form. In the presence of an electrochemical gradient of protons, an uncoupler would become protonated outside of the cells (assuming that the polarity of the gradient were positive and acid outside, and negative and basic inside), and pass through the membrane into the cytosol, where it would become deprotonated because of the difference in pH. The deprotonated compound would then return to the external medium to complete the cycle. In effect, therefore, uncouplers act as ionophores for protons, dissipating the electrochemical gradient. The other effect of uncouplers, the stimulation of respiration, occurs because the electrochemical gradient which is established by the electron transport chain may

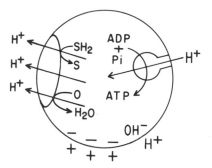

FIG. 12-2 Chemiosmotic coupling of oxidation to phosphorylation. The respiratory chain, indicated on the left, is shown including three "loops", although the figure is not meant to depict the exact stoichiometry of the reactions. Electrogenic extrusion of protons is catalyzed by the respiratory chain, establishing the electrochemical proton gradient with a polarity as indicated at the bottom. Proton uptake is catalyzed by the BF_0F_1, shown on the right, and the energy derived from the gradient is conserved by the formation of ATP.

inhibit its own formation by mass action. Dissipation of the gradient by the uncoupling agent relieves the back pressure, thus increasing the rate of electron flow and proton extrusion.

There is convincing evidence that both the electron transport chain and the ATPase complex are systems which span the membrane and act as electrogenic proton pumps [18,48,53,56,61,74,104,119,211,212]. It should be pointed out that there are details of both oxidation and phosphorylation still to be established [151] and that the chemiosmotic hypothesis as it relates to oxidative phosphorylation has not by any means gained universal acceptance [19,151]. It is not unreasonable to speculate that the hypothesis may well be modified as experimental evidence is gathered, but that, at present, its general tenets are correct and supported by the evidence.

3. Primary Active Transport

As discussed above, primary active transport systems are mechanisms for the conversion of chemical energy into electroosmotic energy. They may consist of enzymes such as the F_0F_1 complex, which converts the energy of the terminal phosphoric anhydride bond of ATP into an electrochemical proton gradient, or they may be a composite of systems, such as the electron transport chain, which couples group translocating dehydrogenases and electron translocating cytochromes into an electrogenic proton pump. The energy utilized to establish an electrochemical gradient can be derived from sources other than chemical energy. For example, the photophosphorylating systems found in chloroplasts and some bacteria convert the electromagnetic energy of light into proton gradients [147]. Likewise, the light-driven proton pump of Halobacterium halobium is a primary active transport system in which the protein bacteriorhodopsin converts the energy of captured photons into an electrochemical gradient of protons [15, 63]. Thus, primary active transport systems are the biological equivalent of fuel cells and solar cells [150]: they convert chemical or electromagnetic energy into electrical and osmotic potentials. And, of course, these systems developed more than a billion years before we "invented" them.

In summary, the establishment and maintenance of the proton motive force across the bacterial membrane is the result of active extrusion of hydrogen ions. As mentioned above, three mechanisms for proton extrusion are currently recognized. (1) For heterotrophs incubated under aerobic conditions, the components of the electron transport chain effect extrusion of protons. The older literature on bacterial oxidative phosphorylation has been summarized comprehensively by Gel'man et al. [49], while Mitchell's chemiosmotic views were first applied to bacterial energy transduction mechanisms by Harold [59, 166]. There is great interest currently in the bacterial system, primarily because of the availability of mutants of E. coli defective in electron transport and in the ATPase complex [29, 188, 196, 207]. (2) Under anaerobic conditions of incubation, the BF_0F_1 complex operates in the direction of ATP hydrolysis (the ATP having) been generated by fermentative pathways), leading to H^+ extrusion. (3) For photosynthetic organisms and for halophiles such as H. halobium, energy for H^+ extrusion is derived from the absorption of photons.

The steady-state level of the pmf is thus maintained at the membrane level. So, the poise of the H^+ gradient is a result of the rate of H^+ extrusion minus the rate of H^+ influx. This latter rate is dependent on the rate of H^+-coupled processes, such as BF_0F_1 activity in the direction of ATP synthesis and H^+-coupled solute transport systems.

4. Secondary Active Transport

By the present definitions, most bacterial active transport systems studied by bacterial physiologists and biochemists are secondary active transport systems, since the transported solutes are species other than protons, e.g., amino acids, sugars, ions, etc. Indeed, until recently there was a sharp demarcation between those workers who investigated "active transport" and those who studied "oxidative phosphorylation." Even though Mitchell pointed out almost two decades ago that the barrier between the fields was artificial, it has only been within the last few years that the equivalence of primary active transport and oxidative phosphorylation has been generally recognized.

Many amino acids, sugars, and ions are taken up by mechanisms in which the coupling to metabolic energy is secondary to the translocation event. The accumulation of metabolites occurs at the expense of a previously formed gradient of another species (usually an ion), thus the term "secondary". The phenomenon of facilitated diffusion (carrier-mediated transport without accumulation of the solute) has often been differentiated from carrier-mediated "up-hill" accumulation; however, these are now recognized to be different aspects of the same theme. Even though one species may appear to be accumulated against its potential gradient via a secondary active transport system, the overall diffusion process always occurs down an overall electrochemical gradient. So, when we observe up-hill transport of a solute, it may be that we are not aware of the overall reaction.

Mitchell has proposed the convenient terms "symport", "antiport" and "uniport" for secondary transport carriers in order to emphasize three basic mechanisms of ion cotransport. Transport is catalyzed by the porter (carrier molecule); no inferences about the molecular mechanisms of translocation should drawn from these terms. Sym-coupled reactions are those in which two different molecules or ions are translocated by the same porter in the same direction at the same time. Anti-coupled reactions utilize the same carrier at the same time for two different solutes which are transported in opposite directions. The porter of a uni-type reaction has only one substrate. The occurrence of each type will be discussed later; we shall at this point consider each in theoretical terms.

First, let us consider some simple thermodynamic aspects of coupled reactions. More detailed discussions are also available [149,194]. In order to transport a substrate, S, against its electrochemical gradient, sufficient energy to balance both the chemical and electrical components of the gradient must be provided according to the expression

$$\Delta G = RT \ln \frac{[S]_i}{[S]_o} - mF\Delta\psi \tag{3}$$

where i and o refer to the inner and outer aqueous phases, respectively, and m is the charge on S. The basic postulate of Mitchell is that the proton motive force, Δp, established by a primary active transport system, provides the driving force for the accumulation of S. Since Δp can be related to the free energy change ΔG, the following relationship can be derived from Eqs. (2) and (3)

$$nF\Delta p = RT \ln \frac{[S]_i}{[S]_o} - mF\Delta\psi \tag{4}$$

where n is the gram ions of protons associated with the transport of a mole of S. From the expanded form of Eq. (2), Eq. (4) can be rewritten as

$$n\Delta\psi - nZ\,\text{pH} = Z\,\log\frac{[S]_i}{[S]_o} - m\Delta\psi \tag{5}$$

Thus, from Eq. (5), it is possible to determine the concentration ratio, $[S]_i : [S]_o$, which a solute can attain for a given value of the proton motive force:

$$\log\frac{[S]_i}{[S]_o} = \frac{(n + m)\Delta\psi - nZ\Delta\text{pH}}{Z} \tag{6}$$

As will be discussed below, Eq. (6) equally well describes symports, uniports, and antiports.

(a) Symports. A symport in simplest terms is a mechanism for translocating two substrates simultaneously in the same direction by means of one carrier. If one substrate is flowing down its potential gradient, the energy derived from that flux can be coupled to the accumulation of the second substrate up its electrochemical gradient. Symports can be electrogenic, with a build-up of charge on one side of the membrane during the translocation process, or electroneutral, if the charge distribution is the same after the transport event as it was before the event. We can consider several examples. If one of the substrates is a proton and the cosubstrate is a monovalent anion, then Eq. (6) reduces to

$$\log\frac{[S]_i}{[S]_o} = -\Delta\text{pH} \tag{7}$$

Thus, the driving force for the accumulation of a monovalent anion by a 1:1 proton symport is the pH gradient alone (Fig. 12-3), and the concentration ratio of anion in:out is equivalent to the antilog of the pH gradient.

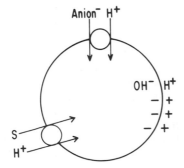

FIG. 12-3 Diagrammatic representation of symport reactions. On the left is shown an electrogenic symport, with a 1:1 cotransport of an uncharged solute (S) with a proton. At the top an electroneutral cotransport reaction is shown, with an anion and a proton transported in a 1:1 ratio.

Note that this consideration does not define the molecular mechanism of the reaction, which cannot be deduced from thermodynamics. The biochemistry of transport reactions remains to be determined.

Continuing with proton-coupled carriers as examples, let us consider an electrogenic symport reaction with an uncharged substrate S. If the ratio of protons to S is again 1:1, then every time a translocation event occurs, one positive charge is accumulated within the cell. In this situation, Eq. (6) reduces to

$$\log \frac{[S]_i}{[S]_o} = \frac{\Delta \psi}{Z} - \Delta pH \tag{8}$$

or

$$\log \frac{[S]_i}{[S]_o} = \frac{\Delta p}{Z} \tag{9}$$

It can be seen, then, that while electroneutral symport reactions utilize only the pH gradient, electrogenic ones use the total proton motive force (Fig. 12-3).

The examples considered above are simple ones. The situation becomes more complicated if $n > 1$ or $m < 1$. It has been suggested that the stoichiometry of proton:solute symports may not be fixed, but may vary with the ionization state of the carrier [89]. Moreover, mutational changes in the carrier may result in altered stoichiometry. For example, the isolation of a mutant strain of E. coli in which the stoichiometry of the proton:alanine symport system changed from 1:1 to 4:1 ($n = 4$) has recently been reported [25]. Finally, while the above cases involve protons as the cosubstrate, the electrochemical gradient of other charged species, such as Na^+, can also be the immediate driving force for certain symports.

(b) Uniports. If a symport is analogous to a multisubstrate enzyme, a uniport is the analog of the single substrate enzymatic reaction. Uniports can be described by Eq. (6), setting $n = 0$:

$$\log \frac{[S]_i}{[S]_o} = \frac{m \Delta \psi}{Z} \tag{10}$$

Equation (10) is a form of the Nernst equation. When the substrate is uncharged ($m = 0$), the uniport is a facilitated diffusion system (Fig. 12-4), that is, a porter which is not coupled to another potential gradient and that cannot, therefore, catalyze the net accumulation of substrate, but only equilibrate it across the membrane. (It should be pointed out that symports may also give the appearance of facilitated diffusion is one is only aware of the substrate which is going down its electrochemical gradient. For

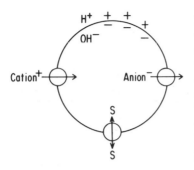

FIG. 12-4 Diagrammatic representation of uniport reactions. Left: A
cation moves into the cell in response
to a membrane potential, positive
outside. Right: An anion moves out
of the cell in response to the membrane potential. Bottom: An uncharged solute (S) moves passively,
but the reaction is facilitated by the
porter.

example, in the absence of an electrochemical proton gradient, a coupled
substrate would flow down its concentration gradient, and would, of course,
bring protons in up their gradient.)

In the more usual situations, m is either greater or less than 0. If
greater, S would be a cation which is accumulated inside of the cell to a
level dependent on the magnitude of the membrane potential. If less, the
substrate would be an anion; in this case the logarithm of the concentration
ratio would have a negative value, that is, the concentration of anion outside of the cell would have to be greater than that inside the cell. Thus,
uniports can be utilized for facilitated diffusion of uncharged substrates, for
accumulation of cations, or for extrusion anions (Fig. 12-4).

(c) Antiports. The third type of coupled reaction is the antiport, in
which two substrates are translocated simultaneously by the carrier in
opposite directions. Thermodynamically such a reaction is equivalent to
a symport in which the two cosubstrates have opposite charges. For
example, a proton:cation antiport in which n and m are both unity is equivalent to a hydroxyl:cation symport, where n = -1. This can be expressed
in the same form as Eq. (7), except that the sign of the reaction is reversed:

$$\log \frac{[S]_i}{[S]_o} = \Delta pH \tag{11}$$

Again, the pH gradient is the force which drives the translocation of the
ionic cosubstrate against its potential gradient, except that the cation is
extruded from the cell instead of concentrated within (Fig. 12-5). In the
example above, the reaction considered was electroneutral. However,
we can conceive of electrogenic antiports with one of the substrates uncharged. This could be the mechanism by which uncharged solutes, such
as the end-products of metabolic pathways, are extruded from the cell, the
driving force being the uptake of a proton. No antiport of this nature has
been described as yet, but it is not an unlikely mechanism for controlling
pathways of catabolism or for osmoregulation.

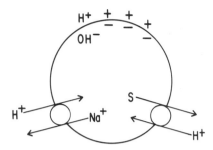

FIG. 12-5 Diagrammatic representation of an antiport reaction. Left:
Electroneutral exchange of a cation
for a proton. Right: Hypothetical
electrogenic exchange of an uncharged
solute (S) for a proton.

B. Other Modes of Energy Coupling

1. Group Translocation

One of the forms of vectorial metabolism suggested by Mitchell [146] involves the transfer of chenical groups rather than whole molecules across
the membrane. Either during or following translocation the particular
chemical group would combine with an acceptor group. The obvious difference between this process of group translocation and secondary solute
transport systems is that the molecule found inside the cell has been modified during passage through the membrane. An example of a group translocation mechanism is the PEP:sugar phosphotransferase system.

In 1964 Kundig et al. [112] reported the discovery of a membrane-
associated enzyme system which could transfer the phosphate of phosphoenolpyruvate to the 6-hydroxyl of hexoses. The reaction sequence as
found in E. coli and S. typhimurium is

(1) PEP + HPr $\xrightleftharpoons{\text{enzyme I}}$ phospho-HPr + pyruvate

(2) Phospho-HPr + sugar $\xrightarrow{\substack{\text{enzyme II}\\ \text{complex}}}$ sugar phosphate + HPr

SUM: PEP + sugar $\longrightarrow$ sugar phosphate + pyruvate

Enzyme I catalyzes the transfer of phosphate from PEP to a histidinyl
residue of the "histidine protein" HPr. Enzyme I and HPr are common to
all sugar phosphotransferase systems. Enzyme II, which catalyzes the
transfer of phosphate from phospho-HPr to a glycosyl unit, is specific for
individual sugars; thus there is a family of enzyme II proteins. In other
bacteria such as S. aureus there is an additional protein, factor III, which
accepts the phosphate from phospho-HPr, again in a phosphohistidinyl
linkage, and donates the phosphate to the sugar.

The role of this system in the group translocation of sugars is now quite clear. The first suggestion that there was a relation came from the work of Kundig et al. [113]. They determined that cells of E. coli subjected to the osmotic shock procedure of Neu and Heppel [157] lost phosphotransferase activity, TMG transport, and α-methylglucoside transport. Simultaneously, HPr was released into the medium. Addition of HPr to shocked cells restored transport, although the experiments were frequently not reproducible. A disturbing feature of those experiments is that it would appear that HPr is acting on the outside of the cell, since it seems unlikely that a protein, even a small one such as HPr, can get into the cells. Does HPr then function from either side of the membrane? Similarly, the studies of Kaback [86] showed that lysozyme-EDTA vesicles of E. coli have phosphotransferase activity, presumably due to HPr and enzyme I within the vesicles. Addition of those two proteins exogenously stimulated sugar phosphorylation, but much of the sugar phosphate was found outside the vesicles. These results led Kaback to propose a model in which enzyme II is the actual translocating element, and that the vectorial nature of the reaction is a function of the location of the phosphate donor [86].

The role of this and other enzyme systems in the group translocation of small molecules is discussed in detail in Chaps. 2 and 3 and will not be repeated here.

2. Transport Systems Linked to
    Phosphate Bond Energy

Until recently there was a consensus that, aside from group translocation systems, there is only one basic mechanism for the energy coupling of solute transport systems. This assumption resulted in part from the widespread use of membrane vesicles for the study of transport and in part from the fact that the model for in vivo transport studies was the lactose permease of E. coli. Since it was clear that solute transport systems in vesicles and lactose transport in vivo depended on some vague "energized membrane state," the implication was that the remaining question was the nature of the "energized state."

On another basis, though, transport systems had been segregated into two classes. In a large number of gram-negative species cold osmotic shock treatment decreases the activity of numerous transport systems, with concomitant release of solute binding proteins, which were shown to be in some manner part of the transport system (see Chap. 11). In nearly every case transport systems could be categorized in one of the two classes: (1) shock sensitive, usually with an associated binding protein and absent in membrane vesicles; and (2) shock resistant, almost always active in vesicles and without an associated binding protein. There was little reason to believe that there were mechanistic differences in the energy coupling of shock-sensitive and shock-resistant systems: both types of transport systems are sensitive to uncouplers and inhibitors of oxidative phosphorylation. Also, while no two systems exhibit identical specificity, there are many examples in which both a shock-sensitive system and a shock-resistant system could transport the same substrate. Again, the most detailed investigations of energy coupling were performed predominately with vesicular systems or, when intact cells were used, the lactose system was analyzed. Thus, the most detailed studies were done with shock-resistant systems.

As a portion of his PhD thesis project, Berger [11] determined that shock-sensitive systems are less sensitive to sulfhydryl reagents than are shock-resistant systems. Because Kaback and his coworkers [10] had demonstrated that transport into membrane vesicles is sensitive to sulfhydryl reagents due to inhibition of the electron transport chain, it occurred to Berger that shock-sensitive systems may not depend on electron transport. That thought led him to perform a series of now classic experiments on the energetics of the shock-sensitive glutamine transport system [12]. His results demonstrated quite convincingly that shock-sensitive systems are coupled to energy in a very different way from shock-resistant systems: shock-resistant systems are coupled to a proton motive force, while shock-sensitive systems are coupled much more directly to phosphate bond energy.

Schairer and Haddock [197] had recently shown that electron transport was not obligatory for the shock-resistant lactose transport system in intact cells. They showed that energy generated by ATP hydrolysis via the BF_0F_1 complex could energize the lactose transport system in the presence of cyanide, an inhibitor of electron transport. Yet the BF_0F_1 is necessary, since cyanide completely eliminated β-galactoside transport in a mutant lacking a functional BF_1. This was quite an important observation in its own right, since at the time most attention was focused on the electron transport coupling hypothesis of Kaback. Berger carried the logic one step further: if shock-resistant systems require not electron transport, but rather a proton motive force, is it possible that shock-sensitive systems require neither? To answer this, Berger performed the simple but elegant experiment of determining whether energy derived from electron transport could energize shock-sensitive systems if none of the energy could be used to generate phosphate bond energy, i.e., in a BF_1-deficient mutant. Berger decided to investigate what exogenous energy sources were capable of driving transport via shock-sensitive systems.

As mentioned previously, a major problem associated with investigations into the energetics of transport in intact cells is the presence of endogenous energy sources. Because they lack such endogenous reserves, vesicle systems had proven a valuable tool for the study of transport. In order to determine the sources of energy for shock-sensitive systems, Berger developed a starvation procedure involving incubation of cells with the uncoupler 2,4-dinitrophenol at 37°C for extended periods of time. This treatment reduced the endogenous rates of transport and created a dependency on exogenous sources of energy. This procedure was a modification of the technique of Koch [109], in which cells are forced to transport α-methylglucoside in the presence of azide until the work of transport depleted energy stores.

Having essentially depleted the endogenous energy supplies of the cells, Berger was able to look into the ability of exogenous sources of energy to drive transport systems. In cells containing a functional BF_1 both glucose and D-lactate were capable of energizing either proline or glutamine transport. In a BF_1-defective mutant, on the other hand, D-lactate oxidation cannot result in the formation of ATP by oxidative phosphorylation, even though it does generate a pmf. In the cells of the mutant, proline transport was stimulated by D-lactate addition, but glutamine transport was stimulated only by glucose and not by D-lactate. (Note that that statement is relative. D-Lactate metabolism via the tricarboxylic acid cycle generates high-energy phosphate bonds by substrate level phosphorylation. It is the difference in effectiveness of glucose and D-lactate, a 50-fold stimulation of transport activity versus a threefold effect, which supports the conclusion that energy derived from electron transport cannot support glutamine transport.) These results suggested that glutamine transport, unlike proline transport, could occur even in the absence of an electrochemical proton gradient. This was demonstrated more conclusively by the effect of various uncouplers and inhibitors on transport by the two systems. Berger [12] found that the effect of cyanide on proline transport in a BF_1-defective mutant was similar to the effect on β-galactoside transport reported by Schairer and Haddock [197]: complete loss of proline transport in the absence of respiratory energy. Glutamine transport, on the other hand, was partially inhibited by cyanide, but a large portion of the transport remained even at concentrations of cyanide sufficient to prevent the formation of a pmf.

Berger used two other types of inhibitors to support his hypothesis. The uncouplers DNP and carbonylcyanide-p-trifluoromethoxyphenylhydrazone (FCCP) uncoupled proline transport in the BF_1-defective mutant, as would be expected for a system dependent on a proton motive force. Glutamine transport was relatively insensitive to the action of uncouplers. Arsenate, which depletes the pools of high energy phosphate compounds, abolished glutamine transport with little effect on proline uptake.

Berger's results also make clear why the different mechanism of coupling of shock-sensitive systems was not discovered earlier. In wild-type strains, cyanide and uncouplers cause depletion of ATP pools, the result of a vain attempt by the cells to maintain a proton gradient. So, both shock-sensitive and shock-resistant systems appear to depend on a pmf when studied in a strain containing a functional BF_0F_1 complex.

C. Measurements of the Components of the Proton Motive Force

The electrochemical gradient of hydrogen ions has been measured by various techniques, outlined below, in both intact cells and in vesicle preparations. Only a brief description of various procedures is given here, primarily to illustrate the rationale of the methods used and to point out potential sources of error.

Because of the extremely small size of bacterial cells, it is impossible to measure the activity of hydrogen ions across the bacterial membrane with microelectrodes, as is done in certain animal cells. Hence other methods have been used. These fall into three classes: (1) the movement of hydrogen ions or potassium ions as detected from the changes in electrochemical activity of the extracellular or extravesicular bulk medium, (2) the distribution of radioactively labeled acids, bases, or permeant ions between the medium and the cells or vesicles, and (3) the changes in fluorescence of dyes whose fluorescence is quenched or enhanced in response to either the electrical potential across the membrane or the hydrogen ion activity.

It is appropriate here to mention that it is common for biochemically oriented investigators to express uptake measurements in terms of radioactive substrate taken up per gram of cell protein or per gram dry weight of cells. This is inadequate for consideration of concentration gradients of solutes; indeed, changes in the intracellular or intravesicular volumes could lead to apparent changes in uptake without actual changes in the concentration ratio attained. The common methods for estimating the intracellular or intravesicular volumes are based on determining the total water space of centrifuged cells either by gravimetric means or from the content of ^3H_2O of the separated cells equilibrated first with ^3H_2O, and by determining the exclusion volume of impermeant radioactively labeled molecules [134]. In the case of bacterial cells, the exclusion volume of large molecules such as inulin or polyethylene glycol is larger than that measured by small impermeant molecules such as sugars. This difference represents the space between the cell wall and the plasma membrane of the cell [96].

1. Measurements of ΔpH

The movement of hydrogen ions into and out of cells and vesicles can be quantitated by measuring the pH of the bulk medium with a glass electrode. The theoretical basis for these methods has been presented by Mitchell and Moyle [154,155]. The method has been applied to mitochondria [155], bacterial cells [24,71,99,187,191,198,199,225-227,229], and membrane vesicles [74]. Briefly, if a suspension of cells is placed in a lightly buffered medium, e.g., 3 mM glycyglycine buffer, then over pH ranges of 0.1 units or less, the change in pH is a reflection of changes in the hydrogen ion concentration. Although pH is a logarithmic function, over small pH ranges pH can be considered directly proportional to proton concentration. The buffering capacity of the cells is measured with standard acid addition. Carbonic anhydrase is often included in the reaction mixture to facilitate the interconversion of carbon dioxide and carbonic acid [154]. In the case of intact cells it is necessary to prevent the reexpulsion of protons; thus for facultaive anaerobes such as E. coli, the cells have to be depleted of endogenous energy stores and strict anaerobiosis must be observed to prevent oxidative metabolism. In the case of anaerobes such as streptococci, anaerobic conditions are not necessary. It is also necessary to dissipate any membrane potential set up from the movement of protons which would impede any further movement of protons. This can be accomplished by adding sufficient permeant ions such as thiocyanate [156]. The proton permeability of the membrane can be calculated by measuring the rate of proton flux back into the cell from the time course of alkalization of the medium. The ΔpH can be calculated from the steady-state values and from the determination of the buffering power of the external phase and the total buffering power of the preparations (and hence that of the inner phase). In an analogous manner the K^+ ratio in:out can be determined [155] with the use of a K^+-specific electrode when the membrane has been rendered permeable to potassium ions by valinomycin. The membrane potential is then calculated using the Nernst equation:

$$\Delta\psi = -\frac{RT}{mF} \ln \frac{[cation]_i}{[cation]_o} \tag{12}$$

where n is replaced by m.

The ΔpH can also be measured from the distribution of weak organic acids or bases [134, 222]. This measurement is dependent on the fact that the undissociated forms of weak organic acids or bases rapidly equilibrate across the cell membrane, while their ionized forms are impermeable; with methylamine, the undissociated form is charged and thus impermeant. In the case of an intracellular pH greater than that of the medium, the accumulation of an acid [e.g., 5,5-dimethyl-2,4-oxazolidinedione (DMO), acetate, propionate, benzoate] is used, while in the case of an interior more acid that the exterior, a base (e.g., methylamine) is used. This is necessary because the total intracellular volume of cells or vesicles in a suspension is usually very much smaller than the total test volume. Hence the accumulation of a radioactive substance is much more readily detectable than its exclusion. For example, [^{14}C]DMO has been used to measure ΔpH in Streptococcus faecalis cells [66] as well as in Streptococcus lactis [100] and E. coli vesicles [163]. The cells are incubated with [^{14}C]DMO and usually separated from the incubation mixture by centrifugation or by filtration through membrane filters. The cells cannot be washed because the labeled compound flows out very rapidly [101]. So, it is necessary to

have an accurate estimate of the contaminating extracellular fluid volume.
This can be measured by including in the reaction mixture ^{3}H-labeled
impermeable molecules which can occupy the external space up to the
plasma membrane. Sucrose is suitable for cells of S. coli. D-Sorbitol
or taurine are suitable for S. lactis cells. Alternately, [3H]inulin has been
used; however, an underestimate of the extracellular volume is obtained,
because inulin can penetrate up to the cell envelope but not to the plasma
membrane [96]. Kaback and coworkers could not show a ΔpH in respiring
E. coli vesicles because they had washed them to remove the extracellular
fluid, which also washes out the radiolabeled DMO, acetate, or propionate;
using a flow dialysis method, described below, they were able to show a
ΔpH in respiring E. coli vesicles [180].

    Another method for separating cells from medium is by centrifugation
of the cells through silicone oil of the appropriate density [96,135]. The
amount of trapped medium is about 1/10 of that obtained by filtration.

    A third technique for the determination of ΔpH in membrane vesicles
is that of flow dialysis, developed by Colowick and Womack [26] and adapted
to bacterial vesicles by Ramos et al. [180]. The apparatus, briefly, consisted of two chambers separated by dialysis tubing impermeable to compounds larger than 6000 to 8000 daltons. The membrane vesicles, suspended
is potassium phosphate buffer plus a radioactive weak acid such as DMO,
acetate, or propionate, were placed in the upper chamber; the same buffer
was pumped through the lower chamber and samples collected at intervals for
counting of radioactivity. Ionophores, electron donors, etc. were added to
the upper chamber at appropriate intervals. The contents of both chambers
were oxygenated and stirred continuously with magnetic stirrers. When the
radioactivity in the fractions from the lower chamber were plotted against
time, there was a slow, steady loss of radioactivity from the system as it was
flushed with buffer. Addition of an electron donor, which was expected to
cause proton efflux from the vesicles, caused a rapid decrease in the radioactivity in the collected fractions, reflecting uptake of the indicator acid
into vesicles; the radioactivity per fraction increased rapidly when proton
influx was effected by addition of the ionophore nigericin, which catalyzes
H^+-K^+ exchange: the decrease in radioactivity per fraction then returned
to the same basal rate as that of untreated vesicles. The value for the
radioactivity per fraction obtained after nigericin treatment was taken to
represent $\Delta pH = 0$.

    In these experiments pH; value were calculated using this value and the
equation of Waddell and Butler [222].

$$pH_i = pK_a' + \log \left\{ \left[\frac{C_t}{C_e} \left(1 + \frac{V_e}{V_i}\right) - \frac{V_e}{V_i} \right] [10^{pH_o - pK_a'} + 1] - 1 \right\} \quad (13)$$

where pH_o is the extracellular pH, pK_a' is the negative logarithm of the
apparent ionization constant of DMO (6.32 at 25°C and 6.21 at 37°C), V_e

and V_i are, respectively, the extracellular and intracellular fluid volumes of the separated cells, C_t is the concentration of DMO in the total sample water, and C_e is the concentration of DMO in the medium.

In the case of weak acids or bases whose pK'a values are very different from the pH values usually encountered, the concentration of the impermeant charged form both inside and outside is very large compared to that of the uncharged form. Thus the ratio in:out of the total base of acid will, for practical purposes, be equal to the ratio of the hydrogen ions across the membrane. In that case, the Nernst relationship [Eq. (12)] may be used to calculate the pH_i.

2. Measurement of $\Delta\psi$

A number of dyes that respond to membrane potential have been developed [223]. Compounds useful for bacterial cells are cationic cyanine dyes whose fluorescence is quenched when there is a membrane potential, negative inside, across a membrane. The mechanism postulated by Waggoner [223] for this quenching is that the cationic dye molecules stack at or near the side of the membrane adjacent to the negative side; such stacking results in the sharing of electrons, thus leading to fluorescence quenching. At specified concentrations of dye and cells, the degree of quenching is proportional to the magnitude of the $\Delta\psi$. The cationic dyes of the cyanine series were first used in red blood cells [81, 209]. In both human red cells and in the giant red cell of Amphiuma, the dye was added to cells incubated in media of varying K^+ concentrations. There was a decrease in the fluorescence upon addition of the ionophore valinomycin, indicating a potassium diffusion potential. As the K^+ concentration of the medium was increased, decreasing $\Delta\psi$, the fluorescence level increased. When the fluorescence intensity was equal to the initial level before addition of valinomycin, the potassium ratio in:out was used to calculate the $\Delta\psi$. In the case of giant red cells of Amphiuma, the value calculated by this method was in excellent agreement with that measured directly with microelectrodes.

Hoffman et al. [81] found empirically that a direct relationship exists between fluorescence level and the logarithm of the external K^+ concentration. This calibration curve is the one used in studies measuring $\Delta\psi$ in S. faecalis cells [118]. Cyanine dyes have also been used to estimate $\Delta\psi$ in H. halobium vesicles [182], as well as in E. coli vesicles [55] and chromatophores of Rhodospirullum rubrum [168]. The determination of $\Delta\psi$ in the absence of valinomycin involves (where the Nernst equation is not valid) generating a calibration curve that relates the fluorescence intensity to a known membrane potential. Thus, the fluorescence intensity is measured as a function of K^+ concentration in valinomycin-treated cells in the presence of dye [96, 100].

Other fluorescent dyes have been used to observe qualitative changes in bacterial membrane energization, including the dyes quinacrine [191] and 9-aminoacridine [193, 195, 210], which are believed to respond to the H^+ concentration at the surface of the membrane [120]. The fluorescent compound 8-aniline-1-naphthalene sulfonate (ANS) has also been used in E. coli vesicles [55, 181]; however, the mechanism of ANS fluorescence quenching is obscure [223]. Schuldiner and Kaback [200] have followed the fluorescence of safranine 0, which qualitatively shows an increase in fluorescence intensity when the membrane is energized, and a decrease upon dissipation of Δp.

A second method for determing $\Delta\psi$ uses the distribution of permeant cations (when the inside is negative compared to the outside; anions would be used for the opposite polarity) and calculation using the Nernst equation. As mentioned before, K^+ itself can be used if the membrane has been rendered permeable by valinomycin treatment. In its absence, a number of artificial permeant cations have been utilized, following their introduction by Skulachev and coworkers [52, 54, 121, 211].

The cation dibenzyldimethylammonium (DDA^+) was first used in bacteria by Harold and Papineau [64] with S. faecalis cells, and later applied to E. coli vesicles. This cation is not entirely satisfactory as a probe for $\Delta\psi$ because its uptake requires the presence of catalytic amounts of the anion tetraphenyl boron (TPB^-) and because of nonspecific adsorption of DDA^+ to cells and membrane filters. Also, K^+-loaded vesicles take up little DDA^+ compared to Na^+-loaded vesicles [5, 75].

Another permeant cation is triphenylmethylphosphonium ($TPMP^+$). Tritiated $TPMP^+$ has been used extensively with E. coli vesicles [54, 180, 200]. The ratio in:out of $TPMP^+$ was used with the Nernst equation to calculate a $\Delta\psi$. At the moment this cation appears to be a useful probe for vesicles; however, intact cells give problems in nonspecific adsorption of cation [137, 200]. Moreover, reports showing that the permeability of lipophilic cations such as DDA^+ or $TPMP^+$ may decrease when the membrane has a large Donnan potential or may increase in the presence of lipophilic anions such as TPB^- or PCB^- indicate that a closer scrutiny of the use of these ions for quantitation of   may be necessary [5, 52, 54, 75, 214].

IV. COUPLING OF PRIMARY SYSTEMS
     TO SECONDARY ACTIVE TRANSPORT

A. Proton/Solute Symports

1. Sugar-effected Proton Movements

The experimental evidence reviewed in this section was obtained primarily with intact bacterial cells depleted of endogenous energy yielding substances.

Hence, we are dealing with carrier-mediated solute translocation without accumulation. Accumulation can be achieved by an imposition of Δp, either "artificial", as with K^+ diffusion potential, or by allowing metabolism of added energy yielding substrates. However, to demonstrate, for example, the cotransport of H^+ and a solute, it is necessary to observe unidirectional fluxes of H^+ in the absence of proton reextrusion, hence the use of energy-depleted systems.

The first experimental evidence for a solute/proton symporter was obtained by West [225] and West and Mitchell [226,227]. Mitchell had suggested in 1963 that carriers such as the lactose carrier in E. coli cells might be β-galactoside/proton symporters catalyzing the tightly coupled translocation of a sugar molecule and a hydrogen ion together across the plasma membrane of the cell. Such a suggested mechanism for solute/ hydrogen ion cotransport is analogous to the solute/sodium ion cotransport that occurs across eucaryotic plasma membranes [204]. In both cases, the driving force is the inwardly directed thermodynamic pressure of the cation moving down its electrochemical gradient [69]. The experiments carried out by West will be described in some detail to illustrate the conceptual and technical framework of this kind of study.

West used cells of E. coli to explore the lactose transport system, which is probably the best understood permeation system for the class of solutes whose accumulation is energized by a proton motive force. To prevent the reextrusion of the hydrogen ions, West treated the cells with inhibitors of glycolysis, such as iodoacetate, and prevented respiration by conducting the experiments under stringently anaerobic conditions. The build-up of a membrane potential, positive inside, resulting from H^+ influx, which might impede further proton entry, was prevented by the addition of the permeant anion thiocyanate. The experiment carried out was to measure the inflow of hydrogen ions that occurred when a β-galactoside was added to the cells. The cells were placed in a weakly buffered medium, and the pH of the medium was measured with a glass electrode. On addition of galactoside, the pH of the bulk medium increased. This alkalinization of the bulk medium was only observed in cells expected to be able to translocate β-galactosides and was absent in cells genetically deficient in the carrier protein (y⁻) or whose carrier was inhibited by N-ethylmaleimide, p-chloromercuribenzoate (pCMB), or formaldehyde.

As one would expect from an uncharged molecule cotranslocated with a proton [n = 1, m = 0, in terms of Eq. (8)], galactoside transport is an electrogenic event resulting in the formation of both a pH gradient and a membrane potential. This was determined by following the rates of H^+ entry and K^+ exit with ion-specific electrodes in cells rendered K^+ permeable by treatment with valinomycin. Equivalent numbers of the cations moved across the membrane, but in opposite directions, indicating the electrogenic nature of the lac transporter. West and Mitchell [226] also showed that a pH gradient, generated by the cells' respiratory chain when a pulse of oxygen was added to the anaerobic reaction mixture, could be dissipated by the addition of galactoside. When a pulse of thiocyanate was added to anaerobic cells to establish a membrane potential, negative inside, the translocation of galactoside stimulated the entry of protons into the cell, again showing proton/sugar cotransport. Furthermore, as expected, the proton ionophore FCCP abolished galactoside-effected proton uptake. West eliminated the possibility of a galactoside/Na^+ symporter, because Na^+ ions added to the system in a range of 0.05 mM to 50 mM had no effect on H^+ influx and because Na^+ influx did not occur when galactosides were added to the cells.

To complicate matters some, many sugars can be transported by more than one membrane carrier, the carriers being distinguishable by their affinity constants for the substrate, association with periplasmic proteins, and genetic map location. This is striking in the case of the various galactose transport systems of E. coli (see Chap. 4). Thus when a high concentration of substrate is required to elicit proton movement, it is not possible to assign unequivocally a H^+ symporter mechanism to a particular carrier, unless the others with similar substrate specificity are absent.

In their studies Henderson and Kornberg [72] have confirmed the association of proton movements and the transport of certain sugars by bacteria. Advantage was taken of the inducible nature of the sugar transport systems and the wide variety of available mutants of E. coli, including ones with deficient or altered carriers and those with an inactive first enzyme in the biochemical sequence of the metabolism of the substrate. Mutants of this last class enabled the use of metabolizable sugars as nonmetabolizable transport substrates. The experiments carrier out were the same as those of West, that is, the pH_0 was measured after addition of substrate to energy-deprived cells incubated anaerobically. Henderson and Kornberg [72] found that protons were taken up when galactose was added to cells that are lactose carrier-deficient and which had been induced with D-fucose to transport galactose. Alkalinization was also observed when D-fucose was added, consonant with this carrier's capacity to transport this sugar. When isopropyl-β-D-thiogalactoside was added, the pH_0 was unaltered, as expected, since these cells lack the lactose transport system. In another strain of E. coli [47], D-fucose-induced cells took up protons when D-galactose or D-fucose were added, but not with L-fucose or L-arabinose, which are not substrates of the induced carrier. Addition of methyl-β-galactoside (the substrate of the mgl system) did not cause proton movement in D-fucose-induced cells; possibly this is due to association of the D-fucose inducible mgl system with a periplasmic binding protein and thus energization by phosphate bond energy rather than by a chemiosmotic mechanism.

The L-arabinose uptake system [162] was associated with proton uptake elicited by L-arabinose addition, as did addition of D-fucose, a nonmetabolizable substrate of the L-arabinose carrier [71,73]. In all these systems, uncoupling agents such as carbonylcyanide-m-chlorophenylhydrazone (CCCP), DNP, tetrachlorosalicylanilide (TCS) or azide abolished the sugar-elicited proton movements.

Galactoside-stimulated H^+ influx was also observed in cells of the gram-positive organism S. lactis 7962 [99]. Unlike E. coli, experiments with this anaerobe could be performed under aerobic conditions, since the cells lack oxidative metabolism and are devoid of stored energy yielding compounds. In a mutant of S. lactis that is incapable of translocating TMG, alkalinization of the medium did not take place when sugar was added.

2. Stoichiometry of Sugar/H^+ Symporters

While qualitative data are relatively abundant, stoichiometric determinations of solute/H^+ transport are as yet scarce. The earliest and clearest work is that of West and Mitchell [227], who determined the stoichiometry of the galactoside/proton symport by measuring the uptake of hydrogen ions and [^{14}C]lactose molecules in parallel reactions. It was necessary to prevent the efflux of the galactoside during the separation of the cells from the medium; thus the carrier was inhibited by addition of PCMB to the reaction immediately following uptake of the sugar. The initial rates of proton and galactoside entry were shown to be equal at first, but with time the H^+ influx decreased more rapidly than the lactose influx. When the ratios of H^+ translocated:lactose translocated were plotted against time, they extrapolated to a 0 time value very close to 1, thus leading West to conclude that in this system the two processes are coupled with a stoichiometry of 1 g-ion H^+ translocated per mole of lactose. As will be discussed below, preliminary results using flow dialysis indicate that the stoichiometry of proton-solute reactions may vary with medium pH [89]. The significance of such a variation is not clear, but one possibility is that amino acid residues of the carrier are being titrated [89].

3. Distinction Between Active Transport and Facilitated Diffusion

A distinction between active transport (accumulation of solute against a concentration gradient) vs facilitated diffusion (carrier-mediated translocation down a concentration gradient) has been emphasized in the older literature [108]. With the present concepts of ion-coupled transport such a distinction may not be necessary. For example, in the β-galactoside transport system of S. lactis, it is simplest to consider the membrane carrier as symmetrical with respect to affinity for sugar and protons at the two faces of the membrane, and capable of translocating sugar and protons equally well in either direction. Accumulation of sugar within the cells occurs when an asymmetry is imposed by a Δp. In E. coli cells the necessity of proton movement together with β-galactoside movement was emphasized by Cecchini and Koch [23], who found that cells extensively depleted of endogenous energy stores were unable to transport o-nitrophenyl-β-galactoside (ONPG) down its concentration gradient unless an energy source was added. Addition of CCCP also restored facilitated diffusion of this lactose analog, in agreement with the idea that in the absence of proton reextrusion, the H^+/solute symporter cannot function.

A different conclusion has been reached by Kaback and collaborators [181, 201-203], based on their finding that the binding of β-galactosides by the lac carrier of E. coli membrane vesicles requires energy in the form of a membrane potential. Using a series of dansyl-β-galactosides, fluorescent analogs of lactose, these workers found that the sugar analogs were unable to enter the vesicles in the absence of an energy source, that is, there was no facilitated diffusion. Although there was some energy-independent binding of the radioactively labeled sugars to the vesicles, the inability of these sugars to induce the lac operon in intact cells, and their capacity to induce in a cell-free extract, are convincing evidence that the sugars, in fact, cannot penetrate the cell and presumably, therefore, the vesicles either. The authors suggested that the lac carrier protein may be negatively charged and that in the absence of a membrane potential (positive outside) the protein is not accessible to the sugar in the medium. Obviously these interesting results require further exploration, as they seem to relate directly to the molecular mechanism of solute translocation.

4. Energization by Artificially Imposed
   Proton Motive Forces

Bacterial cells or vesicles depleted of energy yielding compounds transport solutes only down a concentration gradient and do not accumulate them. For example, in S. lactis 7962 cells equilibration across the membrane of β-galactosides is catalyzed by a membrane carrier, but accumulation against a concentration gradient occurs only when the cells are provided with a fermentable energy source, such as glucose or arginine [97]. Alternatively, energization can be effected by imposition of an "artificial" Δp.

Energization of sugar transport by streptococci was effected by imposition of an electrical potential across the membrane [98]. Cells of S. lactis, like most gram-positives, contain a concentrations of K^+ on the order of 0.3 to 0.4 M; when placed in buffers containing low concentrations of K^+, the cells lose little of this cation, showing that the membrane in the absence of an energy source is poorly permeable to this cation. Indeed, in streptococci the unidirectional flux of K^+ requires an energy source [65]. Since S. lactis is sensitive to the potassium ionophore valinomycin, addition of that compound to cells in the absence of energy yielding substrate caused K^+ to flow down its chemical gradient, resulting in an electrical potential, negative inside. This process came to a halt when the chemical and electrical gradients of K^+ balanced. Hydrogen ions would be expected to be attracted toward the negatively charged inside, and indeed, there was an influx of protons. This resulted in a ΔpH, acid inside, opposite in polarity to that generated by metabolizing cells, where the inside is more alkaline. Since the microbial membrane is poorly permeable to H^+, the entry occurred by a number of routes which are proton-linked membrane carriers, including the galactoside carrier. As Mitchell has pointed out, in the absence of a transport substrate a carrier would not be expected to facilitate the translocation of protons, as such a property would result in an uncoupled system. The result of valinomycin addition was, thus, the temporary accumulation of the sugar. This accumulation process was insensitive to DCCD, showing that the electrical potential could drive transport without prior ATP synthesis, while, as expected, glycolytically driven TMG accumulation was DCCD sensitive. Eventually the accumulated TMG exited from the cells, reaching a concentration equal to that of the medium, because the Δp was dissipated as K^+ exchanged with H^+.

Similarly, a series of experiments with membrane vesicles from E. coli have shown the Δp to be the driving force for the uptake of specific sugars and amino acids. Hirata et al. [75,76], working with vesicles prepared by the method of Kaback [87], showed that imposition of an artificial membrane potential caused the vesicles to temporarily accumulate proline, glycine, lysine, lactose, and TMG. The membrane potential was imposed by adding valinomycin to K^+-loaded vesicles; the K^+-specific ionophore monactin had the same effect as valinomycin, while nigericin, which catalyzes electroneutral exchange of protons and K^+, resulted in no accumulation of proline. The valinomycin-effected uptake of proline was inhibited by the uncoupling agents CCCP, TCS, and also by nigericin, which becomes an uncoupler in the presence of valinomycin. Uptake was insensitive to DCCD or to respiratory chain inhibitors. These data clearly demonstrate that transport carriers need not be tightly coupled to the respiratory chain, but can respond to an electrical difference across the membrane. In addition, intact cells of E. coli accumulate galactoside in response to an artificially imposed ΔpH [37].

Hamilton and his colleagues have studied the uptake of amino acids in response to Δp by energy-depleted cells of S. aureus [159-161]. From their data (outlined below) the authors propose that the carriers for the various amino acids function as follows: basic amino acids are transported by uniports in response to $\Delta\psi$, neutral amino acids by H^+ symport mechanisms in response to Δp, and acidic amino acids by H^+ symports in response to ΔpH (Table 12-1).

Cells of S. aureus depleted of energy yielding reserves under anaerobic conditions demonstrate a membrane potential that is the result of a Donnan distribution of permeant ions [159]. Under these conditions the proton motive force is 0. When [^{14}C]lysine, which bears a positive charge at neutral pH, was added to such resting cells, there was a slow exchange of the labeled amino acid with the lysine of the cell [159]. The ratio of [^{14}C]lysine in:out reached a level equivalent to the equilibrium potential of K^+ ions measured under the same experimental conditions. The authors

TABLE 12-1 Known Symports, Antiports, and Uniports

System	Solute	Co-Ion	Porter	Driving Force	Ref.
E. coli cells	β-galactosides	H^+	Symport	--	71, 225-227
	lactate, alanine serine, glycine	H^+	Symport	--	25
	gluconate	H^+	Symport	ΔpH	186
	galactose	H^+	Symport	--	71
	fucose	H^+	Symport	--	72
	L-aribinose	H^+	Symport	--	73
	glucose-6-phosphate	H^+	Symport	ΔpH	35, 231
	glucuronate	H^+	Symport	ΔpH	114
	Na^+	H^+	Antiport	Δp	228
	glutamate	Na^+	Symport	Na^+ gradient or $\Delta\psi$	236, 237
	melibiose	Na^+	Symport	Na^+ gradient or $\Delta\psi$	239
E. coli vesicles	proline, glycine	H^+	Symport	$\Delta\psi$	75, 76
	lysine	H^+	-	$\Delta\psi$	75, 76
	thiomethyl-β-galacto-side, lactose	H^+	Symport	$\Delta\psi$	75, 76
	Ca^{2+}	H^+	Antiport	ΔpH	219, 221
	glutamate	Na^+	Symport	Na^+ gradient or $\Delta\psi$	238
S. faecalis cells	glycine, alanine serine, threonine	H^+	Symport	$\Delta\psi$	8

	lactate	H^+	Symport	ΔpH	63
	phosphate	OH^-	Antiport	ΔpH	67
	Na^+	H^+	Antiport	ATP	65, 67
S. lactis cells	β-galactosides	H^+	Symport	Δp	99
S. aureus cells	glutamate, aspartate	H^+	Symport	ΔpH	43, 159, 160
	isoleucine, glycine	H^+	Symport	Δp	161
	lysine	--	Uniport	$\Delta\psi$	161
Mycobacterium phlei cells	proline	--	Uniport	$\Delta\psi$	173
P. denitrificans vesicles	sulphate	H^+	Symport	--	21
	phosphate	H^+	Symport	ΔpH	21, 22
Clostridium pasteurianum cells	gluconate	H^+	Symport	ΔpH	17
	galactose	H^+	Symport	$\Delta\psi$ and/or ΔpH	17
H. halobium vesicles	leucine	Na^+	Symport	Na^+ gradient or $\Delta\psi$	132
	glutamate	Na^+	Symport	Na^+ gradient	116, 117
	Ca^{2+}	Na^+	Antiport	Na^+ gradient	242
Azotobacter vinelandii vesicles	Ca^{2+}	H^+	Antiport	ΔpH	14

postulate that lysine is translocated as a cation by a membrane carrier and that it distributes itself like other permeant cations. This distribution of [14C]lysine was unchanged by the addition of the proton ionophore TCS and by valinomycin added after TCS, as expected of a Donnan distribution. Addition of KCl after the ionophores, however, resulted in the efflux of the [14C]lysine from the TCA-soluble pool, presumably because of the decrease in the cation equilibrium potential. Changing the H^+ concentration had a similar effect: intracellular [14C]lysine levels reacted predictably to additions of HCl or NaOH, decreasing with acidification of the medium and increasing with alkalinization. In contrast, the neutral amino acid glycine (and also isoleucine) and the negatively charged amino acid glutamate (and also aspartate) equilibrated across the membrane to ratios in:out of about 1 [159].

The authors then reasoned that experimental manipulation of the Donnan equilibrium should result in the imposition of a temporary proton motive force which would decay to 0 as the ions reequilibrate. They accomplished this by changing the K^+ ratios in:out or the ΔpH. Thus valinomycin treatment stimulated the rate of [14C]lysine uptake, in consonance with a uniport mechanism responsive to $\Delta\psi$; the final lysine level was equal to that of untreated cells. TCS added before valinomycin prevented the increased rate of lysine uptake. Glycine (and isoleucine) uptake was also increased temporarily after valinomycin addition, but TCS reduced the final intracellular level to a ratio in:out less than 1. The rate of equilibration of the anionic amino acid glutamate (and asparate) was not affected by the ionophores valinomycin and TCS.

When the ΔpH was altered by decreasing the extracellular pH from 7 to 5, the uptake of [14C]lysine (that is, the rate at which the amino acid reached a Donnan distribution) was unaffected [160]. In contrast, isoleucine influx was stimulated temporarily by acid addition, and the amino acid effluxed when base was added, confirming a H^+ symport mechanism driven by Δp. The anionic amino acid glutamate was rapidly taken up when the medium pH was decreased by 2 pH units, and it effluxed rapidly when the pH_O was again raised above pH 7.0. This efflux after the increase in pH_O was stimulated by valinomycin treatment, presumably because the membrane was then freely permeable to K^+, which then acted as a counterion. The authors postulated an electroneutral glutamate/H^+ symport mechanism driven by ΔpH. This was confirmed by an experiment in which the polarity of the ΔpH was reversed. The cells were depleted of K^+ by antibiotic treatment and washing; then these cells were placed in high K^+ medium and treated with valinomycin and TCS. Here the K^+ ions could go into the cell down their chemical gradients, driving proton efflux. The resulting ΔpH (alkaline inside) permitted the accumulation of [14C]glutamate. Addition of alkali to the outside reduced the pH, and the amino acid exited.

The transport of several ions has been shown to be coupled to ΔpH [14, 21, 22, 219-221]. Burnell et al. [21] demonstrated that sulfate and phosphate transport in membrane vesicles of <u>Paracoccus denitrificans</u> could be driven by artificially derived pH gradients. The gradients could be established by the addition of KCl in the presence of nigericin or by NH_4Cl pulses. Similarly, Tsuchiya and Rosen [221] found that Ca^{2+} transport in membrane vesicles of <u>E. coli</u> occurred when the vesicles were subjected to a pH shift, in which the outer aqueous phase was made suddenly more basic. Presumably, the anions are transported via a proton symport [21], while a cation/proton antiport has been suggested for calcium transport [14, 221].

The work by Burnell et al. [21] and Tsuchiya and Rosen [221] addresses an interesting and important question: whether transport carriers are bidirectional and what determines the direction of a transport reaction. Both right-side-out and everted membrane vesicles of the two organisms transport sulfate or calcium in response to an artificially imposed ΔpH. These results suggest that symport and antiport carriers function in either direction, with the direction being determined by the polarity of the applied force. Since the force is always applied unidirectionally by biological energy transducers such as electron transport chains, solute fluxes are normally observed in one direction. These conclusions should be considered tentative in light of recent observations of Adler and Rosen [3], which suggest that the absolute orientation of "right-side-out" membrane vesicles of <u>E. coli</u> may not be the same as the orientation of the cytoplasmic membrane in vivo (see Chap. 1).

B. <u>Other Cations Associated with</u>
   <u>Solute Transport</u>

1. Requirement for Specific Monovalent Cations

The cotransport of solutes and Na^+ is the generally accepted mechanism for animal cell plasma membranes [204], while bacterial transport systems generally do not show this requirement. (But see Recent Developments.) A number of systems, however has now been shown to involve Na^+ or K^+ (Table 12-2). There is little evidence in these cases for dependence on cation gradients such as for energy coupling. Rather, it appears that the affinity of carrier for substrate is affected, in some cases, or the capacity of the cell to retain solute, in others. For example, in a marine pseudomonad, MacLeod and his collaborators have described the requirement for Na^+ and K^+ ions for the transport of α-aminoisobutyric (AIB), as well as other amino acids, in both intact cells and membrane vesicles [32, 33, 216, 217].

TABLE 12-2 Effect of Cations on Transport

System	Substrate	Cation	Effect	Ref.
Aerobacter aerogenes	citrate	K^+	requirement	34
Bacillus licheniformis (cells and vesicles)	glutamate	Na^+	stimulation	133
E. coli	glucose-6-phosphate	K^+, Mg^{2+}, Ca^{2+}	stimulation	35
E. coli	proline	Li^+	stimulation	102,103
E. coli (cells and vesicles)	glutamate	K^+ Na^+	stimulation requirement	38,57 95,138, 236-238
Marine pseudomonad	α-aminoiso-butyrate	Na^+,K^+	requirement	32,33 216,217
Micrococcus lysodeikticus	succinate, glucose, valine	Na^+, Li^+	stimulation	7
Mycobacterium pheli (vesicles)	proline	Na^+, Li^+	requirement	77
S. typhimurium	melibiose	Na^+	requirement	213,240
S. aureus	glutamate	K^+	stimulation	31

In their proposed model [216] Na^+ is required for entry and acts by increasing the affinity of the carrier for the solute to be accumulated. A very similar model has been proposed by Halpern and colleagues [57] for the requirement for K^+ and Na^+ for glutamate transport by cells and vesicles of E. coli [38, 62, 138].

2. Na^+/Solute Symports

Bacterial cation/solute systems described so far have typically involved hydrogen ions; in contrast the extreme halophile H. halobium is a bacterial system that utilizes a sodium ion/solute mechanism for amino acid transport. Lanyi and his collaborators have conducted an elegant series of studies employing envelope vesicles prepared from this cell [116, 117, 132, 182]. The membrane vesicles were presumably of the same sidedness as the inner membrane of intact cells and offered the advantage that the salt composition of the intravesicular fluid could be varied by subjecting the vesicles to osmotic shock in media of varying salt composition. It was necessary to maintain the total salt concentration of both the intra- and extravesicular fluids at 3 M salt, with NaCl, KCl, or combinations of the two salts, in order to maintain the integrity of the membrane [116,117]. These vesicles, like the membrane of the intact cell, contain patches of purple membrane which contain only one protein, bacteriorhodopsin. This protein is a light-driven primary active transport system for protons [163]. The vesicles, like the intact cell [82], are capable of accumulating leucine and glutamate when irradiated with light [132].

[14C]Leucine was concentrated several hundred fold without chemical alteration by such vesicles, with light as energy source [132]. The optimum uptake of the amino acid was observed when the vesicles containing 3 M KCl were placed in medium containing 1.5 M KCl. The requirement for external NaCl was absolute, while K^+ ions stimulated amino uptake, but were not essential. Instead of light, the imposition of a Na^+ gradient out/in the dark also resulted in leucine accumulation; this effect was insensitive to the addition of FCCP, suggesting a Na^+/leucine symport rather than a H^+/leucine porter. This leucine accumulation by vesicles was sensitive to addition of the permeant cation $TPMP^+$, indicating the participation of a membrane potential. Indeed, imposition of a $\Delta\psi$, by addition by valinomycin to K^+-rich vesicles in a Na^+ medium, also produced a temporary accumulation of the amino acid; intact starved cells showed a similar effect [16].

The translocation of glutamate was also light energized, but differed from leucine transport in an apparently more stringent requirement for Na^+ ions cis to the transport substrate [116,117]. Irradiated K^+-rich vesicles in Na^+-containing media accumulated glutamate to ratios in:out of over 58,000; exit of glutamate was not observed when the light was turned off or when excess unlabeled glutamate was added to the system. The extremely high ratio in:out of glutamate resulted from the inability of the amino acid to exit, because no Na^+ ions were present in the cells; indeed, efflux of glutamate from preloaded vesicles was rapid when Na^+ ions were present inside the vesicles. In contrast, accumulated [14C]leucine exited from vesicles when the light was turned off or when large amounts of nonradioactive leucine were added. For glutamate translocation, in addition to the Na^+ requirement, there was also a requirement for K^+ on the trans side. The function of K^+ ions trans to the transport substrate is not fully understood: Lanyi speculated that binding of K^+ on the trans side of the membrane accelerates the cycling of the carrier [117], as has also been suggested for other bacterial transport systems [57,216,217]. Like leucine, glutamate accumulation could be elicited in the dark by imposing a sodium gradient out/in. This accumulation was relatively insensitive to FCCP, while the light-energized accumulation was abolished by the uncoupler.

Lanyi and MacDonald [115] have suggested that light energizes proton extrusion by bacteriorhodopsin. This hydrogen ion gradient is utilized for extrusion of Na^+ ions. This was shown by the light-effected expulsion of

^{22}Na from preloaded vesicles by a process that is inhibited by TPMP$^+$ and FCCP [116]. The Na$^+$/H$^+$ exchange appears to be electrogenic by an antiporter with a stoichiometry of H$^+$/Na$^+$ > 1, since the efflux of Na$^+$ from preloaded vesicles was inhibited by TPMP$^+$ while uptake of Na$^+$ by pre-irradiated vesicles was stimulated by TPMP$^+$ and by FCCP. Energization of the membrane due to light adsorption was shown to consist of a $\Delta\psi$ (measured by quenching of the fluorescence of a cyanine dye and by DDA$^+$ accumulation) and a ΔpH [182]. Thus H$^+$ ions are involved indirectly, namely by exchange of an electrochemical hydrogen ion gradient for a sodium ion gradient. The flux of Na$^+$ ions down their concentration gradient then is directly coupled to leucine and glutamate uptake.

The authors point out that leucine bears no net charge at the pH values (about 6.5) of the experiment, thus a leucine-Na$^+$ carrier complex is presumably positively charged and would be expected to respond by moving toward a negatively charged region. Indeed, dissipation of the membrane potential by addition of the permeant cations tetrabutylammonium or TPMP$^+$ resulted in inhibition of leucine uptake in irradiated vesicles, while these compounds had little effect on the concomitant extrusion of protons. In contrast, FCCP abolished both proton extrusion and amino acid uptake. On the other hand, the glutamate molecule at pH 7 bears a net negative charge. Thus a glutamate-Na$^+$ carrier complex would be expected to bear no net charge and not to respond to a negatively charged region. This was demonstrated by the minor effect of TPMP$^+$ on the initial rate of light-energized glutamate uptake. At pH 5 the glutamate molecule is protonated and therefore expected to be more lipid soluble; indeed, when the intravesicular pH was lowered, accumulated glutamate was able to exit from the vesicle.

V. QUANTITATIVE ASPECTS OF PROTON GRADIENT COUPLED SOLUTE TRANSPORT

Is the pmf sufficient to account for the accumulation of the solutes that utilize the chemiosmotic mechanisms discussed above? As yet few quantitative data have been obtained on the relationships between the steady-state levels of accumulated solute and the poise of Δp. The importance of quantitative relationships lies in the ultimate aim to elucidate the mechanisms of these energy coupling systems. For example, if the accumulation of a nonelectrolyte sugar is due to the cotranslocation of protons by a 1:1 symport mechanism, then the level of accumulated sugar should equal the Δp value measured under the specified conditions of incubation (and expressed in the same concentration units). In the same way, a ΔpH-driven system should exhibit a quantitative relationship between solute accumulation and ΔpH and not to $\Delta\psi$.

A beginning has been made by measuring β-galactoside levels within S. lactis cells in relation to Δp, as determined by measurements of both $\Delta \psi$ and ΔpH. Kashket and Wilson [99] utlilized artificially imposed Δp to effect the accumulation of [14C]TMG by treating S. lactis cells suspended in a medium of low K^+ concentration with valinomycin (See Sec. IV.A4). The calculated Δp values ($\Delta \psi$ minus the ΔpH values, in this case) showed a direct 1:1 relationship to the steady-state accumulation of TMG. However, when the TMG ratio in:out was 1, the Δp value was 25 mV. This discrepancy could be due to a number of reasons; one factor may be that the calculated values for $\Delta \psi$ are too high because measured potassium concentrations were used instead of activities. Since the intracellular ionic strength is higher than that of the medium used, the activity coefficient for K^+ is lower inside than outside. A second factor is that the potassium concentration immediately outside the plasma membrane is probably higher than that in the bulk medium. In order to compensate for these presumptive factors, Kashket and Wilson [100] carried out further experiments with cells incubated in media of relatively high ionic strength (by adding NaCl to 0.3 M) and with 10 to 40 mM K^+ added to the medium. In this set of experiments the Δp was generated and maintained by fermentation of glucose or arginine. The $\Delta \psi$ was determined from the fluorescence of 1,1'-dihexyl-2,2'-oxacarbocyanine. Under such conditions, the Δp values and the level of TMG accumulation showed a direct 1:1 relationship, in consonance with an obligatory coupling between proton and sugar uptake. The values obtained for $\Delta \psi$ under conditions of high medium NaCl were lower (about 80 mV) with glucose as energy source (Table 12-3) than those calculated for metabolizing cells in media without added NaCl [96] (Table 12-3). The effect of lowering the Δp is not due to plasmolysis of the cell, (and hence a decrease in the cell water space), but to the dissipation of Δp by the work entailed for the extrusion of Na^+ by the cells [96].

Schuldiner and Kaback [200], using E. coli vesicles, determined the $\Delta \psi$ from the concentration ratio of $TPMP^+$ and related that value to the concentration ratio of lactose. The vesicles were energized by various electron donors and the Δp varied by addition of various concentrations of either CCCP or the electron transport inhibitor 2-heptyl-4-hydroxy-quinoline-N-oxide. These workers found that the log of the concentration ratio of lactose accumulated was related to the log of the concentration ratio of $TPMP^+$. Similar relationships were found with proline, tyrosine, glutamic acid, and glycine. However, the lactose ratios exceeded those of $TPMP^+$; for example, a concentration ratio for lactose of 100, assuming one lactose taken up per H^+, would require a Δp of 120 mV, but a $\Delta \psi$ of 75 mV, which by itself is not sufficient to energize lactose uptake, was observed. Hence the authors postulated that the stoichiometry of the transport system may be one lactose per two H^+ translocated. However, these workers failed to detect a ΔpH because the DMO flowed out when the filtered vesicles were washed. When flow dialysis was used, Ramos et al.

TABLE 12-3 Proton Gradients in Metabolizing Cells and Vesicles

System	Incubation Condition	Methods		Energy Source	$\Delta\psi$ (mV)	$-Z\Delta pH$ (mV)	Δp (mV)	Ref.
		$\Delta\psi$	ΔpH					
E. coli cells (valinomycin treated)	pH$_o$ = 6 pH$_o$ = 7 pH$_o$ = 8 K$^+$ = 1 mM	86Rb distribution	DMO and Methylamine	respiration	74 82 88	109 53 -30 reversed	183 115 50	164
	pH$_o$ = 7	DDA$^+$ Uptake	--	--	74	--	--	103
E. coli vesicles	pH$_o$ = 6.6	DDA$^+$ Uptake	not done	respiration	124	--	--	5,75
E. coli spheroplasts	pH$_o$ = 6.45 to 6.75	K$^+$ flux	H$^+$ flux	respiration	132	98	230	24
S. faecalis cells		DDA$^+$ Uptake	DMO	glycolysis	150-190	48-60	198-250	64,66
		cyanine dye	not done	glycolysis	140	--	--	118

S. lactis cells	high salt medium	cyanine dye	DMO	glycolysis	36	44	80	100
	$pH_o = 7$	cyanine dye	DMO	arginine fermentation	39	0	39	100
	low salt medium $pH_o = 5$	cyanine dye	Benzoate	glycolysis	95	58	153	96
S. aureus cells		K^+ flux	H^+ flux	respiration	134	77	211	24
H. halobium cells	$pH_o = 6.0$	TPMP$^+$ Uptake	DMO	light	113	73	186	137
	$pH_o = 6.7$				126	42	168	
	$pH_o = 8.0$				151	8	159	
H. halobium vesicles	ratios of K^+ and Na^+ in/out varied	cyanine dye	H^+ flux	light	34–120	106–119	153–229	182

[180] were able to measure a ΔpH in E. coli vesicles, of a size and polarity (inside alkaline) consonant with a Δp value greater than $\Delta\psi$ and more in line with the sugar levels observed.

As Ramos et al. [180] have shown, the composition of Δp in energized E. coli vesicles, like that of valinomycin-treated intact cells [164], is dependent on the pH_O (Table 12-3). The Δp decreases from a value of 180 mV at pH_O 5.5 to about 75 mV at pH_O 7.5. The $\Delta\psi$ remained nearly constant, being about 65 mV at pH_O 5 and rising to 75 mV at pH_O 7. Since the pH_i remains essentially at 7.5, the ΔpH values are equivalent to 114 mV at pH_O 5.5 and decrease to 0 at pH 7.5 or above.

To assign a specific component of Δp as the driving force for a particular transport system, Kaback [89] reasoned that if the stoichiometry of solute transport is one solute molecule per proton, then measuring the accumulation of various solutes at different pH_O values should indicate whether $\Delta\psi$, ΔpH, or Δp is the driving force. For example, Δp or ΔpH-driven systems should exhibit maximum accumulation at pH_O values near 5, while $\Delta\psi$-driven systems should show less sensitivity to the medium pH. However, such simple relationships were not found experimentally: the steady-state concentration ratios of the solutes tested (succinate, lysine, proline, lactose, and glucose-6-phosphate) were generally higher at pH_O values of 7 and below than at pH 8. While the Δp at pH_O 5.5 was sufficient for solute accumulation, at higher pH_O values the solutes accumulated to levels higher than the measured Δp. To explain these data, Kaback proposed that the stoichiometry between solute and protons may vary with change in pH_O, namely that the ratio of hydrogen ion translocated per solute molecule may be 1 at pH_O 5.5, but that the ratio may rise to 2 or 3 as the external pH is increased. For organic acids Kaback postulates that at lower pH_O values one H^+ is taken up per molecule of undissociated acid, while at more alkaline pH_O, one proton is taken up in association with the substrate and another is associated with the transport carrier. It is obvious that work needs to be done to resolve these questions, since they bear directly on the mechanisms by which carriers operate.

Variation of stoichiometry of hydrogen ions translocated per solute molecule has indeed been observed by Hamilton and his colleagues [25]. These workers showed that one alanine molecule was taken up by E. coli cells together with one proton; on extended growth in a chemostat with alanine as carbon source, a mutant strain was selected which showed a stoichiometry of one alanine per two protons. Further growth resulted in the selection of a mutant with an alanine:proton relationship of 1:4. The stoichiometries of D-alanine, serine, and glycine increased in parallel to that of L-alanine, but no proton uptake was observed on isoleucine or valine addition. Hence the ala-ser-gly carrier system was affected, but not the binding protein-associated aliphatic hydrophobic amino acid system. The increase in stoichiometry of alanine uptake has an important implication for energy coupling: referring to Eq. (6), if a proton/alanine symporter has a stoichiometry of n = 1, with no net charge on the solute (i.e., m = 0), then for a Δp of about 230 mV alanine could be accumulated with a concentration ratio of about 1000. With a stoichiometry of four protons per alanine (n = 4), a Δp = 230 mV would suffice for an alanine accumulation ratio of 10^{16}. Such a gradient would not occur, in fact, since the actual level of intracellular amino acid attained is the result of the rates of substrate uptake, leakage out of the cells, and metabolism. While the authors were not able to measure the alanine levels achieved by these mutants, they did show that with low alanine concentrations in the medium the mutants grew faster than the parental cells, which is consistent with an increased cellular concentration of alanine.

If a higher ratio of protons to solute increases the concentration ratio as the power of the proton/solute ratio, why has not natural selection for carriers with ever-increasing ratios of cotransport occurred? It is obvious that the larger the portion of the proton motive force which is used for a single transport event, the less energy will be available for other cellular functions. The actual energetic cost to the cell of nutrient transport has yet not been determined. Such studies are of obvious importance for the understanding of the mechanisms of energy coupling, as well as of the general physiology of the cell. Purdy and Koch [174] measured the rate at which ONPG is transported into cells down a concentration gradient in energy–depleted cells of E. coli. The cells were then energized by supplying glucose at low, steady rates. Under these conditions, 86 molecules of ONPG were translocated per molecule of glucose oxidized over a 6.5-fold range in the rate of glucose addition, suggesting that transport is the major consumer of metabolic energy, at least under the conditions of these assays. It is of value to note that this study furnishes independent evidence for a chemiosmotic-type mechanism for β-galactoside transport in E. coli. Assuming a proton: ONPG ratio of 1:1, a value of 76 ONPG per glucose would be predicted. Other proposed mechanisms would yield lower values. Thus, of the proposed mechanisms, a chemiosmotic one is most consistent with the data.

VI. ENERGY COUPLING OF TRANSPORT SYSTEMS SENSITIVE TO OSMOTIC SHOCK

As discussed above, Berger [12] has shown that the shock-sensitive glutamine transport system of E. coli is coupled directly to ATP (or a derivative of ATP) as a source of energy, while the shock-resistant proline transport system is obligatorily coupled to a proton motive force. It was of importance at that time to demonstrate that the difference in energy coupling properties was related to the sensitivity of transport to osmotic shock treatment. Berger and Heppel [13] expanded the number of examples of ATP-linked uptake systems to include the systems for diaminopimelic acid (the cystine-general system), for histidine, for arginine (the arginine-specific system), and for ornithine (the lysine-arginine-ornithine or LAO system). The list of shock-sensitive systems which have been shown to require ATP has since grown (Table 12-4).

However, binding proteins have not been discovered for some of the E. coli systems which are sensitive to osmotic shock. One of the phosphate transport systems in E. coli is listed among those demonstrated to require ATP. However, the reports by Rae and Strickland [178, 179] do not clearly identify ATP as the source of energy, but only eliminate direct participation of a proton motive force.

Of great interest is the observation by Harold and Spitz [67] that two transport systems in the gram-positive organism S. faecalis are not coupled to a proton motive force in any direct manner. The systems transport anions: one for phosphate and arsenate and the other, asparate and glutamate.

TABLE 12-4 Transport Systems Which Require Phosphate Bond Energy

Organism	Transport System	Associated Binding Protein	Ref.
E. coli[a]	arginine specific	yes	13
	cysteine general	yes	13
	galactose specific	no	230
	glutamine	yes	12,13
	glycylglycine	no	27
	histidine	yes	13
	leucine specific	yes	107,232
	leucine-isoleucine-valine (LIV-I)	yes	107,232
	lysine-arginine-ornithine (LAO)	yes	13
	methionine	no	94
	β-methylgalactoside	yes	230
	phosphate	yes	178,179
	potassium (Kdp system)	no	183
	ribose	yes	30
H. halobium	histidine		82
S. faecalis	arsenate (phosphate)		67
	aspartate (glutamate)		67

[a]In the case of E. coli each transport system has been shown to be shock sensitive, but in some cases associated binding proteins have not been found.

Again, the assumption was made that, since a proton motive force is not sufficient for energizing these transport systems, the process requires ATP. However, the data do not demonstrate the direct involvement of ATP, in the sense that no procedures were used which specifically reduced the ATP pools and simultaneously reduced the transport of those anions. Still, a product of glycolysis does appear to be required, and, as the authors pointed out, in anaerobes the ultimate energy donor for transport is ATP.

The original distinction made by Berger was between shock-sensitive and shock-resistant systems in the gram-negative organism E. coli. Defining the energy coupling of transport systems according to which of those two classes a particular system belongs can no longer be considered rigorous enough in view of some recent reports. First is the discovery of ATP-linked systems in gram-positive cells, which lack an outer membrane and hence a periplasmic region as in found in gram-negative cells. Second, several transport systems in cells of E. coli have been shown to be sensitive to osmotic shock, yet those same systems are present in membrane vesicles. For example, the transport of a number of carboxylic acids by intact cells is sensitive to osmotic shock, with the release of a carboxylic acid binding protein; yet, membrane vesicles are capable of transporting those acids, apparently by the same transport system [126-129]. A similar observation has been made by Silhavy et al. [206] with the glycerol-3-phosphate transport system. The system is shock sensitive, yet is retained in membrane vesicles. A protein related to the system is released following osmotic shock treatment. Since glycerol-3-phosphate and carboxylic acids are negatively charged at neutral pH, it has been suggested [206] that the function of those proteins is to facilitate the passage of solutes through the negatively charged outer membrane. Such a role would not be unique, since a number of outer membrane receptors have been reported (see Chap. 9). The fact that the carboxylic acid binding protein has affinity for both succinate and D-lactate, while vesicles transport those two carboxylic acids by separate systems [126-129], suggests that the binding protein is not immediately involved in passage of the solutes through the inner membrane. While the energetics of those systems have not been investigated in detail, it is reasonable to assume that in vesicles the main source of energy is the proton motive force. Of course, another possible explanation for those results is that different systems are measured in intact cells and vesicles.

Those results are not in conflict with those of Berger, but do point out the possible semantic traps into which we may fall. For the present, the designations "ATP-linked" and "pmf-linked" seem sufficient and more exact than "shock-sensitive" and "shock-resistant." When the detailed mechanisms of the coupling reactions are determined, new terminology may be needed.

Aside from the semantic problem, there is the more fundamental question of whether ATP linkage and coupling to the pmf are mutually exclusive. Several preliminary studies have shown that a few transport

systems have characteristics of both types of coupling. Rosen [189] has
found that lysine is transported by at least two different transport systems
in E. coli. One, the LAO, is a classical shock-sensitive, ATP-linked
system. The other, the lysine-specific, is found in membrane vesicles,
is not sensitive to osmotic shock treatment, but is also no more sensitive
to uncouplers and electron transport chain inhibitors than are ATP-linked
system. Moreover, succinate metabolism is unable to drive lysine transport. By these criteria, it appears that the system is ATP-linked. However,
depletion of ATP pools by arsenate treatment does not totally inhibit lysine-
specific transport, as it does ATP-linked systems. Rosen has suggested
[189] that this system can couple to either source of energy. However, a
third lysine transport system cannot be excluded from the data. A third
system would have to be shock resistant but ATP linked to fit the data.

Another anomalous system is the trkA system of E. coli, one of the
three transport systems for potassium ion [184]. This system is shock
resistant, yet is not measurable in membrane vesicles [183]. The system
exhibits the same sensitivity to uncouplers and electron transport chain
inhibitors as do other shock-resistant systems, but is inhibited by arsenate
treatment. Although the trkA system is shock resistant and has no associated binding protein, Rhoads and Epstein [183] have postulated that the
energy for translocation is supplied by ATP, and that a pmf is required to
expose the substrate binding site of the carrier on the external face of the
inner membrane. This mechanism is similar to that proposed by Kaback
and coworkers [201] in which the lactose transport carrier cannot bind
substrate in the absence of a proton motive force.

Another transport system in E. coli which is shock resistant but sensitive to the action of arsenate is the LIV-II, one of the leucine transport
systems [232]. However, the three anomalous systems also show differences. For example, the LIV-II and the lysine-specific systems are relatively insensitive to uncouplers, while the trkA is sensitive . BF_1 mutants
depleted of endogenous energy reserves are unable to utilize succinate for
the transport of lysine and potassium ions, but leucine uptake by the LIV-II
system is able to use the analogous oxidizable compound D-lactate (although
this observation must be qualified by the fact that D-lactate, unlike succinate, generates ATP through substrate-level phosphorylation). The
common features of each of these systems are the resistance to osmotic
shock and the sensitivity to arsenate. It is not clear at this time whether
these are exceptions to Berger's observations, whether they represent
intermediate cases, or whether the results are in some way artifacts.

Recently, Lieberman and Hong [122, 124] have reported the isolation
of mutants of E. coli, called ecf mutants, which are defective in oxidative
phosphorylation and in the transport of a number of substrates by both
shock-sensitive and shock-resistant systems. The exact nature of the
mutation is not known, although it appears that two simultaneous mutations
are necessary for expression of the phenotype [124]. Yet, ATP-linked

systems are defective in this mutant even though ATP pools are not decreased. By both biochemical and genetic criteria the mutation(s) can be differentiated from mutants in the BF_0F_1 complex. For example, the strain has normal proton permeability and Mg^{2+}-ATPase activity. The mutation also allows K^+ to leak into the cell, while other metabolites leak into the extracellular medium. It is not clear whether the defect(s) is primarily related to energy coupling, since a number of other cellular functions, including phospholipid, DNA, and RNA synthesis, are also impaired [123]. The properties of these mutants are similar to those found in wild-type cells after treatment with colicin K, as will be discussed in the next section. As will be seen, the postulated mechanism of action of colicin K is activation or creation of a pore for anions, leading to depolarization of the membrane.

If both the defect in ecf mutants and in colicin K-treated cells are related to the maintenance of the membrane potential, then the question is how the membrane potential is involved in ATP-linked systems. The explanation may be related to the suggestions of Kaback and coworkers [201] and Rhoads and Epstein [183] that the membrane potential is required for binding of substrates to carrier proteins. Thus, ATP may be necessary but not sufficient for energizing some shock-sensitive transport systems. Likewise, some shock-resistant systems may require a membrane potential for purposes other than energizing translocation.

Finally, the question still remains as to the mechanism by which the energy of ATP is coupled to shock-sensitive systems, whether those systems are primary active transport systems, or whether the transport occurs by some other as yet unknown mechanism.

VII. BACTERIOCINS AND ENERGY TRANSDUCTION

A. Mechanisms of Bacteriocin Action

Bacteriocins, high molecular weight proteins produced by certain bacteria, have the property of killing of other bacterial species [130]. Although the function of these proteins is unknown, there is considerable interest in the mechanism by which they produce their lethal effect. Certain bacteriocins, including colicins A, El, Ia, Ib, and K and staphylococcin 1580, interfere specifically with energy metabolism [36, 50, 84, 130, 158, 224]. Thus, studies of the interaction of these molecules with the bacterial membrane produce information about the mechanism of energy transduction, as well as about the mechanism of action of the bacteriocins themselves. In order to illustrate the types of information which can be gained, we will consider primarily the literature concerning the mechanism of action of colicin K, although many of the observations have been made with other bacteriocins as well.

The first event in colicin K action is the binding of that protein to a specific receptor on the outer membrane of an E. coli cell [130,171]. The next series of events can be divided into two stages. In stage I the cell has not yet been killed, for it can be rescued by destruction of the colicin by trypsin treatment. Stage II is characterized by actual death of the cell; trypsin no longer reverses the effects. However, the transition from stage I to II requires oxidative energy, since inhibition of electron transport with cyanide or anaerobiosis, or uncoupling with DNP or CCCP, prevents the irreversible step [169]. Stage II can be delayed, also, by chilling the cells. The energy required is in the form of a proton motive force rather than ATP per se, even though colicin K depletes ATP pools, since a BF_1 mutant treated with colicin K was killed even though intracellular ATP pools increased [83,172]. Moreover, in energy-depleted cultures of both the mutant and parent, killing did not occur until an energy source, either glucose or D-lactate, was added, even though D-lactate oxidation was only capable of regenerating the ATP pools of the mutant to 10% of normal levels. The requirement for a proton motive force to effect the transition from stage I to stage II is reminiscent of the need for a membrane potential in dansylgalactoside binding to the lactose carrier.

This similarity prompted Jetten and Jetten [83] to postulate that a protein of the inner membrane involved in colicin K action can perform its function only when acted on by the pmf. Although not specifically spelled out in their suggestion, we assume that the affinity of that site is increased by generation of a pmf.

How does colicin K kill the cells? That question has been answered in terms of what causes death. Colicin K interferes with oxidative metabolism, and, as a consequence, the membrane of the cell becomes more permeable to small molecules, resulting in the efflux of such solutes as the cations K^+ and Mg^{2+}. In an elegant series of experiments, Kopecky et al. [111] demonstrated that killing could be prevented by the addition of 100 mM KCl and 1 mM Mg^{2+} to the medium, approximately the physiological intracellular concentrations of those ions. Thus, it appears that depletion of those two cations leads to cell death, most likely because those ions are necessary for macromolecule biosynthesis.

Still, the efflux of those cations cannot be the primary event in colicin K action, because the polarity of the membrane potential of the cell is normally positive outside, yet colicin K causes positive ions to flow outwards. The efflux of cations certainly indicates the absence of a positive membrane potential at the time that cation efflux occurs. Thus, a more primary event must be related to an effect of colicin on the pmf. Using 3,3'-dihexyloxacarbocyanine, one of the fluorescent probes which measures membrane potential, Brewer [20] determined that colicin K causes an immediate depolarization of the membrane. The depolarization occurred during stage II, since the fluorescence changes could be prevented by trypsin treatment. In analogy with certain BF_1 mutants which cannot main-

tain membrane potentials because of an increase in the proton permeability
of the membrane [2,3,187], one possibility was that colicin K acted as an
uncoupler. Brewer was able to rule out this possibility on two grounds.
First, the membrane of colicin-treated cells retained normal proton
permeability. Second, the intracellular pH, and hence the chemical gradient of protons, was unaltered by colicin K, as measured by distribution
of the weak organic acid DMO. So, an initial event in the action of colicin
K is dissipation of the pmf by decreasing the magnitude of the membrane
potential alone and not by collapsing the pH gradient. The mechanism of
colicin K action is not similar to that of valinomycin, which produces its
killing in S. faecalis by facilitating K^+ efflux and depolarizing the membrane [66,234]. Brewer was able to eliminate this possibility because, if
colicin K acted as an electrogenic cationophore, the membrane would have
become hyperpolarized rather than depolarized by the efflux of the cations.
In order to depolarize the membrane, a cation has to flow into the cell. So,
addition of high extracellular amounts of K^+ might be expected to increase
the rate or extent of the depolarization in treated cells. Instead, addition
of mono- and divalent cations to the medium had no effect on the depolarization when added before colicin treatment, but repolarized the membrane
when added after the colicin.

An alternate mechanism for the observed depolarization is anion efflux.
Brewer found a slight stimulation of phosphate influx following colicin addition, but this was not considered adequate for the observed depolarization.
However, it may be significant that phosphate uptake was stimulated, whereas most other transport systems were inhibited. A large stimulation of
^{32}P-labeled inorganic phosphate entry would be expected if countertransport occurred during the efflux of intracellular phosphate. However colicin
K may produce a gating of phosphate, and then countertransport might not
occur. Experimentally, this could be determined by the effect of high
external phosphate concentrations on the transition from stage I to stage II,
as measured by killing or membrane depolarization. In any case, whether
the ion is phosphate or some other anion, it seems reasonable to conclude
that one of the immediate events in colicin K action is the creation or activation of anion channels.

It is tempting to speculate that the actual primary event in the process
is the binding of colicin K to a membrane protein, and that the complex
becomes a channel for one specific anion. The mechanism is likely to be
more complex than this, since colicins cause a number of changes in the
membrane. For example, Knepper and Lusk [106] have recently shown that
colicin K treatment causes a number of specific membrane proteins to disappear from the membrane. This may mean that depolarization of the
membrane is a secondary event. In that regard, Cramer and coworkers
[29,70,167] have found that colicin E1 binding produces fluorescence changes similar to those reported by Brewer [20], except that the fluorescent
probe, N-phenyl-1-naphthylamine, is uncharged and therefore would not be

expected to show a direct response to changes in a membrane potential.
The changes may reflect an increase in the microviscosity of the membrane
upon colicin El binding. Moreover, Helgerson and Cramer [70] have shown
that FCCP, colicin El, and colicin K all produce similar changes in the
E. coli cell envelope, changes which have been interpreted as alterations
in the structure of the inner and outer membranes rather than in the membrane potential. Perhaps both interpretations are correct: changes occur
both in the structure of the membrane and in its polarization. In addition,
it is probable that some of the agents used have more than one effect: the
related uncoupler CCCP, for example, is known to cause immediate uncoupling, but also acts as a sulfhydryl reagent after a longer incubation [92].
It is not certain which of those phenomena is being measured in the reports
of Helgerson and Cramer [70]. Thus depolarization of the membrane may
result in structural changes which cause the disappearance of membrane
proteins, or the reverse may be true, namely that structural changes within
the membrane produce depolarization. Finally, there is uncertainty as to
what parameter the changes in fluorescence of the various probes actually
indicate.

B. Active Transport and Bacteriocins

We have seen that the action of colicin K is intimately involved with the
maintenance of the membrane potential. Colicins thus become useful tools
for the study of energy transduction for active transport. Colicins have
been found to cause efflux of intracellular metabolites [131, 224] and to
inhibit the influx of a large number of solutes [36, 50, 84, 131]. In wild-type
cells of E. coli, colicin K has been shown to deplete ATP pools [36, 172],
which could be prevented by prior treatment of cells with DCCD [172].
When a BF_1 mutant was treated with colicin K, ATP levels increased,
leading to the reasonable conclusion that dissipation of the pmf increased
the rate of ATP hydrolysis by the BF_1 [172].

Plate et al. [172] investigated the effect of colicin K on the shock-
sensitive glutamine and shock-resistant proline transport systems. These
workers reported that both transport systems were inhibited by colicin
treatment in the BF_1-positive strain, as would be expected since both the pmf
and ATP levels are reduced. However, in a BF_1 mutant glutamine transport was also inhibited, even though the intracellular concentration of ATP
was actually increased threefold by colicin treatment, compared to the concentration found in the untreated parent. Still, the inhibitory effect on
glutamine transport was less complete than on proline transport. While
Berger's data showed that ATP is necessary for transport by shock-sensitive systems [12, 13], these experiments with colicin K suggest that ATP
may not be sufficient for energization.

Plate [170] recently reported the isolation of a mutant of E. coli defective both in active transport and in response to colicin K. The mutant was isolated by selection for resistance to the antibiotic neomycin, followed by screening for inability to utilize succinate for growth, followed by screening for resistance to colicin K. A similar procedure has been utilized for the selection of proton-permeable BF_1 mutants, which similarly cannot utilize succinate for growth and are defective in the coupling of oxidative energy to active transport [4,187,188,208]. Plate's mutant, however, differed from BF_1 mutants because (1) Mg^{2+}-ATPase activity was unimpaired, (2) the mutant was not abnormally permeable to protons, and (3) treatment with DCCD did not restore transport activity, as had been found for the BF_1 mutants. Plate speculated that the mutation was not a defect in F_0, since he he found normal solute transport activity in a mutant defective in the BF_0, but he was not able to eliminate that possibility since genetic mapping had not been performed. The mutant defective in active transport and in response to colicin K showed normal oxidation of electron transport chain substrates. If the components of the oxidative phosphorylation pathway (the BF_0F_1 and electron transport chain) are normal, why is the mutant unable to utilize succinate for growth? Presumably respiration-coupled phosphorylation is impaired, although this was not measured. One could speculate that if the primary action of colicin K were the activation or formation of an electrogenic anion channel, then mutation of a protein responsible for translocation of that anion could produce the effects observed in the mutant. Plate points out that the mutations may be leaky, and therefore, only partial dissipation of the membrane potential may have occurred, which might explain why the mutation was not lethal. These are admittedly speculations that await experimental proof, but the idea is an attractive one. Proton-permeable mutants are unable to maintain a pmf; why should it not be possible to isolate anion-permeable mutants unable to maintain a membrane potential? The notion that colicins promote increased anion permeability is supported by the findings of Gould et al. [51], who showed that colicins E1 and K caused a large increase in the $H^+:O$ ratio of anaerobic cells upon addition of a pulse of oxygen. This effect was also seen when thiocyanate was added. Thus, the authors concluded that the colicins had increased the permeability of the membrane to an electrically compensatory counter ion.

Interestingly, although the shock-resistant proline and TMG transport systems were defective in Plate's mutant, the shock-sensitive glutamine and arginine transport systems were normal. This is in contrast with the effect of colicin K on shock-sensitive transport systems in the wild-type cell. A temperature-sensitive revertant of the mutant was isolated by ability to utilize succinate at 27 but not at 42°C. Using either glucose or succinate as an energy source, proline transport was only slightly defective when the cells were grown and assayed at the permissive temperature. At

42°C proline transport was reduced considerably with either carbon source.
Glutamine transport in the revertant, on the other hand, was slightly elevated compared to the mutant strain when assayed at the permissive temperature, whether glucose or succinate was used as energy source. However,
at 42°C, the restrictive temperature, succinate was oxidized in the temperature sensitive revertant at only 60% of the rate found in the parent. Under
the same conditions, intracellular ATP levels increased to nearly twofold.
It seems strange that the ATP levels increased if oxidative phosphorylation
were not occurring at the restrictive temperature. However, the cells had
not been depleted of endogenous energy reserves, so that ATP might have
been formed by some other pathway. In any case, a defect was shown to
occur in a shock-sensitive system under conditions where the ATP pools
increased. Again, this suggests that ATP may be necessary but not sufficient for energization of shock-sensitive transport systems, possibly reflecting the need for a membrane potential for solute-carrier interaction.

VIII. AVENUES OF FUTURE STUDY

It is hazardous to venture predictions about the course that these studies
will follow in the next few years. Nevertheless, we seem at the moment to
have solved many of the conceptual problems of energy coupling. The molecular mechanisms of energy coupling and carrier function are less well
understood and will probably be the focus of future study. We predict that
the main areas in which the immediate future of the field lies are (1) quantitation, (2) genetics, (3) isolation of carriers, and (4) elucidation of the
mechanism of ATP-linked systems.

With the reasonable assumption that certain transport systems are
coupled to the pmf, the quantitative relationship between them must be
determined. The two most promising lines of research are the use of
cyanine dyes to measure membrane potentials [223] and the use of flow
dialysis to measure uptake of solutes [26,180]. Both have been useful in
obtaining numbers, the question being whether those numbers are the actual
values of the $\Delta\psi$ and/or ΔpH, or just proportional to those parameters.
Caution must be exercised in the interpretation of the numbers; for example,
it has been assumed that certain organic cations respond to the membrane
potential as predicted by the Nernst equation and therefore can be used to
determine the actual value of $\Delta\psi$. Experiments demonstrating that the
permeability of DDA^+ and TPMP is modified in the presence of an organic
anion or by a Donnan potential suggest that the cations may not be as permeant as assumed [5,52,54,75,214]. If that is the case, then the numbers
which are generated may not directly yield $\Delta\psi$. Such problems are primarily technical and will be solved. With the use of these techniques, it should
be possible to determine the stoichiometry of symport reactions. This in
turn will yield information about the carrier itself, such as the number of
sites for each substrate, the order of addition, and so on. Eventually,

transport reactions will be analyzed in the same ways as enzymatic ones are now: quantitative analysis will put this area on to the same firm ground as enzymology now enjoys.

Genetic analysis of transport systems will certainly provide information on the number of components of individual systems. Already certain components have been shown to be shared by several different transport systems, and even between transport systems and chemoreceptors [1]. Mutations which affect the stoichiometry of symport reactions will undoubtedly prove to be of considerable value in determining the mechanism of carrier function. Two such mutants exist. A strain of E. coli which apparently has altered the lac permease from a 1:1 proton/sugar symport to a sugar uniport has been isolated [229]. It would be interesting to compare an amino acid analysis of the M protein from this strain with that of the wild type to see whether any amino acid residues with an ionizable group have changed. As has been discussed previously, a mutant of E. coli has been isolated in which the proton/alanine symport has a stoichiometry of 4:1, rather than 1:1 as found in the wild type [25]. The amount of information which such genetic tricks will yield will be greatly enhanced when the carriers themselves can be isolated and studied in vitro.

It should be obvious that there is only one way to really learn about the molecular mechanisms involved in transport: to isolate, purify, and characterize the components of a transport system. Proteins related to ATP-linked transport systems have been isolated and purified, but it is still uncertain what role these binding proteins play in the overall transport reaction. Recently several groups of investigators have isolated membrane proteins which appear to be involved with transport of solutes. More importantly, they have been able to incorporate these proteins into artificial membranes in a functional manner. This has been most successful for several primary transport systems, the light-driven H^+-translocating protein bacteriorhodopsin from H. halobium [177] and the H^+-translocating TF_0F_1 from the thermophilic bacterium PS3 [212]. In several reports the F_0F_1 and bacteriorhodopsin were incorporated into the same vesicle, creating a light-driven ATP synthetase [177, 233]. Similar studies are being performed using transport proteins isolated from eucaryotic systems, such as the Na^+/K^+-ATPase [176] and Ca^{2+}-ATPase [175]. Less information has been obtained with secondary transport systems for several reasons. First, the isolation of membrane proteins involved in secondary transport systems is considerably more difficult. Once removed from the membrane, there is no convenient assay. The proteins of the primary transport systems, on the other hand, have enzymatic activity (ATPase) in solution. Second, it is also much more difficult to assay a secondary transport protein than a primary one when incorporated into a vesicle, because the primary system is still coupled to chemical energy, while the secondary system has lost its primary source of energy. Two groups of investigators have recently reported the isolation and insertion into vesicles of alanine carrier proteins. Hayakawa et al. [68] isolated a protein complex from B. subtilis and incorporated it into liposomes of different lipid composition. The vesicles were capable of accumulating alanine with energy supplied by NADH oxidation. The very fact that NADH could energize alanine transport, however, suggests that the protein fraction was heterogeneous. A purified carrier of a secondary transport system would not be expected to exhibit concentration of solute unless it were coupled in some way to another gradient, for example, a potassium gradient. Hirata et al. [78, 79] used this concept to develop an assay system for the alanine carrier of the thermophilic bacterium PS3. The partially purified protein fraction was incorporated into liposomes, which were then loaded with K^+. Addition of valinomycin resulted in uptake of alanine. A similar concept has been used to develop an assay for calcium transport in aggregated vesicles of E. coli [190]. Vesicles were solubilized with deoxycholate, and the supernatant solution was diluted into buffer at pH 5.5. This resulted in the formation of vesicles which lacked energy-dependent calcium transport, but which did accumulate calcium when the pH of the medium was rapidly shifted to pH 8.0. Finally, Amanuma et al. [6] have used an organic solvent to extract a proline carrier from the membrane of E. coli. The protein was functional on incorporation into liposomes. Although these studies are preliminary, the results suggest that it will be possible to define the systems chemically as well as physiologically.

The last conceptual frontier (in our opinion) involves the nature of transport systems which utilize high-energy phosphate bond energy rather than a pmf. At this time, the direct donor of energy is unknown; without that knowledge it is difficult even to predict what the coupling mechanism could be. The function of binding proteins in these systems is unknown. Attempts to restore transport in osmotically shocked cells by addition of purified binding proteins have been irreproducible. Addition of the proteins to membrane vesicles is without effect. It is possible that experiments of the latter type could succeed if the correct energy donor were known and supplied. If ATP were the donor, then it would be necessary to add ATP to the correct side of the membrane (i.e., the side with the BF_0F_1), while adding the binding protein either to that side or the other. The problem could be simplified if it were not necessary to add binding protein. The discovery that phosphate transport in the gram-positive organism S. faecalis is apparently linked to phosphate bond energy may provide such a system [67], since binding proteins have not been found to exist in gram-positive bacteria.

A number of other unresolved questions remain with regard to energy coupling. For example, there are the observations that in some systems the exit of accumulated solute is accelerated by addition of energy donors, for example, the exit of α-methylglucoside from E. coli cells [80] or of

α-aminoisobutyrate from S. lactis cells [215]. Another unresolved question is whether transport systems could couple alternately to either or simultaneously to both phosphate bond energy and a pmf. It may be that a particular system may be coupled to one source of energy, for example phosphate bond energy, and yet be modulated by a pmf. That may be the situation with systems such as the trkA system for K^+ [183].

In conclusion, the field of bacterial transport in general and the particular portion of it related to energy coupling has progressed very rapidly in recent years. There is no reason to believe that the progress will slow, and there are still enough unresolved questions to keep the field interesting and lively for years to come.

RECENT DEVELOPMENTS

In the six months since this chapter was submitted there have been some interesting and important developments. One intriguing finding is the demonstration that a proton motive force is not essential for growth of bacteria under certain conditions. Harold and Van Brunt [235] cultured cells of Streptococcus faecalis with gramicidin and other ioniphores in order to eliminate electrochemical gradients. Growth of the cells occurred when the K^+ concentration in the medium was maintained at 0.3 N, that of amino acids at the millimolar level, the pH above 7.0. We can conclude that a pmf allows the cell to maintain an internal environment conducive to growth when the external environment is hostile, for example, when the concentration of nutrients in the medium is too low to allow for macromolecular synthesis in the absence of concentrative mechanisms. However, we would predict that obligate aerobes and facultative anaerobes could not grow aerobically in the presence of gramicidin. The former, of course, requires a proton motive force for ATP synthesis. Under aerobic growth conditions, the latter would hydrolyze glycolytically-derived ATP via the BF_0F_1 in the presence of gramicidin.

Second, the utilization of an electrochemical gradient of Na^+ to energize active transport has recently been shown to be more general than previously thought. A Na^+/solute symport has been described for glutamate transport in Escherichia coli [236-238]. Similarly, the melibiose transport system of Escherichia coli [239] and Salmonella typhimurium [240] has been shown to be coupled to the electrochemical gradient of Na^+. Tokuda and Kaback [241] have demonstrated elegantly separate functions for $\Delta\psi$ and for the chemical gradient of Na^+ in melibiose transport. The former lowers the affinity of sugar binding sites, while the latter increases the number of available sites. Finally, Belliveau and Lanyi [242] have reported the existence of a Na^+/Ca^{2+} antiport in Halobacterium halobium, presumably responsible for the extrusion of Ca^{2+} from the cell.

REFERENCES

1. J. Adler, Ann. Rev. Biochem., 44:341-356 (1975).
2. L. W. Adler and B. P. Rosen, J. Bacteriol., 128:248-256 (1976).
3. L. W. Adler and B. P. Rosen, J. Bacteriol., 229:959-966 (1977).
4. K. Altendorf, F. M. Harold, and R. D. Simoni, J. Biol. Chem., 249:4587-4593 (1974).
5. K. Altendorf, H. Hirata, and F. M. Harold, J. Biol. Chem., 250: 1405-1412 (1975).
6. H. Amanuma, K. Motojima, A. Yamaguchi, and Y. Anraku, Biochem. Biophys. Res. Commun., 74:366-373 (1977).
7. M. Ariel and N. Grossowicz, Biochim. Biophys. Acta, 352:122-126 (1974).
8. S. S. Asghar, E. Levin, and F. M. Harold, J. Biol. Chem., 248: 5225-5233 (1973).
9. E. M. Barnes and H. R. Kaback, Proc. Natl. Acad. Sci. USA, 66: 1190-1198 (1970).
10. E. M. Barnes and H. R. Kaback, J. Biol. Chem., 246:5518-5522 (1971).
11. E. A. Berger, PhD Thesis, Cornell University, 1973.
12. E. A. Berger, Proc. Natl. Acad. Sci. USA, 70:1514-1518 (1973).
13. E. A. Berger and L. A. Heppel, J. Biol. Chem., 249:7747-7755 (1974).
14. P. Bhattacharyya and E. M. Barnes, Jr., J. Biol. Chem., 251: 5614-5619 (1976).
15. R. A. Bogomolni, R. A. Baker, R. H. Lozier, and W. Stoeckenius, Biochim. Biophys. Acta, 440:68-88 (1976).
16. J. Boonstra, M. T. Huttunen, W. N. Konings, and H. R. Kaback, J. Biol. Chem., 250:6792-6798 (1975).
17. I. R. Booth and J. G. Morris, FEBS Lett., 59:153-157 (1975).
18. D. H. Boxer and R. A. Clegg, FEBS. Lett., 60:54-57 (1975).
19. P. D. Boyer, B. O. Stokes, R. G. Wolcott, and C. Degani, Fed. Proc., 34:1711-1717 (1975).
20. G. J. Brewer, Biochemistry, 15:1387-1392 (1976).
21. J. N. Burnell, P. John, and F. R. Whatley, J. Biochem., 150:527-530 (1975).
22. J. N. Burnell, P. John, and F. R. Whatley, FEBS Lett., 58:215-218 (1975).
23. G. Cecchini and A. L. Koch, J. Bacteriol., 123:187-195 (1975).
24. S. H. Collins and W. A. Hamilton, J. Bacteriol., 126:1224-1231 (1976).
25. S. H. Collins, A. W. Jarvis, R. J. Lindsay, and W. A. Hamilton, J. Bacteriol., 126:1232-1244 (1976).
26. S. P. Colowick and F. C. Womack, J. Biol. Chem., 244:774-777 (1969).

27. J. L. Cowell, J. Bacteriol., 120:139-146 (1974).
28. G. B. Cox and F. Gibson, Biochim. Biophys. Acta, 346:1-25 (1976).
29. W. A. Cramer, S. K. Phillips, and T. W. Keenan, Biochemistry 12:1177-1181 (1973).
30. S. J. Curtis, J. Bacteriol., 120:295-303 (1974).
31. R. Davies, J. P. Folkes, E. F. Gale, and L. C. Bigger, Biochem. J., 54:430-437 (1953).
32. G. R. Drapeau and R. A. MacLeod, Biochem. Biophys. Res. Commun., 12:111-115 (1963).
33. G. R. Drapeau, T. I. Matula, and R. A. MacLeod, J. Bacteriol., 92:63-71 (1966).
34. R. G. Eagon and L. S. Wilkerson, Biochem. Biophys. Res. Commun., 46:1944-1950 (1972).
35. R. C. Essenberg and H. L. Kornberg, J. Biol. Chem., 250:939-945 (1975).
36. K. L. Fields and S. E. Luria, J. Bacteriol., 97:57-63 (1969).
37. J. L. Flagg and T. H. Wilson, J. Bacteriol., 125:1235-1236 (1976).
38. L. Frank and I. Hopkins, J. Bacteriol., 100:329-336 (1969).
39. E. F. Gale, J. Gen. Microbiol., 1:53-76 (1947).
40. E. F. Gale, Biochem. J., 48:286-290 (1951).
41. E. F. Gale, Biochem. J., 48:290-296 (1951).
42. E. F. Gale, Symp. Soc. Exp. Biol., 8:242-253 (1954).
43. E. F. Gale, Biochim. Biophys. Acta, 266:182-205 (1972).
44. E. F. Gale and P. D. Mitchell, J. Gen. Microbiol., 1:299-313 (1947).
45. E. F. Gale and T. F. Paine, Biochem. J., 48:298-301 (1951).
46. E. F. Gale, E. S. Taylor, P. D. Mitchell, and G. R. Crowe, J. Gen. Microbiol., 1:77-85 (1947).
47. A. K. Ganesan and B. Rotman, J. Mol. Biol., 16:42-50.
48. P. B. Garland, J. A. Downie, and B. A. Haddock, Biochem. J., 152:547-559 (1975).
49. N. S. Gel'man, M. A. Lukoyanova, and D. N. Ostrovskii, Bio-membranes, 6:1-261 (1975).
50. M. J. R. Gilchrist and J. Konisky, J. Biol. Chem., 250:2457-2462 (1975).
51. J. M. Gould, W. A. Cramer, and G. van Thienen, Biochem. Biophys. Res. Commun., 72:1519-1525 (1976).
52. L. Grinius, A. Jasaitis, Y. Kadziauskas, E. Liberman, V. Skulachev, V. Topali, L. Tsofina and M. Vladimirova, Biochim. Biophys. Acta, 216:1-12 (1970).
53. B. Griniuviene, V. Chmieliauskaite, and L. Grinius, Biochem. Biophys. Res. Commun., 56:206-213 (1974).
54. B. Griniuviene, V. Chmieliauskaite, V. Melvydas, P. Dzheja, and L. Grinius, Bioenergetics 7:17-28 (1975).
55. B. Griniuviene, P. Dzheja, and L. Grinius, Biochem. Biophys. Res. Commun., 64:790-796 (1975).
56. B. A. Haddock and M. W. Kendall-Tobias, Biochem. J., 152:655-659 (1975).

57. Y. S. Halpern, H. Barash, S. Dover, and K. Druck, J. Bacteriol., 114:53-58 (1973).
58. W. A. Hamilton, Adv. Microb. Physiol., 12:1-53 (1975).
59. F. M. Harold, Bacteriol. Rev., 36:172-230 (1972).
60. F. M. Harold, Curr. Top. Bioenerg., 6:84-149 (1977).
61. F. M. Harold, in, The Bacteria, Vol. VI (J. R. Sokatch and L. N. Ornston, eds.), Academic, New York, in press.
62. F. M. Harold and K. Altendorf, Curr. Top. Membr. Trans., 5:1-50 (1974).
63. F. M. Harold and E. Levin, J. Bacteriol., 117:1141-1148 (1974).
64. F. M. Harold and D. J. Papineau, J. Membr. Biol., 8:27-44 (1972).
65. F. M. Harold and D. J. Papineau, J. Membr. Biol., 8:45-62 (1972).
66. F. M. Harold, E. Pavlasova, and J. R. Baarda, Biochim. Biophys. Acta, 196:235-244 (1970).
67. F. M. Harold and E. Spitz, J. Bacteriol., 122:266-277 (1975).
68. K. Hayakawa, T. Ueda, I. Kusaka, and S. Fukui, Biochem. Biophys. Res. Commun., 72:1548-1553 (1976).
69. E. Heinz, P. Geck, and C. Pietrzyk, Ann. N.Y. Acad. Sci., 264: 428-441 (1975).
70. S. L. Helgerson and W. A. Cramer, Biochemistry 16:4109-4117 (1977).
71. P. J. F. Henderson, in Comparative Biochemistry and Physiology of Transport (L. Bolis, K. Block, S. E. Luria, and F. Lynen, eds.), North-Holland, Amsterdam, 1974, pp. 409-424.
72. P. J. F. Henderson and H. L. Kornberg, in Energy Transformation in Biological Systems (Ciba Foundation Symposium no. 31), Elsevier, Amsterdam, 1975, pp. 243-269.
73. P. J. F. Henderson and A. Skinner, Trans. Biochem. Soc., 2:543-545 (1974).
74. E. L. Hertzberg and P. C. Hinkle, Biochem. Biophys. Res. Commun., 58:178-184 (1974).
75. H. Hirata, K. Altendorf, and F. M. Harold, Proc. Natl. Acad. Sci. USA, 70:1804-1808 (1973).
76. H. Hirata, K. Altendorf, and F. M. Harold, J. Biol. Chem., 249: 2939-2945 (1974).
77. H. Hirata, F. C. Kosmakos, and A. F. Brodie, J. Biol. Chem., 249:6965-6970 (1974).
78. H. Hirata, N. Sone, M. Yoshida, and Y. Kagawa, J. Biochem. (Tokyo), 79:1157-1166 (1976).
79. H. Hirata, N. Sone, M. Yoshida, and Y. Kagawa, Biochem. Biophys. Res. Commun., 69:665-671 (1976).
80. P. Hoffee, E. Englesberg, and F. Lamy, Biochim. Biophys. Acta, 79:337-350 (1964).
81. J. F. Hoffman and P. C. Laris, J. Physiol. (Lond.), 239:519-552 (1974).

82. J. S. Hubbard, C. A. Rinehart, and R. A. Baker, J. Bacteriol., 125:181-190 (1976).

83. A. M. Jetten and M. E. R. Jetten, Biochim. Biophys. Acta, 287:12-22

84. A. M. Jetten and G. D. Vogel, Biochim. Biophys. Acta, 311:483-495 (1973).

85. H. R. Kaback, Fed. Proc., 19:130 (1960).

86. H. R. Kaback, J. Biol. Chem., 243:3711-3724 (1968).

87. H. R. Kaback, Meth. Enzymol., 22:99-120 (1971).

88. H. R. Kaback, Science, 186:882-892 (1974).

89. H. R. Kaback, J. Cell. Physiol., 89:575-593 (1976)

90. H. R. Kaback and E. M. Barnes, J. Biol. Chem., 246:5523-5531 (1971).

91. H. R. Kaback and L. S. Milner, Proc. Natl. Acad. Sci. USA, 66: 1008-1015 (1970).

92. H. R. Kaback, J. P. Reeves, S. A. Short and F. J. Lombardi, Arch. Biochem. Biophys., 160:215-222 (1974).

93. H. R. Kaback and E. R. Stadtman, Proc. Natl. Acad. Sci. USA, 55: 920-927 (1966).

94. R. J. Kadner and H. H. Winkler, J. Bacteriol., 123:985-991 (1975).

95. S. Kahane, M. Marcus, H. Barash, and Y. S. Halpern, FEBS Lett., 56:235-239 (1975).

96. E. K. Kashket and S. L. Barker, J. Bacteriol., 130:1017-1023 (1977).

97. E. R. Kashket and T. H. Wilson, J. Bacteriol., 109:784-789 (1972).

98. E. R. Kashket and T. H. Wilson, Biochem. Biophys. Res. Commun., 49:615-620 (1972).

99. E. R. Kashket and T. H. Wilson, Proc. Natl. Acad. Sci. USA, 70: 2866-2869 (1973).

100. E. R. Kashket and T. H. Wilson, Biochem. Biophys. Res. Commun., 59:879-886 (1974).

101. E. R. Kashket and T. H. Wilson, unpublished observations.

102. T. Kawasaki and Y. Kayama, Biochem. Biophys. Res. Commun., 55:52-59 (1973).

103. Y. Kayama and T. Kawasaki, J. Bacteriol., 128:157-164 (1976).

104. M. B. Kemp, B. A. Haddock, and P. B. Garland, Biochem. J., 148:329-333 (1975).

105. A. Kepes and G. N. Cohen, in The Bacteria, Vol. IV: The Physiology of Growth (I. C. Gunsalus and R. V. Stanier, eds.), Academic, New York, 1962, pp. 179-221.

106. J. E. Knepper and J. E. Lusk, J. Biol. Chem., 251:7577-7580 (1976).

107. H. Kobayashi, E. Kin, and Y. Anraku, J. Biochem. (Tokyo), 76:251-261 (1974).

108. A. L. Koch, Biochim. Biophys. Acta, 79:177-200 (1964).

109. A. L. Koch, J. Mol. Biol., 59:447-459 (1971).

110. W. N. Konings and H. R. Kaback, Proc. Natl. Acad. Sci. USA, 70:
     3376-3381 (1973).
111. A. L. Kopecky, D. P. Copeland, and J. E. Lusk, Proc. Natl. Acad.
     Sci. USA, 72:4631-4634 (1975).
112. W. Kundig, S. Ghosh, and S. Roseman, Proc. Natl. Acad. Sci. USA,
     52:1067-1074 (1964).
113. W. Kundig, F. D. Kundig, B. Anderson, and S. Roseman, J. Biol.
     Chem., 241:3234-3245 (1965).
114. A. E. Lagarde and B. A. Haddock, Biochem. J., 162:183-187 (1977).
115. J. K. Lanyi and R. E. MacDonald, Biochemistry 15:4608-4614
     (1976).
116. J. K. Lanyi, R. Renthal, and R. E. MacDonald, Biochemistry 15:
     1603-1609 (1976).
117. J. K. Lanyi, V. Yearwood-Drayton, and R. E. MacDonald, Biochem-
     istry 15:1595-1602 (1976).
118. P. C. Laris and H. A. Pershadsingh, Biochem. Biophys. Res.
     Commun., 57:620-626 (1974).
119. H. G. Lawford, J. C. Cox, P. B. Garland, and B. A. Haddock,
     FEBS Lett., 64:369-374 (1976).
120. C. P. Lee, Biochemistry 10:4375-4381 (1971).
121. E. A. Lieberman and V. P. Skulachev, Biochim. Biophys. Acta, 216:
     30-42 (1970).
122. M. A. Liberman and J-S. Hong, Proc. Natl. Acad. Sci. USA, 81:
     4395-4399 (1974).
123. M. A. Lieberman and J-S. Hong, J. Bacteriol., 125:1024-1031 (1976).
124. M. A. Lieberman and J-S. Hong, Arch. Biochem. Biophys., 172:312-
     315 (1976).
125. F. Lipmann, Adv. Enzymol., 1:99-162 (1941).
126. T. C. Y. Lo, M. K. Rayman, and B. D. Sanwal, Can. J. Biochem.,
     52:854-866 (1974).
127. T. C. Y. Lo and B. D. Sanwal, Biochem. Biophys. Res. Commun.,
     63:278-285 (1975).
128. T. C. Y. Lo and B. D. Sanwal, J. Biol. Chem., 250:1600-1602
     (1975).
129. T. C. Y. Lo and B. D. Sanwal, Mol. Gen. Genet., 140:303-307
     (1975).
130. S. E. Luria, in Bacterial Membranes and Walls (L. Lieve, ed.),
     Dekker, New York, 1973, pp. 293-320.
131. J. E. Lusk and D. L. Nelson, J. Bacteriol., 112:148-160 (1972).
132. R. E. MacDonald and J. K. Lanyi, Biochemistry 14:2882-2889
     (1975).
133. R. A. MacLeod, P. Thurman, and H. F. Rogers, J. Bacteriol.,
     113:329-340 (1973).
134. P. C. Maloney, E. R. Kashket, and T. H. Wilson, Meth. Membr.
     Biol., 5:1-49 (1975).

135. J. A. Manno and D. Schaechter, J. Biol. Chem., 245:1217-1223 (1970).
136. A. Martonosi, Curr. Top. Membr. Trans., 3:83-197 (1972).
137. H. Michel and D. Oesterhelt, FEBS Lett., 65:175-178 (1976).
138. K. M. Miner and L. Frank, J. Bacteriol., 117:1093-1098 (1974).
139. P. Mitchell, J. Gen. Microbiol., 9:273-287 (1953).
140. P. Mitchell, J. Gen. Microbiol., 11:X (1954).
141. P. Mitchell, J. Gen. Microbiol., 11:73-82 (1954).
142. P. Mitchell, Symp. Soc. Exp. Biol., 8:254-261 (1954).
143. P. Mitchell, Nature, 180:134-136 (1957).
144. P. Mitchell, Nature, 191:144-148 (1961).
145. P. Mitchell, Biochem. J., 81:24P (1961).
146. P. Mitchell, Biochem. Soc. Symp., 22:142-169 (1963).
147. P. Mitchell, Biol. Rev., 41:445-502 (1966).
148. P. Mitchell, Nature, 214:400 (1967).
149. P. Mitchell, Comp. Biochem., 22:167-197 (1967).
150. P. Mitchell, Bioenergetics, 3:5-24 (1972).
151. P. Mitchell, Biochem. Soc. Trans., 4:399-430 (1976).
152. P. Mitchell and J. M. Moyle, J. Gen. Microbiol., 9:257-272 (1953).
153. P. Mitchell and J. Moyle, Farad. Soc. Discuss., 21:258-283 (1956).
154. P. Mitchell and J. Moyle, Biochem. J., 104:588-600 (1967).
155. P. Mitchell and J. Moyle, Eur. J. Biochem., 7:471-484 (1967).
156. P. Mitchell and J. Moyle, Eur. J. Biochem., 9:149-155 (1969).
157. H. C. Neu and L. A. Heppel, J. Biol. Chem., 240:3685-3692 (1965).
158. D. Nieva-Gomez, J. Konisky, and R. B. Gennis, Biochemistry 15: 2747-2753 (1976).
159. D. F. Niven and W. A. Hamilton, FEBS Lett., 37:244-248 (1973).
160. D. F. Niven and W. A. Hamilton, Eur. J. Biochem., 44:517-522 (1974).
161. D. F. Niven, R. E. Jeacocke, and W. A. Hamilton, FEBS Lett., 29:248-252 (1973).
162. C. P. Novotny and E. Englesberg, Biochim. Biophys. Acta, 117: 217-230 (1966).
163. D. Oesterhelt and W. Stoeckenius, Proc. Natl. Acad. Sci. USA, 70:2853-2857 (1973).
164. E. Padan, D. Zilberstein, and H. Rottenberg, Eur. J. Biochem., 63:533-541 (1976).
165. R. Panet and D. R. Sanadi, Curr. Top. Membr. Trans., 8:99-160 (1976).
166. E. Pavlasova and F. M. Harold, J. Bacteriol., 98:198-204 (1969).
167. S. K. Phillips and W. A. Cramer, Biochemistry 12:1170-1176 (1973).
168. U. Pick and M. Avron, Biochim. Biophys. Acta, 440:189-204 (1976).
169. C. A. Plate, Antimicrob. Ag. Chemother., 4:16-24 (1973).
170. C. A. Plate, J. Bacteriol., 125:467-474 (1976).

171. C. A. Plate and S. E. Luria, Proc. Natl. Acad. Sci. USA, 69:2030-2034 (1972).
172. C. A. Plate, J. L. Suit, A. M. Jetten, and S. E. Luria, J. Biol. Chem., 249:6138-6143 (1974).
173. R. Prasad, V. K. Kalra, and A. F. Brodie, J. Biol. Chem., 251:2493-2498 (1976).
174. D. R. Purdy and A. L. Koch, J. Bacteriol., 127:1188-1196 (1976).
175. E. Racker, J. Biol. Chem., 247:8198-8200 (1972).
176. E. Racker and L. W. Fisher, Biochem. Biophys. Res. Commun., 67:1144-1150 (1975).
177. E. Racker and W. Stoeckenius, J. Biol. Chem., 249:662-662 (1974).
178. A. S. Rae and K. P. Strickland, Biochem. Biophys. Res. Commun., 62:568-576 (1975).
179. A. S. Rae and K. P. Strickland, Biochem. Biophys. Acta, 433:564-582 (1976).
180. S. Ramos, S. Schuldiner, and H. R. Kaback, Proc. Natl. Acad. Sci. USA, 73:1892-1896 (1976).
181. J. P. Reeves, F. J. Lombardi, and H. R. Kaback, J. Biol. Chem., 247:6204-6211 (1972).
182. R. Renthal and J. K. Lanyi, Biochemistry 15:2136-2143 (1976).
183. D. B. Rhoads and W. Epstein, Fed. Proc., 35:1168 (1976).
184. D. B. Rhoads, F. B. Waters, and W. Epstein, J. Gen. Physiol., 67:325-342 (1976).
185. R. B. Roberts and I. Z. Roberts, J. Cell Comp. Physiol., 36:15-39 (1950).
186. A. Robin and A. Kepes, FEBS Lett., 36:133-136 (1973).
187. B. P. Rosen, Biochem. Biophys. Res. Commun., 53:1289-1296 (1973).
188. B. P. Rosen, J. Bacteriol., 116:1124-1129 (1973).
189. B. P. Rosen, Fed. Proc., 35:77 (1976).
190. B. P. Rosen, unpublished observations (1976).
191. B. P. Rosen and L. W. Adler, Biochim. Biophys. Acta, 387:23-36 (1975).
192. B. P. Rosen and J. S. McClees, Proc. Natl. Acad. Sci. USA, 71:5042-5046 (1974).
193. H. Rottenberg, Bioenergetics 7:61-74 (1975).
194. H. Rottenberg, FEBS Lett., 66:159-163 (1976).
195. H. Rottenberg and C-P. Lee, Biochemistry, 14:2675-2680 (1975).
196. V. Schairer, P. Friedl, B. I. Schmid, and G. Vogel, Eur. J. Biochem., 66:257-268 (1976).
197. H. V. Schairer and B. A. Haddock, Biochem. Biophys. Res. Commun., 48:544-551 (1972).
198. P. Scholes and P. Mitchell, J. Bioenerg., 1:61-72 (1970).
199. P. Scholes and P. Mitchell, Bioenergetics, 1:309-323 (1970).
200. S. Shuldiner and H. R. Kaback, Biochemistry 14:2675-2680 (1975).

201. S. Schuldiner, G. K. Kewar, H. R. Kaback, and R. Weil, J. Biol. Chem., 250:1361-1370 (1975).
202. S. Schuldiner, H. Kung, H. R. Kaback, and R. Weil, J. Biol. Chem., 250:3679-3682 (1975).
203. S. Schuldiner, R. D. Spencer, G. Weber, R. Weil, and H. R. Kaback, J. Biol. Chem., 250:8893-8896 (1975).
204. S. G. Schultz and P. F. Curran, Physiol. Rev., 50:637-718 (1970).
205. A. Schwartz, G. E. Lindenmayer, and J. C. Allen, Curr. Top. Membr. Trans., 3:1-82 (1972).
206. T. J. Silhavy, I. Hartig-Beecken, and W. Boos, J. Bacteriol., 126:951-958 (1976).
207. R. D. Simoni and P. W. Postma, Ann. Rev. Biochem., 43:523-554 (1975).
208. R. D. Simoni and M. K. Shallenberg, Proc. Natl. Acad. Sci. USA, 69:2663-2667 (1972).
209. P. J. Sims, A. S. Waggoner, C-H. Wang, and J. F. Hoffman, Biochemistry 13:3315-3330 (1974).
210. A. P. Singh and P. D. Bragg, Eur. J. Biochem., 67:177-186 (1976).
211. V. P. Skulachev, Curr. Top. Bioenerg., 4:127-190 (1971).
212. N. Sone, M. Yoshida, H. Hirata, and Y. Kagawa, J. Biol. Chem., 250:7917-7923 (1975).
213. J. Stock and S. Roseman, Biochem. Biophys. Res. Commun., 44:132-138 (1971).
214. S. Szmelcman and J. Adler, Proc. Natl. Acad. Sci. USA, 73:4387-4391 (1976).
215. J. Thompson, J. Bacteriol., 127:719-730 (1976).
216. J. Thompson and R. A. MacLeod, J. Biol. Chem., 246:4066-4074 (1971).
217. J. Thompson and R. A. MacLeod, J. Bacteriol., 120:598-603 (1974).
218. T. Tsuchiya, J. Biol. Chem., 251:5315-5320 (1976).
219. T. Tsuchiya and B. P. Rosen, J. Biol. Chem., 250:7687-7692 (1975).
220. T. Tsuchiya and B. P. Rosen, J. Biol. Chem., 250:8409-8415 (1975).
221. T. Tsuchiya and B. P. Rosen, J. Biol. Chem., 251:962-967 (1976).
222. W. J. Waddell and T. C. Butler, J. Clin. Invest., 38:720-729 (1959).
223. A. Waggoner, J. Membr. Biol., 27:317-334 (1976).
224. L. Wendt, J. Bacteriol., 104:1236-1241 (1970).
225. I. C. West, Biochem. Biophys. Res. Commun., 41:655-661 (1970).
226. I. West and P. Mitchell, Bioenergetics 3:445-462 (1972).
227. I. West and P. Mitchell, Biochem. J., 132:587-592 (1973).
228. I. West and P. Mitchell, Biochem. J., 144:87-90 (1974).
229. I. West and T. H. Wilson, Biochem. Biophys. Res. Commun., 50:551-558 (1973).
230. D. B. Wilson, J. Bacteriol., 120:866-871 (1974).
231. H. H. Winkler, J. Bacteriol., 116:203-209 (1973).
232. J. M. Wood, J. Biol. Chem., 250:4477-4485 (1975).

233. M. Yoshida, N. Sone, H. Hirata, and Y. Kagawa, Biochem. Biophys. Res. Commun., 67:1295-1300 (1975).

234. M. H. Zarlengo and S. G. Schultz, Biochim. Biophys. Acta, 126: 308-320 (1966).

235. F. M. Harold and J. Van Brunt, Science, 197:372-373 (1977).

236. R. E. MacDonald, J. K. Lanyi, and R. V. Greene, Proc. Natl. Acad. Sci. USA, 74:3167-3170 (1977).

237. T. Tsuchiya, S. M. Hasan, and J. J. Raven, J. Bacteriol., 131:848-853 (1977).

238. S. M. Hasan and T. Tsuchiya, Biochem. Biophys. Res. Commun., 78:122-128 (1977).

239. T. Tsuchiya, J. Raven, and T. H. Wilson, Biochem. Biophys. Res. Commun., 76:26-31 (1977).

240. H. Tokuda and H. R. Kaback, Biochemistry, 16:2130-2136 (1977).

241. H. Tokuda and H. R. Kaback, Biochemistry 17:698-705 (1978).

242. J. W. Belliveau and J. K. Lanyi, Arch. Biochem. Biophys., 186: 98-105 (1978).

Numbers in brackets are reference numbers and indicate that an author's work is referred to although his name is not cited in the text. Underlined numbers give the page on which the complete reference is listed.

A

Abelson, P. H., 1[3], 2[3], 5
Abelson, R., 93[127], 102
Abramova, N., 485[139], 492
Abramowitz, A., 188[145], 216
Abrams, A., 236[98], 276[98],
    282, 288[97], 315, 324,
    499[18], 500[18], 504[18,
    47,48], 507[62], 510[76],
    511[18,48], 516[98],
    517[18,47], 518[47],
    519[114], 548, 549, 550,
    551, 552
Abu-Sabe, M., 159[252,253],
    169
Adams, E., 188[152], 217
Addison, J. M., 364[1,2,3,4],
    369
Adler, J., 142[121], 144[135,
    136], 148[129,130],158[136],
    165, 520[119,120], 521[121,
    122], 528[161], 541[220,
    223,224], 542[226], 543[236],
    552, 553, 555, 556, 582[214],

[Adler, J.]
    608[214], 609[1], 612, 619
Adler, L. W., 25, 28[190],
    33[190], 40, 267[1], 279[175],
    308, 320, 578[191], 582[191],
    591, 605[2,3], 612, 618
Adolfsen, R., 504[54,55], 549
Ailhaud, G., 84[105], 102,
    394[47], 410
Aithol, H. N., 32[205], 40
Ajl, S. J., 393[41], 394[41],
    409
Aksamit, R. R., 144[137],
    145[137,147], 159[137,147],
    165, 531[234], 542[234], 556
Albendea, J. A., 504[49], 549
Albert, A., 484[131], 492
Albright, F. R., 424[84], 457
Alderete, J. F., 12[61], 35
Alemohammad, M. M., 451[220],
    462
Alexander, J. K., 58[65], 66[65],
    100, 129[11], 161
Alexandrov, S. L., 195[193],
    209[193], 218
Allen, J. C., 566[205], 619

Altekar, W. W., 405[95], <u>411</u>
Altendorf, K. H., 18[136,141],
    20[136], 23, 27[192], 28[141],
    32, <u>38</u>, <u>40</u>, 137[71], 150[198],
    156[198], <u>163</u>, <u>167</u>, 201[209,
    210], <u>219</u>, 227[90], 231[3],
    232[3,105], 234[90], 235[90],
    236[90], 237[90], 255[90],
    268[90], 274[90], 275[90],
    276[90], 277[90], 279[2,4,90],
    280[90], 296[90], <u>308</u>, <u>314</u>, <u>315</u>,
    516[92,94], 517[94], 519[94],
    526[148], <u>551</u>, <u>553</u>, 563[5],
    582[5,75], 587[75,76], 588[75,
    76], 592[62], 596[5,75], 607[4],
    608[5,75], <u>612</u>, <u>614</u>
Amanuma, H., 13[72], 33[213],
    <u>36</u>, <u>41</u>, 175[42,44], 181[42,
    83,85,92], 182[85], 183[103],
    194[85,92], 195[42], 196[103],
    197[103], 198[44], 199[83,85],
    200[42,83], 201[92], 202[85],
    203[44], 205[85], 206[85],
    207[85,92], 210[42,85,92], <u>213</u>,
    <u>214</u>, <u>215</u>, <u>218</u>, 533[182,185],
    541[185], 547[249], <u>554</u>, <u>556</u>,
    610, <u>612</u>
Amaral, D., 109[40], 116[40],
    <u>124</u>
Amelunxen, R. E., 514[82], <u>551</u>
Ames, B. N., 244[128], 251[128],
    <u>317</u>, 327[5], 332[5], 333[5],
    341[5], 342[5], 357[5], 358,
    367, <u>369</u>, 431[127], 434[127],
    435[142], <u>459</u>, 483[116], <u>491</u>,
Ames, G. F.-L., 9[27,28], 10[27,
    28,48,51], 15, <u>34</u>, <u>35</u>, <u>37</u>,
    146[165], 148[165,172,174],
    <u>166</u>, 189[162,163,164,165,166,
    167,168], 190[167,168],
    196[162], 201[163,164,165,168],
    210, <u>217</u>, 327[5], 332[5,81],

[Ames, G. F.-L.]
    333[5], 341[5], 342[5], 357[5],
    358[5], 367[5], <u>369</u>, <u>374</u>, 416[24,
    25], 420[49], <u>455</u>, <u>456</u>, 528[158],
    530[186], 533[158], 534[186,188,
    189], <u>553</u>, <u>554</u>
Amoore, J. E., 14[91], <u>36</u>
Anand, N., 486[145], <u>492</u>
Andeeva, S., 64[55], 66[55], <u>100</u>
Anderson, A., 159[251], <u>169</u>
Anderson, B., 49, 63[47], <u>99</u>, <u>100</u>,
    172[5], <u>212</u>, 575[113], <u>616</u>
Anderson, J. J., 179[88], 181[88],
    182[88], 203[88], <u>215</u>
Anderson, R., 50[23], 53[23], 65[59,
    60,61], 67, 69[60,76,77,78], 70,
    73[77], 77, 78[60], <u>99</u>, <u>100</u>, <u>101</u>
Andreesen, J. R., 300[5], <u>309</u>
Andreoli, A. J., 479[90], <u>490</u>
Andreu, J. M., 504[49], 514[81],
    <u>549</u>, <u>551</u>
Andry, K., 486[149], <u>492</u>
Angelmaier, D., 85[111], <u>102</u>
Anraku, Y., 9[20,22], 10[38,45],
    11[56], 12[20,22,38], 13[20,22
    72], 14[22,45,80], 16[115],
    20[115], 30[115], 33[213], <u>34</u>, <u>35</u>,
    <u>36</u>, <u>37</u>, 141[112], 143, 146[123],
    <u>164</u>, <u>165</u>, 172[7,8], 174[7,8,35,
    37], 175[42,44], 177[59], 178[72],
    179[72,102], 181[35,42,72,75,78,
    79,83,84,85,90,91,92], 182[72,
    85,90], 183[102,103], 184[102],
    192[171], 193[90,171,172,173,174],
    194[7,8,35,37,85,92,191], 195[42,
    199], 196[75,102,103], 197[103],
    198[174,203,206], 199[75,83,85,
    90,207], 200[42,78], 201[75,90,92,
    211], 202[85,90], 203[44,90],
    205[85], 206[84], 207[35,85,90,
    92], 210[42,90,92,199], <u>212</u>, <u>213</u>,
    <u>214</u>, <u>215</u>, <u>217</u>, <u>218</u>, <u>219</u>, 517[101,

[Anraku, Y.]
102,108], 518[101,102],
530[174,210], 532[174,175],
533[182,185], 539[210],
541[185,225], 547[249], 551,
554, 555, 556, 600[107],
610[6], 612, 615
Antissier, F., 156[233], 168
Antonov, V. K., 195[193],
209[193], 218
Aono, H., 532[170], 554
Archer, E. G., 332[179], 338[179],
343[179], 346[179], 365[179],
381
Ariel, M., 592[7], 612
Arima, K., 484[123,125], 485[136,
137], 492
Armitage, A. K., 486[145], 492
Armstrong, G. B., 175[51], 213
Armstrong, J. B., 133[48],
136[48], 162
Aronson, P. S., 2[10], 5
Artman, M., 486[151], 492
Asano, A., 21[162,164,165],
31[165], 32[205], 39, 40
Asatoor, A. M., 326[115], 376
Asensio, J., 307
Asghar, S. S., 588[8], 612
Ash, D. E., 306
Ashgar, S. S., 187[138], 216
Ashwell, G., 386[6], 408
Atagün, M., 483[113], 491
Atkinson, D. E., [115], 316
Autissier, F., 387[14], 388[14],
408
Avi-Dor, Y., 13[69], 24[69], 36
Avron, M., 581[168], 617
Avtalion, R. R., 485[143], 492
Awazu, S., 183[103], 196[103],
197, 215
Ayling, P. D., 188, 216
Azam, F., 260[6], 301[6], 309

B

Baarda, J. R., 234[91,93,94,96],
235[91], 236[91,96,98], 276[96,
98], 279[95,96], 288[92], 314,
315, 516[98], 551, 567[66],
579[66], 596[66], 605[66], 614
Babcock, D., 293[20], 310
Bachi, B., 387[19], 409
Bachman, E., 483[115], 491
Bachmann, B. F., 426[98], 433[98],
446[98], 458
Bachmann, B. J., 139[90], 155[90],
164, [222], 219, 225[7], 226,
300[7], 309
Bachmann, R., 431[163], 460
Baird, B. A., 508[63], 513[63],
550
Baker, R. A., 569[15], 593[82],
600[82], 612, 615
Baker, R. E., 21[165], 31[165], 39
Ball, E., 340[64], 373
Baltscheffsky, H., 504[56], 550
Baltscheffsky, M., 504[56,57], 550
Barak, Z., 326[7], 327[6,7], 332[7],
333[6,9], 338[7], 341[6,8], 342[6,
8], 347[7], 353[8], 363, 367[6],
369, 421, 457
Barash, H., 146[162], 166, 186,
201[135], 216, 277[113], 302[84],
314, 316, 536[195,196], 555,
592[57,95], 593[57], 614, 615
Barker, H. A., 467[16], 488
Barlow, V., 504[51,52], 549
Barnes, E. M., Jr., 17[122,128],
18[139], 37, 38, 129[17], 136[65,
66], 137[65], 157[241], 161, 163,
169, 175[49,50], 193[50], 213,
269[15], 270, 309, 520, 545[247],
552, 556, 562[9,10,90], 563[9,10,
90], 564[10,90], 576[10], 589[14],
591[14], 612, 615

Baron, C., 503[42], 510[76], 516[98], 519[114], 549, 550, 551, 552

Barran, L. R., 135[57, 58], 162

Barrett, C., 424[79], 457

Barrett, J. T., 386[2], 401[2], 405[2], 408

Bartley, W., 14[90], 36

Bassford, P. J., Jr., [88], 444[189], 445[189, 192, 193], 446[189, 194], 447[192], 448[194, 201, 202], 449[201], 450[194], 453[189, 192], 454[193], 457, 461, 467[14], 488

Baugh, C. M., 473[49], 474[59], 489

Baumann, L., 129[13], 161

Baumann, P., 129[13], 161

Bayer, E. A., 527[155], 553

Bayer, M. E., 414[6], 447[203], 451, 455, 461

Bayliss, M. E., 334[201], 382

Beacham, I. R., 424[82, 83], 457

Beacham, S., 94, 102

Beamer, K. C., 332[103, 214], 338[103, 214], 343[103, 214], 359[215], 375, 383

Beauchamp, R. S., 241[8], 309

Bechet, J., 400[75], 410

Beck, C., 93, 94, 102

Beck, W. S., 467[20, 21], 488

Becker, J. M., 331[69], 332[119], 333[11, , 66, 118], 336[10, 66, 119], 338[66], 341[118], 342[118], 343[66], 344[66], 348[11, 66, 119], 349[10, 119], 350[66, 119], 352[10, 66, 119], 353[10, 11, 66, 118, 119], 355, 357[11, 119], 358[119], 360[11], 362, 366[10], 367[66], 368[119], 369, 373, 376, 377, 480, 491

Beckwith, J. R., 140, 148[176], 164, 166

Beechey, R. B., 502, 517[99], 549, 551

Beever, R. E., 307

Bekkonzjin, A. G., 130[28], 161

Belaich, A., 160[256], 169

Belaich, J.-P., 160[256], 169

Belet, M., 130[24], 161

Bell, R. M., 286[203], 287[203], 321

Belliveau, J. W., 611, 620

Benedetto, J.-P., [29], 456

Benedict, C. D., 263[63], 313

Benemann, J. R., 301[9], 309

Bennett, R. L., 226[10], 284[10], 285[10, 11, 229], 286[10, 11, 229], 287[10, 229], 309, 323, 529[167], 554

Bennett, W., 85, 86[112], 102, 389[46], 394[46], 409

Berg, C. M., 177[64], 178[64], 214

Berg, H. C., 146[151], 165

Berger, A., 334[161, 162], 336[213], 379, 383

Berger, E. A., 4[18], 5, 10, 11, 12, 13[63], 35, 107[30, 31], 120, 123, 147[170], 150[197], 154[217], 156[197], 158[217], 166, 167, 168, 180[115], 184, 194, 215, 218, 232[12], 255[12], 268[12], 286[12], 309, 334[58], 345[58], 372, 520[117], 530[199], 536[199], 537[202], 545[202], 552, 555, 576[12], 577, 589[11], 599, 600[12, 13], 606[12, 13], 612,

Bergstein, P. E., 118[63], 125

Berlin, R., 90, 102

Berman, K., 532[178], 554

Berman, R. H., 129[13], 161

Berman, S. M., 129[13], 161

Berman-Kurtz, M., 131[32], 161

Betz, H., 336[62], 373

Bhattacharyya, P., 227[196], 231[13], 242[196], 256[196], 257[196], 260[13, 14], 269[15], 270, 271[196], 309, 321, 589[14], 591[14], 612

Bhumiratana, A., 69[78], 101

Bielig, H.-J., 301[137], 317

Bigger, L. C., 592[31], 613
Bihler, I., 121, 122[80], 125
Birdsell, D. C., 17, 37
Bisschop, A., 19[143], 23[143],
  30[143,202], 38, 40, 396[71],
  398[71], 399[71], 410
Bjare, U., 334[12], 370
Blackburn, T. H., 333[63], 338[63],
  373
Blakley, R. L., 467[16], [17], 488
Blaurock, A. E., 278[17], 309,
  516[88], 551
Blondin, G., 232[79], 314
Bloom, G. D., 416[19], 421[64],
  455, 457
Boardman, N. K., 280[172], 319
Bockrath, R. C., 486[149], 492
Boekhout, M., 415[16], 455
Bogin, E., 504[58], 550
Bogomolni, R. A., 278[125,127],
  316, 317, 569[15], 612
Boguslavsky, L. I., 279[18], 309
Bolton, E. T., 1[3], 2[3], 5,
  93[127], 102
Boniface, J., 110[48,49], 112, 124
Boonstra, J., 17[129], 30[202], 38,
  40, 193[176], 217, 562[16],
  593[16], 612
Boos, W., 8[8,9], 9[8,9], 10[50,
  52], 14[82], 34, 35, 36, 106[26],
  123, 134[51], 139[83,84],
  142[51], 143[125,126,127,128],
  144[127,128,131,132,140,141,
  254], 145[127,140,149,150],
  146, 147[84], 148[131,132],
  149[140], 150[198,199], 151[153],
  153[153], 154[216], 156[198,199],
  157[240], 158[83,124,125,140,
  153,216], 162, 163, 165, 166,
  167, 168, 169, 172[19], 175[41],
  195[41,194], 201[214,215],
  209[216,217,218], 210[220], 212,
  213, 218, 219, 428[114], 458,
  496[5], 497[5], 528[159],
  530[211], 539[211,212], 540[211,

[Boos, W.]
  213], 541[215,216,218,219,221],
  [213], 545[244], 548, 553, 555,
  556, 601[206], 619
Booth, I. R., 589[17], 612
Borst-Pauwels, G. W. F. H., 196,
  218
Böszörményi, Z., 177[58], 214
Boulouis, D., 504[59], 550
Bourd, G. I., 64, 66[55], 100,
  106[11], 123
Boutet, A., 336[94], 375
Bowen, D. O., 338[216], 342[216],
  343[216], 365[216], 368[216], 383
Boxer, D. H., 294[19], 310, 568[18],
  612
Boyer, P. D., 119[57], 124, 336[13],
  370, 519, 532[178], 552, 554,
  568[19], 612
Bradbeer, C., 9[29], 34, 149[184,
  185], 167, 265[44], 311, 442[176,
  179,180,183], 443[176,179,180,
  183,185], 444[190], 445[185,190,
  193], 454[193], 460, 461, 466[7],
  467[7,11,14], [13], 487, 488,
  544[240], 556
Bradfield, G., 293[20], 310
Bradley, D., 293[20], 310
Bragg, P. D., 25[185], 40, 120, 125,
  423[76], 457, 500[26], 503[41],
  504[26,41], 507[46], 508[26,46],
  510[75], 513, 517[41,100], 518[41,
  100], 519[75], 524[139], 533[46],
  548, 549, 550, 551, 553, 582[210],
  619
Brand, M. D., 14[89], 36
Braun, V., 9[32], 35, 149[187], 167,
  226[85], 244[86], 248[21,85,86,87],
  249[21,85,86,87], 251[87],
  252[85], 310, 314, 414[7], 415[3,
  9,10,12], 422, 423[71], 431[125],
  432[129], 433[136,160,161,162],
  434[136], 438[136,162], 439[136],
  440, 441[171], 455, 457, 459,
  460

Brazenaite, J., [52], 613
Breitman, T. R., 181[73], 214
Brenner, M., 419[37], 456
Breton, A., 308
Brewer, G. J., 604, 605, 612
Bridgeland, E. S., 188, 216
Briggs, T., 389[32], 392[32], 409
Britten, R. J., 1[3], 2[3], 5
Broach, J., 281[22], 310
Brock, M., 415[16], 455
Brock, T. D., 332[15], 338[15],
    343[15], 347[14,15], 370
Brockman, R. W., 24[182], 40
Broda, E., 262[28], 310
Brodie, A. F., 16[114], 19[151],
    21[114], 31[163,165], 32, 37,
    38, 39, 40, 504[58], 547[250],
    550, 556, 589[173], 592[77],
    614, 618
Bromberg, F. G., 106[25], 123,
    139[86], 195[86], 163
Bromwell, K., 188[145,146], 216,
    530[208], 538[208], 539[209],
    555
Bronner, F., 266[24], 267[24],
    268[78], 270, 271, 274[24], 310,
    314
Brooke, M. S., 149[190], 167
Brooks, J. C., 499[16], 548
Brot, N., 246, 310, 432[144],
    435[145], 459
Brown, C. E., 138[74], 144[74],
    146[74], 155[221], 163, 168
Brown, C. M., 294[25], 310
Brown, D. A., 146[151], 165
Brown, J. L., 336[16,67], 370,
    373
Brown, J. P., 473[55], 489
Brown, K. A., 251[26], 310
Brown, K. D., 179[95,96], 183,
    215
Brown, T. D. K., 388[23], 389[23],
    390[23], 393[23], 409
Brucker, H., 334[25], 370
Bruneteau, M., [29], 456

Bryan, L. E., 486[156,157,158],
    487[156,157,158], 493
Bryant, M. P., 330[153], 332[153,
    154], 338[154], 342[154], 343[154],
    379
Bryce, G. F., 246, 310, 432[144],
    435[145], 459
Bucheder, F., 262[28], 310
Budd, K., 263[163], 297[139,140,141,
    142], 298[139,140,141], 317, 318,
    319
Budreau, A., 177[62], 214
Buehring, K. U., 473[56], 489
Bukland, N. F., 333[17], 348[17],
    370
Bulman, 93, 94, 102
Bulmer, G. S., 525[144], 553
Burgess, A. H., 524[140], 553
Burman, L. G., 421[64], 457
Burnell, J. N., 30[203], 40, 289,
    290[29,30], 292[30], 310, 589[21,
    22], 591[21,22], 612
Burns, D. J. W., 307
Burns, R. O., 84[104], 101, 395[48],
    410
Burrous, S. E., 14[78], 36
Burston, D., 364[1,2,3,4], 369
Bursztyn, H., 419[43], 456
Burton, C., 282[83], 283[83], 314
Busuttil, J., 334[84], 374
Butler, T. C., 579[222], 580, 619
Buttin, G. N., 13[68], 15[100],
    24[68], 36, 37, 133[49], 139,
    140[96], 157[49], 158[96], 162,
    163, 164, 172[22], 212
Button, D. K., 69[74], 78[74], 100,
    129[15], 130[15], 161
Button, R., 93[127], 102
Buxbaum, L., 256[66], 258[66],
    259[66], 302[66], 313
Buxton, R. S., 444[186], 461
Byers, B. R., 244[31], 246, 251[31],
    310
Byrave, F. L., 14[99], 36

C

Cabantchik, Z. I., 298[32,179], 299[32,179], 311, 320

Caldwell, P. V., 355[200], 382

Cameron, M., 389[33], 393[33], 396[33], 402[33], 407[33], 409

Campbell, J. J. R., 386[1], 401[1], 408

Canovas, J. L., 129[13], 161

Cardelli, C., 433[165], 439[165], 460

Cardenas, J., 299[33], 311

Carillo-Castaneda, G., 395[56], 396[56], 410

Carls, R. A., 405[97], 411

Carreira, J., 514[81], 551

Carrol, R. C., 32[208], 40

Carson, J., 17[134], 20[134], 38, 415[14], 416[14], 431[14], 455

Carstensen, E. L., 223[132], 317

Carter, J. R., 2[11], 5, 133[45, 46], 134[46], 137[46], 157[46], 160[46], 162, 525[146,147], 553

Casadaban, M. J., 148[176], 166

Cascieri, T., 331[20,21], 332[20], 334[20,21], 336[20,21], 338[20], 340[18,19], 341[20], 342[20], 343[20], 345[20,21], 346[19,20], 349[20,21], 352[20,21], 353, 356[20,21], 357[20,21], 358[20, 21], 359[20,21], 360[20], 361[20,21], 362, 363, 364[20], 366[19,20], 367, 370

Cassidy, P. J., 483[109], 491

Cattell, K. J., 517[99], 551

Cavari, B. Z., 13[69], 24[69], 36

Cecchini, G., 15[104], 37, 151[207], 153[207], 167, 585, 612

Celis, T. F. R., 179[109], 184[109], 215

Center, M. S., 452[215], 462

Cha, Y.-A., 17[131], 38, 402[89], 404[89], 411

Chaiet, L., 150[195], 167

Chalumean, H., 129[7], 131[7], 161

Chance, B., 32[206], 40

Chandler, J. L. R., 479[89], 490

Chang, G. W., 442[177], 460, 466[3], 487

Chang, J. T., 442[177], 460, 466[3], 487

Chapman, V. A., 425[95], 458

Charmella, L. J., III, 177[64], 178[64], 214

Chatterjee, A. K., 420[52], 456

Chavin, S. I., 527[154], 553

Chegwidden, K., 226[177], 285[177], 286[177], 287[177], 289[177], 320

Chen, C., 14[89], 36

Cheng, K.-J., 414[5], 455

Chernyakhovskaya, M. Yu., 326[191], 381

Chipley, J. R., 263[34], 311, 530[228], 542[227,228], 556

Chlebowski, J. F., 306

Chopra, I., 485[140], 492

Chmieliauskaite, V., 568[53], 582[54], 608[54], 613

Chou, G., 174[33], 213

Choules, G. L., 333[22], 335[22], 370

Christensen, H. N., 121, 125, 173[24], 212, 327[23], 330[24], 343[24], 370

Christian, J. H. B., 296, 297[35], 311

Chung, A., 334[195], 382

Chused, T. M., 140[99], 149[99], 164

Cicmanec, J. F., 476[75], 490

Ciferri, O., 419[43], 456

Cintron, N. M., 479[91], 490

Cipolloni, P. B., 332[214], 338[214], 343[214], 383

Cirakoglu, C., 200[208], 218, 533[184], 554

Cirillo, V. P., 49, 50[17], 58[99], 64[16,17], 99, 101, 129[9], 130[9], 161, 223[36], 311

Clark, D., 239[190], 240, 241[190, 193], 302[190], 321

Clarke, P. H., 401[84], 405[84], 411

Claus, D. R., 146[167], 166, 181[74], 182[74], 196[74], 199[74], 200[74], 201[74], 202[74], 203[74], 207[74], 207[74], 214, 532[179], 533[179], 544

Claverie, J. M., 129[7], 131[7], 161

Clegg, R. A., 294[19], 310, 568[18], 612

Cleland, W., 63, 81[43], 100

Cobley, J. G., 265[37], 311

Coddington, A., 293[38], 295[38], 311

Cohen, B. L., 334[25], 370

Cohen, G. N., 2, 5, 15[100], 37, 133[49], 134[52], 136[52], 157[49,52], 162, 172[1,2,22,23], 173[1], 175[1], 181[23], 199, 212, 222[39], 311, 332[59], 337[72], 370, 373, 561[105], 615

Cohen, M. A., 542[229], 556

Cohen, N. S., 21[165], 31[165], 32[205], 39, 40, 547[250], 556

Cohn, M., 154[215], 168, 426[101], 458

Coleman, J. E., 306

Collins, E. B., 402[86], 404[86], 411

Collins, S. H., 479[99], 491, 572[25], 588[25], 596[24], 597[24], 598[25], 600[25], 609[25], 612

Colowick, S. P., 15[107], 37, 580, 608[26], 612

Cooper, B. A., 468[52], 473, 489

Cooper, R. A., 159[251], 169

Cooper, S., 179[101], 183[101], 215

Copeland, D. P., 604[111], 616

Cordaro, C., 46[10], 54[10], 64[48], 65, 98, 100

Cordaro, J. C., 50[22,23], 53[22, 23], 75, 99, 101, 483[113], 491

Corpe, W. A., 130[26], 161

Corwin, L. M., 187, 216

Cosloy, S. D., 177, 178[67], 214

Costerton, J. W., 414[5], 420[52], 455, 456

Costilow, R., 69[78], 101

Cota-Robles, E. H., 17, 37, 451, 461

Cowell, B. S., 433[164], 439[164], 460

Cowell, J. L., 10[41], 12[41], 35, 194[192], 218, 332[27], 338[27], 342, 343[27], 344, 345[29], 370, 600[27], 613

Cowie, D. B., 1[3], 2[3], 5, 93[127], 102

Cox, G. B., 251[149], 252[149], 279[40], 311, 318, 432[149], 434[135], 435[135,139,146], 436[149], 441[149], 459, 460, 519[113], 552, 560[28], 613

Cox, G. S., 193[175], 217

Cox, J. C., 568[119], 616

Cozzarelli, N. R., 150[192], 156[192], 167

Cramer, W. A., 569[29], 605, 606, 607[51], 613, 614, 617

Crane, F. L., 517[103], 518[103], 551

Crawford, E. J., 474[60], 490

Criddle, R. S., 301[107], 315

Cronan, J. E., Jr., 15[102], 37, 286[203], 287[203], 321

Crosby, L. K., 425[95,96], 458, 473[48], 489

Cross, R. L., 519[111], 552

Crowe, G. R., 561[46], 563[46], 613

Crumm, M., 474[60], <u>490</u>
Cseh, E., 177[58], <u>214</u>
Cuppoletti, J., 293[41], <u>311</u>
Curran, P. F., 583[204], 591[204], <u>619</u>
Curtis, S. J., 10[40], 12[40], <u>35</u>, 64, 78, <u>100</u>, 106[22], <u>123</u>, 154[214], 159[214], <u>168</u>, 600[30], <u>613</u>
Curtiss, R., 177[64], 178[64], <u>214</u>

D

Dagley, S., 401[82,83], <u>411</u>
Dahl, M. M., 543[236], <u>556</u>
Dailey, H. A., Jr., <u>306</u>
Dalen, A. B., 364, <u>370</u>
Dalmark, M., 298[80], 299[80], <u>314</u>
Dalrymple, J. A., 364[4], <u>369</u>
Damadian, R., 296, <u>311</u>
D'Ambrosio, S. M., 188[147,148], 196[147,148], <u>216</u>
Dancey, G. F., 523[134], 524[135], <u>552</u>
Dann, L., 432[154], 436[154], <u>460</u>
D'Aoust, J. Y., 135[57,58], <u>162</u>
Datta, A., 502[32], <u>549</u>
David, J., 159[250], <u>169</u>
Davidson, G. E., 473[55], <u>489</u>
Davies, J. K., 420[57], 424[81], 425[57], 431[94], 432[94], 433[94,167], 436[94], 439[94, 167], 441[94], 446[57,94], <u>456</u>, <u>457</u>, <u>458</u>, <u>460</u>, 483[118], <u>491</u>
Davies, M., 227[58], 229[58], 237[58], <u>312</u>
Davies, P. L., 25[185], <u>40</u>, 510[75], 517[100], 518[100], 519[75], <u>550</u>, <u>551</u>
Davies, R., 592[31], <u>613</u>
Davis, B., 537[200], <u>555</u>
Davis, B. B., 466[2], <u>487</u>
Davis, B. D., 149[189], <u>167</u>,

[Davis, B. D.]
177[60], 184, <u>214</u>, <u>215</u>, 442[175], <u>460</u>, 486[145,146,147], <u>492</u>
Davis, R. H., 478[85], 481[85], <u>490</u>
Dawes, W. A., 401[82,83], <u>411</u>
Deans, J., 273[53], <u>312</u>
De Boer, B. L., 509[69], <u>550</u>
Decad, G., 9[34], 13[34], 14[34], 20[34], <u>35</u>
DeCicco, B. T., 281, <u>322</u>
De Felice, M., 178[87], 181[76,87], 182[87], 199[76], <u>214</u>, <u>215</u>, 327[29], 332[29,46,47], 333[29], 338[29], 341[29], 342[29], 343[29], 345, 360, 367[29], <u>371</u>, <u>372</u>
Degani, C., 568[19], <u>612</u>
De Gier, J., 15[112], <u>37</u>
De Jerphanion, M.-B., 509[70], <u>550</u>
De Jong, L., 388[24], 391[24], 397[24], <u>409</u>
DeLaey, P., 326[190,191], <u>381</u>
Del Bene, V. E., 484[133], <u>492</u>
Del Campo, F. F., 109[43], 119[73], 120[43], <u>124</u>, <u>125</u>
Delobbe, A., 129[7], 131[7], <u>161</u>
DeLong, S. S., 425[95], <u>458</u>
DeMoss, J. A., 300[76,202], <u>313</u>, <u>321</u>
DeMoss, R. P., 14[78], <u>36</u>
Dennis, E., 85, <u>102</u>
Der, C.-L., 254[50], 255[50], 256[50], 258[50], <u>312</u>
Deshusses, J., 130[22,23,24], <u>161</u>
Deters, D. W., 506[43], 517[107], 518[107], <u>549</u>, <u>551</u>
Deuel, T. F., 22, <u>39</u>
De Vlieghere, M., 85[107], 86[107], <u>102</u>, 393[38], <u>409</u>
DeVries, W., 17[129], <u>38</u>, 193[176], <u>217</u>
DeZeeuw, J. R., 484[129], <u>492</u>
DeZwaig, R. N., 387[18], 388[16, 18], <u>409</u>
Dhal, M. M., 144[135], <u>165</u>
Dice, J. F., 336[44], <u>371</u>

Dickinson, E. S., 479[92], <u>491</u>

Diddens, H., 483[117], <u>491</u>

Didier-Fichet, M. L., 386[7], <u>408</u>

Diedrich, D. L., [88], <u>457</u>

Diehl, J. F., 338[216], 342[216], 343[216], 365[216], 368[216], <u>383</u>

Dietz, G. W., 130[20], 140, 141[20, 112, 113], 156[20,103,113,116], <u>161</u>, <u>164</u>, 194[191], <u>218</u>, 287[43], <u>311</u>, 545[246], <u>556</u>

Dietzler, D. N., 118[63], <u>125</u>

DiGirolamo, P. M., 442[176,179, 183], 443[176,179,183], <u>460</u>, <u>461</u>, 466[7], 467[7,11,13], 470[23], <u>487</u>, <u>488</u>, 544[240], <u>556</u>

DiMasi, D. R., 149[184], <u>167</u>, 265[44], <u>311</u>, 443[185], 445[185], <u>461</u>

Dingle, S., 56[40], 58[40], 65[40], <u>99</u>

Dittmer, J. C., 19[147], <u>38</u>

Dittmer, K., 347, 352, 353[32], <u>371</u>

Dobbs, F., 473[55], <u>489</u>

Dockter, M. E., 485[134,135], <u>492</u>

Doddema, H., 396[71], 398[71], 399[71], <u>410</u>

Dorn, F., 229[81], 236[81], 275[81], <u>314</u>

Dorpema, J. W., 19[144], <u>38</u>

Doudoroff, M., 173, <u>212</u>

Dover, S., 536[195], <u>555</u>, 592[57], 593[57], <u>614</u>

Dowben, R. M., 499[17], <u>548</u>

Dowd, D., 2[9], <u>5</u>

Downie, J. A., 293[72], 294[72], <u>313</u>, 568[48], <u>613</u>

Drachev, L. A., 279[45], <u>311</u>

Drapeau, G. R., 591[32,33], 592[32,33], <u>613</u>

Dreyfuss, J., 194[185], <u>218</u>, 290[46, 47], 291[46,47,159], 292[159], 299[46], <u>312</u>, <u>319</u>, 528[162], <u>553</u>

Drift, J. A. M.v.d., 510[73], <u>550</u>

Druck, K., 302[84], <u>314</u>, 536[195], <u>555</u>, 592[57], 593[57], <u>614</u>

Drucker, H., 334[30,31], <u>371</u>

Dubin, D. T., 486[146,147], <u>492</u>

Dubler, R. E., 396[78], 401[78], <u>411</u>

Duijts, G. A. H., 452[209,210], <u>462</u>

Dular, U., 282[207], <u>322</u>

Duncombe, G., 88[117], <u>102</u>

Dunlop, P., 283[173], <u>319</u>

Dunn, E., 283[162], <u>319</u>

Dunn, F. W., 332[179], 338[179], 340[64], 341[33,178], 343[179], 346[179], 347, 352, 353[32], 365[179], <u>368</u>, <u>371</u>, <u>373</u>, <u>380</u>, <u>381</u>

DuVigneaud, V., 358[210], <u>383</u>

Dvoraki, H. F., 194[189], <u>218</u>

Dworsky, P., 452[212], <u>462</u>

Dzheja, P., 581[55], 582[54,55], 608[54], <u>613</u>

E

Eagon, R. G., 19[145], <u>38</u>, 129[19], <u>161</u>, 387[21,22], 388[21,22], 402[88], 404[88,91,92], <u>409</u>, <u>411</u>, 592[34] <u>613</u>

Earhart, C. F., 452[211], <u>462</u>

Eaton, M. W., 273[53], <u>312</u>

Eberhard, S., 56[40], 58[40], 65[40], <u>99</u>

Ebner, E., 336[62], <u>373</u>

Ecker, W., 84[102], 85[102], <u>101</u>

Edwards, H. M., Jr., 263[34], <u>311</u>

Egami, F., 303[48], <u>312</u>

Egan, J. B., 50[27,28,29], 69[74, 75], 70[75,80,81], 78[74,75], 79, <u>99</u>, <u>100</u>, <u>101</u>, 129[14,15], 130[15], <u>161</u>

Egorova, V. V., 326[191], <u>381</u>
Ehring, R., 395[49], <u>410</u>
Eidels, L., 141[107], 156[107], <u>164</u>
Eisele, A., 364[36], <u>371</u>
Eisele, K., 363[34,35], 364[36], <u>371</u>
Eisenberg, M. A., 469[78], 476, 477[78,79], <u>490</u>
Eisenberg, M. R., 477[79], <u>490</u>
Eisenstadt, E., 227[49,52,185, 196], 230[49], 233[51], 241[184], 242[52,185,196], 243[184,185], 254[50,185], 255[50], 256[50, 66,185,196], 257[185,196], 258[50,66], 259[66,185], 271[185, 196], 302[66], <u>312</u>, <u>313</u>, <u>320</u>, <u>321</u>, 407[102], <u>411</u>
Elbrink, J., 121, 122[80], <u>125</u>
El Ghazzawi, E., 300[5], <u>309</u>
Ellar, D. J., 273[53], <u>312</u>, 507[61], <u>550</u>
Ellenbogen, L., 470[22], <u>488</u>
Elliker, P., 58[73], 68[73], <u>100</u>
Elliott, B. B., 299[54,55], <u>312</u>
Elliott, W. H., 334[102], <u>375</u>
Ellis, J. H., 146[154], <u>166</u>, 210[219], <u>219</u>
Emery, T., 243[56], 244[56], 246, 251[56], <u>312</u>
Emmens, M., 17[132], <u>38</u>
Engler, M., 50[23], 53[23], <u>99</u>
Englesberg, E., 118, <u>125</u>, 145[148], 146[148], 155[222], 195[148], <u>165</u>, <u>168</u>, 531[230], 542[230], <u>556</u>, 584[162], 610[80], <u>614</u>, <u>617</u>
Englund, P., 419[41], <u>456</u>
Ennis, H. L., 14[84], <u>36</u>, 229[228], <u>323</u>
Enoch, H. G., 273[57], 299[57], 300[57], <u>312</u>, 523[133], <u>552</u>
Epstein, W., 10[44], 12[44], 14[81], <u>35</u>, <u>36</u>, 64, 78, <u>100</u>, 106[16,22], <u>123</u>, 227, 228[171], 229[58,59,

[Epstein, W.]
171], 230[60,61,171], 231[16], 232[171], 236[171], 237, 275[171], 287[224], 296[60,183], <u>309</u>, <u>312</u>, <u>319</u>, <u>320</u>, <u>323</u>, 532[172], <u>554</u>, 600[183], 602[183,184], 603, 611[183], <u>618</u>
Erecinska, M., 32[206], <u>40</u>
Ericcson, L. H., 336[181], <u>381</u>
Eriksson-Grennberg, K. B., 419[41], 425[93], <u>456</u>, <u>458</u>
Esaki, K., 468[29], 470[29], 471[29, 36], <u>488</u>, <u>489</u>, 543[238], <u>556</u>
Essenberg, R. C., 13[76], <u>36</u>, 141[115], 156[115], <u>164</u>, 588[35], 592[35], <u>613</u>
Estroumza, J., 394[47], <u>410</u>
Euffanti, A. A., 130[26], <u>161</u>
Euwe, M. S. T., 452[209], <u>462</u>
Evans, D. J., Jr., 502[30], <u>549</u>
Even, H. L., 283[173], <u>319</u>
Even-Shoshan, A., 146[168], <u>166</u>, 186[132], <u>216</u>
Eytan, G. D., 32, <u>40</u>

F

Fagan, J., 513[78], <u>550</u>
Faik, P., 156[226,227], <u>168</u>, 387[20], 388[20], <u>409</u>
Failla, M. L., 262[62], 263[63,64], 264[64], 265[64], <u>313</u>
Falcoz-Kelly, F., 140[101], <u>164</u>
Fan, C. L., 188[154], <u>217</u>
Farmer, J. L., 473[54], <u>489</u>
Fassenden-Raden, J. M., 22[171], <u>39</u>
Feigelson, P., 188, <u>216</u>
Ferenci, T., 64[50], 65[50], 66[50], <u>100</u>, 141[106], 142[106,120], 154[216], 156[106], 158[216], <u>164</u>, <u>165</u>, <u>168</u>
Ferguson, R. B., 476[70], <u>490</u>
Feucht, B. U., 20[155], <u>39</u>, 58[66], 66[66], <u>100</u>, 106[19,23], 109[45],

[Feucht, B. U.]
  112[23], 113[23], 114[19,45],
  117[62], 119[62], 123, 124,
  129[8], 131[38], 161, 162,
  388[30], 392[30], 409
Fickel, T. E., [37], 333[38],
  357[38], 358[38], 371
Fields, K. L., 603[36], 606[36],
  613
Fiethen, B., 84[106], 85[106], 102,
  389[44], 393[34], 394[44],
  395[44], 409
Fillingame, R. H., 25[186], 40,
  279[65], 313, 516[93], 517[93],
  519[93], 551
Fischer, J., 16[113], 23[113],
  28[113], 37
Fisher, L. W., 609[176], 618
Fisher, R. J., 22[176], 25[176],
  26[176], 39
Fisher, S., 254[50], 255[50],
  256[50,66], 258[50,66], 259[66],
  302[66], 312, 313
Flagg, J. L., 12[60], 35, 587[37],
  613
Fletterick, C. G., 336[174], 380
Florsheim, H. A., 349[164],
  350[164], 352[164], 353[164],
  358[164], 359[164], 362[164],
  363[39], 371, 380
Folkes, A. T., 363[159a], 379
Folkes, J. P., 592[31], 613
Foltz, E. L., 150[195], 167
Forbes, E. C., 283[162], 319
Ford, J. E., 467[9], 488
Forn De Salas, M., 424[87], 457
Foster, S. J., 485[141], 492
Foulds, J., 424[79], 457
Fournier, R. E., 396[69], 398[69],
  399[69], 400[76], 410, 527[152],
  553
Fox, C. F., 2, 5, 10, 32, 35, 40,
  55[37,38], 64[37], 99, 106[12],
  123, 132, 133[46], 134[46],
  137[46], 157[46], 160[46], 162,

[Fox, C. F.]
  172, 174, 175[3], 212, 525[145,
  146,147], 553
Fraenkel, D. G., 64[49], 100, 140,
  164
Frank, L., 146[155], 166, 185[124,
  125], 186[125], 216, 277[67], 313,
  536[194], 555, 592[38,138], 613,
  617
Franklin, T. J., 481, 484[124,126,
  129], 485[138,141], 491, 492
Fredericq, P., 431, 459
Freedberg, W. B., 150[192,193],
  156[192], 167
Freedman, M. L., 231[122], 248[122],
  316
Freer, J. H., 507[61], 550
Freese, E., 27, 40, 58[98], 101,
  193[181], 218
Frerman, F. E., 85, 86[112,113],
  87[115,116], 88[114,117], 102,
  389[46], 393[42], 394[42,46], 409
Freund, T. S., 271[23], 274, 310
Frick, K., 432[157], 433[167],
  436[157], 439[167], 440[157],
  460
Fridovich, I., 306
Friedberg, I., 141, 150[197], 156[197],
  164, 167, 334[58], 345[58], 372,
  537[202], 545[202], 555
Frieden, E., 303[68], 313
Friedkin, M., 474[60], 490
Friedl, P., 569[196], 582[196], 618
Friedman, S., 63[46], 64[46], 66[46],
  67[46], 100
Frost, G. E., 9[33], 35, 149[188],
  167, 226[70], 246[71], 248[70,71],
  249, 252[70], 253[69,71], 313,
  432[147,158], 434[133,158],
  435[133], 436[147], 437, 440[158],
  441[147,158], 450, 453[158], 459,
  460
Frumkin, S., 140[102], 164
Fruton, J. S., 333[173], 337[173],
  347[171,175], 348[171,175],

[Fruton, J. S.]
  354[175], 359[80,173,175],
  360[175], 374, 380
Fu, M. L., 442[183], 443[183],
  461, 467[11], 488
Fujii, K., 470[23], 488
Fujisawa, H., 473[47], 489
Fukui, S., 130[21], 161, 423[75],
  457, 610[68], 614
Fulco, A. J., 395[50], 410
Furlong, C. E., 144[134], 145[134],
  159[134], 165, 180[117,119,126],
  181[82], 185[117,119,126],
  186[126,127], 187, 199[82],
  200[208], 203[82], 214, 215, 216,
  218, 528[157], 530[181,192],
  531[235], 533[181,183,184],
  535[192], 536[198], 542[235],
  547[252], 553, 554, 555, 556
Furmanski, P., 393[41], 394[41],
  409
Futai, M., 13[75], 16[118], 19, 22,
  23, 24, 25[170], 26[118], 27[118],
  28[178], 29[178], 31[118,178],
  32[118], 36, 37, 38, 39,
  193[172,173], 217, 497[8],
  504[44], 508[44], 510[44],
  517[108], 518[44], 521[125],
  522[127], 548, 549, 551, 552

G

Gachelin, G., 74[84,85], 101, 118,
  119, 120, 125
Gaensslen, R. E., 14[94,95], 36
Gale, E. F., 1[2], 2, 5, 104, 122,
  331[40], 332[40], 363[159a],
  371, 379, 561[39,40,41,42,43,
  44,45,46], 563[44,46], 589[43],
  592[31], 613
Galliers, E., 58[98], 101
Galsworthy, P. R., 290[154],
  291[154], 292[154], 318,
  529[165], 554

Gander, J. E., 17[134], 20[134], 38,
  415[14], 416[14], 431[14], 455
Ganesan, A. K., 76[91], 101, 106[24],
  123, 138[73], 140[73], 146[73],
  155[73,80], 157[73], 158[73,80],
  163, 541[217], 555, 584[47], 613
Gárdos, G., 177[58], 214
Garen, A., 441[169,170], 460
Gargus, J., 520[119], 552
Garland, P. B., 293, 294[72], 313,
  568[48,104,119], 613, 615
Garrett, R. H., 295[180,181], 320
Gatmaitan, J. S., 307, 336[54,126,
  127], 372, 377
Gaumont, Y., 474[59], 489
Gay, P., 129[7], 131[7], 161
Gazdar, C., 106[18], 123
Geck, P., 583[69], 614
Gel'man, N. S., 521[124], 552,
  569[49], 613
Gennis, R. B., 603[158], 617
Gerber, N. H., 533[183], 554
Gerdes, R. G., 226[177], 285[73,
  177], 286[73,74,177], 287[74,
  177], 289[177], 290[74], 313,
  320, 532[169], 547[257], 554,
  557
Gerhardt, P., 14[88], 36, 187[40],
  216, 422, 457
Gershanovitch, V. N., 64[55],
  66[55], 100, 106[11], 123
Geuther, R., 452[213], 462
Ghei, O. K., 396[70,72,73], 398[68,
  70,72], 399[70,72], 400[70,72,
  73,77], 410
Gho, D., 160[255], 169
Gholson, R. K., 479[88,90], 490
Ghosh, S., 47[11], 58[11], 98,
  172[4], 212, 564[112], 574[112],
  616
Gibbons, R. J., 332[197], 382
Gibbs, P. A., 330[149], 379
Giberman, E., 333[146], 335[146],
  378

Gibson, F., 226[236], 246[129, 237], 250[236,237], 251[149], 252[149], 279[40], 306, 311, 317, 318, 324, 430[119], 432[140,141,143,147,148,149], 433[148], 434[135], 435[119, 135,138,139,140,141,143,146], 436[147,148,149], 441[147,148, 149], 459, 460, 519[113], 552, 560[28], 613

Gibson, J., 388[28], 392[28], 396[28], 401[28], 409

Giddens, R. A., 139[88], 154[88], 163

Gilchrist, M. J. R., 120, 125, 433[164], 439[164,166], 460, 603[50], 606[50], 613

Gilliland, G. L., 547[256], 557

Gilvarg, C., 326[7,41,143,185], 327[6,7], 330[142], 332[7,143, 185], 333[6,38,43,141,185], 336[185], 337[43,141], 338[7], 341[6,8,143], 342[6,8,141], 346[143], 347[7,143], 348[42,92], 349[42,92], 351[92], 353[8,42, 92], 354, 355[141], 356[141], 357[38,141], 358[38], 361[141], 362[143], 363, 367[6,42,43], 369, 371, 375, 381, 421, 457

Ginzburg, B. Z., 297[75], 313

Ginzburg, M., 297[75], 313

Giroux, J., 135[57], 162

Gladding, C., 483[113], 491

Glaser, D. A., 231[123], 316

Glaser, J. H., 300[76], 313

Glaser, L., 78, 101

Glass, T., 81[97], 101

Glover, G. I., 188[147,148], 196[147,148], 216

Gnirke, H., 415[9], 455

Göhring, W., 483[117], 491

Gold, V., 341[163], 343[163], 349[164], 350[164], 352[164], 353[164], 358[163,164], 359[164], 362[163,164,165], 380

Goldberg, A. L., 336[44], 371

Goldine, H., 420[50], 456

Goldsmith, J., 295[77], 313

Goldthwaite, C., 483[121], 492

Golecki, J. R., 416[21], 417[21,30], 419[21], 451[207], 455, 456, 462

Golub, E. E., 266[24], 267[24], 268[78], 270[24], 271[24], 274[24], 310, 314

Goodman, S. R., 486[154], 493

Gordon, A. S., 139[91], 143[127], 144[127], 145[127], 146[91], 147[91], 164, 165, 195[194], 197, 209[216], 218, 219, 527[156], 539[212], 540[213], [213], 553, 555

Gorneva, G. A., 24, 40, 306

Gosh, S., 3[15], 5

Gottschalk, G., 58[100], 101, 300[5], 309

Gould, J. M., 607, 613

Goy, M. F., 521[121], 552

Grahl-Nielsen, O., 331[45], 371

Gray, W. R., 333[22], 335[22], 370

Green, D. E., 232[79], 314

Greene, R. V., 611[236], 620

Greengard, P., 121[79], 125

Gries, E.-M., 241[227], 252[227], 323, 402[99], 406[99], 411

Griffith, D. D., 354, 356[172], 361[172], 380

Griffith, T. W., 471[35], 476[35], 489, 531[237], 543[237], 556

Grinius, L., 568[53], 581[55], 582[52,54,55], 608[52,54], 613

Griniuviene, B., 568[53], 581[55], 582[54,55], 608[54], 613

Grogan, E., 50[23], 53[23], 99

Gronlund, A. F., 129[5], 160, 187, 188, 216, 217

Gross, J. D., 177[62], 214

Grossowicz, N., 13[69], 24[69], 36, 468[62], 474[62,63], 475[65], 490, 592[7], 612

Grover, W. H., 141, 164

Groves, D. J., 129[5], 160
Gruzdkov, A. A., 326[191], 381
Gryder, R. M., 188[152], 217
Grzelakowska-Sztabert, B., 474[64], 490
Guardiola, J., 178[86,87], 181[76, 86,87], 182[86,87], 199[76], 214, 215, 327[29], 332[29,46, 47,48], 333[29], 338[29], 341[29], 342[29], 343[29], 345[29], 366[29], 367[29], 371, 372
Guest, J. R., 179[97], 183[97], 215, 395[53], 396[53], 410
Guidotti, G., 298[106], 299[106], 315
Guirard, B. M., 331[49], 372
Gulik-Krzywicki, T., 20[156], 39, 419[37], 456
Gunn, R. B., 298[80], 299[80], 314
Günther, Th., 229[81], 236[81], 275[81], 314
Guroff, G., 188[146], 216, 530[208], 538[208], 539[209], 555
Gustafsson, P., 149[178], 166, 419[40], 456
Guterman, S. K., 249[82], 314, 432[153,154,155], 436[153,154, 155], 460
Gutnick, D. L., 486[152], 492, 504[45], 513[45], 517[45], 518[45], 519[112], 549, 552
Gutowski, S. J., 398[67], 399[67], 410
Guymon, L. F., 19[145], 38, 387[21,22], 388[21,22], 409
Guzman, R., 76[91], 101, 106[24], 123, 138[73], 140[73], 146[73, 157], 147[157], 148[157], 155[73], 157[73], 158[73], 163, 166, 544[242], 556

H

Haas, D., 424[83], 457

Haavik, H. I., 306
Hachimori, A., 504[53], 549
Hackette, S. L., 282[83], 283[83], 314
Hacking, A. J., 155[224], 156[224], 168
Haddock, B. A., 265[37], 293[72], 294[72], 311, 313, 568[48,56,104, 119], 576, 577, 588[114], 613, 615, 616, 618
Hagihira, H., 118[71], 125, 140[100], 164
Haguenauer, R., 74[86], 101, 119, 125
Halegoua, S., 415[11], 455
Halfenger, G. M., 486[155], 493
Hall, F. F., 334[49a], 372
Hall, R. E., 143[127], 144[127], 145[127], 165, 195[194], 218, 539[212], 540[212], 555
Haller, I., 418[33], 420[59], 423, 424[59], 456, 457
Halpern, Y. S., 9[14], 14[84], 18[14], 34, 36, 118[70], 125, 146[162, 168], 166, 172[18], 180[134], 186[128,129,130,131,132,133, 134], 187[128,129], 201[135], 210, 212, 216, 277[113], 302[84], 314, 316, 332[50], 372, 536[195, 196,197], 555, 592[57,95], 593[57], 614, 615
Hamelin, O., 136[69], 153[69], 163
Hamilton, W. A., 479[99], 491, 560[58], 567[161], 572[25], 587[159,160,161], 588[25], 589[159,160,161], 590[159,160], 596[24], 597[24], 598, 600[25], 609[25], 612, 614, 617
Hammes, G. G., 508[63], 513[63], 550
Hammond, C., 433[167], 439[167], 460
Hampton, M. L., 27, 40
Hancock, R. E. W., 226, 248[21,85], 249[21,85], 252[85], 310, 314,

[Hancock, R. E. W]
   420[58], 425[58], 431, 433[136,
   160,162], 434[136], 438[136,
   160,162], 439[136,160,162],
   441[171], 446[58], 451[222],
   457, 459, 460, 461, 486[144,
   147], 492
Hane, M. W., 483[119], 491
Hanna, M. L., 442[181,182], 461,
   466[8], 467[8,12], 487, 488,
   531[239], 544[239], 556
Hannig, K., 415[18], 452[18], 455
Hanson, R. L., 502[31], 517[31],
   518[31], 549
Hanson, R. S., 405[97], 411
Hanson, T., 65[59], 69[76], 100,
   101
Hantke, K., 9[32], 35, 149[187],
   167, 226[85], 244[86], 248[21,
   85,86,87], 249[21,85,86,87],
   251[87], 252[85], 310, 314,
   414[7], 423[71], 431, 432[129],
   433[136,160,161], 434[136],
   438[136,160], 439[136,160,161],
   440, 455, 457, 459, 460
Hardy, K. G., 425[89], 458
Hare, J. F., 25, 28, 40, 503[39],
   517[39], 518[39], 549
Harold, F. M., 8[5,6], 9[5,6],
   11[6,53], 18[136], 20[136], 24,
   27[192], 34, 35, 38, 40, 104[2],
   120[2], 122, 175[46,47],
   187[138], 201[209,210], 213,
   216, 219, 227[90], 231[3], 232[3,
   105], 234[88,89,90,91,93,94,
   96], 235[90,91], 236[90,91,96,
   98,100], 237[90], 255[90],
   268[89,90], 269[89], 274[88,89,
   90], 275[90], 276[90,96,98,
   100], 277[88,89,90], 279[2,88,
   89,90,95,96,100], 280[88,89,
   90], 286[89], 287, 288[89,92,
   97,101], 289[101], 296[88,89,
   90], 302[101], 307, 308, 314,
   315, 496[3], 497[3], 498, 499,

[Harold, F. M.]
   505, 516[92,98], 519[3], 520[3],
   548, 551, 560[60,61], 563[5,166],
   564[61], 565[60,61], 567[66],
   568[61], 569[63], 579[66], 582[5,
   75], 586[65], 587[75,76], 588[8,
   63,65,67,75,76], 592[62], 596[5,
   64,66,75], 600[67], 605[66],
   607[4], 608[5,75], 610[67], 611,
   612, 614, 617, 620
Harold, R. L., 236[98], 276[98],
   288[97], 315
Harris, D. A., 517[110], 518[110],
   519[110], 552
Harris, J. I., 333[173], 337[173],
   359[173], 380
Harris, P. E., 129[6], 161, 177[64],
   178[64], 214
Harris, R. D., 233[218], 235[218],
   322
Hartig-Beecken, I., 10[52], 35,
   150[198,199], 156[198,199], 167,
   545[244], 556, 601[206], 619
Hartline, R. A., 396[78], 401[78],
   411
Hartman, P. E., 483[113], 491
Hartmann, A., 248[21], 249[21], 310,
   433[160], 438[160], 439[160], 460
Harvey, R. J., 402[86], 404[86],
   411
Hasan, S. M., 19[150], 38, 503[40],
   517[40], 518[40], 549, 611[237,
   238], 620
Hasegawa, T., 407[103], 411
Hasin, M., 418[35], 456
Hassid, W. Z., 173[21], 212
Hatfield, D., 426[106], 458
Hauschild, A. H. W., 330[52],
   338[51,53], 372
Hayakawa, K., 610, 614
Hayashi, R., 468[31,32], 471[31,32,
   38,41], 472[42], 488, 489
Hayashi, S., 107[27], 123, 130[29],
   150, 156[230], 161, 167, 168
Hayes, W., 447[198]; 461

Hayman, S., 307, 336[54,126, 127], 372, 377

Hays, J. B., 48[12], 49[19], 50[12, 19,25], 51[12,19,30], 52[12,19], 53[19], 63[46], 64[46], 66[46], 67[46], 68[12], 78[12], 79[96], 80, 81[97], 83, 98, 99, 100, 101, 108[38], 124

Haza, J., 330[120], 334[120], 377

Hazelbauer, G. L., 9[30], 34, 144[135,136,138,143,144], 145[143,144,146], 158[136,138], 165, 195, 218, 427, 428, 458, 542[226], 543[236], 547[254], 556, 557

Hearfield, D. A. H., 330[55], 372

Heavy, L. R., 144[145], 165, 547[253], 556

Hedges, A. J., 448[200], 453, 461, 462

Heidrich, H.-G., 415[18], 452[18], 455

Heinen, W., 300, 315

Heiniger, V., 334[56], 372

Heinz, E., 583[69], 614

Heldt, H. W., 14[92], 36

Helenius, O., 52[32], 99

Helgerson, S. L., 605[70], 606, 614

Heller, I., 446, 461

Hellingwerf, K. J., 19, 38

Hemmingsen, B. B., 260[6], 301[6], 309

Henderson, G. B., 468[53], 470[23], 473, 475[69], 488, 489, 490

Henderson, P. J. F., 10[46], 35, 139[87,88], 154[88], 163, 578[71], 584[71,73], 588[71, 72,73], 614

Henderson, R., 278[103,104], 315, 516[89], 551

Hendler, R. W., 524[140], 553

Hendlin, D., 150[195], 167

Hengstenberg, H., 48[13], 66[13], 98

Hengstenberg, W., 49[15], 50[27, 28,29], 51[101], 52[15], 69[74], 78[74], 83, 84, 98, 99, 100, 101, 129[15], 130[15], 161

Henning, U., 415[9], 418[33], 420[59], 423[71], 424[59,86], 446[195], 455, 456, 457, 461

Heppel, L. A., 3[12,14], 5, 9[18, 19,20,21,23,24], 10, 12[20,21, 23,24,59], 13[20,24,63], 18[59], 22[170], 24[182], 25[170], 34, 35, 39, 40, 107[29,31], 120[31], 123, 130[20], 131, 140, 141[20,112], 142[39], 147[169,170], 150[197], 154[217], 156[20,103,116,197], 158[217], 161, 162, 164, 166, 167, 168, 172[8,17], 174[6,8,32, 34,38], 180[115,117,118], 184, 185, 194[8,38,189,191], 209[118], 212, 213, 215, 216, 218, 232[12], 255[12], 268[12], 286[12], 309, 334[57,58], 345[57,58], 372, 498[9], 501[27], 502[27,34], 503[34], 504[44], 508[44], 510[44], 511[27], 512[77], 518[44], 522[129], 528[9], 530[190,199], 535[190], 536[199], 537[202], 545[202,245], 548, 549, 550, 552, 554, 555, 556, 575, 599, 600[13], 606[13], 612, 617

Herbert, A. A., 395[53], 396[53], 410

Herbert, D., 337[58a], 372

Hermann, K. O., 143[126], 165, 541[221], 555

Hermodson, M. A., 542[232], 547[255], 556, 557

Hernandez, S., 150[195], 167

Hernandez-Asensio, M., 109[43], 119, 120[43], 124, 125

Hertzberg, E. L., 16[117], 21, 26[117], 28[117], 37, 568[74], 578[74], 614

Hesse, J., 106[16], 123

Higa, S., 341[163], 343[163],
    349[164], 350[164], 352[164],
    353[164], 358[163,164],
    359[164], 362[163,164,165], 380
Higashi, T., 504[58], 550
Higginson, B., 484[126], 492
Hill, C., 445, 461
Hill, F., 85[111], 102
Hill, K., 50[29], 99
Hinds, T. R., 21[166], 32[205],
    39, 40
Hinkle, P. C., 16[117], 21, 26[117],
    28[117], 33[210], 37, 41,
    278[167], 319, 568[74], 578[74],
    614
Hintze, D. N., 473[54], 489
Hirashima, A., 415[11], 423[70],
    455, 457
Hirata, H., 16[114], 17[133],
    18[136], 20[136], 21[114,162,
    163,164], 25[187], 31[163],
    32[205], 33[212], 37, 38, 39,
    40, 41, 175[43], 193[180],
    197[180], 198[43], 201[209,
    210], 213, 218, 219, 231[3],
    232[3,105], 308, 315, 503[36,
    37,38], 504[37], 507[60],
    514[36,37,60], 515[36,37,38],
    516[60], 526[149], 549, 550,
    553, 563[5], 568[212], 582[5,
    75], 587, 588[75,76], 592[77],
    596[5,75], 608[5,75], 609[212,
    233], 610, 612, 614, 619, 620
Hirosawa, K., 16[115], 20[115],
    30[115], 37, 192[171], 193[171],
    217
Hirose, S., 336[121], 344[121],
    377
Hirsch, M. L., 332[59], 373
Hirshfield, I. N., 331[60], 333[17,
    61], 336[61], 348[17], 354[61],
    356[61], 361[61], 370, 373
Hitchings, G. H., 473, 489
Ho, C., 535[191], 554
Ho, M. K., 298[106], 299[106], 315

Hoberman, H. D., 331[182], 381
Hobson, A. C., 160[255], 169
Hochhauser, S. J., 479[91], 490
Hochstadt, J., 2[9], 5, 92[123], 93,
    94, 95[130], 97, 102
Hochstadt-Ozer, J., 90, 93[126],
    102
Hoehn, B., 446[195], 461
Hoffee, P., 118, 125, 610[80], 614
Hoffman, J. F., 581[81,209], 614,
    619
Hofnung, M., 9[31], 34, 148[171,
    183], 158[171,183], 166, 426[99,
    104,106,107], 427, 429[112],
    454[112], 458
Hofschneider, P. H., 415[18],
    452[18], 455
Hofstädter, L. J., 106[23], 112[23],
    113[23], 123, 131[38], 162
Hofsten, B., 334[12], 370
Hogarth, C., 273[53], 312
Hogg, R. W., 138[74], 144[74,139,
    142], 145[139,142,148], 146[74,
    148], 155[221], 195[139,142,148],
    163, 165, 168, 520[119], 531[230],
    542[230,232], 547[255], 552, 556,
    557
Holby, M., 293[20], 310
Holden, J. T., 189[155,156,157],
    201[213], 217, 219, 538[204],
    555
Holdsworth, E. S., 467[9], 488
Holland, E. M., 451[221], 462
Holland, I. B., 425[91], 445, 451[221],
    458, 461, 462
Holley, R. W., 121, 125
Holme, T., [198], 382
Holzer, H., 336[62], 373
Hong, J.-S., 29[199], 40, 139[220],
    154[220], 168, 602[124], 603[123],
    616
Hopkins, I., 185[124], 216, 277[67],
    313, 536[194], 555, 592[38], 613
Hopkins, J. W., 241[8], 309

Horecker, B. L., 138, 140[101], 151[202], 152[202], 153[78], 163, 164, 167
Hou, C., 423[76], 457, 500[26], 503[41], 504[26,41], 507[46], 508[26,46], 510[75], 513, 517[41], 518[41], 519[75], 533[46], 548, 549, 550
Houghton, R. L., 22, 25, 26[176], 39
Howes, W. V., 426[102], 458
Hubbard, J. S., 593[82], 600[82], 615
Hubbard, S. A., 502[29], 549
Huber, R. E., 292[204], 301[107], 315, 322
Huennekens, F. M., 468[53], 470[23], 473, 475, 488, 489, 490
Huet, J., 509[70], 550
Hullah, W. A., 333[63], 338[63], 373
Humphreys, J., 341[33], 368[33], 371
Hupkes, J. V., 15[112], 37
Hurwitz, C., 486[148], 492
Hurwitz, J., 499[13], 517[13], 548
Hutchings, B. L., 484[127], 492
Hütter, E., 431[163], 460
Hutter, R., 483[115], 491
Huttunen, M. T., 562[16], 593[16], 612

I

Iaccarino, M., 178[86,87], 181[76, 86,87], 182[86,87], 199[76], 214, 215, 327[29], 332[29,46, 47,48], 333[29], 338[29], 341[29], 342[29], 343[29], 345[29], 366[29], 367[29], 371, 372
Iezuitova, N. N., 326[191], 381
Ifland, P. W., 340[64], 373

Ikawa, M., 338[78], 359[78], 374
Imai, I., 407[103], 411
Ingraham, J., 93[124], 94[124], 102
Ingram, J. M., 414[5], 455
Inouye, M., 414, 415[11], 422, 423[70], 455, 457
Irr, G., 155[222], 168
Irvin, R. T., 420[52], 456
Isaacson, P., 486[150], 492
Ishikawa, S., 504[50], 549
Isturiz, T., 156[228,229], 168, 387[18], 388[16,18], 409
Isuchiya, D., 327[5], 332[5], 333[5], 341[5], 342[5], 357[5], 358[5], 367[5], 369
Itoh, J., 198[203], 218
Itol, J., 175[42], 181[42], 195[42], 200[42], 210[42], 213
Iwai, K., 473[47], 489
Iwashima, A., 468[44], 471[33,34, 37], 472[44,45,46], 488, 489
Iwata, K. K., 180[119], 185[119], 216
Iyima, T., 407[103], 411
Izaki, K., 395[54,55,57], 396[54,55], 410, 484[123,125], 485[136,137], 492

J

Jackman, L., 92[123], 102
Jackson, M. B., 150[195], 167, 333[66], 336[66], 338[66], 343[66], 344, 348[66], 350[66], 352, 353, 367[66], 373
Jacob, F., 132[40], 162
Jacobs, A. J., 547[250], 556
Jacobs, M., 233[218], 235[218], 322
Jacobson, D. W., 470[23], 488
Jakes, R., 365[144], 378
Janacek, K., 177[57], 214
Jarvis, A. W., 572[25], 588[25], 598[25], 600[25], 609[25], 612
Jarvis, B. D. W., 335[145,186], 378, 381

Jasaitis, A. A. , 279[45], 311,
    582[52], 608[52], 613
Jasper, P. E. , 444[187],
    461
Jasper, P. L. , 222[109,191],
    227[108], 230[108], 238[108],
    239[108,109], 241[110],
    253[191], 254[110], 255[110,
    191], 259[191], 260[110,191],
    266[110], 303[109], 316, 321
Jeacocke, R. E. , 567[161],
    587[161], 589[161], 617
Jehen, A. M. , 154[219], 168
Jenkins, W. T. , 525[143], 553
Jensen, C. , 504[47,48], 511[48],
    517[47], 518[47], 549
Jensen, R. A. , 188[147,148],
    196[147,148], 216
Jetten, A. M. , 603[84], 604[83,
    172], 606[84,172], 615, 618
Jetten, M. E. R. , 604[83], 615
Jimeno-Abendano, J. , 156[231],
    168, 224[116], 230[116],
    296[116], 316, 386[7,8],
    387[8], 388[8], 408
Johansson, B. C. , 504[56,57], 550
John, P. , 30[203], 40, 289, 290[29,
    30], 292[30], 310, 589[21,22],
    591[21,22], 612
Johnseine, P. , 241[193], 254[192],
    255[192], 302[192], 321
Johnson, A. C. L. , 17[131], 38
Johnson, C. L. , 402[89], 404[89],
    411
Johnson, G. L. , 336[67], 373
Johnson, M. K. , 336[68], 373
Johnson, W. C. , 10[50], 35,
    144[131], 148[131], 165
Jones, J. E. , 331[69], 373
Jones, T. H. D. , 133, 136[47],
    162
Jones-Mortimer, M. C. , 64, 100,
    139[88], 154[88], 163
Jorgensen, S. E. , 140[99], 149[99],
    164

Jorovitz-Kaya, N. V. , 106[11], 123
Jose leau-Petit, D. , 21, 39
Judice, J. J. , 109[45], 114[45], 124,
    388[30], 392[30], 409
Jung, G. , 483[117], 491
Jungermann, K. , 11[54], 35

                          K

Kaback, H. R. , 2, 5, 8[11,12,13],
    15[108,109], 16[113,120], 17[11,
    12,13,120,122,123,126,127],
    18[11,12,13,109,120,135,137,
    139,140,142], 19[142,146,148],
    20[156], 22, 23[12,113,120,140],
    24, 25[140], 26[140], 27, 28[12,
    113,140], 29[140,199,200], 31,
    33, 34, 37, 38, 39, 40, 46[5], 69,
    70[82], 73, 78[94], 91, 98, 101,
    109[41,46], 116[41,59], 117[41,
    59], 119, 124, 134[56], 135[56,
    59,60,61,62,63], 136[61,62,65,
    66], 137[59,62,65,70], 139[91],
    141[114], 146[91,114,158],
    147[91,158], 157[61,62,241,242],
    160[258], 162, 163, 164, 166,
    169, 172[14,15], 175[14,15,40,
    49,50,52,53,56], 181[89], 190[14,
    15,45], 191[89], 192[89], 193[45,
    50,175,182,183], 197[202],
    198[89], 202[89], 203, 212, 213,
    217, 218, 231[111,126], 232[111,
    126], 260, 267, 271, 275[111,126],
    277[113], 289[111], 316, 317, [70,
    71], 373, 387[17], 391[25],
    394[45], 408[17], 409, 479[100],
    481, 491, 496[1,2,6], 497[7],
    520[2,6], 521[126], 522[128],
    527[156], 545[247], 548, 552, 553,
    556, 560[88], 562[9,10,16,86,88,
    90,91,93,110], 563[9,10,90,180,
    200], 564[10,90], 572[89], 575[86],
    576, 580[180], 582[180,181,200],
    585[89], 586, 587, 593[16], 595,
    598[180], 602, 603, 606[92],

[Kaback, H. R.]
608[180], 611[240], <u>612</u>, <u>615</u>,
<u>616</u>, <u>618</u>, <u>619</u>, <u>620</u>
Kaczorowski, G., 16[113], 23[113],
28[113], <u>37</u>
Kadner, R. J., 10[43], 12[43],
14[83,85], <u>35</u>, <u>36</u>, 141[109,
110], 150[110], 156[109], <u>164</u>,
179[99,100], 183, <u>215</u>, 265[112],
<u>316</u>, 442[176], 443[176,184],
444[189], 445[189,192,193],
446[189,194], 447[192],
448[194,201,202], 449[201],
450[194], 453[189,192],
454[193], <u>460</u>, <u>461</u>, 466[4],
[13], 467[14], 483[111], <u>487</u>,
<u>488</u>, <u>491</u>, 600[94], <u>615</u>
Kadziauskas, Y., 582[52], 608[52],
<u>613</u>
Kagawa, Y., 17[133], 25[133,187],
32[207], 33[212], <u>38</u>, <u>40</u>, <u>41</u>,
175[43], 193[180], 197[43,180],
198[43], <u>213</u>, <u>218</u>, 500, 503[36,
37,38], 504[37], 507[60],
514[36,37,60], 515[36,37,38],
516[60], 526[149], <u>548</u>, <u>549</u>,
<u>550</u>, <u>551</u>, <u>553</u>, 568[212],
609[212,233], 610[78,79], <u>614</u>,
<u>619</u>, <u>620</u>
Kahan, F. M., 150[195], <u>167</u>,
483[109,110], <u>491</u>
Kahan, J. S., 483[109], <u>491</u>
Kahana, R., 424[82], <u>457</u>
Kahane, I., 524[138], <u>553</u>
Kahane, S., 186[128,129], 187[128,
129], 210[129], <u>216</u>, 277[113],
<u>316</u>, 536[197], <u>555</u>, 592[95],
<u>615</u>
Kalb, A. J., 134[55], <u>162</u>
Kalckar, H. M., 138[72], 143[72],
146[72,152], 144[132], 148[132],
<u>163</u>, <u>165</u>, <u>166</u>
Kallio, R. F., 386[2], 401[2],
405[2], <u>408</u>
Kalra, V. K., 19[151], 21[168,

[Kalra, V. K.]
169], 32[205], <u>38</u>, <u>39</u>, <u>40</u>, 504[58],
<u>550</u>, 589[173], <u>618</u>
Kamen, M. D., 301[9], <u>309</u>
Kamio, Y., 416[28], 417[28], 418[28,
34], <u>455</u>, <u>456</u>
Kanegasaki, S., 9[16], <u>34</u>
Kanner, B. I., 25[188], <u>40</u>, 486[152],
<u>492</u>, 504[45], 513[45], 517[45],
518[45], 519[112], <u>549</u>, <u>552</u>
Kanzaki, S., 181[75,91], 196[75],
199[75], 201[75], <u>214</u>, <u>215</u>
Karibian, D., 149[190], <u>167</u>
Kasahara, M., 33[210], <u>41</u>, 193[174],
198[174], <u>217</u>
Kashket, E. K., 580[96], 581[96],
595[96], 596[96], <u>615</u>
Kashket, E. R., 11[57], 12[57], <u>35</u>,
77,78,<u>101</u>,134[53], 152[53,208,
210,211,212], 153[208,210],
157[244], <u>162</u>, <u>167</u>, <u>168</u>, <u>169</u>,
578[99,134], 579[100,101,134],
581[100], 585[99], 586[97,98],
589[99], 596[100], <u>615</u>, <u>616</u>
Kashket, S., 467[20,21], <u>488</u>
Kasuga, T., 426[103], <u>458</u>
Katchalski, E., 333[9,146], 334[161,
162], 335[146], 336[10], 347,
348[42], 349[10,42], 352[10],
353[10,42], 366[10], 367[42],
<u>369</u>, <u>371</u>, <u>378</u>
Katzir, Katchalski, E., 332[119],
336[119], 348[119], 349[119],
350[119], 352[119], 353[119],
357[119], 358[119], 368[119],
<u>377</u>
Kaufman, J. T., 467[20], <u>488</u>
Kaulen, A. D., 279[45], <u>311</u>
Kawasaki, T., 468[29], 470[28,29],
471[29,30,36,39,40], 472[43],
<u>488</u>, <u>489</u>, 543[238], <u>556</u>, 592[102,
103], 596[103], <u>615</u>
Kay, W. W., 180[120,121,122], 185,
186, 187, 188, <u>216</u>, <u>217</u>, 389[31,
33], 392[31], 393[31,33], 395[51,

[Kay, W. W]
  52,58], 396[3,51,52,58,70,72,
  73], 398[68,70,72], 399[70,72],
  400[70,72,73,77], 402[33],
  407[33], 409, 410
Kayama, Y., 592[102,103],
  596[103], 615
Kayushin, L. P., 277[114],
  278[114], 316
Keenan, T. W., 569[29], 605[29],
  613
Kelker, N., 65[60], 67, 69[60,76,
  77], 70, 73[77], 77, 78[60],
  100, 101
Kellerman, A., 547[254], 557
Kellermann, O., 130[25], 144[25],
  145[25,146], 147[25], 154[25],
  158[25], 161, 165, 195[195],
  218, 426[105], 428[105], 458,
  528[160], 543[160], 553
Kemp, A., Jr., 517[110], 518[110],
  519[110], 552
Kemp, J. D., [115], 316
Kemp, M. B., 568[104], 615
Kendall-Tobias, M. W., 568[56],
  613
Kennedy, E. P., 2[11], 5, 10,
  13[70], 15[101], 24[70], 25[184],
  28[184], 35, 36, 37, 40, 50[26],
  99, 132[41], 133[45,46,48],
  134[46], 136[47,48], 137[46],
  157[41,46], 160[46], 162, 172,
  174[31], 175[3,51], 212, 213,
  239[130], [131], 241[147,148],
  317, 318, 502[31], 517[31],
  518[31], 525[145,146,147],
  549, 553
Kepes, A., 2, 5, 21, 39, 74[86],
  101, 119, 125, 129[2], 132[43],
  134[52], 136[52], 151[201],
  153[201], 155[225], 156[225,
  231,233], 157[52], 160, 162,
  167,168, 172[2], 173[26], 195,
  212, 218, 224[116], 230[116],
  296[116], 316, 337[72], 373,

[Kepes, A.]
  386[8], 387[8,14], 388[8,14], 408,
  518[104], 540[214], 541[214], 551,
  555, 561[105], 588[186], 615, 618
Kertai, P., 177[58], 214
Kerwar, G. K., 135[61], 136[61],
  139[91], 146[91], 147[91], 157[61],
  163, 164, 175[53], 213, 481[102],
  491
Kessel, D. H., 177[62,63], 179[104],
  181[63], 184[104], 214, 215,
  332[73,74], 333[74], 338[74],
  341[74], 342, 343[74], 344,
  346[74], 354, 356[74], 359,
  360[74], 361[74], 363, 366, 373
Kessler, R., 232[79], 314
Ketchum, P. A., 300[117], 316
Kewar, G. K., 586[201], 602[201],
  603[201], 619
Keysary, A., 333[146], 335[146]
  378
Khalifah, L. I., 149[185], 167,
  444[190], 445[190], 461
Khalil, M., 395[61], 396[61], 397[61],
  410
Kholand, B., 526[148], 553
Kida, S., 286, 316
Kihara, H., 331[75,76,77,79],
  332[75,76,77,79], 338[78], 359,
  373, 374
Kim, B. S., 227[59], 229[59],
  237[59], 312
Kim, I. C., 524[139], 553
Kimura, H., 193[172], 217
Kin, E., 10[38], 11[56], 12[38], 35,
  181[84], 206[84], 214, 517[102],
  518[102], 551, 600[107], 615
King, K., 254[192], 255[192],
  302[192], 321
Kinghorn, J. R., 283[162], 319
Kingman, J., 415[16], 455
Kinscherf, T. G., 254[225], 260[225],
  323
Kiritani, K., 187[141], 216
Kisluik, R. L., 474[59], 489

Kistler, W. S., 523[131], 552
Kistu, S. G., 534[189], 554
Kitihara, K., 467[18], 488
Kitsutani, S., 522[130], 552
Kittredge, J. S., 189[155], 217
Kiuchi, K., 485[136], 492
Klatt, O. A., 331[79], 332[79], 374
Klaus, S., 452[213], 462
Klein, K., 84[106], 85, 102, 389[44], 393[34], 394[44], 395[44], 409
Klein, W. L., 119[57], 124
Kleiner, D., 281, 316
Kleinzeller, A., 173[25], 212
Kline, W. L., 181[94], 215
Klingenberg, M., 14[93], 36, 122[83], 125
Klofat, A., 58[98], 101
Klopotowski, T., 178[87], 181[76, 87], 182[87], 199[76], 214, 215, 332[46,47], 372
Klyutchova, V. V., 106[11], 123
Knauf, P., 298[179], 299[179], 320
Knepper, J. E., 605, 615
Knight, I. G., 517[99], 551
Knowles, C. J., 451[220], 462
Kobashi, M., 473[47], 489
Kobayashi, H., 10[38], 12, 13[72], 35, 36, 181[84,92], 194[92], 198[206], 201[92], 206[84], 207[92], 210[92], 214, 215, 218, 517[101,108], 518[101], 533[185], 541[185], 551, 554, 600[107], 615
Koch, A. L., 11, 15[58,103,104], 35, 37, 109[44], 110[48,49,50], 112, 124, 150[196], 151[206, 207], 153[207], 167, 173[27], 212, 302[120], 316, 577, 585[108], 599, 612, 615, 618
Koch, J. P., 140[99], 149[99], 150[191], 156[230], 164, 167, 168

Kogut, M., 386[3], 401[3], 405[3], 408, 486[150], 492
Kohn, L. D., 18[140], 23[140], 25[140], 26[140], 28[140], 29[140, 200], 38, 40, 391[25], 409, 521[126], 552
Kohn, L. K., 522[128], 552
Kokkonen, H., 479[96], 491
Kolata, G. B., 223[121], 316
Koltushkina, G. G., 326[191], 381
Komatsu, Y., 94, 102
Kon, S. K., 467[9], 488
Kondrashin, A. A., 279[18,45], 309, 311
Konings, W. N., 17[122,129,132], 18[142,143,144], 23, 30[143,201, 202], 37, 38, 40, 175[50], 193[50, 176,181], 213, 217, 218, 388[24, 26], 391[24,26], 395[26], 396[71], 397[24], 398[26,71], 399[71], 400[26], 401[26], 409, 410, 562[16,110], 593[16], 612, 616
Konisky, J., 120, 125, 433[164,165, 167], 439[164,165,166], 460, 603[50,158], 606[50], 613, 617
Kopecky, A. L., 604, 616
Koplow, J., 420[50], 456
Kornberg, A., 13[68], 24[68], 36
Kornberg, H. L., 10[46], 13[76], 35, 36, 46[8], 64[50], 65[50], 66, 76, 83, 98, 100, 101, 109[39, 40], 116[39,40,58], 124, 129[6], 131, 139[35,89], 141[106,108, 115], 142[106], 155[89], 156[106, 108,115,226,227], 161, 164, 168, 180[120,121], 185, 216, 386[4], 387[19,20], 388[20,27], 392[27], 395[51,52], 396[51,52], 408, 409, 410, 584, 588[35,72], 592[35], 613, 614
Kort, E. N., 521[121,122], 552
Korte, T., 52, 99
Koser, S. A., 466[1], 474[1], 487
Koshland, B. Jr., 531[234], 542[234], 556

Koshland, D. E., Jr., 104[4], 122,
    144[137], 145[137,145,147],
    159[137,147], 165, 521[123],
    547[253], 552, 556
Kosmakos, F. C., 16[114], 21[114,
    167], 37, 39, 592[77], 614
Kostellow, A. B., 18[135], 38
Kotyk, A., 173[25], 177[57], 212,
    214
Kozloff, L. M., 425[95,96], 458,
    473[48,49,50], 489
Kozlov, I. A., 279[18], 309
Kraas, E., 483[117], 491
Kralovic, M. L., 254[194], 266,
    321
Kralovic, M. U., 22[172], 39
Krause, A. E., 524[136], 552
Krawitz, T., 485[139], 492
Krehl, W. A., 359[80], 374
Kreishman, G. P., 535[191], 554
Kropp, J., 483[109], 491
Krulwich, T. A., 11[55], 17[130],
    35, 38, 56[39], 58[39,64], 66[63,
    64], 99, 100, 129[12], 161
Kubitschek, H. E., 231[122],
    248[122], 316
Kuenen, J. G., 17[132], 38
Kuhn, J., 179[98], 183[98], 215
Kuhn, T. S., 3[16], 5
Kukumoto, J., 336[95], 375
Kula, M. R., 336[167], 380
Kulpa, C. F., Jr., 452, 462
Kundig, F. D., 172[5], 212,
    575[113], 616
Kundig, W., 3, 5, 46[6], 47,
    49[18], 53, 54[34], 55[33],
    56[6], 58[11], 63[35], 65[33],
    98, 99, 112[56], 113, 124,
    172[4,5], 212, 564[112], 574,
    575, 616
Kung, F.-C., 231[123], 262, 316
Kung, H., 586[202], 619
Kung, S. F., 137[70], 163
Kunkel, H. O., 334[49a], 372
Kuo, T., 419[38], 456

Kuriki, Y., 510[74], 550
Kusaka, I., 610[68], 614
Kusch, M., 134[54], 152[54,208],
    162, 167
Kustu, S. G., 9[27], 10[27], 34,
    189[166,167], 190[167], 217,
    281[22], 310, 332[81], 374
Kuzuya, H., 530[208], 538[208],
    555
Kwok, S. C., 282[207], 322

                            L

Labelle, J. L., 135[58], 162
Ladygina, V. G., 130[28], 161
Lagarde, A. E., 151[204], 152[204],
    157[236,237,239], 167, 168,
    387[9,10], 388[9,10], 391[10],
    408, 588[114], 616
Lakshmanan, S., 332[154], 338[154],
    342[154], 379
Lamberti, A., 327[29], 332[29],
    333[29], 338[29], 341[29], 342[29],
    343[29], 345[29], 366[29], 367[29],
    371
Lambeth, D. O., 499[16], 548
Lamy, F., 118, 125, 610[80], 614
La Nauze, J., 273[53], 312
Lange, R., 396[74], 399[74], 410,
    527[153], 553
Langman, L., 246[237], 250[237],
    306, 324, 432[141,147,148],
    433[148], 435[141], 436[147,148],
    441[147,148], 459, 460
Langridge, R., 529[164], 553
Langworthy, T. A., 333[180], 381
Lankford, C. E., 430[121], 459
Lanyi, J. K., 16[116], 21[116],
    32[116], 37, 193[177,178,179],
    217, 306, 307, 547[251], 556,
    581[182], 589[116,117,132],
    592[116,117,132,182], 593[116,
    117,132], 594[116,182], 597[182],
    611[236], 616, 618, 620
Lardy, H. A., 499[16], 548

Larimore, F. L., 283[173], 307, 319

Laris, P. C., 581[81,118], 596[118], 614, 616

Larsen, S. H., 520[119], 521[121, 122], 552

Larson, R. J., 502[34], 503[34], 509[71], 510[71], 512[77,80], 513[80], 514[80], 549, 550

Lau, C., 425[92], 458

Lauppe, H. F., 48[13], 66[13], 98

Lawford, H. G., 402[94], 405[94], 411, 568[119], 616

Lazdunski, A., 334[82,83,84,117], 374, 376

Lazdunski, C., 334[82,83,84,117], 374, 376

Leach, F. R., 187[136],216, 332[86], 338[85,86], 342[86], 343[85,86], 346[86], 348[85], 359[86], 360[86], 366, 374, 471[35], 476[35], 479[98], 489, 491, 531[237], 543[237], 556

Leal, J., 447[196,197], 461

Leckie, M. P., 118[63], 125

Lecocq, J., 327[5], 332[5], 333[5], 341[5], 342[5], 357[5], 358[5], 367[5], 369

Leder, I. G., 158[248], 169

Lederberg, E. M., 426, 458

Lederberg, J., 173[21], 212, 419[43], 456

Lee, C. P., 32[206], 40, 582[120], [195], 616, 618

Lee, N., 155[222], 168

Lee, S.-H., 504[58], 550

Lehninger, A. L., 14[89,99], 36, 49[20], 99

Leive, L., 13, 35, 149[179], 166, 184, 215, 416[20], 419[44,45], 420[45,46,47,48], 452, 455, 456,462, 537[200], 555

LeMinor, L., 429, 458

Lengeler, J., 66, 100, 129[4], 143[126], 160, 165, 541[221], 555

Lennarz, W. J., 20[158], 39, 415[17], 416[17], 424[84], 455, 457

Leong, J., 251, 252[124], 316, 441, 460

Lessely, B. A., 530[228], 542[228], 556

Lester, R. L., 273[57], 299[57], 300[57], 312, 523[133], 552

Letellier, L., 419[37], 456

Leung, K., 95[131], 102

Leutgeb, W., 415[8], 455

Lever, J. E., 146[160,165], 148[165], 166, 189[164,165], 201[164,165], 217, 528[158], 530[186], 533[158], 534[186], 553, 554

Levin, E., 187[138], 216, 280, 315, 569[63], 588[8,63], 612, 614

Levin, H., 332[122], 377

Levin, Y., 333[43], 337[43], 367[43], 371

Levine, A. E., 523[134], 552

Levine, E. M., 302[195], 321, 332[87,88], 337[88], 338[87,88], 344[89], 346[87,89], 359, 360[89], 366, 374

Levinson, S., 58[64], 66[64], 100

Levy, J. S., 307

Levy, L., 424[82], 457

Levy, S. B., 485[142], 492

Levy, S. R., 420[48], 456

Lewis, A., 278[125], 316

Li, C., 2[9], 5

Lieberman, E. A., 279[45], 311, 582[52,121], 608[52], 613, 616

Lichstein, H. C., 469[71,72,73], 476[70,72,74,75,76], 480, 490, 491

Lie, R. F., 301[9], 309

Lieberman, M. A., 139[220], 154[220], 168, 602[124], 603[123], 616

Liggins, G. L., 265[112], <u>316</u>, 443[184], <u>461</u>, 466[4], <u>487</u>

Lightbrown, J. W., 486[150], <u>492</u>

Lillah, A., 49, 64[16], <u>99</u>

Lilligh, T. T., 20, <u>38</u>

Lin, E. C. C., 8[2], 9[2,15], <u>34</u>, 46[9], 67[70], 68[9], 74[87], <u>98</u>, <u>100</u>, <u>101</u>, 107[27,28], 118[71], <u>123</u>, <u>125</u>, 130[29,30,31,32,33], 131[37], 132[42], 140[100], 149, 150[191,192,193,194], 155[224], 156[192,224,230], 157[37], <u>161</u>, <u>162</u>, <u>164</u>, <u>167</u>, <u>168</u>, 172[13], <u>212</u>, 523[131,132], <u>552</u>

Lin, J. J.-C., 423, <u>457</u>

Lindberg, A. A., 424[85], <u>457</u>

Lindenmayer, G. E., 566[205], <u>619</u>

Linder, R., 524[141], <u>553</u>

Lindop, C. R., 517[99], <u>551</u>

Lindsay, R. J., 572[25], 588[25], 598[25], 600[25], 609[25], <u>612</u>

Linnett, P. E., 502[29], <u>549</u>

Lipmann, F., 564[125], <u>616</u>

Litchfield, C. D., 334[90,91], <u>374</u>, <u>375</u>

Livoni, J. P., 295[77], <u>313</u>

Lo, T. C. Y., 150[200], <u>167</u>, 185[123], <u>216</u>, 395[59,60,61,62], 396[59,61,62,63], 397[59,61,63, 64,65,66], 398[65], 408[65], <u>410</u>, 526[150,151], 547[248], <u>553</u>, <u>556</u>, 601[126,127,128, 129], <u>616</u>

Lombardi, F. J., 18, <u>38</u>, 146[158], 147[158], <u>166</u>, 181[89], 190, 191[89], 192[89], 197[202], 198[89], 202[89], 203, <u>215</u>, <u>218</u>, 231[126], 232[126], 275, <u>317</u>, 497[7], 527[156], <u>548</u>, <u>553</u>, 582[181], 586[181], 606[92], <u>615</u>, <u>618</u>

Long, M. M., 233[218], 235[218], <u>322</u>

Long, R. A., 19[147], <u>38</u>

Long, W. S., 295[199], <u>321</u>

Loomis, W., 64[56], <u>100</u>

Lorand, L., 336[147], <u>378</u>

Losick, R., 348[92], 349[92], 351[92], 353[92], <u>375</u>

Low, K. B., 139[90], 155[90], <u>164</u>, [222], 219, 225[7], 226[7], 300[7], <u>309</u>, 426[98], 433[98], 446[98], <u>458</u>

Lowery-Goldhammer, C., 499[12], <u>548</u>

Lozier, R. H., 278[125,127], <u>316</u>, <u>317</u>, 569[15], <u>612</u>

Lu, C. Y.-H., 295[199], <u>321</u>

Lubin, M., 177[62,63], 179[104], 181[63], 184[104], <u>214</u>, <u>215</u>, 332[73,74], 333[74], 338[74], 341[74], 342, 343[74], 344, 346[74], 354, 356[74], 359, 360[74], 361[74], 363, 366, <u>373</u>

Luckey, M., 244[128], 251[128], <u>317</u>, 431, 432[130], 434, 438, 439, <u>459</u>, 483[116], <u>491</u>

Lugtenberg, E. J. J., 418[32], <u>456</u>

Luke, R. K. J., 246[129,237], 250[237], <u>317</u>, <u>324</u>, 432[140,141, 149], 435[140,141], 436[149], 441[149], <u>459</u>, <u>460</u>

Lukoyanova, M. A., 521[124], <u>552</u>, 569[49], <u>613</u>

Lupo, M., 118[70], <u>125</u>, 186[131], <u>216</u>

Luria, S. E., 154[219], <u>168</u>, 425[90], 436, <u>458</u>, <u>460</u>, 603[36,130], 604[130,171,172], 606[36,172], <u>613</u>, <u>616</u>, <u>618</u>

Lusk, J. E., 226[161], 239[130], 241[161], 242[161], 253[161], 265[161], 303[161], [131], <u>317</u>, <u>319</u>, 604[111], 605, 606[131], <u>615</u>, <u>616</u>

Lute, M., 425[95,96], <u>458</u>, 473[48,49], <u>489</u>

M

Maas, W. K., 177[60,61], 179[107, 108,109], 184[61,107,108,109], 214, 215, 538[206], 555

Mabuchi, I., 525[142], 553

McCann, L., 279[40], 311

McCann, M. T., 20[155], 39

McCarty, R. E., 14, 36, 502[33], 513[78,79], 549, 550

McClees, J. S., 16[119], 22, 28[119], 33[209], 37, 41, 267, 268, 269[176], 270[176], 273, 274[176], 320, 562[192], 618

McClung, J. A., 504[55], 549

McConnell, H. M., 136, 163

MacDonald, R. E., 16[116], 21[116], 32[116], 37, 193[177, 178,179], 217, 307, 589[116, 117,132], 592[116,117,132], 593[116,117,132], 594[116], 611[236], 616, 620

Macdonald-Brown, D. S., 294[25], 310

McDowell, F., 56[40], 58[40], 65[40], 99

McElhaney, R. N., 15[112], 37

McElwee, P. G., 474[61], 490

McGinnis, J. F., 106[21], 107[20], 123

McGowan, E. B., 144[141], 165, 209[217], 219

McHugh, G. L., 333[104], 336[104], 376

McIntosh, M. A., 452[211], 462

McKay, L., 58[73], 68, 100

McKenna, C. E., 301[9], 309

McKillen, M. N., 396[69], 398[69], 399[69], 402[98], 405[98], 410, 411

MacKinnon, K., 333[111], 336[111], 349[111], 352[111], 355[111], 356[111], 361[111], 376

McLaren, J., 479[93], 491

McLellan, W. L., 151[202], 152[202], 167

MacLeod, R. A., 189[161], 217, 277[210], 322, 591[32,33,216, 217], 592[32,33,133,216,217], 593[216,217], 613, 616, 619

McMurry, L., 485[142], 492

MacPhee, D. G., 419[38], 456

Maeda, A., 453, 462

Maeda, M., 198[206], 218, 517[108], 551

Magasanik, B., 64[56], 100, 105[10], 122, 123, 149[190], 167

Magnani, J. L., 118[63], 125

Magnuson, J. A., 485[134,135], 492

Magnusson, R. P., 513[79], 550

Magrath, D. I., 435[138], 459

Mahajan, V. K., 542[233], 556

Maity, B. R., 140[102], 164

Makineni, S., 363[39], 371

Malamy, M. H., 226[230], 284[230], 285[11,229,230,231], 286[11,229, 231], 287[229], 290[230], 309, 323, 529[167], 532[171], 554

Male, C. J., 473[50], 489

Mallette, M. F., 331[20,21], 332[20], 334[20,21], 336[20,21], 338[20], 340[18,19], 341[20], 342[20], 343[20], 345[20,21], 346[19,20], 349[20,21], 352[20,21], 353, 356[20,21], 357[20,21], 358[20, 21], 359[20,21], 360[20], 361[20, 21], 362, 363, 364[20], 366[19, 20], 367, 370

Maloney, P. C., 11[57], 12[57,61], 35, 578[134], 579[134], 616

Mandelbaum-Shavit, F., 468[62], 474[62,63], 475[65,66,67], 490

Mandersloot, J. G., 15[112], 37

Manno, J. A., 580[135], 617

Mäntsälä, P., 479[94,95,96,97], 491

Marcovich, H., 447[196,197], <u>461</u>

Marcus, M., 180[134], 186[128, 129,133,134], 187[128,129], 210[129], <u>216</u>, 277[113], <u>316</u>, 536[197], <u>555</u>, 592[95], <u>615</u>

Marquis, R. E., 14[88], <u>36</u> 187[140], <u>216</u>, 223[132], <u>317</u>

Marrian, D. H., 363[159a], <u>379</u>

Marrs, B. L., 486[154], <u>493</u>

Martin, A., 17[132], <u>38</u>

Martin, H. H., 17[125], <u>37</u>

Martin, W. G., 135[57,58], <u>162</u>

Martonosi, A., 566[136], <u>617</u>

Maryanski, J. H., 58[72], 67, <u>100</u>, 129[10], <u>161</u>

Marzluf, G. A., 292[133,134], 293[135], <u>317</u>, 330[204], 332[203], 334[203,204], 336[203,204,205], 338[204,205], 339[203], 341[203,204,205], 342[205], 343[205], 345, 350[205], 354[205], 367[203, 204,205], 368[203,204,205], <u>382</u>

Masi, D. R. D., 9[29], <u>34</u>

Massey, V., 524[137], <u>553</u>

Masui, M., <u>306</u>

Mata, J. M., 150[195], <u>167</u>

Matheson, A. T., 334[93], 336[94], <u>375</u>

Matill, Ph., 334[56], <u>372</u>

Matin, A., 388[26], 391[26], 395[26], 398[26], 400[26], 401[26], <u>409</u>

Matney, T. S., 479[88,89,90], <u>490</u>

Matsuhashi, M., 149[177], <u>166</u>, 419[42], <u>456</u>

Matsumura, Y., 336[95], <u>375</u>

Matsuura, A., 471[33,37], <u>488</u>, <u>489</u>

Matthews, D. M., 326[96,97,98, 99], 327[97,98], 329[98], 330[99], 331[99,100,101],

[Matthews, D. M.] 332[99], 335[97,98], 338[97,98], 347[97,98], 364[1,2,3,4], <u>369</u>, <u>375</u>

Matula, T. I., 591[33], 592[33], <u>613</u>

May, B. K., 334[102], <u>375</u>

Mayhew, S. G., 524[137], <u>553</u>

Mayo, J., 58[51], 64[51], <u>100</u>

Mayshak, J., 332[103], 338[103], 343[103], <u>375</u>

Meadow, P. M., 401[84], 405[84], <u>411</u>

Measures, J. C., 297[136], <u>317</u>

Medić, M., 427[110], <u>458</u>

Medvecky, N., 529[166], 530[166], 532[168], <u>554</u>

Medveczky, H., 201[212], <u>219</u>

Mee, B., 477[79], <u>490</u>

Meentzen, M., 277[151], 278[151], <u>318</u>

Meersche, J. V., 85[107], 86[107], <u>102</u>, 393[38], <u>409</u>

Meinhart, J. O., 332[105], <u>376</u>

Meisch, H.-U., 301[137], <u>317</u>

Meisler, N., 337[106], 343[106], <u>376</u>

Melton, T., 65, <u>100</u>, 483[113], <u>491</u>

Melvydas, V., 582[54], 608[54], <u>613</u>

Menzel, J., 417[30], 451[207], <u>456</u>, <u>462</u>

Mergenhagen, S. E., 419[45], 420[45], <u>456</u>

Merrick, J. M., 542[229], <u>556</u>

Merrifield, R. B., 331[208,209], 332[208,209], 353[108], 358[108, 210], 368, <u>376</u>, <u>382</u>, <u>383</u>

Mertz, W., 300, <u>317</u>

Messer, A., 134, 195[50], <u>162</u>

Metelsky, S. T., 279[18], <u>309</u>

Metzer, E., 186[128,129], 187[128, 129], 210[129], <u>216</u>, 536[197], <u>555</u>

Meury, J., 224[116], 230[116], 296[116], 316
Mével-Ninio, M., 22, 26[175], 39
Meyn, T., 293[20], 310
Michel, G., [29], 456
Michel, H., 582[137], 597[137], 617
Michels, P. A. M., 19[144], 38
Michoalds, G. E., 532[176], 554
Miki, K., 523[132], 552
Miller, A., 332[109], 376
Miller, A. G., 297[139,140,141, 142], 298[139,140,141], 317, 318
Miller, A. K., 150[195], 167, 483[110], 491
Miller, C. G., 333[104,111], 336[104,110,111], 349[111], 352[111], 355[111], 356[111], 361[111], 376
Miller, D. L., 188[153,154], 217
Miller, H. K., 332[112], 376
Miller, R. C., Jr., 452[214], 462
Miller, T. W., 150[195], 167
Milne, M. D., 326[113,114,115], 376
Milner, L. S., 19[146], 38, 442[174], 460, 562[91], 615
Minamiura, N., 336[95], 375
Mindrich, L., 407[101], 411
Miner, K. M., 146[155], 166, 185[125], 186[125], 216, 592[138], 617
Mingioli, E. S., 442[175], 460, 466[2], 487
Mirsky, R., 504[51,52], 549
Mitchell, A. D., 502[29], 549
Mitchell, P. D., 2, 3, 5, 15[111], 25, 37, 40, 44[1,2], 45, 98, 108[32,33], 123, 128[1], 160, 172[10,11], 175[54,55], 210[17], 212, 213, 223, 268[144], 269, 271[182], 274[144], 275[226],

[Mitchell, P. D.]
277[144], 279[145], 280, 289, 318, 320, 323, 498, 499[22], 501, 516[95,96,97], 548, 551, 560[150], 561[44,46], 563[44,46,139,140, 141,142,143,152,153], 564[139, 140,141,142,143,146,152,153], 565[144,150], 566[148], 567[145, 147], 568[151], 569[147,150], 570[149], 574, 578[155,198,199, 226,227], 579[154,155,156], 583, 585, 588[226,227,228], 613, 617, 618, 619
Miura, T., 20, 39, 415[13], 455
Miyai, K., 74[87], 101, 131[37], 157[37], 162
Miyairi, S., 130[21], 161
Miyata, I., 468[29], 470[28,29], 471[29,30], 488, 543[238], 556
Mizuno, S., 483[120], 492
Mizushima, S., 20, 39, 415[13], 416[26], 423[77], 455, 457
Mlynar, D., 336[212,213], 383
Mocholes, S., 150[195], 167
Moczydlowski, E. G., 109[42], 118[67], 120, 124, 125
Mogelson, J., 227[185], 242[185], 243[185], 254[185], 256[185], 257[185], 259[185], 271[185], 320
Momsen, W., 519[111], 552
Monod, J., 2, 5, 15[100], 37, 104, 105[5], 122, 132[40], 133[49], 138[77,78], 153[78], 157[49], 162, 163, 172[1,22], 173[1], 175[1], 212, 222[39], 311
Monroy, G. C., 509[68], 510[68], 550
Monteil, H., 504[59], 550
Moo-Penn, G., 347[14], 370
Mora, J., 187[137], 216, 482, 491
Mora, W. K., 117[62], 119[62], 124
More, J., 14[87], 36

Morikawa, A., 10[45], 14[45], <u>35</u>, 179[102], 183[102,103], 184[102], 196[102,103], 197[103], <u>215</u>

Morris, D., 504[47,48], 511[48], 517[47], 518[47], <u>549</u>

Morris, J. E., 334[25], <u>370</u>

Morris, J. G., 589[17], <u>612</u>

Morris, R. G., 533[183], <u>554</u>

Morse, M. L., 50[27,28,29], 69[74,75], 70[75,80,81], 78[74, 75], 79, <u>99</u>, <u>100</u>, <u>101</u>, 129[14, 15], 130[15], <u>161</u>

Mortenson, L. E., 282[239], 299[33,54,55,239], 300[239], <u>311</u>, <u>312</u>, <u>324</u>

Motojima, K., 33[213], <u>41</u>, 175[44], 198[44], 203[44], <u>213</u>, 547[249], <u>556</u>, 610[6], <u>612</u>

Moudrianakis, E. N., 504[54,55], <u>549</u>

Moyle, J. M., 271[182], <u>320</u>, 499[22], 516[96], <u>548</u>, <u>551</u>, 563[152,153], 564[152,153], 578[155], 579[154,155,156], <u>617</u>

Mueller, J. H., 363[116], <u>376</u>

Mühlradt, P. F., 416[21], 417[21], 419[21], 451, <u>455</u>, <u>456</u>, <u>462</u>

Muller, C. R., 137[71], <u>163</u>, 526[148], <u>553</u>

Müller-Hill, B., 160[255], <u>169</u>

Mulligan, J. H., 469[82], 477[82], <u>490</u>

Munn, E. A., 502[29], <u>549</u>

Munoz, E., 504[49], 507[61], 509[72], 514[81], <u>549</u>, <u>550</u>, <u>551</u>

Murakawa, S., 395[54,55,57], 396[54,55], <u>410</u>

Muramatsu, N., 504[53], <u>549</u>

Muraoka, S., 198[204], <u>218</u>

Murayama, T., 334[93], <u>375</u>

Murgier, M., 334[82,117], <u>374</u>, <u>376</u>

Murphy, J. C., 300[187], <u>320</u>

N

Nachbar, M. S., 501, <u>548</u>

Nagel, de Zwaig, R., 156[228,229], <u>168</u>

Naider, F., 331[69], 332[119], 333[11,66,118], 336[10,66,119], 338[66], 341[118], 342[118], 343[66], 344[66], 348[11,66,119], 349[10,119], 350[66,119], 352[10, 66,119], 353[10,11,66,118,119], 355, 357[11,119], 358, 360[11], 362, 366[10], 367[66], 368[119], <u>369</u>, <u>373</u>, <u>376</u>

Nakae, T., 9[34,35], 13[34,35], 14[34,35], 20[34,35], <u>35</u>, 149[181, 182], <u>166</u>, 419[36], 421, 422[36], 423[36,78], 430[63], <u>456</u>, <u>457</u>

Nakamura, K., 423[77], <u>457</u>

Nakane, P. K., 9[25], <u>34</u>, 194[188], <u>218</u>

Nakayama, H., 471[38,41], 472[42], <u>489</u>

Nakazawa, A., 48[14], <u>98</u>

Nakazawa, T., 48[12], 49[19], 50[12, 19,25], 51[12,19], 52[12,19], 53[19], 68[12], 78[12], <u>98</u>, <u>99</u>, 108[38], <u>124</u>

Nanninga, N., 418[31], 456, 524[140], <u>553</u>

Naraki, T., 181[75], 196[75], 199[75], 201[75], <u>214</u>

Narconis, R. J., 486[154], <u>493</u>

Nash, W. C., 266[24], 267[24], 270[24], 271[24], 274[24], <u>310</u>

Neal, J. L., 196, <u>218</u>

Neale, S., 179[105], 184[105], <u>215</u>

Neef, V. G., 470[23], <u>488</u>

Neet, K. E., 104[4], <u>122</u>

Neidle, A., 332[109], <u>376</u>

Neilands, J. B., 222[146], 243[146], 244[128], 251[128,222], 252[124], <u>316</u>, <u>317</u>, <u>318</u>, <u>323</u>, 430[117,118, 120,122], 431[127,130], 432[126,

[Neilands, J. B.]
  128,130,157], 434[127],
  435[120,142], 436[157],
  438[130], 439[130], 440[157],
  441, 458, 459, 460, 483[116],
  491
Neiuwenhuis, F. J. R. M., 510[73],
  550
Nekvasolova, K., 330[120],
  334[120], 377
Nellis, L. F., 484[132], 492
Nelson, D. L., 241[147,148], 318,
  606[131], 616
Nelson, H., 506[43], 509[67],
  510[67], 517[107], 518[107],
  549, 550, 551
Nelson, N., 504[45], 506[43],
  509[67], 510[67], 513[45],
  517[45,107], 518[45,107],
  519[112], 549, 550, 551, 552
Nelson, S. O., 188[147], 196[147],
  216
N'Eman, Z., 524[138], 553
Nemecek, I. B., 279[45], 311
Nemos, G., 156[234,235], 168
Neu, H. C., 3[12], 5, 9[18], 34,
  131, 142[39], 150, 162, 172,
  174[6,33], 212, 213, 575, 617
Neuhaus, F. C., 177[66], 178[66],
  181[66,70], 187[139], 214, 216,
  482[108], 491
Neujahr, H. Y., 468[25], 470[25,
  26,27], 488
Neumann, C., 281[22], 310
Neusch, J., 483[115], 491
Nevins, M. P., 442[182], 461
  467[12], 488
Newhard, J., 93[124], 94[124],
  102
Newman, M. J., 59[95], 79, 83,
  101, 117[61], 119[61], 124
Newton, A., 149[186], 167,
  248[221], 249[221], 323,
  432[134,150,151], 434[134],

[Newton, A.]
  436, 440[150], 441[134], 450,
  459, 460
Newton, N. A., 432[149], 436[149],
  441[149], 460
Ng, M. H., 509[72], 550
Ngo, D. T. C., 479[93], 491
Nichoalds, G. E., 9[25], 34, 181[80],
  194[188], 214, 218
Nieuwenhuis, F. R. J. M., 517[106],
  518[106], 551
Nieva-Gomez, D., 603[158], 617
Nikaido, H., 9[34,35], 13[34,35],
  14[34,35], 20[34,35], 35, 149[181],
  166, 414, 416[25,28], 417[1,28],
  418[28,34], 419[36], 420[1,49],
  421, 422[36], 423[36], 430[63],
  455, 456
Nikaido, K., 10[51], 35
Nishi, A., 336[121], 344[121], 377
Nishimune, T., 468[31,32], 471[31,
  32], 488
Nishino, A., 468[44], 471[34],
  472[44], 489
Niven, D. F., 567[161], 587[159,
  160,161], 589[159,160,161],
  590[159,160], 617
Noack, D., 452[213], 462
Nobrega, F. G., 499[14], 548
Nochoalds, G. E., 146[159], 166
Nomura, M., 453, 462
Nonomura, Y., 193[173], 217
Norberg, C. L., 295[77], 313
Nordström, K., 149[178], 166,
  419[40,41], 421[64], 425[93],
  456, 457, 458
Normark, S., 149[178], 166, 416[19],
  419[40], 420[54], 455, 456
Norrell, S. A., 442[181], 461,
  466[8], 467[8], 487, 531[239],
  544[239], 556
Norton, J. E., 525[144], 553
Nose, Y., 468[29,44], 470[28,29],
  471[29,30,33,34,37], 472[44,45,

[Nose, Y.]
  46], 488, 489, 543[238],
  556
Nosoh, Y., 504[53], 549
Nossal, N. G., 9[19], 34,
  174[32], 213
Novak, D., 427[110], 458
Novel, G., 386[7], 408
Novotny, C. P., 584[162],
  617
Nunn, W. D., 15[102], 37
Nurmikko, V., 479[96], 491
Nutter, W. E., 332[166], 338[166],
  341[166], 380

O

O'Barr, T. P., 332[122], 377
O'Brien, I. G., 251[149], 252[149],
  318, 430[119], 432[149], 435[119,
  139,146], 436[149], 441[149],
  459, 460
Ødegaard, P., 331[45], 371
Odom, R., 389[32], 392[32], 409
Oehr, P., 241[227], 252[150,227],
  318, 323, 402[99,100], 406[99,
  100], 411
Oeschger, M. P., 448, 461
Oesterhelt, D., 277[151,153],
  278[151,152,153], 318, 515[86],
  551, 579[163], 582[137],
  593[163], 597[137], 617
O'Farrell, P. H., 10[49], 35
O'Gara, F., 307
Oginsky, E. L., 442, 460, 466,
  487
Ohki, M., 178[72], 179[72],
  181[72], 182[72], 206[72], 214
Ohno, K., 503[38], 515[38], 549
Ohta, N., 290[154], 291[154],
  292[154], 318, 529[165], 554
Oishi, M., 499[14], 548
Okamoto, H., 507[60], 514[60],
  516[60], 550
Olden, K., 25[184], 28[184], 40

Olivera, B. M., 479[93], 491
Olsen, W. L., 415[18], 452[18], 455
Olson, R. E., 486[154], 493
Ondera, K., 424[80], 457
Oppenheim, J. D., 24, 39
Ordal, G. W., 144, 148[129,130],
  165, 521[122], 528[161], 541[220,
  224], 552, 553, 555
Ornston, L. N., 392[29], 409
Ornston, M. K., 392[29], 409
Ortega, M. V., 395[56], 396[56],
  410
Osborn, M. J., 17[134], 20, 38,
  141[107], 151[202], 152[202],
  156[107], 164, 167, 415[14],
  416[14], 431, 455
Ose, D. E., 306
Oshima, R., 528[157], 553
Ostroumov, S. A., 279[45], 311
Ostrovskii, D. N., 521[124], 552,
  569[49], 613
Otsuji, N., 532[170], 554
Ovchinnikov, Yu. A., 306
Overath, P., 84[102,103,106],
  85[102,103,106,109], 86[109],
  101, 102, 136[69], 153[69], 163,
  389[44], 393[34,39], 394[43,44],
  395[44,49], 409, 410, 419, 456
Owens, M. S., 300[117], 316
Oxender, D. L., 3, 5, 8[3,4], 9[3,
  4,25], 10[4], 12[3,4,62], 13[62],
  14[4,77], 34, 35, 36, 138[75],
  146[159,167], 163, 166, 172[16,
  20], 174[39], 177[68], 178[68],
  179[69,88], 181[68,69,74,80,81,
  88,93,94], 182[74,88], 194[39,
  186,188], 196[69,74], 199[69,74,
  186], 200[74,186], 201[74,81,93],
  202[74], 203[74,88], 207[74], 212,
  213, 214, 215, 218, 327[125],
  332[123,124], 337[123,124], 377,
  530[173], 532[173,176,177,179],
  533[179], 554
Oya, N., 478[83], 490

P

Padan, E., 596[164], 598[164], 617

Pai, C. H., 476, 477[77,81], 490

Paigen, K., 105[9], 106[21], 107[20], 123

Paine, T. F., 561[45], 613

Panchenko, L. F., 130[28], 161

Panet, R., 566[165], 617

Papineau, D. J., 236[100], 276[100], 279[100], 315, 582, 586[65], 588[65], 596[64], 614

Parada, J. L., 395[56], 396[56], 410

Pardee, A. B., 9[26], 34, 140[92, 93], 157[92], 158[92], 164, 194[187], 218, 286[157], 290[47,154,160], 291[47,154, 155,156,159,160], 292[154, 159], [158], 312, 318, 319, 396[69], 398[69], 399[69], 400[76], 402[96,98], 405[96, 98], 410, 411, 527[152], 528[162,163], 529[164,165], 530[163], 553, 554

Parisi, E., 17[134], 20[134], 38, 415[14], 416[14], 431[14], 455

Park, M. H., 226[161], 241[161], 242[161], 253[161], 265[161], 303[161], 319

Parnes, J. R., 14[82], 36, 146, 151[153], 153[153], 158[153], 166

Parra, R. J., 432[148], 433[148], 436[148], 441[148], 460

Parsons, R. G., 144[139,142], 145[139,142], 195[139,142], 165

Passetto-Nobrega, M., 499[14], 548

Pastan, I., 106[13,14], 123

Patel, L., 19[148], 38

Pateman, J. A., 283[162], 319

Patni, N. J., 58[65], 66[65], 100, 129[11], 161

Paton, W. H. N., 263[163], 319

Patterson, E. K., 307, 336[54,126, 127], 372, 377

Pauli, G., 84[103], 85[103,109], 86[109], 101, 102, 393[39], 394[43], 395[49], 409, 410

Pavlasova, E., 234[96], 236[96], 276[96], 279[96], 315, 563[166], 567[66], 569[166], 579[66], 596[66], 605[66], 614, 617

Payne, J. W., 326[99,131,138,139, 140,143], 327[140], 330[99,140, 142], 331[99,100,101,133], 332[99,138,140,143], 333[128, 129,130,132,136,137,139,141], 334[139], 335[139], 336[132,133, 134,139,140], 337[131,132,135, 136,141], 338[131], 339[138], 341[128,129,130,132,137,143], 342[129,130,135,136,141], 343[131], 344[131,132,134], 346[131,138,143], 347[138,139, 143], 348[128,129,137], 349[128, 129,137], 350[137], 351[130,137], 353[129,137], 354[138], 355[141], 356[141], 357[141], 360[131], 361[141], 362[143], 363[131,136], 364[1,2,3,4,132], 365[144], 366[128,130,132], 367[128,129, 131], 369, 375, 377, 378, 421, 457

Payne, W. J., 293[164], 294[164], 319

Pearce, L. E., 335[145], 378

Pecht, M., 333[146], 335[146], 378

Pedersen, P. L., 499[15], 504[15], 507[15], 518, 548

Pellissier, C., 334[83], 374

Penefsky, H. S., 499[21], 502[32], 549, 549

Penrose, W. R., 146[159], 166, 181[80,81], 201[81], 214, 532[176,

[Penrose, W. R.]
177], 554
Perlman, G. E., 336[147], 378
Perlman, R., 106[13,14], 123
Perry, J. W., 158[248], 169
Pershadsingh, H. A., 581[118],
596[118], 616
Peterkofsky, A., 106[18], 123
Peters, V. J., 331[155], 332[148],
359, 363[148], 378, 379
Pfaff, E., 14[93], 36
Phelps, D. C., 517[103], 518[103],
551
Phibbs, P. V., 129[19], 161
Phillips, A. W., 330[55,149], 372,
379
Phillips, G. N., Jr., 542[233],
547[256], 556, 557
Phillips, S. K., 569[29], 605[29,
167], 613, 617
Pick, U., 581[168], 617
Pickart, L., 331[150], 379
Pietrzyk, C., 583[69], 614
Pilwat, G., 229[165], 236[165],
275[165], 319
Pine, M. J., 336[151], 379
Pinnock, C., 331[152], 379
Piperno, J. R., 3, 5, 14[77], 36,
146[159], 166, 179[69], 181[69,
80], 194[186], 196[69], 199[69,
186], 200[186], 214, 218,
530[173], 532[173,176], 554
Pittman, K. A., 330[153], 332[153,
154], 338[154], 342[154],
343[154], 379
Plante, L. T., 474[60], 490
Plate, C. A., 154[219], 168,
604[169,171,172], 606[172],
607, 617, 618
Podoski, E. P., 386[3], 401[3],
405[3], 408
Pogell, B. M., 140[102], 164
Pollack, J. R., 244[128], 251[128],
317, 430[120], 431[127],
434[127], 435[120,142], 459,
483[116], 491

Pollock, J. J., 524[141], 553
Portis, A. R., Jr., 14[96,97,98],
36
Postma, P. W., 8[10], 9[10], 24, 25,
26, 34, 40, 46[7], 49[7], 53[7],
54[7], 58[7], 63[7], 64[7], 65[7],
69, 73, 74[7], 75[89], 79[7], 98,
101, 104[3], 120[3], 122, 131[36],
139[36], 162, 175[48], 213, 281,
322, 393[37], 396[79], 401[79,80,
81], 402[93], 404[79,93], 409,
411, 496[4], 497[4], 548, 560[207],
569[207], 619
Potalies, R., 386[7], 408
Potter, A. L., 173[21], 212
Pouysségur, J. M., 156[226],
157[237,238], 168, 387[9,10,11,
12,13,15,20], 388[9,10,11,13,20],
391[10], 408, 409
Power, J., 155[222], 168
Prakash, O., 469[78], 476, 477[78,
79], 490
Prasad, R., 19[151], 21[168,169],
38, 39, 589[173], 618
Prescott, J. M., 331[155], 332[148],
334[49a,90,91,156,196,201],
359[148], 363[148], 372, 374,
375, 378, 379, 382
Pressman, B. C., 306
Prestidge, L. S., 140[92], 157[92],
158[92], 164, 194[184,185], 218,
[158], 291[159], 292[159], 319,
528[162], 553
Preston, Y. A., 442[183], 443[183],
461, 467[11], 488
Price, H. D., 143[127], 144[127],
145[127], 165, 195[194], 218,
539[212], 540[212], 555
Price, M. B., 333[61], 336[61],
354[61], 356[61], 361[61], 373
Puck, T. T., 441[169], 460
Pugsley, A. P., 248[166], 249[166],
306, 319, 432[156], 433[156,159],
436[156], 437[159], 438, 440[156],
460

Pullman, M. E., 502[32], 509[68], 510[68], 549, 550
Purdy, D. R., 599, 618
Putnam, E. W., 173[21], 212

Q

Quay, S. C., 12[62], 13[62], 35, 172[20], 179[88], 181[88,94], 182[88], 203[88], 212, 215
Quinlan, D. C., 2[9], 5
Quiocho, F. A., 542[233], 547[256], 556, 557

R

Rabinowitz, M., 181[73], 214
Racker, E., 25[188], 32[207,208], 33[211], 40, 41, 277[168], 278[167,168], 319, 499[23], 500, 502[32,33], 506[43], 509[67], 510[67], 515, 516[84], 517[107], 518[107], 548, 549, 550, 551, 609[175,176,177], 618
Rader, J. I., 470[23], 488
Rader, R. L., 2[9], 5, 95[130], 97, 102
Radojkovic, J., 138[79], 143, 163
Rae, A. S., 286[169], 319, 532[168], 554, 600[178,179], 618
Rahmanian, M., 146[167], 166, 181[74,93], 182[74], 196[74], 199[74], 200[74], 201[74,93], 202, 203[74], 207, 214, 215, 532[179], 533[179], 554
Ramirez, J. M., 109[43], 119[73], 120[43], 124, 125
Ramos, F., 400[75], 410
Ramos, S., 157[242], 169, 387[17], 408[17], 409, 479[100], 491, 496[6], 520[6], 548, 563[180], 580[180], 582[180], 598[180], 608[180], 618
Randall, L. L., 420[51], 456

Randall-Hazelbauer, L., 158[247], 169, 427[108], 454[108], 458
Rao, M. R. R., 405[95], 411
Rao, N., 425[95], 458
Rapley, L., 14[92], 36
Rashed, I., 144[254], 169
Ratledge, C., 251[26], 310
Raufuss, E., 84[102], 85[102], 101
Raven, J. J., 160[257], 169, 611[237, 239], 620
Ravnio, R. P., 525[143], 553
Ray, L. E., 334[157], 379
Rayman, M. K., 185[123], 216, 395[59], 396[59], 397[59,66], 410, 601[126], 616
Raymond, J., 231[123], 262[123], 316
Razin, S., 58[99], 101, 129[9], 130[9,27], 161, 418[35], 456, 524[138], 553
Reader, R. W., 521[122], 552
Reber, G., 130[22,23,24], 161
Reeve, E. C. R., 483[122], 492
Reeves, H. C., 393[41], 394[41], 409
Reeves, J. P., 24[179], 29[199], 39, 40, 122[81,82], 125, 175[52], 213, 231[126], 232[126], 275[126], 317, 582[181], 586[181], 606[92], 615, 618
Reeves, P., 248[166], 249[166], 306, 319, 420[57,58], 424[81], 425[57, 58], 431[94], 432[94,156], 433[94, 156,159], 436[94,156], 437[159], 438, 439[94], 440[156], 441[94], 446[57,58,94], [88], 451[222], 456, 457, 458, 462
Rehn, K., 415[9,10], 455
Reid, K. G., 189[156,157], 217
Reitz, R. H., 187[139], 216
Renthal, R., 193[179], 217, 581[182], 589[116], 592[116,182], 593[116], 594[116,182], 597[182], 616, 618
Ressler, C., 358[210], 383

Rest, R. F., 129[18], <u>161</u>

Reuser, A. J. J., 401[80,81], <u>411</u>

Reynard, A. M., 484[132], <u>492</u>

Reynolds, H., 332[122], <u>377</u>

Rhoads, D. B., 10[44], 12[44], 14[81], <u>35</u>, <u>36</u>, 227[170,171], 228, 229[171], 230[171], 232[171], 236[171], 237[170, 171], 275[171], <u>319</u>, 532[172], <u>554</u>, 600[183], 602[183,184], 603, 611[183], <u>618</u>

Richards, F. M., 425[92], <u>458</u>

Richardson, J. P., 499[12], <u>548</u>

Richarme, G., 195, <u>218</u>, 540[214], 541[214], <u>555</u>

Richey, D. P., 131[32,33], <u>161</u>

Richman, J., 159[252,253], <u>169</u>

Richmond, M. H., 420[60], <u>457</u>

Rick, P. D., 141[107], 156[107], <u>164</u>

Rickenberg, H. V., 15[100], <u>37</u>, 106[17], <u>123</u>, 133[49], 157[49], <u>162</u>, 172[22,23], 181[23], 199, <u>212</u>, 332, <u>370</u>

Riebling, V., 11[54], <u>35</u>

Rinehart, C. A., 593[82], 600[82], <u>615</u>

Riordan, C., 76, 83, <u>101</u>, 131, 139[35,89], 155[89], <u>161</u>, <u>164</u>

Ritt, E., 14[93], <u>36</u>

Roantree, R. J., 419[38], <u>456</u>

Robbie, J. P., 149[180], <u>166</u>, 420, <u>457</u>

Robbins, A. R., 144[133], 146[156, 157], 147, 148[156,157], <u>165</u>, <u>166</u>, 541[222], 544[241,242], <u>555</u>, <u>556</u>

Robbins, J. C., 177[68], 178[68], 181[68], <u>214</u>

Robert-Baudouy, J., 156[234, 235], <u>618</u>, 386[7], <u>408</u>

Roberts, I. Z., 561[185], <u>618</u>

Roberts, R. B., 1[3], 2[3], <u>5</u>, 93, <u>102</u>, 561[185], <u>618</u>

Robertson, D. C., 129[18], <u>161</u>

Robertson, D. E., 535[191], <u>554</u>

Robertson, R. N., 280[172], <u>319</u>

Robin, A., 155[225], 156[225], <u>168</u>, 224[116], 230[116], 296[116], <u>316</u>, 588[186], <u>618</u>

Rodgers, G. C., 430[117], <u>458</u>

Rodriquez-Navarro, A., <u>307</u>

Rodwell, V. W., 188, <u>217</u>

Rogers, H. F., 592[133], <u>616</u>

Rogers, M., 484[133], <u>492</u>

Rogers, T. O., 469[73], 476[74], <u>490</u>

Roisin, M. P., 518[104], <u>551</u>

Rola, F. H., 499[14], <u>548</u>

Rolfe, B., 424[80], <u>457</u>

Rollin, C. F., 336[94], <u>375</u>

Romano, A., 56[40,41], 58[40], 59[41], 65[40], <u>39</u>

Roncari, G., 335[158], <u>379</u>

Roon, R. J., 283[173], <u>307</u>, <u>319</u>

Rosano, C. L., 486[148], <u>492</u>

Rose, S., 55[38], <u>99</u>

Roseman, S., 3[15], <u>5</u>, 8[7], 9[7], 18[7], <u>34</u>, 46[7], 47[11], 48[12], 49[7,18,19], 50[12,19,21,22,23, 24,25], 51[12,19,30], 52[12,19], 53[7,19,22,23], 54[7,34], 56[24], 58[7,11,66], 63[7,35], 64[7,24], 65[7,35], 66[66], 67[24], 68[12], 69, 70, 73[83], 74[7], 75[89], 78[12], 79[7], 81[44], <u>98</u>, <u>99</u>, <u>100</u>, <u>101</u>, 106[25], 108[34,35,37,38], 110[51,52], 111[52], 112[53,54, 55], 113[51,52,53,54,55], 114[54, 55], <u>123</u>, <u>124</u>, 129[3,8], 139[86], 140, 158[98], 195[86], <u>160</u>, <u>161</u>, <u>163</u>, <u>164</u>, 172[4,5], <u>212</u>, 343[159], <u>379</u>, 483[113], <u>491</u>, 564[112], 574[112], 575[113], 592[213], <u>616</u>, <u>619</u>

Rosen, B. P., 3[14], 5, 9[24], 12[24], 13[24,64,67], 15[67], 16[119], 19[150], 20, 22[173], 25, 27[67], 28[119,190], 33, 34, 35, 37, 38, 39, 40, 41, 107[29], 123, 146[161,163,166], 147[163], 150[197], 156[197], 166, 167, 172[17], 179[110,111], 184[110, 111,112], 189[169,170], 196[170], 212, 215, 217, 267[1], 268, 269[176,214], 270[176], 273, 274[176], 279[174,175], 308, 320, 334[58], 345[58], 372, 486[153], 487[153], 492, 516[90,91], 517[109], 530[187], 203,205], 533[187], 537[201, 202,203], 538[205,207], 545[202], 551, 554, 555, 562[192,219,220], 569[188], 578[187,191], 582[191], 588[210,221], 591[219,220, 221], 602[189], 605[2,3,187], 607[187,188], 610[190], 612, 618, 619
Rosen, O. M., 106[15], 121[78], 123, 125
Rosenberg, H., 9[33], 35, 149[188], 167, 201[212], 219, 226[70,177], 244[178], 246[71, 178], 247[178], 248[70,71], 249[178], 250[178], 252[70, 178], 253[71,178], 285[73,177], 286[73,74,177], 287[74,177], 289[177], 290[74], 313, 320, 398[67], 399[67], 410, 430[123], 432[147,149,158], 434[133, 158], 435[133], 436[147,149], 437, 440[158], 441[147,149, 158], 450, 453[158], 459, 460, 529[166], 530[166], 532[168, 169], 547[257], 554, 557
Rosenberg, T., 151[205], 167, 173[28,30], 213
Rosenbusch, J. P., 423[74], 457

Rosenfeld, H. J., 179[109], 184[109], 188, 215, 216
Rosing, J., 517[110], 518[110], 519[110], 552
Roth, J. H., 189[163], 201[163], 217
Rothfield, L. I., 420[56], 456
Rothman-Denes, L. B., 106[16], 123
Rothstein, A., 298[32,179], 299[32, 179], 311, 320
Rotman, B., 76[91], 101, 106[24], 123, 138[73,79], 140[73], 142, 143, 146[73,154,156,157], 147[157], 148[156,157], 155[73, 80], 157[73], 158[73,80,94], 163, 164, 166, 210[219], 219, 541[217], 544[241,242], 555, 556, 584[47], 613
Rottem, S., 418[35], 456
Rottenberg, H., 582[193], 570[194], 582[195], 596[164], 598[164], 617, 618
Roussel, G., 504[59], 550
Rowland, I., 184[106], 215
Rowlands, D. A., 363, 379
Roy, L. E., 334[195], 382
Roy-Burman, S., 91, 92[122], 93[122], 102
Rubin, C. S., 121[78], 125
Rudnick, G., 134, 135, 137, 162, 175[56], 213, 481[103], 491
Ruiz-Herrera, J., 402[87], 404[87], 411
Rumley, M. K., 13[70], 24[70], 36, 133[48], 136[48], 162, 175[51], 213
Russel, R. P. B., 177[65], 178[65], 214
Ryabova, I. D., 24, 40, 306
Ryden, A. C., 334[12], [160], 370, 379
Ryter, A., 454[218], 462

S

Sabet, S. F., 444, 445, <u>461</u>
Sachan, D. S., 404[90], <u>411</u>
Sacktor, B., 2[10], <u>5</u>
Saier, M. H., Jr., 20, <u>39</u>, 50[21], 58[66], 59[95], 66[66], 73[83], 79, 83, <u>99</u>, <u>100</u>, <u>101</u>, 106[19, 23,25], 108[36], 109[45], 110[51,52], 111[52], 112[23,53, 54,55], 113[23,51,52,53,54,55], 114[19,45,54,55], 116[60], 117[61,62], 118[60], 119[61,62], 121[60], <u>123</u>, <u>124</u>, 129[8], 131[38], 139[86], 195[86], <u>161</u>, <u>162</u>, <u>163</u>, 388[30], 392[30], <u>409</u>
Salaj-Šmic, E., 427[110], <u>458</u>
Salanitro, J. P., 85[108], 86[108], <u>102</u>, 389[40], 393[35,36,40], <u>409</u>
Salton, M. R. J., 9[17], 24, <u>34</u>, <u>39</u>, 501, 507[61], 509[72], 524[141], <u>548</u>, <u>550</u>, <u>553</u>
Samuel, D., 84[105], <u>102</u>, 394[47], <u>410</u>
Sanadi, D. R., 22[176], 25[176], 26[176], <u>39</u>, 566[165], <u>617</u>
Sanborn, D., 333[17], 348[17], <u>370</u>
Sanchez, R. S., 156[228], <u>168</u>, 387[18], 388[18], <u>409</u>
Sander, D. C., 479[98], <u>491</u>
Sanderman, H., 526[148], <u>553</u>
Sandermann, H., Jr., 137[71], <u>163</u>
Sanderson, K. E., 420[52], <u>456</u>
Sandine, W., 58[73], 68[73], <u>100</u>
Sanemori, H., 472[43], <u>489</u>
Sanno, Y., 130, <u>161</u>
Santy, P. A., 200[208], <u>218</u> 533[184], <u>554</u>
Sanwal, B. D., 150[200], <u>167</u>, 185[123], <u>216</u>, 386[5], 395[59, 60,61,62], 396[59,61,62,63], 397[59,61,63,66], <u>408</u>, <u>410</u>,

[Sanwal, B. D.] 526[150,151], <u>553</u>, 601[126,127, 128,129], <u>616</u>
Sarid, S., 333[9], 334, <u>369</u>, <u>379</u>
Sarvas, M. O., 143[125], 158[125], <u>165</u>, 201[214], <u>219</u>, 541[218], <u>555</u>
Sasaki, T., 467[18,19], <u>488</u>
Sato, T., 149[177], <u>166</u>, 419[42], <u>456</u>
Satre, M., 509[70], <u>550</u>
Savagean, M. A., 178[71], 181, <u>214</u>
Sawyer, L. E., 331[60], <u>373</u>
Sawyer, M. H., 129[13], <u>161</u>
Scarborough, G. A., 13[70], 24[70], <u>36</u>, 50[26], <u>99</u>
Schachtele, C., 58[51], 64[51,52], <u>100</u>
Schachter, H., 62, 96[42], <u>99</u>
Schaechter, D., [135], <u>617</u>
Schaedel, P., 157[240], <u>168</u>
Schaeffer, S., 55[36], <u>99</u>
Schaff, R., 524[140], <u>553</u>
Schairer, H. U., 84[103], 85[103], <u>101</u>, 394[43], <u>409</u>, 576, 577, <u>618</u>
Schairer, V., 569[196], 582[196], <u>618</u>
Schaller, K., 431[125], <u>459</u>
Schandell, A., 519[115], <u>552</u>
Schatz, G., 32[208], <u>40</u>
Schechter, E., 20[156], <u>39</u>, 419[37], <u>456</u>
Schellenberg, G. D., 533[183], 536[193], 547[252], <u>554</u>, <u>556</u>
Scherrer, R., 422, <u>457</u>
Schlecht, S., 419[39], <u>456</u>
Schleif, R., 146[164], 155[223], 195[164], <u>166</u>, <u>168</u>, 531[231], 542[231], <u>556</u>
Schloemer, R. H., 295[180,181], <u>320</u>
Schmid, B. I., 569[196], 582[196], <u>618</u>
Schmid, K., 159[249], <u>169</u>

Schmidt, G., 419[39], <u>456</u>
Schmitges, C. J., 424[86], <u>457</u>
Schmitt, R., 140[97], 158[97], 159[249], <u>164</u>, <u>169</u>
Schnaitman, C., 424[87], <u>457</u>
Schnaitman, C. A., 9[29], 20[158], <u>34</u>, <u>39</u>, 149[184], <u>167</u>, 265[44], <u>311</u>, 415[15,17], 416[15,17,22, 23,27], 423[22,23], 443[185], 444, 445[185,192,193], 447[192], 448[201], 449[201], 450[22], [88], 453[192], 454[193], <u>455</u>, <u>457</u>, <u>461</u>, 467[14], 484[130], <u>488</u>, <u>492</u>
Schnebli, H. P., 507[62], <u>550</u>
Schneider, D. L., 32[207], <u>40</u>
Schneider, H., 135[57,58], <u>162</u>
Scholes, P., 271[182], <u>320</u>, 578[198,199], <u>618</u>
Schor, M. T., 509[72], <u>550</u>
Schrecker, O., 48[13], 66[13], <u>98</u>
Schuhmann, L., 277[151], 278[151], <u>318</u>
Schuldiner, S., 15[108,109], 18[109], 19[148], <u>37</u>, <u>38</u>, 109[46], <u>124</u>, 135[62,63], 136[61], 137[62,70], 157[61, 62,242], <u>163</u>, <u>169</u>, 175[53,56], <u>213</u>, 479[100], 481[102], <u>491</u>, 496[6], 520[6], <u>548</u>, 563[180, 200], 580[180], 582[180,200], 586[201,202,203], 595, 598[180], 602[201], 603[201], 608[180], <u>618</u>, <u>619</u>
Schultz, S. G., 230[60,61], 287[224], 296[60,183], <u>312</u>, <u>320</u>, <u>323</u>, 583[204], 591[204], 605[234], <u>619</u>, <u>620</u>
Schvo, Y., 349[164], 350[164], 352[164], 353[164], 358[164], 359[164], 362[164], <u>380</u>
Schwartz, Z., 566[205], <u>619</u>
Schwartz, A. C., 524[136], <u>552</u>
Schwartz, J. H., 177[61], 184[61], <u>214</u>, 538[206], <u>555</u>

Schwartz, M., 144[140], 145[140, 150], 149[140], 154[216], 158[140, 216,245,246,247], <u>165</u>, <u>168</u>, <u>169</u>, 175[41], 195[41], 209[218], <u>213</u>, <u>219</u>, 426[100,106,107], 427[108, 109], 428[114], 429, 454[108,218], <u>458</u>, <u>462</u>, 541[216], 547[254], <u>555</u>, <u>557</u>
Schwarz, H., 136[69], 153[69], <u>163</u>
Schwarz, V., 415[8], <u>455</u>
Scott, J. M., 473[55], 474[61], <u>489</u>, <u>490</u>
Scribner, H. E., 227[185,196], 241[184], 242[185,196], 243[184, 185], 254[185], 256[185,196], 257[185,196], 259[185], 266[197], 269[197], 270[197], 271[185,196], 274[197], <u>320</u>, <u>321</u>
Segel, I. H., 282[83], 283[83], 292[216], 293[20,41,216], 295[77], 300[216], 301[107], <u>310</u>, <u>311</u>, <u>313</u>, <u>314</u>, <u>315</u>, <u>322</u>
Sekizawa, J., 423[75], <u>457</u>
Semenov, A. Y., 279[45], <u>311</u>
Senior, A. E., 499[16,20], <u>548</u>
Serrano, R., 25[188], <u>40</u>
Sgaramella, V., 419[43], <u>456</u>
Shabolenko, V. P., 64[55], 66[55], <u>100</u>, 106[11], <u>123</u>
Shadur, C. A., 177[66], 178[66], 181[66,70], <u>214</u>, 482[107,108], <u>491</u>
Shahrabdi, M. S., 486[156], 487[156], <u>493</u>
Shallenberg, M. K., 607[208], <u>619</u>
Shandell, A., 279[198], <u>321</u>
Shane, B., 331[152], <u>379</u>, 468[58], 469[84], 473, 474[58], 475, 478, 480[84], <u>489</u>, <u>490</u>
Shankman, S., 341, 343[163], 349[164], 350[164], 352, 353, 358[163], 359[164], 362[163, 164], 363[39], <u>371</u>, <u>380</u>
Shanmugam, K. T., 282[186], <u>307</u>, <u>320</u>

Shannon, R., 453, 462
Shapiro, B. M., 523[134],
    524[135], 552
Shapiro, M., 84[104], 101,
    395[48], 410
Shapiro, S., 140[102], 164
Shavronskaya, A. G., 106[11],
    123
Shaw, J., 422[69], 457
Shcherbakov, G. G., 326[191],
    381
Shechter, E., 175[52], 213
Shelton, D. C., 332[103,166,214],
    338[103,166,214], 341[166],
    343[103,214], 359[215], 375,
    380, 383
Shen, C., 422[69], 457
Shertzer, H. G., 33[211], 41
Shida, H., 429, 458
Shimada, K., 525[142], 553
Shin, Y. S., 473[56,57], 489
Shinagawa, H., 529[164], 553
Shinozawa, T., 429, 458
Shive, W., 340[64], 341[33],
    368[33], 371, 373
Short, S. A., 16[113], 17[123],
    18, 23[140], 25, 26, 28[113,
    140], 29[140,200], 37, 38, 40,
    193[182,183], 218, 391[25],
    409, 522[128], 552, 606[92], 615
Shovlin, V. K., 419[45], 420[45],
    456
Shrecker, O., 51[101], 101
Shum, A. C., 300[187], 320
Shuman, H. A., 144[254], 148[176],
    166, 169, 454[218], 462
Sidlo, J., 330[120], 334[120],
    377
Siepen, D., 336[167], 380
Silhavy, T. J., 10[50,52], 35,
    139[83,84], 141[131], 144[132,
    140,141], 145[140,149,150],
    147[84,173], 148[131,132,173,
    176], 149[140], 150[199],
    156[199], 158[83,140], 163,

[Silhavy, T. J.]
    165, 166, 167, 175[41], 195,
    209[217,218], 213, 218, 219,
    428[114], 458, 541[215,216],
    545[244], 555, 556, 601[206],
    619
Silver, S., 22[172], 39, 222[109,
    189,191], 227[52,185,196], 231[16,
    122], 233[51], 239[109,188,190],
    240, 241[8,110,184,190,193],
    242[52,185,196], 243[184,185],
    248[122], 253[191], 254[50,110,
    185,192,194,225], 255[50,110,
    191,192], 256[50,66,185,196],
    257[185,196], 258[50,66], 259[66,
    185,191], 260[110,191,225],
    266[110,189,197], 269, 270,
    271[185,189,196], 272[189],
    273[189], 274[197], 302[66,190,
    192,195], 303[109], 307, 309,
    312, 313, 316, 320, 321, 323,
    444[187], 461
Silverman, M. P., 306
Silverstein, S. C., 28[113], 37
Simkins, R., 69[77], 73[77], 101
Simmonds, S., 326[169,170], 332[87,
    88,105], 333[173], 334[176,192],
    336[169,170,174], 337[88,106,
    169,170,173], 338[87,88],
    343[106,168], 344[89,168,169,
    170], 346[87,89,169,170],
    347[171,175], 348[171,175],
    354, 356[172], 359[173], 360[89,
    175], 361[172], 366, 374, 376,
    380, 381
Simon, E. J., 177[61], 184[61],
    214, 538[206], 555
Simoni, R. D., 8[10], 9[10], 27[192],
    34, 40, 48[12], 49, 50[12,19,21,
    24,25], 51[12,19,30], 52[12,19],
    53[19], 56[24], 63, 64[24], 67[24],
    68[12], 70, 73, 78[12], 81[44],
    98, 99, 100, 104[3], 108[37,38],
    110[51,52], 111[52], 113[51,52],
    116[60], 118[60], 120[3], 122,

[Simoni, R. D.]
  124, 175[48], 213, 279[2,198],
  308, 321, 496[4], 497[4],
  516[92], 519[115], 548, 551,
  552, 560[207], 569[207],
  607[4,208], 612, 619
Simonpietri, P., 160[256], 169
Simons, K., 52[32], 99
Sims, P. J., 581[209], 619
Singer, S. J., 104[1], 122,
  210[221], 219, 327, 380
Singh, A. P., 120, 125, 582[210],
  619
Singleton, R., 514[82], 551
Sips, H. H., 30[202], 40
Sistrom, W. R., 13[73], 15[73],
  36
Sithold, B., 415[16], 455
Skinner, A., 584[73], 588[73],
  614
Skipper, N. A., 335[145,186],
  378, 381
Skulachev, V. P., 277[114],
  278[114], 279[18,45], 309,
  311, 316, 568[211], 582[52,
  121,211], 608[52], 613, 616,
  619
Skye, G. E., 282[83], 283[83],
  314
Slade, H. D., 187[139], 216
Slater, E. C., 517[110], 518[110],
  519[110], 552
Slayman, C. L., 295[199], 321
Slayman, C. W., 200, 321
Sleisenger, M. H., 364[4], 369
Smarda, J., 431, 459
Smilowitz, H., 450[219], 462
Smit, J., 416[28], 417[28],
  418[28], 455
Smith, D., 424[87], 457
Smith, E. L., 467[10], 488
Smith, I., 483[121], 492
Smith, J., 64, 100, 141[106,108],
  142[106], 156[106,108], 164,
  388[27], 392[27], 409

Smith, J. B., 499[18], 500[18],
  501[27], 502[27,34], 503[34],
  504[18,47], 505[64], 506[64],
  508[64,65,66], 509[71], 510[64,
  65,66,71], 511[18,27,64], 512[64,
  65,66,77,80], 513[64,66,80],
  514[80], 517[18,47], 518[47,64,
  66,105], 519[105,114], 548, 549,
  550, 551, 552
Smith, M., 50[24], 56[24], 64[24],
  67[24], 99
Smith, M. R., 333[180], 381
Smith, P. F., 333[180], 381
Smith, P. H., 466, 487
Smith, R. L., 332[179], 338[179],
  341[178], 343[179], 346[179],
  365[179], 380, 381
Smulson, M. E., 181, 214
Snell, E. E., 14[87], 36, 187[136,137],
  216, 331[49,75,76,77,79,155],
  332[75,76,77,79,86,148], 338[78,
  85,86], 342[86], 343[85,86], 346[86],
  348[85], 359[78,86,148], 360[86],
  363[148], 366, 372, 373, 374,
  378, 379, 469[82,84], 477[82],
  478, 480[84], 482, 490, 491
Snow, G. A., 246, 321, 483[114],
  491
Sobel, M. E., 11[55], 17[130], 35,
  38, 56[39], 58[39], 66[39], 99,
  129[12], 161
Sokatch, I. R., 525[144], 553
Solomon, E., 67, 74, 100, 101,
  131[37], 157[37], 162
Somerfield, P., 293[20], 310
Somerville, R. L., 179[98], 183[98],
  215
Sompolinsky, D., 485[139,143],
  492
Soncek, S., 433[167], 439[167],
  460
Sone, N., 17[133], 25[133,187],
  33[212], 38, 40, 41, 175[43],
  193[180], 197[43,180], 198[43],
  213, 218, 503[36,37,38], 504[37],

[Sone, N.]
    507[60], 514[36,37,60], 515[36,
    37,38], 516[60], 522[130],
    526[149], 549, 550, 552, 553,
    568[212], 609[212,233], 610[78,
    79], 614, 619, 620
Soon-Ho Lee, 547[250], 556
Southard, J. H., 232[79], 314
Spencer, R. D., 135[63], 163,
    586[203], 619
Sperl, G. T., 300[202], 321
Speth, V., 417[30], 451[207], 456,
    462
Spicer, S. S., 194[189], 218
Spielman, P. M., 302[195], 321
Spitz, E., 288[101], 289[101],
    302[101], 315, 588[67], 600[67],
    610[67], 614
Spitzen, J., 17[124], 37
Spoonhower, J., 278[125], 316
Sprague, G. F., Jr., 286[203],
    287[203], 321
Springer, M. S., 521[122], 552
Springer, S. E., 292[204], 322
Sprott, G., 189[161], 217
Spudich, E. N., 9[28], 10[28],
    34, 148[172], 166, 189[168],
    190[168], 201[168], 210, 217,
    420[49], 456, 534[188], 554
Squires, R., 341[163], 343[163],
    358[163], 362[163], 380
Sramek, S., 87[115,116], 88[114],
    102
Stadtman, E. R., 17[126], 37,
    90, 102, 172, 175, 212, 562[93],
    615
Stadtman, T. C., 300[205], 322
Staehelin, L. A., 18[141], 23,
    28[141], 32, 38
Stainer, R. Y., 499[19], 548
Stanley, S. O., 294[25], 310
Stapley, E. O., 150[195], 167,
    483[110], 491
Stein, R., 48[13], 66[13], 98
Stein, W. D., 8[1], 34

Steinberg, R., 84[106], 85[106],
    102, 389[44], 393[34], 394[44],
    395[44], 409
Steinfeld, A. S., 333[66], 336[66],
    338[66], 343[66], 344[66], 348[66],
    350[66], 352[66], 353[66], 367[66],
    373
Steinhart, R., 279[220], 322, 503,
    507, 512[35], 549
Stephenson, M., 1[1], 5
Stern, J. R., 17[131], 38, 402[89],
    404[89,90], 407[104], 411
Sternweis, P. C., 22[170], 25[170],
    39, 501[27], 502[27,34], 503[34],
    504[44], 505[64], 506[64], 508[44,
    64,65,66], 509[71], 510[44,64,65,
    66,71], 511[27,64], 512[64,65,66,
    77], 513[64,66], 518[44,64,66],
    548, 549, 550
Stevenson, J., 189[160], 217
Stevenson, R., 307
Stiles, C. D., 108[36], 123
Stimler, N. P., 141[107], 156[107],
    164
Stinnett, J. D., 19[145], 38, 387[21],
    388[21], 409
Stinson, M. W., 542[229], 556
Stock, J., 108[34], 123, 140, 158[98],
    164, 592[213], 619
Stock, R., 13[74], 36
Stocker, B. A. D., 419[38], 456
Stoeber, F. R., 151[204], 152[204],
    156[232,234,235], 157[236,238,
    239], 167, 168, 386[7], 387[11,12,
    13], 388[11,13], 408
Stoeckenius, W., 277[153,168],
    278[125,127,152,153,168], 307,
    316, 317, 318, 319, 515[85,86],
    516[84,88], 551, 569[15], 579[163],
    593[163], 609[177], 612, 617, 618
Stoffel, W., 84[102], 85[102], 101
Stokes, B. O., 568[19], 612
Stokes, F. N., 386[1], 401[1], 408
Stokstad, E. L. R., 331[152], 379,
    468[58], 473[56,57], 474[58], 475,
    489

Stoll, E., 336[181], 381
Stone, D., 331[182], 381
Strange, P. G., 144[145], 165, 547[253], 556
Stratis, J. P., 483[113], 491
Straus, L. D., 525[143], 553
Strenkoski, L. F., 281, 322
Strickland, K. P., 286[74,169], 287[74], 290[74], 313, 319, 532[168], 547[257], 554, 557, 600[178,179], 618
Strominger, J. L., 482, 491
Stroobant, P., 27, 28, 40
Struve, W. G., 136, 163
Subramanian, K. N., 478[85], 481[85], 490
Sughrue, M. J., 118[63], 125
Suit, J. L., 154[219], 168, 604[172], 606[172], 618
Sun, I. L., 517[103], 518[103], 551
Sundaram, T. K., 479[92], 491
Surdin-Kerjan, Y., 308
Sussman, A. J., 326[185], 332[185], 333[184,185], 336[183,184,185], 381
Sussman, M., 63, 79[96], 80, 81[97], 83, 100, 101
Suzuki, H., 10[45], 14[45], 35, 179[102], 183[102], 184[102], 196[102], 215
Suzuki, I., 282[207], 322
Switzer, R. L., 259[208], 265[208], 322
Sykes, R. B., 420[60], 457
Szeto, K. W., 336[174], 380
Szmelcman, S., 9[31], 34, 130[25], 144[25,140], 145[25,140,146, 150], 147[25], 148[183], 149[140], 154[25,216], 158[25, 140,183,216], 161, 165, 166, 168, 175[41], 195[41,195], 209[218], 213, 218, 219, 426[105], 427, 428[105,114],

[Szmelcman, S.]
429[112], 454[112], 458, 528[160], 541[216], 543[160], 547[254], 553, 555, 557, 582[214], 608[214], 619

T

Taber, H., 486[155], 493
Takahashi, H., 395[54,55,57], 396[54,55], 410
Takeuchi, Y., 503[38], 515[38], 549
Tamaki, S., 149[177], 166, 419[42], 456
Tamura, T., 473[57], 489
Tanaka, S., 67[70], 100
Tanaka, Y., 11[56], 23, 28[178], 29[178], 31[178], 35, 39, 193[172], 217
Tarshis, M. A., 130[28], 161
Tatum, E. L., 347[175], 348[175], 354[175], 359[175], 360[175], 380
Taylor, A. C., 139[90], 155[90], 164
Taylor, A. L., [222], 219, 225[7], 226[7], 300[7], 309, 390, 411, 426[98], 433[98], 446[98], 458
Taylor, E. S., 561[46], 563[46], 613
Taylor, R. T., 442[181,182], 461, 466[8], 467[8,12], 487, 488, 531[239], 544[239], 556
Teather, E. M., 136[69], 153[69], 163
Tempest, D. W., 337[185a], 381
Templeton, B. A., 178[71], 181, 214
Terada, H., 198[204,205], 218
Thaler, M. M., 331[150], 379
Thauer, R. K., 11[54], 35
Thipayathasana, P., 279[209], 322, 519[116], 552
Thirim, J. P., 426[99], 458
Thomas, A. A. M., 517[106], 518[106], 551

Thomas, E. L., 18[137], 38
Thomas, J., 138[77,78], 153[78], 163
Thomas, T. D., 335[186], 381
Thomassen, E., 93[124], 94[124], 102
Thompson, T. E., 503[42], 549
Thompson, J., 277[210], 322, 591[216,217], 592[216,217], 593[216,217], 610[215], 619
Thorne, G. M., 187, 216
Thurman, P., 592[133], 616
Timmis, K., 448[200], 461
Timofeeva, N. M., 326[191], 381
Tokita, F., 332[187], 381
Tokuda, H., 160[258], 169, 611[240], 620
Topali, V., 582[52], 608[52], 613
Torriani, A., 262[211], 263, 322
Toscano, W. A., Jr., 396[78], 401[78], 411
Tosteson, D. C., 297[75], 298[80, 212], 299[80], 313, 314, 322
Toth, K., 227[196], 242[196], 256[66,196], 257[196], 258[66], 259[66], 266[197], 269[197], 270[197], 271[196], 274[197], 302[66], 313, 321
Toye, N. O., 334[176], 380
Traylor, T. G., 301[9], 309
Trifore, J., 56[41], 59[41], 99
Tristram, H., 179[105], 184[105, 106], 215
Tritsch, G. L., 331[45], 371
Tritz, G. J., 479[88,89,90], 490
Trotter, C. D., 390, 411
Tsofina, L., 582[52], 608[52], 613
Tsuchiya, T., 20, 22[173], 26[191], 28, 33[209], 39, 40, 41, 160[257], 169, 267, 268, 269[214], 322, 517[109], 551, 562[218,219,220], 588[219, 221], 591[219,220,221], 611[237,238,239], 619, 620

Tsukagoshi, N., 32, 40
Tsukamura, M., 483[120], 492
Tsvetkova, V. A., 326[191], 381
Tucker, A. N., 20, 38
Tulyaganova, E. Ch., 326[191], 381
Tweedie, J. W., 292[216], 293[216], 300[216], 322
Tyler, B., 64, 100, 105[10], 123

U

Ueda, T., 610[68], 614
Uemura, J., 416[26], 455
Ugolev, A. M., 326[188,189,190, 191], 329[188], 381
Ullah, A., 49[17], 50[17], 64[17], 99
Umbarger, H. E., 186[130], 216, 333[194], 337[194,199], 342, 366[194], 382
Unsöld, H. J., 143[126], 165, 541[221], 555
Unwin, P. N. T., 278[104], 315, 516[89], 551
Urry, D. W., 233[217,218], 235[217, 218], 322
Utech, N. M., 189[156,157], 217

V

Valentine, R. C., 282[186], 320
van Balgooy, J. N. A., 189[155], 217
van Beek, W. P., 15[112], 37
Van Brunt, J., 307, 611, 620
van Dam, K., 198[205], 218, 396[79], 401[79], 404[79], 411, 509[69], 510[73], 517[106], 518[106], 550, 551
Van Den Elzen, H. M., 486[156,157, 158], 487[156,157,158], 493
Vanderwinkel, E., 85[107], 86[107], 102, 393[38,41], 394[41], 409
Van de Stadt, R. J., 509[69], 517[110], 518[110], 519[110], 550, 552

Van Gool, A. P., 418[31], 456
van Heerikhuizen, H., 415[6], 455
Van Knippenberg, P. H., 452[209, 210], 462
Van Lenten, E. J., 334[192], 381
van Thienen, G., 24, 25, 26, 40, 607[51], 613
Vasington, F. D., 146[161], 166, 189[169,170], 196[170], 217, 530[187], 533[187], 554
Vatter, A. E., 507[62], 550
Venkateswaran, P. S., 141[111], 150[111], 156[111], 164, 423[70], 457, 483[112], 491
Verkleij, A.J., 418[32], 456
Verneulen, C. A., 19[143], 23[143], 30[143], 38
Verses, C., 425[96], 458
Ververgaert, P. H. J. Th., 418[32], 456
Vignais, P. V., 509[70], 550
Villafranca, J. J., 306
Villarreal-Moguel, E. I., 402[87], 404[87], 411
Visentin, L. P., 336[94], 375
Visser, A. S., 281, 322, 393[37], 409
Visser, D., 91, 92[122], 93[122], 95[131], 102
Viswanatha, T., 527[155], 553
Vladimirova, M., 582[52], 608[52], 613
Voenhuis, M., 19[143], 23[143], 30[143], 38
Voet, A. B., 510[73], 550
Vogel, G. D., 279[220], 322, 503, 507, 512[35], 549, 569[196], 582[196], 603[84], 606[84], 615, 618
Vogell, W., 14[93], 36
Vogt, V. M., [193], 382
Volcani, B. E., 260[6], 301[6], 309
Volkov, A. G., 279[18], 309

Voll, M. J., 420[47], 456
Von der Haar, R. A., 333[194], 337[194], 342[194], 366[194], 382
von Hugo, H., 58[100], 101
Von Meyenburg, K., 423, 457
Vorisek, J., 129[2], 160
Vorotyntseva, T. I., 195[193], 209[193], 218

W

Wachsman, J. T., 13[74], 36
Wada, S., 306
Waddell, W. J., 579[222], 580, 619
Waelsch, H., 332[109,112], 376
Waggoner, A. S., 581[209,223], 582[223], 608[223], 619
Wagner, C., 389[32], 392[32], 409
Wagner, F. W., 334[156,157,195, 196], 379, 382
Wahren, A., 332[197], [198], 382
Wakabayashi, Y., 472[45], 489
Wakil, S. J., 84[104], 101, 395[48], 410
Wallenfels, K., 157[240], 168
Waller, J. R., 469[71,72], 476[72, 76], 490
Wallick, H., 150[195], 167, 483[110], 491
Walsh, C. T., 16[113], 23[113], 28[113], 37
Walter, L., 58[73], 68[73], 100
Walter, R., 65[61], 100
Waltho, J. A., 296, 297[35], 311
Wang, C. C., 149[186], 167, 248[221], 249[221], 323, 432[134, 150,151], 434[134], 436, 440[150], 441[134], 450, 459, 460
Wang, C-H., 581[209], 619
Ward, J., 78, 101
Wargel, R. J., 177[66], 178[66], 181[66,70], 214, 482[107,108], 491
Wasmuth, J. J., 337[199], 342[199], 382

Watanabe, K., 9[26], 34, 194[187], 218, 290[160], 291[160], 319

Waters, F. B., 14[81], 36, 227[171], 228[171], 229[171], 230[171], 232[171], 236[171], 237[171], 275[171], 319, 532[172], 554, 602[184], 618

Waygood, B., 48[14], 98

Wayne, P. K., 106[15], 123

Wayne, R., 244[128], 251[128,222], 317, 323, 431[127,130], 432[126, 128,130,157], 434[127], 436, 438[130], 439[130], 440, 459, 460, 483[116], 491

Webb, M., 262[223], 323

Weber, G., 135[63], 163, 586[203], 619

Wecksler, M., 156[229], 168, 388[16], 409

Wedler, F. C., 306

Weeks, G., 84[104], 101, 395[48], 410

Wegener, W. S., 85[108], 86[108], 102, 389[40], 393[35,36,40], 409

Weidel, W., 441[172], 460

Weiden, P. L., 287[224], 323

Weigand, R. A., 420[56], 456

Weigel, N., 48[14], 49[18], 98, 99

Weil, R., 15[108], 37, 109[46], 124, 134[56], 135[56,59,60,61, 62,63], 136[61,62], 137[59,62, 70], 157[61,62], 162, 163, 175[52,53,56], 213, 481[102, 103], 491, 586[201,202,203], 602[201], 603[201], 619

Weinberg, E. D., 263[63,64], 264[64], 265[64], 313

Weiner, J. H., 12[59], 13, 18[59], 27[193], 29, 35, 40, 147[169], 150[197], 156[197], 166, 167, 180[117,118], 181[82], 185[117, 118], 199[82], 200, 203[82], 209[118], 214, 215, 216, 334[58], 345[58], 372, 522[129],

[Weiner, J. H.] 530[181,190], 533[181], 535[190], 537[202], 545[202], 552, 554, 555

Weiss, A. A., 254[225], 260[225], 323

Weiss, B., 479[91], 490

Weiss, R. L., 478[85,86,87], 481[85,86,87], 490

Weissbach, H., 18[137], 38, 193[175], 217, 442[174], 460

Weller, R., 432[144], 459

Weltman, J. K., 499[17], 548

Wendt, L., 603[224], 606[224], 619

Wenzel, D., 50[23], 53[23], 99, 109[45], 114[45], 124, 388[30], 392[30], 409

Werkheiser, W. C., 14[90], 36

West, I. C., 13[71], 15[110,111], 24[71], 25, 30[110], 36, 37, 40, 108[32,33], 123, 152[209], 154[243], 167, 169, 175[54,55], 213, 275[226], 323, 578[225,226, 227,229], 583, 585, 588[225,226, 227,228], 609[229], 619

Wethamer, S., 486[151], 492

Wetzel, B. K., 194[189], 218

Whatley, F. R., 30[203], 40, 289, 290[29,30], 292[30], 310, 589[21, 22], 591[21,22], 612

Whipple, M. B., 194[185], 218, 291[159], 292[159], 319, 528[162], 553

White, B. J., 479[91], 490

White, D. A., 20, 39, 415[17], 416[17], 424[84], 455, 457

White, D. C., 193[182,183], 218

White, J. C., 9[29], 34, 149[184], 167, 265[44], 311, 442[183], 443[183,185], 445[185], 461, 467[11], 488

Whitfield, C., 442[174], 460

Whitney, E., 241[193], 321, 444[187], 461

Whittaker, J. R., 355[200], 382

Wiame, J. M., 400[75], 410

Wickner, S., 499[13], 517[13], 548
Wickner, W., 28, 40
Widdas, W. F., 173[29], 213
Wiesmeyer, H., 154[215], 159[250], 168, 169, 426[101], 458
Wieth, J. O., 298[80], 299[80], 314
Wilbrandt, W., 151[205], 167, 173[28,30], 213
Wilchek, M., 527[155], 553
Wilkerson, L. S., 402[88], 404[88,91,92], 411, 592[34], 613
Wilkes, S. H., 334[156,196,201], 379, 382
Wilkinson, B. J., 273[53], 312
Wilkinson, S., 364[1,2,3,4], 369
Wille, W., 407[102], 411
Williams, A. M., 401[85], 411
Williams, B., 105[9], 123
Williams, G. R., 402[94], 405[94], 411
Williams, M. A., 473[57], 489
Williams, R. J. P., [131], 317
Willicke, K., 241[227], 252[150, 227], 318, 323, 396[74], 399[74], 402[96,98,99,100], 405[96,98], 406[99,100], 407[101.102], 410, 411, 527[153], 553
Willis, D. B., 229[228], 323
Willis, R. C., 144[134], 145[134], 159[134], 165, 180[116,119, 126], 185[119,126], 186[126, 127], 187, 200[208], 215, 216, 218, 528[157], 530[192], 531[235], 533[183,184], 535[192], 536[198], 542[235], 553, 554, 555, 556
Willsky, G. R., 226[230], 284[230], 285[229,230,231], 286[229,231], 287[229], 290[230], 308, 323, 529[167], 532[171], 554

Wilson, D. B., 10[39], 12[39], 14[86], 35, 36, 109[42], 118[67], 120, 124, 125, 138[76], 139, 143[76], 151[203], 153[203], 154[218], 155[76,218], 158[218], 163, 167, 168, 302[232], 323, 544[243], 545[243], 556, 600[230], 619
Wilson, D. M., 12[61], 35
Wilson, F., 77, 78, 101
Wilson, G., 55[37], 64[37], 99, 106[12], 123
Wilson, K. J., 334[156], 379
Wilson, N. L., 296[183], 320
Wilson, O. H., 201[213], 219
Wilson, P. W., 401[85], 411
Wilson, Q. H., 538[204], 555
Wilson, T. H., 11[57], 12[57,60, 61], 14[79], 35, 36, 109, 112, 118[71], 124, 125, 130[31], 134[53, 54], 136, 140[100], 149[180], 151[67,68], 152[53,54,67,208, 209,210,211,212], 153[67,68, 208,210], 157[67,244], 160[257], 161, 162, 163, 164, 166, 167, 168, 169, 172[12], 173, 175[12], 212, 302[234], 303[234], 323, 420, 457, 578[99,134,229], 579[100,101,134], 581[100], 585[99], 586[97,98], 587[37], 589[99], 595, 596[100], 609[229], 611[239], 613, 615, 616, 619, 620
Winkler, H. H., 10[43], 12[43], 14[79], 35, 36, 109, 112, 118, 124, 125, 129[16], 136, 141[105, 110], 142[117], 150[110], 151[67], 152[67], 153[67], 156[105,117], 157[67], 161, 163, 164, 172[12], 173, 175[12], 212, 285[233], 287[233], 302[233,234], 303[234], 323, 483[111], 491, 588[231], 600[94], 615, 619
Wiseman, G., 326[202], 382
Wishnow, R., 64[56], 100

Wittenberger, C. L., 58[72], 67, 100, 129[10], 161
Wojdani, A., 485[143], 492
Wolcott, R. G., 568[19], 612
Wolf, F. J., 150[195], 167
Wolff, H., 415[12], 455
Wolfinbarger, L., 330[204], 332[203], 334[203, 204], 336[203, 204, 205], 338[204, 205], 339[203], 341[203, 204, 205], 342[205], 343[205], 345, 350[205], 354[205], 367[203, 204, 205], 368[203, 204, 205], 382
Wolfson, E. B., 11[55], 17[130], 35, 38, 56, 58[39], 66[39], 99
Wolfson, P., 66[63], 100
Wolf-Watz, H., 416[19], 455
Womack, F. C., 15[107], 37, 580, 608[26], 612
Wong, B. B., 226[161], 241[161], 242[161], 253[161], 265[161], 303[161], 319
Wong, P. T. S., 134[53], 136, 151[68], 152[53], 153[68], 157[244], 162, 163, 169
Wood, J. M., 10[42], 12[42], 35, 181[77], 196[77], 199[77], 202, 203, 205, 207, 214, 532[180], 554, 600[232], 602[232], 619
Wood, R. C., 473, 489
Wood, T. H., 483[119], 491
Woodrow, G. C., 226[236], 246[235], 250[235, 236], 306, 323, 324, 432[143], 435[143], 459
Woodrow, M. L., 149[185], 167, 442[180], 443[180], 444[190], 445[190], 461, 467, 488
Woodruff, H. B., 150[195], 167, 483[110], 491
Woods, S. L., 448, 461

Wooley, S. O., 332[15], 338[15], 343[15], 347[15], 370
Woolfolk, C. A., 180[116], 185, 215
Woolley, D. W., 331[208, 209], 332[208, 209], 349[207], 352[207], 353[108, 201, 207], 358[108, 210], 376, 382, 383
Wright, A., 9[16], 34
Wright, D. W., 338[211], 383
Wrischer, M., 427[110], 458
Wu, H. C. P., 138[82], 139[82], 141[111], 150[111], 153[213], 156[111], 163, 164, 168, 420[55], 423[70], 456, 457, 483[112], 491
Wynants, J., 400[75], 410

Y

Yabu, K., 189[158, 159], 217
Yagil, E., 94, 102, 424[82, 83], 457
Yamada, K., 471[39, 40], 472[43], 489
Yamada, T., 483[118], 491, 525[142], 553
Yamaguchi, A., 13[72], 33[213], 36, 41, 175[44], 181[92], 194[92], 198[44], 201[92], 203[44], 207[92], 210[92], 213, 215, 533[185], 541[185], 547[249], 554, 556, 610[6], 612
Yamamoto, T., 22, 26[175], 39, 336[95], 375
Yamasaki, H., 472[43], 489
Yamato, I., 16[115], 20, 30[115], 37, 178[72], 179[72], 181[85, 90], 182[72, 85, 90], 192, 193[90, 171, 172, 173], 194[85], 199[85, 90], 200, 201[90], 202[85, 90], 203[90], 205, 206[72, 85, 90], 207[90], 210[85, 90], 214, 215, 217
Yariv, J., 134, 162, 333[146], 335[146], 378

Yariv, M., 134[55], 162
Yaron, A., 336[212,213], 383
Yearwood-Drayton, V., 193[178],
217, 307, 589[117], 592[117],
593[117], 616
Yeh, F. M., 331[60], 373
Yoder, O. C., 332[103,214],
338[103,214], 343[103,214],
359[215], 375, 383
Yokota, T., 426[103], 458
Yoshida, M., 17[133], 25[133,
187], 33[212], 38, 40, 41,
175[43], 193[180], 197[43,180],
198[43], 213, 218, 503[36,37,
38], 504[37], 507[60], 514[36,
37,60], 515[36,37,38], 516[60],
526[149], 549, 550, 553,
568[212], 609[212,233], 610[78,
79], 614, 619, 620
Yoshimura, F., 510[74], 550
Young, E. A., 338[216], 342[216],
343[216], 365[216], 368, 383
Young, I. G., 226[236], 244[178],
246[178,237], 247[178],
249[178], 250[178,236,237],
252[178], 253[178], 306, 320,
324, 430[123], 432[141,143,
147], 434[135], 435[135,141,
143], 436[147], 441[147], 459,
460
Young, I. L., 432[148], 433[148],
436[148], 441[148], 460
Young, J. D., 327[5], 332[5],
333[5], 341[5], 342[5], 357[5],

[Young, J. D.]
358[5], 367[5], 369
Young, W., 73[83], 101
Yu, P. H., 336[167], 380

Z

Zähner, H., 431[163], 460, 483[115,
117], 491
Zaidenzaig, Y., 485[139], 492
Zakrzewski, S. F., 474[64], 490
Zand, R., 181[81], 201[81], 214,
532[177], 554
Zarlengo, M. H., 282, 324, 605[234],
620
Zevely, E. M., 475[68,69], 490
Zgaga, V., 427[110], 458
Zilberstein, D., 596[164], 598[164],
617
Zimmerman, U., 229[165], 236[165],
275[165], 319
Ziska, P., 330[217,218], 383
Zitzmann, W., 279[4], 308,
516[94], 517[94], 519[94], 551
Zuber, H., 335[158], 336[181],
379, 381
Zukin, R. S., 144, 165, 521[123],
547[253], 552, 556
Zumft, W. G., 282[239], 299[239],
300[239], 324
Zwaig, N., 156[228,229], 168,
387[18], 388[16,18], 409

SUBJECT INDEX

A

Acetate (see also pH gradient,
measurement of) transport,
392–393
Acetyl CoA:acetoacetate CoA-
transferase
in fatty acid transport in
E. coli, 86
Acetyl CoA:butyrate CoA-
transferase
in fatty acid transport in
E. coli, 86
Acids, weak organic (see also)
pH gradient, measurement
of), 579
Aconitate transport, 405, 407
Actinomycin D, transport in
whole cells, 13
Active transport (see Transport,
active)
Acylation, vectorial (see also
Group translocation)
Acyl CoA-synthetase, role in fatty
acid transport in E. coli, 85
Adenine phosphoribosyltrans-
ferase, 90
Adenine transport, 90–91
Adenosine transport, 93
Adenosine-5'-diphosphate, in
preparation of membrane
vesicles, 19–20

Adenosine-5'-triphosphatase
antibody against, in determination
of membrane orientation,
24, 25
bound nucleotides, 517–519
inhibitor, 509
latent activity, 509
in membrane vesicles, 22
mutants, 10, 279, 519, 576, 577,
607
and orientation of membrane
vesicles, 24–27
oxidative phosphorylation by,
566–568
proton transport by, 279–280
role in transport, 520, 569, 576
subunits, 506–517
F_0 subunits, 515–517
F_1 subunits, 506–514
Adenosine-5'-triphosphate
as energy source for transport,
22, 520, 569, 576
formation, 566
intracellular concentrations of, 11
Adenosine-5'-triphosphate-linked
transport systems, 4, 10,
575–577, 599–603, 608, 610
ADP (see Adenosine-5'-diphosphate)
Aerobacter aerogenes, active transport in whole cells, 13
Aerobactin, role in iron transport,
435

Argobacterium tumefaciens, transport of sugars, 130
Alanine carrier protein, 33, 526
Alkaligens (Hydrogenomonas) eutropha, transport of ammonia, 281
L-Alloisoleucine, 200
Allosteric activation, 104
Allosteric inhibition, 104, 109
α subunit of F_1 (see also Adenosine-5'-triphosphatase; F_1), 506
Amethopterin, 473
Amino acid transport (see also Transport, amino acids), 171-212
9-Aminoacridine (see also, pH gradient, measurement of), 582
L-α-Aminobutyrate, 200
9-Amino-6-chloro-2-methoxy-acridine, 26
Aminoglycoside transport, 486-487
α-Aminoisobutyrate, association of cations, 591
transport, 591
Ammonia transport, 14, 281-284
AMP (see Adenosine-5'-mono-phosphate)
Aniline, 14
8-Aniline-1-naphthalene sulfonate, (see also Membrane potential, measurement of), 592
Anions (see also Transport, anions) permeant (see also Membrane potential, measurement of), 582
transport, 284-301
Antibiotic transport (see also) Transport, antibiotics), 481-487
Antiport, 570, 573-574
APG (see 2-Nitro-4-azidophenyl-1-thio-β-D-galactoside)
Aprotic solvents, 137
Arabinose transport, 138, 146, 584

Arabinose binding protein, 144-145, 147, 542, 547
D-Arabinose-5-phosphate, 141
Arginine transport, 147, 184, 193
Arginine binding protein, 538
argP gene, relationship to arginine transport in E. coli, 184
aroP gene, relationship to aromatic amino acid transport in E. coli, 183
Arsenate
effect on active transport systems, 119, 577
transport in E. coli, 284-288
Arthrobacter pyridinolis, transport of amino acids in membrane vesicles, 11, 193
Asparagine transport, 185, 193
Aspartate transport
in E. coli, 185, 395
in membrane vesicles of B. subtilis and S. aureus, 193
in Pseudomonas, 401
ATP (see Adenosine-5'-triphosphate)
ATPase (see Adenosine-5'-triphosphatase)
Autoradiography, use of in determination of membrane vesicle orientation, 23
L-Azetidine-2-carboxylate, 184
Azidophenylgalactosides, as photoaffinity labels, 135
Azotobacter vinelandii
ammonia transport, 281
calcium transport, 269-270
membrane vesicles, 17
sugar transport, 129

B

Bacillus megaterium
aminoglycoside transport, 486
calcium transport, 266
glycine-alanine transport, 187

[Bacillus megaterium]
  membrane vesicles, amino acid
    transport, 193
Bacillus subtilis
  arsenate resistant mutants,
    400
  calcium transport, 266
  magnesium transport, 242-243
  manganese transport, 253,
    255-260
  membrane vesicles, 17
  amino acid transport, 193
  phenylalanine-tyrosine-tryp-
    tophan transport, 182-188
  sugar transport, 131
Bacillus thuringiensis, molyb-
    date transport, 300
Bacteria
  aerobic, 11
  anaerobic, 11, 129
  photosynthetic, 129
  thermophilic, 17
Bacteriocins, energy trans-
    duction, 603-608 (see also
    Colicins)
Bacteriophage M13
  coat protein of, 28
  orientation of membrane
    vesicles, 28
Bacteriorhodopsin, 21, 277-278,
    515-517, 569
Bases, weak organic (see also
    pH gradient, measurement
    of), 579
Basic amino acid transport, 184
Benzoate (see also pH gradient,
    measurement of), 579
β subunit of F_1 (see also
    Adenosine-5'-triphosphatase;
    F_1), 506
BF_0F_1 complex (see Adenosine-
    5'-triphosphatase)
BF_1 mutants (see Adenosine-5'-
    triphosphatase)
bfe gene product, functional
    properties of, 447-451

Binding proteins (see also
    Osmotic shock procedure)
  for amino acids, 172, 532-539
    arginine, 538
    cystine, 536-537
    glutamate-aspartate, 186,
      535-536
    glutamine, 535
    histidine, 189, 533-534
    leucine-isoleucine-valine,
      199-200, 532-533
    lysine-arginine-ornithine,
      537-538
    phenylalanine-tryptophan-
      tyrosine, 538
  for carbohydrates, 144-146,
      539-543
    arabinose, 542
    galactose, 143-147, 539-542
    glucose, 542
    maltose, 543
    ribose, 542
  for carboxylic acids, 395,
      397-398
  for inorganic ions, 528-529,
      532
    phosphate, 286, 529, 532
    potassium, 233
    sulfate, 291-292, 528-529
  retention effect of, 145-146,
      195
  role in active transport, 146-147,
      207-210, 544-545
  role in chemotaxis, 142, 144, 541
  for vitamins, 543-544
    cyanocobalamin, 467, 544
    thiamine, 471, 543
Biotin transport, 475-477
brnR gene, relation to leucine-
    isoleucine-valine transport
    in E. coli, 182
N-Bromoacetyl-β-D-galactosamine,
    134
Bromosuccinate, 398
Brucella abortus, sugar transport,
    129

C

Calcium transport
  binding factors, 275
  in cells, 14, 266-267
  driven by an artificially-imposed
    proton motive force, 591
  in membrane vesicles, 22, 25,
    267-271
  orientation of membrane
    vesicles, 25
  during sporulation, 271-274
Canavanine, 184
Candida utilis, transport of zinc,
    263-265
Carbohydrate transport (see also
    Phosphoenolpyruvate-sugar
    phosphotransferase system;
    Transport, Carbohydrates),
    46-84, 105-122, 127-160
Carbonylcyanide-m-chloro-
    phenylhydrazone (see also
    Uncouplers), 151, 254, 266,
    288, 584, 587, 595, 604, 606
Carbonylcyanide-p-trifluoro-
    methoxypheynlhydrazone
    (see also Uncouplers), 266,
    577, 593, 594, 606
CO_2, evolution of as assay for
    transport in intact cells, 15
Carrier proteins, isolation
  alanine, 33, 525, 610
  bacteriorhodopsin, 515-517, 609
  dicarboxylate, 395-397, 525
  lactose, 132-133, 137, 525
  proline, 33, 197-198, 527, 610
  proton, 515-517
Catabolite repression, 105
Cations (see also Transport,
    cations)
  associated with solute transport,
    591-594
  colicin action, 604
  efflux, 266, 274, 277
  permeant (see also Membrane
    potential, measurement of), 582

[Cations]
  transport of, 226-284
CCCP (see Carbonylcyanide-m-
    chlorophenylhydrazone)
Cells, assays for transport in,
    8-16
Chemiosmotic hypothesis (see
    also Proton motive force),
    3-4, 277, 498, 563-573
Chemotaxis
  and adenosine-5'-triphosphatase,
    520-521
  and periplasmic binding proteins,
    142, 144, 541
Chloride transport, 295-299
Chromium transport, 300
Citrate transport, 401, 404-407
  and iron transport, 252-253, 434
  and magnesium transport, 405-407
Clostridium pasteurianum, molyb-
    date transport, 299-300
Clostridium perfringens, 11
Cobalt transport, 241, 265
Colicins
  and energy transduction, 606-608
  mechanism of action, 603-606
  receptors, 424-425
  and enterochelin system, 436,
    439-440
  and outer membrane adhesion
    sites, 453
Counterflow, 151, 152
crr gene, relationship to regulation
    of carbohydrate transport,
    113
Cyanine dyes (see also Membrane
    potential, measurement of),
    581, 595, 604, 608
Cyanocobalamin (Vitamin B12)
    transport, 9, 149, 442-446,
    466-470, 544
cycA gene, relationship to amino
    acid transport, 177
Cyclic AMP (see Cyclic adenosine-
    5-monophosphate)
Cyclic adenosine-5-monophosphate

[Cyclic adenosine-5-monophosphate]
activation of protein kinases, 121
and carbohydrate transport, 106
in membrane vesicles, 19
Cyclic electron flow, and amino
acid uptake, 19
Cycloleucine, inhibition of leucine
transport, 199
D-Cycloserine transport, 177,
482
Cystine transport, 13, 147, 184
Cytoplasmic membranes, (see
Membrane, cytoplasmic)

D

Dansylgalactosides, 135-137,
586
DCCD (see N, N'-dicyclohexyl-
carbodiimide)
DDA (see dibenzyldimethylammonium ion)
δ subunit of F_1 (see also Adeno-
sine-5'-triphosphatase; F_1),
506, 510
Deoxyadenosine transport, 93
2-Deoxy-D-glucose-6-phosphate,
140
dhuA gene, relation to histidine
transport in S. typhimurium,
189
Diaminopimelic acid transport,
147, 184
Diauxie, 104, 106
5-Diazo-4-oxo-L-norvaline, and
asparagine transport in
E. coli, 185
Dibenzyldimethylammonium ion
(see also Membrane potential,
measurement of), 582, 608
Dicarboxylic acid transport (see
also Transport, carboxylic
acids), 150, 398, 401, 527
N, N'-dicyclohexylcarbodiimide,
13, 19, 235, 267, 279, 288,
503, 515-517, 519, 566, 587,
606

Diffusion pore, and outer membrane, 422-424
3, 3'-Dihexyloxacarbocyanine (see
also Cyanine dyes; membrane
potential, measurement of),
604
5, 5-Dimethyl-2, 4-oxazolidinedione
(see also pH Gradient, measurement of), 579-581
2, 4-Dinitrophenol (see also
Uncouplers), 12, 254, 584,
604
Dislocated enzymes, and vesicle
orientation, 29
DMO (see 5, 5-Dimethyl-2, 4-
oxazolidinedione)
DNP (see 2, 4-Dinitrophenol)
Donnan equilibrium, 590

E

ecf mutants, relationship to
energy coupling, 602-603
Electrical potential (see Membrane
potential)
Electrochemical proton gradient
(see Proton motive force)
Electron microscopy
freeze fracture, 23, 418
and orientation of membrane
vesicles, 23
Electron transfer particles from
M. phlei, 21
Electron transfer systems, 564
Electroosmotic energy (see also
Proton motive force), 568
Electrophoretic fractionation of
membrane vesicles, 21
Enterochelin, and iron transport,
246-250, 430, 435-440
Enzyme I (see also Phosphoenolpyruvate-sugar phosphotransferase system), 47-49, 61,
108, 110, 574
Enzyme II (see also Phosphoenolpyruvate-sugar phosphotransferase system), 47-48, 52-56,

[Enzyme II]
    60-80, 108, 116, 117, 119,
    574
Enzyme III (see also Phosphoenol-
    pyruvate-sugar phosphotrans-
    ferase system), 47-48, 50-54
    57, 60-67, 70, 74, 77, 80-84,
    108, 574
ε subunit of F_1 (see also Adeno-
    sine-5'-triphosphatase; F_1),
    506, 508
Equilibrium dialysis, 145, 195
Escherichia coli
  membrane vesicles, 21
    everted, 21-22
    orientation, 23-30
    preparation of, 16-18, 20-21
    transport in
      of amino acids, 177-187
      of anions, 284-288, 292-294,
        296
      of antibiotics, 482-487
      of carbohydrates, 128-160
      of carboxylic acids, 386-398
      of cations, 227-234, 239-255,
        262-263, 266-269, 275, 279,
        283-284
      of fatty acids, 84-89, 392-398
      of nucleosides, 93-96
      of peptides, 347-350, 358,
        362-364, 366-369
      of purines, 90-92
      of pyrimidines, 92-93
      of vitamins, 466-467, 470-473,
        476-477
Exit reaction, of carbohydrate
    transport, 150-154
Exogenous induction of glucose-
    6-phosphate transport,
    141-142

                    F

Facilitated diffusion, 8, 73, 107,
    130-131, 136, 572, 585-586
Fatty acid transport, 84-89,
    392-395

FCCP (see Carbonylcyanide-p-
    trifluoromethylphenylhydrazone)
Ferric-enterochelin receptor,
    437-439
Ferrichrome system, 431-434
Ferricyanide, and sidedness of
    membrane vesicles, 27
Flow dialysis, 15, 580
Fluorocitrate, and citrate trans-
    port, 404-407
3-Fluoro-L-erythromalate, and
    dicarboxylic acid transport,
    399
F_0 (see also Adenosine-5'-triphos-
    phatase), 500
  N,N'-dicyclohexylcarbodiimide-
    binding protein, 517
  mutations, 519
  and oxidative phosphorylation,
    566
  proton translocation by, 279-280
  subunits, 515-517
F_0F_1 (see Adenosine-5'-tri-
    phosphatase)
Folate transport, 472-475
F_1 (see also Adenosine-5'-tri-
    phosphatase)
  localization in membrane
    vesicles, 22, 24-27
  mutations, 519
  and oxidative phosphorylation,
    566
  subunits, 506-514
Formate dehydrogenase, 523
Fosfomycin (see Phosphonomycin)
French pressure chamber, use in
    preparation of membrane
    vesicles, 21-22, 31, 267
Fructose transport, 65, 116, 129
Fructose-1-phosphate transport,
    141
Fructose-6-phosphate transport,
    140-142
D-Fucose
  and galactose transport, 139
  and proton movement, 584

Fumarate transport, 395
Fungi, peptide transport in,
    353-354, 362-363, 368

G

Galactinol, induction of meli-
    biose transport, 140
Galactitol transport, 66, 129
Galactose
    binding protein, 143-147, 153,
        195, 539-542
    exit reaction, 151
    genetics of transport, 138-139,
        147, 541
    phosphoenolpyruvate:sugar
        phosphotransferase system,
        68
    and proton movements, 584
    transport, 131, 137-142, 143,
        151, 203
β-Galactoside (see also Lactose
    transport, M protein)
    dansylgalactosides, 136
    photoaffinity label, 134-135,
        137
    and proton movements, 582-584
    spin label, 136
    stoichiometry with protons,
        585, 595
    sulfhydryl label, 132-134, 137
    transport, 132-137
D-Galacturonate transport, 386
γ subunit of F_1 (see also Adeno-
    sine-5'-triphosphatase, F_1),
    506, 512
General aromatic amino acid
    transport system, 183
Ghost fraction, from M. phlei,
    21
gltR Gene, relation to glutamate
    transport in E. coli, 186
gltS Gene, relation to glutamate
    transport in E. coli, 186
Glucitol transport, 129
Gluconate transport, 387

Glucose
    binding protein, 542
    regulatory effect, 104
    transport, 64, 129
Glucose-1-phosphate transport,
    130, 141
Glucose-6-phosphate transport,
    140-142
α-Glucosides, and regulation of
    carbohydrate transport,
    110, 112
β-Glucoside transport, 55
D-Glucuronate transport, 386
Glutamate
    binding protein, 535, 547
    proton symport, 590
    sodium symport, 593, 611
    transport, 146, 185-187, 192
Glutamine
    and ATP-linked transport systems,
        576-577
    binding protein, 534
    transport, 13, 147, 185
Gramicidin, 611
α-Glutamylhydroxamate, and
    glutamine transport, 185
α-Glutamylhydrazide, and gluta-
    mine transport, 185
Glycerate transport, 391
Glycerol
    facilitated diffusion of, 130
    regulation of transport, 112
α-Glycerol phosphate (Glycerol-3-
    phosphate) dehydrogenase, 522
    and membrane vesicle orientation,
        27, 29
    transport, 149-150
Glycerylgalactoside transport,
    139, 143
Glycine-alanine transport, 177-181,
    187, 191, 193
Glycolate transport, 391-392
Glyoxylate transport, 391
Group translocation (see also
    Phosphoenolpyruvate-sugar
    phosphotransferase system), 8,

[Group translocation]
44-98, 497, 574-575
of carbohydrates, 46-84
of fatty acids, 84-89, 393-394
of nitrogenous bases and
nucleosides, 89-98
Guanine transport, 91
Guanidine-HC1, 19

H

Halobacterium halobium
amino acid transport, 592-594
and bacteriorhodopsin, 277-279,
515-516, 569
calcium transport, 611
measurement of membrane
potentials, 581
preparation of membrane
vesicles, 21
Halobacterium salinarium,
glutamate-aspartate transport in, 189
Hexose phosphate transport, 116,
140-142
hisJ protein, relationship to
histidine transport in S.
typhimurium, 189-190
hisP gene, relationship to histidine transport in S. typhimurium, 189, 190
Histidine
binding protein, 189-190,
533-534
transport, 148, 189-190, 193,
600
HPr (see also Phosphoenolpyruvate-sugar phosphotransferase system), 47, 49-50,
60, 108, 574-575
hrbA gene, relationship to
leucine-isoleucine-valine
transport, 182, 206
hrbR gene, relationship to leucine-
isoleucine-valine transport,
182

Hydrogen ion translocation (see
Proton transport)
Hydrolases, and peptide utilization,
333-336
Hydroxamate-iron transport
system, 251-252
β-Hydroxyamyl-L-aspartate
transport, 185
β-hydroxyaspartate transport,
185
3-Hydroxy-3-butynoate transport,
391
Hydroxycitrate transport, 404
β-Hydroxyglutamate transport,
186
Hypoxanthine transport, 91

I

ilvP gene, relationship to leucine-
isoleucine-valine transport,
187
Inducer exclusion, 105, 110
Iron transport, 9, 149, 243-253,
430-431
Isocitrate transport, 404-405
Isoleucine transport (see Leucine)
Isopropyl-β-thiogalactoside, 106

K

Kaback vesicles (see Membrane
vesicles)
3-Keto-3-deoxygluconate transport,
152, 387
α-Ketoglutarate transport, 399,
401
Klebsiella aerogenes, 17, 130

L

D-Lactate
dehydrogenase, 521-522
coupling to transport carriers,
562-563
and membrane vesicle
orientation, 23-24

[D-Lactate]
  as energy source for transport,
    17, 577
  transport, 390-391, 395
L-Lactate transport, 390-391
Lactobacillus
  cyanocobalamin transport, 466
  folate transport, 473
  peptide transport, 352, 353,
    358, 359, 362, 363, 366,
    368
Lactobacillus casei, transport
  of glycine-alanine, 187
Lactobacillus plantarum
  transport of aspartate-gluta-
    mate, 189
  transport of biotin, 476
Lactose transport, 2, 132-137,
    210, 525
  energetics, 583-587, 595,
    598, 599
  exit reaction, 151-153
  by phosphoenolpyruvate: sugar
    phosphotransferase system,
    68
  regulation of, 109
lacY protein, relationship to
    lactose transport, 173-174
Lambda phage receptor, 429
Leucine-isoleucine-valine
  binding proteins for, 195,
    199-203, 206-207, 532-533
  transport, 146, 181-182, 187,
    192, 193, 593
livR gene, relationship to leucine-
    isoleucine-valine transport,
    182
Lipoic acid transport, 479
Lipopolysaccharide, 149, 417,
    451
Liposomes, 33, 609-610
lstR gene, relationship to leucine-
    isoleucine-valine transport,
    182
Lysine-arginine-ornithine
  binding protein, 537-538

[Lysine-arginine-ornithine]
  transport, 13, 146, 184, 188-189,
    192-193, 587, 590
Lysostaphin, 17

                    M

Magnesium transport, 239-243,
    405-406
Malate transport, 395, 398
Malonate transport, 401
Maltose
  binding protein, 144-145, 195,
    428, 543
  and λ receptor, 428
  by phosphoenolpyruvate:sugar
    phosphotransferase system,
    68
  regulation, 112
  transport, 9, 147-149, 154, 426
Maltotriose, 428
Manganese transport, 253-261
Mannitol transport, 66, 129, 131
Mannose transport, 129
Mannose-6-phosphate transport,
    141
Marine pseudomonad, transport
  in, 193, 591
Melibiose transport, 112, 134,
    137, 140, 611
Membrane
  adhesion sites, 451-454
  chromatophore, 271
  cytoplasmic, 8, 192-192, 201-202,
    206
  outer, 9, 149-150, 193, 413-454,
    545-546, 601
  proteins, 495-547
  vesicles, 21
    everted, 22, 267-269, 591
    Kaback, 22, 267-269, 562-563,
      575
  orientation, 21-32
  preparation, 16-23, 192-193
  reconstitution with membrane
    proteins, 29, 137, 522

Membrane potential (see also
   Chemiosmotic hypothesis;
   proton motive force), 18,
   108, 109, 119, 152
 artificially-imposed, 586-587
 as component of the proton
   motive force, 565-566
 as driving force for transport,
   223-224, 570-573
 measurement of, 581-582,
   595-596
 Nernst equation, 223-224, 572
metD gene, relationship to
   methionine transport, 183
Methionine transport, 183, 188,
   193
D-Methionine transport, 183
Methylamine transport (see also
   pH gradients, measurement
   of), 281-282, 579
β-Methylgalactoside transport,
   138-139, 142-144, 146-148,
   152-154, 584
α-Methylglucoside transport,
   11, 110, 113, 118-120
α-Methylglutamate transport,
   186
γ-Methylglutamate transport,
   186
α-Methyl-D, L-methionine transport, 188
5-N-Methylphenazonium methylsulfate, and orientation of
   membrane vesicles, 30
metP gene, relationship to
   methionine transport, 183,
   188
Micrococcus denitrificans (see
   Paracoccus denitrificans)
Micrococcus lysodeikticus
 adenosine-5'-triphosphatase
   from, 501
 membrane vesicle orientation,
   24

Mitchell hypothesis (see
   Chemiosmotic hypothesis;
   proton motive force)
Molybdate transport, 299-300
Monactin, 587
Monocarboxylate transport (see
   also Transport, carboxylic
   acids), 386-395
Mosaic structure of membrane
   vesicles, 28
M protein (see also Lactose
   transport)
mRNA, lac specific, 106
Mycobacterium avium, glutamate-
   aspartate transport, 189
Mycobacterium phlei, 19, 193
Mycobacterium smegmatis,
   glutamate-aspartate transport,
   189
Mycoplasma, 130

                        N

NADH (see Nicotinamide adenine
   dinucleotide)
Nectin, 510
Neocosmospora vasinfecta
 chloride transport, 297
 zinc transport, 263
Nernst equation (see also Membrane
   potential), 223-224, 572, 579,
   581
Niacin transport, 479
Nicotinamide adenine dinucleotide
 dehydrogenase, 25, 27-28,
   523-524
 and membrane vesicle orientation,
   25, 27-28, 32
 as energy source for transport,
   25, 30, 267
 generation inside of membrane
   vesicles, 19-20
Nigericin, 580, 587
Nitrate transport, 293-295

Nitrite transport, 293-295
2-Nitro-4-azidophenyl-1-thio-β-
    D-galactoside, as photo-
    affinity label, 134-137
o-Nitrophenyl-β-D-galactoside
    transport, 15, 134, 586,
    599
L-Norleucine transport, 200, 206
Nucleoside transport (see also
    Group translocation), 89-98
Nucleotides, F_1-bound, 517-519

O

Oleate transport, 394
ONPG (see o-Nitrophenyl-β-D-
    galactoside)
Osmotic shock procedure, 3, 9,
    131, 142, 172, 174, 194-195,
    575
Outer membrane (see Membrane,
    outer)
Oxaloacetate transport, 404
Oxidative phosphorylation,
    19-20, 31, 565-568

P

Palmitate transport, 395
Pantothenate transport, 479
Paracoccus denitrificans
  artificially-imposed proton
    motive forces in, 591
  membrane vesicles, 17, 30,
    193
  phosphate transport, 289-290,
    591
  sulfate transport, 292, 591
Penicillin, 17
Peptidase activity, role in
    peptide utilization, 333-336
Peptide transport (see also
    Transport, peptides),
    325-369
Peptidoglycan, 415, 422
Phage receptors, 424, 427

Phenazine methosulfate, 20
Phenylalanine binding protein, 538
Phenylalaline-tyrosine-tryptophan
    transport, 183, 187-188,
    192, 193
pH gradient (see also Chemiosmotic
    hypothesis; proton motive
    force)
  artificially imposed, 587, 590,
    591
  as component of the proton
    motive force, 565-566
  as driving force for transport,
    571-573
  measurement of, 578-581
Phosphate binding protein, 529-547
Phosphate bond energy-linked
    transport systems (see Adeno-
    sine-5'-triphosphate-linked
    transport systems)
Phosphate transport, 284-290,
    400, 605
Phosphatidylserine, 18
Phosphoenolpyruvate
  as energy source for the phos-
    phoenolpyruvate:sugar
    phosphotransferase system,
    47
  preparation of membrane
    vesicles containing, 18
  regulation, 114
  transport, 392
Phosphoenolpyruvate:sugar phos-
    photransferase system (see
    also Group translocation;
    transport, carbohydrates),
    3, 9, 46-84, 105-120, 129,
    131, 172, 574-575
  biological distribution, 56
  comparative physiology, 57-59
  enzymology, 48-56
  function, 69-84
  nomenclature, 47-48
  regulation, 105-120
Phosphonomycin, 141, 150, 483

2-Phosphoglycerate transport, 397

3-Phosphoglycerate transport, 114, 392

D, L-Pipecolic acid, 188

Polyglutamate, 473

Potassium transport, 226-239

Proline transport, 33, 183-184, 188, 193, 577

Propionate (see also pH gradient, measurement of transport), 392-393

Protease activity, role in peptide, utilization, 333-336

Protein kinases, 116, 121

Proteus mirabilis, 193, 301

Proton motive force (see also Chemiosmotic hypothesis; membrane potential; pH gradient), 3-4, 18, 21, 152, 154, 556-575, 578-599, 605

Proton-translocating ATPase (see Adenosine-5'-triphosphatase)

Proton transport, 277-280, 515-517, 565-568
    cotransport with carbohydrates, 152, 582-585
    and orientation of membrane vesicles, 25

Protoplasts, 17

Pseudomonas, transport of peptides, 352-354, 358-359, 362-363, 366-368

Pseudomonas acidovorans, 188

Pseudomonas aeruginosa
    amino acid transport, 187-188
    carbohydrate transport, 129
    preparation of membrane vesicles, 19

Pseudomonas fluorescens, 130

Pseudomonas putida
    amino acid transport, 188-189, 193

[Pseudomonas putida]
    preparation of membrane vesicles, 17

Purine transport, 89-98

Pyridoxine transport, 577-578

Pyrimidine transport, 89-98

Pyruvate transport, 391

Q

Quinacrine (see also pH gradient, measurement of), 582

R

Regulation of carbohydrate transport, 105-122

Rhodopseudomonas capsulata, 227-239

Rhodopseudomonas spheroides, 19

Rhodospirullum rubrum, 581

Ribose binding protein, 144-148, 542

RPr, and regulation of carbohydrate transport, 115-116

S

Safranine O (see also Membrane potential, measurement of), 582

Salmonella, peptide transport, 352-354, 358, 367

Salmonella enteritidis, zinc transport in, 263

Salmonella typhimurium
    amino acid transport, 187-193
    antibiotic transport, 483
    group translocation, 574
    phosphate transport, 290-293
    pyridoxine transport, 477
    sulfate transport, 194

Selenocystine, 184

Selenium transport, 300

Selenomethionine, 183

D-serine, 177

Sideromycins, 483
Siderophores, 430, 431, 434-435, 440-442
Silicate transport, 301
Silicone oil, centrifugation through as assay for transport, 14, 580
Sodium
  antiport with calcium, 611
  solute symports, 592-594, 611
  transport, 274-277
Solute transport, proton gradient-coupled; quantitative aspect of, 594-599
Sonication, use of in preparation of membrane vesicles, 31-32
Sorbitol transport, 66
Sorbose transport, 69, 129
Spheroplasts, 13, 174
Sporulation, calcium accumulation during, 271-274
Staphylococcin, 603
Staphylococcus aureus
  amino acid transport, 193, 561, 587, 590
  carbohydrate transport, 129, 130
  group translocation, 574
  magnesium transport, 260-261
  membrane vesicles, 17
  phosphate transport, 289
  phosphoenolpyruvate:sugar phosphotransferase system, 50, 57, 60, 63, 66, 70, 73, 79
  tetracycline transport, 484
Staphylococcus oranienburg, 296
Streptococcus faecalis
  amino acid transport, 187, 189
  calcium transport, 266
  measurement of membrane potential, 582

[Streptococcus faecalis]
  measurement of pH gradient, 579
  phosphate transport, 288-289
  potassium transport, 234-237
  proton transport, 280
  sodium transport, 276
Streptococcus lactis
  artificially-induced proton motive force, 586
  exit reaction, 152
  measurement of pH gradient, 579
  quantitative aspects of solute transport, 595
Streptomycin transport, 486
Submitochondrial particles, 32
Succinate dehydrogenase, 524
Succinate transport, 395, 401, 526, 547, 601
Sucrose transport, 69
Sulfate binding protein, 291, 528
Sulfate transport, 290-293
Symport, 570-572, 587, 611

T

D, L-Tartarate transport, 399
meso-Tartarate transport, 400
TCS (see Tetrachlorosalicylanilide)
Tetrachlorosalicylanilide, 584, 587, 590
Tetracycline transport, 484
Tetraphenyl boron (see also, Membrane potential, measurement of), 582
Thermophilic bacterium PS-3, 197 503, 526
Thiamine
  binding protein, 471, 543
  kinase, 470
  pyrophosphate, 470
  transport, 470-472
Thiobacillus neapolitanus, 17
Thiocyanate (see also pH gradient, measurement of), 579, 583

Thiodigalactoside, 132–137
Thiolase, 88
Thiomethyl-β-D-galactoside
    transport (see also Lactose
    transport), 133, 137, 140,
    142, 143, 152, 595
D, L-Threo-β-hydroxyaspartic
    acid, 185
Threonine-serine transport, 181,
    191, 193
TMG (see Thiomethyl-β-D-
    galactoside)
Toluene, 25
tonB gene, 436–437, 440–442,
    445–451
TPB (see Tetraphenyl boron)
Transcobalamin II, 470
Transient repression, 105
Transport
  amino acids, 171–211
  anions, 284–304
  antibiotics, 481–487
  carbohydrates, 46–84, 106–122,
    128–160
  carboxylic acids, 386, 408
  cations, 226–284
  energetics, 560–611
  fatty acids, 84–89
  methods of assay, 8–33
  nitrogenous bases, 90–93
  nucleosides, 93–98
  and outer membrane, 414–454
  peptides, 326–369
  primary, 546–569
  proteins, 496–547
  regulation, 106–122
  secondary, 569–573
  vitamins, 466–481

Trehalose transport, 69
Tricarboxylic acid transport
    (see also Transport, car-
    boxylic acids), 401–407
Trifluoroleucine, 200
TPMP (see Triphenylmethyl-
    phosphonium ion)
Triphenylmethylphosphonium ion
    (see also Membrane potential,
    measurement of), 582, 595,
    608
Triton X-100, 25
tryP gene, relationship to
    pheynlalanine-tyrosine-
    tryptophane transport, 183

U

Ubiquinone, 27
Uncouplers, 153, 154, 563, 567,
    568, 577, 587
Uniport, 570, 572–573, 587
Uracil transport, 92
Uridine transport, 94–96

V

Valine transport (see Leucine)
Valinomycin, 231, 579, 581
Vectorial metabolism, 564
Veillonella alcalescens, 17, 193
Vinylglycolate, 16
Vitamin B12 (see Cyanocobalamin)

X

Xanthine transport, 91

Z

Zinc transport, 262–265